Biology

Biology

Harvey D. Goodman
Chairperson, Biology Department
Grover Cleveland High School
Ridgewood, New York

Thomas C. Emmel
Professor, Department of Zoology
University of Florida
Gainesville, Florida

Linda E. Graham
Associate Professor, Department of Botany
University of Wisconsin
Madison, Wisconsin

Frances M. Slowiczek
Science Department Chairperson and
  Life Sciences Teacher
Kearny High School
San Diego, California

Yaakov Shechter
Professor, Department of Biological Sciences, and
  Chairperson, Natural Sciences Core Curriculum
Herbert H. Lehman College of the City University
  of New York
Lecturer, College of Physicians and Surgeons,
  Columbia University
New York, New York

Harcourt Brace Jovanovich, Publishers
Orlando   New York   Chicago   San Diego   Atlanta   Dallas

Acknowledgments

CONTENT SPECIALISTS

Thomas Armentano, Ph.D.
Director, Biotic Resources Program
Holcomb Research Institute
Butler University
Indianapolis, Indiana

Richard K. Boohar, Ph.D.
Associate Professor and Chief Biological Sciences Advisor
Department of Biological Science
University of Nebraska
Lincoln, Nebraska

Dorothy E. Croall, Ph.D.
Instructor of Physiology
University of Texas Health Science Center
Dallas, Texas

Beth DiDomenico, Ph.D.
Department of Molecular Genetics and Cell Biology
University of Chicago
Chicago, Illinois

Thomas Kantz, Ph.D.
Professor
Department of Biological Sciences
California State University
Sacramento, California

Georgia E. Lesh-Laurie, Ph.D.
Professor of Biology and Dean, College of Graduate Studies
Cleveland State University
Cleveland, Ohio

Diane T. Merner, Ph.D.
Department of Biology
University of South Florida
Tampa, Florida

Henry A. Robitaille, Ph.D.
The Land, EPCOT Center
Lake Buena Vista, Florida

Ethel Sloane, Ph.D.
Chair, Department of Biological Sciences
University of Wisconsin
Milwaukee, Wisconsin

Curtis Williams, Ph.D.
Professor of Biology
State University of New York
Purchase, New York

Tommy E. Wynn, Ph.D.
Associate Professor of Botany
North Carolina State University
Raleigh, North Carolina

Alfred Zweidler, Ph.D.
Fox Chase Cancer Center
Philadelphia, Pennsylvania

CURRICULUM SPECIALISTS

Delmous R. Ingram
Former Biology Teacher and Science Department Chairperson
Needham B. Broughton High School
Raleigh, North Carolina

Janis W. Lariviere
Biology–Chemistry Teacher
Anderson High School
Austin, Texas

Robert L. Lehrman
Science Teacher
Roslyn High School
Roslyn, New York

Karen M. Nein
Science Consultant
St. Charles, Missouri

Barney Parker
Biology Teacher
Developmental Research School
Florida State University
Tallahassee, Florida

Daniel S. Sheldon, Ph.D.
Associate Professor
Science Education Center
University of Iowa
Iowa City, Iowa

READING SPECIALIST

Judy Nichols Mitchell, Ph.D.
Associate Professor of Reading
College of Education
University of Arizona
Tucson, Arizona

FEATURE WRITERS AND CONTRIBUTORS

Robert V. Blystone, Ph.D.
Professor of Biology
Department of Biology
Trinity University
San Antonio, Texas

William R. Collien
Biology and Botany Instructor
Triton College
River Grove, Illinois

Daniel H. Franck, Ph.D.
Former Botany Professor
University of Wisconsin
Madison, Wisconsin

Harvey D. Goodman
Chairperson, Biology
Grover Cleveland High
  School
Ridgewood, New York

Mary B. Harbeck
Supervising Director,
  Science
District of Columbia Public
  Schools
Washington, D.C.

Vicki Werner Hoffman
Former Biology Teacher
Coral Gables Senior High
  School
Coral Gables, Florida

Joanne Ingwall, Ph.D.
Associate Professor of
  Physiology and Biophysics
Department of Medicine
Harvard Medical School and
  Brigham and Women's
  Hospital
Boston, Massachusetts

Mary E. Kayusa
Biology Teacher
North Fort Myers High
  School
North Fort Myers, Florida

Irving Kent Loh, M.D., F.A.C.C.
Cardiologist
Los Robles Regional Medical
  Center
Thousand Oaks, California

Glenn K. Leto
Biology Teacher
Barrington High School
Barrington, Illinois

Kenneth Nelson
Biology Teacher
Lyons Township High
  School, South Campus
Western Springs, Illinois

James D. Oilschlager
Biology Teacher
Libertyville High School
Libertyville, Illinois

Henry A. Robitaille, Ph.D.
The Land, EPCOT Center
Lake Buena Vista, Florida

Yaakov Shechter, Ph.D.
Professor, Department of
  Biological Sciences
Herbert H. Lehman College
New York, New York

Kenneth W. Weidlich
Biology Teacher
Hillsborough High School
Belle Mead, New Jersey

FIELD TEST TEACHERS AND SUPERVISORS

Duane R. Ashenfalder,
  Teacher
Parkland High School
Orefield, Pennsylvania

William E. Ayers, Science
  Coordinator
Allentown School District
Allentown, Pennsylvania

William E. Beggs, Secondary
  Science Supervisor
Curriculum and Instruction
  Center
Largo, Florida

Connie Brekke, Teacher
Culver City High School
Culver City, California

Melani Brewer, Teacher
Western High School
Fort Lauderdale, Florida

Edward Davis, Assistant
  Director of Curriculum
  Services
Parkland School District
Allentown, Pennsylvania

Joseph S. Elias, Teacher
William Allen High School
Allentown, Pennsylvania

Helen Fleck, Teacher
North High School
Evansville, Indiana

Robert W. Jackson, Teacher
Snider High School
Fort Wayne, Indiana

Irwin N. Jaeger, Science
  Supervisor
Union High School
Union, New Jersey

Thomas D. Jones,
  Curriculum Coordinator
Maine–Endwell School
  District
Endicott, New York

Major H. Kirby, Jr., Science
  Department Chairperson
Culver City High School
Culver City, California

Max R. Lake, Director
START Math/Science
Fort Wayne, Indiana

Gerald Love, Science
  Supervisor
Bethpage High School
Bethpage, New York

Michael M. McGuyer,
  Teacher
North High School
Evansville, Indiana

Mary W. Pace, Teacher
Coconut Creek High School
Coconut Creek, Florida

Lou Palmeri, Teacher
Bethpage High School
Bethpage, New York

Raleigh Philp, Science
  Department Chairperson
Rowland High School
Rowland Heights, California

Nancy Romance, Ed.D.,
  Science Supervisor
Division of Instruction
Davie, Florida

John J. Sheridan, Teacher
Boca Ciega High School
St. Petersburg, Florida

William V. Soranno, Teacher
Union High School
Union, New Jersey

Robert E. Wright, Teacher
Maine–Endwell High School
Endwell, New York

William K. Wright, Teacher
Lakewood High School
St. Petersburg, Florida

Contents

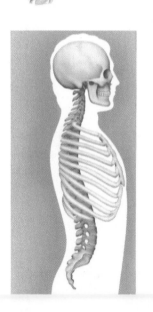

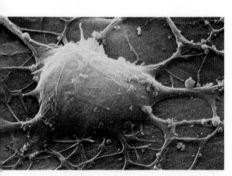

Highlight on Careers

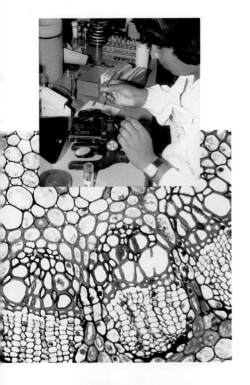

Spotlight on Biologists

Investigations

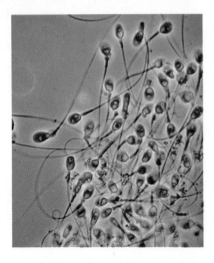

BioTech

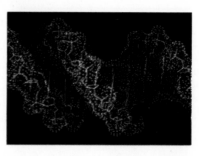

Introducing the Book

The study of biology involves learning about the parts of living things and their functions. Like living things, textbooks are also made up of a variety of parts. This section introduces and explains the function of each of the major parts of this book. Careful reading will prepare you to use the book effectively as you learn about the exciting world of living things.

U N I T

3 Continuity of Life

10 **Fundamentals of Genetics**
The Origin of Genetics • Genetics and Probability • Extending Mendel's Principles

11 **Chromosomes and Genetics**
The Chromosome Theory • The Sex Chromosomes • The Behavior of Genes

12 **Chemical Basis of Genetic**
The Discovery of DNA • DNA and Protein • Changes in the Genetic Code

13 **Human Genetics**
The Study of Human Genetics • The Inhe Human Traits • Detecting Genetic Disorde

14 **Applied Genetics**
Controlled Breeding • Other Methods of I Organisms

Unit Openers

Each of the eleven units opens with a bold black page that features a dramatic photograph and lists the chapters and major topics covered in the unit.

C H A P T E R

6 **Cell Structure and Function**

Outline

The Discovery of Cells
Early Microscopic Observations
The Cell Theory
Size and Shape of Cells

Basic Parts of the Cell
The Nucleus
Cell Membrane and Cell Wall
The Cytoplasm and Organelles

Differences in Cells
Cells Without a True Nucleus
Cells with a Nucleus
Animal and Plant Cells

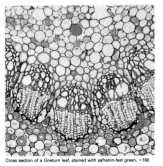

Cross section of a *Gnetum* leaf, stained with safranin-fast green, ×160

Introduction

Organisms can be as large and complex as whales or as small and simple as bacteria. Large or small, each organism is made up of atoms and molecules. The massive whale and the microscopic bacteria are each essentially a combination of chemicals. How then can something be termed alive? At what point do the atoms and molecules cease to be just a collection of chemicals and take on the properties of living organisms?

The discovery over 300 years ago that all living things are composed of cells began a period of study that continues today. Cells are more than a collection of chemicals. They possess a unique organization that gives them all the qualities of life. The study of cells and how they are organized is, then, the study of what constitutes life.

78

Chapter Openers

The chapter opening page presents a striking photograph and a brief introduction to the chapter. The outline in the left margin provides an overview of key topics.

The Nature of Matter

Ask a scientist what makes up the universe and you are likely to get a very simple answer—matter and energy. **Matter** is anything that takes up space. *Energy* is the capacity to move matter. Before learning about how energy affects matter, you will need to know the basic properties and composition of matter.

4.1 Properties of Matter

Every object contains a certain amount of matter. The measure of the amount of matter in an object is its **mass**. The *weight* of an object measures the pull of gravity on its mass. The more mass something has, the more it weighs.

Each kind of matter has specific properties, or characteristics, that distinguish it from every other kind of matter. Scientists generally divide the properties of matter into two classes. **Physical properties** are characteristics that can be determined without changing the basic makeup of the substance. Physical properties include the size, shape, texture, and color of a substance. **Chemical properties** describe how a substance acts when it combines with other substances to form entirely different kinds of matter.

Physical Properties and Physical Change The physical properties of a substance include its odor, its hardness, and its ability to conduct heat and electricity. The temperature at which a substance melts, boils, or freezes is also a physical property.

A substance can undergo changes in its physical properties without changing its chemical properties. One simple kind of physical change occurs when a substance changes in size or shape. Quartz boulders on a beach are ground into tiny grains of sand by the waves. The sand has all the chemical properties the boulders had, even though some of the physical properties have changed.

A more complex physical change occurs when matter changes state. The three common **states of matter** are solid, liquid, and gas. In all three states, matter consists of tiny particles. The particles in a solid are held tightly together. A solid therefore has a definite shape. It also has a definite *volume*—a given amount of a solid takes up a definite amount of space.

The particles in a liquid are held together less tightly than in a solid. They are free to tumble over one another, and so a liquid assumes the shape of any container. Like a solid, however, a given amount of a liquid maintains a constant volume.

Section Objectives

- *Define* the term *matter.*
- *Explain* the difference between physical and chemical properties.
- *List* three types of particles in the atom and describe their arrangement.
- *Distinguish* elements from compounds.
- *Name* the three basic kinds of mixtures important to living things.

Figure 4–1. [Matter, exists...] The solid state... [gas]eous state... [liq]uid water is a... part of all org[anisms]...

Bas[...]

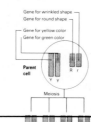

Gene for wrinkled shape
Gene for round shape

Gene for yellow color
Gene for green color

Parent cell

Meiosis

Y R   Y r   y R   y r

Possible gametes

Figure 10–5. The principle of independent assortment states that two or more gene pairs separate independently. Thus gametes may contain a combination of dominant and recessive genes.

generation. However, the F₂ generation also included round, green seeds and yellow, wrinkled seeds. From this experiment, Mendel realized that two traits produced by recessive genes did not have to appear in the same offspring. For example, green color, a recessive trait, could appear with round seeds, a dominant trait. Mendel formulated the **principle of independent assortment** to explain this finding. *The principle of independent assortment states that two or more pairs of genes segregate independently of one another during the formation of gametes.* For instance, the segregation of the genes for seed color does not affect the segregation of genes for seed shape.

Today it is known that most gene pairs segregate independently only if they are located on different chromosomes. Traits determined by two genes on the same chromosome tend to be inherited together. Mendel, however, was able to choose seven contrasting traits, each determined by a gene pair on a different pair of chromosomes.

10.6 Other Genetic Terminology

Since the time of Mendel, the language of genetics has become more precise. As you know, scientists use the term *gene* instead of *factor* to describe the unit of heredity. They also use the term **allele** (uh LEEL) to refer to either member of a pair of genes that determines a single trait. For example, the dominant allele for seed color in peas (Y) produces yellow seeds. The recessive allele (y) produces green seeds.

The pairs of alleles in the cells of an organism make up its **genotype** (JEE nuh typ). These pairs of genes are represented with capital and lowercase letters, such as YY, Yy, and yy. A trait that is actually expressed in an organism is called a **phenotype** (FEE nuh typ). Although environment also affects many visible traits, phenotypes are largely determined by an organism's genotype. For example, a pea plant with the genotype YY will have the phenotype of yellow seeds. What other genotype can produce the same phenotype of yellow seeds?

Reviewing the Section

1. Why did Mendel experiment with pea plants?
2. What pattern of inheritance led Mendel to suggest the existence of factors?
3. How is Mendel's principle of segregation different from his principle of independent assortment?
4. Why was Mendel's work not accepted at first?

Section Objectives

Clearly stated objectives at the beginning of the section provide guidelines for your reading.

Section Heads

Bold headlines help you identify major sections.

Primary Subheads

Major subheads are numbered for easy reference.

Secondary Subheads

Smaller boldface headings point out subcategories.

Key Sentences

Major concepts in each section are highlighted in boldface italic type.

BioTerms

Important biology vocabulary terms are highlighted in boldface type and defined the first time you encounter them. Hard-to-pronounce terms are respelled phonetically.

Reviewing the Section

Each section ends with questions to help you review the main points covered.

Special Features

Readable magazine-style features relate scientific concepts presented in the chapter to other concerns such as career choices or technological developments. Many features use photographs and diagrams to clarify ideas and help bring science to life.

▶ Genes: Facts and Riddles

Scientists have learned a great deal about genes since the work of Watson and Crick in 1953. Most important, scientists now know that a gene is a sequence of n... an active str... that codes fo... protein. As y... the code of e... transcribed i... of mRNA. Th... ecule travels... plasm and p... blueprint for... acids into a ... *polypeptide*... teins consist... polypeptide,... consist of tw... chains. Thus... protein may ... mation from...

genes, one for each polypeptide.

Although scientists have been able to understand how protein synthesis works, many other riddles

acid causes the shape of the red blood cells to change from round to bent, or sickle-shaped.

Another unsolved mystery concerns nucleotide

Highlight on Careers: Laboratory Technician

General Description
The laboratory technician plays a vital role in scientific problem solving. Under directions from a scientist, the technician performs tests and analyzes sample materials. The technician also keeps records of test results and compiles facts for the scientist testing a theory.

The duties of a laboratory technician vary. For example, a medical laboratory technician may work in a hospital, preparing slides from tissue samples and analyzing them under a microscope. A food tester may perform chemi-

cal analyses on samples of yellow dye used to color the food.

Career Requirements
Many technicians work for industry, government, hospitals, and universities. Most positions require two years of college with courses in biology, chemistry, or physics. Some positions are open to graduates of technical training courses.

Technicians enjoy great flexibility. They can start work with a modest amount of higher education and go back to school later if they wish to move

complicated molecules. In a ... puters to help them analyze ... make sense of hundreds or e... information. For example, a ... wing size affects the flight s... lects data by measuring the v... ual hawks. The scientist then ... data into a computer. The co... cal chunks of information, su... of flight speeds. Then the sci... sions from the information a...

Reviewing the Sec...

1. How does a centrifuge w...
2. What is the purpose of n...
3. What tweezerlike tool re...
4. When in the scientific m...

190 Chapter

42 Chapter 3

Spotlight on Biologists: Barbara McClintock

Born: Hartford, Connecticut, 1902
Degree: Ph.D., Cornell University

When geneticist Barbara McClintock first proposed her theory of "jumping genes" in 1947, the scientific community largely ignored her. Now her revolutionary findings that genes can move from one chromosome to another is hailed as one of the most important discoveries of this century. McClintock's lifetime of research with maize plants was acknowledged in 1983 with the Nobel Prize for Medicine or Physiology.

When McClintock began her research, most scientists thought genes maintained a fixed position, like beads in a row. McClintock crossed maize plants and

noted color changes in the leaves and kernels that did not follow a hereditary pattern. She concluded that such mutations resulted from the movement of genes.

Later experiments by molecular biologists working with bacteria confirmed McClintock's theories. These experiments further concluded that the genes move during the replication phase of the cell cycle, prior to cell division. These findings explain the origin of some bacteria strains that are resistant to antibiotics. The "jumping genes" also have significance in evolution, as a source of genetic variability.

McClintock has often had to resist popular opinion to pursue her own interests. Her mother at first op-

posed her desire to attend college. After McClintock studied biology on her own while working in an employment agency, her parents gave in and allowed her to attend Cornell University. For years she had difficulty in securing good teaching and research positions because she was a woman. Since 1941 McClintock has pursued her research at Cold Spring Harbor Laboratory in New York state.

chromosome mutations are harmful or even lethal to organisms that inherit them.

Figure 12–11 shows three major kinds of chromosome mutations. One type of mutation involves a single chromosome. During mitosis or meiosis, a chromosome may break, and part of it may be lost. The loss of a chromosome segment is called a *deletion*. Its effect in animals is usually lethal. Occasionally, the middle section of a chromosome may break away, turn over, and recombine with the same chromosome in reverse order. This type of mutation is called an *inversion*. Inversions may not harm organisms since all the same genes are present on the chromosome.

192 Chapter 12

▶ Features

Each chapter includes features that focus on key scientific concepts and revolutionary discoveries.

Highlight on Careers

More than 30 chapters include descriptions of varied careers in biology. These range from jobs that require vocational training to those demanding advanced degrees.

Spotlight on Biologists

Twenty chapters contain biographical sketches of scientists who have made—or are making—significant discoveries.

4.9 Chemical Reactions

By filling outer energy levels, chemical bonds between atoms make compounds relatively stable. However, adding energy to a compound can break the bonds. The atoms may then form new bonds with other atoms to make new compounds. The process of breaking existing chemical bonds and forming new bonds is called a **chemical reaction.**

Because bonds break only when sufficient energy is supplied, every chemical reaction requires an input of energy to get started. The energy required to start a chemical reaction is called **activation energy.** Heat is the most common form of activation energy. For example, when sugar is exposed to high heat it bubbles, turns black, and gives off water vapor and carbon dioxide. The chemical explanation for this change is that the heat energy causes the sugar molecules to collide faster and faster. These collisions break the existing bonds between the carbon, hydrogen, and oxygen atoms in the sugar. As the bonds break, the atoms form new bonds which result in carbon dioxide, water vapor, and pure carbon. Energy is also released in the forms of heat and light from this activity. No energy is lost; it only changes form.

Atoms do not change in a chemical reaction; they are only rearranged into different combinations. The *products* of the reaction described above—water vapor, carbon dioxide, and pure carbon—contain the same total number of atoms as the sugar originally contained. This fact illustrates one of the basic laws of chemistry: All the atoms that enter into a chemical reaction will be present in the products of the reaction.

As new bonds form, some amount of energy is released. In some cases, this energy helps keep a reaction going. Chemical reactions that release more energy than they use up are called

Q/A

Q: *How do fireflies glow?*

A: Chemical reactions within the firefly's abdomen release energy in the form of light. This phenomenon, which also occurs in fish and many other animals, is called *bioluminescence.*

Figure 4–11. A spark, a flame, and a fire are the results of exergonic chemical reactions in which energy is released as heat and light. The exergonic chemical reactions that occur in cells release energy far more gradually than the exergonic reactions shown below.

Q/A

The Q/A features in the margins offer fascinating details of biological study.

BioTech

In every unit, two colorful pages show you developments, such as gene therapy and space agriculture, that are on the cutting edge of biological research.

BIO**TECH**
Mapping Ancient Climates

Traces of ancient human settlements have been found in the Sahara. How could people have survived in such a dry, barren place? The answer is that the climate of the Sahara, now hot and extremely arid, was once far more hospitable. Over time, climate changes. As a result of technological advances, the nature of these changes in past climates can be studied in detail.

The prevailing weather patterns known as climate result from complex interactions of the atmosphere with the earth's oceans, ice cover, and land masses. Scientists called *paleoclimatologists* study the climate of the earth's past. In many cases, they can

provide a description of the climate of a particular place at a particular time in history, such as the ancient climate of the Sahara.

How do paleoclimatologists gather their data? Obviously they cannot measure the temperature or precipitation at a particular place thousands of years in the past. Likewise, they cannot measure how much of the earth's surface was covered by ice 5,000 or 10,000 years ago. Paleoclimatologists must rely instead on indirect sources for information.

One major source of information lies in layers of sediment buried deep beneath the floors of oceans and lakes. Within the sediment are microfossils, the fossils of microscopic organisms. These microfossils can be removed and studied through the use of a variety of technological tools. The microfossils provide clues to the climatic conditions under which the organisms that produced them must have lived.

Core-sampling device

Microfossils provide clues to the climatic conditions under which the organisms that produced them must have lived.

Ocean Core

Age of core segment in millions of years
Temperature of surface
Present 5°C
1 million 10°C
5 million 15°C
15 million 20°C
25 million 25°C

Computer-generated maps of sea surface temperature from 18,000 years ago (left) and from present (right)

For example, marine organisms called foraminifera live in waters near the surface of oceans. Different varieties of foraminifera, with different shell structures, flourish at different temperatures. The fossil shells of foraminifera record their shapes and thus indicate the temperature of the ocean water in which the creatures once lived.

Researchers examine microfossils under a scanning electron microscope to discover details of texture and structure necessary for identification. Next, they determine the age of microfossils with another technological tool: radiocarbon dating. Evidence from the scanning electron microscope suggests what climatic conditions

Scanning electron microscope

supported the organisms. The radiocarbon dating indicates at what point in the past that climate prevailed.

Using these and other techniques, paleoclimatologists have painstakingly collected information about ancient precipitation patterns and temperatures from all over the world. To interpret their data, they feed it into powerful computers that integrate the data with information about current climate.

268

269

Investigation 8: Photosynthesis

Purpose
To determine the rate of photosynthesis in a plant

Materials
Elodea or any aquatic plant, glass funnel, 1000-mL beaker, test tube, sodium bicarbonate, watch or clock with second hand, wooden splint, water, matches

Procedure
1. Pour water into the beaker until it is half full. Allow the water to sit overnight, which allows chlorine and any other additives to evaporate. Dissolve 3g of sodium bicarbonate in the water.
2. Place an elodea plant in the bottom of the beaker. Put a funnel over the plant, as shown in the diagram.

Test tube
Funnel
Water
Plant
Beaker

3. Fill the test tube with water. Placing your thumb securely over the mouth of the test tube, invert the tube and place it on top of the funnel.
4. Set the beaker in direct sunlight and count the number of gas bubbles that appear in the test tube. Write down the number of bubbles you count after 30, 60, 90, 120, 150,

Analyses and Conclusions
1. Make a grid like the following one and graph your results.

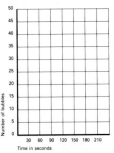

Number of bubbles (y-axis: 0 to 50)
Time in seconds (x-axis: 30 60 90 120 150 180 210)

2. How are your results related to the rate of photosynthesis? What conclusion can you draw from your graph?
3. Compare your results with those of your classmates. Why is it important that only one plant be used in each experiment?

Going Further
· Use the same setup as in the original experiment, but use warm water and test the effect of heat on the rate of photosynthesis.

Chapter 2 Review

Summary

Science can be characterized in two ways. First, it is a body of knowledge about the world around us. Second, it is a logical, organized process of inquiry. This process, called the scientific method, includes the following steps: defining the problem, collecting background information, formulating a hypothesis, conducting a controlled experiment, observing and recording data, and formulating a conclusion. This method of study sets science apart from all other areas of study.

The abiogenesis-biogenesis debate illustrates the scientific method at work. Prior to the 1800s, many people believed in abiogenesis, or the ability of nonliving things to pro-

duce certain forms of life. In the seventeenth century, Francesco Redi's experiments challenged abiogenesis by showing that maggots arose not from decaying meat but from eggs laid by flies. In the eighteenth century, John Needham designed an experiment to support abiogenesis, but Lazzaro Spallanzani's work uncovered flaws in Needham's experiment. The theory of abiogenesis was finally disproved after Louis Pasteur proved in 1864 that microorganisms carried by dust made new microorganisms in flasks of broth. Today the principle of biogenesis is being challenged as scientists study the origin of life on Earth.

BioTerms

abiogenesis (28)	experiment (26)	scientific principle (27)
applied science (27)	experimental group (26)	spontaneous generation (28)
biogenesis (28)	hypothesis (25)	statistics (26)
control group (26)	pure science (27)	theory (27)
controlled experiment (26)	science (24)	variable (26)
data (26)	scientific method (25)	

BioQuiz   (Write all answers on a separate sheet of paper.)

I. Completion

1. The _____ is an organized, logical method of inquiry.
2. The test of a hypothesis is a scientific _____.
3. _____ science puts scientific knowledge to practical use.
4. The theory of _____ states that life arises from living things.
5. _____ are scientific facts collected during an experiment.
6. A _____ is a scientific explanation of known facts.

II. Modified True and False

Mark each statement TRUE or FALSE. If false, change the underlined term to make the statement true.

7. A hypothesis is based on the data that is collected in a controlled experiment.
8. The control group is used to test the variable in an experiment.
9. Statistics is a mathematical method of evaluating numerical data.
10. Maggots in decaying meat are produced by an active principle.

III. Multiple Choice

11. A hypothesis is a) a proposed answer to a scientific question. b) the factor tested in an experiment. c) a scientific explanation of known facts. d) an established scientific truth.
12. Every controlled experiment must have a) two variables. b) two control groups. c) two groups of subjects. d) two hypotheses.
13. A scientific law is a) a theory that has not been proven false. b) a statement of fact. c) an untested hypothesis. d) a tested hypothesis.
14. Redi showed that the presence or absence of air did not affect the presence of maggots on meat by using a) open flasks. b) sealed flasks. c) open jars. d) mesh covered jars.

15. Experiments conducted by Francesco Redi and Lazzaro Spallanzani supported a) biogenesis. b) active principles. c) abiogenesis. d) superstition.

IV. Essay

16. How is an experiment used to test a hypothesis?
17. Why must scientists keep complete records of their experiments?
18. What is the relationship between a hypothesis and a theory?
19. How was Francesco Redi's second experiment different from his first?
20. In what two ways did Louis Pasteur show that microorganisms were carried on dust particles?

Applying and Extending Concepts

1. Many important scientific discoveries begin with ideas, but some begin with lucky accidents. Research the work of Sir Alexander Fleming. What accident led to his discovery of penicillin?
2. Design an experiment to test one of the following hypotheses:
   a. Grass seed A germinates faster at low temperatures than at high temperatures.
   b. Lettuce seed B germinates faster in darkness than in light.
3. Imagine that you are studying the causes of heart disease. Narrow this topic into two questions you can investigate with the scientific method.

4. A biologist feels that trees treated with fertilizer 1 produce more apples than those treated with fertilizer 2. To test this hypothesis, the biologist spreads fertilizer 1 around a red delicious apple tree and an equal amount of fertilizer 2 around a Jonathan apple tree. At the end of the season the biologist counts the number of apples in each tree. The tree fertilized by fertilizer 2 produced 86 apples, while the one treated with fertilizer 1 produced only 45. Evaluate the experiment used to test this hypothesis. What conclusion should be drawn from this experiment?

Related Readings

Milton, J. *Controversy: Science in Conflict.* New York: Messner, 1980. This book highlights four conflicts in biology and shows how scientists using the scientific method can arrive at different conclusions.

Moravcsik, M. J. *How to Grow Science.* New York: Universe Books, 1980. Written by a research physicist, this wonderfully readable book explains what science is and how science is done.

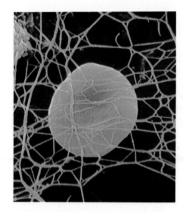

Visual Materials

Many key concepts of biology are easier to learn if you can visualize how they actually occur in living things. Throughout the book, photographs, illustrations, and tables have been carefully selected to clarify and supplement the written text.

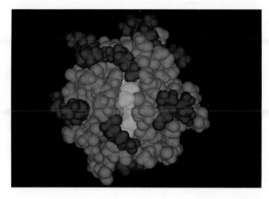

Photographs

Unusual photographs, current photomicrographs, and computer-generated illustrations help you visualize the complex and beautiful biological world.

Illustrations

Bright, precise diagrams and illustrations bring unseen processes and microscopic structures into sharp focus. Many illustrations include detailed insets.

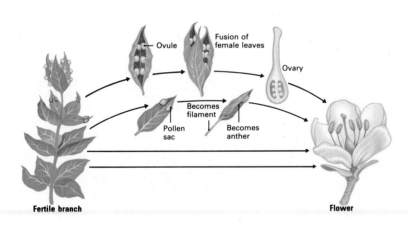

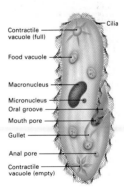

Tables

Crisp, easy-to-read tables make use of color coding, illustrations, and clear organization to help you grasp large amounts of detailed information.

Table 38–1: Four Common Bird Types

Type	Common Examples	Types of Feet	Types of Beaks
Flightless Birds	Penguins, rheas, ostriches	Adapted for running	(Beaks vary)
Water Birds	Ducks, swans, geese	Webbed	Broad and flat for filtering / Long and pointed for fishing
Perching Birds	Sparrows, robins, other songbirds	Toes cling to branches	Short, thick, strong (seed eaters) / Long and slender for probing (insect eaters)
Birds of Prey	Hawks, eagles, owls	Sharp, curving claws	Tearing beaks

Reference Section

At the end of the book, a 56-page Reference Section pulls together valuable information and reference aids that will help you use the book effectively.

Five-Kingdom Classification of Organisms

Kingdom Monera

Prokaryotes (cells lack a true nucleus and membrane-bound organelles); mostly unicellular; some occur in filaments or clusters.

Phylum Schizophyta: Bacteria; about 2,500 species including eubacteria (true bacteria), rickettsias, mycoplasmas, and spirochetes; mostly heterotrophs; some photosynthetic and chemosynthetic autotrophs; reproduction usually asexual by binary fission.

Phylum Cyanophyta: Blue-green algae or cyanobacteria; about 200 species of photosynthetic autotrophs with chlorophyll *a* and accessory pigments; no chloroplasts; mostly filamentous; some unicellular; reproduction asexual by binary fission or fragmentation.

Phylum Prochlorophyta: Protosynthetic autotrophs; contain chlorophyll *a* and *b*, xanthophylls, and carotenes.

Kingdom Protista

Diverse group of unicellular and simple multicellular eukaryotes (cells have a true nucleus and membrane-bound organelles).

Phylum Euglenophyta: Euglenoids; about 800 species of unicellular, photosynthetic/heterotrophic organisms with chlorophyll *a* and *b*; usually having a single flagellum; reproduction asexual.

Phylum Mastigophora: Flagellates; about 2,500 species; mostly parasitic; includes *Trypanosoma* and *Trichonympha*.

Phylum Sarcodina: Sarcodines; about 11,500 species that move by means of pseudopodia; includes amoebas.

Phylum Ciliophora: Ciliates; about 7,200 species; locomotion by cilia, or sessile; includes paramecia and stentors.

Phylum Sporozoa: Sporozoans; about 6,000 species of nonmotile parasites; includes *Plasmodia*, the cause of malaria.

Phylum Chrysophyta: Golden algae; about 12,000 species of photosynthetic autotrophs with chlorophylls *a* and *c* and carotenes, xanthophylls, and fucoxanthins; most are unicellular and aquatic; includes diatoms.

Phylum Pyrrophyta: Fire algae; about 1,100 photosynthetic spec[...] thophyll; unic[...] phytoplankton

Phylum Chloro[...] tosynthetic sp[...] carotenoids; [...] multicellular [...] land plants.

Phylum Phaeo[...] tosynthetic sp[...] fucoxanthin; [...] (612)

Phylum Rhodo[...] synthetic spec[...] tenes, and ph[...] ticellular seav[...]

Kingdom [...]

Eukaryotic heter[...] tion; includes sap[...] ticellular, compo[...] cell wall of mos[...]

Phylum Myxom[...] species; body [...] that creeps b[...] separates int[...] duction; form[...]

Phylum Eumyco[...] cies; mostly f[...] reproduction [...] stages.

Glossary

Pronunciation Key

Symbol	As In	Phonetic Respelling	Symbol	As In	Phonetic Respelling
a	bat	a (bat)	ô	dog	aw (dawg)
ā	face	ay (fays)	oi	foil	oy (foyl)
â	careful	ai (CAIR fuhl)	ou	mountain	ow (MOWN tuhn)
ä	argue	ah (AHR gyoo)	s	sit	s (siht)
ch	chapel	ch (CHAP uhl)	sh	sheep	sh (sheep)
e					
e					
ér					
i					
ī					
k					
o					
ō					

Reference Material

A detailed classification scheme, timeline, laboratory procedures, and safety guidelines are easy to use.

Index

Boldface numbers refer to an illustration on that page.

Glossary

All BioTerms in the book are listed, defined, and referenced to the text.

Index

The fully cross-referenced index tells you where to look for topics, tables, or illustrations.

The Study of Living Things

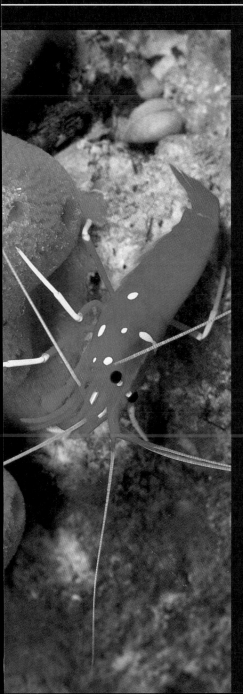

1

Biology: The Study of Life

Wildlife near a waterhole in Africa

Introduction

Many dictionaries define *biology* as the "study of life." The word *life* itself is hard to define, however, because it is so abstract. So you might modify this definition and say that **biology** is the study of living things.

People generally have no trouble telling the difference between things that are alive and things that are not. However, living things vary greatly in size, appearance, and behavior. Consider the great variety of living things on the plains of Africa, for example. Each has its own unique appearance and way of life, but all share certain characteristics that distinguish them from nonliving things. You will begin your study of biology by learning some basic information about these characteristics of living things.

Characteristics of Living Things

The scientific term for a complete, individual living thing is **organism.** Butterflies, trees, bacteria, sponges, and elephants are all organisms. The characteristics shared by these and other organisms define what it means to say that something is alive.

1.1 Organisms Are Made of Cells

One of the most important discoveries biologists have made is that all organisms are made up of the same basic kind of building blocks. These building blocks are called **cells.** Most cells are so small that they are invisible to the unaided eye. You can see some cells, however, if they are magnified. If you gently scrape the moist skin inside your mouth with a toothpick and look at the scraping through a microscope, you can see one type of skin cell. If you look at a drop of your blood under the microscope, you will see red blood cells suspended in the fluid.

Because all organisms are made of cells, cells are called the *structural* units of living things. Cells are not just structural units, however. They are also the *functional* units of organisms. In other words, cells are the smallest units that can carry on the activities of life. All the things that an organism can do are made possible by what its cells can do.

Some organisms, called **unicellular** (yoo nuh SEHL yoo luhr) **organisms,** are single cells. You will study many organisms of this type later in this book. Among those you may already be familiar with are bacteria. Other living things, called **multicellular** (muhl tih SEHL yoo luhr) **organisms,** are made up

Section Objectives

- *List* the major characteristics shared by organisms.
- *Describe* why organisms need energy.
- *State* why it is important for organisms to adapt to their environments.
- *Define* homeostasis and *explain* why it is important to the maintenance of life.

Figure 1–1. The cells of a plant (left), a rat (center), and a human being (right) appear to be quite different. In fact, they have many similarities, and each plays an important role in the structure and function of the organism.

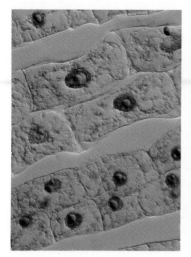
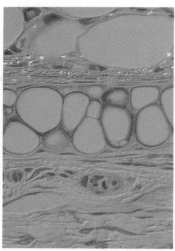

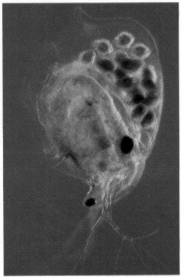

Figure 1–2. A unicellular paramecium (left), a simple, multicellular water flea (center), and a complex animal like a whale (right) are all composed of cells. The more complex an organism, the more diverse are the structures and functions of its cells.

of more than one cell. Some multicellular organisms have only a few cells. Others, such as whales or humans, consist of trillions of cells. Most cells in more complex organisms are highly specialized. That is, each of the many types of cells has its own special functions to perform.

1.2 Organisms Are Highly Organized

Every living cell is a highly complex structural and chemical system. It consists of thousands of substances, many of which have never been made in the laboratory. Organisms make these substances from simpler substances in their surroundings.

Consider a green plant. Starting from a tiny seed, it may eventually grow to be a huge tree. All this living matter, however, was produced from simple, nonliving materials: carbon dioxide from the air, plus water and a few minerals from the soil. In fact, all organisms consist of the same chemical raw materials as nonliving things. The only difference lies in the way in which these raw materials are organized into more complex substances.

Inside a living cell are many complex structures that the cell uses to stay alive. These structures perform many functions. For example, they enable the cell to manufacture substances it needs. Each cell is, in fact, a complete chemical factory. Unlike other factories, however, a cell constructs all its own tools and machinery—even its own building. A cell conducts all its own repairs and even generates its own power. In short, a cell controls and regulates all its own activities.

1.3 Organisms Use Energy

No work of any kind can be done without **energy**, or power. No machine, for example, will run without some source of energy. To get an air conditioner to function, you must plug it in so that it can draw electrical energy. To get a car to run, you must supply gasoline, a source of chemical energy.

Like an automobile engine, you too run on chemical energy. That energy is provided by the food you eat. In the cells of your body, foods are chemically "burned" to release their energy. Cells use the energy to carry out their activities.

What is true of you is true of all organisms. *All living things need energy.* You need energy to walk, to talk—even to think. A bird needs energy to fly, a spider needs energy to spin its web, and a plant needs energy to produce a flower.

Unlike an engine, organisms need energy constantly. An engine that runs out of gas will stop, but it will not be damaged. An organism, however, will die if it goes without food for very long, because it must do a great deal of work just to stay alive. Even while you sleep, your heart beats, you breathe, and your brain sends messages to all parts of your body. In fact, every cell of your body is working and using energy.

Every living thing is constantly building the substances that it needs. Generally, such chemical building requires energy. However, cells are also constantly breaking down other substances, and this process releases energy. The sum of all this

Q/A

Q: *Where does all the energy organisms use come from in the first place?*

A: Ultimately, it comes from the sun. Green plants capture the energy of sunlight in making their own food. Animals get their energy by eating plants or other animals that have eaten plants.

▶ It Takes Energy to Stay Organized

You may be wondering why organisms must use so much energy even when they are not doing anything in particular. Energy is needed simply because organisms are such highly organized systems. A basic law of physics says that no highly organized state can be maintained without using up a lot of energy. You know from experience that it takes energy to keep anything well organized: a committee, a factory, even a notebook. Things don't need any help getting messed up, however. They seem to do that by themselves. In fact, the more complex something is, the more easily it tends to fall apart.

If you do not tune an engine from time to time, it starts to lose its efficiency. If librarians did not constantly expend energy cataloging and shelving books, the library would quickly become so disorganized that it could not be used. The same is true of any living thing. If it did not expend energy to maintain itself, it too would become disorganized. In fact, it would become so disorganized that it would die.

chemical building up and breaking down is **metabolism** (muh TAB uh lihz uhm). You can also look at metabolism as the sum of all the ways in which an organism gets and uses energy.

1.4 Organisms Grow and Develop

A glance at the family photo album reminds you that you were once much smaller than you are now. One of the main ways organisms use food is for growth. The growth of an organism is of a special kind, however. Some nonliving things, such as a fire or a salt crystal, can also grow. This type of growth is just "more of the same." You can grow a salt crystal as large as you want, but except for size it will never change as it grows.

An organism, on the other hand, grows by using materials from its surroundings to make more of itself. A baby does not grow by adding little bits of baby to itself. It grows by drinking milk. Substances in the milk are broken down chemically and made into other substances. Eventually, they become part of new cells of the growing baby.

In addition, the growth of organisms does not occur at a constant rate. Most growth takes place during certain parts of an organism's life. A baby generally doubles its weight during the first five months of life. You wouldn't want to double your weight in the next five months, and fortunately you won't. You are probably now nearly as tall as you are likely to get and also nearly as heavy, unless you overeat. Like you, most animals reach a full adult size and then stay about that size.

Most organisms do not become adults just by growing larger. An adult human being does not look like a giant baby. You can compare the growth of an organism with the growth of a university. When a university has to expand, it does not simply increase the size of all its buildings by 10 percent. The growth is planned and orderly. One year a new science building may be constructed. Later a new gym may be built, and the old one converted to a new cafeteria.

Organisms grow in much the same way. *Growth always takes place in a specific way, which is different for each kind of organism.* Different parts grow at different rates. Sometimes

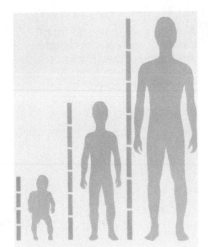

Infant   Six-year-old   Adult

Figure 1–3. Humans grow more rapidly during the first two years of life than at any other time.

Figure 1–4. Three months after beginning life as a single cell, a human fetus is more than 7.5 cm (3 in.) long and weighs nearly 28 g (1 oz.).

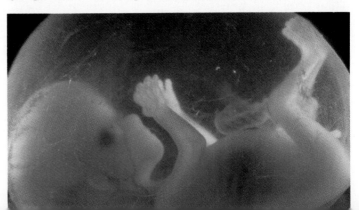

new structures appear, and old ones take on new functions. In short, organisms do not simply grow—they develop. The **development** of a living thing includes all the changes that it undergoes as it matures.

Some developmental changes are dramatic, as when a caterpillar becomes a butterfly or when a tadpole becomes a frog. You may think these cases are unusual, but actually they are not. Consider your own existence before birth. You began as a single cell. Now your body consists of trillions of cells, most of them highly specialized. A liver cell, for example, cannot do the work of a brain cell. The taking on of special functions by different groups of cells is an important part of development.

Figure 1–5. Among the most spectacular of all developmental changes are those that transform this caterpillar into a spicebush swallowtail butterfly.

1.5 Organisms Have a Life Span

Development does not stop when an organism reaches its adult form. Even after growth has stopped, repair, rebuilding, and replacement go on continuously. No organism, however, can keep renewing itself forever. In time the repair processes become less efficient. The organism deteriorates and eventually it dies.

Most organisms have a fixed length of life. The average length of life for an organism is its **life span.** In the late 1700s, the average human life span in most parts of the world was less than 40 years. Today it is about 70 years in the developed nations. The life spans of other organisms vary greatly. Certain

Q/A

Q: *Will further improvement in medicine and nutrition enable people to live 150 or even 200 years?*

A: Probably not. Research indicates that living more than 100 years is becoming more probable, but there seems to be a natural limit to human life.

122
72
60
23
21
15
13.5
11
3

Figure 1–6. The life spans of organisms vary. Some turtles live to be 122 years old, whereas most snakes live only about 12 years. Humans average about 72 years.

insects live only for a single day as adults. At the opposite extreme, one bristlecone pine tree in California is about 4,600 years old. For some other examples, see Figure 1–6.

1.6 Organisms Reproduce Themselves

One of the most vital activities of living things is the production of offspring. The process of producing offspring, or new individuals, is called **reproduction.** Since no organism lives forever, reproduction is necesssary for the continued existence of a **species,** or kind of living thing.

New individuals are always like their parents in *kind* but different in *detail*. A baby may or may not inherit its father's brown hair or its mother's blue eyes. Nevertheless, it is sure to have a human nose, not a trunk like an elephant. It will have two eyes, not eight like some spiders.

1.7 Organisms Respond to Stimuli

stimulus (plural, *stimuli*)

All organisms respond to conditions in their surroundings. Any condition to which an organism can react is called a **stimulus.** What the organism does as a result of the stimulus is a **response.** *The ability to respond to stimuli is typical of all living organisms.* Biologists call this property **irritability.**

Human beings and other higher organisms detect stimuli through senses that include sight, hearing, and smell. What organisms detect affects their behavior. A baby who hears a loud noise may respond by crying. A dog that smells food may sit up, wag its tail, drool, or perhaps run to the place where it expects to be fed. Far simpler organisms also respond to stimuli. Tiny, single-celled creatures can "taste" chemicals in the water in which they live. They will swim toward particles of food and away from harmful substances.

1.8 Organisms Adjust to Their Environments

To survive, an organism must adjust to changes in its environment. The **environment** includes everything in an organism's surroundings that affects it in any way. The environment of a trout, for example, includes the temperature of the water, the amount of dissolved oxygen, and the composition of the bottom (sand, mud, or rock). The environment also includes the organisms that the trout might eat, such as insects or worms, as well as any animals that might eat the trout, such as birds or otters.

Any environment can support only a limited number of each type of organism. Hundreds of birch seedlings may sprout in a meadow, but not all will live long enough to become trees. Some may be destroyed by hungry insects or other animals. Harsh weather may kill others. The remaining seedlings will have to compete with one another for moisture, sunlight, and space to grow. The survivors will be the individuals best adjusted to their environment.

Environmental conditions change, however. Some changes are sudden and dramatic: a flood, a drought, a forest fire, a volcanic eruption. Other, much slower changes can be even more important in the long run. Several times, great sheets of ice have covered large parts of Europe and North America. Mountains have been pushed up by tremendous forces with the earth and then worn down by erosion over millions of years. To survive, organisms must adjust, or *adapt*, to such changes. Any change in an organism that makes it better suited to its environment is called an **adaptation.**

Individual organisms can adapt to many short-term changes in their environments. They can cope with changes in the weather or with the passage of the seasons. As winter comes on, for example, the brown fur of an arctic hare is replaced by white fur. The white coat makes it difficult for enemies to spot the hare in a snow-covered landscape. Any response like this one that increases an individual's chances of surviving is called an *adaptive response*.

Groups of organisms can also adapt to long-term changes. This process occurs over many generations and does not depend on changes by individual organisms during their lifetimes. It depends entirely on the **traits,** or characteristics, received by new individuals from their parents. These traits are **inherited,** or passed from generation to generation.

Figure 1–7. The short-tailed weasel has an adaptive response to winter: Its coat changes color from brown to white.

▶ Homeostasis: Changing to Stay the Same

You can jump from a sauna at 70°C (160°F.) into a cold lake at 15°C (60°F.). Yet your body temperature will change by only a degree or so from its normal 37°C (98.6°F.). You can visit La Paz, Bolivia (altitude 3,630 m, or 11,900 ft.), where the air has 30 percent less oxygen than at sea level. You may feel unusually tired for a few days, but your cells will still get enough oxygen to function normally.

These examples illustrate an important property of living things. Although their surroundings may change greatly, organisms maintain a nearly constant internal environment. It is vital that they do so, because the cells of a multicellular organism are extremely delicate. They need a continuous supply of nutrients and oxygen, and waste products must be removed constantly. Cells cannot tolerate much change in temperature. The concentration of chemicals in the surrounding fluid also cannot change much. If it does, the cells will shrivel up like raisins, or swell and burst.

The environment within a complex organism can be compared to the controlled environment in a greenhouse. Outside the weather changes, and the

Some **variation,** or set of differences, generally exists among individuals in a group of organisms. In a pack of wolves, for example, one animal may have longer legs, another a warmer coat, and another keener hearing. Many such differences are inherited.

If the environment changes, certain traits may take on special importance. They may give the individual that has them an edge in competition with others of its kind. If the climate is growing colder, for example, wolves with exceptionally thick coats will be most likely to survive. Such animals may at first be few in number. In each generation, however, a high proportion of the thin-coated animals will die before they have a chance to reproduce. Many of the thick-coated individuals will survive and pass this trait on to their offspring. The number of thin-coated individuals will be very few. Thus the valuable trait will become more common in each new generation. Eventually it seasons come and go. Inside the greenhouse, the temperature and humidity are always maintained at the ideal level for the plants growing there. The plants have just the right kind of soil, the right amount of water and light, and precise doses of fertilizer. Of course, maintaining such a changeless environment takes work. Similarly, your body works hard to keep its internal environment constant.

Adjusting to the external environment is only part of the problem. An organism must also adjust its life functions to fit its activities at a given moment. If you are running a race, for instance, your muscles are working at top capacity. They need extra fuel and oxygen. They are also producing wastes at a high rate. As a result, your heart and lungs must work harder than usual. Since fuel is being used up so rapidly, much heat is also being released. Your body must get rid of this heat, or your cells will "roast" themselves. One method of temperature control is sweating. The evaporation of moisture from your skin helps cool the body.

To keep its internal environment stable, an organism must maintain a delicate balance between its life functions and its activities and environment. This self-adjusting balance of life functions, environment, and activities is known as **homeostasis** (hoh mee oh STAY sihs).

You may already have recognized an important link between homeostasis and adaptation. Organisms constantly monitor both their internal and external environments. Any change in the external environment that threatens the organism calls forth a response. The organism adapts and so fits better into its changing environment. A change in the internal environment also calls forth a response. Through homeostasis, the organism changes its life functions and so restores the original conditions.

will be the rule rather than the exception. Over a long period, the group of organisms will have adapted to the new, colder conditions.

Reviewing the Section

1. What is meant by the term *biology*?
2. What is the basic structural unit of living things?
3. Why do organisms need energy at all times?
4. What is the difference between an organism's growth and an organism's development?
5. What do biologists mean when they say that an organism's response to a stimulus is adaptive?
6. What is homeostasis, and why is it vital to the health of an organism?

Organization of Living Things

Biology is the study of living things, but biologists do not limit their study to complete, individual organisms. To understand how a car functions, you must study its parts, such as the carburetor, the engine, and the transmission. You may even have to study the parts of the parts. Similarly, to understand organisms, biologists study the workings of organisms, of their parts, and of the parts of their parts.

1.9 Atoms and Molecules

The building blocks of all matter—living and nonliving alike—are called **atoms.** Atoms themselves are composed of still smaller parts. In organisms, however, atoms are never broken down. Instead, they are only rearranged into new combinations.

Ninety-two kinds of atoms are found naturally on the earth. They correspond to the ninety-two natural **elements,** substances that cannot be broken down chemically into simpler substances. Six elements are especially important to life: carbon, hydrogen, oxygen, nitrogen, sulfur, and phosphorus. About twenty others play lesser roles.

How can complex living cells be made up of only about two dozen kinds of atoms? To help answer this question, ask yourself another one: How is it possible that so many different pieces of music—from the classics to the latest pop songs—could have been composed using the same few dozen notes? In each case, a few simple parts can be put together in many different patterns.

Atoms join together and form larger structures called **molecules.** The molecules of many elements consist of a single atom. However, many molecules in living things are made up of millions of atoms.

1.10 Organelles and Cells

Organelles are structures that perform specific functions within living cells. Each organelle is made up of many large molecules. Some organelles serve as the cell's waste disposal units. They store and break down harmful substances produced by the cell. Other organelles serve as the cell's power plants, providing energy for the cell. Still others are chemical factories that make many of the molecules of life.

Organelles can perform some but not all life functions. They can exist only as part of a living cell. Thus the cell is truly

Q/A

Q: *How many kinds of molecules are present in the human body?*

A: No one knows exactly, but 10,000 to 15,000 is probably a good estimate.

the "frontier of life." Cells exhibit all the properties of life discussed earlier in this chapter. Indeed, some cells are complete organisms.

1.11 Tissues, Organs, and Systems

In most multicellular organisms, cells are organized into tissues. A **tissue** is a group of similar cells that perform a common function. Your body contains many kinds of tissues. Nervous tissue consists of cells specialized for carrying messages. Muscle tissue is made up of cells that contract. This specialization of cells makes possible much *division of labor*. Like people with different jobs, the cells of the body depend on one another.

Tissues are organized into organs. An **organ** is a structure composed of a number of tissues that work together to perform a specific task. The eye, for example, is an organ. Muscle tissue controls its movement and focus. Nervous tissue responds to light and sends messages to the brain. Other types of tissue protect and nourish the eyeball. All these tissues are necessary if the eye is to perform its special function of seeing.

In a complex multicellular organism, many tasks are too great for a single organ. They require a **system**—a group of organs that cooperate in a series of related functions. For example, the digestive system carries out the complex job of digesting food. Human beings eat a wide variety of foods. These foods require many kinds of special processing before they can be used by the body. Each organ of the digestive system has its own part to play in the process of digestion. They all must function together to perform the function.

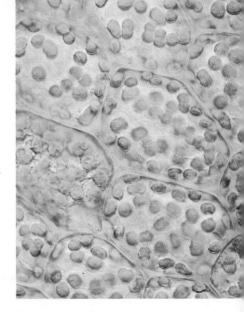

Figure 1–8. The cells that make up a leaf, shown here magnified 400 times, are themselves made up of many parts. Each organelle helps the cell function.

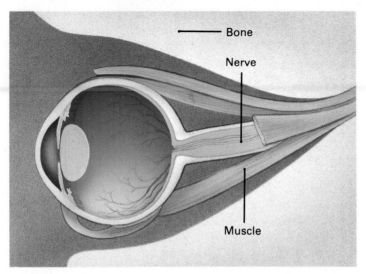

Bone

Nerve

Muscle

Figure 1–9. The eye is an organ of the nervous system. It is made of many kinds of tissues, such as muscle and nerve tissue. Each of these tissues, in turn, consists of various kinds of cells.

1.12 Organisms

Most multicellular organisms have several systems. Keep in mind, though, that many organisms are unicellular. They depend on organelles to perform specialized functions that tissues, organs, and systems perform in multicellular organisms.

Some organisms, called **colonial organisms,** live in groups that resemble multicellular organisms. In reality, however, each cell is a separate organism. Like the residents of an apartment house, they are merely neighbors who get certain advantages from living near one another.

1.13 Groups of Organisms

Biologists are also interested in how living things interact with one another and with their environments. For this reason, biologists study organisms in a series of larger and larger groups.

Populations A **population** is a group of organisms of the same species living in a particular place at a particular time. The place may be large or small, and so may the size of the population. A population may consist of a billion bacterial cells in a tiny puddle of water. In contrast, a few dozen elephants in a huge African game preserve may also be a population.

Biologists often limit the meaning of the term *population* somewhat. They define a population as a group of organisms that **interbreed**—that is, that mate within the population.

Figure 1–10. Similar organisms living in the same area and sharing the same habitat make up a population. Some populations, such as this penguin colony in Antarctica, are composed of hundreds or thousands of individuals.

Highlight on Careers: Exobiologist

General Description

An exobiologist is a scientist who studies the nature and distribution of life in the universe. The main focus of an exobiologist's work is the search for extraterrestrial life, or life elsewhere in the universe. Some scientists now predict that life could be widespread in the universe. The discovery of extraterrestrial life would be one of the most significant events in history.

Since no one yet has any evidence of extraterrestrial life, exobiologists study the way in which life began on the earth. They believe that conditions similar to those that produced life on the earth have occurred elsewhere in the universe and produced other forms of life.

To understand how life began on the earth, exobiologists search for answers to the following questions: As our galaxy came into being, what caused the formation of the six elements needed for life—carbon, hydrogen, oxygen, nitrogen, sulfur, and phosphorus? How were the molecules that make up living things formed from these six elements? What chemical events led to the first self-reproducing systems?

To answer these questions, some exobiologists use existing knowledge to work out theories of how life began. Others conduct complex experiments, analyze the chemicals found in ancient rocks, or use special telescopes to study the chemical composition of comets and asteroids.

Some exobiologists search for intelligent life elsewhere in the universe. They use radio telescopes to listen for radio signals from possible civilizations out in the galaxy.

Career Requirements

Exobiology draws on the knowledge and skills of many fields, including astronomy, biology, geology, and biochemistry. An individual who runs a research project must have a Ph.D. in his or her chosen field. Persons with M.S. or B.A. degrees may serve on research teams.

For Additional Information

Extraterrestrial Research Division, Code LX
NASA Ames Research Center
Moffett Field, CA 94035

Communities Different populations that live in the same area and interact with one another make up a **community.** Like populations, communities can be large or small. The community of organisms in and around a pond, for example, would probably include populations of many kinds of plants, fish, insects, shellfish, and amphibians.

Populations in a community interact in many ways. The most obvious form of interaction is eating or being eaten. Each animal population depends on other plant or animal populations for its food. Populations may also depend on one another for shelter. Many animals live in hollow trees, and birds use twigs and grass to build their nests.

Ecosystems A community of living things and its physical environment make up an **ecosystem.** A biologist might study a forest, a prairie, or a coral reef as an ecosystem. Living things, naturally, are greatly affected by their environments. However, the physical environment is also affected by the organisms that inhabit it. Beavers dam streams, creating lakes and ponds. Lakes gradually fill up with decaying matter from dead plants and animals. Eventually the lakes become meadows. Dead organisms and animal wastes enrich the soil, making it more fertile for new plants.

Over long periods, organisms can produce great environmental changes. Evidence indicates that billions of years ago the earth's atmosphere had no oxygen. Today it is one-fifth oxygen. All the oxygen was released by green plants over millions of years. Only after this oxygen was released could animal life develop.

Biomes Ecosystems can be grouped together into still larger units called biomes. A **biome** is a large geographic area that has the same major forms of life. The nature of a biome is shaped

Figure 1–11. Plants and animals found in deserts (left), mountains (top right), and grasslands (bottom right) vary because the geography and climate differ from region to region.

largely by the geography of the region and its climate, especially the temperature and amount of rainfall. These features determine the types of plant life that can flourish in the region. The kinds of plants, in turn, largely determine what kinds of animals will be found in the area.

Biomes are generally identified by their climate, their vegetation, or both. For example, the most familiar of the major kinds of biomes include tropical rain forest, grassland, and desert.

The Biosphere   All the ecosystems on earth together make up the **biosphere.** *The biosphere includes all the life-supporting environments on this planet, together with all the organisms that inhabit them.*

The biosphere may seem vast to you, but from a spacecraft you would see that the realm of life is actually very limited. A few birds have been observed at altitudes of 8,000 m (26,000 ft.) above sea level. A few organisms inhabit depths of the oceans, about 10,000 m (33,000 ft.) below the surface. Seen from space, however, the entire region between these two extremes is only a tiny fraction of the 6,400–km (4,000–mi.) radius of the earth. Proportionally, it is no thicker than the peel of an apple.

Most life is actually confined to a far narrower region. The vast majority of organisms live in the top 100 m (330 ft.) of the ocean, the bottom 100 m of the atmosphere, on the earth's solid surface, or a few meters deep in the soil. This 200–m "skin" on the surface of the planet is home for about 5 million kinds of organisms. Whether the earth's biosphere is the only home for living things in the universe is one of today's great unanswered questions.

Figure 1–12. The earth's biosphere is home for about 5 million kinds of organisms.

Reviewing the Section

1. Which six elements are the most important ones in living things?
2. What are some of the functions performed by organelles within cells?
3. How does a community differ from a population?
4. In what way does the specialization of cells benefit an organism?
5. What might you conclude about the climates in two regions that have similar plant and animal life?
6. Why do scientists study groups of organisms such as populations and ecosystems?

Section Objectives

- *List* some reasons why people study biology.
- *Explain* how the study of biology helps promote human welfare.
- *Name* some major fields of modern biology and describe the subject matter of each.

The Science of Biology

Now that you have some basic knowledge about living things, you are ready to consider the study of living things—biology. What is the value of biology? Why do people study biology at all?

1.14 Biology and Human Welfare

People study biology for many reasons. Some people choose to study biology chiefly as a way of helping humans to live longer and healthier lives. Biological research has laid the foundations of modern medicine. It has brought about the conquest of many diseases. It is basic to our knowledge of nutrition. It has helped us to raise more and better food. In these and many other ways, biology has made a tremendous contribution to human well-being.

1.15 The Diversity of Life

Some people study biology simply for the pleasure of learning about the world of living things. You may think that by now biologists must know everything there is to know about living things. In fact, millions of species of organisms have not even been named yet, let alone studied. Those that are well known show us that life is far from predictable. Nothing that you can read in works of fantasy or science fiction is as strange as some of the creatures that actually live on this planet. Consider the following examples:

- A plant, a variety of bamboo, flowers only once every 120 years or so. All plants of this species flower at exactly the same time, whether they are growing in Japan, Great Britain, or the United States.
- A certain type of bird may fly for three years, apparently without alighting. These birds cover nearly a million miles—eating, drinking, sleeping, and mating in flight—before they settle down to raise a family.
- A frog that must remain moist at all times lives in the bone-dry deserts of Australia. On the rare occasions when it rains, this frog soaks up water through its skin like a sponge. It then buries its swollen body in the earth and seals itself up in a waterproof, plasticlike material of its own making. It can live for two years or more until the rains come again.

Figure 1–13. This Australian frog encases itself in a waterproof membrane that prevents water-loss during dry weather.

1.16 The Human Machine

Some people study biology because they are curious about themselves. They wish to know more about the human body and the human brain. How do these amazing "machines" work? What keeps the heart beating? How can people remember thousands of names and faces? Why do people sleep? Are there really certain foods you can eat to live longer? Biology is the science that seeks answers to questions such as these.

1.17 The Web of Life

Another reason for studying biology is especially important for everyone today. Biology helps us understand our place in the living world. *All organisms in the biosphere are interrelated and affect one another in many ways.* The phrase "the web of life" expresses this idea well. This web cannot be torn without harming the human species. People too are part of the biosphere and depend for their day-to-day survival on countless other living things.

Living things supply all food. They provide important raw materials, such as wood, cotton, and wool. Plants replenish the life-giving oxygen in the air, and they also help prevent erosion and control floods. Many medicines and drugs were first obtained from living things, and hundreds of others almost certainly remain to be discovered. Human fate is tied up with the fate of all these organisms.

1.18 Biology and the Future

An extremely important reason for studying biology is to help people understand, and perhaps control, the future. In the near future, new medical techniques may wipe out many diseases, while disease caused by polluted environments may increase sharply. Scientists may soon be able to create entirely new organisms in the laboratory, but many thousands of existing species are faced with extinction. New agricultural technologies may make it possible to raise more food, but the number of people to be fed is rising rapidly. People may establish settlements in space or on other planets, but the environment on the earth is seriously threatened by pollution and the wasteful use of natural resources.

As citizens of this world of the future, you will have to understand both its promises and its problems. A knowledge of biology is essential to this understanding.

Figure 1–14. Destruction of natural habitats, like this rainforest in New Guinea, threatens the existence of many kinds of plants and animals. Such damage to the environment could ultimately threaten human existence.

1.19 Some Fields of Modern Biology

Biology is a large, complex, and rapidly growing science. A couple of centuries ago, there were only two main areas of biology: **zoology,** the study of animals; and **botany,** the study of plants. Today new specialties are constantly emerging. Table 1–1 lists some of the major divisions of biology.

Table 1–1: Some Major Fields of Biology

Anatomy	The study of the external and internal structure of organisms
Biochemistry	The study of the chemical makeup and processes of organisms
Botany	The study of plants
Cell biology	The study of the structure and activities of living cells
Ecology	The study of how organisms interact with one another and with their environments
Evolutionary biology	The study of how organisms have changed through time
Genetics	The study of heredity, or how traits are transmitted from generation to generation
Microbiology	The study of organisms too small to be seen without a microscope
Physiology	The study of how organisms carry on their life processes and how various parts of the organisms perform their special functions
Zoology	The study of animals

The various fields of modern biology are closely interrelated. For example, a thorough knowledge of biochemistry is important for anyone who is studying cell biology, genetics, or physiology.

Reviewing the Section

1. What are some reasons why people study biology?
2. How does the study of biology help promote human welfare?
3. Why is it important to understand the place of human beings in the web of life?
4. What were originally the two main areas of biology?
5. In which field of biology would you be most likely to study communities, ecosystems, and biomes? Explain your answer.

Investigation 1: Identifying Life

Purpose
To identify signs of life

Materials
Lens paper, microscope, glass slide, coverslip, medicine dropper, pond water, toothpick, petroleum jelly

Procedure
1. Use lens paper to clean the microscope lens mirror, glass slide, and coverslip. *Why should each microscope investigation begin in this manner?*
2. Place a drop of pond water on the clean glass slide.
3. Using a toothpick, make a ring of petroleum jelly around the drop of pond water. Surround but do not touch the pond water.

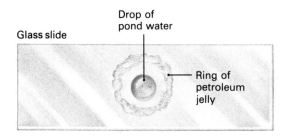

Glass slide

Drop of pond water

Ring of petroleum jelly

4. Carefully place the coverslip on top of the petroleum jelly. *What is the purpose for steps 3 and 4?*
5. Make sure the microscope is level, and place the prepared slide on the microscope stage. Observe the slide under low power. Follow the procedure for using a microscope found on page 830. *What might occur if the microscope were tilted?*
6. Illustrate all the different organisms that you observe under the microscope.
7. List the signs of life that you observe.
8. Compare your illustrations and list with those of other students in the class. *Why is comparison of results an important step in all scientific investigations?*
9. As a class, prepare a composite list of the signs of life recorded.

Analyses and Conclusions
1. What characteristics distinguished the living from the nonliving things on the slide?
2. What life processes were not evident on the slide? Explain why.
3. If you were to observe your slide on the following day, what changes might have occurred? Explain.

Going Further
- Place a live earthworm in a culture dish. Observe and explain how the earthworm reacts to each of the following: light, a weak acid, a weak base, and temperature changes.
- Plant several bean seeds and observe them once a day for about 10 days. What life processes do you observe?

Chapter 1 Review

Summary

The millions of kinds of organisms on the earth share certain characteristics. All are composed of complex structural and chemical units called cells. Organisms begin life as a single cell, develop into adults, and eventually deteriorate and die. They all use energy to perform important functions, including reproducing, responding to stimuli, and maintaining a constant internal environment. Organisms that survive are those well suited to their environments.

To understand organisms, biologists study parts of organisms, including organelles, cells, tissues, organs, and systems. They also study organisms in larger and larger groups to learn how organisms relate to one another.

Dozens of fields of biology have emerged from botany and zoology, the original areas of biological study. Today specialists discover new species, improve human life, and may also hold the key to solving future problems.

BioTerms

adaptation (9)	energy (5)	organism (3)
atom (12)	environment (9)	population (14)
biology (2)	homeostasis (11)	reproduction (8)
biome (16)	inherited (9)	response (8)
biosphere (17)	interbreed (14)	species (8)
botany (20)	irritability (8)	stimulus (8)
cell (3)	life span (7)	system (13)
colonial organism (14)	metabolism (6)	tissue (13)
community (15)	molecule (12)	trait (9)
development (7)	multicellular organism (3)	unicellular organism (3)
ecosystem (16)	organ (13)	variation (10)
element (12)	organelle (12)	zoology (20)

BioQuiz   (Write all answers on a separate sheet of paper.)

I. Completion

1. The sum of all the ways an organism gets and uses energy is its _____ .
2. A _____ is a large geographic area that has the same major forms of life.
3. Offspring are always like their parents in kind but different in _____ .
4. Tissue specialized for contraction is _____ .
5. The study of plants is called _____ .
6. Ninety-two kinds of _____ are found in nature.

II. Modified True and False

Mark each statement TRUE or FALSE. If false, change the underlined term to make the statement true.

7. A <u>tissue</u> is a group of similar cells that perform a common function.
8. The <u>biosphere</u> includes all life-supporting environments on the earth and the organisms that live in them.
9. An <u>organ</u> is a structure within a cell that performs a specific function.

III. Multiple Choice

10. The structural and functional units of living things are a) organelles. b) atoms. c) cells. d) tissues.
11. A community is made up of several a) ecosystems. b) biomes. c) environments. d) populations.
12. The study of the structure of organisms is called a) anatomy. b) physiology. c) ecology. d) biochemistry.
13. Organs are found only in organisms that are a) multicellular. b) unicellular. c) colonial. d) microscopic.
14. The ability to respond to stimuli is a) homeostasis. b) reproduction. c) biogenesis. d) irritability.
15. Organisms of the same species living in the same place make up a a) community. b) population. c) biome. d) ecosystem.

IV. Essay

16. How does your growth differ from that of a salt crystal?
17. What is meant by "the web of life"?
18. What do biologists mean when they say that an organism's response is adaptive?
19. Why must an organism maintain a constant internal environment?
20. How do colonial organisms differ from unicellular and multicellular organisms?

Applying and Extending Concepts

1. When an octopus encounters a threatening organism, it squirts out a cloud of black fluid. In what way is the octopus's response to danger an adaptive response?
2. During the 1970s the *Viking* landers examined soil samples on Mars for signs of life. Find out about these experiments. Then write a brief description of one of them. Be certain to explain what characteristic of living things the experiment was designed to detect.
3. Read about recent developments in one of the fields of modern biology listed in Table 1–1. Explain how one of these developments could help promote human welfare or help people to better understand the significance of their place in the web of life.
4. Exercising on a hot day may make you sweaty and thirsty. Explain both the sweating and the thirst in terms of the mechanism of homeostasis.

Related Readings

Attenborough, D. *Life on Earth: A Natural History*. Boston: Little, Brown, 1981. This beautifully illustrated book surveys the earth's life forms and their amazing ability to adapt.

Campbell, P. N., ed. *Biology in Profile: An Introduction to the Many Branches of Biology*. Elmsford, N. Y.: Pergamon, 1980. Each essay in this collection covers one field of biology, reviewing its past contributions and present concerns.

Capra, F. "The Dance of Life." *Science Digest* 90 (April 1982): 30–31. A research physicist presents his observations on the systems of life, from the cell through the biosphere.

Weitz, C. A. "Weathering Heights." *Natural History* 90 (November 1981): 72–82. This article describes the adaptations that enable people to live in the Andes and Himalaya Mountains.

2

Science and Problem Solving

A scientist collecting data in a marsh

Introduction

What causes cancer? Why do children resemble their parents? What happened to the dinosaurs? For centuries people have asked questions such as these about themselves and the world around them. And over those hundreds of years, people have found answers to such questions by developing a special method of inquiry.

Today the word **science** is used to describe both the body of knowledge that exists about the world and the method of study used to arrive at that knowledge. The individual pictured above is doing scientific work. Collecting information is an important part of the scientific process. This information, and the generalizations that may follow from it, will become part of the body of knowledge called science.

The Scientific Method

Section Objectives

The goal of science is to establish principles and thereby to acquire knowledge about the natural world. Scientists establish principles through a logical, organized method of study called the **scientific method.** Many different procedures are part of the scientific method, but all of them draw on the following series of logical steps.

- *Name* the five steps in the scientific method.
- *Explain* the function of a hypothesis.
- *Define* the elements of a controlled experiment.
- *Distinguish* between pure and applied science.

2.1 Defining the Problem

The first step in the scientific method is to identify the problem. For example, a scientist might be interested in acid rain. This pollutant forms when chemicals released by cars and factories mix with moisture in the air and fall as rain. Acidity of rain is represented on a scale of 1 to 7, where 1 represents high acidity and 7 represents no acidity.

A scientist might be curious about the effect of acid rain on wildlife. However, trying to investigate all the animals that make up "wildlife" would be impossible. Instead, the scientist would focus on a smaller group of organisms. For example, the researcher might pose the question, "Does acid rain affect the development of salamanders?"

2.2 Collecting Background Information

After stating the problem as a clear question, the scientist collects information about the problem. The scientist studying the effects of acid rain would need to understand normal salamander development and the characteristics of areas affected by acid rain. He or she would also want to know whether anyone had studied this question or a related one.

The scientist could find information in books and scientific journals. He or she could also use a computer to search the scientific information published each year. In this way, the scientist could avoid duplicating the work of others and benefit from recent discoveries.

2.3 Formulating a Hypothesis

Next the scientist offers a **hypothesis** (hy PAHTH uh sihs), which is a proposed answer to the question. The hypothesis can be based on information available to the scientist. It can also be an educated guess. *The hypothesis is a statement that can be tested.* The scientist studying salamanders might state the

Figure 2–1. To understand the biology of a salamander, a scientist first collects information on its environment. The scientist studies the geographical area where the animal lives and its habitat within that area.

following hypothesis: "Salamanders that develop under acid rain conditions show a greater number of developmental abnormalities than salamanders that develop in unpolluted waters."

2.4 Testing the Hypothesis

Next the scientist tests the hypothesis to see whether it is supported by evidence. The test of a hypothesis is an **experiment.** The purpose of an experiment is to test only the condition that varies in the hypothesis. This condition is called the **variable.** In the case of the salamanders, the variable is the acidity of the water where salamanders develop. The effect of this variable is tested in a **controlled experiment,** or an experiment in which all the conditions are alike except for the condition being tested.

To carry out a controlled experiment, a scientist uses two identical groups of subjects. The group that is exposed to the variable is called the **experimental group.** The other group is the **control group.** In the sample experiment then, the experimental group of salamanders would be raised in water with the same acidity as a polluted lake or pond. The control group would be raised in water with normal acidity.

To be valid, this or any other experiment must be designed to show that only the variable being tested produces changes in the subject. The experiment must also be designed in such a way that other scientists can repeat it.

2.5 Making and Recording Observations

In order for an experiment to be reproduced by other researchers, a scientist must keep careful records. These records must state how the experiment was planned, how it was carried out, what equipment was used, and how long it took. In addition, the scientist must record all the observations made during the experiment. Such information may include drawings, tables, graphs, diagrams, written observations, photographs—even sound recordings.

2.6 Drawing Conclusions

The answer to a scientific question is formulated by drawing a conclusion based on **data,** which are scientific facts collected during the experiment. Often, scientists form their conclusions with the help of **statistics** (stuh TIHS tihks), a mathematical method of evaluating numerical data. Statistical tests help determine whether important differences exist between data obtained

Figure 2–2. This scientist is using a computer to analyze the data he has collected through his experiments.

Today the work of scientists can be divided into at least two major categories. **Pure science** involves the search for new knowledge. **Applied science** puts the findings of pure science to practical use.

Science, whether pure or applied, includes many fields of study. Yet, scientists in all fields use the scientific method to solve problems. The scientific method, however, is a creative thought process—not a checklist of steps. Each scientist develops a way to satisfy the unique requirements of his or her area of study.

Although the aim of all scientists is to discover knowledge, no theory or principle can be considered final. New information may cause theories or principles to be revised. For instance, people once thought that atoms were the smallest bits of matter in living and nonliving things. Later, research revealed smaller particles called protons, electrons, and neutrons. Protons were considered the smallest units until still more research showed they can be divided into even tinier particles. Thus science finds answers to the questions at hand, but may also raise new, exciting questions.

from the experimental and control groups. This evidence either supports or does not support the hypothesis. In the salamander study, the scientist found that of 1,000 eggs raised in normal water, 6 were abnormal. Of 1,000 eggs raised in acidic water, 440 were abnormal. From this, the scientist could conclude that more salamanders develop abnormally in acidic water.

Before presenting a conclusion to the scientific community, a scientist retests the hypothesis several times. Later other scientists repeat the experiment until the hypothesis and the conclusion are supported or rejected. When a hypothesis explains how an event occurs, it becomes a **scientific principle** or **law.** When a hypothesis explains why events occur, it becomes a **theory.** At all times, however, theories and principles are subject to revision or replacement by a new theory that provides a better or more complete explanation.

Reviewing the Section

1. What are five main steps in the scientific method?
2. What is a controlled experiment?
3. How does a hypothesis differ from a theory?
4. How do pure science and applied science differ?

- *Compare* the theory of biogenesis and that of abiogenesis.
- *List* Redi's hypotheses and explain how he tested them.
- *Compare* Needham's and Spallanzani's experiments.
- *List* the steps in Pasteur's experiment.

Using the Scientific Method

Through the ages people have developed theories that explained a great many natural occurrences. The following investigation shows how one long-accepted scientific theory was discredited by scientific experimentation and how the theory that replaced it is being modified.

2.7 The Question of Spontaneous Generation

Today the theory of **biogenesis** (by oh JEHN uh sihs) is part of the definition of living things. *The theory of biogenesis states that all living things arise from other living things.*

The theory of biogenesis may seem obvious to you, but it was a very controversial issue until only about 100 years ago. As recently as the late 1800s, many people believed that some organisms form from nonliving materials. This concept is referred to as **spontaneous generation,** or **abiogenesis** (ay by oh JEHN uh sihs).

The idea of abiogenesis can be traced to the Greek philosopher Aristotle, whose major work was done during the 300s B.C. Aristotle stated that some fish were produced by mud at the bottom of rivers and oceans. Obviously, he did not see the release and eventual development of fish eggs. Over the centuries people explained other events through abiogenesis. In the 1600s, for instance, the Belgian physician Jean van Helmont stated that mice arose from a dirty shirt and a few grains of wheat placed in a dark corner. According to another belief common in van Helmont's time, decaying meat produced tiny white wormlike creatures called maggots. This belief persisted because no one saw flies laying eggs on decaying food or the eggs hatching into maggots.

Figure 2–3. This engraving, done in 1552, shows "barnacle geese" being hatched from trees. Scientists quickly dispelled this and other myths, but struggled for centuries with the question of spontaneous generation.

2.8 Testing the Theory of Abiogenesis

The theory of abiogenesis states that some organisms arise from nonliving materials. It took nearly 200 years of experimentation to replace abiogenesis with a more accurate theory.

The Debate Begins In 1668 an Italian physician named Francesco Redi challenged the belief that decaying meat will eventually turn into flies. He began with the following hypothesis: "Flies come from eggs laid by other flies on decaying flesh. The decaying flesh serves as a source of food for the developing flies."

To test this hypothesis, Redi filled two sets of four jars with chunks of veal, snake, fish, or eel. He sealed one set of jars and left the other set open to the air. ***The variable in Redi's experiment was whether or not the jars were sealed.*** The sealed set of jars composed the experimental group. The control group consisted of jars left open to the air.

During the next few days, Redi observed flies entering and leaving the open jars. Several days later, the open jars contained rotting or decaying flesh and maggots. The covered jars also contained decaying flesh but no maggots. Based on these observations, Redi concluded that maggots do not arise through spontaneous generation but that they come from eggs laid by flies on rotting meat.

Other scientists of Redi's day did not agree. They believed that Redi had prevented spontaneous generation by keeping air out of the jars. They felt there was an "active principle" present in air needed for abiogenesis.

Redi answered his critics with a second experiment that began with the following hypothesis: "Flies arise from eggs, not rotten flesh, and the presence or absence of air is not a deciding factor." In this second experiment Redi did not seal the experimental group of jars. Instead, he covered them with fine mesh. The mesh allowed air to enter the jars but kept the flies out.

As before, Redi observed that flies entered the open jars and that maggots developed in the rotting flesh. He also observed that flies landed on the mesh covering the experimental jars but

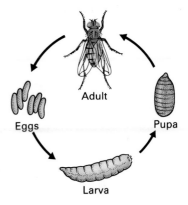

Figure 2–4. Redi showed that eggs and maggots, natural parts of a fly's life cycle (above), were not spontaneously generated in jars of meat. By covering jars with a screen (below), he excluded flies and proved that maggots did not arise from a mixture of air and meat.

Redi's first experiment

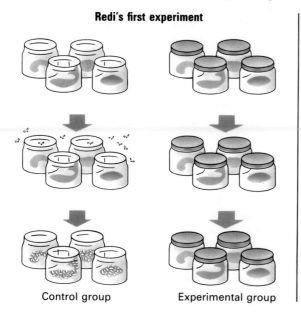

Control group   Experimental group

Redi's second experiment

Control group   Experimental group

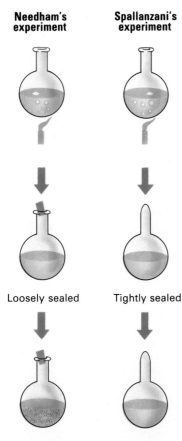

Needham's experiment **Spallanzani's experiment**

Loosely sealed   Tightly sealed

Figure 2–5. Needham's experiment (above left) seemed to show that microorganisms could spontaneously generate after the broth in a corked flask had been boiled. Spallanzani's experiments (above right) showed that Needham had not sealed his flasks correctly. Spallanzani sealed his flasks completely, and no organisms grew in the broth.

that no maggots appeared on the meat inside. Again, Redi concluded that maggots did not arise by spontaneous generation but hatched from eggs laid by flies.

Most scientists of the time could not deny Redi's experimental data, yet some clung to the principle of abiogenesis. These scientists pointed out that what was true for flies was not necessarily true for all organisms.

Needham and Spallanzani By the 1700s many scientists were using the microscope to study bacteria and other tiny organisms. In the mid-1700s, an English scientist named John Needham used microscopic observations to support the theory of abiogenesis.

To test the theory, Needham boiled meat broth for several minutes in loosely sealed flasks. Immediately after boiling, Needham examined the broth under a microscope and saw no living things. Then Needham used cork stoppers to loosely reseal the flasks and allowed them to cool.

After a few days, Needham reexamined the broth and found it was teeming with microorganisms. He concluded that the microorganisms had spontaneously generated from the nonliving materials of the broth.

About 25 years later, an Italian priest and biologist named Lazzaro Spallanzani challenged Needham's work. Spallanzani felt that Needham's experiment was flawed in two ways. First, Spallanzani believed that Needham had not boiled the broth long enough to kill all the life it contained. Thus new organisms could have been produced by organisms that survived the boiling. Secondly, Spallanzani thought that fresh microorganisms could have entered the flasks through the loose seals and reproduced once inside.

Spallanzani designed an experiment to disprove abiogenesis for microorganisms. He boiled seeds in water for one hour to produce a broth. Then he sealed the flasks by melting their glass necks closed. Finally, he placed the flasks in boiling water for several hours and then left them to rest.

Several days later Spallanzani broke the necks of the flasks and examined the broth under a microscope. He found no signs of life. Spallanzani concluded that microorganisms do not arise spontaneously. The supporters of abiogenesis were quick to disagree. They argued that the long period of boiling had destroyed the ''active principle'' in the broth.

In response to the critics, Spallanzani tested a new hypothesis: ''If boiling destroys some active principle, longer boiling will destroy more active principle.'' In his experiment, Spallanzani filled flasks with broth and covered them with loose seals. Each flask was boiled for a different period of time, from 30 minutes to two hours. Then Spallanzani allowed the sealed flasks to sit undisturbed.

After eight days Spallanzani examined the broth in the flasks with a microscope and found living organisms in each one. He also discovered that the broth boiled the longest had the most organisms. Spallanzani concluded that boiling did not destroy the "active principle" in broth. He also incorrectly concluded that longer boiling actually made the broth more supportive for microorganisms.

Pasteur Settles the Question

The abiogenesis debate continued until 1864 when Louis Pasteur, a French chemist, began his investigations. As a result of earlier studies, Pasteur hypothesized that microorganisms are carried on dust particles in the air.

Pasteur developed a two-part experiment to test this hypothesis. First, he sealed flasks filled with broth and boiled them long enough to kill all the microorganisms present. Then he took the flasks to places with varying amounts of dust, such as mountain meadows and country roads. At each place Pasteur exposed a different flask to the air.

Several days later Pasteur examined each flask and found microorganisms in each one. The flasks that were exposed to dustier areas contained the most microorganisms. Pasteur concluded that microorganisms were carried in the air in differing amounts depending on the area.

Next Pasteur retested his hypothesis in the laboratory. First he filled a series of flasks with broth, melted the neck of each one, and bent it in an S-shaped curve. In this way, air could move in and out of the flasks, but dust was caught in the curves. Then Pasteur boiled the flasks, forcing the air out and killing microorganisms in the broth.

As the broth cooled, air flowed into the flasks but the curves in the neck prevented dust from reaching the broth. After several days, Pasteur saw no microorganisms in the flasks, although the broth in each one was exposed to the air. Then Pasteur tipped a flask, allowing some broth to come into contact with dust trapped in the S-shaped curve. In a few days, Pasteur saw microorganisms in the broth.

Pasteur's experiment showed that although air was allowed to enter the flasks, it produced no life. At the same time, Pasteur proved that boiling broth did not affect its ability to support life. *Instead, Pasteur showed that the microorganisms in the flasks came from microorganisms carried on dust particles, not from the air itself.*

Pasteur's experiment

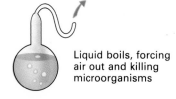

Liquid boils, forcing air out and killing microorganisms

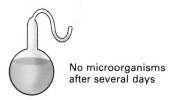

Liquid cools, drawing air and dust in and trapping dust

No microorganisms after several days

One flask tipped, mixing dust with liquid

Liquid contaminated

Figure 2–6. To silence critics who claimed that air was necessary for spontaneous generation, Pasteur heated broth in an open, swan-necked flask. No microorganisms grew in the broth. By tilting the flask, Pasteur then proved that microorganisms in the neck of the flask, and in the air, could contaminate the broth.

Spotlight on Biologists: Lewis Thomas

Born: New York, 1915
Degree: M.D., Harvard University

Lewis Thomas is one of America's foremost medical researchers in immunology, microbiology, and pathology. However, he is perhaps best known to the public as the author of *Lives of a Cell, Medusa and the Snail,* and an array of other fascinating books and essays.

Thomas's books have wide appeal because they translate complex scientific subjects, such as DNA, into language that nonscientists can understand. Thomas instills in readers some of his own curiosity and sense of wonder about life.

Thomas draws on his vast experience in medical science to write his thought-provoking essays. In the last 50 years, Thomas has served as a researcher, as a doctor, and as a professor at several prestigious American universities. He now serves as president of Memorial Sloan-Kettering Cancer Center in New York.

Thomas's interest in medical science began at an early age, nurtured by his mother, a nurse, and by his father, a physician.

Thomas has always specialized in medical research, studying everything from bacteria to gene compatibility. As medical science has advanced, Thomas has turned to analyzing how new technologies affect the doctor-patient relationship. These observations keep Thomas in the forefront of medical science.

The work of Redi, Spallanzani, Pasteur, and others provided enough evidence to convince scientists that organisms do not arise from nonliving things. By discrediting the theory of abiogenesis, they contributed to the development of the theory of biogenesis.

Today, however, the principle of biogenesis may have to be modified. When considering the origin of life on Earth, some scientists have hypothesized that the first cells arose from nonliving materials. In Chapter 15 you will read more about this hypothesis as well as the experiments that have been performed to test it.

Reviewing the Section

1. How do the theories of abiogenesis and biogenesis differ?
2. How did Spallanzani improve Needham's experiment?
3. How did Pasteur disprove abiogenesis?

Investigation 2: Designing an Experiment

Purpose
To design and carry out an experiment that uses the scientific method

Materials
Lima bean seeds, glass beakers, humus or potting soil, sand, gravel, powdered clay, water

Procedure
1. Using the materials listed, design an experiment to determine the best soil type for the germination and growth of lima bean seeds. *What factors do you need to control in the experiment?*
2. Carry out the experiment. Make a table similar to the following one, extending it to 14 days. Record your observations each day for the next two weeks.

Observations on Germination and Growth of Seeds

Soil type	Day 1	Day 2	Day 3	Day 4	Day 5
Humus					
Sand					
Gravel					
Clay					

Analyses and Conclusions
1. Which soil type was best for germination? Which was best for growth? How could you determine why one soil type was better than another for germination or growth?
2. How do your results compare with those of your classmates? Account for any differences.
3. Explain why water is both a necessity and a possible problem in seed germination. Design an experiment to test your explanation.

Going Further
• Repeat the soil experiment using a different type of seed, such as radish seed. How do your results compare with those in the original experiment?

• Prepare six containers with the same type of soil and at least three seeds in each. Place three containers in a closet and the other three on a window sill in a brightly lighted area. Make daily observations for two weeks. Then reverse the containers and observe them for two more weeks. State a conclusion based on your observations.
• Using the following diagram, design and carry out an experiment testing the growth of plants exposed to different colored lights. How do different colors of light affect the growth of the plants?

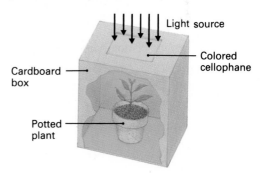

Light source

Colored cellophane

Cardboard box

Potted plant

Chapter 2 Review

Summary

Science can be characterized in two ways. First, it is a body of knowledge about the world around us. Second, it is a logical, organized process of inquiry. This process, called the scientific method, includes the following steps: defining the problem, collecting background information, formulating a hypothesis, conducting a controlled experiment, observing and recording data, and formulating a conclusion. This method of study sets science apart from all other areas of study.

The abiogenesis-biogenesis debate illustrates the scientific method at work. Prior to the 1800s, many people believed in abiogenesis, or the ability of nonliving things to pro-duce certain forms of life. In the seventeenth century, Francesco Redi's experiments challenged abiogenesis by showing that maggots arose not from decaying meat but from eggs laid by flies. In the eighteenth century, John Needham designed an experiment to support abiogenesis, but Lazzaro Spallanzani's work uncovered flaws in Needham's experiment. The theory of abiogenesis was finally disproved after Louis Pasteur proved in 1864 that microorganisms carried by dust produced new microorganisms in flasks of broth. Today the principle of biogenesis is being challenged as scientists study the origin of life on Earth.

BioTerms

abiogenesis (**28**)
applied science (**27**)
biogenesis (**28**)
control group (**26**)
controlled experiment (**26**)
data (**26**)

experiment (**26**)
experimental group (**26**)
hypothesis (**25**)
pure science (**27**)
science (**24**)
scientific method (**25**)

scientific principle (**27**)
spontaneous generation (**28**)
statistics (**26**)
theory (**27**)
variable (**26**)

BioQuiz   (Write all answers on a separate sheet of paper.)

I. Completion

1. The _____ is an organized, logical method of inquiry.
2. The test of a hypothesis is a scientific _____.
3. _____ science puts scientific knowledge to practical use.
4. The theory of _____ states that life arises from living things.
5. _____ are scientific facts collected during an experiment.
6. A _____ is a scientific explanation of known facts.

II. Modified True and False

Mark each statement TRUE or FALSE. If false, change the underlined term to make the statement true.

7. A <u>hypothesis</u> is based on the data that is collected in a controlled experiment.
8. The <u>control group</u> is used to test the variable in an experiment.
9. <u>Statistics</u> is a mathematical method of evaluating numerical data.
10. Maggots in decaying meat are produced by an <u>active principle</u>.

III. Multiple Choice

11. A hypothesis is a) a proposed answer to a scientific question. b) the factor tested in an experiment. c) a scientific explanation of known facts. d) an established scientific truth.
12. Every controlled experiment must have a) two variables. b) two control groups. c) two groups of subjects. d) two hypotheses.
13. A scientific law is a) a theory that has not been proven false. b) a statement of fact. c) an untested hypothesis. d) a tested hypothesis.
14. Redi showed that the presence or absence of air did not affect the presence of maggots on meat by using a) open flasks. b) sealed flasks. c) open jars. d) mesh covered jars.
15. Experiments conducted by Francesco Redi and Lazzaro Spallanzani supported a) biogenesis. b) active principles. c) abiogenesis. d) superstition.

IV. Essay

16. How is an experiment used to test a hypothesis?
17. Why must scientists keep complete records of their experiments?
18. What is the relationship between a hypothesis and a theory?
19. How was Francesco Redi's second experiment different from his first?
20. In what two ways did Louis Pasteur show that microorganisms were carried on dust particles?

Applying and Extending Concepts

1. Many important scientific discoveries begin with ideas, but some begin with lucky accidents. Research the work of Sir Alexander Fleming. What accident led to his discovery of penicillin?
2. Design an experiment to test one of the following hypotheses:
   a. Grass seed A germinates faster at low temperatures than at high temperatures.
   b. Lettuce seed B germinates faster in darkness than in light.
3. Imagine that you are studying the causes of heart disease. Narrow this topic into two questions you can investigate with the scientific method.
4. A biologist feels that trees treated with fertilizer 1 produce more apples than those treated with fertilizer 2. To test this hypothesis, the biologist spreads fertilizer 1 around a red delicious apple tree and an equal amount of fertilizer 2 around a Jonathan apple tree. At the end of the season the biologist counts the number of apples in each tree. The tree fertilized by fertilizer 2 produced 86 apples, while the one treated with fertilizer 1 produced only 45. Evaluate the experiment used to test this hypothesis. What conclusion should be drawn from this experiment?

Related Readings

Milton, J. *Controversy: Science in Conflict*. New York: Messner, 1980. This book highlights four conflicts in biology and shows how scientists using the scientific method can arrive at different conclusions.

Moravcsik, M. J. *How to Grow Science*. New York: Universe Books, 1980. Written by a research physicist, this wonderfully readable book explains what science is and how science is done.

Tools and Techniques of a Biologist

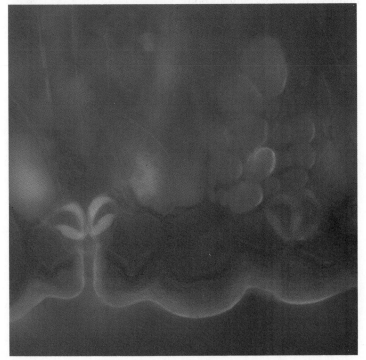

Cross section of a *Welwitschia* leaf, with fluorescent staining, ×120

Introduction

The scientific method is based on the observation of objects and events. Early biologists were limited to studying organisms large enough to be easily visible. Most cells, however, are far too small to be seen by the unaided eye.

Powerful microscopes and other highly specialized tools and techniques now enable scientists to study otherwise invisible cells and their structures, like the food-producing parts of the leaf cell shown above. Various tools help scientists determine the function of the structures they have discovered. Biologists are also taking advantage of advances in computer technology. Computers process the data scientists collect, including the information provided by microscopes and other sophisticated tools.

The Microscope

Without the help of a magnifying glass, your eyes see only a limited amount of detail. For example, two dots less than 0.1 mm (0.004 in.) apart blur into a single fuzzy dot. When you consider that many of your body cells are one-fourth the size of the smallest dot you can see, the importance of the microscope becomes clear. *By allowing scientists to see what the unaided eye cannot see, the microscope greatly increases the amount of data available for scientific inquiry.*

3.1 The Light Microscope

The microscope most often used in biological research today is the *light microscope*. This microscope uses light to form an enlarged image of the *specimen,* or object being viewed.

The magnifying glass, called a **simple microscope,** is the most basic light microscope. It is a single *lens,* or curved piece of glass. The lens bends light rays as they pass through it, causing the specimen to appear between 2 and 20 times its actual size. This apparent increase in the object's size is called **magnification.** An increase in visible detail is called **resolution.** The reason for magnifying an object is to allow a viewer to distinguish more details.

The most commonly used light microscope is the **compound light microscope,** which contains two kinds of lenses. The **ocular** lens set is positioned near the viewer's eye. It forms part of the **eyepiece** of the microscope. The **objective** lens set is positioned near the specimen. As Figure 3–1 shows, light travels through the specimen and the lenses and into the eye of the viewer.

The lenses of the compound light microscope determine its degree of magnification. Each lens is marked with a number and the symbol ×, which stands for *times.* Thus a lens marked 20× magnifies an object 20 times. To calculate the total magnification, multiply the power of the objective by the power of the ocular. A 43× objective and a 10× ocular can therefore magnify the image of a specimen 430 times. Compound light microscopes are used to view living organisms as well as preserved cells mounted on glass slides.

Biologists also use a variety of other light microscopes. The **stereomicroscope,** used to study large specimens, has an ocular lens and an objective lens for each eye. This arrangement of lenses provides a three-dimensional view of the specimen's surface magnified 5 to 60 times. The **phase contrast microscope**

- *Distinguish* between a simple and a compound light microscope.
- *State* the major events in the development of the microscope.
- *Compare* transmission and scanning electron microscopes.
- *Describe* the preparation of specimens for viewing under light microscopes and electron microscopes.

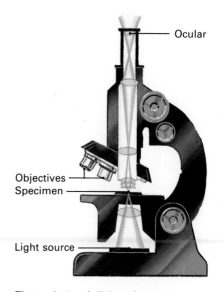

Figure 3–1. A light microscope uses mirrors to focus light and create a magnified image of a specimen. Objects to be viewed under a light microscope usually require special preparation.

clarifies features inside living cells. As the beam of light passes through the specimen, the edges of cell structures bend the waves of light. The bent light waves cross, or interfere with, the unbent light waves. The special lenses of the phase contrast microscope use this interference to reveal boundaries between cell parts, which appear brighter to the viewer.

Light microscopes are important tools, but they have one major drawback. These microscopes can magnify an object by any desired amount, but eventually the details of the object become fuzzy. Why does this occur? The resolving power, or the ability of a microscope to provide clear details, depends on the objective lens. The best objective lens can distinguish objects as close together as 0.2 micrometers (μm). A **micrometer** is a unit of measure that equals one-millionth of a meter (0.000039 in.). By combining this objective with the right ocu-

▶ A Look at Early Microscopes

When you think of microscopes, probably the last thing that comes to mind is the Middle Ages. Yet historians believe the first microscopes came into use during the mid-1400s. These microscopes, the familiar magnifying glasses now called simple microscopes, were used by scientists to study insects.

About 140 years after the first simple microscopes were introduced, several inventors in Europe discovered that one magnifying lens could be used to enlarge the image produced by another lens. Two of these inventors— Dutch eyeglass makers named Hans and Zacharias Janssen—are credited with using this knowledge to develop the first compound light microscope in 1590. Their compound microscope was very basic, with a lens at each end of a tube.

Although early compound microscopes provided greater magnification than simple microscopes, they produced distorted images. For this reason, most scientists preferred high quality single lenses. The finest such lenses were produced by a Dutch merchant named Anton van Leeuwenhoek in the 1670s and 1680s. During that time, Leeuwenhoek ground more than 400 lenses with magnification powers of 50 to 300 times. Leeuwenhoek designed each of his

lenses for a specific purpose. For example, Leeuwenhoek used the lens shown in the picture to view pond water where he saw bacteria and other tiny organisms he called "cavorting beasties." Leeuwenhoek was the first person to produce drawings of microscopic organisms. His work eventually led to *microbiology,* the study of microscopic life.

lar lens, the viewer can get useful magnification up to 2,000 times. Resolution at higher magnification is not possible with any type of light microscope.

3.2 The Electron Microscope

Physicists who wanted to solve the problem of achieving higher resolution knew they could do so only by using an energy beam with a shorter wavelength than that of light. Certain atomic particles called *electrons* have such a wavelength. The **electron microscope** creates enlarged images with a beam of electrons instead of a beam of light. Scientists introduced the first commercial electron microscope in 1935.

Modern electron microscopes produce both high magnification and high resolution, but they too have a major limitation. Living things cannot be viewed under an electron microscope because they cannot survive the techniques used to prepare them for viewing. Even if they could, they would die in the airless interior of the microscope. The air is removed because electrons cannot travel very far in air.

Today scientists use two kinds of electron microscopes. The **transmission electron microscope (TEM)** sends a beam of electrons through the specimen. The beam creates a clear, detailed image on a televisionlike screen magnified 200,000 times or more.

The **scanning electron microscope (SEM)** sends a beam of electrons across the specimen from left to right, a process called *scanning*. As the beam moves, electrons bounce off the specimen in different directions. These electrons produce a three-

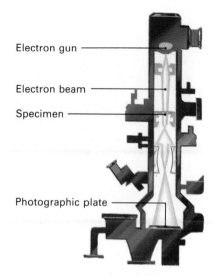

Electron gun

Electron beam

Specimen

Photographic plate

Figure 3–2. A transmission electron microscope uses magnets to focus an electron beam.

Figure 3–3. A diatom looks different under a light microscope (left, ×200), a transmission electron microscope (center, ×4,500), and a scanning electron microscope (right, ×325).

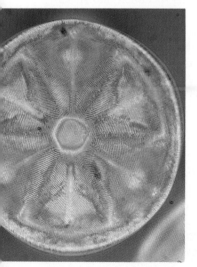
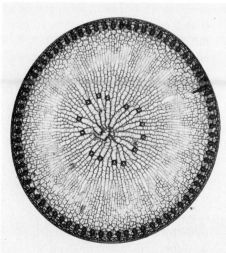

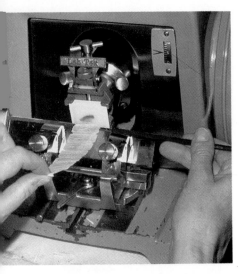

Figure 3–4. A microtome like this one is used to cut thin slices of a specimen to be viewed under a light microscope.

dimensional view of the specimen's surface on a televisionlike screen. The scanning electron microscope magnifies objects as much as 100,000 times—not as much as the transmission electron microscope. The advantage of the SEM lies in its three-dimensional views of the object's surface features.

3.3 Preparation of Specimens

For study under the light microscope, cells must be thin enough for light to pass through them. However, cell parts would remain invisible unless they were made to block some of the light. For this reason, most cells must be prepared for viewing. The first step in this process is to treat cells with substances that fix their parts in place. Thick tissues are sometimes embedded in wax and sliced into very thin pieces with a cutting instrument called a **microtome** (MY kruh tohm). To ensure that structures in the cell will block light and become visible, biologists color them with dyes called **stains.** For instance, *safranin* colors some tissues red. *Crystal violet* colors certain tissues blue. Because certain cell parts can absorb only certain stains, biologists can highlight specific structures without staining the entire cell.

Obviously these processes kill cells or disturb their contents, possibly giving scientists a distorted picture of organisms and their structures. To avoid this problem, scientists have developed a few **vital stains,** which are dyes that highlight structures in living tissues.

Specimens to be viewed under an electron microscope are prepared differently. First the specimen is dried and embedded in hard plastic. Then a microtome cuts the specimen into slices no more than one micrometer thick. Finally, the slices are coated with a thin layer of metal molecules, such as lead. These molecules block electrons just as stains block light, causing certain cell parts to stand out from the others. Sometimes whole specimens are coated with metal and then treated with chemicals. Then the specimen is removed and the biologist views only a replica of the organism's surface.

Reviewing the Section

1. How do simple and compound light microscopes differ?
2. What is the total magnification of a 20× objective paired with a 10× ocular?
3. Why do biologists stain cells?
4. Why is magnification useless without resolution?

Other Tools and Techniques

Microscopes enable scientists to see minute cells and organisms. To understand how these units of life work, biologists use other tools and techniques. Among the most important are centrifugation, microdissection, and the use of computers.

3.4 Centrifugation

Cells and microscopic organisms are very small, but they are made up of many substances. Scientists can separate the substances that make up cells by spinning cell parts at high speeds, a process called **centrifugation** (sehn trihf yuh GAY shuhn). Cells, which may be suspended in a sugar solution, are broken apart in a blender. The resulting liquid is placed in a tube. A machine called a *centrifuge* spins the tube at speeds of up to 20,000 revolutions per minute. This rotation forces the cell parts to settle in layers, with the heaviest parts at the bottom of the tube and the lightest ones at the top. Then scientists can remove each layer and study its contents to determine what materials make up cells.

3.5 Microdissection

To learn what the structures within cells do, biologists remove or add structures to cells through a kind of surgery called **microdissection.** By removing the nucleus from one organism and replacing it with a nucleus from another organism, for example, biologists can learn what functions the nucleus performs.

The cell surgeon uses special tools to perform surgery on a single cell. One of these tools is the *micromanipulator,* a special machine that translates large movements of the hand into microscopic movements. A variety of tools may be attached to the micromanipulator. *Microelectrodes* measure electric currents within the cell. A tweezerlike instrument called a *microforceps* is used to remove materials from the cell or to insert new material. Together, these instruments form an operating room the size of your thumbnail.

3.6 Computers

The computer has many applications in biology today. Some biologists, for example, use computers to determine how atoms are arranged within complex molecules that are necessary for life. Computers are also used to create visual models of these

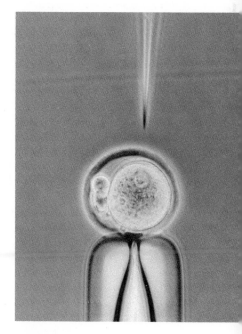

Figure 3–5. Using an extremely fine syringe, a biologist injects nucleic acid into this mouse cell. This procedure allows scientists to study the biochemistry of the mouse cell nucleus.

Highlight on Careers: Laboratory Technician

General Description

The laboratory technician plays a vital role in scientific problem solving. Under directions from a scientist, the technician performs tests and analyzes sample materials. The technician also keeps records of test results and compiles facts for the scientist testing a theory.

The duties of a laboratory technician vary. For example, a medical laboratory technician may work in a hospital, preparing slides from tissue samples and analyzing them under a microscope. A food tester may perform chemi-

cal analyses on samples of yellow dye used to color the food.

Career Requirements

Many technicians work for industry, government, hospitals, and universities. Most positions require two years of college with courses in biology, chemistry, or physics. Some positions are open to graduates of technical training courses.

Technicians enjoy great flexibility. They can start work with a modest amount of higher education and go back to school later if they wish to move

up in the field. They can also be mobile, since there is a great demand for technicians all over the country.

For Additional Information
American Chemical
  Society
1155 Sixteenth Street, N.W.
Washington, DC 20036

complicated molecules. In addition, many scientists use computers to help them analyze data. A computer helps a scientist make sense of hundreds or even thousands of individual bits of information. For example, a biologist may want to learn how wing size affects the flight speed of hawks. The scientist collects data by measuring the wings and speeds of many individual hawks. The scientist then enters this raw, or unprocessed, data into a computer. The computer processes the data into logical chunks of information, such as average wing size and range of flight speeds. Then the scientist can more easily draw conclusions from the information available.

Reviewing the Section

1. How does a centrifuge work?
2. What is the purpose of microdissection?
3. What tweezerlike tool removes or inserts cell parts?
4. When in the scientific method are computers most useful?

Investigation 3: Staining Biological Specimens

Purpose
To develop the skill of staining specimens for study under a microscope

Materials
Microscope, lens paper, glass slide, coverslip, distilled water, medicine dropper, forceps, onion, Lugol's iodine, absorbent paper

Procedure
1. Clean the microscope lens and mirror, glass slide, and coverslip with lens paper.
2. The bulb of an onion is actually made up of modified leaves. Obtain an onion leaf from your teacher. Remove a piece of the thin "skin" from the concave side of the onion leaf.

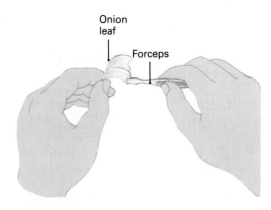

Onion leaf
Forceps

3. Prepare a wet mount of a thin piece of onion skin. Follow the procedure for making a wet mount described on page 831.
4. Observe the specimen under the microscope, first with the low power objective and then with the high power objective. Follow the steps described on page 830 when switching from low to high power. Draw a detailed diagram of the specimen at each magnification and label the diagrams.
5. Position a piece of absorbent paper at one edge of the coverslip. Place a drop of Lugol's iodine at the opposite edge of the

coverslip. The stain will be drawn through the specimen by capillary action.
6. Return the slide to the microscope stage and observe the specimen at low and high power. Draw diagrams and label them.

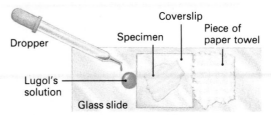

Dropper
Lugol's solution
Specimen
Coverslip
Piece of paper towel
Glass slide

Analyses and Conclusions
1. Summarize the similarities and the differences between the unstained specimen and the stained one.
2. State a possible cause for each of the following problems:
   - specimen is too dark
   - specimen is too light
   - slide keeps slipping off the microscope stage
   - coverslip cracks
   - specimen is unclear under high power
3. State an advantage of using a biological stain. What might be a disadvantage?

Going Further
Many research microscopes are equipped with an oil immersion lens that can magnify objects up to 900 times. Research how the oil immersion lens works. What role does the oil play?

Chapter 3 Review

Summary

Microscopes are among the most important biological tools. The microscope biologists most often use is the compound light microscope. This instrument has two sets of lenses—the objective near the specimen and the ocular, or eyepiece, near the eye. Together, these lenses can magnify images up to 2,000 times.

Other light microscopes include the single-lens simple microscope, or magnifying glass; the stereomicroscope with lenses for both eyes; and the phase contrast microscope for defining features inside cells. Electron microscopes use beams of electrons to greatly enlarge images. Transmission electron microscopes magnify specimens up to 200,000 times. Scanning electron microscopes produce three-dimensional images of a specimen's surface.

Most specimens viewed under a microscope are not alive. They are prepared by processes such as staining and slicing. However, vital stains enable biologists to study living cells under a light microscope.

Biologists use a wide variety of other tools and techniques. Through centrifugation, biologists learn about the substances that compose cells. Microdissection enables biologists to study the function of various cell parts. Computers perform a wide range of jobs, from drawing images of complex molecules to analyzing data.

BioTerms

centrifugation (**41**)

compound light
   microscope (**37**)

electron microscope (**39**)

eyepiece (**37**)

magnification (**37**)

microdissection (**41**)

micrometer (**38**)

microtome (**40**)

objective (**37**)

ocular (**37**)

phase contrast
   microscope (**37**)

resolution (**37**)

scanning electron
   microscope (**39**)

simple microscope (**37**)

stain (**40**)

stereomicroscope (**37**)

transmission electron
   microscope (**39**)

vital stain (**40**)

BioQuiz   (Write all answers on a separate sheet of paper.)

I. Completion

1. _____ is an increase in visible detail.
2. _____ help biologists analyze data.
3. Tissues embedded in wax or plastic are sliced thin with a _____ .
4. _____ stains do not harm organisms.
5. _____ involves performing surgery on an individual cell.
6. _____ microscopes are used by scientists to view the surface features of large specimens.

II. Modified True and False

Mark each statement TRUE or FALSE. If false, change the underlined term to make the statement true.

7. Light microscopes magnify up to 5000×.
8. Phase contrast microscopes use a beam of light to magnify objects.
9. The first microscopes came into use in the Middle Ages.
10. Specimens viewed under electron microscopes are coated with vital stains.

III. Multiple Choice

11. To view a cell magnified 200,000×, you should use a a) stereomicroscope. b) compound light microscope. c) scanning electron microscope. d) transmission electron microscope.
12. A microscope with a 10× objective and a 50× ocular magnifies a) 60 times. b) 150 times. c) 500 times. d) 40 times.
13. Visual models of molecules are produced by a) simple microscopes. b) centrifuges. c) computers. d) vital stains.
14. Microdissection enables scientists to learn a) what materials make up a cell. b) what cell parts do. c) how cells are shaped. d) where cell parts are located.
15. Centrifugation involves a) cell surgery. b) staining cells. c) drawing cell parts. d) separating cell parts.

IV. Essay

16. How do microscopes help biologists understand the world around us?
17. How does a phase contrast microscope work?
18. In what way are a stereomicroscope and a scanning electron microscope similar?
19. Why can an electron microscope achieve greater resolution than a light microscope can?
20. How are cells prepared differently for viewing under an electron microscope than for a light microscope?

Applying and Extending Concepts

1. Use a pint food jar to collect a sample of water from a pond or a puddle. Observe drops of the sample under a compound light microscope, and make drawings of the organisms you see. Do library research to find the names of the three most common organisms you have drawn.
2. Research the way in which the electron microscope works. Then prepare a poster illustrating the means by which electron microscopes produce images and present your findings to the class.
3. Two other techniques that biologists use to study the makeup of living things are paper chromatography and tissue culture. Research one of these techniques to learn what steps it involves and what information it reveals.
4. A scientist must always guard against producing artifacts. A scientific *artifact* is data produced by experimental manipulation. For example, an air bubble trapped in a glass slide is an artifact, which may be mistaken for the nucleus of a cell. Though artifacts can occur in any realm of scientific inquiry, the chances are high in doing cell research. How does the preparation of cells for viewing increase the dangers of producing artifacts? What are some ways in which to avoid them?

Related Readings

Bowser, H. "Micromasterpieces." *Science Digest* 89 (December 1981): 46–55. This article compares electron microscope photographs with early drawings.
Miller, J. A., "Cell-E-Vision." *Science News* 119 (April 11, 1981): 234–235. This article explains how standard video equipment permits the study of living cells.
Scharf, D. "Electron Eye Catchers." *Science 81* (September 1981): 62–65. This article describes techniques used to view living organisms with an electron microscope.

Barn swallow drinking

Introduction

One of the most basic characteristics of living things is that they change and cause changes in the world around them. A swallow, for example, converts food and water into muscle, bone, and feathers. Other chemical reactions produce energy that the bird uses to fly, sing, and reproduce. Chemical changes involve tiny particles of matter invisible to the unaided eye.

How can you best understand these changes? This chapter begins by introducing the fundamental principles of chemistry, which explain how substances are formed and how they interact. The chapter will focus chiefly on how those principles apply to nonliving matter. Chapter 5 will apply them further to help you understand how chemical changes affect the functioning of organisms.

The Nature of Matter

Ask a scientist what makes up the universe and you are likely to get a very simple answer—matter and energy. **Matter** is anything that takes up space. *Energy* is the capacity to move matter. Before learning about how energy affects matter, you will need to know the basic properties and composition of matter.

4.1 Properties of Matter

Every object contains a certain amount of matter. The measure of the amount of matter in an object is its **mass.** The *weight* of an object measures the pull of gravity on its mass. The more mass something has, the more it weighs.

Each kind of matter has specific properties, or characteristics, that distinguish it from every other kind of matter. Scientists generally divide the properties of matter into two classes. **Physical properties** are characteristics that can be determined without changing the basic makeup of the substance. Physical properties include the size, shape, texture, and color of a substance. **Chemical properties** describe how a substance acts when it combines with other substances to form entirely different kinds of matter.

Physical Properties and Physical Change The physical properties of a substance include its odor, its hardness, and its ability to conduct heat and electricity. The temperature at which a substance melts, boils, or freezes is also a physical property.

A substance can undergo changes in its physical properties without changing its chemical properties. One simple kind of physical change occurs when a substance changes in size or shape. Quartz boulders on a beach are ground into tiny grains of sand by the waves. The sand has all the chemical properties the boulders had, even though some of the physical properties have changed.

A more complex physical change occurs when matter changes state. The three common **states of matter** are solid, liquid, and gas. In all three states, matter consists of tiny particles. The particles in a solid are held tightly together. A solid therefore has a definite shape. It also has a definite *volume*—a given amount of a solid takes up a definite amount of space.

The particles in a liquid are held together less tightly than in a solid. They are free to tumble over one another, and so a liquid assumes the shape of any container. Like a solid, however, a given amount of a liquid maintains a constant volume.

Section Objectives

- *Define* the term *matter.*
- *Explain* the difference between physical and chemical properties.
- *List* three types of particles in the atom and describe their arrangement.
- *Distinguish* elements from compounds.
- *Name* the three basic kinds of mixtures important to living things.

Figure 4–1. Water, like all matter, exists in three states. The solid state is ice; the gaseous state is steam. Liquid water is an important part of all organisms.

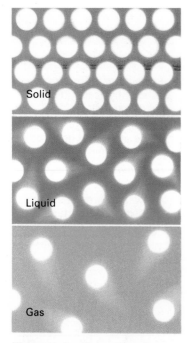

Figure 4–2. The atoms in a solid move slowly and are regularly arranged. In a liquid, they move more quickly. Atoms in a gas move most rapidly of all.

The particles in a gas are held so loosely that they disperse to fill any given volume. A gas can also be compressed into a smaller space. That is why a small tank of compressed helium can hold enough gas to fill hundreds of balloons.

Matter commonly changes from one state to another. Such a change of state usually results from a change in temperature. When you remove ice from the freezer, heat from the surrounding air causes the frozen water to melt, or change into its liquid state. When you heat that water in a pan on the stove, the liquid changes into a gas called water vapor. Heating gives particles more energy, causing them to move faster and spread farther apart. Cooling causes particles to move more slowly and so to move closer together.

Chemical Properties and Chemical Change Although the physical properties of a substance are relatively simple to study, determining its chemical properties requires complex observations. When chemists study chemical properties, they try to determine how one substance breaks down into other substances. They also experiment to see how different substances combine.

When a substance undergoes a chemical change, it is changed into one or more different substances. For example, an explosive chemical change occurs when the metal sodium comes in contact with water, converting these two substances into hydrogen gas and lye.

You experience chemical changes every day, although you may be unaware of them. For example, every time you eat, the food you consume goes through complex chemical changes. These changes convert the food into energy your body can use to run, think, or produce new cells. The clues to how these changes occur lie in the basic composition of matter.

4.2 Atoms

The basic building blocks of all matter are **atoms.** They are so small that only the most powerful electron microscopes can detect individual atoms. The largest known naturally occurring atom, the uranium atom, is just over one nanometer (one ten-billionth of a meter) wide.

Atoms are made of still smaller components called **subatomic particles.** The atom's central core, called the **nucleus,** consists of two types of subatomic particles— protons and neutrons. **Protons** carry a positive electrical charge (+1). **Neutrons,** as their name suggests, are electrically neutral. The nucleus thus has an overall positive charge.

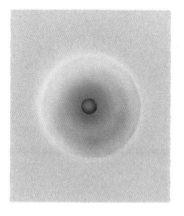

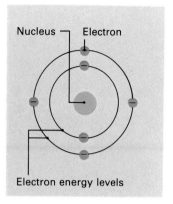

Nucleus ⌐ Electron

Electron energy levels

Negatively charged particles called **electrons** move around the nucleus at tremendous speeds. They do not follow well-defined paths, and so they are said to form a cloud around the nucleus. Each electron has a negative electrical charge (-1) equal to a proton's positive charge. In most atoms, the number of electrons equals the number of protons. Thus the atom is electrically neutral.

Under certain conditions, an atom may gain or lose electrons. An atom that has gained or lost one or more electrons is called an **ion.** When an atom loses electrons, it becomes an ion with a positive charge. When an atom gains electrons, it becomes an ion with a negative charge.

4.3 Elements

Each basic type of atom makes up an **element,** a substance that cannot be changed into a simpler substance by chemical means. For example, carbon is made entirely of carbon atoms. If you try to break down a lump of carbon—by heating it, for example—you will find that the carbon does not turn into any other substance. Ninety-two elements occur naturally on the earth. At least 16 others have been artificially created in laboratories and may exist in the cores of certain stars.

About 25 of the 92 naturally occurring elements are found in living things. Most organisms are made chiefly of the elements carbon, hydrogen, oxygen, nitrogen, sulfur, and phosphorus. Iron, iodine, and other *trace elements* make up less than 0.1 percent of the human body but must be present for the body to function normally.

Some of the basic information about elements appears in the periodic table on pages 50 and 51. Scientists use the periodic table as a tool to classify elements and their properties and to predict the behavior of elements. Notice in the table that each

Figure 4–3. Many types of illustrations are used to represent an atom. One (left) shows the nucleus surrounded by a "cloud" of electrons. Another (center) portrays the nucleus, electrons, and electron orbits. The Bohr model (right) shows electrons occupying distinct energy levels around the nucleus.

Figure 4–4. Atoms can be seen only as tiny bumps even when magnified over 500,000 times as in this electron microscope image.

element is represented by a symbol, usually the first letter or two of its common name. For example, the letter H stands for hydrogen, and Ca stands for calcium. The symbols for some elements come from other languages. For example, the symbol for iron is Fe from the Latin *ferrum,* and the symbol for sodium is Na from Latin *natrium.*

Elements in the periodic table are arranged in rows in order of their **atomic number,** which is the number of protons in the nucleus of one atom of the element. A nitrogen atom (N) has seven protons, so nitrogen has an atomic number of seven. What is the atomic number of iron?

The periodic table also shows each element's **mass number,** which equals the number of protons plus the number of neutrons in an atom. This number tells you the approximate mass of an atom relative to other kinds of atoms. Protons and neutrons are roughly equal in mass. The mass of an electron is so slight that it is not figured into the total mass of the atom. The mass number of ordinary oxygen, therefore, is 16 (8 protons + 8 neutrons).

Table 4–1: Periodic Table of the Elements

Common Elements in Living Things	
Name	Symbol
Calcium	Ca
Carbon	C
Chlorine	Cl
Chromium	Cr
Cobalt	Co
Copper	Cu
Fluorine	F
Hydrogen	H
Iodine	I
Iron	Fe
Magnesium	Mg
Manganese	Mn
Molybdenum	Mo
Nitrogen	N
Oxygen	O
Phosphorus	P
Potassium	K
Selenium	Se
Silicon	Si
Sodium	Na
Sulfur	S
Tin	Sn
Vanadium	V
Zinc	Zn

4.4 Compounds and Molecules

Two or more elements that are chemically combined form a **compound.** Although thousands of compounds occur in the nonliving world, most are made by living things. The basic materials of living tissues are complex compounds.

Each compound has its own special properties, which differ from the properties of the individual elements in that compound. Recall, for example, that sodium is a metal that reacts explosively with water. Chlorine is a poisonous green gas. Yet when sodium and chlorine combine chemically, they form a compound called sodium chloride—ordinary table salt.

Elements vary greatly in their ability to form compounds. Sodium, chlorine, and other elements in groups IA and VIIA of the periodic table are extremely reactive elements. They have such a strong tendency to combine that it is difficult to keep them as pure elements. In contrast, helium and other elements in group VIIIA ordinarily do not form compounds. Most of these elements are gases under normal conditions.

| | Elements making up 99.3% of living things |
| Elements making up about 0.7% of living things |
| Trace elements making up less than 0.01% of living things |

						IIIA	IVA	VA	VIA	VIIA	VIIIA
											2 **He** 4
						5 **B** 11	6 **C** 12	7 **N** 14	8 **O** 16	9 **F** 19	10 **Ne** 20
VIIIB	VIIIB	VIIIB	IB	IIB		13 **Al** 27	14 **Si** 28	15 **P** 31	16 **S** 32	17 **Cl** 35	18 **Ar** 40
26 **Fe** 56	27 **Co** 59	28 **Ni** 59	29 **Cu** 64	30 **Zn** 65	31 **Ga** 70	32 **Ge** 73	33 **As** 76	34 **Se** 79	35 **Br** 80	36 **Kr** 84
44 **Ru** 101	45 **Rh** 103	46 **Pd** 106	47 **Ag** 108	48 **Cd** 112	49 **In** 115	50 **Sn** 119	51 **Sb** 122	52 **Te** 128	53 **I** 127	54 **Xe** 131
76 **Os** 190	77 **Ir** 192	78 **Pt** 195	79 **Au** 197	80 **Hg** 200	81 **Tl** 204	82 **Pb** 207	83 **Bi** 209	84 **Po** (209)	85 **At** (210)	86 **Rn** (222)
108 •	109 •									

• Elements synthesized, but not officially named

61 **Pm** (145)	62 **Sm** 150	63 **Eu** 152	64 **Gd** 157	65 **Tb** 159	66 **Dy** 163	67 **Ho** 165	68 **Er** 167	69 **Tm** 169	70 **Yb** 173	71 **Lu** 175
93 **Np** 237	94 **Pu** (244)	95 **Am** (243)	96 **Cm** (247)	97 **Bk** (247)	98 **Cf** (251)	99 **Es** (252)	100 **Fm** (257)	101 **Md** (258)	102 **No** (259)	103 **Lr** (260)

▶ Isotopes

All the atoms of an element have the same number of protons. The number of neutrons, however, can vary. Atoms of an element with different numbers of neutrons are **isotopes** (EYE suh tohps).

Elements naturally occur as mixtures of isotopes. Because each isotope has a different number of neutrons, an isotope is identified by its mass number. Tin, for example, has ten isotopes ranging from tin-112 to tin-124.

An element's most common isotope is usually its most stable form. The less stable isotopes release small amounts of excess energy called *radioactive emissions,* which can be detected easily by radiation devices.

Radioactive isotopes, or *radioisotopes,* are used widely in scientific research. In a process called "labeling," a small amount of a radioisotope is injected into a living organism. The radioisotope can be traced as it circulates through the organism.

In agriculture, the radioisotope phosphorus-32 has been used to test how well plants absorb fertilizer. In this way, scientists showed that applying nutrients directly onto leaves can be more effective than feeding plants through the roots.

Doctors use iodine-131 to diagnose the condition of the thyroid gland. Since iodine naturally concentrates in the thyroid gland, the radioactive tracer also accumulates in the gland. Radiation scanners can then detect the gland's shape and size, and can monitor its activity.

The smallest particle of a compound or element that can have stable, independent existence is called a **molecule.** A molecule of an element may consist of one, two, or more atoms of that element. Each molecule of a compound contains two or more different atoms.

The number of atoms of each element in a molecule is indicated by a **molecular formula.** For example, the molecular formula for hydrogen is H_2. The number 2 in the formula is a subscript. It tells you that two *atoms* of hydrogen make up one *molecule* of hydrogen. Any molecule that contains exactly two atoms is called a *diatomic molecule*. CO is the molecular formula for another diatomic molecule, carbon monoxide.

A molecular formula also shows the proportions of each kind of atom found in a compound. ***Atoms combine in specific ratios to form molecules.*** For example, a water molecule (H_2O) always contains hydrogen and oxygen in a ratio of two hydrogen atoms to one oxygen atom. Add one more atom of oxygen and you have a molecule of hydrogen peroxide (H_2O_2), a syrupy, corrosive compound used in bleaches, antiseptics, and rocket fuel. You could say that the proportion of atoms in a compound gives that compound its distinctive chemical personality.

Using a molecular formula, you can also determine a molecule's molecular weight. The **molecular weight** equals the sum of the mass numbers of all the atoms in the molecule. Glucose is a simple sugar with the molecular formula $C_6H_{12}O_6$. What is the molecular weight of this molecule?

4.5 Mixtures

In nature, elements and compounds rarely exist in a pure state. Instead they are normally found mixed together. In a **mixture,** the molecules of different substances mingle without combining chemically.

Each substance in a mixture retains all its chemical properties. The components of a mixture also retain their physical properties, which makes it possible to separate them physically. For example, you can use a paper filter to separate smoke from air. Mixtures can contain solids, liquids, and gases. Unlike compounds, mixtures can be composed of substances in any ratio. The three kinds of mixtures most important to living things are solutions, suspensions, and colloids.

Solutions When you stir sugar into a glass of water, the sugar granules break up into individual molecules that disperse throughout the water. The resulting mixture is called a solution. A **solution** forms when the particles of one substance are dispersed in another substance to make a uniform mixture. Solutions can be mixtures of gases, mixtures of liquids, or mixtures of gases and liquids. For example, fish get the oxygen they need from the oxygen dissolved in water.

A substance that can dissolve other substances is called a **solvent.** Water is often called the "universal solvent" because it can dissolve a great variety of substances. Most of the chemical processes of living things take place in water solutions. In fact, life on earth would not be possible without water. The substance that dissolves in the solvent is called the **solute.**

When an ionic solute such as salt dissolves, the individual ions of the solute separate from one another and then disperse in the solvent. The separation of ions in solution is called *dissociation.* The sodium and chloride ions in salt dissociate when the salt dissolves in water.

The amount of solute dissolved in a given amount of solvent is called the **concentration** of the solution. For example, the normal salt solution that doctors give patients to replace lost body fluids has a concentration of 0.9 percent, or 9 parts of salt per 1,000 parts of water. Every solution has a maximum possible concentration. The solution becomes *saturated* when the

Q/A

Q: *What makes soda pop fizz?*

A: Carbonated drinks contain carbon dioxide that is put into solution under pressure. When you open the can or bottle, the pressure is released, and the carbon dioxide is free to escape as gas bubbles.

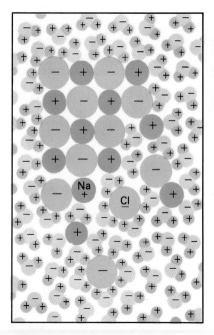

Figure 4–5. When the solute sodium chloride (NaCl) is dissolved in the solvent water (H_2O), it dissociates and forms the ions Na^+ and Cl^-.

Figure 4–6. A suspension is formed when a river picks up millions of dirt particles on its way to the sea.

solvent holds all the solute it can. A glass of iced tea, for example, can hold only a certain amount of sugar in solution. Additional sugar settles to the bottom of the glass without dissolving.

Suspensions   A mixture in which particles are temporarily mixed together is called a **suspension.** You can distinguish a suspension from a solution by its appearance. Suspensions look cloudy; solutions are clear.

The particles in a suspension are larger than molecules. For example, fine dirt mixed in water forms a suspension. The water will appear murky and discolored while dirt particles are suspended in it. If you let the water stand for several minutes, the dirt particles will gradually settle to the bottom of the glass. You can also separate the particles from a suspension by passing the mixture through a filter.

Suspensions are important to living things. For example, blood cells are suspended in the liquid portion of blood. The bloodstream's constant circulation keeps the blood cells from settling out in any one location.

Colloids   A **colloid** is a mixture in which the suspended particles are smaller than those in an ordinary suspension but larger than the particles of solute in a solution. Colloidal particles cannot be filtered out through paper, and they do not settle to the bottom of a container. Unlike most suspensions, a colloid will last indefinitely if left undisturbed.

Powdered gelatin forms a colloid when it is mixed with warm water. As long as the particles of gelatin remain uniformly dispersed in the warm water, the colloid is said to be in the *sol* state. A sol behaves much like a liquid. When the gelatin mixture is chilled, the gelatin particles join together to form a tangled network. The water becomes trapped within this network, giving the colloid a semisolid consistency called the *gel* state.

Colloids are important to living organisms. The interior of a living cell is a colloid. The functioning parts of the cell, as well as particles of food, remain dispersed throughout the cell rather than settling in any one area.

Reviewing the Section

1. How are subatomic particles arranged within an atom?
2. What distinguishes an element from a compound?
3. What does a molecular formula tell about a molecule?
4. How does a colloid differ from an ordinary suspension?

Energy and Chemical Change

All living things must constantly use energy to stay alive. Organisms obtain the energy they need by converting food into various forms of energy within their cells. To understand how these conversions occur, you will first need to understand the basic nature of energy.

4.6 Forms of Energy

Both food energy and the energy of a falling object demonstrate one of the basic principles of energy. *Any process of change involves the conversion of one form of energy into another.* A book resting on the edge of a shelf has a form of energy called potential energy. **Potential energy** is energy an object possesses because of its position or its composition. When the book falls off the shelf, its potential energy is converted into **kinetic energy,** the energy of motion.

Water backed up behind a dam has a great deal of potential energy because of the force of gravity acting on the water. As the water flows down through the dam, its potential energy is converted into kinetic energy. The dam's turbines in turn convert this kinetic energy into electricity. The potential chemical energy in wood, gasoline, and other fuels is converted into heat and other forms of kinetic energy as the fuel burns.

4.7 Energy Levels in the Atom

The energy used by living things is chemical potential energy. This chemical energy is actually the energy of electrons within atoms. Electrons speed around the nucleus within specific regions of space. These regions are called **energy levels** because the electrons in each region have a specific amount of energy. Electrons in the first level—that is, in the level closest to the nucleus—have the least amount of energy. The more energy an electron has, the farther it is from the nucleus.

Each energy level can hold only a certain number of electrons. The first level can hold only 2 electrons. The second level can hold up to 8 electrons, and the third can hold up to 18 electrons. Atoms may have over 6 energy levels.

Atoms fill their energy levels from the inside out. For example, the first level must be filled before any electrons will occupy the second level. Neon, with 10 protons and 10 electrons, has 2 electrons in the first level and 8 in the second. Both levels are filled to capacity.

- *Distinguish* between two kinds of energy.
- *Explain* how stability affects the chemical behavior of an atom.
- *State* the difference between ionic and covalent bonds.
- *Summarize* what happens to chemical bonds in a chemical reaction.
- *Balance* a chemical equation.

Figure 4–7. When moving down an incline, the slider's energy of position is converted into kinetic energy, the energy of action.

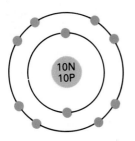

Figure 4–8. A Bohr model shows that a neon atom has 10 electrons. The first energy level contains 2 of these electrons. The second energy level contains the other eight. The presence of 8 electrons in the outer energy level makes neon extremely stable. Under normal circumstances, neon rarely forms molecules.

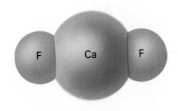

Figure 4–9. When regularly arranged, millions of molecules of calcium fluoride (top) form large crystals of fluorite (bottom).

Like neon, all elements in group VIIIA of the periodic table have their outer energy level filled. Generally, an atom is most stable when its outer energy level is filled to capacity. This fact explains why atoms of neon and other similar elements rarely combine with other atoms. Atoms with three or more energy levels are most stable with 8 electrons in the outer level. Atoms tend to interact in ways that result in stable, complete energy levels.

4.8 Formation of Chemical Bonds

Atoms in any molecule or compound are held together by forces called **chemical bonds.** Chemical bonds result from the interaction of electrons in the outer energy levels of atoms. Two different kinds of interaction between atoms may produce chemical bonds. In one kind of interaction, an atom with "extra" electrons in its outer energy level may transfer them to an atom that needs electrons in its outer level. In another kind of interaction, atoms that need electrons in their outer energy levels may share the electrons available.

A transfer of electrons makes each atom an ion and so gives it a stable outer energy level. The electrical attraction between the positively and negatively charged ion forms a bond called an **ionic bond.** Sharing electrons also creates stable outer levels. Two atoms that share electrons are held together by a **covalent bond.** Both kinds of interactions produce stable outer layers and bind atoms together.

Ionic Bonds   The interaction of sodium and chlorine illustrates how ionic bonds are formed. Both sodium and chlorine are unstable elements because their outer energy levels are incomplete. A sodium atom has one electron in its outer level. A chlorine atom has seven electrons in its outer level. Sodium tends to enter chemical reactions in which it will lose its outer electron, while chlorine tends to enter reactions in which it will gain an electron. When sodium and chlorine atoms combine, as shown in Figure 4–10, the sodium atom transfers its single outer electron to the chlorine atom. This transfer gives each atom a stable outer energy level of eight electrons.

The loss of an electron makes the sodium atom a positively charged ion, symbolized as Na^+. At the same time, the gain of an electron makes the chlorine atom a negatively charged ion, Cl^-. The electrical attraction between the oppositely charged ions forms an ionic bond between them. The resulting compound, NaCl, or table salt, is called an *ionic compound* because it consists of ions held together by an ionic bond.

Covalent Bonds   When salt dissolves in water, the sodium and chlorine ions dissociate and disperse through the water. This dissociation indicates that some ionic bonds can be easily broken. Covalent bonds are relatively strong bonds. Covalent bonds do not break easily because the bonded atoms share electrons.

Water is the most common example of a covalent compound. A water molecule is composed of an oxygen atom and two hydrogen atoms that share electrons. An oxygen atom has six electrons in its outer energy level. It needs two more for a stable outer level. In a water molecule, each of the two hydrogen atoms shares its single electron with the oxygen atom, completing the oxygen's outer level. At the same time, the oxygen shares one of its electrons with each of the hydrogen atoms, completing their outer levels. Altogether, four electrons are shared in a water molecule—one contributed by each of the two hydrogen atoms and two contributed by the oxygen atom.

Figure 4–10.   A carbon atom (top) forms a covalent bond with each of four hydrogen atoms to form methane (CH_4). When sodium donates an electron to chlorine (bottom), the resulting ions, Na^+ and Cl^-, attract each other and form an ionic bond.

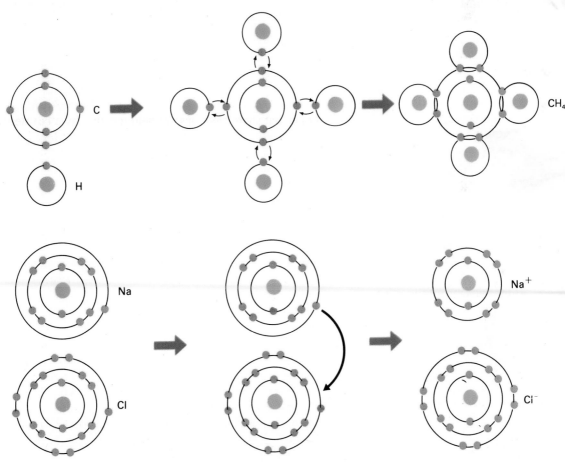

Carbon has four outer electrons available for sharing. In the gas methane, each carbon atom completes its outer energy level by bonding with four separate hydrogen atoms, as Figure 4–10 shows. In this bond, each hydrogen atom also completes its own outer shell. Like water, methane is a compound with *single bonds*. That is, each bond consists of one pair of electrons. Atoms may share more than one pair of electrons. A covalent bond in which atoms share two pairs of electrons is called a *double bond*.

Highlight on Careers: Crime Laboratory Technician

General Description

A crime laboratory technician collects and analyzes evidence relating to a criminal or civil offense. The facts assembled by a crime lab technician are then used in court to help determine the guilt or innocence of a person who has been accused of committing the crime.

Crime lab technicians are called to the scene of murders, bank robberies, and other serious crimes. There, the technicians collect fingerprints, take photographs, and gather other kinds of evidence about the crime.

In the laboratory the technicians analyze the evidence they have collected. They may run chemical tests to determine the victim's blood type. They may analyze handwriting samples, tool marks, or the explosives used in a crime.

Many crimes are solved because of such work by crime lab technicians. Many cases are settled out of court, saving the public time and money. Most importantly, crime laboratory technicians may discover evidence that frees a person who has been wrongly accused of committing a crime.

Most crime lab technicians are employed by city and state governments or federal agencies such as the Federal Bureau of Investigation. Other crime laboratory technicians are often employed by private laboratories.

Career Requirements

Most crime laboratory technicians have a B.S. degree in chemistry, biology, or physics. A two-year community college course qualifies students for entry-level jobs in this field.

Since the future of another person may rest on a crime laboratory technician's work, these professionals must produce thorough, accurate reports. They must also be able to work with people and give testimony in court. Because of the detailed nature of their work, they must also be careful and methodical.

For Additional Information
American Academy of
  Forensic Scientists
225 South Academy
  Boulevard
Colorado Springs, CO
  80910

4.9 Chemical Reactions

By filling outer energy levels, chemical bonds between atoms make compounds relatively stable. However, adding energy to a compound can break the bonds. The atoms may then form new bonds with other atoms to make new compounds. The process of breaking existing chemical bonds and forming new bonds is called a **chemical reaction.**

Because bonds break only when sufficient energy is supplied, every chemical reaction requires an input of energy to get started. The energy required to start a chemical reaction is called **activation energy.** Heat is the most common form of activation energy. For example, when sugar is exposed to high heat it bubbles, turns black, and gives off water vapor and carbon dioxide. The chemical explanation for this change is that the heat energy causes the sugar molecules to collide faster and faster. These collisions break the existing bonds between the carbon, hydrogen, and oxygen atoms in the sugar. As the bonds break, the atoms form new bonds which result in carbon dioxide, water vapor, and pure carbon. Energy is also released in the forms of heat and light from this activity. No energy is lost; it only changes form.

Atoms do not change in a chemical reaction; they are only rearranged into different combinations. The *products* of the reaction described above—water vapor, carbon dioxide, and pure carbon—contain the same total number of atoms as the sugar originally contained. This fact illustrates one of the basic laws of chemistry: All the atoms that enter into a chemical reaction will be present in the products of the reaction.

As new bonds form, some amount of energy is released. In some cases, this energy helps keep a reaction going. Chemical reactions that release more energy than they use up are called

Q: *How do fireflies glow?*

A: Chemical reactions within the firefly's abdomen release energy in the form of light. This phenomenon, which also occurs in fish and many other animals, is called *bioluminescence.*

Figure 4–11. A spark, a flame, and a fire are the results of exergonic chemical reactions in which energy is released as heat and light. The exergonic chemical reactions that occur in cells release energy far more gradually than the exergonic reactions shown below.

▶ Chemical Equations

Scientists use a special shorthand called **chemical equations** to describe what happens in chemical reactions. The substances entering the reaction, called *reactants,* are symbolized on the left side of the equation. An arrow points in the direction of the products that result from the reaction, symbolized on the right side.

A chemical equation also shows the relative amount of each substance that will take part in the reaction. For example, the equation

$$H + H \rightarrow H_2$$

states that one atom of hydrogen combines with another atom of hydrogen to make one molecule of hydrogen.

Notice that the same number of hydrogen atoms are present on both sides of the equation. An equation is *balanced* if the numbers of each kind of atom are equal on both sides of the equation. Balanced equations demonstrate the **law of conservation** of mass, which states that matter cannot be created or destroyed.

Now look at the equation below, which shows hydrogen and oxygen combining to form water. The equation

$$H_2 + O_2 \rightarrow H_2O$$

is not balanced—only one atom of oxygen appears in the product. You cannot add a subscript to the product because subscripts are fixed numbers that tell the number of atoms of an element in one molecule of a substance. Instead, scientists use numbers called *coefficients* to show the correct proportion of substances in the reaction. The balanced equation for water reads

$$2H_2 + O_2 \rightarrow 2H_2O$$

The coefficient, 2, multiplies all the numbers that follow it. The balanced equation shows four hydrogen atoms and two oxygen atoms on each side.

exergonic (ehk suhr GAHN ihk) **reactions.** Wood burns in an exergonic reaction that both sustains itself and gives off heat and light energy. Reactions that use up more energy than they release are called **endergonic** (ehn duhr GAHN ihk) **reactions.** Many cellular reactions are endergonic. Living things need a regular supply of nourishment to stay alive because the endergonic reactions within their cells use more energy than they release.

Reviewing the Section

1. What are energy levels in the atom?
2. How do covalent bonds form?
3. Why is activation energy needed in chemical reactions?
4. The element potassium has 19 electrons. Is it more likely to form ionic or covalent bonds? Why?
5. How would you balance the equation $Al + O_2 \rightarrow Al_2O_3$?

Investigation 4: Comparing Mixtures and Compounds

Purpose
To identify the differences between a mixture and a compound

Materials
Iron filings, sulfur, bar magnet, three heat-resistant test tubes, test tube rack, test tube holder, paper, spatula, safety goggles, lab apron, Bunsen burner, matches

Procedure
1. Make a table similar to the following one, adding any other properties you wish. Complete the table, listing the observable physical properties of iron and sulfur. Use the magnet to determine magnetism. In separate test tubes, mix small amounts of each element with water to determine the solubility of each.

Characteristics of Iron and Sulfur

Element	Iron	Sulfur
Color		
Texture		
Density		
Magnetism		
Solubility		

2. Using the tip of a spatula, mix small amounts of the iron and the sulfur on a piece of white paper. Determine a procedure for separating the two elements.
3. **CAUTION: Put on goggles and a lab apron.** Again using the tip of the spatula, place a pinch of iron and sulfur in a 50-mL or larger heat-resistant test tube. Heat the tube carefully over a Bunsen burner in a well-ventilated room or under a ventilation hood.
4. Remove the test tube from the heat as soon as the sulfur has melted and has apparently combined with the iron. Compare the newly formed substance with the original products and with the substance formed in step 2. *Could you easily separate the newly formed substance into its original elements?*

Analyses and Conclusions
1. How did the substance formed in step 2 differ from the substance formed in step 3? How did the method of combining the elements differ in each step? Which substance is a mixture, and which is a compound?
2. What is the relationship between a physical or a chemical change and a mixture or a compound?
3. How would the final product differ if you used a greater quantity of iron than sulfur in step 3? A greater quantity of sulfur than iron?

Going Further
- A student accidentally spilled the contents of a container of salt on top of some sand. Design a procedure that would allow the student to separate the two substances.
- A scientist discovers a new compound with properties that would make it valuable for space missions. However, to produce the compound, the scientist must identify its component elements. What procedures would you use to find the component elements?
- The labels of chlorine bleach products always have a printed warning that advises the user not to combine the product with any other product containing ammonia. Why does this warning appear? Use chemical equations to support your explanation.

Chapter 4 Review

Summary

The physical and chemical properties of matter are determined by the composition of each substance. Atoms, which make up all matter, consist of protons and neutrons in the nucleus and electrons in energy levels around the nucleus. Each chemically different kind of atom corresponds to a different element.

Elements combine in fixed ratios to form compounds. A compound has different properties than its component elements. Molecules are the smallest particles of an element or compound that can exist independently. The molecules in a mixture retain all their properties and so can be physically separated. Solutions, suspensions, and colloids are the basic kinds of mixtures important to living things.

Energy is the capacity to move matter from one location to another. Energy transfer in the atom involves electrons as they interact to form chemical bonds. Ionic bonds form when electrons are transferred from one atom to another. Covalent bonds form between atoms that share electrons. In a chemical reaction, existing bonds between atoms are broken and new bonds form. Activation energy is necessary to start a chemical reaction.

BioTerms

activation energy (**59**)
atom (**48**)
atomic number (**50**)
chemical bond (**56**)
chemical equation (**60**)
chemical property (**47**)
chemical reaction (**59**)
colloid (**54**)
compound (**51**)
concentration (**53**)
covalent bond (**56**)
electron (**49**)
element (**49**)

endergonic reaction (**60**)
energy level (**55**)
exergonic reaction (**60**)
ion (**49**)
ionic bond (**56**)
isotope (**52**)
kinetic energy (**55**)
law of conservation (**60**)
mass (**47**)
mass number (**50**)
matter (**47**)
mixture (**53**)
molecular formula (**52**)

molecular weight (**53**)
molecule (**52**)
neutron (**48**)
nucleus (**48**)
physical property (**47**)
potential energy (**55**)
proton (**48**)
solute (**53**)
solution (**53**)
solvent (**53**)
states of matter (**47**)
subatomic particle (**48**)
suspension (**54**)

BioQuiz (Write all answers on a separate sheet of paper.)

I. Completion

1. Atoms of the same element with different numbers of neutrons are _____ .
2. The measure of the amount of matter in an object is its _____ .
3. An _____ chemical reaction releases more heat than it uses.
4. Most of the chemical processes in cells take place in water _____ .
5. Covalent bonds are created when two or more atoms _____ electrons.
6. A _____ becomes saturated when the solvent holds all the solute it can.
7. _____ bonds are easily broken.

II. Multiple Choice

8. In order for a chemical reaction to start, there must be available a) ionic bonds. b) activation energy. c) potential energy. d) covalent bonds.
9. Substances that cannot be changed chemically into simpler substances are a) elements. b) compounds. c) molecules. d) atoms.
10. The subatomic particle involved in forming chemical bonds is the a) neutron. b) electron. c) proton. d) nucleus.
11. When an atom loses an electron, the atom becomes a positively charged a) compound. b) molecule. c) element. d) ion.
12. A chemical equation is balanced when each side has the same number of a) atoms. b) compounds. c) subscripts. d) coefficients.

13. Atomic interactions tend to result in a) a change of state. b) stable outer energy levels. c) extra neutrons. d) a solution.
14. The gel state is a form of a a) compound. b) solution. c) colloid. d) mixture.

III. Essay

15. How do radioactive emissions occur?
16. What do chemical equations illustrate about the nature of matter and chemical reactions?
17. What happens to chemical bonds in a chemical reaction?
18. What is the difference between kinetic and potential energy?
19. Do particles have most kinetic energy in a solid, a liquid, or a gas?
20. How do electrons determine the chemical properties of an element?

Applying and Extending Concepts

1. Use your school or public library to research a specific use of radioisotopes in medicine, agriculture, or environmental studies. Then write a report about the discoveries the isotope has made possible.
2. Crude oil comes from the ground as a mixture of many different compounds called *grades* of oil. Heat is used to separate the grades and purify them into gasoline, kerosene, and other petroleum products. What physical property must therefore vary between the grades of oil? Write a brief paragraph to explain your answer.
3. Water is a very unusual compound. Most compounds contract as they freeze; water expands. As a result, ice weighs less than water. What would happen to a lake and the organisms in it if ice were heavier than liquid water?

Related Readings

Asimov, Isaac. *Asimov on Chemistry*. New York: Doubleday, 1974. This book lays a good foundation for understanding basic chemistry and its applications in many branches of science.

Brady, J. E., and G. E. Humiston. *General Chemistry: Principles and Structures*, 2d ed. New York: Wiley, 1980. This basic text clearly explains the fundamental principles of chemistry.

Stine, W. R. *Applied Chemistry*, 2d ed. Boston: Allyn & Bacon, 1981. This nontechnical book discusses the applications of chemistry in daily life.

BIO*TECH*

Biosterometrics

Computer image of cell

Fossil skull (below right) imbedded in rock may be viewed clearly using computerized tomography (below).

A biologist taps the keys of a computer keyboard, bringing an image of the structure of a skin cell to the computer screen. A paleontologist examines the computer-generated image of a fossil skull. A physician studies a computer image of a small section of a patient's artery. All three scientists are using new technological processes of **biostereometrics**—the study of biological form and function in three dimensions.

Biosterometrics allows biologists in many fields of study to see things in entirely new ways. The cell, for example, can be examined in ways never before possible. Electron micrographs taken at different angles are assembled into a three-dimensional image on a computer screen. The image can be viewed from any angle and sliced in any direction. Cell structures can be colored to emphasize relative size, location, and distribution. These graphic representations of the tiny internal structures of cells lead to greater understanding of how cells function.

Paleontologists use *computerized tomography,* which combines X rays with computer imaging to study fossil bones

Biostereometrics allows biologists to see living things in new ways.

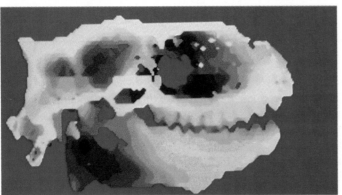

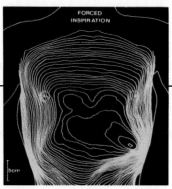

FORCED INSPIRATION

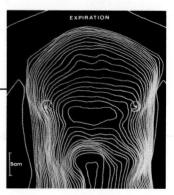

EXPIRATION

that cannot be easily separated from rock. The X ray distinguishes differences in density between rock and fossil. The computer then processes this information to produce an image of the fossil alone.

Nuclear magnetic resonance (NMR) imaging is among the newest of the biostereometric techniques. NMR provides a way in which to obtain cross-sectional pictures that show differences among tissues on a chemical level. Through

the use of NMR, medical researchers can tell the difference between living and dead tissue and between healthy and diseased tissue. They can also differentiate between arteries with normal blood flow and arteries that are obstructing blood flow. Researchers are also finding NMR useful in analyzing how

Surface maps of a human torso indicate changes in body shape during inhalation (left) and exhalation (right).

well drugs work. NMR images taken of affected tissues or body parts before and after medication show clearly which drugs work and which do not.

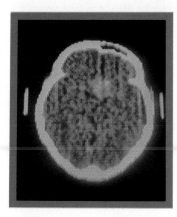

NMR imaging magnet consists of four rings that generate a magnetic field. The pulse from the radio-frequency coil provides data for a computer image, such as the brain outline above.

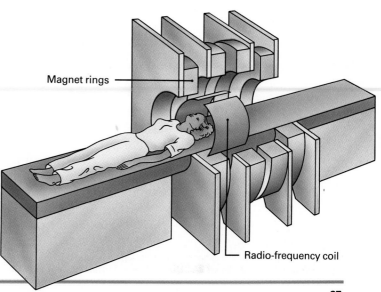

Magnet rings

Radio-frequency coil

5

The Chemistry of Living Things

Mushrooms producing biological light

Introduction

By the beginning of the nineteenth century, scientists had begun to apply their growing knowledge of chemistry to living things. They wondered how organisms like fireflies managed to convert food to produce energy and light. They wondered how plants and animals could perform the same chemical reactions in their bodies that in the laboratory produced enough heat to kill the organisms. They also wondered how the chemistry of living things could help explain the great diversity of organisms on the earth.

The answers to these and similar questions formed the basis of the science of *biochemistry*. Biochemistry explores how the properties of chemicals make life possible and how these chemicals help determine the characteristics of living things.

Inorganic Compounds and Life

One of the earliest discoveries about the chemistry of living things was that most compounds made by organisms contain carbon. Scientists therefore came to classify most of the thousands of carbon compounds as **organic compounds. Inorganic compounds** are those not made by living things.

Many inorganic compounds are essential to life. Nitrogen compounds and minerals such as sodium, potassium, and iron provide elements that all organisms need. Water is also an inorganic compound.

5.1 Water

Water is the most important inorganic compound for living organisms. Most cellular activities take place in water solutions. Water is important to living things because it is an excellent solvent and has a high *heat capacity*. Water has a high heat capacity because it can absorb and release a great deal of heat energy before changing temperature. This property of water protects organisms from overheating or freezing.

The chemical structure of a water molecule (Figure 5–1) explains its effectiveness as a solvent and its high heat capacity. The single oxygen atom in a water molecule strongly attracts the electrons of the two hydrogen atoms. As a result, the oxygen atom in a water molecule has a slight negative charge, while each of the two hydrogen atoms has a slight positive charge. The opposite charges found at either end of the molecule make water a **polar compound.** Polarity makes water an excellent solvent. Its polar ends attract the ends of other polar molecules as well as ions such as the sodium and chloride ions in salt.

Water has a high heat capacity because water molecules also attract each other. When water is heated, most of the heat energy is used first in breaking the bonds between water molecules. Then only a relatively small amount of heat energy is available to increase the movement of the molecules and raise the water's temperature.

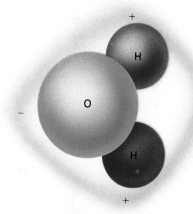

Figure 5–1. The unequal distribution of positive and negative charges around a water molecule causes water to be polarized. Its polarity makes water an excellent solvent.

5.2 Acids and Bases

The vast majority of water molecules always remain as H_2O. However, the attraction between water molecules causes a few molecules to *dissociate,* or separate, into a positively charged hydrogen ion (H^+) and a negatively charged hydroxide ion (OH^-). In pure water the number of hydrogen ions equals the

Spotlight on Biologists: Linus Pauling

Born: Portland, Oregon, 1901
Degree: Ph.D., California Institute of Technology

The American chemist Linus Pauling made major breakthroughs in understanding the molecular structure of matter. His spiral staircase model for certain proteins has had wide biochemical application.

By applying to proteins his knowledge of chemical bonding, Pauling was able to suggest explanations for some diseases. He first proposed, for example, that sickle-cell anemia was due to deficiency of oxygen in the blood.

Pauling's familiarity with chemistry came at an early age from his father, who was a pharmacist. When he was 13 years old, Pauling began experiments with a friend's home chemistry laboratory. From these beginnings he launched a career that eventually led to his winning the Nobel Prize for Chemistry in 1954.

After World War II Pauling became a strong advocate for nuclear disarmament and for a ban on nuclear testing. In 1962 he was awarded a second Nobel prize—for peace.

Q/A

Q: *Why do bottles of drain cleaner warn against getting any of the product near your eyes or mouth?*

A: Drain cleaners contain powerful bases that cause severe burns. The bottles also instruct users to treat burn victims immediately with lemon juice, vinegar, or other acidic solutions that will neutralize the base.

number of hydroxide ions. Certain compounds called **acids,** however, release hydrogen ions in water and so increase the concentration of H^+ in a solution. **Bases** are the chemical opposite of acids. They release hydroxide ions in solution or *accept* hydrogen ions.

The relative concentration of hydrogen ions in a substance is measured by the **pH scale.** The strongest acids have a pH near 0, the strongest bases a pH near 14, and a neutral solution a pH of 7. Pure water has a pH of 7. Each kind of organism has particular pH balances that must be maintained at all times. Human blood, for example, has a pH of 7.4; stomach acid a pH of 2.0. Fluctuations in pH can change the rate and nature of internal chemical reactions and can seriously endanger an organism's life.

Reviewing the Section

1. What is an organic compound?
2. What is the difference between an acid and a base?
3. How does the chemical structure of water explain its high heat capacity?

Organic Compounds

Although many inorganic substances are essential to life, the vast majority of substances in living things are organic compounds. *Carbon forms the structural backbone of all organic molecules.* Of all the elements, only carbon is versatile and stable enough to make up the tremendous variety of molecules found in living things. A carbon atom can bond with up to four atoms at once. Carbon atoms can also bond with one another to form rings or long chains.

Most organic molecules are constructed of basic units that repeat over and over. These units are called **monomers.** When two monomers combine chemically, a new compound is formed. The reaction often releases two hydrogen atoms and one oxygen atom that unite to form a molecule of water. A reaction that produces water in this way is called a **condensation reaction,** or a *dehydration synthesis.* The condensation of many monomers produces a complex molecule called a **polymer.** Many organic molecules are polymers.

The most common organic compounds in living things are classified in four major groups. These groups are carbohydrates, lipids, proteins, and nucleic acids.

5.3 Carbohydrates

Carbohydrates are organic compounds that contain carbon and hydrogen and oxygen in the same ratio as in water—two hydrogen atoms for each oxygen atom. Familiar carbohydrates include sugars and starches. Some carbohydrates, such as cellulose, are used as structural materials. Others, such as sugars, provide quick energy or store energy in cells.

The monomers that make up all carbohydrates are single-sugar molecules. These simple molecules are called **monosaccharides.** Two common monosaccharides, glucose and fructose, are shown in Figure 5–2. They both have the same molecular formula, $C_6H_{12}O_6$, but differ in the arrangement of their atoms. Compounds such as these, in which the same atoms are arranged differently, are called **isomers.** Most monosaccharides are five- or six-carbon sugars. Each molecule is built on a chain of five or six carbon atoms. Both plants and animals use monosaccharides for energy. These compounds, however, may enter the body not as monosaccharides but as compounds formed from the condensation of monosaccharides.

The condensation of two monosaccharides produces a **disaccharide,** or double-sugar molecule. Maltose, known as malt

Section Objectives

- *Identify* the chemical similarities and differences among the four main kinds of organic compounds.
- *Summarize* what happens in a condensation reaction.
- *Explain* how chemists use structural formulas.
- *Describe* the basic structure of an amino acid and explain how amino acids form proteins.
- *Summarize* the interaction between an enzyme and its substrate.

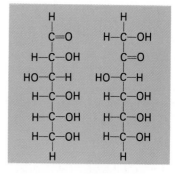

Figure 5–2. Glucose (left) and fructose (right) are isomers. They have the same structural formula, $C_6H_{12}O_6$, but their atoms are arranged differently and they have different chemical properties.

▶ Structural Formulas

If you have ever drawn stick figures to represent people, then you already know how chemists represent molecules. Just as a stick figure shows the basic parts of a human body, so **structural formulas** show atoms bonded together in a molecule.

The lines between letters represent the bonds between atoms. Below is the formula for a water molecule. Each line represents a single covalent bond between a hydrogen atom and the oxygen atom.

enable a scientist to picture the differences between isomers. Compare the formulas for glucose and fructose (a). Many organic molecules are more accurately drawn as a ring. This method

shows the structures of glucose and fructose even more clearly (b).

Structural formulas can be used to show the condensation of two glucose molecules to form a maltose molecule (c).

Double bonds, like those in a molecule of carbon dioxide,

$$O=C=O$$

are represented by two lines ($=$). Three lines ($\equiv$) represent triple bonds. One of the values of structural formulas is that they

sugar, is made of two condensed glucose molecules. Lactose, or milk sugar, is a molecule of glucose combined with a molecule of galactose. Both human milk and cow's milk contain lactose, composed of glucose and galactose. Sucrose, common table sugar, is made of glucose joined to fructose.

The largest carbohydrates are **polysaccharides.** These molecules may consist of thousands of monomers. Plants store food in the form of *starch*, a polysaccharide that is a polymer of glucose (Figure 5–3). Animals store excess sugars as *glycogen*, another polymer of glucose. When you are active, the glycogen stored in your liver and muscles is broken down to release glucose for quick energy. **Hydrolysis,** the reaction that breaks down complex molecules, is the reverse of a condensation reaction. In hydrolysis, water molecules combine with parts of the long glycogen molecules to form molecules of glucose.

Changing just a few atoms in a large molecule can dramatically alter the molecule's chemical behavior. Both starch and cellulose have almost the same kinds of atoms. Both are polymers of glucose, but the atoms are arranged differently in each. Because of this, the chemical behavior of the two polymers is different. Starch is a storage molecule that most organisms can easily break down into glucose. Most organisms cannot break down cellulose. Instead, cellulose forms the tough, fibrous tissues that support plant stems and transport water in plants.

5.4 Lipids

Lipids are a chemically diverse group of substances that include fats, oils, and waxes. Beef fat, butter, and olive oil are examples of lipids. Like carbohydrates, lipids contain carbon, hydrogen, and oxygen. Lipids are classified together because they are all insoluble in water. In living things, lipids serve mainly as storage of energy. Lipids are also part of cell membranes and thus help regulate what enters and leaves cells.

Lipid molecules are somewhat more complex than carbohydrate molecules. The backbone of many lipids is a three-carbon molecule called *glycerol,* to which three *fatty acids* are attached. Fatty acids are chainlike molecules that typically contain between 14 and 22 carbon atoms. Each chain bonds with the glycerol molecule through a condensation reaction.

5.5 Proteins

Egg whites, gelatin, hair, and muscle among many other materials are made of proteins. **Proteins** are the basic building materials of all living things. Protein molecules contain carbon, hydrogen, and oxygen. Unlike lipids and carbohydrates, they also contain nitrogen as well as sulfur and other elements. All proteins are made of monomers called **amino acids.**

Amino Acids All amino acids have the same basic structure. As shown in Figure 5–5, in an amino acid four groups of atoms are bonded to a central carbon atom. Each group has distinctive chemical characteristics. The *acid group,* COOH, tends to give up a hydrogen ion. The *amino group,* NH_2, acts as a base

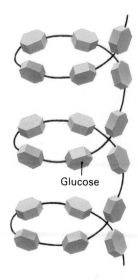

Figure 5–3. Starch is a polymer, a polysaccharide formed from many glucose units. When glucose is needed by cells, starch molecules are broken down by hydrolysis.

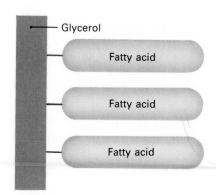

Figure 5–4. Many lipids consist of a single glycerol molecule and three fatty acids.

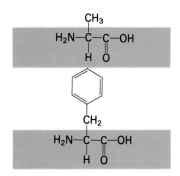

Figure 5–5. The amino acids alanine (top) and phenylalanine (bottom) have similar stuctures. The tinted portion is the same in almost all amino acids.

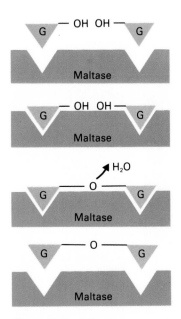

Figure 5–6. The enzyme maltase catalyzes the reaction that unites two glucose molecules to form the disaccharide maltose.

because it tends to combine with hydrogen ions. The third group consists of a single hydrogen atom.

The three groups listed so far are identical in all amino acids. Variety is introduced in the fourth group, called the *R group*. The R group can be a single hydrogen atom or a complex chain of atoms. Each of the 20 amino acids has its own specific R group.

Protein Structure   A protein molecule is formed through a series of condensation reactions. In each reaction, an acid group combines with an amino group from another amino acid. The bond that holds these two groups together is called a **peptide bond.** A molecule formed by two bonded amino acids is a *dipeptide*. Three or more amino acids bonded together form a *polypeptide*. All proteins consist of polypeptides.

The sequence of amino acids and the resulting shape of the protein molecule give each protein its unique characteristics. In *fibrous proteins,* for example, the polypeptide chains are long and stretched out. They may wrap around each other like strands of a rope, or they may be interwoven to form flexible layers. Fibrous proteins form muscles, spider webs, wool, nails, hooves, horns, and beaks. *Globular proteins* twist and fold into complex patterns because of the attraction between the electrically charged groups along the polypeptide chain. Hemoglobin and other globular proteins make up 90 percent of the solid matter in blood.

Enzymes   Some proteins act as **catalysts**—that is, they speed up chemical reactions in cells. Protein catalysts called **enzymes** control the rate of reactions without themselves being affected by the reactions. Some reactions occur up to one million times faster with enzymes than without them.

You can think of an enzyme as a key that fits only one lock. The "lock" is a specific molecule, which is called the enzyme's **substrate.**

When an enzyme and its substrate come into contact, chemical bonds form between them at a special place on the enzyme called the **active site.** Bonding of the enzyme to the substrate at the active site weakens, or "unlocks," some of the bonds in the substrate. As a result the substrate can more easily enter into chemical reactions. When two or more substrates are involved in the reaction, the enzyme holds the molecules together at the point where the reacting molecules will connect. Some reactions also require the presence of small, nonprotein molecules called *coenzymes*. Coenzymes usually help the enzyme bind to the substrate. Many vitamins act as coenzymes. Vitamins are very complex molecules that are smaller than proteins. Some are soluble in water and others are not. Vitamins are classified as members of groups with similar chemical characteristics. Vitamins of the B group are water soluble. Vitamin A, found in carrots and cod liver oil, is not water soluble.

Substances called *inhibitors* act to regulate enzyme activity. Inhibitors may block the active site of an enzyme. They may also distort the enzyme's shape by bonding with it elsewhere. Some poisons act as enzyme inhibitors.

Enzymes are marvels of biological efficiency. Most chemical reactions require a large amount of energy to get started. The action of enzymes lowers the energy required to start reactions so that the reactions can take place at normal body temperature. Enzymes also speed up reactions to the rates necessary to sustain life. Enzymes are also biologically efficient because they are not used up during chemical reactions. They can be used repeatedly. Thus the cell does not have to use extra energy to make new enzymes each time an enzyme is involved in a reaction.

5.6 Nucleic Acids

How do cells know which proteins, lipids, or carbohydrates to manufacture at any given time? Considering the huge number of different kinds of cells, the information must be extremely detailed and precise. Yet it must also be passed along to each new cell and each new organism, so it must be stored in compact form.

All instructions for cellular activity are carried by a class of organic compounds called **nucleic acids.** There are two kinds of nucleic acid. **Deoxyribonucleic acid,** or **DNA,** records the instructions and transmits them from generation to generation. **Ribonucleic acid,** or **RNA,** "reads" the instructions and carries them out. Both DNA and RNA are made of complex monomers called **nucleotides.** DNA is found primarily in the nucleus. RNA is found in both the nucleus and the cytoplasm. The nature and function of nucleic acids are discussed in Chapter 12.

Q/A

Q: *What is the advantage of storing energy in large molecules?*

A: Large molecules of condensed monomers take up less space than an equivalent amount of the monomer. Also, large molecules are usually insoluble in water and not easily broken down. Starch and glycogen are therefore more efficient storage units than glucose.

Reviewing the Section

1. What happens in a condensation reaction?
2. Why are fructose and galactose referred to as isomers of glucose?
3. What is the structure of a lipid molecule?
4. How do amino acids form proteins?
5. Why are enzymes necessary to living things?

Investigation 5: Nutrients in Foods

Purpose
To identify nutrients in foods by performing chemical tests

Materials
Safety goggles, whole milk and various other foods, Benedict's solution, hot plate, large beaker, test tubes, test tube holder, test tube rack, crucible, Lugol's iodine, brown wrapping paper or paper bag, Biuret reagent

Procedure
1. **CAUTION: Put on safety goggles and a lab apron. Leave them on for the entire investigation.** Test small samples of each food for the presence of water, sugar, starch, protein, and fat using the following chemical tests. Thoroughly wash and dry the crucible or test tube after each test.

Test for Water
Place the food sample in a crucible. Using a test tube holder, hold an inverted test tube over the crucible. Heat the food until only a residue remains. Any fluid that condenses on the glass surface is water.

Test for Sugar
CAUTION: Do not get Benedict's solution on your skin. If you do, wash it off immediately. Add about 5 mL (10 drops) of Benedict's solution to a test tube containing a food sample. Heat gently in a boiling water bath. The solution will turn green to brick red, depending on the amount of sugar.

Test for Starch
Place a drop of Lugol's iodine on a food sample. The drop will turn blue-black if starch is present.

Test for Fat
Rub a sample of a solid food or place a few drops of a liquid food on brown paper. Hold the paper to light. If the food contains fat, you will see a translucent spot that will not disappear even when dry.

Test for Protein
1. Add 2–3 drops of Biuret reagent to a test tube containing a food sample. A color change from pink to purple indicates the presence of protein.
2. Make a table like the one below and record the results of your test.

Nutrients Present in Foods

Food			
Water			
Sugar			
Starch			
Fat			
Protein			

Analyses and Conclusions
1. Compare your results with those of other students. Explain any differences.
2. Why do you think milk has long been considered the almost perfect food?
3. If you extended this experiment, how could you use the data to help plan a balanced diet?

Going Further
- Test a plain soda cracker for the presence of sugar. Chew another piece of the cracker for two to three minutes but do not swallow it. Place the chewed cracker in a test tube and test it for sugar. Explain your results.
- Some food products are labeled "protein enriched" or "protein fortified." Test a few of these to verify the accuracy of the claims.
- Pasteurized food products must have certain nutrients replaced after the pasteurization process. Research the effects of pasteurization. Suggest alternatives to this process.

Chapter 5 Review

Summary

Biochemistry is the study of of how the properties of atoms and molecules make life possible. For living things, water is the most important inorganic compound. Certain properties of water are essential for life. The polarity of water makes it an excellent solvent and gives water a high heat capacity. Acids and bases are also important to living things because of the part they play in the body's chemical balance.

All organic compounds contain carbon. Organic molecules typically form when two or more monomers join in a condensation reaction. Monosaccharides, the monomers that form carbohydrates, contain carbon, hydrogen, and oxygen. Cellulose and starch are polysaccharides.

Lipids are complex compounds that are insoluble in water. Fats and oils are lipids that store concentrated energy and protect cells.

Proteins are highly complex molecules that contain nitrogen. They are built of amino acids, which can combine in thousands of ways. Enzymes are proteins that catalyze chemical reactions within organisms. The instructions for all cellular activity are carried by nucleic acids.

BioTerms

acid (**68**)
active site (**72**)
amino acid (**71**)
base (**68**)
carbohydrate (**69**)
catalyst (**72**)
condensation reaction (**69**)
deoxyribonucleic acid (DNA) (**73**)
disaccharide (**69**)

enzyme (**72**)
hydrolysis (**71**)
inorganic compound (**67**)
isomer (**69**)
lipid (**71**)
monomer (**69**)
monosaccharide (**69**)
nucleic acid (**73**)
nucleotide (**73**)
organic compound (**67**)

peptide bond (**72**)
pH scale (**68**)
polar compound (**67**)
polymer (**69**)
polysaccharide (**70**)
protein (**71**)
ribonucleic acid (RNA) (**73**)
structural formulas (**70**)
substrate (**72**)

BioQuiz   (Write all answers on a separate sheet of paper.)

I. Completion

1. A condensation reaction produces a new compound and releases a molecule of _____.

2. Disaccharides are formed from two _____.

3. Water dissolves ionic compounds because it is a _____ compound.

4. _____ acids consist of four chemical groups around a central carbon atom.

II. Modified True and False

Mark each statement TRUE or FALSE. If false, change the underlined term to make the statement true.

5. Bases release <u>hydroxide</u> ions in water.

6. <u>Amino acids</u> carry all instructions for cellular activity.

7. Enzymes <u>slow down</u> reactions.

8. <u>Hydrolysis</u> breaks down complex sugars into their monomers.

III. Multiple Choice

9. Chemical bonds form between an enzyme and its substrate at the a) polar compound. b) active site. c) monomer. d) condensation reaction.
10. Water dissociates into one hydrogen ion and one a) carbon atom. b) hydroxide ion. c) sodium ion. d) oxygen ion.
11. A strong acid might have a pH of a) 0–3. b) 4–7. c) 8–11. d) 12–14.
12. All organic compounds contain a) hydrogen. b) nitrogen. c) phosphorus. d) carbon.
13. Glucose is a monosaccharide used for a) storing energy. b) quick energy. c) building cells. d) catalyzing reactions.

14. Isomers have the same a) bonds. b) structure. c) atoms. d) traits.

IV. Essay

15. How do lipids function in the body?
16. How do enzymes lower the amount of energy needed for chemical reactions?
17. Why must living things maintain a fairly constant pH balance?
18. Which chemical elements or traits can be used to identify carbohydrates, lipids, and proteins?
19. What gives each kind of protein its unique characteristics?
20. Why is water ideally suited to support the life of organisms?

Applying and Extending Concepts

1. Honeycombs are made of wax, a kind of lipid. What might happen if a honeycomb were made of a carbohydrate such as sucrose?
2. When proteins are cooked, their structure is permanently changed and their enzymes cease to function. This process is called *denaturation*. Why, do you suppose, is denaturation used to preserve foods? Be sure to use examples of particular foods in your answer.
3. Two French chemists, Anselme Payen and Jean F. Persoz, performed early experiments that led to the discovery of digestive enzymes. Use your school or public library to research the work of these men and write a brief report summarizing their experiments. If time permits, present your report to the class.
4. Fats and oils can store up to six times more energy per gram than can carbohydrates. How might the differences in plant and animal life explain why plants store energy as starch and animals store energy as fats and oils? Why do you suppose arctic animals such as whales and seals store much of their food as oil rather than as fat?

Related Readings

Adler, I. *How Life Began*. New York: Harper & Row, 1977. This introductory book explains biochemical processes at the molecular level.

Berman, W. *Beginning Biochemistry*, rev. ed. New York: Arco, 1980. Written with verve, this handbook gives a comprehensive overview of basic biochemistry.

Lehmann, P. L. "Doling Out Dietary Minerals." *Sciquest* 53 (December 1980): 6–10. This article lists sources of minerals and describes dangers of overdoses.

Patrusky, B. "Why Do We Cry Tears?" *Science 81* 2 (July–August 1981): 104. The article answers its question by explaining the biochemistry of tears.

The Cell

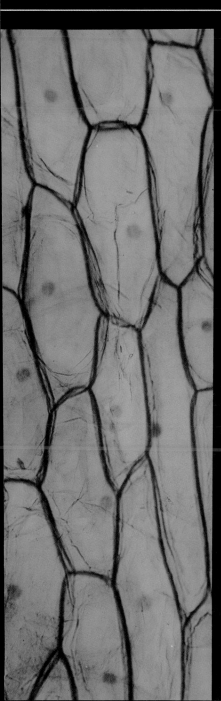

Cell Structure and Function

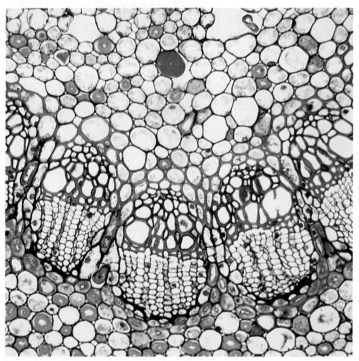

Cross section of a *Gnetum* leaf, stained with safranin-fast green, ×160

Introduction

Organisms can be as large and complex as whales or as small and simple as bacteria. Large or small, each organism is made up of atoms and molecules. The massive whale and the microscopic bacteria are each essentially a combination of chemicals. How then can something be termed alive? At what point do the atoms and molecules cease to be just a collection of chemicals and take on the properties of living organisms?

The discovery over 300 years ago that all living things are composed of cells began a period of study that continues today. Cells are more than a collection of chemicals. They possess a unique organization that gives them all the qualities of life. The study of cells and how they are organized is, then, the study of what constitutes life.

The Discovery of Cells

Most cells are too small to be seen with the unaided eye. The development in the seventeenth century of the compound light microscope gave biologists the tool they needed to study cells. The first compound light microscopes magnified an object 270 times. Today's light microscopes can magnify objects up to 2,000 times. Electron microscopes can magnify objects more than 200,000 times. Even with these modern tools, however, some details of cell structures are much too small to be seen.

6.1 Early Microscopic Observations

One of the first records of observations using the compound light microscope was made by the English scientist Robert Hooke. During the 1660s Hooke made a simple but significant discovery. Hooke carefully shaved a very thin section of cork from a plant stem and looked at it under his microscope. In his book *Micrographia,* Hooke recorded, ''I could . . . plainly perceive it to be perforated and porous, much like a honeycomb, but the pores were not regular. . . .'' ***Hooke named the structures that he saw under a microscope ''cells.''*** These small boxlike cavities reminded him of the small rooms of a monastery where monks lived. Hooke was not looking at living cells. He was looking at the nonliving outer walls of what once had been living cork cells. Nonetheless, this was a significant discovery about the structure of living things.

During the next 160 years, microscopes and their lenses gradually improved. In the 1820s, a French botanist named René Dutrochet became one of the first to make a generalized statement about the structure of living things. He examined parts from many animals and plants under the microscope and concluded that various parts of organisms are composed of cells. To rephrase Dutrochet, the leaves of a tree and the skin on your hand are each made of cells.

The 1830s brought a flurry of activity in cell study. Pieces of all sorts of organisms were studied under the microscope. In Scotland, Robert Brown announced that a cell contains a large, central part, or **nucleus.** Another French scientist, Felix Dujardin, reported that cells are not hollow, empty structures. He described them as full of a clear, jellylike fluid.

A great wealth of information was accumulating. What did these seemingly isolated facts about cells mean? By the late 1830s, two German scientists working separately began to formulate a general theory about cells. After extensive research

Section Objectives

- *List* the contributions of Hooke, Schleiden, Schwann, and Virchow to the development of the cell theory.
- *State* the cell theory.
- *Give* examples of cells in which structure is well-suited to the function of the cell.
- *Name* a major factor that limits the size of cells.

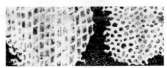

Figure 6–1. In the 1660s Robert Hooke used this microscope (top) to view the microscopic world for the first time. He observed and illustrated thinly sliced cork (bottom) and called the pores he saw ''cells.''

Q: *Who first suggested the use of magnifying lenses to look at tiny objects?*

A: Leonardo DaVinci was the first to suggest the use of magnifying lenses—more than 100 years before Hooke made his discovery.

on plants, Matthias Schleiden concluded that all plants and plant parts are composed of living cells. Theodore Schwann came to the same conclusion about animals. The separate conclusions of Schleiden and Schwann, when considered together, meant that all living things are composed of cells. Cells are the basic building blocks of living things, just as bricks are the basic structural units of many buildings.

The last bit of information required to complete the general theory of cells was supplied by German physician, Rudolph Virchow. Virchow advanced the idea that existing cells give rise to new cells. He stated, ''All cells come from living cells.'' Life is an unbroken chain of cells going all the way back to the first living thing.

6.2 The Cell Theory

The conclusions of Schleiden, Schwann, and Virchow are summarized in the **cell theory,** which has three parts:

1. *All organisms are composed of cells.*
2. *Cells are the basic units of structure and function in organisms.*
3. *All cells come from preexisting cells.*

Like most theories, this one has some exceptions. If every cell comes from a parent cell, where did the first cell come from? Billions of years ago the environmental conditions on the earth were much different than they are today. Could conditions at that time have enabled the first living cell to form from non-living materials?

Another exception to the cell theory is the virus. A virus is a package of nucleic acid wrapped in a protein coating. It possesses only a few of the structures of a cell. In fact, a virus must invade and occupy a cell in order to reproduce. Without a host cell, a virus is as lifeless as a grain of sand. A virus is not a cell. Is a virus alive, or is it somewhere between life and nonlife?

6.3 Size and Shape of Cells

Cells exist in a great variety of sizes and shapes. The smallest cells are bacterialike organisms called *mycoplasmas.* Some of these organisms cause diseases that infect the respiratory system. Mycoplasmas are about 0.1 to 0.3 micrometer (μm) in diameter. One micrometer equals one-millionth of a meter (0.000039 in.). Mycoplasmas can be seen only with an electron microscope. The smallest cells that can be seen with a light microscope are larger bacteria, from 1 to 5 μm in diameter. At

Human egg cell = 14 red blood cells

Red blood cell = 35 microorganisms

Figure 6–2. One human egg cell has a diameter equal to that of 14 human red blood cells. The diameter of one red blood cell equals that of 35 microorganisms called rickettsias.

▶ The Size of Cells

Although the range of cell size from mycoplasma to an ostrich egg is great, these two extreme examples are not typical of cells in general. Most living cells are between 2 and 200 μm in length. Can it be mere coincidence that almost all cells are exceedingly small? What factor limits cells to their microscopic size?

One of the most important factors in determining cell size is the relationship between surface area and volume. Two cube-shaped cells can be used to illustrate this relationship. The smaller cube is one unit on a side, and the larger cell is twice that size, or two units on a side.

The volume of a cube equals length times width times height. The surface area is calculated by multiplying width times length

times six—the number of sides of the cube. Thus doubling the width of a side of the cell increases the volume eight times. The surface area, however, increases only four times.

What does this mean to the cell? The volume reflects the amount of living material in the cell. The chemicals required to keep this material alive must enter or leave the cell's surface. It follows that a cell must have a sufficient amount of surface area to keep the living material adequately supplied.

The smaller of the two cells has six units of surface area for each unit of volume. When the width of the cell is doubled, the ratio of surface area to volume drops to 3:1. If the cell grew to a cube four units on a side, the problem would get worse. The

ratio of surface area to volume would then drop to 1.5:1.

A problem has developed. As the cell increases in size, it has proportionately less surface area than volume. Its need for materials increases faster than its ability to supply those materials.

An ostrich egg can grow as large as it does because its food, the yolk, is already inside the cell. In bird's eggs almost the entire volume of the egg cell is given over to food. The rest of the cell's apparatus appears as only a small patch on the surface of the yolk.

The relationship between surface area and volume limits the size of living things. It is an important factor whether determining the size of a cell or the size of a gorilla.

the opposite extreme is the ostrich egg cell, which measures about 100 mm (4 in.) in diameter. The volume of an ostrich egg cell is approximately one million trillion times greater than that of the smallest mycoplasma. In everyday terms, this is like comparing the sizes of a whale and a flea.

Cells exhibit an even greater variety in shape than in size. Some *unicellular,* or one-celled organisms, such as a paramecium or a euglena, have a definite shape. Others, such as amoebas, are constantly changing in shape. Even within the human body, cells exhibit a great variety of shapes. Cells near the surface of the skin are flattened, irregularly shaped discs a few

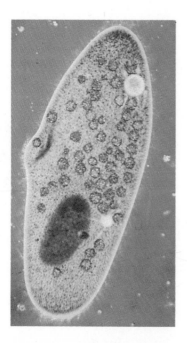

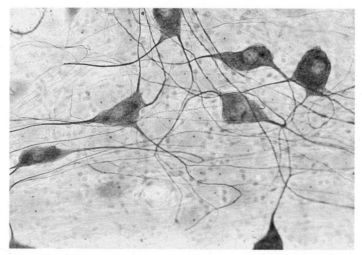

Figure 6–3. Cells are often modified in ways that meet the special needs of the organism. The cell of a unicellular organism, such as a paramecium (left), is adapted to meet all the organism's needs. Human nerve cells have unusual shapes related to the task they perform.

micrometers across. In contrast, nerve cells are long, thin, and stringlike. Some human nerve cells are over a meter in length.

Why is there such diversity in the size and shape of cells? You might as well ask why balls are round and tennis rackets flat. Most things in the everyday world are designed to perform a function. Their sizes and shapes relate to their functions. This same principle holds true for cells. Skin cells function as a covering for the body just as shingles on a roof cover a house. The similarity of the broad, flat shape of skin cells and shingles is related to their similarity in function. Is it just a coincidence that nerve cells are long and thin? Nerve cells function as transmission lines throughout the body. Shape is related to function in nerve cells just as it is in telephone wires.

Biological structure is closely related to the function that is performed by that structure. This basic principle holds true for the structures composed of cells, the cells themselves, and the parts that make up the cells.

Reviewing the Section

1. What contributions did Schleiden, Schwann, and Virchow each make to the development of the cell theory?
2. Describe what Robert Hooke actually saw when he looked at cork shavings through the microscope.
3. State the cell theory.
4. How do you account for the variety in the shapes of cells?
5. Why is the surface area an important factor in limiting a cell's growth?

Basic Parts of the Cell

A "typical" cell cannot be described any more successfully than can a "typical" animal. Nevertheless, most cells do have certain characteristics in common. The structure of most cells can be divided into three basic parts: the nucleus, or control center; the cell membrane, or outer boundary of the cell; and the cytoplasm, or everything between the nucleus and the membrane.

The nucleus and its component parts form the control center of the cell. The **cytoplasm** (SYT uh plaz uhm) is the material between the nucleus and the outer boundary. Within the cytoplasm are found cellular **organelles**—tiny structures that perform specialized functions in the cell. Organelles function in the cell in much the same way as organs function in the human body. Each organelle has a special task that helps maintain the life of the cell.

You can think of the cell as a microscopic factory. The cell takes in raw materials and manufactures products much as factories do in human society. Consider an automobile factory, for example. Such a factory can be very large and spread out over a square kilometer or more. To produce a car, a great many smaller tasks must be performed. One part of the factory may make the engine and another part the body. A large factory also has departments that do not directly contribute to the final product. The maintenance department keeps the factory itself in good working order. The janitorial service sees that the factory is clean. Management makes sure that everything is done correctly and on time.

6.4 The Nucleus

Using the analogy of the factory, the cell nucleus represents management. In many cells the nucleus is the most prominent internal structure. Many nuclei are spherical, but some are cylindrical and others disclike.

The nucleus performs two important functions for the cell. *First, the nucleus controls most activities that take place in the cell.* The nucleus provides the instructions for building proteins. It determines not only how they will be made but which proteins will be made and when. Proteins, in turn, regulate most of the other chemical processes in the cell.

Second, the nucleus transmits hereditary information. Each generation of cells must have the appropriate directions for maintaining all life's functions. The nucleus is responsible for

- *Identify* the three basic parts of most cells.
- *Compare* the structure of the cell membrane with the cell wall.
- *State* the function of each organelle.
- *Give* an example of a cytoplasmic inclusion.

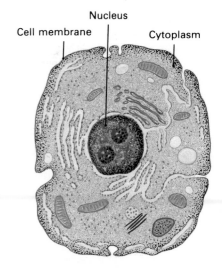

Figure 6-4. A typical animal cell is composed of a cell membrane, cytoplasm, a nucleus, and various organelles. Each component plays an important role in the cell's biology.

Cell Structure and Function **83**

Figure 6–5. The large, spherical object that occupies the center of this electron photomicrograph is a cell nucleus magnified 16,000 times. The nucleus directs cell activities and is often called the control center of the cell.

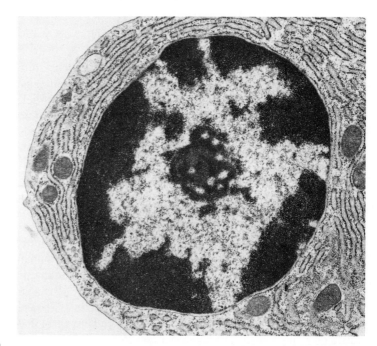

nucleoli (singular, *nucleolus*)

the orderly transfer of that information to the next generation of cells. The nucleus contains the hereditary material. This material consists of the nucleic acid DNA and, with proteins, is formed into rod-shaped or rope-shaped bodies called **chromosomes.** The chromosomes are clearly visible only when the cell is dividing. In between cell divisions the hereditary material is no longer rod-shaped. In this more diffuse state the material is referred to as **chromatin.** *Chromosomes carry the hereditary information from one generation of cells to the next.*

The nucleus of most cells has one or more spherical bodies called **nucleoli** (noo KLEE uh ly). A nucleolus is most prominent in cells that actively make proteins. Nucleoli are made of DNA, another type of nucleic acid called RNA, and proteins.

The chromosomes and nucleoli are surrounded by two nuclear membranes, or the **nuclear envelope,** that forms a boundary with the cytoplasm. This envelope is a double layer of lipids and proteins with openings, or **nuclear pores,** scattered throughout. The nuclear pores allow materials to pass between the cytoplasm and the nucleus.

6.5 Cell Membrane and Cell Wall

A fence and security system enclose the outer boundary of many large factories. The security department regulates what enters and leaves the factory. The cell, like the factory, has an outer

Highlight on Careers: Cell Biologist

General Description

A cell biologist is a scientist who examines the cell and its organelles in detail. Using a small tissue sample, the cell biologist isolates cells and studies them under a microscope.

Cell biologists, also known as cytologists, tend to specialize in either plant or animal cells. In either case, they are looking for answers to the same key questions: How do cells reproduce? Why do some cells become diseased and die? How are cells affected by physical and chemical changes?

Many cell biologists teach in universities while continuing to research basic questions. A growing number work for privately owned research and development laboratories or industrial companies.

Many cell biologists who work for private laboratories focus on how human cells pass on hereditary information. How and why normal cells become cancerous is another area of concern. Aging is also an important topic of investigation for the cell biologist, who attempts to track every stage in the life of a cell.

Basic research in cell biology may yield important results for many areas of human health. This is a rapidly growing field.

Career Requirements

An entry-level job as a technician in an industrial, government, or private laboratory requires a B.S. degree in biology. To advance in the field, however, a cell biologist must earn an M.S. or Ph.D. degree in cell biology.

A Ph.D. is also required for college teaching positions. College or university work offers the scientist not only the potential for advancement but also the opportunity to do independent self-directed research.

Personal requirements for the cell biologist include a keen interest in scientific experiments and abilities in chemistry and mathematics. Patience and an aptitude for detail are also important assets.

For Additional Information
American Society for Cell Biology
4316 Montgomery Avenue
Bethesda, MD 20014

boundary composed of lipids and proteins, called the **cell membrane.** The membrane, sometimes called the *plasma membrane*, holds the cell together. It protects the contents of the cell and helps give the cell a shape. ***The cell membrane regulates what enters and leaves the cell.***

The cells of plants, algae, fungi, and some bacteria possess a **cell wall** in addition to a cell membrane. ***The cell wall aids in the protection and support of the cell.*** The wall is very porous and so allows water and dissolved substances to pass through

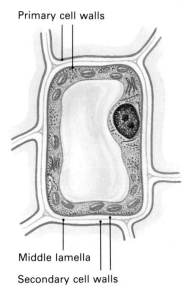

Primary cell walls

Middle lamella

Secondary cell walls

Figure 6–6. Unlike an animal cell, a typical plant cell is surrounded by a multilayered cell wall. Adjacent plant cells are usually joined by a sticky substance called a middle lamella.

easily. As Figure 6–6 shows, the cell wall lies outside the cell membrane.

Cell walls of plants are built in a series of steps. Two newly formed cells build a partition between themselves called the **middle lamella** (luh MEHL uh). This structure contains a gluey substance, *pectin*, that helps hold the cells together.

Each of the cells then forms a *primary cell wall* on its side of the middle lamella. This structure is composed of *cellulose*, a fibrous material. The elasticity of the cellulose fibers allows the wall to stretch as the cell grows.

When the cell is completely grown, plants that have woody stems add a *secondary cell wall*. This wall is composed of cellulose and *lignin*, a substance that stiffens the cellulose. Wood is a material that consists mainly of rigid secondary cell walls.

6.6 The Cytoplasm and Organelles

The term *cytoplasm* is generally used to describe everything within the cell except the nucleus. Today, scientists know that this material is more than the clear jellylike fluid first observed by Dujardin. With powerful electron microscopes, scientists have discovered that cytoplasm is a highly complex material. It is not as uniform in consistency as jelly. The cytoplasm contains numerous organelles, many of them *membrane-bound,* or enclosed by a membrane.

If you look at living cells through a microscope, you will see that the cytoplasm is in constant motion. This flow of cytoplasm is called **cytoplasmic streaming.** The cytoplasm is more than just a place where things happen. It is part of the living material of the cell. Cytoplasmic streaming is one mechanism that moves materials throughout the cell.

Endoplasmic Reticulum   A series of canals or channels called the **endoplasmic reticulum** (ehn duh PLAZ mihk rih TIHK yuh luhm), or **ER,** winds through the cytoplasm. This network of interconnecting, flattened sacs and tubes is lined with a thin, delicate membrane. *The endoplasmic reticulum is the cell's internal transport system.* The ER connects with the nuclear envelope. It serves as a transportation route for materials moving between various parts of the cytoplasm and the nucleus of the cell.

The amount of ER within a cell varies with the cell's function. Cells that produce large quantities of proteins for use outside the cell have the most ER. For example, cells that line the stomach and those that produce regulatory hormones possess a large, complex ER.

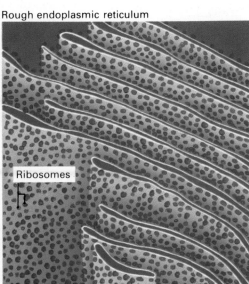

Rough endoplasmic reticulum

Ribosomes

Cells contain two types of endoplasmic reticulum—*smooth ER* and *rough ER*. Smooth ER plays an important role in building the lipids that will be used in the plasma membrane of the cell. Smooth ER also helps neutralize poisons. Drugs such as amphetamines and morphine are broken down within the smooth endoplasmic reticulum. Rough ER gets its name from the presence of ribosomes attached to it.

Ribosomes Tiny, knoblike organelles called **ribosomes** (RY buh sohmz) are manufacturing centers of the cell. They are not enclosed by a membrane. ***Ribosomes are sites of protein synthesis.*** Proteins used within the cell are formed on "free" ribosomes scattered throughout the cytoplasm. Proteins exported out of the cell are synthesized on the "bound" ribosomes attached to the membrane of the rough ER. Because these manufacturing units lie along the transport system, the products can be more easily collected for relocation.

Proteins are vital to life—all cells must produce them. Ribosomes, accordingly, are the most numerous of all the organelles. The proteins produced by the rough ER collect in large fluid-filled sacs. These sacs are like warehouses along the transport systems. When a quantity of protein has accumulated, it may be released from the cell or transferred to a packaging area.

Figure 6–7. Magnified 20,000 times, endoplasmic reticulum resembles folded strings winding through the cytoplasm (left). The illustration shows that ER consists of sheets of membranes (right). Rough ER is covered with ribosomes; smooth ER is not.

Q/A

Q: *Is hair made of cells?*

A: No. Hair is made of protein that is secreted by hair follicle cells. The hair follicle surrounds the root of the hair.

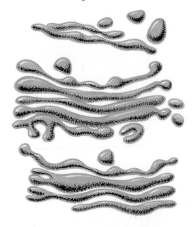

Golgi bodies

Figure 6–8. Golgi bodies are stacks of membranes that float in the cytoplasm. The small, membrane-bound sacs around the edge of each Golgi body store chemicals.

Figure 6–9. The three-dimensional nature of the membranes of the mitochondrion are illustrated below. These same structures are magnified 70,800 times in the photograph below right.

Golgi Bodies   As chemicals collect in the ER, small saclike pieces of the membrane are pinched off. These tiny sacs gradually combine to form **Golgi** (GOHL jee) **bodies,** which appear to be stacks of tubes with membranous sacs at the ends. As Figure 6–8 shows, they look like tiny flattened balloons in the cytoplasm. Golgi bodies get their name from their discoverer, the Italian biologist Camillo Golgi. They are also known as the *Golgi complex* or *Golgi apparatus.*

Golgi bodies are areas for the storage and packaging of chemicals. The accumulated chemicals are secreted from the cells. As with the ER, the number and size of the Golgi bodies are greater in cells that produce large quantities of chemicals. Cells that make saliva and other materials that aid digestion have large numbers of Golgi bodies. Mucus, the substance that coats the nasal passages, is packaged and stored in the Golgi bodies of the mucus-producing cells.

Mitochondria   The organelles that release energy from the nutrients taken into the cell are **mitochondria** (myt uh KAHN dree uh). The mitochondria are the cell's powerhouses. They vary in shape from almost spherical to sausagelike. As Figure 6–9 shows, mitochondria are composed of two membranes. The structure of each membrane is similar to that of the cell membrane. The outer membrane is a smooth enclosure, while the inner membrane has many folds. This folding of the membrane greatly increases the internal surface area. A complex series of energy-releasing chemical reactions takes place on the surface of this inner membrane. *Chemical activity in the mitochondria provides energy for the cell.*

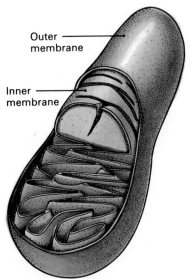

Outer membrane

Inner membrane

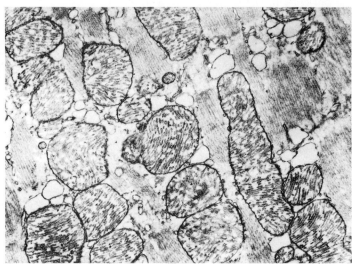

Mitochondria are most common in active cells. A liver cell may have as many as 2,500 mitochondria. A heart muscle cell has even more. An interesting aspect of mitochondria is their ability to reproduce. Mitochondria contain DNA and ribosomes. They reproduce within the cytoplasm from materials already in the cytoplasm or imported through the cell membrane.

Plastids   Another group of specialized organelles that can reproduce themselves are **plastids.** They are found only in plants and in some algae. *Some plastids contain food; others contain pigments.* Like mitochondria, each type is composed of an outer membrane surrounding a complex system of inner membranes.

Leucoplasts (LOO kuh plasts) are colorless plastids that store food. The cells of roots and stems may contain many leucoplasts. *Chromoplasts* contain the pigments responsible for red, orange, and yellow color in fruit, flowers, and autumn leaves. *Chloroplasts* contain the green pigment chlorophyll. Chloroplasts are the site of food production in plants and algae.

Vacuoles   A factory has a warehouse or a storeroom. Some cells, especially plant cells, require a storage area for certain substances. **Vacuoles** are bubblelike structures that store water, other liquids, waste materials, or food particles. A membrane keeps the contents of the vacuole separate from the cytoplasm. This membrane functions in much the same way as the cell

Figure 6–10.   The photograph below left shows a plant cell with a large central vacuole magnified 610 times. Water in the central vacuole exerts pressure on the cell that helps it retain its shape (below right). The central vacuole is also the site of waste storage for plant cells.

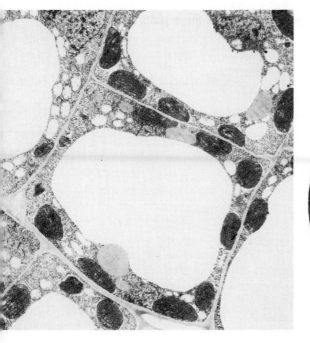

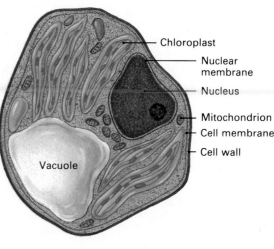

Chloroplast

Nuclear membrane

Nucleus

Mitochondrion

Cell membrane

Cell wall

Vacuole

membrane. It regulates the passage of materials between the cytoplasm and the contents of the vacuole.

The interior of many plant cells contains one large water-filled vacuole, as Figure 6–10 shows. The pressure exerted by the water in the vacuole helps maintain the shape of the cell. When the plant is deprived of water, the vacuoles collapse and the plant wilts. Animal cells have only small vacuoles or none at all.

Lysosomes Membrane-bound organelles that are formed in the Golgi bodies are **lysosomes** (LY suh sohmz). Loaded with strong destructive enzymes, lysosomes digest large particles found in the cell.

One example of lysosomes at work can be seen in the action of white blood cells. These cells engulf bacteria and other foreign objects in the blood. When the cell encircles a bacterium, the cell membrane becomes a vacuole around it. Lysosomes fuse with the vacuole membrane and dump their chemical contents inside. The powerful enzymes destroy the bacterium.

Other Organelles Cells are made more rigid by long, slender tubes called **microtubules** (my kroh TOOB yoolz). These hollow cylinders of protein often lie just beneath the cell membrane. They help support the cell and maintain its shape. **Spindle fibers** are microtubules that appear during cell division. These temporary structures help move chromosomes through the cytoplasm. Microtubules are used during cell division to construct **centrioles** (SEHN tree ohlz), small dark bodies located outside

▶ Cytoplasmic Inclusions

Each of the cellular structures discussed in this chapter is an active participant in the functioning of a cell. Each has a specific task to perform to maintain everyday life in the cell.

Cells also include materials that do not participate in the everyday functions of cell life. Much smaller than the smallest organelle, these materials are called *cytoplasmic inclusions*. Lipid droplets, starch grains, and yolk granules are examples of cytoplasmic inclusions that function as a cellular food supply.

Crystals of a substance called *guanine* are often found as inclusions in the skin cells of fish and amphibians. These crystals give fish their silvery appearance. Guanine crystals are also present in the eyes of animals such as cats. At night, the crystals reflect light, enabling cats to see in dim light. The crystals also cause cat's eyes to appear to glow in the dark.

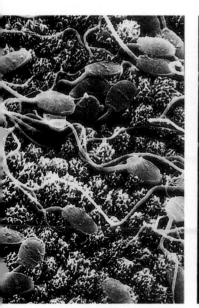
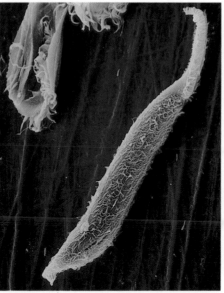
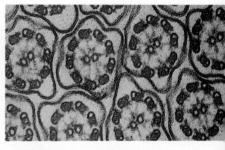
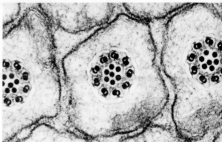

the nucleus in many cells. These cylindrical structures generally exist in pairs and perform a function only during cell division.

Cilia (SIHL ee uh) and **flagella** (fluh JEHL uh) are hairlike projections that stick out from the surface of the cell. Cilia are short, threadlike organelles. When present, they are generally numerous. Flagella are longer and less numerous. Flagella aid in the locomotion of unicellular organisms. Cilia aid in locomotion as well as in the movement of substances along the cell's surface.

Cilia and flagella are similar in structure. Each of these organelles contains nine pairs of microtubules that form a cylinder. In the center of the cylinder are two unpaired microtubules. Protein links also connect the outer microtubules with the inner ones, and the nine outer pairs with one another.

Figure 6–11. The flagellum of spermatozoa (left) and the cilia of *Euplotis* (center) enable the organisms to move. Cross sections of cilia (top right) and a flagellum (bottom right) magnified 20,000 times show microtubules arranged in some variation of the "9 + 2" scheme.

Reviewing the Section

1. What are the three basic units of all cells?
2. What is the main function of the cell membrane?
3. What is the function of each of the three structures that contribute to the support of a plant cell?
4. How do lysosomes function to destroy bacteria?
5. How do the products of cytoplasmic ribosomes differ from those of ribosomes located on rough ER?
6. What is a cytoplasmic inclusion? Give an example.

Differences in Cells

- *Distinguish* between prokaryotes and eukaryotes.
- *Give* examples of prokaryotic organisms and eukaryotic organisms.
- *State* the differences in structure between plant and animal cells.

Not all of the organelles discussed thus far are present in all cells. A cell's structure is closely related to the function it performs. The organelles found in a cell reflect its function. Muscle cells, for example, have large concentrations of mitochondria because of the high energy requirements of muscles.

Cells can be grouped according to their similarities and differences. All cells can be divided into two large categories—**eukaryotes** (yoo KAR ee ohts), or cells with a nucleus, and **prokaryotes** (proh KAR ee ohts), or cells without a true nucleus.

6.7 Cells Without a True Nucleus

Prokaryotes have nuclear material that is not surrounded by a nuclear membrane. The DNA, or hereditary material, of prokaryotic cells is not arranged into chromosomes. Rather, the DNA is a single circular molecule. The earliest cells found in the fossil record did not have a nucleus surrounded by a nuclear membrane. Today, prokaryotes include bacteria and blue-green algae.

Figure 6–12. Prokaryotic cells, such as the cells of bacteria and blue-green algae, lack a true nucleus. They also lack the mitochondria, Golgi bodies, lysosomes, and other components that characterize eukaryotic cells.

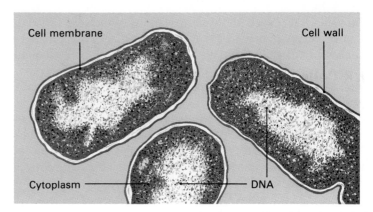

Prokaryotic cells differ from eukaryotic cells in several other ways. They have no mitochondria, chloroplasts, or endoplasmic reticulum. They do not possess Golgi bodies, lysosomes, and vacuoles. Thus they do not have any membrane-bound organelles. Prokaryotic cells do have a cell membrane and a cell wall, although these walls differ from the walls of plant cells. The flagella of bacteria are also quite different in structure and mode of action, though their purpose is the same as for eukaryotes. Prokaryotic cells also have ribosomes and these are much like those of the highest eukaryotic organisms.

Table 6–1: Differences Between Prokaryotes and Eukaryotes

Characteristics	Prokaryotes	Eukaryotes
Nuclear membrane	Absent	Present
Chromosomes	Single circular molecule of nucleic acid	Multiple, composed of nucleic acid and protein
Membrane-bound organelles	Absent	Present
Cell membrane	Present	Present
Cell wall	Present, contains muramic acid	When present, does not contain muramic acid
Ribosomes	Small	Large
Chlorophyll	When present, not contained in chloroplasts	When present, contained in chloroplasts
Flagella	Lack 9+2 microtubular structure	Have 9+2 microtubular structure
Cytoplasmic streaming	Does not occur	Occurs

6.8 Cells with a Nucleus

All other cells are eukaryotes. *Eukaryotes are cells that possess a well-defined nucleus surrounded by a nuclear membrane.* The complex chromosomes of eukaryotic cells can be seen during cell division. All of the membrane-bound organelles can be found in eukaryotic cells. Each organelle is specialized to perform its task well. The many organelles of eukaryotic cells allow for a greater division of labor. The Golgi bodies package chemicals, for example, and the ER forms a route to transport the materials. The greater efficiency that results makes it possible for eukaryotic organisms to be multicellular and to grow larger than prokaryotic organisms.

6.9 Animal and Plant Cells

If you were to compare a tree with a cat, you would find them very different. When the individual cells of these organisms are compared, however, they have many characteristics in common. Cells are the basic structural and functional units of both

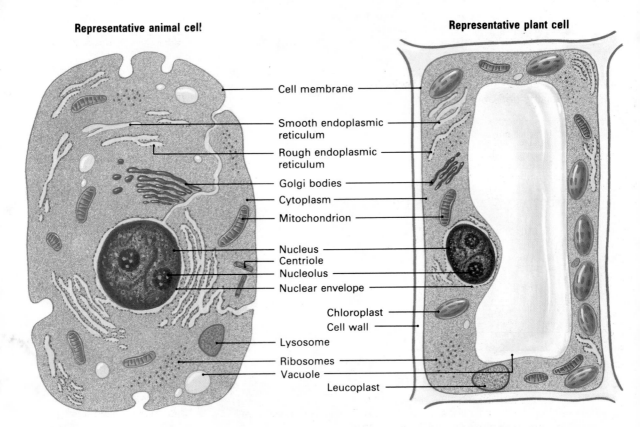

Representative animal cell

Representative plant cell

Cell membrane

Smooth endoplasmic reticulum

Rough endoplasmic reticulum

Golgi bodies

Cytoplasm

Mitochondrion

Nucleus
Centriole
Nucleolus
Nuclear envelope

Chloroplast
Cell wall

Lysosome

Ribosomes
Vacuole
Leucoplast

Figure 6–13. Animal and plant cells are similar but have some structural differences. Plant cells have cell walls, large central vacuoles, and chloroplasts. Animals lack these structures.

plants and animals. They share many characteristics: both are multicellular organisms; all plant and animal cells are eukaryotic; and most organelles are present in both types of cells. Some characteristics, however, are unique to the cells of each type of organism.

Plant cells may contain three structures not found in animal cells. Cell walls, large central vacuoles, and plastids are characteristic of plant cells. These structures are not found in animal cells. Centrioles are found in some but not all types of plant and animal cells.

Reviewing the Section

1. In what ways do eukaryotes and prokaryotes differ?
2. What allows eukaryotes to be multicellular and to grow larger than prokaryotes?
3. What characteristics of plant cells are not found in animal cells?

Investigation 6: Comparing Cells

Purpose
To compare different types of cells from various plants and animals

Materials

Microscope, glass slides, coverslips, paper towels, lens paper, variety of plant and animal specimens (such as leaves, roots, fish scales, fish muscle tissue), forceps, scalpel, medicine dropper, Lugol's iodine, methylene blue

3. Observe each stained slide under a microscope. *How do the stained specimens differ from the unstained?*
4. Draw diagrams of the types of cells. Label the parts of the cells. *What major differences do you observe among the cells?*
5. Make a table like the one shown. Fill in the table by indicating the appropriate structures that you were able to observe in the specimens you examined.

Specimen name	Plant or animal	Cell wall	Cell membrane	Nucleus	Other
			(present/ absent)		

Procedure
1. As your teacher directs, bring to class a variety of plant and animal specimens. Following the procedures listed on page 831, prepare a wet mount of tissue from each specimen.
2. Observe each specimen under a microscope. If necessary to make cell details easier to observe, use a stain, either Lugol's iodine or methylene blue. The following diagram shows how to stain a slide.

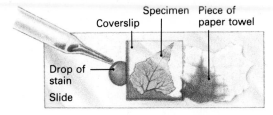

Analyses and Conclusions
1. Study your diagrams and compare the different types of cells in terms of size and organelles.
2. Drawing on your experiences, describe common problems in preparing slides and how to overcome them.

Going Further
Prepare slides to compare living and nonliving cells as well as stained and unstained cells.

Chapter 6 Review

Summary

Improvements in microscopes led nineteenth century scientists to formulate the cell theory. This theory states that all living things are composed of cells, cells are the basic unit of structure and function in living things, and cells come only from preexisting cells.

Most cells have a nucleus, cytoplasm, and a cell membrane. The nucleus acts as the control center of the cell. The cell membrane regulates what materials pass in and out of a cell. The chemical reactions required to maintain life occur in the cytoplasm. Numerous organelles in the cytoplasm perform specific functions within the cell.

Prokaryotes are cells without a true nucleus. Their nuclear material is not surrounded by a nuclear membrane. Prokaryotic cells have ribosomes and cell walls, but lack the other organelles found in eukaryotic cells.

Eukaryotes are cells with a true nucleus, surrounded by a double membrane. Organelles in eukaryotic cells include mitochondria, lysosomes, endoplasmic reticulum, Golgi bodies, ribosomes, and lysosomes. Plant cells also have plastids, large central vacuoles, and cell walls. Animal cells lack these organelles characteristic of plants. Certain types of plant and animal cells have centrioles.

BioTerms

cell membrane (**85**)
cell theory (**80**)
cell wall (**85**)
centriole (**90**)
chromatin (**84**)
chromosome (**84**)
cilia (**91**)
cytoplasm (**83**)
cytoplasmic streaming (**86**)

endoplasmic reticulum (**86**)
eukaryote (**92**)
flagella (**91**)
Golgi body (**88**)
lysosome (**90**)
microtubule (**90**)
middle lamella (**86**)
mitochondria (**88**)
nuclear envelope (**84**)

nuclear pore (**84**)
nucleolus (**84**)
nucleus (**79**)
organelle (**83**)
plastid (**89**)
prokaryote (**92**)
ribosome (**87**)
spindle fiber (**90**)
vacuole (**89**)

BioQuiz   (Write all answers on a separate sheet of paper.)

I. Completion

1. The basic units of all living things are called _____ .
2. Nuclear pores permit the passage of materials between the nucleus and the _____ .
3. Two organelles that project outside the cell are cilia and _____ .
4. Structures used in cell division and made up of microtubules are _____ .

II. Modified True and False

Mark each statement TRUE or FALSE. If false, change the underlined term to make the statement true.

5. Cell membranes are made of <u>proteins</u> and lipids.
6. Bacterial cells lack nuclear membranes and so are classified as <u>eukaryotes</u>.
7. According to the cell theory, the basic units of all living things are <u>organelles</u>.

III. Multiple Choice

8. Vacuoles function to a) destroy foreign objects in the blood. b) store fluids in plant cells. c) transport materials between the nucleus and membrane. d) manufacture proteins.

9. Chromosomes carry a) food and oxygen to the nucleus. b) hereditary information between cell generations. c) proteins from the ribosomes. d) waste materials to the lysosomes.

10. Ribosomes function in cells as a) sites of protein synthesis. b) control centers. c) transport routes. d) enzyme packages.

11. Rough ER is rough because of the presence of attached a) microtubules. b) secondary cell walls. c) ribosomes. d) lysosomes.

12. Pectin holds together the a) cell membrane. b) primary cell wall. c) middle lamella. d) vacuole.

13. An exception to the cell theory is represented by a) protozoa. b) viruses. c) yeasts. d) green algae.

14. A structure common to plant cells but not found in animal cells is the a) Golgi apparatus. b) nucleus. c) cell membrane. d) cell wall.

IV. Essay

15. What factors determine the size of the cell? Why is this so?

16. In what ways are plant and animal cells alike? In what ways are they different?

17. How is the shape of a skin cell related to its function?

18. What are the functions of the cell nucleus?

19. What are the steps involved in building a plant cell wall?

20. How does the structure of the membranes in mitochondria aid in the energy production of the cell?

Applying and Extending Concepts

1. How would you determine from a small sample whether a material is plant or animal in origin?

2. Glands secrete hormones, which are composed of lipids. What organelles would you expect to find in gland cells?

3. Would you expect to find more mitochondria in a muscle cell or in a skin cell? Why?

4. Which organelle is most directly responsible for the production of hair?

5. New technologies are making it possible to introduce new characteristics into cells. If an animal's cells were given the ability to produce cell walls, how might this affect the animal's survival?

Related Readings

Berns, M. W. *Cells*. New York: Holt, Rinehart and Winston, 1977. This book provides a comprehensive introduction to the study of cells.

Frankel, E. *DNA: The Ladder of Life*. 2d ed. New York: McGraw-Hill, 1978. This book provides an introduction to cell structure and chemistry.

Sheeler, P. and D. E. Bianchi. *Cell Biology: Structure, Biochemistry, and Function*. New York: Wiley, 1983. Cells, organelles, and other constituents are investigated in this up-dated edition. The structure, biochemistry, and function of these parts is discussed with emphasis on genetics and recent developments.

7

Cellular Transport

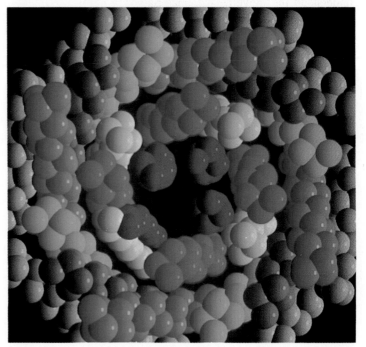

Computer-generated image of sodium transport molecule

Introduction

A cell is a single unit of life. Whether it is a unicellular organism or whether it is just one small part of a multicellular organism, a cell must carry on life processes. Every cell, for example, must meet certain needs, such as nutrition, transport, respiration, and excretion. A cell must import the materials necessary for these processes from its environment. A cell also must export the waste products resulting from these processes to avoid polluting itself.

A nation has a customs service that controls the flow of materials across its borders. Similarly, a cell has a cell membrane that regulates the passage of materials across its outer boundary. As with nations, what moves in and out of a cell affects the cell's health and well-being.

The Movement of Materials

Whether part of a living system or not, all molecules are governed by the same physical laws. A knowledge of the physical properties that affect molecular movement will help you better understand the movement of materials in a living cell.

7.1 Molecular Movement

In solid and liquid matter, molecules are packed closely together. In a gas, however, molecules are relatively far apart. In all substances—solids, liquids, and gases—molecules are in constant motion. Even in solids, molecules vibrate in a fixed space.

In liquids and gases, molecular motion is totally random. Imagine a room full of table-tennis balls that are in constant motion. A ball moves in one direction until it collides with another ball, then each flies off in a new direction. In the same way, the movement of molecules in liquids and gases has no plan or design.

This constant random movement of molecules is referred to as *Brownian movement,* after Robert Brown, the nineteenth-century Scottish scientist who first described it. If you look through a microscope at a particle of dust suspended in a drop of water, you can see the results of Brownian movement.

The study of cells is primarily concerned with the movement of molecules in the liquid state. All the substances important to life are most often part of a **solution,** a mixture in which the molecules of one substance are evenly dispersed in another. The substance that makes up the greater part of the solution is called the **solvent,** and the substance dissolved in the solvent is called the **solute.** *Water is the solvent of most solutions involved in cell activities.*

7.2 Diffusion

Consider what happens if you place a drop of food coloring into a beaker of water. All of the molecules are in constant, random motion. As they move about, the food coloring molecules spread farther and farther apart until they are evenly spaced throughout the beaker of water.

Diffusion is the process by which molecules of a substance move from areas of higher concentration of that substance to areas of lower concentration. The dispersal of a drop of food coloring through the beaker of water is an example of diffusion

- *Describe* the movement of molecules in solids, liquids, and gases.
- *Define* the terms *concentration gradient* and *diffusion.*
- *Describe* the effects of temperature, pressure, and concentration on diffusion rate.

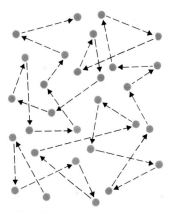

Figure 7–1. Molecules undergoing Brownian motion move in random paths. This natural movement of molecules allows solutes to disperse throughout solvents and create solutions.

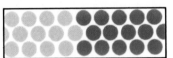
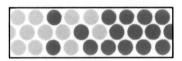
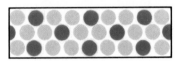

Figure 7–2. When food coloring is placed in water (left), it starts to diffuse. It moves with the concentration gradient (center), diffusing from high to low concentration until it is evenly distributed throughout the water (right). The illustrations show diffusion at the molecular level.

Q/A

Q: *When you smell the aroma of turkey on Thanksgiving Day, what is actually entering your nose?*

A: The heat of the oven causes tiny molecules of fat to break free of the cooking turkey. They diffuse through the air and some stimulate the sense receptors in your nose.

in liquids. Diffusion continues until the molecules of food coloring are evenly distributed throughout the molecules of water in the container. The net, or overall, movement of the molecules results in a uniform concentration of the food coloring through the substance. Molecular movement continues, but the random motion of molecules will not change their overall distribution. ***Diffusion is one of the major mechanisms of molecular transport in cells.*** Many materials move into, out of, and through cells by the process of diffusion.

The difference in concentration of molecules of a substance from the highest to the lowest number of molecules is called the **concentration gradient.** Molecules of a substance that are moving from areas of high concentration of that substance to areas of low concentration are moving with the concentration gradient.

Diffusion only takes place from areas of high concentration to areas of lower concentration. The steeper the grade from high to low, the more rapid the diffusion rate.

Concentration is not the only factor that affects the rate of diffusion. When the temperature of a liquid or gas is raised, the molecules move faster and rebound farther after collisions. An increase in the temperature of a substance thus increases the rate of diffusion. Consider how much faster a spoonful of sugar dissolves in a cup of hot tea than in iced tea.

An increase in pressure also results in an increase in the rate of diffusion. The molecules of substances under high pressure

Highlight on Careers: Biophysicist

General Description

A biophysicist is a scientist who uses the knowledge and techniques of the physical sciences to answer biological questions. For example, a biophysicist might study the ways in which a cell "recognizes" that a substance is dangerous and prevents that substance from crossing the cell membrane.

Biophysicists specialize in a wide variety of areas other than cell transport, however. Some use X rays to discover the structure of molecules. Others study how nerve cells conduct impulses.

Most biophysicists work on research projects in government laboratories or in universities. A few are employed by industrial laboratories.

Biophysics is responsible for many important scientific advances. Biophysicists, for instance, worked on the computer scanning equipment used to detect tumors and other diseases.

Career Requirements

Biophysicists who direct research projects must have a Ph.D. and knowledge of both biology and physics. Persons with a B.A. or M.S. degree may assist a biophysicist on a research project. Laboratory technicians, for example, are generally individuals with a B.A. degree.

Biophysicists must be

patient, methodical, and creative. They must also have mathematical ability, since many biological systems are explained in mathematical terms.

For Additional Information

Biophysical Society
9650 Rockville Pike
Bethesda, MD 20814

are squeezed more closely together than those under low pressure. Under high pressure a molecule has less empty space through which it can move before a collision sends it in a new direction. Conversely, molecules under low pressure diffuse slowly because there is less chance of collision.

Reviewing the Section

1. Compare the movement of molecules in solids, liquids, and gases.
2. What is the meaning of the terms *concentration gradient* and *diffusion?*
3. Would you expect diffusion to occur more rapidly in a system at high temperature and low pressure, or a system at low temperature and low concentration? Why?

Diffusion Through Membranes

Section Objectives

- *Describe* the process of diffusion through a selectively permeable membrane.
- *Define* the terms *isotonic, hypotonic,* and *hypertonic.*
- *Describe* the behavior of cells in isotonic, hypotonic, and hypertonic solutions.
- *Discuss* the cause of turgor pressure and its effect on plant tissues.
- *Explain* how cytolysis is used to study cells.

The cell and many organelles within the cell are surrounded by a membrane. Any material that moves into or out of the cell or these organelles must pass through the membrane.

7.3 Characteristics of Cell Membranes

The cell membrane acts as a barrier, isolating the cell from its environment. This barrier must allow some materials to pass through, however, or the cell would die. ***Membranes control the passage of materials into and out of the cell.*** Cell membranes are **selectively permeable**—that is, they allow only certain substances to pass through them. Water, for example, passes through the membrane, but the substances dissolved in the water may or may not pass through.

The cell membrane is an active, living part of the cell, visible only through an electron microscope. The membrane's structure is generally described as a double layer of lipid molecules with proteins scattered through the lipid layers. As Figure 7–3 suggests, some proteins are partially embedded in one layer of the lipids like icebergs. Other proteins rest on the surface of the lipids or poke completely through the double layer. The lipids are fluid and some of the proteins are free to move about, forming different patterns. For this reason, this model of the cell membrane is called the *fluid-mosaic model.*

Oxygen and carbon dioxide can dissolve in lipids and so pass right through the cell membrane. Water molecules, which do not dissolve in lipids, and some other small molecules pass into the cell through openings formed by proteins in the lipid layers. Although water molecules diffuse no differently than any other molecules, the diffusion of water through a membrane is called **osmosis** (ahz MOH suhs). ***Water diffuses into cells by osmosis.***

Figure 7–3. The cell membrane is composed of a double layer of lipids in which proteins are embedded. Because both lipids and proteins are in motion, the membrane is called fluid-mosaic.

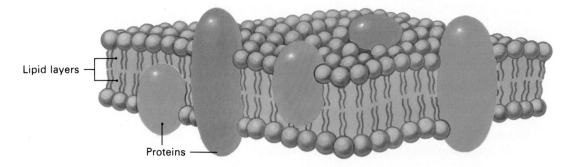

Lipid layers —

Proteins —

7.4 Osmosis and Living Cells

Water makes up 70 to 95 percent of a living cell. Since water is the most abundant substance in cells, its movement into and out of cells is of vital importance. The cell has no control over osmosis. Water will flow in and out of the cell until the concentration of water molecules is equal on each side of the membrane. When the concentration of two solutions separated by a membrane is the same, the two solutions are in a state of **equilibrium.** *Water will continue to diffuse back and forth across a cell membrane until an equilibrium is reached.* Even after equilibrium is reached, movement of molecules continues. However, the number of molecules present on both sides of the membrane remains equal.

The movement of water across a cell membrane depends on the concentration gradient of the water across that membrane. The concentration of water on each side of the membrane, in turn, is determined by the concentration of solutes in that water solution. In an **isotonic** (eye suh TAHN ihk) **solution** the concentration of solutes outside the cell is the same as that inside a cell. In a **hypotonic solution** the concentration of solutes outside the cell is lower than that inside the cell. In a **hypertonic solution** the concentration of the solutes outside the cell is greater than that inside the cell. An easy way to remember the differences among these three solutions is to recall that *iso-* means ''equal,'' *hypo-* means ''less than,'' and *hyper-* means ''more than.''

Isotonic Solutions If a cell is placed in an isotonic solution, the rate of osmosis into the cell is exactly the same as the rate of osmosis out of the cell. As a result, no net movement of water takes place. Isotonic solutions are important to living organisms. Plasma, the liquid portion of whole blood, is an isotonic solution with respect to blood cells. Accident victims who have lost large amounts of blood often receive transfusions of plasma instead of whole blood. This increases the volume of the victims' blood without upsetting the balance between the plasma and the blood cells.

Hypotonic Solutions Because the concentration of solutes in a hypotonic solution is lower, the concentration of water is relatively higher in the solution than in·a cell placed in the solution. Water moves from the solution into the cell—that is, from an area of higher concentration of water to an area of lower concentration of water.

Freshwater plants often exist in hypotonic solutions. As water diffuses into the cell, the cell swells and increases its

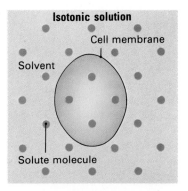

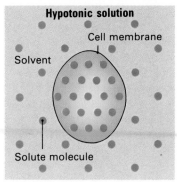

Figure 7–4. A cell placed in an isotonic solution has the same solute concentration inside and outside the cell. A cell placed in a hypotonic solution has a higher solute concentration than that of the surrounding medium.

Using Cytolysis to Study Cell Membranes

Distilled water is the ultimate hypotonic solution. It is composed of 100 percent water with no dissolved solutes. When red blood cells are placed in distilled water, they swell and burst. The breakdown of cells, particularly for study purposes, is called *cytolysis.* The cytolysis of red blood cells is known as *hemolysis.* When the red blood cell swells, the membrane ruptures like an over-inflated balloon. The contents of the cell are expelled, leaving behind a membrane "ghost."

The membrane ghosts are heavier than either the water or the cell contents. Through the use of a centrifuge—a machine that rapidly whirls the fluids—the membranes can be separated out from materials of different densities.

This procedure produces nearly pure samples of cell membranes for study.

internal pressure. The pressure that builds in a plant cell as a result of osmosis is called **turgor** (TUHR guhr), or *turgor pressure.* The excess water entering a plant cell is often stored in a large central vacuole. Increase of the turgor pressure forces the cytoplasm and the cell membrane against the plant cell wall, causing the cell to become stiff. The cell wall prevents the cell from bursting. In this way, turgor makes the soft tissues of stems and leaves more rigid. Water will continue to diffuse into the plant until the solutions inside and outside the cell are in equilibrium.

Animal cells do not have a cell wall and therefore cannot reach equilibrium in a hypotonic solution. As water flows in, the cell swells and bursts. Clearly, then, it is very important for a cell to be able to remove excess water efficiently. Organisms have developed a number of mechanisms to remove excess water. As a general rule, animals use energy to pump excess water from cells before any damage results. Unicellular organisms living in fresh water have **contractile** (kuhn TRAK tuhl) **vacuoles,** which actively pump excess water out of the cell. Freshwater fish and other gill-breathing animals remove excess water from their bodies through excretion.

Hypertonic Solutions   In a hypertonic solution, the concentration of solutes is higher than that inside a cell placed in the solution. As a result, the concentration of water is lower in the solution surrounding the cell than it is inside the cell itself.

Figure 7–5. A cell placed in a hypertonic solution has a lower solute concentration than that of the surrounding medium. Water diffuses out of the cell until equilibrium is reached.

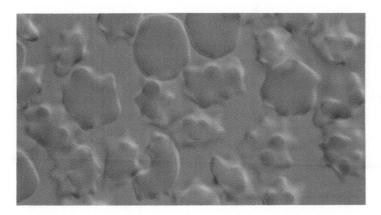

Figure 7–6. Red blood cells placed in a hypertonic solution shrink as water diffuses out of the cell.

Cells placed in a hypertonic solution shrivel up and lose their shape because more water flows out of the cells than into the cells.

Drinking sea water is dangerous because the salt water of the ocean is hypertonic relative to human tissues. If people drank sea water, their cells would lose more water through osmosis than the cells would take in. Some animals, however, have developed ways of living part of their lives in salt water and part in fresh water. Salmon, for example, regulate the level of salt in their body cells. These fish actively excrete salt through their gills when in salt water. In fresh water, salmon excrete urine, or waste fluids, with a low concentration of salt.

The flow of water out of the cells in a hypertonic solution can cause the loss of turgor pressure. In northern climates, salt is often spread on roads, driveways, and sidewalks to melt ice in winter. As the ice melts, a hypertonic solution of salt and water forms and is carried away as runoff. This solution does not disappear harmlessly. Instead, it collects at the side of the road or along the edge of the driveway. If the concentration of salt collected in the soil is high enough, the plants in the affected area will wilt and die.

Reviewing the Section

1. What is a selectively permeable membrane?
2. Summarize the effects of hypotonic, hypertonic, and isotonic solutions on animal cells.
3. How does the concentration of water molecules in a solution affect turgor pressure?
4. Compare the effects of a hypotonic solution on plant cells and on animal cells. How do you account for the differences?

- *Compare* the processes of simple diffusion and facilitated diffusion.
- *Compare* active transport and passive transport.
- *Define* the terms *endocytosis* and *exocytosis.*
- *Describe* phagocytosis and pinocytosis.

Other Means of Transport

Osmosis and diffusion take place without any use of energy by cells. For this reason, osmosis and diffusion are considered *passive* processes. These processes are not the only ways materials move across the cell membrane, however. The transport of some materials involves the active participation of the cell and in some cases the use of cell energy. Transport methods that involve work by the cell are considered *active* processes.

7.5 Carrier Transport

Carrier molecules are proteins in the cell membrane that transport large molecules or molecules that cannot dissolve in the lipids that make up the cell membrane. Carrier molecules function like moving vans. They pick up other molecules on one side of the membrane, carry them across, and deposit them on the other side of the membrane.

Each carrier molecule is highly specific in its function. A carrier may transport one type of molecule and refuse another almost identical one. Although biologists do not completely understand the details of this transport method, they do know that transport by carrier molecules occurs in two ways.

Facilitated diffusion involves the use of a carrier molecule but follows the rules of simple diffusion. *In facilitated diffusion, substances move with the concentration gradient from high to low, but carrier molecules speed up the movement of diffusing substances.* The cell does not expend energy in this process. Glucose enters most cells through facilitated diffusion.

Active transport is a second transport method using carrier molecules. *Active transport involves the movement of materials against the concentration gradient.* In active transport, molecules are moved from regions of low concentration to

Figure 7–7. The diagrams below show the active transport of a sodium ion across a membrane. The sodium ion first attaches to a carrier protein (left). The protein then changes shape (center) and delivers the ion across the membrane (right).

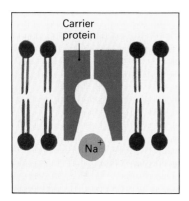

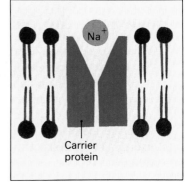

regions of higher concentration. Moving a molecule against the concentration gradient requires the use of energy. Liver cells store glucose and have a higher concentration than the surrounding bloodstream has. Thus active transport is required to move glucose into liver cells.

Many models have been proposed to show how active transport actually works. The currently preferred model suggests that a solute such as a sodium ion enters an opening in a carrier molecule located in the membrane. The carrier molecule changes its shape, channeling the sodium to the other side of the membrane, where the sodium ion is released.

7.6 Bulk Transport

Often materials that cannot pass through the membrane need to be transported into or out of the cell. The molecules may be so large or they may be present in such great quantities that movement through the membrane is not possible. *Bulk transport* methods allow droplets of fluid, particles of food, or globules of protein to move across the cell boundary without actually passing through the membrane. **Endocytosis** (ehn doh sy TOH sihs) is the bulk transport of substances *into* the cell. *Endo-* means "into." **Exocytosis** (ehk soh sy TOH sihs) is the bulk transport of substances *out of* the cell. *Exo-* means "out of." These methods require cells to use energy.

Endocytosis Endocytosis begins as the cell membrane encloses a substance or particle, forming a pouch. The pouch is drawn into the cell and then pinched free of the cell membrane. **Phagocytosis** (fag oh sy TOH sihs) is the term used to describe the movement of solids or large particles into the cell. Amoebas use this method of taking in food. After the food particle is surrounded, the cell membrane fuses and forms a vacuole within

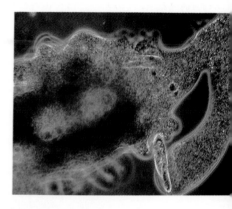

Figure 7–8. When an amoeba engulfs a paramecium (top), it does so by endocytosis. The illustration below shows the steps in this process. The cell membrane first forms a pocket (left), then pinches off (center), and food is taken into the cell (right).

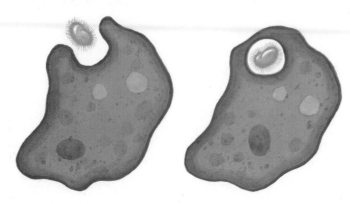

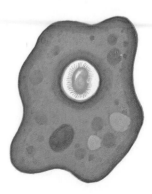

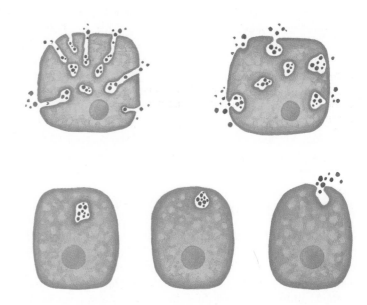

Figure 7–9. During pinocytosis (top), liquids and small particles are taken into the cell in small membrane packets. In exocytosis (bottom), materials leaving the cell are packaged in membrane sacs, transported to the cell membrane, and dumped outside the cell.

Q/A

Q: *How does a cell membrane increase in size?*

A: A cell membrane increases in size in a process similar to exocytosis. A membrane-enclosed vacuole manufactured by the Golgi bodies moves to the surface of the cell, where it fuses with the cell membrane.

the cell. Once the food is inside the cell, lysosomes fuse with the vacuole and dissolve its contents. **Pinocytosis** (pihn oh sy TOH sihs) refers to the movement of liquids with solutes and small particles into the cell. A small region of the cell membrane is pulled inward forming a pocket or channel. The liquid flows into this pocket. The membrane near the entrance of the pocket comes together, leaving the membrane-enclosed vacuole inside the cell.

Exocytosis   Exocytosis is essentially the reverse of endocytosis. Substances to be removed from the cell are enclosed in membrane vacuoles. The vacuoles then move to the surface of the cell, where the membrane of the vacuole fuses with the cell membrane. The contents of the vacuole are then expelled from the cell through the opening in the membrane.

Reviewing the Section

1. What is a carrier molecule?
2. How is facilitated diffusion similar to simple diffusion? How is it different from active transport?
3. How are endocytosis and exocytosis different, and how are they the same?
4. Distinguish between the processes of phagocytosis and pinocytosis.

Investigation 7: Enzymes and Diffusion

Purpose

To investigate the relationship between enzymes and the diffusion of molecules through a membrane

Materials

150-mL beaker, dialysis tubing, string, starch solution, amylase or diastase, test tubes, safety goggles, large beaker, hot plate, Benedict's solution, medicine dropper, Lugol's iodine

Procedure

1. Tie one end of a length of dialysis tubing with string. Pour the starch solution into the tube until it is two-thirds full. Tie the other end of the tubing with string.
2. Place the dialysis bag in a 150-mL beaker of water as shown in the diagram.

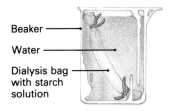

Beaker —
Water —
Dialysis bag with starch solution —

3. Wait 10 minutes and test samples of the water in the beaker for the presence of sugar and starch. **CAUTION: Put on safety goggles. Do not get Benedict's solution on your skin. If you do, wash it off immediately.** To test for sugar, add about 5 mL (10 drops) of Benedict's solution to a test tube containing a sample of the water from the beaker. Heat gently in a boiling water bath. The solution will turn green to brick red, depending on the amount of sugar present. To test for starch, place a drop of iodine in a test tube containing a sample of the water. The iodine will turn blue-black if starch is present. Record your results. *What is the purpose of waiting 10 minutes? What do your results indicate?*

4. Replace the water in the beaker. Untie one end of the dialysis bag. Add 10 drops of amylase or diastase solution to the dialysis bag. Again place the bag in the beaker.
5. Wait 10 minutes and test the water in the beaker for the presence of sugar and starch. Record your results. *How do the results in this step differ from those in step 3?*

Analyses and Conclusions

1. What is the difference between the results in step 3 and those in step 5? What accounts for this difference?
2. Explain how the enzyme affected the diffusion of molecules through the membrane.
3. Explain how the dialysis membrane demonstrated selective permeability.

Going Further

Create your own dialysis membranes using cellophane, plastic wrapping, and other materials. Summarize the differences in permeability among the membranes.

Chapter 7 Review

Summary

Both the speed of molecules and the distance between them increase as matter changes from solid to liquid to gas. The motion of molecules in liquids and gases is random.

Molecules diffuse from areas of higher concentration to areas of lower concentration. A rise in temperature, pressure, or concentration increases the rate of diffusion.

Osmosis is the diffusion of water. Diffusion and osmosis do not require the use of energy. An isotonic solution has the same concentration of solutes as that of a cell immersed in the solution. No net movement of solutes or water takes place across the cell membrane. A hypotonic solution has a lower concentration of solutes than the concentration of those solutes inside the cell. As a result, water moves into the cell. A hypertonic solution has a higher concentration of solutes than the cell, causing water to move out of the cell.

Turgor pressure increases as water flows into a plant cell. Turgor stiffens plant tissues.

Some substances are transported across the cell membrane by carrier molecules. Facilitated diffusion does not require energy; active transport does. Large molecules are moved in and out of the cell by bulk transport.

BioTerms

active transport (**106**)
carrier molecule (**106**)
concentration gradient (**100**)
contractile vacuole (**104**)
diffusion (**99**)
endocytosis (**107**)
equilibrium (**103**)

exocytosis (**107**)
facilitated diffusion (**106**)
hypertonic solution (**103**)
hypotonic solution (**103**)
isotonic solution (**103**)
osmosis (**102**)
phagocytosis (**107**)

pinocytosis (**108**)
selectively
 permeable (**102**)
solute (**99**)
solution (**99**)
solvent (**99**)
turgor (**104**)

BioQuiz   (Write all answers on a separate sheet of paper.)

I. Completion

1. Temperature, pressure, and _____ affect the rate of diffusion.
2. Some proteins in the cell membrane function as _____ molecules.
3. Turgor pressure builds up in a plant cell as a result of _____.
4. _____ moves substances against the concentration gradient.
5. A _____ membrane allows some substances to pass through but prevents others from doing so.

II. Modified True and False

Mark each statement TRUE or FALSE. If false, change the underlined term to make the statement true.

6. Osmosis does not require an active expenditure of cell energy.
7. Cells swell in a hypotonic solution.
8. Random movement results in an even dispersion of molecules.
9. Contractile vacuoles in some organisms pump water out of the cell.
10. Endocytosis moves wastes out of the cell.

III. Multiple Choice

11. Diffusion occurs only a) with active transport. b) with a concentration gradient. c) against a concentration gradient. d) in an isotonic solution.
12. The dissolved substance in a solution is the a) turgor. b) solute. c) solvent. d) gradient.
13. Solutions stop diffusing when a) the cell runs out of energy. b) solutes are concentrated. c) equilibrium is reached. d) temperatures are constant.
14. Pinocytosis a) digests lipids. b) moves water across a membrane. c) moves liquids into a cell. d) moves solids out of a cell.
15. Water is vital to cellular transport because it a) contains necessary oxygen. b) is lightweight. c) is the most common solvent. d) is isotonic.

IV. Essay

16. How does the structure of the cell membrane allow some substances to pass through?
17. Will a teaspoon of salt dissolve more quickly in a cup of fresh water or a cup of salt water? Why?
18. How do freshwater fish rid themselves of excess water?
19. Why do raisins plump up more quickly in hot water than in cold water?
20. How do scientists prepare samples of cell membranes for study?

Applying and Extending Concepts

1. List methods used to transport materials across the cell membrane. Indicate which methods require cell energy and which methods do not.
2. Cells taken from a frog and a human are placed in an 0.8 percent salt solution. The frog cells swell and burst, while the human cells shrink. Explain these results.
3. Why might overfertilization result in wilted plants?
4. Chemicals that dissolve easily in fat pass through the cell membrane more quickly than fat-insoluble molecules of similar size and weight. Suggest an explanation for this phenomenon.
5. Determine the concentration of salt that is isotonic to plant cells, such as elodea, by placing leaves of the plant in solutions that vary in salt concentration. What should you look for?

Related Readings

Black, I. "The Sea Within Us." *Science Digest* 88 (November/December 1980): 62–65. The article gives an explanation of diffusion, osmosis, and concentration gradients. As its title suggests, the article discusses their importance for human cells and fluids.

Brown, M. S. and J. L. Goldstein. "Receiving Windows for the Cells." *Science Year 1980*. Chicago: World Book, 1979. The text provides an interesting discussion of receiver sites on the cell surface.

Cook, R. E. "The Cold Facts of Winter Wheat." *Natural History* 91 (November 1982): 24–28. The text explains how changes in the cell membrane enable winter wheat varieties to withstand freezing temperatures.

Cells and Energy

Trees capture energy from sunlight

Introduction

Take a moment to consider the cell as a machine. Like any machine, the cell does work. The cell's work may be simply to stay alive, or the cell may perform an important function as part of a multicellular organism. Every machine, whether natural or artificial, requires energy to function. Most machines require energy in specialized forms—farm machinery uses gasoline, toasters use electricity, windmills use the wind. The cell too requires a specialized form of energy.

In this chapter you will investigate the cell's source of energy and how the cell uses the energy. You will learn about the specialized fuel required by all living things and the unique way some living things convert the light energy of the sun into chemical energy.

Energy for Living Cells

Cells require energy to undertake the many tasks necessary for life. This energy is stored in the form of chemical bonds in food. However, food molecules cannot deliver energy directly to living systems. The energy stored in food's chemical bonds must first be transferred to molecules capable of providing energy where it is needed.

8.1 Energy Transfer in Cells

Like all chemical reactions, the reactions that take place within cells require energy to get started. This extra energy, called *activation energy,* enables molecules to collide with enough force to break existing chemical bonds. Within cells, proteins called *enzymes* lower the amount of activation energy needed to start reactions. Enzymes allow reactions to occur that normally could not take place in a cell. Enzymes come in many different shapes. As Figure 8–1 shows, enzymes combine briefly with the reacting molecules, then are released unchanged. Enzymes themselves are not used up in a reaction.

Many reactions within cells are *endergonic.* That means that in addition to activation energy, these reactions require a steady input of energy to keep them going. In most cases, this energy is supplied by a molecule called **ATP** (adenosine triphosphate). ATP consists of a base, a sugar, and a chain of three phosphates. The base is *adenine* and the sugar is *ribose.* Together they form *adenosine.* Attached to the adenosine are three phosphates. Adenosine with one phosphate, represented as ⓟ, is called AMP (adenosine monophosphate). Adenosine with two phosphates is called ADP (adenosine diphosphate).

The bond that joins the second and third phosphates in ATP is easily broken. Enzymes allow ATP to readily transfer the third phosphate to another molecule. ***By transferring a phosphate, ATP provides energy that can be used to drive endergonic reactions in the cell.*** ATP is the transfer molecule that allows the energy in food molecules to be delivered to cells wherever needed.

To understand this process more clearly, consider a hypothetical reaction:

$$W + X \rightarrow Y + Z$$

If the products of this reaction, Y and Z, have more energy in their chemical bonds than do the reactants, W and X, then this reaction is an endergonic one. It will not take place without the

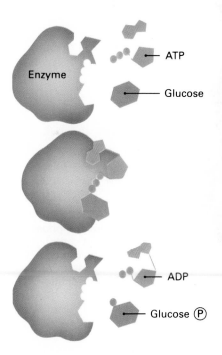

Figure 8–1. A glucose molecule is activated when it receives a phosphate group from ATP. This reaction requires an enzyme that brings the molecules close to one another.

Figure 8–2. ATP, ADP, and AMP differ only in the number of phosphate groups that are attached to the adenosine molecule. ATP has three phosphate groups, ADP has two, and AMP has one.

input of energy. This energy is supplied by the transfer of a phosphate from ATP to one of the reactants:

$$\text{ATP} + \text{W} \rightarrow \text{ADP} + \text{W} - \textcircled{P}$$

W has formed a new bond with $\textcircled{P}$, so W $-$ $\textcircled{P}$ has more energy than W. In fact, W $-$ $\textcircled{P}$ and X have more energy than Y and Z. As a result, the reaction can now proceed as follows:

$$\text{W} - \textcircled{P} + \text{X} \rightarrow \text{Y} + \text{Z} + \textcircled{P}$$

8.2 ATP-ADP Cycle

For ATP to be an effective energy transfer molecule, it must lose its final phosphate group. The phosphate group is returned to ATP by adding a $\textcircled{P}$ to ADP. The series of reactions between ATP and ADP form a cycle; the products of the first reaction are used for the second. In terms of energy, the ATP-ADP cycle can be compared to a battery that continually recharges itself. The reforming of ATP recharges the battery.

The phosphate group is returned to ATP by adding a $\textcircled{P}$ to ADP during the process of **cellular respiration.** Glucose is broken down and the energy in its chemical bonds transferred to the energy bonds of ATP. The glucose used is produced through **photosynthesis,** a process in which green plants convert the energy from sunlight into chemical energy.

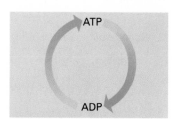

Figure 8–3. ATP becomes ADP when it loses a phosphate group as it contributes energy to cell reactions. ATP is then reformed from ADP during cellular respiration.

Reviewing the Section

1. What role does ATP play in cells?
2. Why do cellular reactions require enzymes?
3. Why is the ATP-ADP relationship called a cycle?

Capturing Energy

The ultimate source of the energy that powers cells is the sun. Green plants and certain other organisms capture the light energy of the sun through the process of photosynthesis.

8.3 Requirements for Photosynthesis

Photosynthesis requires light, chlorophyll, and raw materials. Enzymes are needed for the reactions to proceed. The chlorophyll present in plants traps the light of the sun. Carbon dioxide drawn from the air, and water absorbed through the roots are the usual raw materials for photosynthesis. Glucose, a six-carbon sugar, is the end product of photosynthesis. Oxygen and water are byproducts. The basic equation for the process of photosynthesis can be written:

$$6CO_2 + 12H_2O \xrightarrow{\text{light, enzymes, chlorophyll}} C_6H_{12}O_6 + 6O_2 + 6H_2O$$

This equation summarizes the overall process of photosynthesis. It does not show the many individual reactions.

The rate of photosynthesis depends on factors such as the availability of the raw materials, the intensity of the sunlight, and the temperature. If water becomes scarce, as when a severe drought occurs, photosynthesis may stop altogether. The greater the intensity of sunlight, in general, the higher the rate of photosynthesis. A temperature range of between 20°C (68°F.) and 35°C (95°F.) is best. Above 35°C and below 0°C, the activity of the enzymes required for chemical reactions is lessened and the rate of photosynthesis decreases.

Light Light provides the energy for photosynthesis. To understand how this energy is converted into chemical energy, you must understand something about the nature of light.

The visible portion of the sun's light is but a small part of the radiation emitted by the sun. When white light passes through a prism, the light spreads out into an array of colors called the **visible spectrum.** This effect can also be observed when water droplets in the atmosphere act as a prism, producing a rainbow.

Light energy comes in packets, or units, called **photons.** The energy is not the same for all kinds of light. A photon of violet light, for example, has almost twice the energy of a photon of red light.

Section Objectives

- *State* the function of each of the pigments involved in photosynthesis.
- *List* the requirements for photosynthesis to occur and its products.
- *Summarize* the main events of the light and dark reactions of photosynthesis.

Q/A

Q: *Does photosynthesis always use carbon dioxide and water as raw materials?*

A: No. Some forms of bacteria use hydrogen sulfide in place of water to complete photosynthesis.

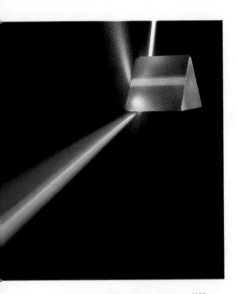

Figure 8–4. The many different wavelengths that make up sunlight are revealed when light passes through a prism. Green plants reflect some wavelengths but use others for processes such as photosynthesis.

The color of an object seen by the eye is the color of light reflected by the object. A red dress, for example, absorbs all of the visible spectrum except red, which is reflected. Green plants reflect the green portion of the spectrum, while absorbing other colors. The violet, blue, and red portions of the spectrum provide the most energy for photosynthesis.

Chlorophyll and Other Pigments Substances that absorb light are called **pigments.** Early investigators demonstrated that only the green parts of plants performed photosynthesis. **Chlorophyll** (KLAWR uh fihl), the green pigment present in plants, is necessary for photosynthesis to begin. Chorophyll absorbs energy from all but the green portion of the visible spectrum. Several types of chlorophyll are active in the presence of light. Plants contain *chlorophyll a* and *chlorophyll b*; other types of chlorophyll are found in other photosynthetic organisms.

Chlorophyll acts as a "light trap" during photosynthesis. When a photon strikes a chlorophyll molecule and is absorbed, the photon's energy is transferred to an electron of the chlorophyll molecule. The energized electrons are like stretched rubber bands—they cannot remain for long in this "excited" state. The electron is raised to a higher energy level. As the electron returns to its original energy level, it releases the absorbed energy, which is then used in chemical reactions.

Xanthophylls (ZAN thuh fihlz) are yellow pigments that absorb light energy in other parts of the spectrum and pass it on to chlorophyll. **Carotenes** are orange pigments that perform the same function. These two pigments are present in most green plants but in lesser amounts than chlorophyll, which usually masks their presence. In the fall when many leaves stop producing chlorophyll, xanthophyll and carotene become more prominent. As a result, the foliage changes to its autumn colors of red, yellow, and orange. The red leaves of brown algae and plants such as coleus and begonias also are due to the presence of accessory pigments, which mask the presence of chlorophyll.

In blue-green algae, which are prokaryotes, pigments involved in photosynthesis are part of a membrane system found throughout the cytoplasm. Green plants, however, are eukaryotes. Their chlorophyll and other photosynthetic pigments are found in chloroplasts.

Figure 8–5 shows that a chloroplast consists of flattened structures enclosed by a double membrane. Photosynthesis begins in the **grana,** stacks of tiny disklike sacs. These stacks look like dark dots when seen under an electron microscope. A dense fluid, **stroma,** fills the space between the grana and the outer membranes.

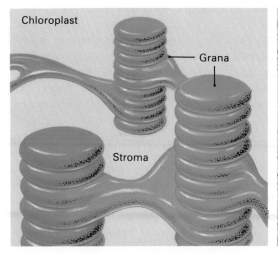

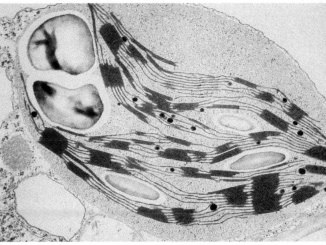

Carbon Dioxide and Water Carbon dioxide (CO_2) is the source of the carbon and oxygen atoms used in constructing glucose. Plants use water (H_2O) as the source of the hydrogen atoms needed in making glucose. Plants release oxygen from the water molecule into the air. This process is the source of most of the oxygen found in the atmosphere.

8.4 Process of Photosynthesis

Biochemists divide the process of photosynthesis into two phases. The first phase is called the **light reactions,** or light phase. As the name implies, this phase uses light energy. *Light reactions involve the trapping of light energy and the formation of materials required in the next phase of the process.* Light reactions can proceed only in the presence of light.

The **dark reactions** comprise the second phase of photosynthesis. *The dark reactions use the products from the light reactions to form glucose.* Thus the light reactions must take place for the dark reactions to proceed. The two phases together form one continuous process. The reactions of the dark phase are not part of the light-trapping process. The dark reactions can occur with or without light; they do not require light.

The Light Reactions The light reactions of photosynthesis use some of the trapped energy to convert ADP into ATP, which stores the energy for later use. Some energy is also used to split water molecules into hydrogen and oxygen.

The light reactions can be described as a series of steps. Some of these steps happen simultaneously. Follow these steps in Figure 8–6 on the next page as you read.

Figure 8–5. Disklike membranes inside the chloroplasts are stacked atop one another to form grana (above left). Magnified 44,000 times (above), grana appear as dense, membranous layers embedded in the stroma. The light reactions of photosynthesis take place in the grana, the dark reactions in the stroma.

1. The chlorophyll molecules in the grana absorb photons of light.
2. The energy from the photons boosts electrons (e^-) from the chlorophyll molecules to a higher energy level.
3. The energized electrons move from one molecule to another in a series of reactions called an **electron transport chain.** Each time a transfer is made, some energy is released.
4. The energy released from the electrons as they move down the electron transport chain is used ultimately to form ATP molecules by uniting ADP molecules and phosphates. Both ADP and phosphates are readily available in the stroma of the chloroplast.
5. The electrons lost from the chlorophyll molecule are replaced by electrons from a water molecule. This process splits the water molecule into hydrogen ions and oxygen gas. The hydrogen combines with a *hydrogen acceptor* molecule, a molecule that readily accepts hydrogen ions. The oxygen escapes into the atmosphere.

The release of energy as electrons are transferred from molecule to molecule in the electron transport chain can be compared to water falling from a height. The water can fall free in a waterfall, its energy unharnessed. The falling water can also be directed so that it hits the paddles of a wheel. The water hitting the paddles turns the wheel. Each time the water hits another

Figure 8–6. The light reactions of photosynthesis take place in the parts of the plant that contain chlorophyll. Light energy is used to split water into hydrogen ions and oxygen gas. In the process, electrons are released and ATP is formed. ATP and the hydrogen ions are then used in the dark reactions.

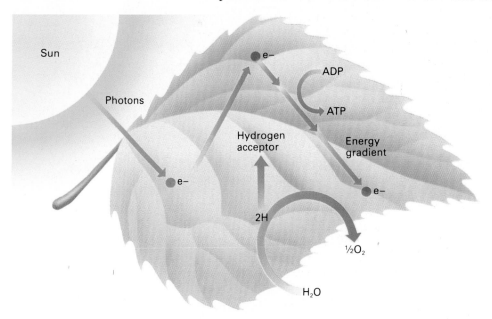

Sun

Photons

e–

ADP

ATP

Hydrogen acceptor

Energy gradient

e–

e–

2H

½O₂

H₂O

paddle, it holds less energy. By the time the water reaches the bottom of the wheel, the water is almost still. Its original energy has been captured to do the work of turning the wheel. In the same way, the energy of electrons is captured for the work of the cell.

The Dark Reactions The second phase of photosynthesis uses the energy stored in ATP and the hydrogen locked into the hydrogen acceptor to form glucose. This phase also uses carbon

Highlight on Careers: Plant Physiologist

General Description

A plant physiologist is a scientist who specializes in the structure and function of plants. These specialists use electron microscopes to examine plant cells and to track each stage of photosynthesis. The scientists also study how plant tissues are affected by changes in light, air, water, soil, and other environmental factors.

Agricultural sciences and any environmental sciences involving plants depend on plant physiologists for accurate information on how plants grow. Plant physiologists also examine how plants respond to weather conditions, fertilizers, and conditions of the soil.

Today many plant physiologists work on genetic engineering experiments. They are trying to find ways to alter a plant's heredity so that the plant will be hardier and produce more offspring.

The maximum production of renewable resources is the ultimate concern of the plant physiologist. For the growing world population, more food must be produced from the limited and precious environment.

Some plant physiologists work for large agricultural chemical companies. Others work in university laboratories or for federal and state environmental protection agencies. Environmental protection workers may do field work, taking samples of plants and soils. Many plant physiologists combine part-time teaching with laboratory work.

Career Requirements

To be a plant physiologist, a student needs a strong interest in chemistry, biochemistry, physics, and

mathematics as well as in biology. A capacity to do detailed work is essential.

Entry level positions in plant physiology are available for those with a B.S. or M.S. in biology or botany and some coursework in plant physiology. Teaching positions at universities generally require a Ph.D. in the subject.

For Additional Information

American Society of Plant
  Physiologists
P.O. Box 1688
Rockville, MD 20850

dioxide as a source of carbon for the glucose. The dark phase requires several enzymes and forms several byproducts. This second phase of photosynthesis takes place in the stroma of the chloroplasts.

The dark phase is also known as the *Calvin cycle*. This name recognizes the work of Melvin Calvin, the American scientist who first identified the process in the 1950s. The cycle begins and ends with a five-carbon sugar, **RDP** (ribulose diphosphate), which is abundant in chloroplasts. There are four major steps in the cycle. As you read, follow these steps in Figure 8–7.

1. Carbon dioxide from the atmosphere combines with RDP in a series of reactions. These reactions use ATP as the energy source and form a substance called *PGA* (phosphoglyceric acid), a molecule containing three carbon atoms.
2. PGA reacts with hydrogen from the light reactions to form **PGAL** (phosphoglyceraldehyde).
3. Most of the PGAL formed during the dark reactions is used to make more RDP. This RDP then unites with more carbon dioxide, beginning another cycle of dark reactions.
4. Some of the PGAL is combined to form glucose. Two three-carbon molecules of PGAL are required to form one molecule of glucose ($C_6H_{12}O_6$).

Figure 8–7. The dark reactions begin when carbon dioxide from the atmosphere unites with the five-carbon sugar ribulose diphosphate (RDP). The resulting six-carbon molecule is then broken down into phosphoglyceraldehyde (PGAL). Some PGAL is used to synthesize glucose—food for plants and animals.

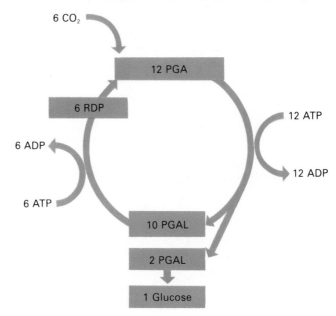

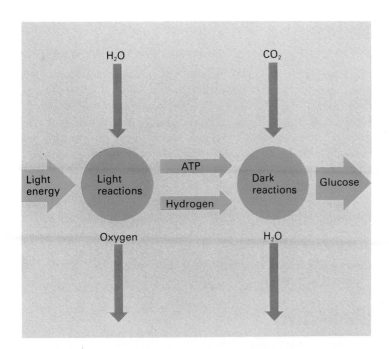

Figure 8–8. Photosynthesis consists of two sets of reactions. The light reactions are driven by the energy of sunlight, which is captured by chlorophyll. The dark reactions use ATP and hydrogen ions from the light reactions to convert atmospheric carbon dioxide into glucose.

Glucose formed in photosynthesis is often not immediately used by the plant. Excess glucose is stored by plants in the form of starch, a polysaccharide made of thousands of glucose molecules. When glucose is needed, the cell breaks down starch and releases just as much glucose as is demanded. In this way plants regulate the amount of glucose available to the cells. If large amounts of starch must be stored, the plant often forms a special storage organ. A potato, for example, is a plant organ used to hold excess amounts of starch.

Plants also combine glucose molecules to form other carbohydrates, including sucrose, the commonly used table sugar. Cellulose, the material that makes up plant cell walls, is also formed from the union of many glucose molecules.

Reviewing the Section

1. What factors affect the rate of photosynthesis?
2. What are the functions of chlorophyll, xanthophyll, and carotene in photosynthesis?
3. Where does photosynthesis occur in eukaryotes?
4. What must be accomplished during the light reactions before the dark reactions can proceed?
5. How is PGAL formed during the dark reactions, and how is it used?

Releasing Energy

Q/A

Q: *Is dieting related to cellular respiration?*

A: Yes. A person with a high metabolic, or respiration, rate burns up energy much faster than a person with a low metabolic rate. That is why some people never have to diet, but seem to be able to eat anything without gaining weight. Studies indicate that exercise increases a person's metabolic rate.

Although respiration and breathing are often thought of as the same, they are in fact two different processes. Breathing is the exchange of gases between an organism and its external environment. Respiration occurs within all living cells. *Cellular respiration involves breaking the chemical bonds of organic food molecules and releasing energy that can be used by the cells.* These food molecules are produced in plants during the process of photosynthesis.

Cellular respiration may be compared in some ways to the burning of a log in a fireplace. Both release energy. Burning a log, however, is an uncontrolled process that gives off large amounts of heat energy. Cellular respiration is an enzyme-controlled process in which the energy given off is trapped in molecules of ATP.

8.5 Glycolysis

The first step in cellular respiration is a process called **glycolysis** (gly KAHL uh sihs). In glycolysis, a glucose molecule is broken in half to form two three-carbon molecules of a substance called **pyruvic** (py ROO vihk) **acid.**

Glycolysis takes place in the cytoplasm of the cell. The process involves a series of nine enzyme-controlled reactions. Two reactions early in the process are endergonic; each requires the input of one molecule of ATP. The later reactions release enough energy to combine four molecules of ADP and four molecules of free phosphate to get four molecules of ATP. Thus, glycolysis produces an overall gain of two molecules of ATP for each molecule of glucose.

In addition to producing two three-carbon molecules of pyruvic acid and two ATPs, glycolysis also releases four hydrogen atoms. These hydrogens then combine with a hydrogen acceptor, as in the light reactions of photosynthesis. The following equation summarizes the overall reaction:

$$C_6H_{12}O_6 \rightarrow 2C_2H_3OCOOH + 4H$$
glucose   pyruvic acid

Glycolysis is an **anaerobic** (an ehr OH bihk) process—that is, no oxygen is required for the process to take place. Glycolysis is followed by one of two processes. If oxygen is present in the cell, the pyruvic acid is broken down further through the process of **aerobic** (ehr OH bihk) **respiration.** This process results in an additional gain of ATP molecules. Without oxygen,

▶ Producers and Consumers

All life on earth is locked together in a never-ending cycle of construction and destruction. Green plants use the energy of the sun to turn carbon dioxide and water into glucose and oxygen. Both plants and animals consume the glucose and oxygen and in turn give off carbon dioxide and water. The plants use these products to continue the cycle.

Green plants and other organisms that can use inorganic molecules to produce organic food molecules are called **autotrophs** (AWT uh trahfs). Many organisms cannot produce their own food molecules from inorganic substances. They are called

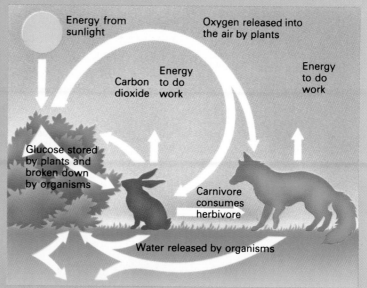

heterotrophs (HEHT uh uh trahfs). Heterotrophic organisms must rely on autotrophs for energy.

All life depends on autotrophs. Animals are hetero-

trophic organisms. People may grow their own food crops, but autotrophic plants are responsible for the conversion of light energy into food.

the pyruvic acid cannot be used to release more energy. However, the hydrogen acceptor molecules must release the hydrogen atoms, or else glycolysis could not continue. Hydrogen atoms are removed through **fermentation,** an anaerobic process that breaks down pyruvic acid into ethyl alcohol or lactic acid. Together, the processes of glycolysis and fermentation make up **anaerobic respiration.**

8.6 Fermentation

Fermentation occurs in some of the less complex organisms, such as some bacteria and yeasts. Most microorganisms convert pyruvic acid to ethyl alcohol, while some animal cells and other microorganisms produce lactic acid.

Alcoholic fermentation combines the hydrogen and the pyruvic acid formed during glycolysis to produce ethyl alcohol.

Figure 8–9. This diagram summarizes the reactions of cellular respiration. First, two ATPs are produced as glucose is broken down by glycolysis to pyruvic acid. The pyruvic acid is then further broken down by fermentation, or aerobic respiration. Aerobic respiration is highly efficient, producing 36 additional ATPs for every glucose molecule used.

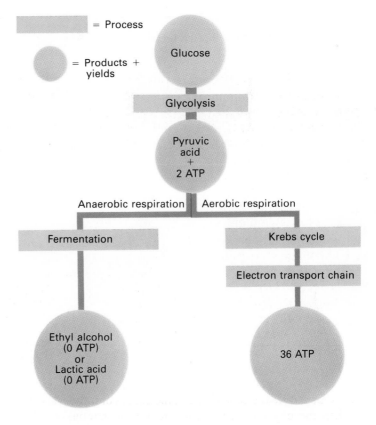

Carbon dioxide is also given off as a byproduct. The general equation of the process can be written:

$$2C_2H_3OCOOH + 4H \rightarrow 2C_2H_5OH + 2CO_2$$
pyruvic acid   ethyl alcohol

Most of the energy originally stored in the glucose remains in the bonds of the ethyl alcohol molecule. For this reason, alcohol is a good fuel.

Lactic acid fermentation combines pyruvic acid and hydrogen from glycolysis to form lactic acid. The general equation for this process can be written:

$$2C_2H_3OCOOH + 4H \rightarrow 2CH_3CHOHCOOH$$
pyruvic acid   lactic acid

Lactic acid fermentation occurs in animal muscle cells. When oxygen is available, these cells carry on aerobic respiration. During strenuous exercise, oxygen concentration diminishes, and muscle cells are forced to use lactic acid fermentation. This process uses the hydrogen stored by the hydrogen acceptor molecule and makes the acceptor molecule available

The Taste of Anaerobic Respiration

For centuries people in various parts of the world have known how to put the waste products of the respiratory process to good use. The microorganisms that carry on anaerobic respiration are used today in many commercial processes. Anaerobic respiration provides food, drink, and even fuel for automobiles.

Yeasts are tiny organisms used as "fermentation factories" to produce both carbon dioxide and alcohol. The carbon dioxide generated during anaerobic respiration is trapped in the baker's flour and water mixture, making the dough rise. The alcohol found in wine, beer, and gasohol is produced as a result of the fermentation of sugars carried on by yeast.

Lactic acid fermentation is also an important industrial process. Certain bacteria ferment milk, producing carbon dioxide and lactic acid. The sour flavor of yogurt, buttermilk, cottage cheese, and sour cream results from the lactic acid produced by bacteria.

In the 1800s the gold miners in the western United States discovered that they could make bread without yeast. They used bacterial fermentation to produce the carbon dioxide needed to make their dough rise. The lactic acid produced by these bacteria gave their sourdough bread a unique and characteristic flavor.

for reuse in glycolysis to obtain more energy. The accumulation of lactic acid in the cells is one of the causes of muscle soreness. When the oxygen concentration returns to normal, the lactic acid is converted back to pyruvic acid, aerobic respiration is begun again, and the soreness fades.

8.7 Aerobic Respiration

Like fermentation, aerobic respiration also begins with the pyruvic acid produced through glycolysis. Although glycolysis takes place in the cytoplasm of the cell, aerobic respiration takes place on the folded membranes inside the mitochondria. Associated with these folded membranes are all the enzymes and coenzymes needed in the reactions that make up the process of aerobic respiration.

The equation for aerobic respiration is essentially the reverse of that for photosynthesis:

$$\underset{\text{glucose}}{C_6H_{12}O_6} + 6O_2 \overset{\text{enzymes}}{\longrightarrow} 6CO_2 + 6H_2O + \text{energy}$$

Aerobic respiration results in a maximum energy gain of 38 molecules of ATP from each molecule of glucose—that is, 36 in addition to the 2 gained from glycolysis. Some cells, because of a high metabolic rate, must produce a greater amount of energy. Brain cells, muscle cells, and insect flight cells, for example, contain large numbers of mitochondria.

Most species of living things carry on aerobic respiration because of its high energy yields. The highly active and complex organisms that inhabit the earth today could not have evolved without the large amount of energy provided by aerobic respiration.

Pyruvic Acid Conversion   The first step of aerobic respiration breaks down pyruvic acid into carbon dioxide, hydrogen, and a two-carbon acetyl group. The carbon dioxide is released as a waste product, and the hydrogen combines with a hydrogen acceptor. The acetyl group then combines with a molecule called *coenzyme A* (CoA) and forms **acetyl CoA.** The function of coenzyme A is to carry the acetyl group into the next stage of respiration, the Krebs cycle.

Krebs Cycle   The acetyl CoA produced from pyruvic acid transfers its acetyl group to a four-carbon molecule to form a six-carbon molecule called **citric acid.** The citric acid then enters a series of reactions called the **Krebs cycle,** or *citric acid cycle*. This reaction series involves many enzymes and coenzymes found within the mitochondria. The Krebs cycle occurs in a series of steps that begin and end with the same substance, citric acid. In other words, some of the molecules required to begin the cycle are re-formed by the reactions that end the cycle.

The reactions of the Krebs cycle involve breaking the chemical bonds between carbon and hydrogen atoms and forming

Figure 8–10. The Krebs cycle is the heart of aerobic respiration. Pyruvic acid from glycolysis is converted to an acetyl group that enters the cycle after combining with coenzyme A. As the cycle turns, hydrogen ions are released. These ions are picked up by hydrogen acceptors and carried to the electron transport chain.

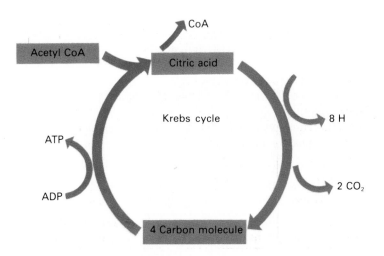

new compounds. As Figure 8–10 shows, one turn of the cycle produces two molecules of carbon dioxide as a waste product. The hydrogen atoms removed during the cycle combine with hydrogen acceptors. Some of the energy released during one turn of the cycle is used to convert one molecule of ADP to ATP. Most of the rest of the energy is in the form of electrons in the hydrogen atoms transferred to the hydrogen acceptors.

Electron Transport Chain By the end of the Krebs cycle, energy from the original glucose molecule has been used to produce four molecules of ATP—two during glycolysis and two during the Krebs cycle. Most of the energy of the glucose, however, remains in the hydrogen atoms transferred to acceptor molecules. There are 24 of these hydrogens—4 from glycolysis, 4 from the conversion of pyruvic acid, and 16 from the Krebs cycle.

In the final stage of aerobic respiration, energy from the 24 hydrogen atoms is released and used to produce ATP. This process occurs in an electron transport chain similar to the one in photosynthesis. The process begins as the 24 hydrogen atoms form 12 pairs of hydrogen ions and 12 pairs of electrons. The electrons are then passed in pairs from molecule to molecule in the electron transport chain. Each transfer is accompanied by a small loss in energy for the electron. The energy released is used to form ATP.

At the end of the chain, the electrons combine with hydrogen ions and oxygen atoms to form water. The oxygen comes from the air and is the final hydrogen acceptor in the process of aerobic respiration.

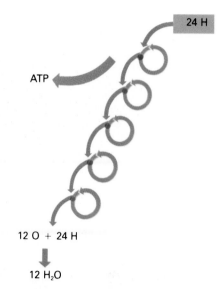

Figure 8–11. The electron transport chain consists of a series of molecules that accept and donate hydrogen ions arriving from the Krebs cycle. As ions pass down the chain they release energy and ATP is formed.

Table 8–1: A Comparison of Respiration and Photosynthesis

	Respiration	**Photosynthesis**
Where	In all living cells	In chlorophyll-containing cells
When	All the time	In light only
Input	Carbon compounds and oxygen	Carbon dioxide and water
Output	Carbon dioxide and water	Carbon compounds, oxygen, water
Energy source	Chemical bonds	Light
Energy results	Energy released (36 ATP)	Energy stored
Chemical reactions	Oxidation of carbon compound	Reduction of carbon compound

Figure 8–12. During the processes of glycolysis and aerobic respiration, the energy present in the chemical bonds of one glucose molecule is transferred to the energy in the phosphate bonds of 38 molecules of ATP. The processes are structured so that energy is released a little at a time.

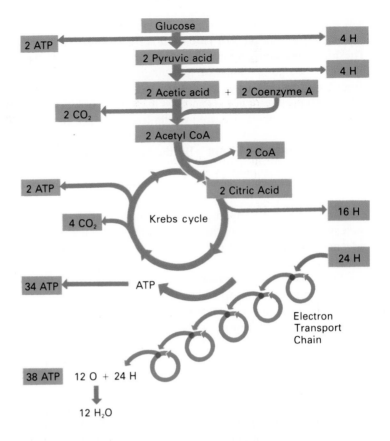

Summary of ATP Produced

As Figure 8–12 shows, *the complete breakdown of one molecule of glucose results in a maximum net yield of 38 molecules of ATP.* Glycolysis and the Krebs cycle each yield a net gain of 2 ATP. The electron transport chain produces 34 ATP molecules.

Reviewing the Section

1. Why do all cells need to perform some type of cellular respiration?
2. Name three uses of anaerobic respiration in the commercial food industry.
3. What is the net gain of ATP molecules from one glucose molecule by each of the following: glycolysis, conversion of pyruvic acid, Krebs cycle, electron transport chain?
4. If fermentation does not produce any additional ATP, why is it necessary for the functioning of some cells?
5. Describe how heterotrophs depend on autotrophs.

Investigation 8: Photosynthesis

Purpose
To determine the rate of photosynthesis in a plant

Materials
Elodea or any aquatic plant, glass funnel, 1000-mL beaker, test tube, sodium bicarbonate, watch or clock with second hand, wooden splint, water, matches

Procedure
1. Pour water into the beaker until it is half full. Allow the water to sit overnight, which allows chlorine and any other additives to evaporate. Dissolve 3g of sodium bicarbonate in the water.
2. Place an elodea plant in the bottom of the beaker. Put a funnel over the plant, as shown in the diagram.

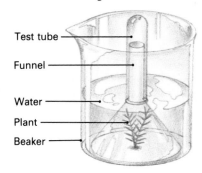

Test tube —
Funnel —
Water —
Plant —
Beaker —

3. Fill the test tube with water. Placing your thumb securely over the mouth of the test tube, invert the tube and place it on top of the funnel.
4. Set the beaker in direct sunlight and count the number of gas bubbles that appear in the test tube. Write down the number of bubbles you count after 30, 60, 90, 120, 150, 180, and 210 seconds.
5. Place your thumb over the mouth of the test tube and remove it from the water. Turn the test tube right side up and insert a glowing splint to test for the presence of oxygen in the tube.

Analyses and Conclusions
1. Make a grid like the following one and graph your results.

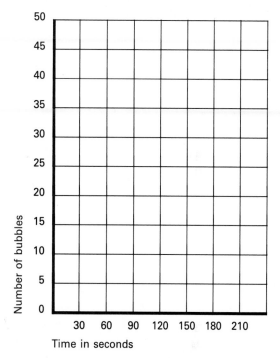

2. How are your results related to the rate of photosynthesis? What conclusion can you draw from your graph?
3. Compare your results with those of your classmates. Why is it important that only one plant be used in each experiment?

Going Further
- Use the same setup as in the original experiment, but use warm water and test the effect of heat on the rate of photosynthesis.
- Compare the rate of photosynthesis under different conditions of light, such as bright and cloudy days.
- Increase the amount of sodium bicarbonate added to the water, and compare the results with those in the original experiment.

Chapter 8 Review

Summary

Cell processes depend on the transfer of energy. A special energy-carrying molecule, ATP, stores energy and releases it as needed in the cell.

All energy used by cells originally comes from sunlight. Through photosynthesis, green plants trap light energy and convert it into chemical energy stored in the bonds of glucose. The light reactions of photosynthesis use light energy to produce ATP and to split water into hydrogen ions, oxygen, and electrons. The dark reactions use the ATP and hydrogen to form glucose.

Cellular respiration releases the ATP energy stored in the glucose produced during photosynthesis. Anaerobic respiration takes place without the presence of free oxygen. The process begins with glycolysis, which produces 2 molecules of ATP by converting glucose to pyruvic acid. During fermentation, the pyruvic acid is converted to alcohol or lactic acid. This process does not produce any additional ATP, but is required for glycolysis to continue. Aerobic respiration requires free oxygen. The pyruvic acid produced during glycolysis is broken down further by the Krebs cycle and the electron transport chain. The final product is an additional 36, or a total of 38, molecules of ATP.

BioTerms

acetyl CoA (**126**)
aerobic respiration (**122**)
anaerobic (**122**)
anaerobic respiration (**123**)
ATP (**113**)
autotroph (**123**)
carotene (**116**)
cellular respiration (**114**)
chlorophyll (**116**)
citric acid (**126**)

dark reactions (**117**)
electron transport
  chain (**118**)
fermentation (**123**)
glycolysis (**122**)
grana (**116**)
heterotroph (**123**)
Krebs cycle (**126**)
light reactions (**117**)
PGAL (**120**)

photon (**115**)
photosynthesis (**114**)
pigment (**116**)
pyruvic acid (**122**)
RDP (**120**)
stroma (**116**)
visible spectrum (**115**)
xanthophyll (**116**)

BioQuiz   (Write all answers on a separate sheet of paper.)

I. Completion

1. During _____ glucose is broken down into two pyruvic acid molecules.
2. Photosynthesis produces glucose and releases water and _____.
3. _____ converts pyruvic acid to lactic acid.
4. _____ absorbs energy from all the visible spectrum except green.
5. Two turns of the _____ cycle utilize one glucose molecule.
6. _____ lower the activation energy of some chemical reactions.
7. The Krebs cycle begins and ends with _____.
8. _____ absorb light.

II. Multiple Choice

9. The electron transport chain allows electrons to a) move from the grana to the stroma. b) move from sunlight to chlorophyll. c) convert into carotene. d) release energy.
10. Energy for the cell's use is released when ATP gives up a phosphate, forming a) ATP and (P). b) pyruvic acid. c) ADP and (P). d) citric acid.
11. Photosynthesis begins in the a) mitochondria. b) cytoplasm. c) grana. d) nucleus.
12. To begin the Krebs cycle, an acetyl group and a four-carbon molecule form a) citric acid. b) acetic acid. c) adenine. d) ribose.
13. During the dark reactions, some PGAL is used to form a) alcohol. b) ATP.

c) carbon dioxide. d) glucose.

14. Cellular respiration cannot take place without a) light. b) enzymes. c) chlorophyll. d) carbon dioxide.

III. Essay

15. Why might muscles become sore during strenuous exercise?
16. After ATP releases a (P), how is energy returned to it for future use?
17. How do electrons in chlorophyll become energized?
18. What is the importance of RDP in the dark reactions of photosynthesis?
19. Through what three processes do organisms break down pyruvic acid?
20. How do plants and animals compare as consumers and producers?

Applying and Extending Concepts

1. If four molecules of ATP are produced from the glycolysis of one glucose molecule, why is the net gain only two ATP molecules?
2. Are vegetarians the only heterotrophs that get all their energy from plants? Explain your answer.
3. Make a table comparing aerobic respiration, alcoholic fermentation, and lactic acid fermentation after glycolysis. Include the materials that each process begins

with, what is produced, and the net gain of ATP molecules.
4. When light enters water, different parts of its spectrum penetrate to different depths. Blue-green algae have mechanisms that allow the algae to rise and fall in the water. What advantage is this to the algae?
5. Why are ATP-ADP conversion, dark reactions, and the Krebs reactions all described as cyclical processes? Be specific in your answer.

Related Readings

Arnon, D. I. "Sunlight, Earth Life: The Grand Design of Photosynthesis." *The Sciences* 22 (October 1982): 22–27. This article outlines the course of scientific inquiry into photosynthesis.

Calvin, M. "Probing the Plant Cell's Secrets." *Science Digest* 85 (April 1979): 76–79. The author, a Nobel prize winner, speculates on the possibility of creating artificial photosynthetic membranes. Such membranes could convert the energy of sunlight directly into usable fuels.

Philips, E. J. "Biological Sources of Energy from the Sea." *Sea Frontiers* 28 (January–February 1982). The author explores the use of marine organisms to meet human energy needs.

BIOTECH

Uncovering the World of the Cell

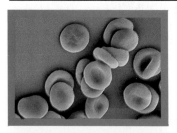

Most cells are very small. An average cell is about 40 micrometers (μm) in diameter. Even smaller are organelles, the structures inside a cell. A mitochondrion is about 1 μm in diameter. If one could flatten the membranes of 10 million mitochondria and spread them out, they would cover a surface area equivalent to that of a postage stamp. A ribosome is about 0.025 μm in diameter. Relatively speaking, an average ribosome inside an average cell is like a marble inside a very large high school auditorium.

Technological advances have made it possible for cell biologists to study organelles as small as ribosomes. In fact, electron microscopes can be used to view objects 0.0002 μm in diameter. The electron microscope magnifies cell structure so well that it can show only a very limited area at one time. Imagine using the electron microscope to photograph the surface of one side of a penny. It would take about 200,000 7.5-by-10-cm prints to assemble the entire image.

Electron microscope technology has dramatically changed biologists' view of cells.

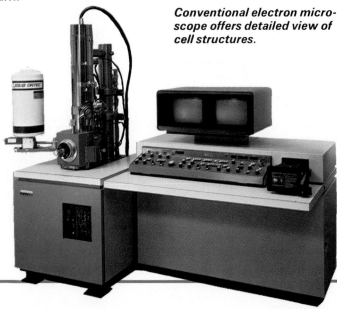

Conventional electron microscope offers detailed view of cell structures.

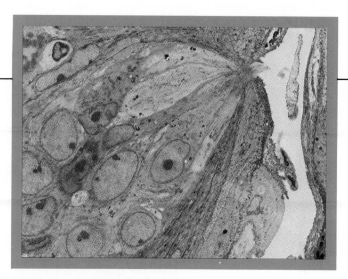

HVEM reveals detailed cross-section of mouse taste bud.

To study a structure like a mitochondrion, a cell biologist uses a machine called an *ultra-microtome.* This device can slice material very fine, to less than 0.02 μm in thickness.

One problem with the slicing operation is that organelles may be distorted as the diamond blade cuts through or

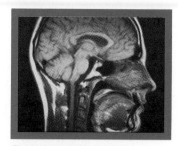

Brain image generated from HVEM.

past them. A recent technological development, the *high-voltage electron microscope* (HVEM), can handle slices as thick as 1μm. The HVEM differs from the conventional electron microscope in that a million-volt beam of electrons is used instead of a 100-kilovolt beam.

A new technological application of electron microscopy is chemical analysis. When electrons in an electron microscope interact with a specimen, they produce X rays, which are collected and analyzed by a device called an *energy dispersive spectrometer.* In combination with a computer, the device can be used to obtain information about chemical structure while a sample is being studied under the electron microscope. The kinds and amounts of various elements can be determined for samples as small as 1 μm in diameter.

Cell Reproduction

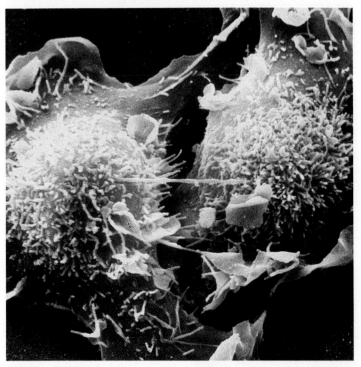

Animal cell dividing

Introduction

Reproduction is a characteristic of all living things. The ability to reproduce is not vital to the life of the individual, but it is crucial to the continuation of the species. Organisms reproduce in a variety of ways, but all these methods are possible only because individual cells can reproduce themselves. Reproduction at the cellular level occurs when one cell, the *parent cell*, divides and forms new cells called the *daughter cells*.

The reproduction of the individual organism depends on the reproduction of the cell. Unicellular organisms reproduce by cell division. In multicellular organisms, the processes of growth and repair depend upon cell division. In addition, a special type of cell division produces the sex cells that combine and form new organisms.

Cell Division

The materials a cell needs for maintenance and growth move into the cell through the cell membrane. Waste materials leave through the cell membrane. As the cell grows, its volume increases at a greater rate than its surface area. If growth were unchecked, the surface area would become too small to accommodate the transfer of materials in and out of the cell. *To maintain a workable ratio of volume to surface area, a cell must divide or stop growing.*

When a cell reaches a certain size, it divides into daughter cells. The daughter cells in turn grow and increase in size until they too divide. The daughter cells produced during cell division are similar in structure to the parent cell. The daughter cells receive portions of the cytoplasm and organelles of the parent cell. Each daughter cell also receives a copy of the hereditary information possessed by the parent cell.

9.1 Growth and Repair Require Cell Division

Multicellular organisms grow by increasing the number of their cells through cell division. Cell division also produces new cells that replace worn out or damaged cells. Thus human bodies rely on cell division to heal cuts, repair broken bones, and replace cells that have a short life span.

The frequency of cell division varies greatly among organisms and among cells within an organism. Bacteria may divide every 20 minutes, while many human cells require 18 to 22 hours to divide. Cells found in the skin and in the lining of the human intestine continue to divide throughout the life of the individual. Many cells in the human body, on the other hand, do not divide.

9.2 Reproduction Depends on Cell Division

Cell division is necessary for reproduction of a single cell or of an entire multicellular organism. Organisms reproduce in two basic ways—asexually and sexually. **Asexual reproduction** is the production of offspring from one parent. Offspring formed asexually have half the cytoplasm and organelles of the parent cell and genetic material identical to the parent. **Sexual reproduction** is the formation of a new individual from the union of two cells. In almost all cases, sexual reproduction requires two parents, and the offspring usually show some characteristics of each parent.

Q/A

Q: *How many cells are produced by the human body in a day?*

A: About 2 trillion additional cells are produced by the body of an adult human during one day, or about 25 million new cells per second.

Spotlight on Biologists: Alfred Lloyd Goldson

Born: New York, 1946
Degree: M.D., Howard University

Howard University radiologist Alfred Goldson is a pioneer in the use of an innovative treatment for cancer. The procedure, which combines surgery and radiation for deep-seated or inoperable tumors, is now used by most major cancer centers. In recognition of his work, the American Cancer Society in 1978 named Goldson among the country's top 50 scientists.

Prior to the 1970s, most physicians in the United States delivered radiation from outside the body, through the skin, after surgery was completed. The interoperative radiation

therapy (IORT) used by Goldson focuses light radiation beams on the site of a cancer during surgery, before suturing takes place. This procedure arrests any malignant cells remaining, without harming adjacent areas.

The IORT procedure was first reported in 1915, but was largely ignored in favor of new, powerful radiation machines that allowed penetration deep into the body. Goldson and some other researchers recognized that these machines also increased the danger of harming healthy surrounding tissues.

Goldson, the son of a physician, began his radiotherapy career at Howard under the late Dr. Ulrich K.

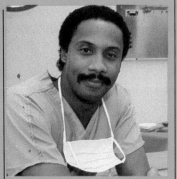

Henschke. Eventually Goldson succeeded Henschke as chairman of Howard's radiotherapy department.

Henschke and Goldson first equipped a Howard University Hospital operating room with the necessary radiation machine, called a linear accelerator. Then, in 1976, Goldson performed the first IORT procedure in the United States.

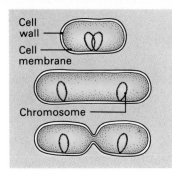

Figure 9–1. Three stages in the process of binary fission are shown in the illustration above.

Asexual Reproduction   The simplest form of asexual reproduction is **binary fission,** in which the cell splits in two. Binary fission is the primary means of reproduction in prokaryotes, which are organisms that lack a true nucleus. First the single, circular chromosome in a prokaryote duplicates, a process called **replication.** As Figure 9–1 shows, both chromosomes then attach themselves to sites on the cell membrane. As the cell grows, a new cell membrane forms between the attachment sites and the two chromosomes are forced apart. Then the cell membrane constricts in the center, ultimately separating the cell into two identical parts.

A second method of asexual reproduction is through the production of tiny, asexual reproductive cells called **spores.** Like binary fission, spore formation begins with the replication of the chromosomes. A protective wall then forms around each set of chromosomes to create the spores. After a spore leaves the parent cell, the spore can remain inactive until environmental conditions are favorable for the growth of the organism. The spore will then germinate. Molds and other fungi reproduce through the formation of spores.

Some organisms, such as yeast, reproduce asexually through a process called **budding.** The cell nucleus and cytoplasm divide into two cells of unequal size. The bud, the daughter cell, pinches off to become a new individual.

Some plants produce organisms that are initially attached to the parent plant and then separate to become individual plants. This process is called **vegetative propagation.** Plants such as strawberries send out ''runners,'' or horizontal above-ground stems. The runners root and develop into new individual plants when they contact fertile soil. Tuber plants such as potatoes propagate through underground runners that develop from potato ''eyes.'' Some plants form tiny plants on the edges of their leaves. The new plants drop off the leaves and root in the soil.

Some animals have the ability to develop lost body parts or even to form new individuals from a single fragment. The development of a new animal from its parts is called **regeneration.** For example, consider what happens if a starfish is cut into several pieces and thrown back into the water. Each of the pieces that contains a portion of the central body will regenerate into a new complete starfish.

Sexual Reproduction   Sexual reproduction results from the joining of two specialized sex cells called **gametes** (GAM eets). The male gamete is called a **sperm** cell and the female gamete an **ovum,** or egg cell. In the process of **fertilization,** a sperm and an ovum combine to form a cell called a **zygote** (ZY goht). Because both parents contribute chromosomes to the zygote, the offspring is usually not identical to either parent but may have some characteristics of each.

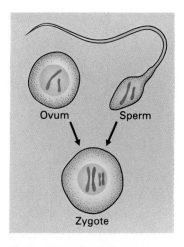

Figure 9–2. Sexual reproduction occurs when gametes, such as a sperm cell and an ovum, unite and form a zygote.

Reviewing the Section

1. What limits the growth of cells?
2. How do newly formed daughter cells compare in structure and function with their parent cell?
3. Which form of reproduction results in the greatest variety of offspring? Why?
4. Compare and contrast spore formation, budding, and binary fission.

- *State* the relationship between the number of chromosomes in a parent cell and the number of chromosomes in the daughter cells produced through mitosis.
- *Outline* the events that take place in each phase of mitosis.
- *Compare* cell division in plants and in animals.
- *List* in sequence the four stages of the cell cycle.

Mitosis

A prokaryotic cell reproduces by replicating its single chromosome and then simply dividing in two. The reproduction of eukaryotic cells is more complex because it involves replication of a greater amount of genetic material and the division of the nucleus. Cell division of eukaryotic body cells involves a process of nuclear division called **mitosis** (my TOH sihs). *As a result of mitosis, each daughter cell receives an exact copy of the chromosomes present in the parent cell.* Before mitosis, the chromosomes in the nucleus of the parent cell replicate. They then divide into identical sets. Mitosis is generally followed by division of the cytoplasm.

Each kind of eukaryotic organism has a specific number of chromosomes in its body cells. Body cells are called **somatic** (soh MAT ihk) **cells.** Every somatic cell within the body of a multicellular organism contains the same number of chromosomes. Somatic cells of humans, for example, possess 46 chromosomes. Somatic cells of mosquitos have 6 chromosomes; those of corn plants have 20; those of goldfish have 94.

9.3 Preparation for Mitosis

Cell division is not a single process but a complex series of events. It involves the division of the nucleus as well as the division of the cytoplasm and its component parts. While some human cells may require 20 hours to go through an entire cycle of growth and division, the actual division of the nucleus and cytoplasm may occur in less than an hour. The time between the formation of a cell through mitosis and the beginning of the next mitosis is called **interphase.** During interphase the cell prepares for division by replicating genetic material, by producing the necessary number of organelles, and by assembling the structures needed for mitosis. Interphase itself is not part of mitosis, but is a necessary phase in the reproduction of the cell.

During interphase chromosomes are not distinguishable under a light microscope. The hereditary material—the nucleic acid DNA and protein—appears as *chromatin,* dense patches within the nucleus. At the start of mitosis, the chromatin coils up and condenses into short, thick rods that become visible under a microscope. Each chromosome, as it is now called, consists of two joined strands called **chromatids** (KROH muh tihdz). Each chromatid is a duplicate of its partner. The point at which the chromatids are held together is called the **centromere** (SEHN truh mihr).

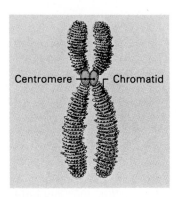

Centromere — Chromatid

Figure 9–3. A chromosome consists of two coiled chromatids that are joined at the centromere.

9.4 Phases of Mitosis

Biologists divide mitosis into four phases: prophase, metaphase, anaphase, and telophase. *Actually, mitosis is a continuous process in which each phase merges into the next.* Here the process is described as it occurs in an animal cell.

Prophase   The first stage of mitosis is called **prophase.** This stage takes up about 60 percent of the total time required for mitosis. Prophase can be divided into early, middle, and late stages. During early prophase, the chromosomes begin to coil up into short rods. The nucleoli break down and begin to disappear. Two pairs of centrioles appear to one side of the nucleus, outside the nuclear membrane.

The centrioles move apart, signaling the beginning of middle prophase. Protein tubes called *spindle fibers* form between

▶ The Cell Cycle

Night follows day, winter follows summer, and high tide follows low tide. Many events in the natural world follow a regular, recurring pattern. One cycle basic to all living organisms is the cycle of the cell.

The **cell cycle** is the sequence of cell growth and division that occurs in a cell between the beginning of one mitosis and the beginning of the next mitosis. The cell cycle has four stages. The first stage includes mitosis and the division of the cytoplasm, called **cytokinesis** (syt oh kih NEE sihs). The remaining stages make up interphase. In a multicellular organism, many cells are permanently in interphase.

Following mitosis, the cell enters a period of intense cellular activity and growth known as the G_1 phase. The cell doubles in size, and organelles such as the ribosomes and mitochondria double in number. Enzyme production is at a high level to accommodate the increased chemical activity.

Those cells that stop growing remain in the G_1 phase. Cells that go on to divide next enter the S, or synthesis, phase. During this period the chromosomes replicate. The S phase is followed by a second period of growth, G_2. During G_2, the structures used in mitosis and cytokinesis are assembled.

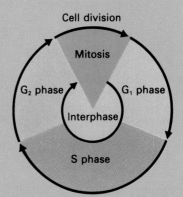

Temperature and other external environmental factors can affect the duration of the cell cycle, which lasts from several hours to several days. Bean cells, for example, complete a cycle in 19 hours. Of this time, 7 hours are spent in S phase, 5 hours in each of the two growth phases, and 2 hours in mitosis.

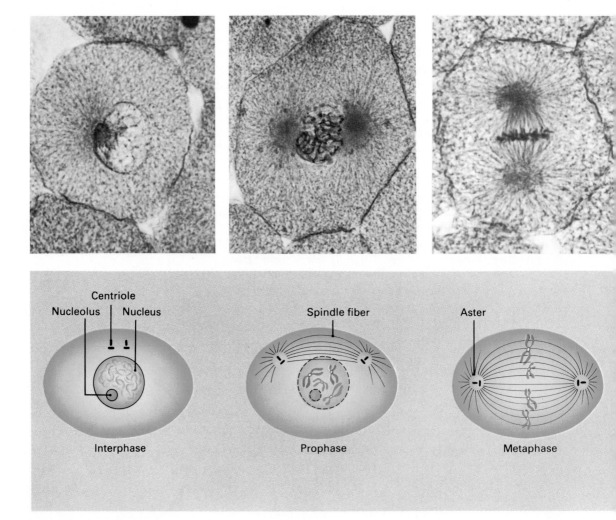

Centriole

Nucleolus | Nucleus

Interphase

Spindle fiber

Prophase

Aster

Metaphase

Figure 9–4. The process of mitosis creates two cells that have the same number of chromosomes as the original cell. Six phases in the process of mitotic cell division are shown above: interphase, prophase, metaphase, anaphase, telophase, and cytokinesis. Although interphase and cytokinesis are not considered true phases of mitosis, they are necessary for cell division.

the centrioles. Additional fibers radiating outward from each centriole form the **aster,** visible in Figure 9–4. The exact function of the aster is not yet known. Plant cells develop spindle fibers, but they lack centrioles and asters. At this stage the nuclear membrane has broken down and disappeared.

By late prophase, the centriole pairs are at opposite ends of the cell. Each centriole is fully formed into a three-dimensional, football-shaped structure. The chromosomes are attached to the centrioles by some of the spindle fibers. Other spindle fibers stretch across the cell from one centriole to the other. Animal cells have centrioles; most plant cells do not.

Metaphase   During **metaphase,** the chromosomes are pushed and pulled by the spindle fibers. Finally they become arranged

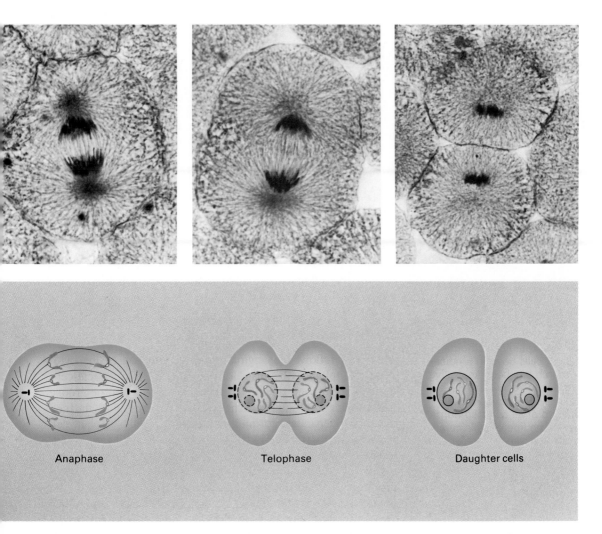

Anaphase

Telophase

Daughter cells

along the cell's midplane, called the *equator*. The centromere of each chromosome is attached to a separate spindle fiber.

Anaphase   Anaphase begins with the separation of the chromatids in each pair. The spindle fibers appear to shorten, pulling the chromatids apart at the centromere. Each chromatid is now called a chromosome. As Figure 9–4 shows, the two sets of separated chromosomes then move through the cytoplasm to opposite ends, or poles, of the cell.

Telophase   The last stage of mitosis is **telophase.** After the individual chromosome strands have reached the opposite poles of the cell, the spindle disappears. A nuclear membrane forms around each set of chromosomes as the chromosomes once more

Figure 9–5. Plant cells do not undergo the same kind of cytokinesis as animal cells. Plants form a cell plate, a structure made from membranes synthesized by Golgi bodies.

Q/A

Q: *Can human cells survive and divide outside the body?*

A: Yes. One line of human cancer cells used in medical research has been kept alive almost 40 years. The HeLa cells are the most widely studied of all cultured human cells. The line of cells came from the body of cancer patient Henrietta Lachs, for whom they were named. HeLa cells now grow in laboratories all over the world.

return to a threadlike mass. The centrioles duplicate; two centrioles are formed in each daughter cell. The nucleoli also re-form within each newly formed nucleus.

9.5 Cytokinesis

Mitosis is followed by cytokinesis, the division of the cytoplasm. Cytokinesis begins during telophase. In an animal cell, the cell membrane pinches together, and a furrow or groove forms along the equator. The groove deepens until the cell membrane separates, forming two daughter cells. A plant cell has a relatively rigid cell wall that prevents the cell from dividing by pinching. In plant cells, a **cell plate** is formed in the middle of the dividing cell from membrane vacuoles produced by the cell's Golgi bodies. Just before cytokinesis, the Golgi bodies migrate to the area where the cell plate will form. Then the membrane vacuoles formed by the Golgi bodies fuse into a membrane that separates the cells. The cell plate extends outward until it separates the two daughter cells. Each of the new cells then forms a cell wall on its side of the cell plate.

Reviewing the Section

1. What is the function of mitosis?
2. Distinguish between chromosomes, chromatids, and centromeres.
3. How does cytokinesis differ in plants and in animals?
4. At what stage of the cell cycle do the chromosomes replicate?
5. Describe chromosome activity during prophase.

Meiosis

All cells produced through mitosis have the same number of chromosomes as their parent cells. Consider what would happen if two cells formed through mitosis combined in sexual reproduction. The offspring would have twice as many chromosomes as its parents. As this process continued, each succeeding generation would have double the chromosome number of its parents. Sexual reproduction does not increase the number of chromosomes, because gametes have only half the number of chromosomes found in somatic cells.

The chromosomes in somatic cells occur in pairs called **homologous** (hoh MAHL uh guhs) **chromosomes.** The chromosomes in a pair are alike in appearance and in the type of genetic information they carry. The 46 chromosomes in human cells, for example, form 23 homologous pairs. Cells that have homologous chromosomes are said to have the **diploid** (DIHP loyd) **number (2n)** of chromosomes. A gamete, however, has only one member from each pair of homologous chromosomes of its parent cell. Thus gametes have half the diploid number, or the **haploid number (n).** Gametes are formed by a type of nuclear division called **meiosis** (my OH sihs). *Meiosis reduces the number of chromosomes to half the number in somatic cells.*

- *Explain* why meiosis is essential for sexual reproduction.
- *Compare* the terms *diploid* and *haploid.*
- *List* the major events in meiosis.

9.6 Phases of Meiosis

The process by which haploid daughter cells are formed from a diploid parent cell requires two successive cell divisions. During the first division, meiosis I, the homologous chromosomes separate. During meiosis II, the chromatids of each chromosome separate.

Meiosis I   As in mitosis, meiosis is preceded by the replication of the DNA that forms the chromosomes. During meiosis I, the homologous chromosomes come together, a process called **synapsis** (sih NAP sihs). The chromosomes twist around each

Table 9–1: Comparison of Mitosis and Meiosis

Nuclear division	No. of daughter cells	Parent cell type	Daughter cell type	Genetic likeness of daughter cells to parent
Mitosis				
1	2	diploid	diploid	identical
Meiosis				
2	4	diploid	haploid	different

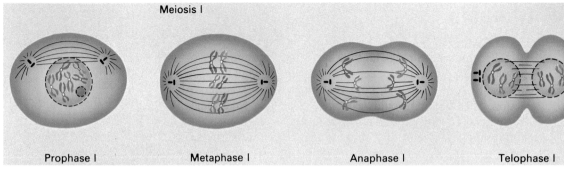

Meiosis I

| Prophase I | Metaphase I | Anaphase I | Telophase I |

Figure 9–6. In the first meiotic division, homologous chromosomes separate. The second division results in four cells, each with half the number of chromosomes present in the original cell.

other, forming a structure called a **tetrad** (TEHT rad). Meiosis I can be divided into the same four phases as mitosis:

- *Prophase I:* The chromatin begins to coil into short rods, and homologous chromosomes are formed. The spindle appears and the nucleoli break down. By the end of prophase I, the nuclear membrane has completely dissolved and, as Figure 9–6 shows, the tetrads are visible.
- *Metaphase I:* The tetrads line up along the equator of the cell. Each tetrad becomes attached to spindle fibers.
- *Anaphase I:* The homologous chromosomes that form each tetrad are pulled apart in pairs. One pair goes to one end of the cell, and the other pair moves to the other end of the cell.
- *Telophase I:* The chromosomes reach the ends of the cell. The cell divides into two daughter cells.

At the end of meiosis I, each daughter cell contains half the number of chromosomes found in the parent cell. That is, one chromosome of each homologous pair is present in each daughter cell. *Meiosis I is reductive division; it reduces the number of chromosomes from the diploid (2n) to the haploid (n) number.*

Meiosis II Each daughter cell produced in meiosis I undergoes another nuclear division in meiosis II. *Meiosis II is similar to mitosis but is not preceded by the replication of DNA.* Meiosis II also has four stages:

- *Prophase II:* Telophase I leads directly into prophase II. A new spindle forms around the paired chromatids.
- *Metaphase II:* The chromosomes line up along the equator, attached at their centromeres to spindle fibers.
- *Anaphase II:* The centromeres duplicate and the chromatids separate. The resulting single chromatids move toward opposite poles of the cell. The chromatids are now called chromosomes.

Q/A

Q: *When does meiosis begin to form egg cells within the human female?*

A: The cells that produce human egg cells begin meiosis during fetal development. These cells remain in prophase I for many years. The completion of meiosis I and meiosis II takes place after sexual maturity.

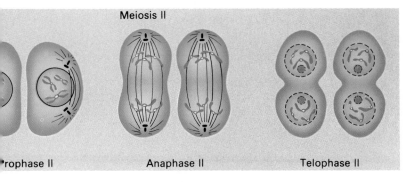

Meiosis II

Prophase II   Anaphase II   Telophase II

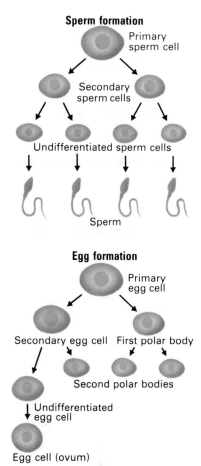

Sperm formation

Primary sperm cell

Secondary sperm cells

Undifferentiated sperm cells

Sperm

Egg formation

Primary egg cell

Secondary egg cell   First polar body

Second polar bodies

Undifferentiated egg cell

Egg cell (ovum)

Figure 9–7. Meiosis of a primary sperm cell (top) yields four haploid sperm. Meiosis of a primary egg cell (bottom) results in only one functional haploid egg cell.

• *Telophase II:* The nuclear membrane forms around each set of chromosomes. The spindle breaks down and the cell undergoes cytokinesis.

Each of the daughter cells formed in meiosis I has divided in two, resulting in a total of four daughter cells produced in meiosis II. ***Each of the daughter cells produced in meiosis II is haploid.*** The cell in Figure 9–6 began with a diploid number of six. It contained three homologous pairs of chromosomes. As a result of meiosis, four haploid cells have been produced. Each cell has three chromosomes, one from each of the homologous pairs of the parent cell.

9.7 Meiosis in Males and Females

Meiosis in male animals results in four cells that *differentiate,* or change, into sperm cells. Meiosis in female animals results in four cells, only one of which becomes an egg. During meiosis I in females, the cytoplasm divides unequally. The smaller of the two cells, called the **first polar body,** may divide again but its cells will not survive. In meiosis II, the division of the egg cell is again unequal. The smaller cell, the **second polar body,** dies. Because of its larger share of cytoplasm, the mature ovum has a rich storehouse of nutrients. These nutrients nourish the young organism that may develop if the ovum is fertilized.

Reviewing the Section

1. Why is meiosis necessary for sexual reproduction?
2. Distinguish between diploid cells and haploid cells.
3. How does anaphase I in meiosis differ from anaphase in mitosis?
4. How does mitosis differ from meiosis in terms of the number and type of daughter cells produced?

Investigation 9: Mitosis

Purpose
To observe the stages of mitosis in a plant root by examining the meristematic area of an onion root tip

Materials

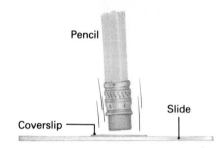

Toothpicks, onion bulb, beaker, scalpel, acetocarmine stain, safety goggles, Bunsen burner, slide, coverslip, pencil, paper towels, microscope

Procedure
1. Insert four or five toothpicks into an onion bulb. Suspend the onion bulb in a beaker of water so that the bottom of the bulb is in the water. The roots should grow freely in the water, as shown in the diagram.

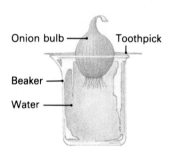

2. When the roots are about 1 cm long, remove their tips by slicing off the end of the root with a scalpel. *Why is the tip of the root being used?*
3. Place the root tips in a beaker with acetocarmine stain. **CAUTION: Put on safety goggles.** Heat gently over a Bunsen burner. Do not allow the stain to boil. *What is the purpose of heating the onion root tip?*
4. Remove the root tips from the stain. Cut a thin slice of each tip. Prepare a wet mount of one or more of the root tips by following the procedure described on page 831.
5. Using the eraser end of a pencil, apply pressure to the coverslip and squash the root tips on the slide. This is most easily accomplished by tapping lightly on the top of the

coverslip. Be sure that you do not tap too hard or the coverslip will break. Blot any excess stain with a paper towel.

6. Observe the slide under a microscope. The acetocarmine should have stained the chromosomes pink or red. Look for cells in the various stages of mitosis. Many of the cells should show some mitotic activity.
7. Draw the cells and label each stage of mitosis. Be sure to locate all phases of the cell division.

Analyses and Conclusions
1. Study your diagrams and describe what is occurring at each stage of mitosis.
2. Describe the position and shape of chromosomes during metaphase.
3. Explain the significance of mitosis. What kinds of cells divide by mitosis?

Going Further
Repeat the experiment using other parts of plants, such as the staminate hairs of a *Tradescantia* or the pollen-producing structures of other flowers. Do the phases of mitosis in these organs correspond to those you have just observed in onion roots?

Chapter 9 Review

Summary

Cells divide to replace worn-out cells, to increase the size of the organism, and to reproduce the organism. Asexual reproduction produces genetically identical offspring from one parent. Binary fission, spore formation, vegetative propagation, and fragmentation are forms of asexual reproduction. Sexual reproduction involves the fusion of gametes supplied by two parents. It produces offspring that are not identical to either parent.

Mitosis is the process of nuclear division that results in the reproduction of body cells and in asexual reproduction in eukaryotes.

Mitosis is preceded by replication of the chromosomes and is divided into prophase, metaphase, anaphase, and telophase.

Meiosis forms the gametes necessary for sexual reproduction. Meiosis reduces the number of chromosomes in gamete cells through two successive nuclear divisions.

Meiosis in a male animal produces four haploid sperm cells, while meiosis in a female produces one haploid egg cell. Fertilization results in the formation of a diploid zygote that may then develop into a new organism.

BioTerms

anaphase (**141**)
asexual reproduction (**135**)
aster (**140**)
binary fission (**136**)
budding (**137**)
cell cycle (**139**)
cell plate (**142**)
centromere (**138**)
chromatid (**138**)
cytokinesis (**139**)
diploid number (**143**)
fertilization (**137**)

first polar body (**145**)
gamete (**137**)
haploid number (**143**)
homologous
    chromosome (**143**)
interphase (**138**)
meiosis (**143**)
metaphase (**140**)
mitosis (**138**)
ovum (**137**)
prophase (**139**)
regeneration (**137**)

replication (**136**)
second polar body (**145**)
sexual reproduction (**135**)
somatic cell (**138**)
sperm (**137**)
spore (**136**)
synapsis (**143**)
telophase (**141**)
tetrad (**144**)
vegetative propagation (**137**)
zygote (**137**)

BioQuiz   (Write all answers on a separate sheet of paper.)

I. Completion

1. Offspring identical to the parent are produced by _____ reproduction.
2. In cell division, the parent cell splits to form two _____.
3. The union of a male and female gamete produces a _____.
4. Two chromatids join at the _____.
5. _____ results in the production of sperm cells and egg cells.
6. Binary fission and regeneration are both forms of _____ reproduction.
7. During _____ the chromosomes become arranged along the cell equator.
8. A _____ forms during the cytokinesis of plant cells.

II. Multiple Choice

9. During the G_1 phase of the cell cycle, the cell a) structures for mitosis are assembled. b) nucleus divides. c) size doubles. d) chromosomes replicate.
10. If an organism could not carry out meiosis, it would a) stop growing. b) fail to produce gametes. c) grow faster. d) shrivel up and die.
11. In diploid cells, homologous chromosomes are a) harmful mutations. b) matching pairs of chromosomes. c) evidence of division. d) found only in humans.
12. Cytokinesis always results in a) two separate cells. b) formation of a zygote. c) two separate tetrads. d) meiosis II.
13. If a cell did not complete interphase, a) the cell would die. b) DNA would not replicate. c) the nuclear membrane would break up. d) the nucleolus would take over the nucleus.
14. Chromosomes form into tetrads during a) cytokinesis. b) spore formation. c) meiosis. d) fertilization.

III. Essay

15. How do molds, yeasts, and plants such as strawberries reproduce?
16. What is the function of mitosis in most animals?
17. Why is meiosis sometimes called a "reduction division"?
18. What are the major events that occur during each stage of mitosis? Describe them.
19. Why must cells divide?
20. What happens during the interphase of the cell cycle?

Applying and Extending Concepts

1. Illustrate the phases of meiosis in a male animal through a series of labeled drawings. Represent the process as it would appear in a parent cell with a diploid number of eight.
2. Human cells normally have 46 chromosomes. Some human hereditary illnesses are a result of abnormal numbers of chromosomes. Explain how it might happen that a human offspring with 47 chromosomes could be produced.
3. The human egg is much larger than the human sperm. Explain how it is possible that a child inherits equally from its mother and father.
4. A horse and a donkey can mate to produce a mule, but the mule is almost always sterile. Horses have a diploid number of 60, and donkeys a diploid number of 66. Suggest an explanation for the fact that the mule is sterile.

Related Readings

Mazia, D. "The Cell Cycle." *Scientific American* 230 (January 1974): 54–64. This classic article describes the events that transpire between the birth of a cell and its division.

Prescott, D. M. *Reproduction of Eucaryotic Cells.* Burlington, N.C.: Carolina Biological, 1978. A leading biologist provides an informative introduction to cell division in eukaryotes.

"Protein Trigger to Cell Proliferation." *Science News* 121 (February 27, 1982). The article discusses a protein that controls cell division and examines its applications for cancer research.

Fundamentals of Genetics

Three generations of a family

Introduction

The fact that organisms reproduce new organisms like themselves may seem obvious to you. You know that horses produce horses, roses produce roses, and people produce people. You also know that organisms possess certain distinguishing characteristics, or traits, exhibited by their parents. Family members, for example, may have similar facial structures, hair texture, and eye color.

Only in the last century have scientists learned how offspring inherit, or receive, traits from their parents. These traits are transmitted by means of information stored in molecules called *DNA*. The passing of traits from parents to offspring is called **heredity.** The study of heredity is one of the most active fields of modern biology.

The Origin of Genetics

The scientific study of heredity is called **genetics.** Modern genetics is based on the knowledge that traits are transmitted by means of **chromosomes,** rod-shaped structures within the nucleus of a cell. Offspring resemble parents because the chromosomes in sperm cells and egg cells contain units of hereditary information. These units are called **genes.** As an individual formed by a sperm cell and an egg cell grows into an adult, the genes influence its development. The genes cause it to resemble the parents who supplied the chromosomes.

Long before people understood the basis of heredity, they bred animals and plants for certain desirable traits. Many breeds of dogs, for example, were developed during the Middle Ages. In the 1800s biologists began to study heredity through scientific experiments. Discoveries made by one of these biologists, Gregor Mendel, became the basis of modern genetics.

10.1 Gregor Mendel: A Pioneer of Genetics

Gregor Mendel was born July 22, 1822, in a small Austrian village. Mendel was an outstanding student and eventually entered a monastery to become a high school teacher. As part of his education, he spent two years at the University of Vienna, where he developed an interest in plant-breeding experiments. When Mendel returned to the monastery, he devoted much of his time to plant-breeding research. Through his experiments, Mendel discovered the basic principles of heredity.

Although scientists before Mendel had performed breeding experiments, none had unlocked the secrets of heredity. Why did Mendel succeed where other scientists had failed? *One important factor in Mendel's success was his choice of garden pea plants for his experiments.* The seeds of these plants were readily available, and the plants could be cultivated quickly in the small garden at the monastery. The large number of offspring produced provided abundant data for Mendel to analyze.

The way in which pea plants reproduce also made them ideal for experimentation. The flower of the garden pea contains *stamens* that produce pollen. The pollen contains the sperm cells—the male **gametes,** or reproductive cells. The same flower also has a *pistil* that contains the eggs, or female gametes. The petals of the flower trap the pollen, which then falls on the pistil and fertilizes the eggs. This process is called **self-pollination.** After many generations of self-pollination, offspring show the same traits as their parents.

Section Objectives

- *Explain* why Mendel succeeded in discovering the principles of heredity when other biologists had failed.
- *Describe* the procedure Mendel followed in his experiments.
- *State* each of Mendel's principles of heredity and give examples of each one.
- *Tell* how Mendel's work was rediscovered.

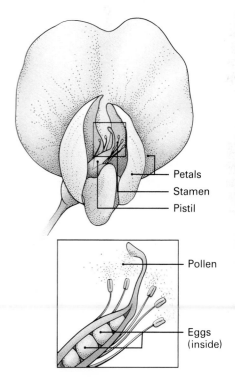

Figure 10–1. The petals of the garden pea plant completely enclose the stamen and pistil, protecting them from foreign pollen. The petals do not open until the plant has self-pollinated.

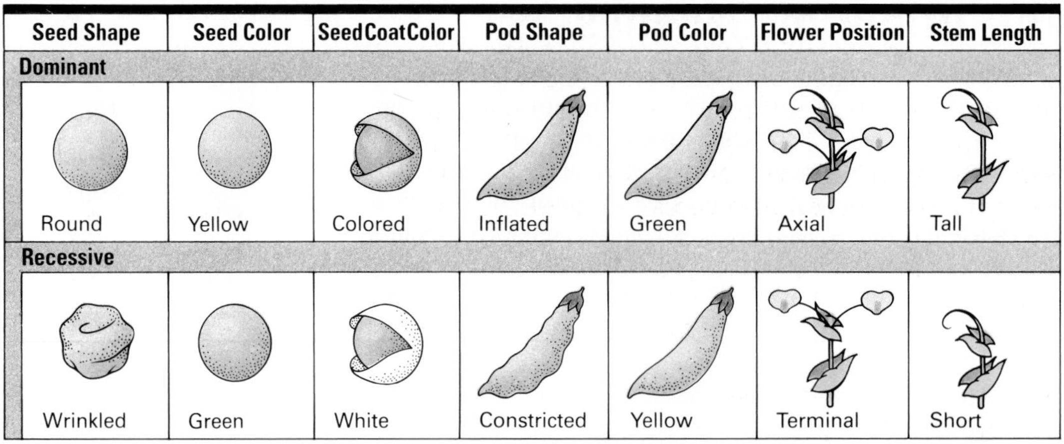

Seed Shape	Seed Color	Seed Coat Color	Pod Shape	Pod Color	Flower Position	Stem Length
Dominant						
Round	Yellow	Colored	Inflated	Green	Axial	Tall
Recessive						
Wrinkled	Green	White	Constricted	Yellow	Terminal	Short

Figure 10–2. Of the many characteristics in the garden pea plant, Mendel chose to study these seven contrasting pairs of traits.

The pea plants were also a good choice because they displayed several traits in one of two contrasting forms. For example, the seeds of the pea plant were either round or wrinkled, never a blend of the two. The plants themselves were either tall or short, never of medium height. Figure 10–2 shows the seven contrasting traits that Mendel studied in pea plants. The choice of plants with distinct traits enabled Mendel to discover how traits are passed from generation to generation. Earlier scientists had studied traits with several intermediate forms and had falsely concluded that all inheritance involved blending of traits.

In addition, Mendel's success resulted from his logical experimental methods and his careful record keeping. First Mendel studied the inheritance of only one trait, then of two traits, and finally of three traits at a time. He also studied the appearance of these traits in three generations. Furthermore, Mendel counted the results of experiments and kept accurate records of them. Later Mendel used these records to calculate the mathematical ratios in which traits appeared. Unlike scientists before him, Mendel suspected that inherited traits would appear in some mathematical pattern.

10.2 Mendel's Experiments

Mendel knew that after many generations of self-pollination pea plants produce offspring identical to themselves. Short plants always produce short offspring. Yellow-seeded plants always produce yellow-seeded offspring.

Mendel wondered what would happen if he crossed two pure-breeding plants with contrasting traits, such as tallness and shortness. To find out he crossed two pea plants. Mendel called these plants the **parental,** or **P, generation.** He performed the parental cross by means of **cross-pollination**, or taking pollen from one plant and dusting it on the pistil of another plant. The results of the parental cross appeared in the first generation of offspring, called the **first filial** (FIHL ee uhl), or **F₁, generation.**

Three possible traits could have appeared in the F₁ generation offspring. The plants could have been tall, short, or of medium height. Mendel soon discovered, however, that all of the F₁ generation plants were tall. The contrasting trait— shortness—seemed to have disappeared.

Mendel continued making parental crosses for each of the seven pairs of contrasting traits. A cross between yellow-seeded plants and green-seeded plants produced F₁ generation seeds that were all yellow. Similarly, a cross between round-seeded plants and wrinkled-seeded plants produced F₁ generation seeds that were all round. *After testing all seven pairs of traits, Mendel discovered that one trait in each pair showed up in the F₁ generation. He noticed that the other trait in each pair seemed to have disappeared.*

Next Mendel allowed the members of the F₁ generation to self-pollinate, producing the **second filial, or F₂, generation.** The results were striking. About three-fourths of the plants in the F₂ generation were tall. However, about one-fourth were short—the trait that had vanished in the F₁ generation. Expressed in mathematical terms, the ratio of tall to short plants was about $3:1$. When Mendel checked the remaining six traits in the F₂ generation, he found that every pair of traits appeared in roughly the same $3:1$ ratio.

As a result of his findings, Mendel proposed that traits that had "disappeared" in the F₁ generation were not lost. Instead, each one was somehow prevented from being expressed. Based on this idea, Mendel reasoned that a pair of elements, which he called *factors,* governs the expression of traits in each individual. Today these factors are called *genes*. According to this idea, every F₁ individual has a pair of genes for each contrasting trait. For example, the pair of genes that determines height in pea plants includes one gene for tallness and one gene for shortness. For some reason, however, only one of each pair of traits appears in the F₁ generation.

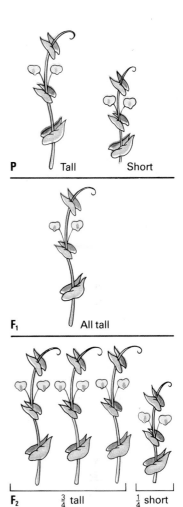

Figure 10–3. When Mendel crossed pea plants with contrasting traits, one trait in each pair disappeared in the F₁ generation and reappeared in the F₂.

10.3 The Principle of Dominance

From his idea of factors, or genes, Mendel developed the **principle of dominance.** *The principle of dominance states that one factor (gene) in a pair may prevent the other factor (gene) in the pair from being expressed.* For example, an F₁ generation seed produced by a cross between a yellow-seeded plant

Q: *The basenji is a breed of dog noted for its inability to bark. Why can't a basenji bark?*

A: The ability to bark is a dominant trait in dogs. All basenjis have two recessive genes for this trait, and so they cannot bark.

and a green-seeded plant contains both a gene for yellow seed color and a gene for green seed color. The gene for yellow seed color prevents the gene for green seed color from being expressed. According to Mendel's principle, a gene that masks the other gene in a pair is **dominant.** The gene that is hidden by the dominant gene in the F_1 generation is **recessive.** The recessive gene may reappear, however, in the F_2 generation.

Today geneticists use symbols to represent genes. A dominant gene is symbolized by a capital letter. For example, Y represents the gene for yellow seed color. The recessive gene in a pair is indicated with the same letter in lowercase. Thus y represents the recessive gene for green seed color. When the two genes in a pair are identical—YY or yy—the individual is **homozygous** (hoh moh ZY guhs), or **purebred.** If both genes are dominant (YY), the individual is **homozygous dominant.** If both genes are recessive (yy), the individual is **homozygous recessive.** Individuals with a dominant and a recessive gene (Yy) are **heterozygous** (heht uhr oh ZY guhs), or **hybrid,** for that trait.

▶ The Fate of Mendel's Ideas

In 1865 Gregor Mendel presented his findings to a meeting of scientists, but they showed little interest in his work. A report of his experiments was published in 1866 and gathered dust on library shelves for the next 35 years.

There were many reasons why the scientific community neglected Mendel's discoveries. One reason was Mendel's mathematical approach. The idea of combining biology and mathematics was not a common one at that time. In fact, scientists of the day believed that biological processes were too complex to be explained by mathematics.

Another reason for the neglect of Mendel's findings was that biologists had not yet discovered the physical basis of heredity. In 1866 there was no workable theory of the cell and its functions. Chromosomes had not yet been observed. Thus it was the fate of Mendel's ideas to await a time when biologists were better able to understand and accept them.

The year 1900 was such a time. Within the brief period of four months, Mendel's work was rediscovered three times. The botanists Hugo DeVries, Carl Correns, and Erich Tschermak all came across Mendel's paper while working independently on problems like those Mendel had solved 35 years earlier. Because of the advances that had taken place in biology during those 35 years, the three scientists immediately recognized the importance of Mendel's discoveries. After these men brought his experiments to the attention of the world, the science of genetics truly got under way.

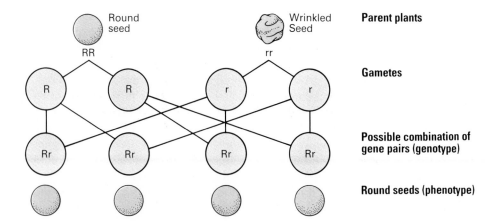

10.4 The Principle of Segregation

Mendel's experiments revealed that a parental trait, such as shortness, can disappear in the F_1 generation. His experiments also showed that the same trait can reappear in the F_2 generation in roughly a 3:1 ratio. To explain how traits can disappear and reappear in a certain pattern from generation to generation, Mendel proposed the **principle of segregation.** *The principle of segregation states that the members of each pair of genes separate, or segregate, when gametes are formed.*

Today biologists know that the principle of segregation describes what happens during *meiosis,* or the formation of gametes. As a result of meiosis, each gamete receives one member of each pair of chromosomes found in body cells. When gametes from two parents unite, new pairs of chromosomes are formed. Mendel, however, did not have the benefit of this knowledge. Chromosomes had not yet been discovered, and the process of meiosis was not observed until about 30 years after Mendel completed his research.

10.5 The Principle of Independent Assortment

Mendel developed his first two principles through experiments involving the inheritance of a single pair of traits. He arrived at his third principle by crossing pea plants with two or more pairs of contrasting traits. For example, he crossed a purebred plant with yellow, round seeds and a purebred plant with green, wrinkled seeds. Seeds produced from this cross were all yellow and round. This result illustrated the principle of dominance.

When these seeds grew into plants and self-pollinated, they produced four types of F_2 generation seeds. The yellow, round seeds and the green, wrinkled seeds resembled the seeds of the P

Figure 10–4. According to the principle of segregation, the two genes in a pair separate when gametes form. One gene goes to one gamete and the other gene goes to a different gamete. The uniting of two gametes in fertilization creates a new gene pair.

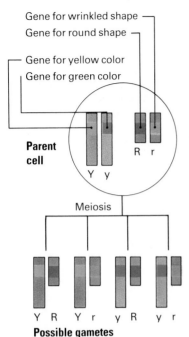

Gene for wrinkled shape
Gene for round shape

Gene for yellow color
Gene for green color

Parent cell

R   r

Y   y

Meiosis

Y R   Y r   y R   y r

Possible gametes

Figure 10–5. The principle of independent assortment states that two or more gene pairs separate independently. Thus gametes may contain a combination of dominant and recessive genes.

generation. However, the F_2 generation also included round, green seeds and yellow, wrinkled seeds. From this experiment, Mendel realized that two traits produced by recessive genes did not have to appear in the same offspring. For example, green color, a recessive trait, could appear with round seeds, a dominant trait. Mendel formulated the **principle of independent assortment** to explain this finding. *The principle of independent assortment states that two or more pairs of genes segregate independently of one another during the formation of gametes.* For instance, the segregation of the genes for seed color does not affect the segregation of genes for seed shape.

Today it is known that most gene pairs segregate independently only if they are located on different chromosomes. Traits determined by two genes on the same chromosome tend to be inherited together. Mendel, however, was able to choose seven contrasting traits, each determined by a gene pair on a different pair of chromosomes.

10.6 Other Genetic Terminology

Since the time of Mendel, the language of genetics has become more precise. As you know, scientists use the term *gene* instead of *factor* to describe the unit of heredity. They also use the term **allele** (uh LEEL) to refer to either member of a pair of genes that determines a single trait. For example, the dominant allele for seed color in peas (Y) produces yellow seeds. The recessive allele (y) produces green seeds.

The pairs of alleles in the cells of an organism make up its **genotype** (JEE nuh typ). These pairs of genes are represented with capital and lowercase letters, such as YY, Yy, and yy. A trait that is actually expressed in an organism is called a **phenotype** (FEE nuh typ). Although environment also affects many visible traits, phenotypes are largely determined by an organism's genotype. For example, a pea plant with the genotype YY will have the phenotype of yellow seeds. What other genotype can produce the same phenotype of yellow seeds?

Reviewing the Section

1. Why did Mendel experiment with pea plants?
2. What pattern of inheritance led Mendel to suggest the existence of factors?
3. How is Mendel's principle of segregation different from his principle of independent assortment?
4. Why was Mendel's work not accepted at first?

watermelons; striped, short watermelons; and striped, long watermelons. The phenotypic ratio of the F_2 generation can be expressed as $9:3:3:1$.

10.12 Codominance

As genetics has progressed, geneticists have learned that the principle of dominance does not hold in all cases. *Not all phenotypes result from dominant or recessive genes. In many cases both alleles for a trait are expressed.* Such alleles show **incomplete dominance,** or **codominance,** a situation in which neither allele is completely dominant or recessive. Codominant alleles produce an intermediate phenotype between the dominant and recessive phenotype. Codominance is a modification of Mendel's principle of dominance.

Codominance may be observed in short-tailed cats. These cats have two alleles for tail length—one for long tail and one for no tail. A Manx cat is homozygous for no tail ($T^N T^N$). Figure 10–12 shows a cross between a Manx cat and a cat homozygous for long tail ($T^L T^L$). Such a cross produces short-tailed offspring. The short-tailed cat is an intermediate phenotype having alleles for both long tail and no tail. Figure 10–12 also shows what happens when two short-tailed cats are crossed. There is a 25 percent chance the offspring will have no tail; a 25 percent chance for a long tail; and a 50 percent chance for a short tail. In this case, both the genotypic and phenotypic ratios are $1:2:1$, because neither allele is dominant.

Q/A

Q: *Does each type of flower have a dominant color?*

A: No. The color of many flowers is determined by codominance. For example, a cross between a white and a red snapdragon produces some pink offspring.

Figure 10–12. In cats, alleles for long tail and no tail are codominant and result in a short-tailed cat. The offspring of two short-tailed cats may include cats with no tail, short-tailed cats, and long-tailed cats in a ratio of $1:2:1$.

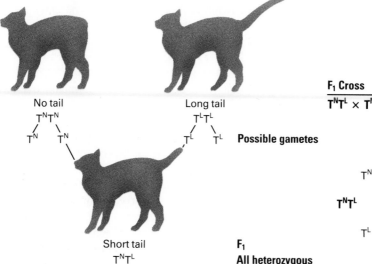

No tail
$T^N T^N$
T^N T^N

Long tail
$T^L T^L$
T^L T^L **Possible gametes**

Short tail
$T^N T^L$

F_1
All heterozygous

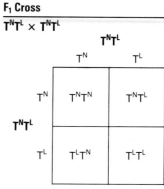

F_1 **Cross**

$T^N T^L \times T^N T^L$

	$T^N T^L$	
	T^N	T^L
T^N	$T^N T^N$	$T^N T^L$
T^L	$T^L T^N$	$T^L T^L$

$T^N T^L$

Figure 10–13. Both the Andalusian chicken and the roan cow are intermediate phenotypes that result from codominance.

As you may have noticed in Figure 10–12, codominant alleles are represented differently from dominant and recessive alleles. In cats the symbol T^N represents the allele for no tail. The symbol T^L represents the allele for a long tail. The genotype $T^N T^N$ produces a cat with no tail, and the genotype $T^L T^L$ produces a cat with a long tail. The short-tailed cat, therefore, has the genotype $T^N T^L$. The raised, small capital letter in these symbols is called a *superscript*. Why, do you think, lowercase letters are not used to represent codominant alleles?

Codominance also shows up in chickens. The Andalusian breed has one allele for black feathers and one for white feathers. Because neither allele is dominant, an Andalusian chicken has both black and white feathers. This combination causes the chicken to appear blue. Breeding two blue Andalusians results in offspring that are black, blue, or white in a ratio of $1:2:1$. Another example of codominance is the reddish-brown, or roan, color in short-horned beef cattle. A roan calf is the offspring of homozygous white and homozygous red parents.

Reviewing the Section

1. In pea plants, red flowers (R) are dominant over white flowers (r). Tall plants (T) are dominant over short plants (t). Diagram a dihybrid cross between a plant with the genotype Rrtt and a plant with the genotype rrTt. How many genotypes and phenotypes are found in the offspring?
2. Use the product rule to predict the results of a dihybrid cross between a plant with the genotype YyRR and a plant with the genotype Yyrr.
3. Can a cattle breeder produce a herd consisting only of roan cattle by crossing roan bulls and white cows? Explain your answer.

Investigation 10: Probability and Genetics

Purpose
To investigate how the principles of heredity are related to the laws of probability

Materials
Part A: 2 different coins
Part B: 100 red beans, 100 white beans, 2 empty coffee cans

Procedure
Part A
1. Flip one of the coins. *What are the chances of the coin turning up heads? tails?*
2. Make a table like Table 1. Use the product rule to predict the results of flipping the two coins simultaneously 20 times. Record the predicted results in the table. Flip both coins simultaneously 20 times. Record the actual results in the table. *How close were your results to the expected results? What might account for any differences?*
3. Repeat step 2, but this time flip both coins 100 times. Record your results in the table. *How do the data from this step differ from those in step 2?*

Table 1	Heads/ Heads	Heads/ Tails	Tails/ Tails
Predicted Results			
Actual Results — 20 Trials			
Actual Results — 100 Trials			

Part B
1. Place 50 red beans and 50 white beans in each coffee can. Assume the beans represent alleles for flower color in a certain plant. Red is dominant over white. Assume one of the cans represents the female parent; the other can represents the male parent.
2. Without looking into the cans, remove one bean from each can. Place the pair of beans into one of three separate groups: red/red,

red/white, white/white. Continue this process until all the beans are removed from the cans. Count the number of pairs in each group. Record the totals in a table like Table 2. Record the genotypes as well. *How many genotypes have resulted from this exercise? What are they? What is the genotypic ratio? How many phenotypes have resulted? Describe them. What is the phenotypic ratio?*

Table 2	Red/ Red	Red/ White	White/ White
Number of Pairs			
Genotype			

Analyses and Conclusions
1. Make a generalized statement relating the number of trials of a chance event and the predicted outcome based on probability.
2. Based on the laws of probability, what results would you expect in part B? How do your actual results compare with the expected results? Account for any differences.
3. What difference in results would you expect if you used 100 beans of each color in each can? Explain your answer.
4. Assume the red beans and white beans represent alleles for a trait that exhibits codominance. What changes in genotype would result? What changes in phenotype would result?

Going Further
Design an investigation similar to Part B that will demonstrate the inheritance of two traits. Perform the investigation and record your results. Explain any discrepancies between your results and those predicted by probability.

Chapter 10 Review

Summary

Gregor Mendel established the fundamentals of genetics through research with garden peas. Mendel found that each trait is influenced by a pair of factors, which are now called genes. Mendel's principle of dominance states that the dominant gene in a pair may mask the expression of the recessive gene. He further found that the genes in a pair separate during gamete formation. Mendel also discovered that the way in which one pair of genes segregates does not affect the way that any other pair of genes segregates.

The two genes in a pair are called alleles. If the alleles are the same, the individual is homozygous. If the pair consists of contrasting alleles, the individual is heterozygous.

Geneticists use probability to predict the results of genetic crosses. Punnett squares are used to display the results of genetic crosses.

A cross involving one trait is called a monohybrid cross. A cross involving two traits is a dihybrid cross. For some traits, neither allele is dominant, and the result of such a cross is an intermediate trait.

BioTerms

allele (**156**)
chromosome (**151**)
codominance (**163**)
cross-pollination (**153**)
dihybrid cross (**161**)
dominant (**154**)
first filial generation (**153**)
gamete (**151**)
gene (**151**)
genetics (**151**)
genotype (**156**)

heredity (**150**)
heterozygous (**154**)
homozygous (**154**)
homozygous dominant (**154**)
homozygous recessive (**154**)
hybrid (**154**)
incomplete dominance (**163**)
monohybrid cross (**160**)
parental generation (**152**)
phenotype (**156**)
principle of dominance (**153**)

principle of independent assortment (**156**)
principle of segregation (**155**)
probability (**157**)
product rule (**158**)
Punnett square (**159**)
purebred (**154**)
recessive (**154**)
second filial generation (**153**)
self-pollination (**151**)
testcross (**159**)

BioQuiz   (Write all answers on a separate sheet of paper.)

I. Completion

1. The combination of all genes in an organism's cells is its _____.
2. A gene that masks the expression of its contrasting allele is _____.
3. In _____, both alleles are expressed and produce an intermediate trait.
4. A cross involving two pairs of traits is called a _____ cross.
5. The appearance of an organism caused by its genotype is called the _____.

II. Modified True and False

Mark each statement either TRUE or FALSE. If false, change the underlined term to make the statement true.

6. A plant of genotype Rr is <u>homozygous</u>.
7. The two genes in a pair that determine a single trait are <u>gametes</u>.
8. A <u>recessive</u> trait can appear only in a homozygous individual.
9. The <u>first filial</u> generation results from a parental cross.

III. Multiple Choice

10. The formation of an equal number of R and r gametes from an Rr individual demonstrates a) independent assortment. b) dominance. c) segregation. d) codominance.
11. The probability of a coin turning up tails five times in a row is a) $\frac{1}{4}$. b) $\frac{1}{8}$. c) $\frac{1}{16}$. d) $\frac{1}{32}$.
12. All individuals that show a recessive trait are a) heterozygous. b) hybrid. c) homozygous. d) self-pollinating.
13. The probable phenotypic ratio resulting from the cross Rr × Rr is a) 1:1. b) 2:2. c) 4:0. d) 3:1.
14. A cross between a short-tailed cat and a Manx cat would produce short-tailed and no-tailed offspring in the ratio a) 1:1. b) 3:1. c) 2:1. d) 4:0.

IV. Essay

15. What is a dihybrid cross?
16. Why were Mendel's studies of breeding successful while those of other scientists were not?
17. What assumptions do geneticists make in applying probability to heredity?
18. How did Mendel explain why one of the traits in a monohybrid cross disappears in the F_1 generation?
19. What is the product rule?
20. Why is it impossible to produce a pure-breeding flock of blue Andalusians?

Applying and Extending Concepts

1. Some Leghorn chickens have colored feathers and some have white feathers. Describe an experiment that would determine which characteristic is produced by a dominant gene. Begin with chickens that are homozygous for colored feathers and homozygous for white feathers.
2. In cattle a recessive gene is responsible for the appearance of horns. The dominant allele causes a hornless condition. One farmer wants to produce only hornless cattle, and another farmer wants to produce only horned cattle. Which farmer would have less difficulty establishing a pure-breeding stock? Explain your answer.
3. One dihybrid cross involves two pairs of *complementary genes* that are inherited separately but influence the same trait. Do library research on complementary genes that produce purple color in sweet pea flowers. Then use Punnett squares to determine the phenotypic ratio in the F_2 generation. How does this ratio differ from the phenotypic ratio of the F_2 generation in a cross where one allele is dominant?

Related Readings

Bornstein, J. and S. Bornstein. *What Is Genetics?* New York: Messner, 1979. This introduction to genetics includes many helpful charts and diagrams.

Iltis, H. *Life of Mendel.* New York: Norton, 1932. This book is the standard biography on the founder of the science of genetics.

Mendel, G. *Experiments in Plant Hybridisation.* Cambridge, Mass.: Harvard University Press, 1965. This book is a translation of Mendel's paper published in 1866.

Truxal, J. G. "Knowing the Odds." In *Science Year 1982,* 180–193. Chicago: World Book, 1981. This introduction to probability includes a discussion of its role in Mendelian inheritance.

Chromosomes and Genetics

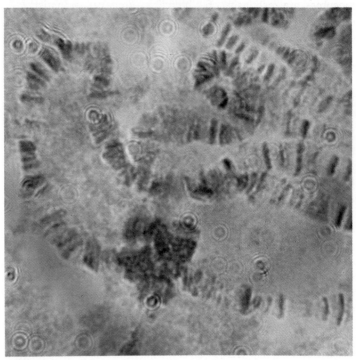

Stained chromosomes

Introduction

Through careful plant-breeding experiments, Gregor Mendel was able to discover the basic principles of heredity. However, Mendel never observed chromosomes or genes. He also never witnessed *mitosis* or *meiosis*, the processes of cell division. For some time, the nature of the factors responsible for heredity remained a mystery.

In the late 1800s, improvements in microscopy allowed scientists to observe chromosomes and the processes of mitosis and meiosis. With this knowledge, and with the rediscovery of Mendel's work in 1900, geneticists made rapid progress in unlocking the mysteries of heredity. Soon it became clear to scientists that Mendel's unseen factors actually do exist as physical parts of the cell.

The Chromosome Theory

In 1882 the German scientist Walther Flemming first observed chromosomes during mitosis. About ten years later, August Weismann, another German scientist, hypothesized that chromosomes are responsible for heredity. However, scientists did not accept the fact that chromosomes are the carriers of heredity until the rediscovery of Gregor Mendel's work in 1900.

11.1 The Work of Walter Sutton

In 1902 Walter S. Sutton was a graduate student at Columbia University in New York City. During his studies, Sutton noticed a strong similarity between Mendel's principles and his own observations of meiosis in grasshoppers. These similarities led Sutton to propose the **chromosome theory**. *The chromosome theory states that hereditary factors, or genes, are carried on chromosomes.*

What observations led Sutton to state the chromosome theory? First Sutton noticed that the chromosomes in each grasshopper cell line up in pairs before meiosis takes place. He also observed that the members of each chromosome pair are homologous, or alike in shape and size. Thus Sutton noted that chromosomes in body cells occur in pairs, corresponding to Mendel's pairs of factors.

Next Sutton observed that homologous chromosomes segregate during meiosis. As a result, each gamete receives one half of the chromosomes—one member from each pair. In addition, Sutton saw that the way in which members of one pair segregate seems to have no effect on how members of another pair segregate. Finally Sutton observed that a sperm and an egg cell, each

- *Describe* the observations that led Walter Sutton to propose the chromosome theory.
- *State* the chromosome theory.
- *Describe* the relationship Sutton believed existed between a gene and a chromosome.

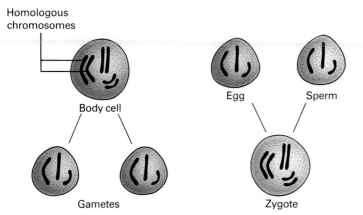

Homologous chromosomes

Body cell

Gametes

Egg   Sperm

Zygote

Figure 11–1. As Walter Sutton noted, the separation of homologous chromosomes into different gametes follows Mendel's principle of segregation. The formation of new homologous pairs in the zygote is consistent with the principle of independent assortment.

Highlight on Careers: Statistician

General Description

A statistician collects and analyzes numerical information, such as the number of offspring from a genetic cross showing a certain trait. Then the statistician uses this information, called data, to help solve problems.

Biometrics is the name of the field that applies mathematical and statistical techniques to biological studies. Statisticians in this field may study the effects of chemicals on genetic material in the cell. They may also investigate the causes of diseases by using public records, or study the rise and fall of animal populations.

Since statisticians are employed by any organization that needs to collect and analyze information, job opportunities are varied. Federal, state, and local government agencies employ statisticians. They are also employed by universities, research institutions, and a variety of industries.

Career Requirements

Individuals who enjoy collecting biological information for analysis need a B.S. degree with strong training in mathematics and the biological sciences. A graduate degree is required for individuals who prefer to analyze data, to develop theoretical methods of solving problems, or to consult with industry and government agencies.

All statisticians, regardless of their specialties, must be comfortable with detail work, skilled in working with numbers, and interested in analyzing information. Although statistics might sound dry, individuals in this field are helping to solve some of the major problems of our

day. By analyzing records of the effects of air pollution and radiation, for instance, statisticians can help determine policies that will improve life for everyone.

For Additional Information
American Statistical
   Association
806 15th Street, N.W.
Washington, DC 20005

carrying half the number of chromosomes found in body cells, unite during fertilization. Therefore, the resulting zygote has a complete set of homologous chromosomes.

After Sutton observed meiosis, he realized that chromosomes behave according to Mendel's principles of heredity. This fact, along with certain other observations Sutton made, contributed to his belief that chromosomes carry the information of heredity. The major points that supported Sutton's hypothesis include the following:

1. Egg and sperm cells provide the only physical link between one generation and the next. For this reason, the hereditary material must be carried in these cells.
2. Hereditary material is probably located in a cell's nucleus rather than in its cytoplasm. Sutton knew that both parents contribute equally to the genetic makeup of offspring. With this idea in mind, Sutton noted that sperm cells have far less cytoplasm than egg cells have. However, the nuclei of the two cell types are about the same size.
3. During meiosis chromosomes tend to behave according to Mendel's principles. Following Mendel's principle of segregation, each pair of homologous chromosomes separates independently of one another.

Interestingly, a German scientist named Theodore Boveri arrived at similar conclusions at about the same time as Sutton. Sutton's findings, however, were published first and were more convincing than Boveri's. Thus Sutton generally receives credit for establishing the chromosome theory.

11.2 Genes and Chromosomes

Scientists quickly realized that organisms have many more traits than they have chromosomes. For example, in the grasshoppers that Sutton observed, only 12 pairs of chromosomes seemed to be responsible for producing hundreds of traits. Sutton reasoned that each chromosome must carry hundreds, perhaps thousands, of smaller particles that contain hereditary information. If such particles were present, they would confirm the existence of Mendel's factors.

In 1909 a Danish biologist named Wilhelm Johannsen first used the word *gene* to describe these tiny particles of inheritance. Evidence for the existence of genes, however, did not come until several years later.

Reviewing the Section

1. What similarity did Sutton observe between meiosis and Mendel's principles?
2. What is the chromosome theory?
3. What led Sutton to think that each chromosome carried many factors?
4. Why was the fact that Mendel published his findings important to the progress of genetic research?

- *Explain* how chromosomes determine the sex of an individual.
- *State* what is meant by a sex-linked trait.
- *Diagram* a cross using a Punnett square to show sex-linked inheritance.

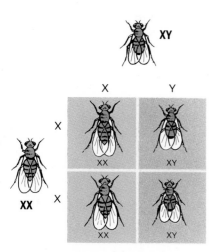

Figure 11–2. This Punnett square shows that a female fruit fly can contribute only an X chromosome to all her offspring. Therefore, sex is determined by the male gamete, which contributes either an X or a Y chromosome.

The Sex Chromosomes

The chromosome theory suggested that genes are located on chromosomes. Some scientists, however, refused to accept Sutton's idea until Thomas Hunt Morgan, an American zoologist, demonstrated that Sutton was correct.

11.3 Chromosomes and Sex Determination

In the early 1900s, Morgan began genetic research at Columbia University using the fruit fly, *Drosophila melanogaster*. Morgan selected *Drosophila* because they are easy to maintain and breed. Just 3 mm (0.1 in.) long, *Drosophila* are so tiny that hundreds can be kept in a jar. In addition, *Drosophila* have a life span of just 10 to 15 days. Therefore, they can produce many generations of offspring in a matter of weeks.

Morgan and his colleagues found another important characteristic: *Drosophila* have only four pairs of chromosomes. The researchers further observed that one pair of chromosomes in males is different from the corresponding pair in females. In males, the two chromosomes in this pair are not the same size or shape. One chromosome is large and rod-shaped; the other is small and hook-shaped. The rod-shaped chromosome is called an *X chromosome*. The hook-shaped one is called a *Y chromosome*. The corresponding pair of chromosomes in females consists of two X chromosomes. Morgan and his colleagues had found the **sex chromosomes**—the chromosomes that determine the sex of an individual. All other chromosomes in an individual are called **autosomes**.

Figure 11–2 shows how sex is determined in *Drosophila*. Following Mendel's principle of segregation, the sex chromosomes separate during meiosis. Each egg has only an X chromosome. Each sperm has either an X or a Y chromosome. When a sperm carrying an X chromosome fertilizes an egg, the offspring will be female (XX). If the sperm that fertilizes the egg carries a Y chromosome, the offspring will be male (XY). Roughly equal numbers of male and female offspring will be produced. ***In most organisms, including fruit flies and humans, sex is determined by gametes from the male parent.***

11.4 Sex-Linked Traits

Morgan's observation of sex chromosomes was the first of several important findings. Another discovery resulted from crosses that Morgan made with red-eyed fruit flies and white-eyed fruit flies. Red is the normal eye color in wild-type *Drosophila*. At this time, Morgan and his associates assumed that the allele for red eye color, R, was dominant over the allele for white eye color, r.

Morgan crossed the white-eyed male with a red-eyed female. Following Mendel's principle of dominance, all members of the F_1 generation had red eyes. Then Morgan allowed the F_1 generation to mate. The F_2 generation also followed the principle of dominance, showing about a 3:1 ratio of red-eyed flies to white-eyed flies. Surprisingly, however, all the F_2 generation flies with white eyes were males. Why were there no females with white eyes?

Morgan concluded that the alleles for eye color are carried only on the X chromosome. Thus, eye color in fruit flies is an example of a **sex-linked trait.** A sex-linked trait is one that is determined by alleles carried only on an X chromosome. A sex-linked trait has no alleles on the Y chromosome.

Punnett squares illustrate how sex-linked traits appear in offspring. Figure 11–3 shows Morgan's parental cross between a white-eyed male and a red-eyed female. Each female gamete carries a dominant allele (X^R) for eye color. Each male gamete, however, has either an X chromosome carrying a recessive allele (X^r) for eye color or a Y chromosome with no allele for eye color. The F_1 generation, then, consists of females heterozygous for red eyes and males with one dominant allele for red eyes. Figure 11–3 also shows the results of the F_1 generation cross. About one-fourth of the offspring are male flies with white eyes. Each of these white-eyed males resulted from the union of a female gamete carrying the recessive allele and a male gamete carrying only a Y chromosome.

Morgan's experiments confirmed Sutton's hypothesis that genes are found on chromosomes. In addition, the discovery of sex-linked traits explained why some characteristics caused by recessive genes appear far more often in males. Since males have only one X chromosome, any recessive allele present on that chromosome will be expressed. Females, with two X chromosomes, are more likely to carry a dominant allele that masks the expression of a recessive allele.

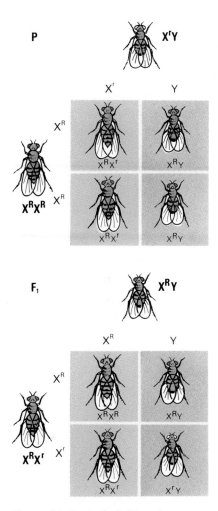

Figure 11–3. In fruit flies, the alleles for red eyes (R) and for white eyes (r) are carried only on the X chromosome.

Reviewing the Section

1. How is the sex of most organisms determined?
2. How did Morgan discover sex-linked traits?
3. Use a Punnett square to show how a white-eyed female *Drosophila* could be produced.

Section Objectives

- *Explain* how the idea of gene linkage developed.
- *Describe* the process of crossing over.
- *Explain* how chromosome maps are made.
- *Discuss* gene mapping.

Thomas Hunt Morgan's work revealed that genes for certain traits are carried on the X chromosomes, and that chromosomes, not genes, segregate during meiosis. Later, Morgan demonstrated that genes are also carried on the autosomes. In addition, his research showed that genes may be exchanged between homologous chromosomes.

11.5 Gene Linkage

In later experiments, Morgan studied other traits in fruit flies. One of his most important experiments involved two traits produced by dominant alleles—gray body color (G) and long wings (L). The corresponding recessive alleles determine black body color (g) and short wings (l).

Morgan crossed flies homozygous for gray bodies and long wings (GGLL) with flies homozygous for black bodies and short wings (ggll). As predicted by Mendel's principles, all of the F_1 flies were heterozygous and had gray bodies and long wings.

According to Mendel's principle of independent assortment, the F_2 generation phenotypes should have shown a 9:3:3:1 ratio. However, when Morgan crossed flies from the F_1 generation, the F_2 generation phenotypes appeared in roughly a 3:1 ratio. About three-fourths had gray bodies and long wings, and about one-fourth had black bodies and short wings. From these results, Morgan concluded that the alleles for gray body and long wings were located on the same chromosome. The Punnett squares in Figure 11–4 illustrate the results of Morgan's crosses.

As a result of Morgan's experiments and the work of certain other scientists, it became clear that genes are linked on all chromosomes. The situation in which two or more genes occur on the same chromosome is called **gene linkage**. Genes that occur together on a chromosome represent a **linkage group**. Fruit flies have 4 pairs of chromosomes and therefore have 4 linkage groups. Humans have 23 pairs of chromosomes and thus 23 linkage groups.

11.6 Crossing Over

When Morgan studied the results of his F_1 cross, he found several offspring with a mixture of dominant and recessive traits. A few flies had gray bodies and short wings. A few others had

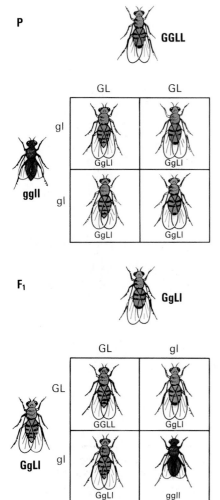

Figure 11–4. In *Drosophila*, alleles for body color and wing length are linked.

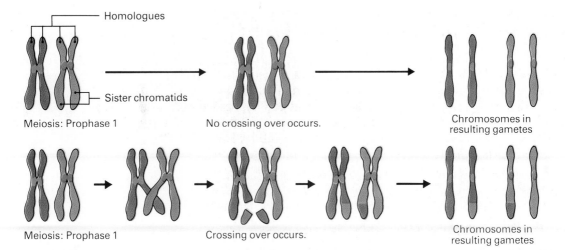

Homologues

Sister chromatids

Meiosis: Prophase 1

No crossing over occurs.

Chromosomes in resulting gametes

Meiosis: Prophase 1

Crossing over occurs.

Chromosomes in resulting gametes

black bodies and long wings. Morgan knew that these individuals could not have been produced if the alleles G and L were always linked. Morgan proposed that alleles from two homologous chromosomes had exchanged places. This process by which alleles exchange places is called **crossing over.** Later experiments by Morgan and others confirmed that crossing over does regularly take place in almost all organisms.

Figure 11–5 illustrates how crossing over occurs. During prophase I of meiosis, the duplicated chromosomes pair up with their homologues. Each duplicated chromosome is made up of two identical sister chromatids. Together, the two chromosomes consist of four chromatids. At this time, breaks may occur in the same spot on two nonsister chromatids. The broken ends of one chromatid then fuse with those of the other chromatid. As a result, each chromatid contains parts from its homologue.

11.7 Chromosome Mapping

The discovery of crossing over revealed that genes are found at certain fixed positions on the chromosomes. The alleles for eye color in fruit flies, for instance, always occupy the same position on homologous chromosomes.

Alfred H. Sturtevant, an undergraduate student in Morgan's laboratory, proposed that genes are located on the chromosomes in a line, like beads on a string. Sturtevant suggested that the frequency of crossing over would be affected by the distance between two genes on a chromosome. He reasoned that widely spaced genes would be more likely to cross over than closely spaced genes. Sturtevant proposed that this assumption could be used to develop a chromosome map like that in Figure 11–6.

Figure 11–5. Crossing over occurs after homologous chromosomes have paired during meiosis. A segment of one chromatid is exchanged with a segment from a nonsister chromatid.

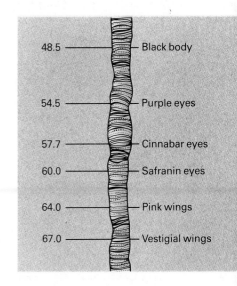

48.5	Black body
54.5	Purple eyes
57.7	Cinnabar eyes
60.0	Safranin eyes
64.0	Pink wings
67.0	Vestigial wings

Figure 11–6. Chromosome mapping was used to determine the position of genes on this chromosome taken from the salivary gland of *Drosophila.*

▶ Giant Chromosomes in Fruit Flies

Chromosomes in most organisms are so small that microscopes can reveal only a few of their details. In the 1930s, however, scientists discovered that the immature form, or larva, of the fruit fly has giant chromosomes in the cells of its salivary glands. These chromosomes are more than 1,000 times larger than the same chromosomes in the cells of the adult fruit fly. They are extremely large because the chromosomes duplicate many times but do not separate. The duplicates line up close to one

another, forming the giant chromosome you see in the photograph.

Giant chromosomes, easily visible under a compound light microscope, gave scientists an opportunity to learn more about chromosomes. Researchers found, for example, that applying stain to these chromosomes produced sequences of dark and light bands. Gene mapping revealed that specific genes were located in specific bands. This discovery supported Sturtevant's belief that genes are arranged in a line on the

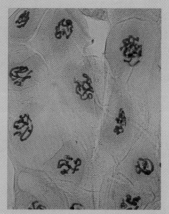

chromosome. Giant chromosomes also enabled scientists to learn more about minute changes in chromosomes that produce severe problems in offspring.

Q/A

Q: *Is crossing over a genetic "mistake"?*

A: No. Crossing over happens regularly, with a constant frequency, in almost all organisms. The process of crossing over is a normal part of meiosis and produces new genetic combinations.

A **chromosome map** is a graphic device that shows where genes are located on a chromosome.

Sturtevant determined the location of genes on a chromosome by calculating the percentage of offspring showing the crossing over of genes. For example, if the two dominant traits produced by linked genes A and B appeared together in 85 percent of the offspring of a testcross (F_1 × aabb), then crossing over must have occurred in the other 15 percent. According to Sturtevant's system, genes A and B would then be considered 15 units apart. Sturtevant and other scientists used many such percentages to develop a chromosome map of *Drosophila*.

Reviewing the Section

1. Explain what led Morgan to suggest the idea of gene linkage.
2. How are chromosome maps made?
3. Explain how crossing over might occur on homologous chromosomes and yet not be detected in an offspring.

Investigation 11: Gene Linkage and Crossing Over

Purpose
To determine patterns of heredity using experimental results

Materials
Paper, pencil

Procedure
Study the following chart, which shows some characteristics of a fictitious insect. Then study the results of the crosses described to answer the questions.

Trait	Possible Phenotypes
Body Color	Blue White
Eye Color	Green Brown
Antennae	Long Short
Bristles	Straight Curly
Number of Chromosomes	2 pairs

Cross #1   white-bodied ♂ × blue-bodied ♀

F_1 All blue-bodied

F_2 62 blue-bodied
    21 white-bodied

**Cross #2   blue-bodied, green-eyed ♂ ×
            blue-bodied, green-eyed ♀**

F_1 21 blue, green
     7 white, green
     8 blue, brown
     2 white, brown

**Cross #3   blue-bodied, short antennaed ♂ ×
            white-bodied, long-antennaed ♀**

F_1 All blue-bodied, short-antennaed

F_2 66 blue, short

19 white, long
 1 blue, long
 1 white, short

**Cross #4   curly-bristled,   green-eyed   ♂ ×
            straight-bristled, brown-eyed ♀**

F_1 All curly-bristled, green-eyed

F_2 75 curly, green
     26 straight, brown
      1 curly, brown
      2 straight, green

Analyses and Conclusions
1. Which traits are dominant in this insect?
2. How many linkage groups are there in this insect?
3. Which genes are linked? What evidence led you to this conclusion?
4. Which of Mendel's principles are demonstrated by the first cross? the second cross?
5. Are any of the listed traits sex-linked? Explain your answer.
6. The following cross-over percentages were determined:
   - 23% between the gene for bristles (C or c) and the gene for eye color (G or g)
   - 16% between the gene for body color (B or b) and the gene for antennae (S or s)
   Use this data to draw chromosome maps for both pairs of chromosomes in a completely heterozygous individual.

Going Further
Use the symbols given in question 6 to diagram each of the above crosses using a Punnett square.

Chapter 11 Review

Summary

Great strides were made in genetic research following the improvement of the compound light microscope and the rediscovery of Mendel's work in 1900. In 1902 Walter Sutton observed that during meiosis chromosomes behave according to Mendel's principles. These observations led Sutton to propose the chromosome theory. This theory states that hereditary information is carried from generation to generation by hereditary factors located on chromosomes.

Thomas Hunt Morgan furthered the study of genetics using the fruit fly. His early observations revealed one pair of chromosomes in male flies that do not match a pair of chromosomes in female flies. These chromosomes, called the sex chromosomes, are designated XY in the male and XX in the female. Morgan showed that the sex chromosomes carry genes for sex-linked traits. Nonsex chromosomes are called autosomes.

Morgan also demonstrated that genes are linked on chromosomes. In addition, he discovered that during meiosis homologous chromosomes regularly exchange genes in a process called crossing over. Alfred Sturtevant determined that the percentage of times genes cross over indicates the distance between genes on a chromosome. Sturtevant used these percentages to draw chromosome maps showing the location of genes for different traits.

BioTerms

autosome (172)
chromosome map (176)
chromosome theory (169)
crossing over (175)
gene linkage (174)
linkage group (174)
sex chromosome (172)
sex-linked trait (173)

BioQuiz   (Write all answers on a separate sheet of paper.)

I. Completion

1. Eye color in *Drosophila* is an example of a _____ trait.
2. A _____ shows the positions of genes on a chromosome.
3. The word _____ is now used to refer to Mendel's factors.
4. Sex of offspring is generally determined by the _____ parent.
5. Geneticists study the frequency of _____ to determine the location of genes on a chromosome.
6. The behavior of _____ during meiosis follows Mendel's principle of segregation.

II. Modified True and False

Mark each statement TRUE or FALSE. If false, change the underlined term to make the statement true.

7. Genes that are found on the same chromosome form a <u>linkage group</u>.
8. The <u>cell</u> theory states that genes are located on chromosomes.
9. In most organisms, if the sperm that fertilizes an egg carries a Y chromosome, the offspring will be <u>male</u>.
10. Chromosomes that determine the sex of an individual are called <u>autosomes</u>.

III. Multiple Choice

11. Crossing over occurs during a) mitosis.
    b) meiosis. c) fertilization. d) replication.
12. Walter Sutton concluded that hereditary material must be located in the
    a) nucleus. b) cytoplasm. c) cell membrane. d) ribosomes.
13. Thomas Hunt Morgan's genetic discoveries came from the study of
    a) grasshoppers. b) pea plants.
    c) bacteria. d) fruit flies.
14. Most genes that determine sex-linked traits are found on a) autosomes.
    b) Y chromosomes. c) X chromosomes.
    d) homologous chromosomes.
15. The number of homologous pairs of chromosomes in an organism determines the number of a) linkage groups.
    b) linked traits. c) genes. d) gametes.

IV. Essay

16. What observation led Walter Sutton to conclude that chromosomes carry the hereditary information?
17. What characteristics make *Drosophila* well suited for genetic study?
18. How did the discovery of sex-linked traits support the chromosome theory?
19. How did Thomas Hunt Morgan explain the fact that linked genes are sometimes inherited separately?
20. Why are genes that are far apart on a chromosome more likely to cross over than genes that are closer together?

Applying and Extending Concepts

1. Explain how the concept of linked genes varies from Mendel's principle of independent assortment. Then restate Mendel's principle to account for linkage.
2. Two dominant traits in fruit flies are long legs (L) and long antennae bristles (B). The recessive traits are short legs (l) and short bristles (b). Design an experiment to find out whether the genes for these traits are linked. Use Punnett squares to determine the phenotypic ratio that would occur if the traits were not linked and the ratio that might indicate linkage.
3. Use the following information about frequency of crossing over to position alleles on a chromosome map. The frequency of crossing over for genes A and B is 16 percent; for genes A and C, 24 percent; for genes B and C, 40 percent.
4. In cats the X chromosome carries the genes for coat color. The allele for yellow coat (Y) is dominant over the allele for black coat (y). A cross between a yellow male and a black female produces three male kittens. What color are the kittens? How do you know?
5. In 1961 geneticist Mary Lyon proposed that one X chromosome in each cell of a female mammal becomes inactive early in the organism's development. Research the Lyon hypothesis. Then use it and the information from question 4 above to explain why almost all calico, or yellow and black, cats are female.

Related Readings

''Does X Mark the Spot?'' *Health* 14 (May 1982): 23–24. This article discusses the role of so-called *fragile X*, an abnormal form of the sex chromosome, in producing mental retardation in males.

Ford, E. B. *Understanding Genetics*. New York: Pica Press, 1979. This introductory book discusses genetics in depth without requiring advanced knowledge of mathematics or chemistry.

12

Chemical Basis of Genetics

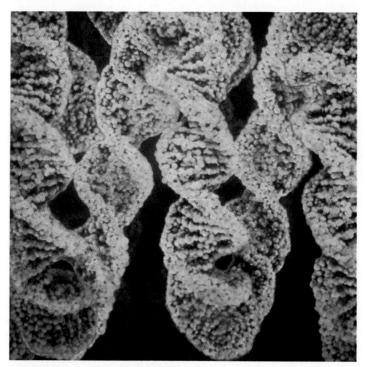

Computer-generated image of DNA

Introduction

Soon after Thomas Hunt Morgan's experiments with fruit flies in 1910, geneticists agreed that genes carried by chromosomes are responsible for heredity. Yet two important questions still puzzled these scientists: Exactly what are genes? How do genes work?

During the next 40 years, answers to these fundamental questions began to emerge from the field of *molecular genetics,* the study of the chemistry of genes. Scientists discovered the chemical makeup of genes as well as the complicated structure that allows them to carry enormous amounts of information. In addition, scientists began to understand how genes affect both the physical structure and the chemical processes of all living things.

The Discovery of DNA

Biochemists provided some early insights into the nature of genes by analyzing the chemical makeup of chromosomes. They found that chromosomes are composed of two different substances. One substance is protein. The other substance is **DNA,** or **deoxyribonucleic** (dee AHK sih ry boh noo KLEE ihk) **acid,** a complex compound classified as a nucleic acid. For years, scientists believed that the protein in chromosomes was the genetic material. Experiments in the 1940s and 1950s, however, showed that the genetic material is DNA.

12.1 Early Proof of DNA's Role

The first evidence that DNA is the genetic material came in 1944 from the research of Oswald Avery, Colin MacLeod, and Maclyn McCarty. These scientists were working at Rockefeller Institute in New York City. Their work was based on a puzzling situation discovered earlier by the English bacteriologist Frank Griffith.

In 1928 Griffith was experimenting with two strains of *Pneumococcus,* a type of bacterium. One of these strains (called S, or smooth) is surrounded by a protective capsule and causes fatal cases of pneumonia in mice. The other strain (called R, or rough) has no protective capsule and does not produce pneumonia in mice. In his experiments, Griffith killed the deadly strain S bacteria with heat. When the killed strain S bacteria were injected into mice, the bacteria did not cause pneumonia. However, mice injected with both heat-killed strain S and living

Section Objectives

- *List* the contributions that various researchers made in discovering that DNA is the hereditary material.
- *Describe* the structure of a DNA molecule.
- *Identify* the parts of a DNA nucleotide.
- *Summarize* the process of DNA replication.

Figure 12–1. The illustrations below show the results of Griffith's experiments. Why did mice die when infected with a mixture of heat-killed S-strain bacteria and live R-strain bacteria?

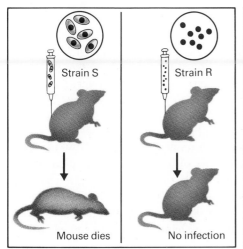

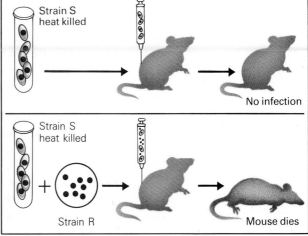

strain R bacteria developed pneumonia and died. What had happened? The strain S bacteria could not have come back to life.

After 10 years of complicated chemical studies, Avery and his associates found the answer to Griffith's question. In their most important experiment, they managed to remove DNA from strain S bacteria and place it in a culture of strain R bacteria. Inside the cells of the strain R bacteria, DNA of the strain S apparently took over and caused the strain R bacteria to develop capsules. Furthermore, the strain R bacteria passed on the genetic instructions of strain S DNA to their offspring. The offspring of the changed strain R bacteria developed protective capsules as well. *As a result of their tests, Avery and his associates concluded that DNA is the genetic material.*

In 1952 biologist Alfred Hershey and his laboratory assistant Martha Chase conducted experiments that supported Avery's findings. Their research showed that a certain type of virus could inject a substance into bacteria and that minutes later, new viruses would appear. Hershey and Chase further showed that the substance injected into the bacteria was DNA. From this observation, Hershey and Chase concluded that DNA was the genetic material because it took control of the bacterial cell and forced it to make new viruses.

12.2 Structure of DNA

The next major breakthrough in genetic research took place in 1953 at Cambridge University in Great Britain. This breakthrough was one of the most important of all scientific discoveries. *James D. Watson, an American biologist, and Francis H. C. Crick, a British biophysicist, discovered the structure of DNA.* Watson and Crick also developed a model of DNA. Their now famous model explains how DNA works.

Existing Knowledge of DNA Watson and Crick had substantial information to draw on as they pieced together the structure of DNA. First, they knew that DNA is a very long, thin molecule. From the work of the American biochemist Phoebus A. Levene in the 1920s, they also knew the chemical makeup of DNA. DNA contains four nitrogen-carrying bases. These four bases are called **adenine** (AD uh neen), **guanine** (GWAH neen), **thymine** (THY meen), and **cytosine** (SYT uh seen). Adenine and guanine are compounds called *purines* (PYOOR eenz). Thymine and cytosine are compounds called *pyrimidines* (pih RIHM uh deenz). In Figure 12–2 you can see that purines have a double ring of carbon and nitrogen atoms; pyrimidines have a single ring of carbon and nitrogen atoms.

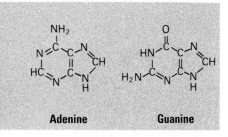

Figure 12–2. Thymine and cytosine are pyrimidines. Each contains a single ring of carbon and nitrogen atoms. Adenine and guanine are purines, which have a double carbon-nitrogen ring.

DNA also contains a phosphate group and a five-carbon sugar called **deoxyribose** (dee ahk sih RY bohs), from which DNA gets its name. Each nitrogen-carrying base is attached to a sugar molecule and a phosphate group. A unit made up of a nitrogen-carrying base, a sugar molecule, and a phosphate group is called a **nucleotide.**

In addition to the clues provided by Levene's research, Watson and Crick had information about the bases in DNA. Erwin Chargaff of Columbia University had shown through chemical analyses that the amount of guanine (G) always equals the amount of cytosine (C). The amount of adenine (A) always equals the amount of thymine (T).

Finally, Watson and Crick had the images of DNA made in 1951 by two British biophysicists, Maurice Wilkins and Rosalind Franklin. These images indicated that the shape of DNA is a *helix* (HEE lihks), or spiral.

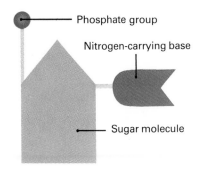

Figure 12–3. A nucleotide consists of a phosphate, a sugar, and a nitrogen-carrying base.

▶ Watson and Crick's Major Clue

The DNA model created by Watson and Crick was based on the work of many scientists. In 1869 Friedrich Meischer, a German scientist, isolated a substance from cell nuclei and called it *nucleic acid*. This substance is now known as DNA. In 1914 Robert Feulgen, another German scientist, discovered that DNA could be stained inside a cell with a red dye, *fuchsin*. This discovery helped scientists learn that DNA is located in chromosomes.

Watson and Crick's major clue was revealed in 1951. That was the year when Maurice Wilkins and Rosalind Franklin of King's College in London, England,

used *X-ray crystallography* to produce images of DNA. Using this technique, they crystallized the DNA and then directed a beam of X rays at the crystal. Some X rays passed through the crystal. Others were scattered onto photographic film. The scattered rays produced a pattern of dark images on the film. As you can see in the photograph, the images cross in the middle. The diagram shows how a helix would produce such intersecting images. Wilkins and Franklin proposed that DNA is a molecule in the shape of a helix. Watson and Crick used this information to explain in detail the structure of DNA.

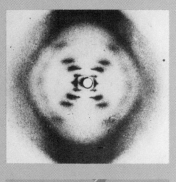

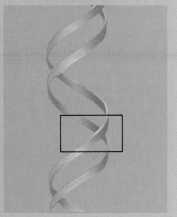

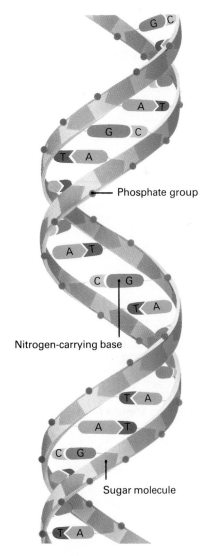

Phosphate group

Nitrogen-carrying base

Sugar molecule

Figure 12–4. A DNA molecule resembles a twisted ladder. Its sides are formed from long chains of sugars and phosphates. The 'rungs' are pairs of nitrogen-carrying bases.

The Watson-Crick Model By unifying the existing information on DNA, Watson and Crick concluded that the DNA molecule is shaped like a *double helix*. As Figure 12–4 shows, a double helix somewhat resembles a spiral staircase, or a twisted ladder. Furthermore, Watson and Crick determined that the sides of the "ladder" are composed of alternating phosphate groups and sugar molecules. The "rungs" on the staircase consist of pairs of nitrogen-carrying bases. The two bases that make up each rung are joined to one another by weak chemical bonds.

A close look at the rungs of the DNA ladder shows that the nitrogen-carrying bases always pair up in a specific pattern. Based on Chargaff's chemical analysis of DNA, Watson and Crick reasoned that a purine and a pyrimidine must pair with each other to make a rung that is the right width. Two purines would produce a rung that is too wide. Two pyrimidines would produce a rung that is not wide enough to connect the sides of the ladder. Adenine (A) pairs only with thymine (T). Likewise, guanine (G) pairs only with cytosine (C).

Notice in Figure 12–4 that each base, or half rung, is attached to a sugar molecule and a phosphate group. A single DNA molecule may be composed of many thousands of such nucleotides. The DNA molecule in the smallest known virus contains about 5,000 nucleotides. Together, all 46 chromosomes in human cells contain more than 5 billion nucleotides.

The pairs of nucleotides forming the DNA ladder can appear in any order. *The sequence of the nucleotide pairs is the code that controls the production of all the proteins of an organism. In fact, a gene is a sequence of nucleotides that controls the production of one type of protein.* You will study genes in more detail later in this chapter. However, you may already realize that the number of genes, and thus the amount of information carried in DNA is staggering. Scientists estimate that information stored in DNA from one human cell equals the amount of information in one thousand 500-page books.

12.3 Replication of DNA

How would you like to copy all of the information in 1,000 books, each 500 pages long? Could you do it in one evening? Of course not. The DNA molecules in your cells, however, can make copies of themselves—and therefore of all the information they contain—in about six hours. The process by which DNA copies itself is called **replication** (rehp luh KAY shuhn).

DNA replicates itself so that every new cell receives a complete copy of the genetic code. Thus DNA replicates before mitosis, when new cells are produced for growth and repair.

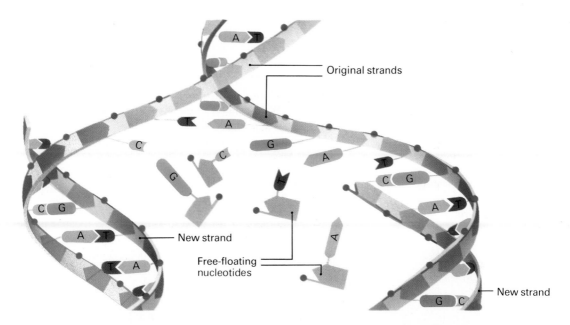

Original strands

New strand

Free-floating
nucleotides

New strand

DNA also replicates before the first division of meiosis so that each gamete receives genetic information from the parent.

The Watson-Crick model made clear how DNA copies itself exactly. Figure 12–5 shows what happens during DNA replication. First, the chemical bonds connecting the bases break in several places and the molecule separates down the middle. As the molecule splits into separate strands, special enzymes cause the proper nucleotides to pair with complementary nucleotides on each single strand. Other enzymes then link the new nucleotides into one long strand. Each original strand serves as a *template,* or pattern, for the creation of the new strand. Every T (thymine) nucleotide pairs with a nucleotide containing an A (adenine). Likewise, every G (guanine) nucleotide pairs with a nucleotide containing a C (cytosine). Notice in Figure 12–5 that each completed DNA molecule contains one old and one new strand. The entire process is powered by energy from ATP and the action of enzymes.

Figure 12–5. In replication, the two strands of a DNA molecule separate as the chemical bonds connecting the bases break. Two complementary strands form, each using one of the single DNA strands as a template.

Q/A

Q: *How many DNA molecules make up one chromosome?*

A: Each chromosome contains one molecule of DNA.

Reviewing the Section

1. What discovery led Avery and his associates to conclude that DNA is the hereditary substance?
2. Describe the shape of a DNA molecule.
3. What are the components of a DNA nucleotide?
4. Why is DNA replication necessary to life?

DNA and Protein Synthesis

- *Contrast* the structures of DNA and RNA.
- *Explain* what is meant by the term *triplet codon*.
- *List* the main steps in the process of transcription.
- *Name* the functions performed by each type of RNA during translation.
- *Discuss* directions for future genetic research.

Once scientists understood the structure of DNA, they were able to understand better how DNA works. Scientists discovered that DNA controls **protein synthesis**, the process by which proteins are made from amino acids. Some proteins are part of the structure of each organism, and other proteins are *enzymes* that control most chemical reactions. The characteristics of any organism are determined by its proteins and, ultimately, by its DNA.

12.4 DNA and RNA

DNA, with its blueprint for protein synthesis, is located in the cell nucleus. Yet the manufacture of protein molecules takes place in the cytoplasm of the cell on structures called *ribosomes*. DNA molecules do not leave the nucleus to control the production of protein. Instead, another type of nucleic acid acts as a messenger between DNA and ribosomes and carries out protein synthesis. This nucleic acid is called **RNA,** or **ribonucleic** (ry boh noo KLEE ihk) **acid.**

DNA and RNA are similar in many ways. For example, both are nucleic acids and are made up of nucleotides arranged in a certain sequence. However, RNA differs from DNA in three important ways. First, RNA nucleotides contain the sugar **ribose** instead of deoxyribose. Second, the pyrimidine base **uracil** (YOOR uh sihl) substitutes for thymine. Like thymine, uracil (U) pairs only with adenine. Third, RNA is primarily a single-stranded molecule, while DNA is usually double-stranded.

There are three kinds of RNA. One kind, called **messenger RNA,** or **mRNA,** carries sequences of nucleotides that code for protein from the nucleus to the ribosomes. A second kind of RNA, called **transfer RNA,** or **tRNA,** picks up individual amino acids in the cytoplasm and carries them to the ribosomes. There the amino acids are joined together in proper order to make a protein. Ribosomes contain the third kind of RNA, called **ribosomal RNA,** or **rRNA.** Ribosomal RNA helps bind mRNA and tRNA together during one step of protein synthesis.

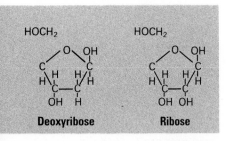

Deoxyribose   **Ribose**

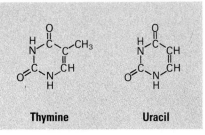

Thymine   **Uracil**

Figure 12–6. RNA differs from DNA by having ribose instead of deoxyribose and uracil instead of thymine. How does the chemical structure of uracil differ from that of thymine?

12.5 Transcription

The process by which mRNA is copied from DNA molecules is called **transcription.** In this process, part of the genetic code of DNA is transferred to molecules of mRNA. First a segment of DNA separates, exposing the two strands. Then one of these strands serves as a template, or pattern, to make a molecule of mRNA. This strand and the molecule of mRNA it produces may be 1,000 to 10,000 nucleotides long. The other strand of DNA does not take part in transcription.

Next mRNA nucleotides in the nucleus bind to their complementary nucleotides on the active strand of DNA. Thus the sequence of bases in mRNA is determined by the order of bases in DNA. For instance, if DNA has the base sequence AGCTGA, the complementary bases in mRNA would then be UCGACU. Recall that in RNA uracil is the complement of adenine. The sequence of bases in mRNA is a code that enables the mRNA to collect the right amino acids and assemble them in the correct sequence to synthesize a particular protein. The mRNA code is a series of three-letter "words," with each base standing for one "letter." Every combination of three "letters," or bases, is called a **codon** (KOH dahn).

How do scientists know that a codon is made up of three bases? Because 20 different amino acids are used to produce proteins, mRNA needs at least 20 codons, one for each amino acid. If a codon consisted of one base, the total number of codons in mRNA would be four—A, C, G, and U. Thus mRNA would be able to collect and assemble only four of the 20 amino acids. With codons two bases long, mRNA could assemble 16 amino acids, because four bases can be arranged in 16 different pairs. Codons with three bases, however, can be arranged in 64 different combinations, more than enough to code for all 20 amino acids. Experiments have confirmed that amino acids are coded by three-base codons, or *triplet codons*.

Figure 12–8 shows the triplet codon for each amino acid. As you can see, several different codons can stand for the same

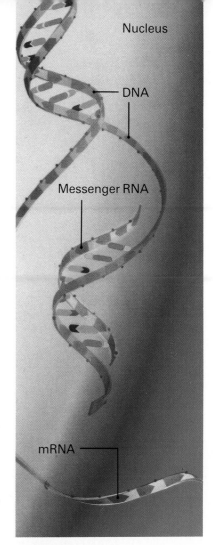

Figure 12–7. During transcription a single strand of DNA serves as a template for the assembly of messenger RNA from free nucleotides. Once formed, mRNA leaves the nucleus and enters the cytoplasm.

Figure 12–8. Each amino acid is coded for by three bases arranged in a specific sequence. According to the chart on the left, what are the triplet codons for alanine? for proline?

Second Base in Codon

		U	C	A	G	
First Base in Codon	**U**	Phenylalanine	Serine	Tyrosine	Cysteine	U
		Phenylalanine	Serine	Tyrosine	Cysteine	C
		Leucine	Serine	Stop Codon	Stop Codon	A
		Leucine	Serine	Stop Codon	Tryptophan	G
	C	Leucine	Proline	Histidine	Arginine	U
		Leucine	Proline	Histidine	Arginine	C
		Leucine	Proline	Glutamine	Arginine	A
		Leucine	Proline	Glutamine	Arginine	G
	A	Isoleucine	Threonine	Asparagine	Serine	U
		Isoleucine	Threonine	Asparagine	Serine	C
		Isoleucine	Threonine	Lysine	Arginine	A
		Methionine	Threonine	Lysine	Arginine	G
	G	Valine	Alanine	Aspartic Acid	Glycine	U
		Valine	Alanine	Aspartic Acid	Glycine	C
		Valine	Alanine	Glutamic Acid	Glycine	A
		Valine	Alanine	Glutamic Acid	Glycine	G

(Third Base in Codon)

amino acid. Other codons do not stand for any amino acid; they are *stop,* or *terminator, codons* that tell RNA to stop synthesizing proteins. The codon AUG, which specifies the amino acid methionine, is the start signal. All protein synthesis begins at an AUG codon.

12.6 Translation

As each section of the genetic code on DNA is transcribed to mRNA, the two strands of DNA rejoin. Then the mRNA moves into the cytoplasm through a pore in the nuclear membrane. In the cytoplasm, ribosomes attach to the mRNA to carry out the formation of a protein in a process called **translation.** This process translates the RNA base sequence into the amino acid sequence of protein. Several ribosomes are involved in translation, thus enabling the cell to use a single mRNA molecule to make many protein molecules at once.

As Figure 12–9 shows, a ribosome is made up of two parts. The smaller part attaches to the mRNA. The larger part contains an enzyme that helps link the amino acids to form a protein. Figure 12–9 also shows that mRNA and rRNA do not carry out translation by themselves. They rely on tRNA, the small molecules that pick up specific amino acids and carry them to the

Figure 12–9. During translation a codon on mRNA links up with an anticodon on tRNA while both are on a ribosome. Amino acids are thus assembled one after another to form a functional protein. The nucleus and the ribosomes are not drawn to scale.

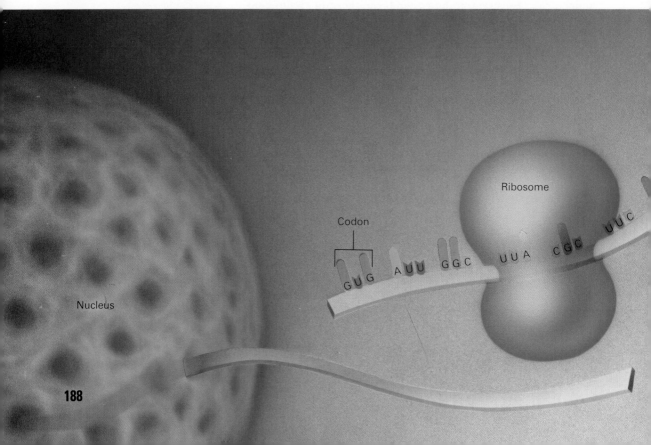

Codon

Ribosome

GUG AUU GGC   UUA CGC UUC

Nucleus

188

mRNA. The cytoplasm contains more than 20 kinds of tRNA, at least one for each of the 20 amino acids needed to synthesize proteins.

Notice in Figure 12–9 that tRNA is about 80 nucleotides long. The three loops of the tRNA molecule give it a shape somewhat like a cloverleaf. One end of tRNA attaches to its amino acid with the help of an enzyme and ATP. The other end of tRNA contains a sequence of three bases that complement the triplet code on mRNA. The sequence of three bases on tRNA is called an **anticodon** (AN tee koh dahn). Since there is at least one codon for each of the 20 amino acids, there are at least 20 different anticodons.

Translation begins at an AUG codon. The ribosomes move along the strand of mRNA and indicate each codon to approaching molecules of tRNA carrying their amino acids. As each codon is indicated, the tRNA with the complementary anticodon binds to the mRNA. In this way amino acids are placed in the proper sequence to make a certain protein. Once the tRNA has bound to the mRNA, an enzyme in the ribosome links the new amino acid to the neighboring amino acid by means of a *peptide bond*. After the amino acids are linked, the tRNA is released and returns to the cytoplasm to collect another amino acid. As each ribosome moves along the mRNA, a chain of amino acids

Q/A

Q: *How does tRNA recognize its specific amino acid?*

A: Transfer RNA recognizes its specific amino acids with the help of activating enzymes in the cytoplasm. Each of the 20 amino acids has a corresponding activating enzyme. With energy provided by ATP, an activating enzyme attaches its amino acid to the correct tRNA molecule.

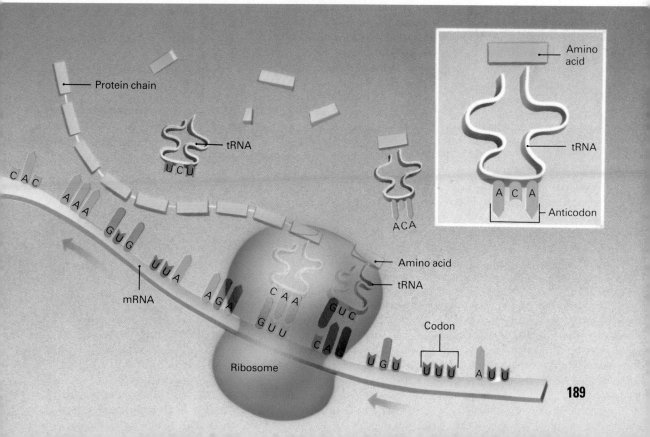

Protein chain

tRNA

UCU

CAC

AAA

GUG

UUA

AGA

CAA

GUU

GUC

CAG

mRNA

Ribosome

UGU

UUU

AUU

Codon

Amino acid

tRNA

ACA

Amino acid

tRNA

ACA

Anticodon

189

▶ Genes: Facts and Riddles

Scientists have learned a great deal about genes since the work of Watson and Crick in 1953. Most important, scientists now know that a gene is a sequence of nucleotides on an active strand of DNA that codes for a certain protein. As you have read, the code of each gene is transcribed into a molecule of mRNA. This mRNA molecule travels into the cytoplasm and provides a blueprint for linking amino acids into a chain called a *polypeptide*. Some proteins consist of a single polypeptide, but others consist of two or more chains. Thus, building a protein may require information from several genes, one for each polypeptide.

Although scientists have been able to understand how protein synthesis works, many other riddles about DNA and genes remain unsolved. One such riddle concerns the relationship between genes and the traits they determine. In only a few cases have scientists been able to understand how a gene-produced protein causes a trait to appear. Scientists do know that in humans a change in one nucleotide causes a change in an amino acid of *hemoglobin*, a protein in red blood cells that combines with oxygen. The change in the amino acid causes the shape of the red blood cells to change from round to bent, or sickle-shaped.

Another unsolved mystery concerns nucleotide sequences that may determine when a gene will become active. Consider for a moment the fact that every cell in your body contains exactly the same genes. How is it then that your brain cells came to be so different from the cells that make up your blood or your skin? The answer lies in the fact that different genes are active in different cells. Just how genes in higher organisms are turned on and off is the subject of much current research.

is assembled in the proper sequence to form a certain protein. Eventually, each ribosome reaches the termination codon on the strand of mRNA. There is no tRNA specifying an anticodon for the termination codon, so no amino acid is added. Protein synthesis stops, and the protein is released from the ribosome. After translation is complete and all the ribosomes have come off, the mRNA breaks down into individual nucleotides.

Reviewing the Section

1. How is RNA different from DNA?
2. How do scientists know that three nitrogen-carrying bases are needed to code for one amino acid?
3. How is RNA manufactured from DNA?
4. Why are tRNA and mRNA needed for protein synthesis?

Changes in the Genetic Code

Each time a cell divides, its DNA replicates so that the genetic code is passed on to the new cell. Usually the DNA copies itself exactly. *Occasionally, replication mistakes or environmental factors cause a change in the genetic code.* A change in a gene or chromosome is called a **mutation.** An organism in which a mutation is expressed is called a **mutant.**

Many mutations have little or no apparent effect on an organism or its offspring. However, some mutations have harmful effects. Mutations that cause death in offspring are called *lethal mutations.*

Section Objectives

- *State* the difference between gene mutations and chromosome mutations.
- *Name* three types of chromosome mutations.
- *Differentiate* between germ mutations and somatic mutations.
- *List* several causes of mutations.

12.7 Gene Mutations

The most common type of mutation involves a change in a single gene. Such a mutation is called a **gene mutation,** or **point mutation.** In this type of change, one nitrogen-carrying base may be substituted for another. In other cases, a base may be lost or added.

A change in a single nucleotide may seem minor. However, when you consider the importance of the triplet code in DNA, you realize that changing even one base in the genetic code can have major consequences. You can compare a change in the triplet code to the change of one digit in the area code of a phone number. By dialing 213 instead of 212, for example, you reach Los Angeles instead of New York City. Likewise, changing one base in the DNA triplet code may mean that a different amino acid is placed in the protein chain. As a result, the protein itself is different from the one originally called for by the code.

The condition called *albinism* is caused by a gene mutation. Because of the change in a nucleotide base, organisms with this condition cannot produce the enzyme responsible for pigment synthesis. The changed base sequence produces a protein that cannot function as an enzyme, so no pigment is produced. Animals with this condition have white hair and pinkish eyes and skin. Some plants with albinism have no chlorophyll and cannot carry out photosynthesis.

12.8 Chromosome Mutations

Chromosome mutations involve changes in many genes. In some cases several genes may be lost, added, or moved to different chromosomes. In other cases, entire chromosomes may be lost or gained. Because so many genes are involved, many

Figure 12–10. Albinism occurs both in animals and in plants. The koala and the corn plants above are white because a gene mutation has resulted in an inability to produce normal pigments.

Spotlight on Biologists: Barbara McClintock

Born: Hartford, Connecticut, 1902
Degree: Ph.D., Cornell University

When geneticist Barbara McClintock first proposed her theory of "jumping genes" in 1947, the scientific community largely ignored her. Now her revolutionary findings that genes can move from one chromosome to another is hailed as one of the most important discoveries of this century. McClintock's lifetime of research with maize plants was acknowledged in 1983 with the Nobel Prize for Medicine or Physiology.

When McClintock began her research, most scientists thought genes maintained a fixed position, like beads in a row. McClintock crossed maize plants and noted color changes in the leaves and kernels that did not follow a hereditary pattern. She concluded that such mutations resulted from the movement of genes.

Later experiments by molecular biologists working with bacteria confirmed McClintock's theories. These experiments further concluded that the genes move during the replication phase of the cell cycle, prior to cell division. These findings explain the origin of some bacteria strains that are resistant to antibiotics. The "jumping genes" also have significance in evolution, as a source of genetic variability.

McClintock has often had to resist popular opinion to pursue her own interests. Her mother at first op-

posed her desire to attend college. After McClintock studied biology on her own while working in an employment agency, her parents gave in and allowed her to attend Cornell University. For years she had difficulty in securing good teaching and research positions because she was a woman. Since 1941 McClintock has pursued her research at Cold Spring Harbor Laboratory in New York state.

chromosome mutations are harmful or even lethal to organisms that inherit them.

Figure 12–11 shows three major kinds of chromosome mutations. One type of mutation involves a single chromosome. During mitosis or meiosis, a chromosome may break, and part of it may be lost. The loss of a chromosome segment is called a *deletion*. Its effect in animals is usually lethal. Occasionally, the middle section of a chromosome may break away, turn over, and recombine with the same chromosome in reverse order. This type of mutation is called an *inversion*. Inversions may not harm organisms since all the same genes are present on the chromosome.

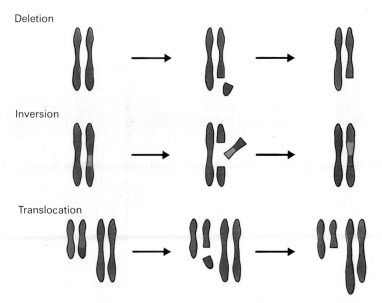

Deletion

Inversion

Translocation

Figure 12–11. Deletions, inversions, and translocations are mutations involving parts of chromosomes. A deletion occurs when part of a chromosome is lost. An inversion results when a segment of DNA is turned around within the chromosome. Translocation occurs when a segment of one chromosome becomes attached to a nonhomologous chromosome.

Another type of chromosome mutation involves changes in two chromosomes. For example, in *translocation,* a fragment of one chromosome may become attached to a nonhomologous chromosome. Translocation involving certain chromosomes may produce serious retardation in humans.

12.9 Somatic and Germ Mutations

Changes in the makeup of genes and chromosomes are the two major kinds of mutations. Mutations can also be categorized according to the type of cell in which they occur.

Germ mutations are mutations that occur in reproductive cells. These mutations can occur during meiosis or after the gamete is mature. Although germ mutations do not affect the individual in which they occur, they can be transmitted to offspring. A mutation is transferred when egg and sperm unite to form a new individual. Some germ mutations may not be expressed for several generations. Because any sort of mutation may occur within a reproductive cell, the effects of germ mutations may range from harmless to lethal.

Mutations that occur in body cells are called **somatic mutations.** Somatic mutations are not passed on to offspring. Yet cells produced by a mutant cell will also contain the mutation. Thus a somatic mutation may affect the individual in which it occurs. Scientists have found evidence that somatic mutations can change genes that control cell reproduction. If such a mutation occurs, cells may reproduce uncontrollably, resulting in a cancerous growth.

Q: *What is a lethal gene?*

A: A lethal gene is one that causes early death in most or all of the individuals possessing it. Since the individual usually dies before reproducing, the gene is seldom passed on. A lethal gene may be expressed shortly after fertilization, causing the death of an embryo. It may also be expressed at any time during development or after birth.

12.10 Effects and Frequencies of Mutations

Many mutations produce genes that are recessive, and so these mutations go undetected. In order for a mutated recessive allele to express itself, it must combine with a similarly mutated allele. Occasionally, however, a mutation produces a dominant gene that is expressed in an organism. For example, one form of dwarfism that occurs among human beings is the result of a dominant mutation.

Nearly all chromosome mutations in animals are harmful because they result in deformities and other traits that make it difficult for the organism to survive in its environment. Occasionally, however, mutations help an organism survive when its environment changes. For example, mutations enabled certain mosquitoes to resist insecticides after these poisons were introduced into their environment. The mutant mosquitoes thus survived and reproduced, creating more insects with resistance to insecticides.

Because many mutations are recessive and thus do not appear in organisms, it is difficult to calculate the exact rate of mutation. Based on studies of dominant and codominant mutations, however, scientists believe the rate of mutation is very low. An estimated rate for spontaneous mutations is only one or two in every 100,000 genes per generation.

Although the rate of spontaneous mutation is low, environmental factors can influence mutation rates. Anything that increases the rate of mutations in cells is called a **mutagen.** Ultraviolet light, for instance, is known to increase mutation rates in bacteria. Scientists believe that ultraviolet rays from the sun may cause skin cancer. X rays and other forms of radiation are also known to be mutagens.

Other mutagens include tars in tobacco smoke, smog, certain viruses, and various chemicals and drugs. The chemical mustard gas, for example, removes guanine from DNA. Nitrous acid, another chemical, removes nitrogen from DNA bases. Both chemicals produce a variety of harmful results.

Reviewing the Section

1. What is a gene mutation?
2. What are three types of chromosome mutations?
3. How are germ mutations and somatic mutations different from each other?
4. Why could the deletion of a chromosome segment have a lethal effect?

Investigation 12: Extracting Nucleic Acid from Cells

Purpose
To extract nucleic acid (DNA and RNA) from liver cells

Materials

Fresh liver homogenate, graduated cylinder, three 50-mL beakers, 100-mL beaker, hot plate, tongs, 15% trichloroacetic acid, 10% sodium chloride solution, 95% ethyl alcohol, stirring rod, funnel, filter paper, glass slide, coverslip, compound microscope, safety goggles, tongs

Procedure
1. **CAUTION: Put on safety goggles and leave them on through Step 12.** Measure 10 mL of liver homogenate into a graduated cylinder.
2. Pour the homogenate into a 100-mL beaker.
3. Add 10 mL of 15% trichloroacetic acid and mix well with a stirring rod.
4. Filter and transfer the material on the filter paper to a clean 50-mL beaker.
5. Add 15 mL of 10% sodium chloride solution and mix well.
6. **CAUTION: Be careful when using any heat source.** Half fill the 100-mL beaker with water. Prepare a boiling-water bath by placing the beaker on a hot plate. Heat the water to a boil.

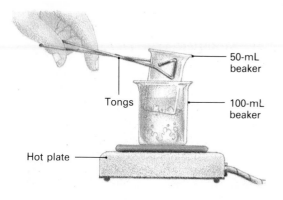

Tongs — 50-mL beaker

Tongs — 100-mL beaker

Hot plate

7. Using tongs, place the 50-mL beaker in the 100-mL boiling-water bath as shown. Heat for 10 minutes.
8. Again using tongs, remove the beaker containing the mixture and allow it to cool.
9. Filter the mixture into a clean 50-mL beaker.
10. Add refrigerated 95% ethyl alcohol slowly to the filtrate in the beaker. The beaker should be tilted so that the alcohol can flow down the side of the beaker.
11. Observe the milky precipitate that forms. This is the nucleic acid of the liver cells.
12. Collect the strands of nucleic acid with a stirring rod and place some of them on a clean slide.
13. Observe the strands of nucleic acid under the microscope under low power, then under high power.

Analyses and Conclusions
1. Describe the appearance of the nucleic acid strands.
2. Explain the procedure you could use to extract the nucleic acid from a fresh steak.

Going Further
You can identify the type of nucleic acid strand (DNA or RNA) by adding the nucleic acid concentration to an equal volume of concentrated hydrochloric acid. Place the mixture in a boiling water bath for 10 minutes. Add several drops of diphenylamine reagent and boil in the water bath for another five minutes. Look for a color change. If the nucleic acid strands turn greenish, RNA is abundant. If the strands turn purplish, DNA is abundant.

Chapter 12 Review

Summary

Since the 1940s geneticists have focused on understanding what genes are and how they work. In 1944 Oswald Avery and his associates showed that DNA is the genetic material in chromosomes. Later Watson and Crick discovered the structure of DNA. They also explained replication, the process by which DNA copies itself. During replication the DNA molecule splits into two strands. Nucleotides in the nucleus bind with complementary nucleotides on each strand of DNA. In this way DNA molecules are produced for each new cell.

Genes are segments of DNA that control protein synthesis. Through transcription, the DNA code is transferred to messenger RNA, which carries the code to the ribosomes. During translation, transfer RNA molecules carry amino acids to mRNA where they are linked in the proper sequence to produce a specific protein.

Mutations, or changes in the genetic code, involve genes or chromosomes. Most major changes in the code are harmful or lethal. Both natural mistakes in replication and environmental factors may produce mutations.

BioTerms

adenine (**182**)
anticodon (**189**)
chromosome mutation (**191**)
codon (**187**)
cytosine (**182**)
deoxyribonucleic acid (DNA) (**181**)
deoxyribose (**183**)
gene mutation (**191**)
germ mutation (**193**)

guanine (**182**)
messenger RNA (**186**)
mutagen (**194**)
mutant (**191**)
mutation (**191**)
nucleotide (**183**)
point mutation (**191**)
protein synthesis (**186**)
replication (**184**)
ribonucleic acid (RNA) (**186**)

ribose (**186**)
ribosomal RNA (**186**)
somatic mutation (**193**)
thymine (**182**)
transcription (**186**)
transfer RNA (**186**)
translation (**188**)
uracil (**186**)

BioQuiz   (Write all answers on a separate sheet of paper.)

I. Completion

1. DNA and RNA belong to a class of compounds called _____.
2. The DNA molecule is in the shape of a _____.
3. Cells contain exactly _____ kinds of amino acids used to build proteins.
4. The three bases on tRNA that attach to mRNA are called the _____.
5. Mutations may involve chromosomes or _____.

II. Modified True and False

Mark each statement TRUE or FALSE. If false, change the underlined term to make the statement true.

6. Cells contain more than <u>50</u> kinds of transfer RNA.
7. RNA contains <u>adenine</u> instead of thymine.
8. RNA nucleotides contain <u>ribose</u>.
9. The triplet code in DNA consists of <u>16</u> three-letter "words."

III. Multiple Choice

10. A DNA nucleotide does not contain
    a) adenine. b) ribose. c) phosphate.
    d) cytosine.
11. In a DNA molecule, adenine joins only
    with a) uracil. b) cytosine.
    c) thymine. d) guanine.
12. If a base sequence in DNA is ATG, the
    complementary sequence in mRNA
    would be a) ATG. b) TAC. c) UAC.
    d) AUC.
13. DNA and RNA a) are nucleic acids.
    b) both undergo replication. c) are
    double-stranded molecules. d) have the
    same nitrogen-carrying bases.
14. A mutation that is passed on to
    offspring is called a) inversion.
    b) somatic mutation. c) translocation.
    d) germ mutation.
15. A sequence of nucleotides that controls
    the synthesis of one protein is a a) gene.
    b) chromosome. c) codon. d) mutagen.

IV. Essay

16. How did Oswald Avery and his
    associates determine that DNA was the
    genetic material in cells?
17. What evidence did Watson and Crick
    use to construct their model of DNA?
18. When does DNA replication occur in
    cells? Explain why this must be so.
19. Why are DNA codons made up of three
    bases?
20. How do cells manufacture proteins?

Applying and Extending Concepts

1. Phoebus A. Levene determined the chemical makeup of the DNA molecule. Based on this information he reached an incorrect hypothesis about the structure of DNA. Do library research to learn what Levene's hypothesis was, and how another scientist, Erwin Chargaff, corrected his error.
2. Messenger RNA is broken down after it has synthesized the necessary protein(s). Explain what problems might occur if the mRNA was not broken down.
3. Oxytocin is a hormone that causes a female mammal to secrete milk for her young. The following set of triplet codons specifies the sequence of amino acids that make up oxytocin: UGU UAC AUC CAA AAC UGC CCA CUA GGA. Translate this code into a sequence of amino acids.
4. Researchers have discovered the existence of cancer genes in humans. Do library research to learn how translocation of these genes may produce cancer.

Related Readings

Gould, S. J. "The Ultimate Parasite." *Natural History* 90 (November 1981): 7–14. Most of the DNA present in our cells does not code for proteins, and this essay explains (and explodes) some of the hypotheses regarding the presence of this "extra" DNA.

Hoagland, M. *Discovery: The Search for DNA's Secrets*. Boston: Houghton Mifflin, 1981. This entertaining history of DNA research explains the chemistry of heredity.

Miller, J. A. "The Gene Idea." *Science News* 121 (March 13, 1982): 180–182. This article traces the development of the concept of the gene from Muller's work on mutations in the 1920s to the discovery of the molecular nature of DNA.

Watson, J. D. *The Double Helix*. New York: Atheneum, 1968. Watson provides a stimulating, day-to-day account of how he, Crick, and their coworkers developed the double helix model of DNA.

13

Human Genetics

Diversity of traits

Introduction

The principles of heredity operate in humans just as they do in all other organisms. Each human cell contains DNA molecules that govern the production of proteins. Because of these proteins, children possess traits similar to those of their parents as well as traits that are unique. As in all other organisms, DNA in humans is transmitted to new cells through mitosis and meiosis. Furthermore, humans show evidence of dominant and recessive genes, sex-linked genes, and mutations.

Geneticists are still probing many mysteries of human genetics, such as the mechanisms that control the expression of genes and the effects of environment on the actions of genes. However, geneticists have identified the sources of many traits and several genetic diseases.

The Study of Human Genetics

Geneticists cannot study people in the same way in which they study bacteria or fruit flies. Since humans choose their own mates, geneticists cannot conduct breeding experiments to determine how human traits are inherited. In addition, humans have a long life span. As a result, it takes many years, not just days or weeks, to produce several generations. Finally, most human families have small numbers of offspring—too few to verify the outcomes of crosses predicted by probability.

To overcome these problems, geneticists use special techniques to study human heredity. Three of the most common methods used to study human genetics are pedigree analysis, population sampling, and twin studies. Each method reveals a different kind of information about human genetics.

13.1 Pedigree Analysis

Geneticists study family trees by observing the occurrence of a trait over several generations. This study helps them understand dominance, sex-linkage, and other facts about genes in humans. Knowing the genetic background of a person is often medically useful.

One study involved people with a streak of white hair near the crown. This trait is called the white forelock trait, or *piebaldness*. Researchers studying people with this trait made a **pedigree,** which is a record that shows how a trait is inherited over several generations. The pedigree in Figure 13–1 shows that piebaldness is inherited in a pattern typical of a dominant trait, controlled by a single dominant-recessive gene pair. Notice in the second generation that some children had white forelocks (W) and some had normal hair (w). This pattern indicates that the parent with the trait is heterozygous (Ww). If the parent were homozygous dominant, all the children would have white forelocks. In addition, the children of heterozygous individuals (Ww) married to homozygous recessive individuals (ww) were about half piebald and half normal. This is the pattern you would predict using probability.

13.2 Population Sampling

Another important way of learning about human genetics is to study the traits that appear in a certain *population,* or an interbreeding group living in a certain area. Obviously, it is impossible to study every person in a population as large as that of the

Section Objectives

- *List* three problems geneticists face in studying human genetics.
- *Name* the kinds of information revealed by the analysis of genealogies.
- *List* two goals of population sampling.
- *Explain* the process called twinning.

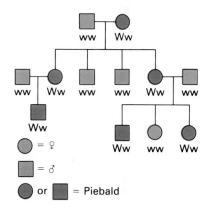

Figure 13–1. This pedigree traces the appearance of the trait of piebaldness through three generations of a single family.

United States. Instead, geneticists use **population sampling,** which involves determining how often a trait appears in a small, randomly selected group and then projecting the results to the population as a whole. To measure the frequency of certain traits, geneticists use questionnaires or tests. Information that scientists collect in this way reveals how frequently certain genes appear in a population or what the rates of mutation are.

Table 13–1: Some Human Characteristics Determined by a Single Gene

Dominant	Recessive
Webbed fingers or toes	Abnormally small head
Very short fingers or toes	Dry, thick skin on palms
Extra fingers or toes	Some forms of deafness
Drooping eyelids	Abnormal fat metabolism
Some kinds of dwarfism	

13.3 Twin Studies

Heredity affects the traits people possess, but factors in the environment may influence the expression of human traits. For example, genes play an important part in determining skin color, yet a person's skin may tan from exposure to the sun. Likewise, individuals who inherit musical or athletic ability may differ in performance due partly to the training they receive. Scientists do not fully understand how heredity and environment affect human traits, but the study of twins has provided some insights into the relationship of their effects.

Like many other organisms, humans can produce two types of twins. **Fraternal twins** develop from two eggs in the mother that are fertilized by two different sperm. These twins are as genetically different as any brothers and sisters. Geneticists are more interested in **identical twins,** who develop from a single fertilized egg that separates into two halves early in development. Barring mutations, identical twins have exactly the same genes and therefore identical traits. For instance, identical twins are always the same sex. Their appearances are so similar that these twins are often difficult to tell apart. Because identical twins have the same genetic makeup, geneticists can attribute differences between them to the environment. For this reason, geneticists study identical twins raised apart and by different families to determine how heredity and environment influence human traits. The results are far from conclusive but have yielded some insight into the genetic influence on human lives.

Two eggs are fertilized by two sperm and divide

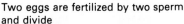
Two masses of cells develop separately and each has a fetal sac

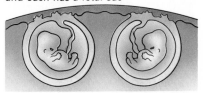

Figure 13–2. Fraternal twins develop when two eggs are fertilized by different sperm (above). Identical twins develop from a single, fertilized egg that splits (below).

Fertilized egg divides into two cells

Two cell masses develop nearly identically in one fetal sac

▶ The Study of Twins

As you might suspect, the study of identical twins raised apart has revealed some amazing things. When the twins in the photograph were reunited after 39 years, they found they both were named Jim and had sons named James Allan. Both enjoyed wood working, disliked baseball, and drove the same model of car.

On the surface it might seem that the twins' identical genes determined all these similarities. But scientists who study twins know the answer is not that simple, especially when trying to explain similarities in behavior.

Consider the fact that identical twins raised apart are more alike than those raised together. You might predict just the opposite— that twins with the same genes and the same environment are the ones most alike. Scientists have learned, however, that a process called *twinning* takes place between twins raised together. Part of the time, the twins enjoy being alike. At other times, the twins attempt to be different so that they will stand out as individuals.

The process of twinning is just one of the problems scientists must take into account when studying twins. For example, although it is assumed that identical twins are genetically alike, a mutation may

alter one twin's genetic code, making that individual different from his or her twin.

At present, no one can explain similarities such as the Jims' choice of car or child's name. It is still impossible to determine whether such similarities are inherited or are merely coincidental.

By studying identical twins, geneticists have learned that genes seem to have a greater influence than the environment on such traits as height, weight, blood pressure, speech patterns, and gestures. They have also discovered that genes play a role in some medical problems once thought to be caused only by environmental factors. For instance, genes can cause a susceptibility to diseases such as diabetes and certain types of cancer.

Reviewing the Section

1. Why is human heredity difficult to study?
2. What information can be learned by studying pedigrees?
3. What can population sampling reveal about human genetics?
4. Why are identical twins used to study human genetics?

Inheritance of Human Traits

- *Identify* some human traits determined by single pairs of genes.
- *Distinguish* between polygenic traits and traits determined by multiple alleles.
- *Compare* sex-linked and sex-influenced traits.
- *List* several consequences of nondisjunction.
- *Describe* what is meant by a karyotype and how such a device is useful.

At the present time, scientists have identified the genetic basis for just a fraction of all human traits, most of them abnormalities. Scientists study abnormalities because most are easy to identify in the population and because many require medical treatment. *Today scientists know that single genes, groups of genes, sex-linked genes, hormones, and chromosome abnormalities all influence human traits.* Although many of the traits described here are disorders, they illustrate the way in which normal traits are inherited. Molecular biologists are now probing the mechanism of action of specific genes to determine how they cause disease. The recognition of dominant and recessive traits in humans is an important step in curing genetic disorders.

13.4 Traits Determined by Single Genes

Scientists have discovered more than 300 human traits determined by single dominant alleles. Some familiar ones are cleft chins, freckles, and free ear lobes. Other traits produced by dominant alleles are less familiar. Individuals with **polydactyly** (pahl ee DAK tih lee), for example, have extra fingers or toes. Persons with **Huntington disease** develop a serious nervous system disorder after the age of 30. This disease begins with a loss of muscle control, then progresses to mental deterioration and death. No cure or treatment for Huntington disease has yet been found.

More than 250 traits, many of them genetic disorders, are known to occur when an individual inherits two recessive alleles for a trait. **PKU**, or **phenylketonuria** (fehn ihl keet uh NYOOR ee uh), is a biochemical disorder produced by two recessive alleles. Individuals with PKU are unable to synthesize the enzyme that breaks down the amino acid phenylalanine. Gradually, phenylalanine accumulates in the blood and destroys brain cells. Children who inherit PKU develop severe mental retardation. Fortunately, testing the blood of newborns reveals the presence of this disease. Most states, in fact, require PKU testing of newborn babies. If PKU is present, children are placed on a diet low in phenylalanine for the first six years of life. This usually prevents brain damage.

Another homozygous recessive condition causes a biochemical disorder called **sickle-cell disease.** In this disease red blood cells become sickle-shaped. Red blood cells contain a protein called *hemoglobin* that carries oxygen to body tissues. In sickle-cell disease, recessive alleles cause the substitution of

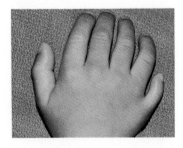

Figure 13–3. Red hair and freckles (above) and a six-fingered hand (below) are traits determined by single gene pairs.

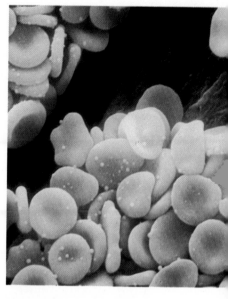

one amino acid in hemoglobin. *Valine* is substituted for *glutamic acid*. When the concentration of oxygen in the blood is low, the abnormal hemoglobin molecules stick together and cause the red blood cells to bend into a sickle shape. These sickle-shaped cells clog small blood vessels, thus depriving tissues of needed oxygen and causing severe pain. Gradually, vital organs are destroyed, and death sometimes occurs within the first 20 years of life. Sickle-cell disease occurs mostly in persons of African descent.

Figure 13–4. The scanning electron microscope image (right) shows normal red blood cells magnified 2,000 times. Sickle-shaped blood cells are shown magnified 5,555 times (left).

13.5 Traits Determined by Multiple Alleles

Several human traits are determined by **multiple alleles,** which are three or more alleles of a gene that can occur at one location on a chromosome and affect a single trait. Blood type, for example, is a trait determined by three different alleles. Of course, each individual carries only two of the three alleles, one from each parent. The two alleles determine a person's blood type. In the whole population, however, there are three alleles. The frequency of those alleles varies from one population to another.

Blood types were discovered in 1900 by Karl Landsteiner, an Austrian scientist. Landsteiner found that when red blood cells from different persons are mixed, some intermingle and others clump together, or *agglutinate*. Through further research, Landsteiner determined that special proteins called *antigens* on the cell membranes of red blood cells cause the clumping. The presence or absence of these antigens determines four different types of red blood cells, which Landsteiner labeled A, B, AB, and O. Type A blood cells have an A antigen on their membranes. Similarly, type B blood cells have a B antigen. Type AB blood cells have both A and B antigens, while type O blood cells have neither A nor B antigens. The type of antigen on a person's blood cells affects the type of blood he or she can receive in a

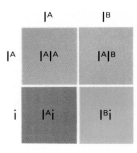

	I^A	I^B
I^A	I^A I^A	I^A I^B
i	I^A i	I^B i

Figure 13–5. A Punnett square shows that children with any of four different blood genotypes can be born to a mother with an I^A i genotype and a father with an I^A I^B genotype. How many phenotypes are possible?

transfusion. For example, a person with type A blood can safely receive type A blood. A transfusion of type B blood, however, will produce clumping, causing the blood to clot in the blood vessels.

The alleles for these four blood types are represented with three symbols. The letter I represents the dominant alleles, while the letter i stands for the recessive alleles. Because the alleles that determine blood types A and B are both dominant, they are written I^A and I^B. The allele that determines type O blood is recessive and is written i. A person with type A blood, then, may possess the genotype $I^A I^A$ or $I^A i$. Similarly, a person with type B blood may have the genotype $I^B I^B$ or $I^B i$. The alleles for types A and B show codominance, so when both are present they both have an effect. The genotype for AB blood, then, is written $I^A I^B$. Since the allele for type O blood is recessive, the genotype for type O blood is ii. Figure 13–5 shows the possible blood types in offspring produced by a mother with an $I^A i$ genotype and a father with an $I^A I^B$ genotype.

13.6 Polygenic Traits

Some traits are **polygenic,** or determined by several genes. In this method of inheritance, none of the genes are dominant. Instead, each gene consists of an *active allele* that has a small additive effect on the phenotype and an *inactive allele* that has no effect. Since each of the genes involved adds something to the phenotype, a continuous range of phenotypes is possible. Skin color, eye color, and height are all polygenic traits.

Figure 13–6. These members of a high school class exhibit a wide range of skin colors. Skin tone is determined by several genes, each of which has only a small effect on the overall color.

13.7 Sex-Linked Traits

Sex in humans, as in many other organisms, is determined by X and Y chromosomes. Females have two X chromosomes while males have an X and a Y chromosome. Sex-linked traits in humans are determined by genes carried only on the X chromosome with no alleles on the Y chromosome. Recessive sex-linked traits rarely appear in females because a dominant allele on one X chromosome will mask the effect of a recessive allele on the other X chromosome. Since males have only one X chromosome, however, recessive genes on that chromosome are always expressed.

Colorblindness In humans the gene for color vision is carried on the X chromosome. The recessive allele of the normal gene may produce **colorblindness,** or the inability to distinguish certain colors. Persons with red-green colorblindness, for example, cannot distinguish between red and green.

Since the gene for color vision is located only on the X chromosome, the normal allele is designated X^C. The recessive allele that determines colorblindness is designated X^c. Homozygous (X^CX^C) and heterozygous (X^CX^c) females have normal color vision. A female who is heterozygous for colorblindness is said to be a **carrier** because she carries the recessive allele but does not express it. Only homozygous recessive females (X^cX^c) are colorblind. Since males have only one X chromosome, they are either colorblind (X^cY) or have normal color vision (X^CY).

Hemophilia Another recessive gene on the X chromosome produces a disorder called **hemophilia** (hee muh FIHL ee uh). People with hemophilia cannot produce the protein needed for normal blood clotting. As a result, small cuts can cause severe bleeding. Hemophiliacs can bleed to death from minor wounds.

Hemophilia appeared in European royal families in the nineteenth century, and a mutant allele in Queen Victoria is believed to have been the source. Since Queen Victoria was heterozygous for the condition, she did not have hemophilia. However, she passed the allele on to one son and two daughters. Through marriage these children carried the disease to royal families in Russia, Germany, and Spain.

13.8 Sex-Influenced Traits

Certain traits appear more often in one sex than the other, but they are not sex-linked. A trait that is generally associated with one sex but is produced by genes carried on autosomes is called

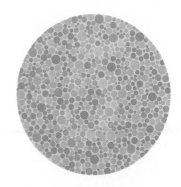

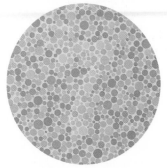

Figure 13–7. The photos above are used to test colorblindness. A person with red-green colorblindness will not be able to read the numbers. What numbers do you read?

Q/A

Q: *Are there other sex-linked traits besides colorblindness and hemophilia?*

A: Yes. There are many known sex-linked traits, and about 50 of them are common. Among these are gout, caused by high levels of uric acid in the blood, and Duchenne muscular dystrophy, a gradual deterioration of the muscles.

Karyotyping Human Chromosomes

Karyotypes are pictures of paired human chromosomes arranged by size. They are used in identifying chromosome abnormalities in fetuses and in infants with abnormal features.

The first step in producing a karyotype is to take a sample of the amniotic fluid surrounding a fetus or a blood sample from a child. The cells in the fluid or white cells in the blood are kept in a tissue culture that nurtures their growth. When the cells begin to divide, *colchicine* is added to the culture. This chemical stops cell division during metaphase.

Next, the cells are placed in a solution that ruptures their membranes, freeing the chromosomes. The chromosomes are then stained and photographed. The photograph is enlarged and cut into pieces containing one chromosome each. These pieces are arranged on a sheet of paper in numbered homologous pairs according to shape, size, and staining bands. As the photograph shows, the chromosomes can easily be studied for abnormalities.

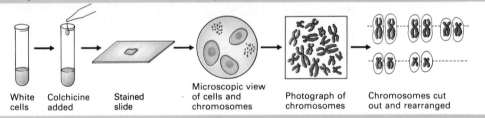

White cells | Colchicine added | Stained slide | Microscopic view of cells and chromosomes | Photograph of chromosomes | Chromosomes cut out and rearranged

a **sex-influenced trait.** Male and female hormones influence the expression of these genes.

Baldness is one example of a sex-influenced trait. In humans, the allele coding for baldness is H^B. This allele is dominant in males but recessive in females. The allele that codes for normal hair, H^N, is recessive in males but dominant in females. Both males and females with genotype $H^B H^B$ are likely to lose their hair. Neither males nor females with genotype $H^N H^N$ are likely to go bald. However, men having $H^B H^N$ usually go bald, but females with this genotype do not. Baldness is affected by the presence of male and female hormones.

13.9 Chromosomal Abnormalities

Some human genetic disorders are determined by abnormal chromosomes. Several of these disorders result from **nondisjunction** (nahn dihs JUHNGK shuhn). Nondisjunction is the failure of a chromosome pair to separate during meiosis or mitosis. When nondisjunction occurs during the first division in meiosis, half of the gametes produced lack one chromosome. The other half have an extra chromosome. When a gamete with one less chromosome combines with a normal gamete, the resulting zygote has 45 chromosomes instead of 46. A gamete with an extra chromosome that combines with a normal gamete produces a zygote with 47 instead of 46 chromosomes. Individuals with 45 chromosomes have the condition called **monosomy** (MAHN uh soh mee). Those with 47 chromosomes have the condition called **trisomy** (try SOH mee). A zygote with more than 47 chromosomes usually does not survive to form an embryo.

A variety of serious problems result from monosomy and trisomy. Children with an extra chromosome 21, for example, have **Down syndrome.** Individuals with Down syndrome have a number of distinctive features, including almond-shaped eyes, short limbs, and thick tongues. They are also mentally retarded in varying degrees.

Males with an extra X chromosome suffer from **Klinefelter syndrome.** This condition results from nondisjunction in the female parent, which produces an egg carrying two X chromosomes. Such an egg fertilized by a Y-carrying sperm results in a zygote with the chromosome configuration XXY. Most males with Klinefelter syndrome are sterile. Many show some degree of mental retardation. Some females who inherit an extra X chromosome develop normally. However, many of these females are sterile.

Individuals with **Turner syndrome** have only one X chromosome and no other sex chromosome. Although these individuals are females (XO), they do not develop normally and they are sterile. In addition, these females are short and have thick, webbed necks. Some are mildly retarded.

20   21   22   23

Figure 13–8. Persons with Down syndrome have distinctive facial characteristics and suffer from mental retardation. Through education such persons are able to make positive contributions to society and many live independently and are self-sufficient.

Reviewing the Section

1. What are four traits determined by single pairs of genes?
2. What are the genetic mechanisms that determine blood types and skin color?
3. What genetic disorders result from nondisjunction?
4. Why is baldness considered a sex-influenced trait?

- *Explain* the purpose of genetic counseling.
- *Name* four disorders that can be detected through amniocentesis.
- *Compare* the processes of ultrasound testing and fetoscopy.
- *List* the steps in chorionic villi biopsy.

Detecting Genetic Disorders

Every year about 250,000 babies in the United States are born with some sort of genetic disorder. ***Physicians and other specialists have developed a variety of tests to identify genetic disorders.*** Since these disorders are produced by abnormal genes or chromosomes, the ultimate cure involves replacing those genes and chromosomes. Although modern techniques such as genetic engineering show promise along these lines, no such cures currently exist.

13.10 Genetic Counseling

A procedure used to inform couples about their chances of passing a harmful trait on to their children is called **genetic counseling.** To determine whether prospective parents carry genes for a disorder, sample cells are tested for certain proteins, and karyotypes are made of their chromosomes. These tests may reveal recessive alleles or abnormal chromosomes that may result in various disorders. Genetic counselors use this information as well as other data, such as pedigree analysis, to explain a couple's chances of transmitting genetic abnormalities to their offspring. As a result of counseling, couples may decide not to have children or to adopt children instead of having their own.

13.11 Tests During Pregnancy

Figure 13–9. Amniocentesis involves withdrawing some of the amniotic fluid that surrounds a fetus (left), and then analyzing the fluid after placing it in a laboratory dish (right).

Nearly 200 genetic disorders can be discovered during pregnancy through the use of various tests. One of these tests, called **amniocentesis** (am nee oh sehn TEE sihs), involves the use of a long needle to withdraw fluid surrounding the fetus. This fluid contains skin cells shed by the fetus as well as fetal wastes

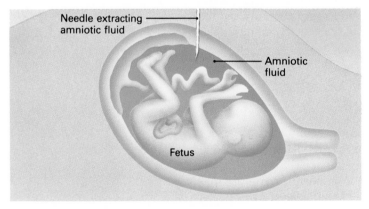

Needle extracting amniotic fluid

Amniotic fluid

Fetus

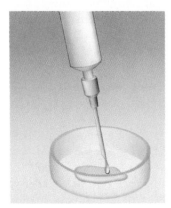

containing various proteins. Chemical tests of the cells and proteins can detect sickle-cell disease and other disorders. Karyotypes produced from fetal cells reveal such problems as Down syndrome and Turner syndrome. Although it is useful in many cases, amniocentesis is still regarded as controversial by some people.

Ultrasound testing is a technique that involves sending high frequency sound waves through the mother's abdomen. Some of the sound waves bounce off the tissues of the fetus and

Highlight on Careers: Genetic Counselor

General Description
A genetic counselor works with a medical team, advising a variety of individuals about genetic disorders. Often, these counselors advise couples about their chances of having children with disorders. Counselors also work with many pregnant women over the age of 35. Such women are more likely than younger women to have babies with Down syndrome or other chromosomal abnormalities. Genetic counselors also advise individuals who think they may have inherited a genetic disorder.

Although individual cases may vary, genetic counselors follow a similar procedure with all of their clients. First they assemble a detailed personal and medical history and construct a family pedigree. If necessary, blood is drawn

for chemical tests performed in the laboratory. Physicians use the pedigrees, clinical examinations, and lab results to make a diagnosis. Then the genetic counselor explains the diagnosis, discusses the probable risks prospective parents face, and outlines the options for dealing with those risks. Genetic counselors also help individuals cope with their fears about genetic disorders. Much of this work is reassurance. Genetic counselors reassure many patients that they do not have the disease they are concerned about. They also help other patients understand that their conditions may be treated because of early detection.

Career Requirements
Genetic counselors need an M.S. degree in human genetics with extensive

clinical experience. After earning the degree, counselors take an examination to be certified by the American Board of Medical Genetics, which also certifies physicians in this area. Desirable qualities for genetic counselors are an interest in helping others and a high degree of empathy, the ability to understand others' feelings.

For Additional Information
American Genetic
  Association
818 18th Street, N.W.
Washington, DC 20060

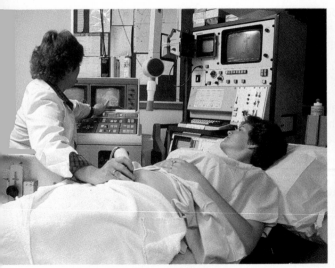
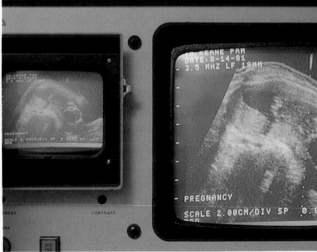

Figure 13–10. Ultrasound waves that penetrate the mother's abdomen (left) are reflected from fetal tissue. The reflected waves form an image of the fetus (right).

produce an image called a *sonogram*. The sonogram can be used to detect an abnormal fetus or one that has died. Ultrasound testing is routinely used to locate the fetus before performing amniocentesis. In this way, the needle can be inserted with less chance of harming the fetus.

A technique called **fetoscopy** (fee TAH skuh pee) allows a physician to actually view the developing fetus. This technique is performed by inserting a slender, hollow needle into the womb through a small incision in the mother's abdomen. The needle is connected to a special microscope. The physician can view the fetus through the needle and determine whether any physical defects exist.

Chorionic villi biopsy is one of the newest means of discovering problems in unborn children. This technique involves the insertion of a thin, hollow tube into the womb with the help of ultrasound equipment. Then a sample is taken of *chorionic villi*, tiny protrusions from a membrane called the *chorion* that surrounds the fetus. These tissues have the same makeup as the fetus and can reveal PKU, sickle-cell disease, and Down syndrome. Chorionic villi biopsy can be done earlier in pregnancy than amniocentesis and provides faster results.

Reviewing the Section

1. Why is genetic counseling important?
2. How is amniocentesis used to detect genetic disorders?
3. What are three uses of ultrasound testing?
4. What is fetoscopy?

Investigation 13: Human Inheritance

Purpose
To observe the occurrence of certain traits

Materials
Mirror

Procedure
1. Make a table like the one shown. Examine the second or middle joint of each finger for hair. If hair is present, circle the letter *H* in the Phenotype column in the Data Table. If no hair is present, circle the letter *h*.
2. Stick out your tongue and try to roll it up at the edges. Look in the mirror to examine the results. If it can be rolled, circle the letter *R*. If it cannot be rolled, circle the letter *r*.
3. Use the mirror to examine your earlobes. *Are they attached directly to your head?* If they are attached, circle the letter *f*. If they hang free, circle the letter *F*.
4. Once again, look in the mirror. *Do you have dimples? A cleft chin?* If you have dimples, circle the letter *D*. If you have a cleft chin, circle *C*; if not, circle *c*.

7. Transfer the data about yourself from your Data Table to the chalkboard where your teacher has prepared a chart to collect this information.
8. Place the class numbers for each of the traits in the appropriate column in the Data Table.
9. Calculate the percentages of each of the contrasting genes for the class.
10. If possible, collect the data for all classes completing this investigation. Compare the percentages for your class with data from other classes.

Analyses and Conclusions
1. What was the ratio of the calculated percentages for each of the traits?
2. What advantage, if any, is there in collecting data from more than one person or class?

Going Further
- Develop an investigation to study a family tree for one of the traits.
- Explain how inherited traits have helped certain organisms to survive.

Trait	Phenotype	Class %	Phenotype	Class %
Finger hair	H		h	
Tongue rolling	R		r	
Earlobes	F		f	
Shape of hairline	V		v	
Dimples	D		d	
Cleft chin	C		c	
Freckles	M		m	

5. *Do you have freckles?* If so, circle *M*; if not, circle *m*.
6. Examine the hairline at the front of your head. If it comes to a point and seems to form a "V," circle the letter *V*. If it is straight across or curved, circle the letter *v*.

Chapter 13 Review

Summary

Geneticists cannot experiment with humans as they can with other organisms. Humans choose their own mates, so breeding cannot be controlled. In addition, humans produce relatively few offspring and take years to produce several generations. As a result, geneticists use pedigrees to study facts about genes, and they use population sampling to measure the frequency of genes in a population. They study identical twins raised apart to understand the influences of heredity and environment.

Through their studies geneticists have learned that cleft chins and polydactyly are each produced by one dominant allele, while PKU and sickle-cell disease each result from two recessive alleles. Multiple alleles determine blood type, and several genes determine skin color. Other traits, such as color blindness, result from genes on the X chromosome. Nondisjunction of chromosomes can result in Down, Turner, and Klinefelter syndromes.

Genetic counselors advise couples of their chances of having a child with a genetic disorder. Amniocentesis, ultrasound testing, fetoscopy, and chorionic villi biopsy may detect disorders in the unborn.

BioTerms

amniocentesis (**208**)
carrier (**205**)
colorblindness (**205**)
chorionic villi biopsy (**210**)
Down syndrome (**207**)
fetoscopy (**210**)
fraternal twins (**200**)
genetic counseling (**208**)
hemophilia (**205**)

Huntington disease (**202**)
identical twins (**200**)
karyotype (**206**)
Klinefelter syndrome (**207**)
monosomy (**207**)
multiple alleles (**203**)
nondisjunction (**207**)
pedigree (**199**)
phenylketonuria (PKU) (**202**)

polydactyly (**202**)
polygenic trait (**204**)
population sampling (**200**)
sex-influenced trait (**206**)
sickle-cell disease (**202**)
trisomy (**207**)
Turner syndrome (**207**)
ultrasound testing (**209**)

BioQuiz   (Write all answers on a separate sheet of paper.)

I. Completion

1. A _____ traces trait inheritance, in chart form, over several generations.
2. Twins produced by the splitting of one fertilized egg are _____ twins.
3. A person with blood type O has the genotype _____ .
4. Recessive _____ traits appear more often in males than in females.
5. _____ are pictures of chromosomes arranged in numbered pairs.

II. Modified True and False

Mark each statement TRUE or FALSE. If false, change the underlined term to make the statement true.

6. Multiple alleles are three or more alleles that can occur on one pair of homologous chromosomes and affect one trait.
7. Height is believed to be a polygenic trait.
8. Several active alleles determine the color of skin.

III. Multiple Choice

9. Klinefelter syndrome is a disorder that results from a) monosomy. b) multiple alleles. c) trisomy. d) antigens.
10. An inherited disease in which phenylalanine accumulates and destroys brain cells is a) PKU. b) sickle-cell disease. c) Down syndrome. d) Huntington disease.
11. Individuals called carriers are a) homozygous dominant. b) heterozygous. c) homozygous recessive. d) mutant.
12. Nondisjunction can cause a) dimples. b) hemophilia. c) baldness. d) trisomy.
13. ABO blood type is determined by a) sex-linked genes. b) a dominant allele. c) multiple alleles. d) several genes.
14. Fluid surrounding a fetus is withdrawn in order to carry out a) fetoscopy. b) ultrasound testing. c) karyotyping. d) amniocentesis.

IV. Essay

15. What information can geneticists learn through population sampling?
16. Why do geneticists use different techniques for studying human heredity than for studying other organisms?
17. Why do scientists study identical twins to learn about the influences of heredity and environment?
18. What is the difference between polygenic inheritance and inheritance by multiple alleles?
19. Why are there so few colorblind females?
20. How can monosomy be detected in a fetus?

Applying and Extending Concepts

1. A couple's daughter is colorblind. Does the father have normal color vision or is he colorblind? How do you know?
2. Suppose that scientists develop techniques for letting parents choose the sex of their child. How might such techniques help prevent genetic disorders?
3. The following pedigree traces the sickle-cell trait through three generations. A circle and a square connected by a horizontal line represent parents. The first line below the parents carries information about their offspring. Dark circles indicate females with sickle-cell disease; dark squares, males with sickle-cell disease. Use this pedigree to determine probable genotypes.

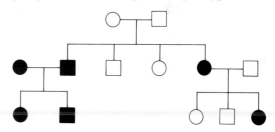

Related Readings

Greenberg, J. "Inheriting Mental Illness." *Science News* 117 (January 5, 1980): 10–12. Evidence of a genetic role in certain mental illnesses is explored in this article.

Halas, D. "The Lethal Legacy." *Science Digest* 90 (August 1982): 28–29. This article discusses genetic errors that cause diseases, including examples of family disorders traceable to one common ancestor.

Powledge, T. A. "Windows on the Womb." *Psychology Today* 17 (May 1983): 36–42. This article discusses procedures that enable physicians to diagnose genetic disorders before birth.

BIO*TECH*

Gene Therapy

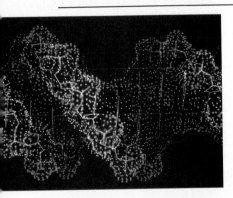

Computer-generated image of DNA illustrates its double helix structure (above). Glowing bands of electrophoresis gel indicate components of DNA (below).

Electrophoresis process reveals the sequence of genes in DNA.

A single defective gene is one of nature's tiniest mistakes, but it can cause a life-threatening problem for a person with a genetic disease. Most genetic diseases cannot be treated medically. Even the few that can be treated, such as hemophilia and sickle-cell disease, present serious problems. Painful and expensive treatment can keep patients alive, but it cannot cure them.

So what can be done about genetic diseases? At present, the only way to cure them is through **gene therapy**—a technique for replacing a defective gene with a normal one.

In simple terms, gene therapy involves isolating the "good" gene from normal cells and cloning, or making many copies of, the "good" gene. These steps are already routine techniques of recombinant DNA technology. Mastering of the steps described below has presented major challenges in perfecting gene therapy.

Genetic engineers think that a promising new technology using viruses may be the key to making gene therapy possible. In this technology, a "good" gene is first inserted into a virus. The virus then invades the patient's cells, serving as a *vector*—a vehicle that transfers the "good" gene into the patient's cells.

New technology using viruses may provide the key to gene therapy.

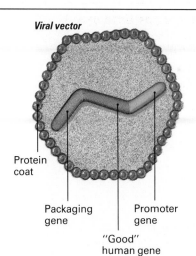

Viral vector

Protein coat

Packaging gene

Promoter gene

"Good" human gene

The viruses used to build vectors consist of a core of RNA surrounded by a protein coat. To build a vector, geneticists first extract just a few of the viral genes, specifically the "promoter" and "packaging" genes. The promoter genes switch on adjacent genes and make them active. The packaging genes direct the entry of viral RNA into the protein coat. The researchers discard the rest of the genes, including those for making the protein coat.

Genetic engineers next splice the "good" human gene in place between the promoter genes and packaging genes. Then many copies of this hybrid DNA are produced through cloning. The cloned DNA is added to a culture of animal cells, which produce hybrid RNA.

The last step in producing the vector is the packaging of this hybrid RNA in a protein coat. Without such a coat, it could not leave the animal cells and invade human cells. This problem is solved by adding to the cell culture an RNA virus from which the packaging genes have been removed. Without these genes, the modified virus can produce protein coats but cannot use them. The hybrid RNA, however, does have packaging genes. Thus it enters the coat and becomes the vector.

When the vector is given to a patient, it injects the hybrid RNA into the cells while leaving the protein coat outside. Once inside the patient's cells, the RNA can never leave because it lacks the coat genes.

Through a process of reverse transcription, the cell produces a DNA copy of the hybrid RNA that becomes incorporated into one of the patient's chromosomes. Thus the "good" gene becomes a permanent part of the patient's genetic material.

Researchers have already produced vectors and used them in gene therapy in animals, but the process is not yet considered safe for humans. However, rapid progress in genetic technology should soon make gene therapy a reality.

Chromatography column is used to separate the components of a nucleic acid.

14

Applied Genetics

Morgan horses, products of special breeding

Introduction

Today's rapidly expanding knowledge of how genes work is creating opportunities for **applied genetics,** the practical use of genetic knowledge. For example, plant and animal breeders use information about genetics to develop improved or entirely new varieties of organisms. One such organism is the Morgan horse shown in the picture.

The future of applied genetics is especially exciting because scientists have found ways to produce improved organisms by manipulating genes. The most revolutionary technique involves transferring genes from one organism to another. The resulting organisms can serve many purposes—from making substances that fight many human diseases to producing new, less expensive sources of energy.

Controlled Breeding

Section Objectives

- *List* several ways in which breeders have adapted plants and animals to human needs.
- *Describe* the process of mass selection.
- *Compare* inbreeding and hybridization.

The effort to improve plants and animals is much older than the science of genetics. In fact, the earliest attempts at breeding probably occurred about 10,000 years ago when people first domesticated plants and animals. At that time, people must have noticed that some of the offspring of strong or productive animals had the same desirable traits as their parents. Breeders picked the animals with the traits they wanted and mated them. Using the same reasoning, farmers saved seeds from the hardiest and most productive plants and planted them for the next year's crop.

The process of selecting individuals with desirable traits to produce the next generation is called **controlled breeding.** In recent decades breeders have applied the knowledge of genetics to controlled breeding. In this way, they have been able to adapt plants and animals to many human needs in a much shorter period of time than was previously possible.

14.1 Goals of Breeders

People who breed plants and animals have many goals. For example, breeders working to increase the food supply have developed animals that produce more offspring. Plant varieties have been developed that produce more seeds and can better withstand disease and harsh weather. These breeders have also developed more nutritious plants, such as corn that contains more protein.

Other breeders have bred plants that are better suited to the mechanized methods of modern farming. One such plant is a variety of tomato that has a thick skin to resist bruising by harvesting machines. This fruit is also uniform in shape so that more of the product will fit into a packing crate. Other breeders develop plants and animals that are bred for qualities that have nothing to do with food. These include fast race horses, purebred dogs, and new strains of ornamental plants.

14.2 Techniques of Controlled Breeding

Modern breeders use combinations of several methods to develop plants and animals with desired traits. *The most important methods of controlled breeding are mass selection, hybridization, and inbreeding.* Each of these traditional methods has been made more effective through the application of genetic principles.

Figure 14–1. This wheat was developed from wild wheat through the process of controlled breeding.

Figure 14-2. Luther Burbank became famous for his success in breeding new plant varieties by mass selection. The Shasta daisy is one of his best-known accomplishments.

Mass Selection   The process of raising a great many plants and animals and selecting the best in each generation for further breeding is called **mass selection.** Luther Burbank, an American plant breeder of the late 1800s and early 1900s, used mass selection to develop more than 800 new or improved fruits, vegetables, flowers, and grains. Among his famous plant varieties are the Shasta daisy and the Burbank potato. Today large corporations and government agencies fund enormous breeding stations so that mass selection is carried out efficiently.

Inbreeding   Frequently breeders want to establish pure lines, or populations of plants or animals made up of genetically similar individuals. Such lines usually *breed true* for certain traits, which means that offspring are almost identical to their parents in these traits. By developing pure lines, breeders preserve desirable traits. The pure lines also serve as known quantities in breeding experiments.

Pure lines are established by following mass selection with **inbreeding,** a method that involves mating genetically similar individuals. With animals, close relatives such as brothers or

Spotlight on Biologists: Severo Ochoa

Born: Luarca, Spain, 1905
Degree: M.D., University of Madrid

By discovering how to synthesize RNA, biologist Severo Ochoa has contributed greatly to scientists' knowledge of genes. The far-reaching effects of Ochoa's basic research have influenced all areas of genetics.

Although a Spaniard by birth, Ochoa has carried out most of his research in the United States. Ochoa was trained in Spain and in Germany. He came to the United States during

World War II and shortly thereafter joined the faculty of the New York University College of Medicine. While engaged in other research, he accidentally discovered a bacterial enzyme that synthesizes RNA. With that information he became the first scientist to produce RNA in the laboratory.

Armed with his synthetic RNA, Ochoa helped to break the genetic code. He investigated the order of RNA nucleotides, then figured out how different nucleotide sequences direct the order in which

amino acids are lined up in molecules.

For his outstanding discoveries in the field of RNA research, Ochoa shared the 1959 Nobel Prize for Medicine or Physiology with Arthur Kornberg. Ochoa is currently affiliated with the Roche Institute of Molecular Biology in Nutley, New Jersey.

sisters are mated over several generations. With plants, inbred varieties are produced by self-pollination. After many generations, inbreeding produces individuals that are homozygous for most traits. Some of these traits are desirable. However, others may be harmful, caused by homozygous recessive genes that are not expressed in a heterozygous individual. For instance, a gene causing deafness has become common in Dalmatians—an inbred pure line variety of dog.

Hybridization The opposite of inbreeding is **hybridization,** a method of crossing two different species, breeds, or varieties. When inbred varieties are crossed, the resulting hybrids may show every possible combination of traits of the parent species. Thus, some hybrid offspring inherit the best traits of each parent and are larger, hardier, and more productive than either parent. This improvement in quality as a result of hybridization is known as **hybrid vigor.**

Over the years some of the greatest breeding advances have come about through hybridization. Santa Gertrudis cattle were developed by mating shorthorn beef cattle with heat- and insect-resistant Brahman cattle from India. The Santa Gertrudis displays the best traits of each parent. Hybridization has also produced new crops such as *triticale,* a cross between wheat and rye. Triticale is more resistant to drought and more nutritious than either wheat or rye.

Hybridization has its drawbacks, however. In some cases offspring inherit the worst traits of each parent. Other hybrids, such as mules, are sterile. Even in hybrids that can reproduce, hybrid vigor may disappear if hybrids are crossed over many generations.

Breeders may combine hybridization with inbreeding and mass selection to produce new true-breeding varieties. On the other hand, hybrid plants can be reproduced asexually. Roses and orchard fruits, for example, are sometimes propagated by *grafting,* or joining a branch of a new variety to the stem and roots of an existing plant. All growth from this branch has the traits of the new variety. In other instances, such as hybrid corn, the original crosses are repeated to produce new hybrid seeds for each year's crop.

Figure 14–3. Santa Gertrudis cattle are the result of successful hybridization. They are resistant to heat and insects and also produce high quality beef.

Reviewing the Section

1. In what ways have plants been improved by breeders?
2. What is mass selection?
3. How do inbreeding and hybridization differ?

Other Genetic Techniques

- *List* the steps involved in cloning plants from a single cell.
- *Explain* how polyploidy occurs in plants.
- *Name* the major steps in cell fusion.
- *Summarize* the methods and potentials of genetic engineering.

Controlled breeding is just one means of developing organisms adapted to human needs. Biologists have also learned to produce identical copies of desirable organisms. ***In the last 30 years, biologists have found new ways to change the genetic makeup of an organism or its offspring by artificial means.*** Among these techniques is a process of transferring genes from one organism to another, a revolutionary technology promising major changes in many fields.

14.3 Cloning

Biologists can now duplicate certain organisms by means of **cloning,** which is the production of organisms with identical genes. Some methods of cloning are simple. Just cutting leaves from a plant, rooting them in water, and planting them results in **clones**—organisms or groups of cells developed from one parent and genetically identical to one another. Grafting is another form of cloning.

A more complex kind of cloning involves growing a complete plant from one *somatic* cell. This is any cell not normally involved in reproduction. First tissues from a parent plant are kept alive in a nutrient solution. Then a single cell is removed and treated with substances that cause it to grow and divide. Eventually, a new plant identical to the parent develops.

Scientists have also cloned a few animals. Frogs have been cloned by replacing the nuclei of frog eggs with nuclei from tadpoles of another kind of frog. Rabbits and mice have also been cloned. Such experiments suggest that cloning many vertebrates is possible. However, that development is not expected soon, in part because cells with implanted nuclei from adult organisms do not grow into normal adults.

Figure 14–4. In one type of cloning, new banana plants are produced by culturing tissue from a single banana plant.

14.4 Polyploidy

Artificially changing the quantity of DNA is another way to improve organisms. One such change is called **polyploidy** (PAHL ih ploy dee), a condition in which an organism has more than two complete sets of chromosomes. A diploid organism has two sets of chromosomes—one set from each parent. A polyploid organism may have three, four, or more sets. For example, diploid strains of wheat have two sets of 7 chromosomes, or a total of 14 chromosomes. Durum wheat, which is used to make pasta, has four sets, or a total of 28 chromosomes.

▶ Building a Better Plant

A method of manipulating genes developed in the 1970s may make it possible to develop hybrids from distantly related plants. This process, called **cell fusion,** joins cells from two very different kinds of plants.

The key to this process is removing the cell walls from the cells that will be fused. These cells, when stripped of their cell walls, are called *protoplasts.* The protoplasts from two species of plants, such as

potatoes and tomatoes, can be combined as shown in the photograph. Then the fused cells are treated with chemicals that cause a new cell wall to

develop. Finally, the cell is treated with substances that stimulate growth. The cells of the new plant contain two complete sets of chromosomes from each parent, making it a hybrid polyploid.

Laboratories are performing many similar experiments that may result in better plants. For instance, the protoplasts in the photograph may result in a potato plant with the tomato's natural resistance to disease.

Polyploidy does not occur in animals since major changes in the number of chromosomes are fatal. However, polyploidy is common in plants and often produces varieties that are larger, hardier, and sometimes more productive than diploid varieties of the same species.

Polyploid plants develop when duplicated chromosomes fail to separate during meiosis. One daughter cell then receives two sets of chromosomes. If this cell is fertilized by another diploid gamete, the fertilized cell will have four sets of chromosomes instead of the normal two.

Biologists produce polyploidy by using *colchicine,* a chemical that prevents cells from dividing after the chromosomes duplicate. The result is a cell with twice the diploid number of chromosomes. Today, breeders use colchicine to grow polyploid cabbages, blueberries, and other plants.

14.5 Recombinant DNA

The most revolutionary means of changing DNA is through **genetic engineering,** the process of transferring DNA segments from one organism into the DNA of another species. The new strand of DNA containing DNA from both species is called **recombinant DNA.** Organisms containing recombinant DNA

Figure 14–5. The large apple in the photograph is an example of fruit produced by a polyploid plant.

have already produced medicines to combat certain human diseases. In the future, recombinant DNA may dramatically change agriculture, food processing, and energy production.

Producing Recombinant DNA

The organism most often used in recombinant DNA technology is *Escherichia coli,* or *E. coli,* a bacterium normally found in the human intestine. Because this bacterium divides about once every 20 minutes, many new bacteria containing recombinant DNA can be produced quickly. These bacteria will then produce the protein coded for by the transplanted DNA. In this way, *E. coli* can be made to generate proteins normally produced by other organisms.

As Figure 14–7 shows, the first step in producing recombinant DNA is to extract DNA from a donor organism. Then enzymes called *restriction endonucleases* are used to remove a gene from the DNA. These enzymes cut the DNA molecule only at certain points and leave a few unpaired bases at one end of each DNA strand. These unpaired bases are called "sticky ends" because they will bond to complementary unpaired bases at the end of another DNA strand.

Next, small circles of DNA called **plasmids** (PLAZ mihdz) are extracted from the *E. coli* cells. The plasmids are broken open with the same enzymes used to remove the gene from the donor DNA. Consequently, the ends of the plasmid DNA and

Figure 14–6. This illustration shows the major steps involved in recombinant DNA technology. The small triangles indicate the points at which enzymes are used to cut the strands of DNA.

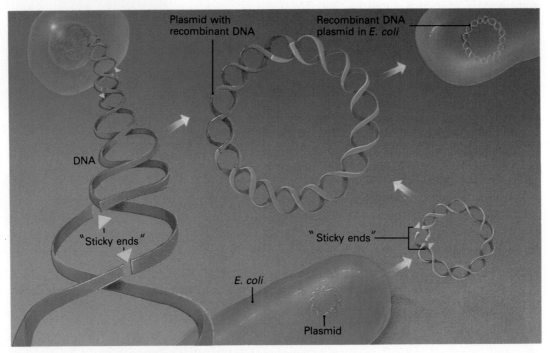

Plasmid with recombinant DNA

Recombinant DNA plasmid in *E. coli*

DNA

"Sticky ends"

"Sticky ends"

E. coli

Plasmid

the donor DNA are complementary. For example, the sequence adenine-cytosine-thymine on the DNA segment will bond to the sequence thymine-guanine-adenine on the plasmid. Recombination occurs when the pieces of donor DNA bond to the ends of the plasmid strands. With the addition of another kind of enzyme, the plasmids with new sections of DNA inserted are made to seal themselves into rings.

Finally, the plasmids are mixed with *E. coli* cells, which are chemically treated so the plasmids travel through their cell membranes. Not every bacterium picks up a plasmid containing recombinant DNA, but only those that have picked up recombinant DNA are allowed to multiply. When the bacteria divide, their plasmids are passed on to daughter cells. Soon a large colony of bacteria with recombinant DNA results. The new genes produce proteins coded for by the new section of DNA.

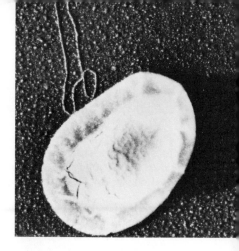

Figure 14–7. This photograph shows a plasmid entering an *E. coli* bacterium magnified 100,000 times.

The Importance of Recombinant DNA When recombinant DNA experiments began in the early 1970s, several scientists asked that such experiments stop until the potential dangers of this technique were evaluated. These scientists feared that cells containing harmful genes might escape from laboratories and cause widespread disease. In 1976 the United States government issued safety guidelines for laboratories developing recombinant DNA. These guidelines required special equipment and methods to prevent problems from occurring.

Since 1976 scientists have made remarkable advances in recombinant DNA technology. In 1978 they first used bacteria to make human *insulin*, a hormone that controls the level of sugar in the blood. Insulin is used to treat individuals with *diabetes mellitus*, a disease in which not enough insulin is produced so the blood sugar rises to dangerous levels. In 1980 scientists produced a bacterium that made human *interferon*, a protein that prevents the multiplication of viruses. Experiments indicate that interferon stops the growth of some cancerous tumors.

In the future, bacteria containing recombinant DNA may produce many substances. This technology may provide other amazing results, such as wheat plants that produce their own nitrogen fertilizer and cures for genetic diseases.

Q/A

Q: *How do experimenters know which bacteria contain recombinant DNA so they can allow only those bacteria to multiply?*

A: Along with segments of DNA, the experimenters attach marker genes to the plasmids. These genes provide resistance to *streptomycin,* an antibiotic that kills certain bacteria. If the recipient bacteria are grown on a medium containing streptomycin, the colonies that survive must contain marker genes and therefore recombinant DNA.

Reviewing the Section

1. Explain how a tree can be grown from a single somatic cell.
2. Why is polyploidy not found among animals?
3. What is recombinant DNA?

Investigation 14: Developing a New Breed of Dog

Purpose
To simulate a cross between two varieties of dogs that results in a new breed having the best characteristics of each variety

Materials
Paper, pencil

Procedure
1. Study the two tables below.

Going Further
- Research some of the breeds of dogs that have been developed through selective breeding techniques such as those described in this investigation.
- Compile a list of sporting dogs and the characteristics that determine each breed.
- Research some other organisms that have been developed through selective breeding techniques.

Characteristic	Phenotype	Genotype
Dog A		
Coat color	Red	RR
Hunting instinct	Does not point at birds	hh
Size	Tall	S^TS^T
Dog B		
Coat color	White	rr
Hunting instinct	Points at birds	HH
Size	Small	S^SS^S

2. Using a Punnett square, diagram the result of a cross between Dog A and Dog B. Show the genotypes and phenotypes of the F_1 generation.
3. Using a Punnett square, diagram a cross between two of the F_1 generation dogs. List all of the possible phenotypes of this cross.

Analyses and Conclusions
1. Is it possible to produce a dog having the following characteristics: red coat, hunting instinct, and medium size?
2. Will there be offspring in the F_2 generation of the same genotype as the F_1 generation? Explain your answer.
3. Explain why the F_2 generation will have fewer white dogs than red dogs.
4. Would it be possible to determine if a red, medium-sized hunting dog is purebred for these traits?

Chapter 14 Review

Summary

Progress in the science of genetics has produced not just new knowledge but also applications that benefit people. Plant and animal breeders now apply genetic information to methods of controlled breeding which have long been used. Mass selection involves raising large numbers of plants or animals and mating the best in each generation. Inbreeding is the mating of genetically similar individuals to produce organisms that are homozygous for desirable traits. In hybridization, two different species or varieties are crossed to produce heterozygous offspring with the best traits of each parent. By using genetic information, breeders can now more quickly achieve results from these methods.

Biologists have also learned to produce identical copies of organisms and to make artificial changes in DNA. Use of the chemical colchicine, for example, produces polyploid plants which have more chromosomes than the normal diploid number. These plants are often larger, hardier, and more productive than diploid varieties. Through genetic engineering, scientists have also created recombinant DNA, which contains DNA from two species. Bacteria containing recombinant DNA have already produced human insulin and human interferon. One day recombinant DNA may be used to manufacture other drugs as well as to provide new sources of food.

BioTerms

applied genetics (216)
cell fusion (221)
clone (220)
cloning (220)
controlled breeding (217)

genetic engineering (221)
hybridization (219)
hybrid vigor (219)
inbreeding (218)
mass selection (218)

plasmid (222)
polyploidy (220)
recombinant DNA (221)

BioQuiz   (Write all answers on a separate sheet of paper.)

I. Completion

1. Plants with more than the usual number of sets of chromosomes are said to be _____.

2. Interferon prevents the multiplication of _____.

3. The technique of transferring DNA from one organism to another organism is known as _____.

4. Hybrid polyploids are created by means of _____.

5. _____ DNA contains DNA from two species.

II. Modified True and False

Mark each statement TRUE or FALSE. If false, change the underlined term to make the statement true.

6. Inbreeding is considered the opposite of cloning.

7. Hybridization can be combined with inbreeding to produce new true-breeding varieties.

8. The plasmids found in bacteria are circles of DNA.

9. Hybridization is a form of controlled breeding.

III. Multiple Choice

10. Using cloning techniques, a plant can be grown from a single a) somatic cell.
    b) reproductive cell. c) gene.
    d) haploid cell.
11. Mass selection is a form of a) cloning.
    b) controlled breeding. c) genetic engineering. d) mutation.
12. Hybrid vigor results from a
    a) heterozygous genotype.
    b) homozygous genotype. c) polyploid genotype. d) recombinant DNA genotype.
13. Genetically identical organisms result from a) grafting. b) cloning.
    c) colchicine. d) hybridization.
14. A section of DNA with the unpaired bases ATC would bond with the plasmid base sequence a) TAG. b) TAC.
    c) GCT. d) CGA.
15. Human interferon has been produced by
    a) cloning. b) mass selection.
    c) colchicine. d) recombinant DNA technology.

IV. Essay

16. What is the purpose of inbreeding?
17. What are two of the drawbacks of hybridization?
18. How do breeders induce polyploidy in plants?
19. What enables a section of DNA to bond with a broken plasmid?
20. Why is recombinant DNA technology considered revolutionary?

Applying and Extending Concepts

1. Imagine that you are a plant breeder who has been asked to develop the ideal plant for space missions, including space stations that are limited in area. Name some traits that this ideal plant might be engineered to have.
2. Seedless grapes, oranges, and watermelons have an odd number of sets of chromosomes. Do library research to learn how plant breeders produce fruits that have no seeds. Describe this process to the class.
3. Citrus growers are at the mercy of cold spells that sometimes severely damage crops. Imagine that you are a breeder searching for a tree that can withstand the cold and still produce large, flavorful oranges. You have some seeds from a hardy species that produces small, bitter fruit. Explain how you could develop a tree with desirable features using the methods of controlled breeding and recombinant DNA technology. List any problems or benefits associated with each method.

Related Readings

McAuliffe, K. and S. McAuliffe. "Keeping Up with the Genetic Revolution." *The New York Times Magazine* (November 6, 1983): 40–97. This article describes current discoveries in genetics and their application to the cure of human diseases.

Rensberger, B. "Tinkering with Life." *Science 81* 2 (November 1981): 44–49. This article discusses developments in the field of molecular biology and the legal and ethical aspects of gene manipulation.

Solomon, S. "Green Genes: Bioengineering New Foods." *Science Digest* 91 (January 1983): 54–59. This article explains how foreign DNA is inserted into plant cells and outlines how plants that are genetically engineered may soon produce drugs and petroleum.

History and Diversity of Life

15

Changes Through Time

Spiral galaxy

Introduction

How was the universe formed? How did life begin on Earth? The answers to these questions remain a mystery. In the last 200 years, however, scientists have collected a great deal of evidence that at least suggests answers. For example, evidence indicates that Earth was formed about 4.6 billion years ago and has undergone many changes since then. The evidence also indicates that living things have changed during the 3.5 billion years they have inhabited Earth.

In this chapter, you will study the current scientific theories for the origin of the universe, Earth, and life on Earth. You will also learn about the variety of evidence indicating that living things have changed over time, a process that is summarized in the theory of **evolution.**

The Beginning of Life

A basic principle of biology states that cells arise only from existing cells. But where did the first cell come from? It is not possible to travel back in time to learn the answer. So scientists have developed theories concerning the beginning of life—and the universe—based on evidence gained from observation and laboratory work.

15.1 Origin of the Universe

Several theories exist to explain the origin of the universe. However, most scientists today favor the big-bang theory proposed in 1927 by Belgian astronomer Georges Lemaitre. The big-bang theory states that the universe began about 15 billion years ago as a dense concentration of matter smaller than a speck of dust. For reasons that today's physicists do not completely understand, this concentration of matter exploded violently. Energy and new forms of matter created by this explosion spread into space. Eventually, gravitational attraction drew these bits and pieces of matter together, forming stars and planets.

15.2 Origin of Earth

Billions of years ago, clouds of molecules, mostly composed of hydrogen gas, began to condense to form stars. One of these stars was the sun. Earth and the other planets of this solar system developed about 4.6 billion years ago out of gases and dust orbiting the sun. At first the dust cloud was very cool and gaseous. Soon gravity caused local accumulations of dust to form planets. The solid matter then formed layers according to density. Heavy elements, such as iron, collected into a dense core. Lighter elements formed a solid crust at the earth's surface. Radioactive decay deep in the earth raised the temperature of the core, and molten rock called *magma* rose to the earth's surface. Volcanic eruptions forced the hot gases such as nitrogen and water vapor out of the magma. These gases, in which there was no free oxygen, then formed an atmosphere.

15.3 Origin of Life

Although scientists will never know exactly how life began, they have gathered general evidence and developed a model of how it could have happened. Scientists agree that the Earth of 4.5 billion years ago was far different from the Earth of today.

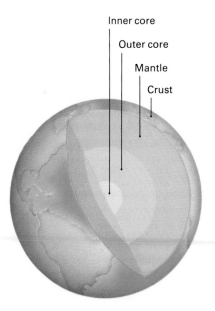

Inner core
Outer core
Mantle
Crust

Figure 15–1. The cross section above shows the layers of the earth. The inner core is hot, solid metal; the outer core is molten metal; the mantle is solid rock; and the outer layer, the crust, is also rock.

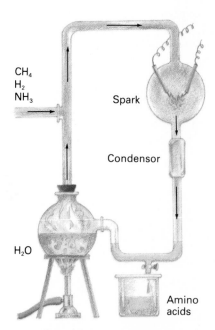

CH₄
H₂
NH₃

Spark

Condensor

H₂O

Amino
acids

Figure 15–2. The diagram above shows how Stanley Miller demonstrated that organic compounds could be formed from inorganic materials.

Q/A

Q: *Are there hypotheses other than Oparin's to explain the origin of simple organic molecules?*

A: Yes. Some scientists estimate that millions of tons of organic molecules may have been formed on comets and deposited when they collided with Earth.

They think much of the surface was covered with hot seas, kept hot by the molten earth. Steam from these seas rose into the atmosphere and formed huge clouds. The water vapor in these heavy clouds slowly cooled, condensed, and fell to Earth in violent rainstorms, accompanied by lightning. On land, active volcanoes spouted gas, steam, and magma.

In 1936 the Russian scientist Alexander Oparin developed a model of how life might have begun on Earth. According to Oparin's hypothesis, the early atmosphere and oceans probably contained ammonia (NH_3), water vapor (H_2O), methane (CH_4), and hydrogen (H_2). Lightning, heat from the earth, and ultraviolet light from the sun provided the energy to split these molecules into atoms. The atoms bonded together to form small *organic* compounds, carbon-containing molecules characteristic of living things. These molecules accumulated in the oceans and formed a kind of "organic soup." After a great length of time, they combined into globules of molecules that could reproduce themselves. These molecules were the first life on Earth.

In 1953 an American biochemist named Stanley Miller tested Oparin's hypothesis. He filled the apparatus shown in Figure 15-2 with carbon dioxide, water vapor, methane, and ammonia. Then he passed a continuous electric spark through the mixture. A liquid trap collected any molecules that might have formed in the apparatus. After seven days, Miller analyzed the liquid and found amino acids, the building blocks of proteins. Miller's experiments showed that amino acids could have formed in the ancient Earth. But life is more than just amino acids. It took hundreds of millions of years after the origin of the earth for cells to form.

The first cells probably resembled anaerobic bacteria, since no free oxygen existed in the atmosphere. They were probably also heterotrophs living off organic compounds in their environments. Cells capable of photosynthesis, such as blue-green algae and certain bacteria, developed later, about 3 billion years ago. These organisms not only made their own food but also gave off oxygen. This oxygen gradually accumulated in the atmosphere and paved the way for more complex forms of life that were capable of utilizing oxygen for their energy needs.

Reviewing the Section

1. What is the big-bang theory?
2. According to Oparin's model, where did the first life forms come from?
3. Describe what the first cells were probably like.

Evidence of Evolution

Organisms have changed dramatically since they appeared on Earth about 3.5 billion years ago. The first simple cells gave rise to more varied and complex organisms, changing in size and structure in response to environmental pressures. Scientists have found evidence of such changes in fossils and in the characteristics of today's living things.

15.4 The Fossil Record

A **fossil** is any preserved part or trace of an organism that once lived. You have probably seen fossils, such as dinosaur skeletons, in museums. Fossils are important sources of evidence that organisms have changed over time.

Fossil Evidence A fossil is formed when all or part of an organism is buried before it can be eaten or before it decays. Most fossils are found in **sedimentary** rock. Sedimentary rocks are formed from *sediments* such as mud, silt, and sand that have been deposited in layers on top of one another. Particles in these sediments are subjected to great pressure. They become cemented together to form sandstone, limestone, shale, and other types of sedimentary rocks.

   Organisms buried in sediments may become fossilized. Soft body structures, such as feathers or leaves, may form **imprints**—that is, impressions in the developing rock. Teeth, shells, bones, and other hard body parts may form **molds**, which are depressions in the rock shaped like the organism

Figure 15–3. The photographs below show three types of fossils: a bee in amber (left), a cast of a trilobite (center), and an imprint of a leaf (right).

Q/A

Q: *How well preserved were the woolly mammoths found frozen in Siberia?*

A: The woolly mammoths were so well preserved that some of their meat was served for dinner at a meeting of paleontologists.

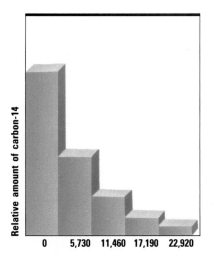

Figure 15–4. The bar graph above shows the rate of decay of carbon-14. What percentage of the atoms have become stable after 11,460 years?

parts. Sometimes the original material decomposes and the resulting molds are filled in with another material, forming **casts.** Other times, hard parts of an organism are replaced by minerals, creating fossils called **petrified fossils.**

Occasionally entire organisms are preserved intact. Many small organisms have been trapped in *amber,* the sap of trees that hardens into a transparent covering. Other, larger organisms have been trapped in tar, which prevented decay.

Determining the Age of Fossils Fossils tell us what kinds of life existed in the past. But knowing the age of fossils enables scientists to understand the sequence in which organisms appeared and how they changed over time.

The age of fossils can be determined by their positions in sedimentary rock. Sediments are built up in layers, the newer layers on top of the older layers. In undisturbed sedimentary rock, the most recent fossils are found in upper layers, and older fossils in lower layers. By comparing the sediment position of fossils, scientists can determine their relative ages.

Scientists can determine the age of fossils more exactly by using *isotopes.* Isotopes are those atoms of an element that have different *atomic masses.* For example, a nucleus of carbon with six protons and six neutrons is the isotope called carbon-12, or ^{12}C. Carbon-13 has six protons and seven neutrons; ^{14}C has six protons and eight neutrons. The isotopes of most value to scientists are **radioactive isotopes.** These isotopes have unstable nuclei that gradually break down into stable elements called **decay elements.** For example, uranium-238 decays into lead-206. Carbon-14 decays into nitrogen-14. Radioactive isotopes are useful because they decay at a constant, known rate. The time it takes for one-half of the isotopes in radioactive material to decay is called the **half-life** of the isotope. The half-life of ^{14}C is 5,730 years.

Carbon-14 is often used to date fossils because it is incorporated into the molecules of the organisms while they are alive. When organisms die, they stop taking in both ^{14}C and ^{12}C. The ^{14}C in their bodies decays over time, but the ^{12}C remains stable. As a result, scientists can determine the age of a fossil by comparing the ^{14}C and ^{12}C in it. The ratio of ^{14}C to ^{12}C in living tissue of today and in living tissue of the past is assumed to be the same. Therefore, a fossil containing half the amount of ^{14}C of the organism's living tissue would be 5,730 years old. A fossil with one-fourth the ^{14}C of living tissue would be 11,460 years old (5,730 plus 5,730). Carbon-14 is used to date relatively young fossils. Uranium-238, with a half-life of 4.5 billion years, is used to date older fossils.

Highlight on Careers: Paleontologist

General Description

A paleontologist is a scientist who studies fossils and other remains in the earth's rock layers. By studying these clues, paleontologists assemble evidence of changes in a species or in an area over thousands or millions of years. The paleontologist deals with great spans of time beyond normal human understanding. Thus, part of the job of a paleontologist is helping people realize how old the earth actually is.

Many paleontologists work for universities, teaching and acquiring rock samples. Paleontologists study these samples for visual evidence of the rock's age and the environment in which it was formed.

Vertebrate paleontologists use fossil bones to make drawings of how extinct animals may have looked. Paleontologists who work for oil companies examine rock samples to determine if certain sites could contain oil and gas deposits.

Career Requirements

Most jobs in universities require a Ph.D. degree in paleontology, geology, botany, or zoology. Entry-level positions in the oil industry are sometimes available to those with an

M.S. degree in geology and related course work or field experience in paleontology.

For Additional Information
Paleontological Research
  Institution
1259 Trumansberg Road
Ithaca, NY 14850

Interpreting the Fossil Record Scientists in various fields interpret the fossil record. **Paleontologists** (pay lee ahn TAHL uh jihsts) look for and study fossils. **Geologists** use fossils to explain the history of Earth. **Physical anthropologists** (an thruh PAHL uh jihsts) study human fossils.

The work of these scientists has revealed many changes in Earth during its long history. For example, fossils of palm trees have been found in the Antarctic, suggesting that the climate there was once warm. Fossils have also revealed tremendous changes in organisms.

Based on the information collected by dating fossils, scientists have developed a geologic time scale of the history of Earth (Table 15–1 on page 234). This scale is a calendar that allows scientists to communicate about events that have occurred since Earth was formed. The scale for events since life began is divided into four **eras,** vast spans of time millions of years long. The eras are divided into periods and **epochs** (EHP uhks).

Table 15–1: Geologic Time Scale

Era	Period	Epoch	Began (millions of years ago)	Significant Events
Cenozoic	Quaternary	Recent	0.025	Complex human societies arise.
		Pleistocene	1.75	The Ice Ages begin. Mammals dominate.
	Tertiary	Pliocene	14	Mammals, birds, modern sea life appear.
		Miocene	26	Mammals diversify on land. Grasslands spread.
		Oligocene	40	Primitive apes appear. Elephant, horse, and camel develop.
		Eocene	55	Large mammals appear. Grasslands and forests present. Fruits and grains develop.
		Paleocene	65	Small mammals begin to take over the land.
Mesozoic	Cretaceous		130	Flowering plants and trees appear. Dinosaurs die out at end.
	Jurassic		180	Dinosaurs abundant. First feathered birds and mammals appear.
	Triassic		225	Insects and cone-bearing trees plentiful. Giant reptiles appear.
Paleozoic	Permian		275	First seed plants appear. Fish, reptiles, and amphibians are plentiful.
	Carboniferous	Pennsylvanian	310	Age of amphibians. Large fern trees, swampy forests; reptiles appear.
		Mississippian	345	Coral reefs formed. Extensive land forests develop.
	Devonian		405	Fish are common. First swampy forests grow. Amphibians and insects appear.
	Silurian		435	First land plants develop. Fish and shell-forming sea animals appear.
	Ordovician		480	Algae and shelled animals are common. First vertebrates appear in sea.
	Cambrian		600	Clams, snails appear. Algae are common.
Precambrian			4.5 billion	Few fossils. Bacteria and algae predominate.

History of Life on Earth The earliest traces of life suggest the presence of microorganisms about 3.5 billion years ago during the earliest era, the Precambrian. The first cells with nuclei appeared about 1.5 billion years ago. These evolved into simple forms of life that gave rise to more complex forms. Land plants, amphibians, and insects came into existence between 435 million and 225 million years ago, during the Paleozoic era. Dinosaurs dominated the Mesozoic era, which began about 225 million years ago. The most recent era—the Cenozoic—began about 65 million years ago. It was not until the last several million years of this era that human beings appeared on Earth.

15.5 Other Evidence of Evolution

Fossils are one major source of evidence supporting evolution. However, the study of living organisms has also revealed important evidence of relationships among organisms.

▶ Biochemical Evidence of Evolution

The most recent evidence supporting evolution has come from **comparative biochemistry,** the study of molecules that make up different living things. Biochemists have found that all living things share certain substances made of proteins. These proteins are composed of various combinations of smaller substances called amino acids. The order in which amino acids are assembled into proteins is determined by DNA.

Cytochrome c, a protein used in aerobic respiration, is one of the proteins researchers studied in many different organisms. The computer-generated image

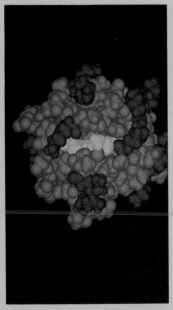

above shows cytochrome c. Researchers paid special attention to the sequences of amino acids in cyto-

chrome c from different organisms. They were able to calculate the number of DNA segments, called nucleotides, needed to account for the differing sequences. They found, for instance, that in human beings and monkeys only 1 nucleotide differed. In human beings and turtles, however, 19 nucleotides were different. Such information suggests that organisms with few differences in DNA nucleotides and therefore in cytochrome c have a close evolutionary relationship. Those organisms with many different nucleotides, therefore, have a more distant evolutionary relationship.

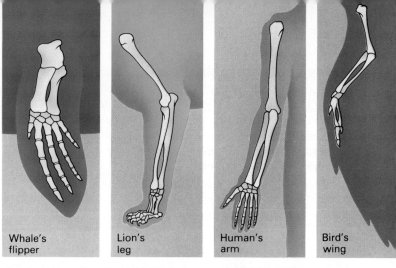

Figure 15–5. Vertebrate limbs are homologous structures. The flipper of a whale, leg of a lion, hand and arm of a human, and wing of a bird differ in size and shape, but they are alike in the number and arrangement of bones.

Whale's flipper | Lion's leg | Human's arm | Bird's wing

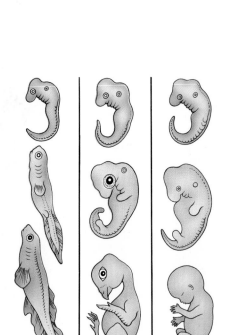

Fish   Chicken   Human

Figure 15–6. Similarities in the development of fish, chicken, and human embryos suggest an evolutionary relationship.

Anatomical Evidence   Body parts with the same basic structure are called **homologous** (hoh MAHL uh guhs) **structures.** *Homologous structures found in different organisms suggest that these organisms have a common ancestry.* In homologous structures, the size and shape of each limb is different, but the number and arrangement of bones is similar.

Body parts that are similar in function but not in basic structure, such as the wings of birds and the wings of insects, are called **analogous** (uh NAL uh guhs) **structures.** These body parts do not indicate an evolutionary relationship.

Other body structures that do provide evidence of evolution are **vestigial** (vehs TIHJ ee uhl) **structures.** These are structures that are reduced in size and appear to have no function. The tiny hip bones in some snakes have no apparent purpose, but they suggest that snakes evolved from ancestors with hips.

Embryological Evidence   Organisms in the early stages of development are called *embryos.* The study of **comparative embryology,** which compares embryos of different species, has found similarities that support the theory of evolution. Most biologists believe that these vertebrates share common genetic instructions for embryo development and, therefore, a common ancestor.

Reviewing the Section

1. How are most fossils formed?
2. How is carbon-14 used to date fossils?
3. What are the major divisions of the geologic time scale?
4. How are homologous and vestigial structures different?
5. How do embryological and biochemical similarities among different animals support evolution?

Investigation 15: Coacervates—Ancestors of Cells?

Purpose
To investigate the conditions under which the first cells may have evolved

Materials
Gelatin solution, test tube, three medicine droppers, gum arabic solution, pH paper, slides, coverslips, compound microscope, safety goggles, hydrochloric acid (HCl)

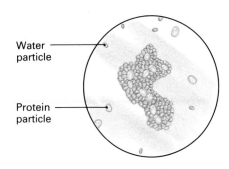

Water particle

Protein particle

Procedure
1. Pour about 5 mL of gelatin solution into a clean test tube.
2. Add about 3 mL of gum arabic solution and mix gently.
3. Test the pH of the gelatin/gum arabic mixture. *What is the pH value?*
4. Make a wet mount of the gelatin/gum arabic mixture. Observe the mount first under the low power of the microscope. Then observe it under the high powers. Record your observations.
5. **CAUTION: Be sure to put on safety goggles. Hydrochloric acid burns the skin and can damage clothing. If you get HCl on your skin or clothes, immediately wash with running water.** Add a drop of HCl to the test tube containing the gelatin/gum arabic mixture. Determine the pH of this mixture. Record the pH.
6. Make a wet mount of the hydrochloric acid mixture. Examine it under the low power of the microscope. Then switch to the high powers, as before. Record your observations.
7. Repeat steps 5 and 6 as many times as necessary until you are able to observe coacervates. The coacervates should appear somewhat like the illustration at the top of the next column. *At what pH did the coacervates first appear?*
8. Repeat steps 5 and 6 until you are unable to observe any coacervates under either power of the microscope.

Analyses and Conclusions
1. Make a general statement about the appearance of the gelatin/gum arabic mixture when coacervates were absent versus when they were present.
2. What kind of living things do coacervates resemble?
3. Gelatin is a protein; gum arabic is a carbohydrate. Explain how a mixture of these materials simulates conditions thought to have been present on the early earth.

Chapter 15 Review

Summary

In the last 200 years scientists have developed several theories explaining the formation of the universe and Earth and the beginning of life. The big-bang theory states that a huge explosion formed the universe about 15 billion years ago. Gases and dust from this explosion eventually formed stars and planets.

The Russian scientist Alexander Oparin suggested that life arose spontaneously from the compounds ammonia, water vapor, methane, and hydrogen. According to Oparin, these compounds bonded together to form the organic compounds from which life arose.

Scientists have also found evidence that life forms have evolved, or changed, since they appeared on Earth. Fossils, or traces of organisms that lived in the past, are one such source of evidence. Carbon-14 and other radioactive isotopes are used to establish an approximate date for these fossils. Other evidence of change is derived from the study of homologous structures and from embryology. Homologous and vestigial body structures as well as similarities in embryos and body chemicals of different organisms reveal evolutionary relationships.

BioTerms

analogous structures (236)

cast (232)

comparative biochemistry (235)

comparative embryology (236)

decay element (232)

epoch (233)

era (233)

evolution (228)

fossil (231)

geologist (233)

half-life (232)

homologous structure (236)

imprint (231)

mold (231)

paleontologist (233)

petrified fossil (232)

physical anthropologist (233)

radioactive isotope (232)

sedimentary (231)

vestigial structures (236)

BioQuiz   (Write all answers on a separate sheet of paper.)

I. Completion

1. Alexander Oparin guessed that life on Earth was generated _____.
2. The _____ states that the universe began with an explosion.
3. _____ is the process by which living things change over time.
4. The _____ is the time needed for half of the isotopes in a sample of radioactive material to decay.
5. _____ fossils are created when hard body parts are replaced with minerals.

II. Modified True and False

Mark each statement TRUE or FALSE. If false, change the underlined term to make the statement true.

6. The longest time spans on the geologic time scale are <u>epochs</u>.
7. <u>Homologous</u> structures are reduced in size and seem to have no function.
8. <u>Imprint</u> fossils are often produced by soft body parts.
9. A footprint preserved in rock is an example of a <u>fossil</u>.

III. Multiple Choice

10. The era during which the earliest traces of life appeared was the a) Precambrian. b) Paleozoic. c) Mesozoic. d) Cenozoic.
11. Scientific evidence indicates that life appeared on Earth about a) 15 billion years ago. b) 3.5 billion years ago. c) 4.6 billion years ago. d) 3 billion years ago.
12. Most fossils are found in a) tar pits. b) amber. c) ice fields. d) sedimentary rock.
13. Uranium-238 is used to date fossils that are a) more than 700 million years old. b) less than 50,000 years old. c) only in sedimentary rocks. d) only volcanic in origin.
14. Unicellular life first arose during the a) Paleozoic era. b) Mesozoic era. c) Precambrian era. d) Cenozoic era.
15. The wings of bats and butterflies are a) analogous structures. b) vestigial structures. c) homologous structures. d) comparative structures.

IV. Essay

16. How do scientists think stars and planets were formed?
17. What conditions probably existed immediately before life appeared on Earth?
18. How is a mold fossil different from a cast fossil?
19. Why do scientists think that organisms have changed since life began?
20. What evidence in living things suggests that many organisms have a common ancestor?

Applying and Extending Concepts

1. Fossil records of horses have been found in different forms. Do library research to learn about these early horses. Create a chart and include the name of each period in which a particular form of the horse existed along with a drawing of each type of horse, its name, and details about its size and characteristics.
2. The eye of a bird and the eye of an insect are both organs used for sight. However, they do not arise from the same parts of the embryo and are not structurally similar. Use your school or public library to compare these analogous organs. Report on their similarities and differences.
3. A fossil is analyzed through carbon-14 dating and found to contain about one-third as much carbon-14 as a comparable amount of living tissue. What is the approximate age of this fossil?
4. The American scientist Sidney Fox extensively studied structures called *proteinoid microspheres*. Do library research to learn the importance of these structures to the scientific explanation of how life began. Present your findings to the class.

Related Readings

Gurin, J. "In the Beginning." *Science 80* 1 (July–August 1980): 44–51. This article describes theories of earliest life.

Park, E. "A Remarkable Tower of Time Tells the Story of Evolution." *Smithsonian* 12 (December 1981): 99–114. This article includes a reproduction of a mural showing 700 million years of life.

Time-Life, Editors. *Life Before Man.* New York: Time-Life Books, 1972. Data about early life, fossils, dinosaurs, and hominids is summarized in this book.

Theories of Evolution

Pink flower mantis resting on leaf

Introduction

In the last 125 years, scientists have made great progress in understanding how organisms *evolve,* or change over time. The most important contribution to this knowledge is a theory of evolution outlined in 1859 by the British naturalist Charles Darwin. This theory explains how the environment molds groups of organisms in such a way that they acquire traits that help them survive. The theory of evolution has been called the most important of all biological ideas.

In this chapter you will learn about Darwin's theory and the various ideas and observations that shaped his thinking. You will also study the mechanisms of evolution, as well as current theories that seek to explain the appearance of new groups of organisms.

The Theory of Evolution

By the early 1800s, scientists had begun to speculate that organisms changed over time, and that new groups of organisms were being formed continuously. Scientists believed this change affected groups of organisms called species. A **species** is a biological group whose members resemble one another and have the capacity to mate and produce fertile offspring. If it was true that species changed, then there had to exist a natural process that caused the origin and change of species. But what was this process? Some scientists, such as Lamarck and Darwin, developed theories to explain how new species arose.

Section Objectives

- *State* Lamarck's theory of evolution.
- *Summarize* Darwin's theory of evolution.
- *Tell* how Darwin's observation of finches of the Galapagos Islands affected his thinking.
- *Explain* how variations arise in species.

16.1 Lamarck's Theory

In 1809 a French biologist named Jean Baptiste de Lamarck presented an explanation of the origin of species in his book *Zoological Philosophy*. Lamarck developed a theory of evolution based on his belief in two biological processes:

- *The use and disuse of organs*. According to Lamarck, organisms respond to changes in their environment by developing new organs or changing the structure and function of old organs. Use is actually an adaptation to the environment; the new organ is an *acquired trait*. Disuse is the response to environmental change by which an organ disappears because it is no longer needed.
- *Inheritance of acquired traits*. Lamarck believed that acquired traits were passed on to the organism's offspring. He called this phenomenon the inheritance of acquired traits. In this way, he said, new generations benefit from useful structures developed by their parents.

Lamarck's theory can be illustrated by explaining how his ideas account for the long necks of giraffes. According to Lamarck's thinking, the earliest giraffes might have had short necks suitable for reaching grass. If grass became scarce, the giraffes would have had to stretch their necks to reach leaves in the trees. The more they stretched, the longer their necks became. Giraffes that acquired the useful trait of long necks then passed the trait on to their offspring. In this way, organisms change as the environment changes.

Lamarck was a respected biologist and philosopher in his day, and his ideas were, at first, widely accepted. However, he could not support his theory of evolution with actual data. Later he suffered severe criticism for his ideas.

Figure 16–1. Fifty years separated the theories of evolution presented by Jean Baptiste de Lamarck (top) and Charles Darwin (bottom).

Figure 16–2. The similar yet different South American rhea (top) and the Australian emu (bottom) are examples of unrelated species that adapted in similar ways to like environments.

16.2 Darwin's Theory

Fifty years after Lamarck presented his theory of evolution, Charles Darwin, a British naturalist, published the *Origin of Species*. In this landmark work, Darwin presented evidence that demonstrated that all living things on Earth evolved from other living things.

The Voyage of the *Beagle* In 1831 Darwin was a young man studying to become a minister. He was not a serious student, preferring to ride horses and collect beetles rather than study. By the time he was 20, he had learned much about nature. When he was offered a position on the *H.M.S. Beagle* to go on a voyage of exploration around the world, he accepted. Thus, he became the ship's naturalist.

At the time the voyage began, Darwin did not accept the idea that species change. He accepted instead two of the prevailing ideas of his time. The first was that the earth was 6,000 years old and had remained unchanged except for the effects of floods and other catastrophes. The second was that organisms were designed especially for certain habitats and appeared on the earth in their present forms.

Early in the voyage, after having read the works of several geologists, Darwin began to change his ideas. He saw evidence that the earth was very old. In South America he witnessed an earthquake that lifted the land level several feet. He realized that mountains could be built by the action of earthquakes over millions of years. He found fossils of marine animals in high mountains, and realized that the rocks must have been lifted out of the ocean.

Darwin also studied animals and plants. On the Galapagos Islands he found animals that were like those of the South American continent, but not exactly alike. He realized that they must have come to the islands from the mainland, and then changed into new species. He also observed the animals and plants of South America, oceanic islands, and the Far East. He saw many examples that indicated that animals in similar environments did not always look exactly alike. For example, the emus of Australia and the rheas of South America look alike, but not exactly alike, and yet occupied the same kind of habitat. If animals were formed for a specific habitat, why would different species be found in similar habitats?

By the time Darwin returned to England, he was convinced that all living things arose by evolution. Over the next 20 years, working in his country house, he gathered evidence for his new theory. His masterful book was published in 1859.

► The Galapagos Islands

The Galapagos Islands, formed about 1 million years ago, straddle the equator 950 km (600 mi.) west of South America. On his famous journey, Darwin saw unique life forms there, including marine lizards and birds called *flightless cormorants.* He also saw giant land tortoises, called *galapagos* in Spanish. The islands were named for these tortoises, which had disappeared everywhere else.

Darwin also observed 13 species of finches. The birds Darwin saw resembled their counterparts on the mainland in general appearance and behavior, but Darwin noticed that each species had a distinctive shape and size of

beak. Moreover, each type of beak was well suited for obtaining a certain kind of food. The finch in the photo at the left, for example, uses its large, heavy beak to break open seeds. The finch at the right has a small, sharp beak for hunting insects.

Darwin saw the 13 species of finches as evidence of the process of evolution. He speculated that the different species

evolved from mainland finches that had come to the islands. The variations in their beaks allowed some of these finches to feed more effectively than others in their new environment. Certain birds, then, were more likely to survive and to pass their useful traits on to their offspring. As the finch populations slowly adapted, a new species was established on each island.

Darwin's Theory of Evolution   Based on his observations and studies, Darwin developed a new theory of evolution. Darwin's ideas, though modified by new knowledge, still form the cornerstone of modern evolutionary thought.

- *Variation exists within species.* Traits vary among individuals of the same species. For example, some gorillas have longer arms than others; some red-tailed hawks have sharper claws than other red-tailed hawks.
- *All organisms compete for limited natural resources.* Organisms compete for food and other necessities of life. These resources are limited. As a result, some organisms will get more of the resources; others will get less.
- *Organisms produce more offspring than can survive.* The number of young that parents can produce is greater than the resources available to support these individuals.

Q/A

Q: *Did other scientists of Darwin's time believe the earth was older than 6,000 years?*

A: Yes. Primary among them was Charles Lyell, a British geologist. Darwin read Lyell's book *Principles of Geology* aboard the *Beagle* and was profoundly influenced by its contention that the earth was millions of years old.

- *The environment selects organisms with beneficial traits.* Darwin believed that organisms with traits well suited to the environment survive and reproduce at a greater rate than organisms poorly suited to the environment. They thus pass desirable traits to their offspring. He called this process **natural selection** since the environment acts to preserve, or select, fit individuals. ''Fitness'' is measured by the number of fertile offspring produced. Some people call natural selection *survival of the fittest,* but ''survival of the fit'' is more accurate.

According to Darwin, a natural force such as bitter cold would favor animals with thick fur. Animals with thick fur would survive the cold temperatures and reproduce in greater numbers to pass on the trait of thick fur. Thick fur is an **adaptation,** a trait that gives the organism an advantage in its particular environment.

16.3 The Origins of Variations

If natural selection is always weeding out the less fit, why is it that individuals in a species vary and do not all look alike? A key to the solution of the origins of variations came from the work of the German biologist August Weismann. Weismann showed that two kinds of variations exist. One is variation produced by the environment and the other is variation produced by changes in what biologists would later call *genes.* Weismann showed that variations caused by the environment—acquired traits—could not be passed to offspring. *Only genetic variations are passed on from generation to generation.* Variation, therefore, could only arise in the organism's genes.

Biologists now know that two fundamental sources of genetic variations exist in species. The first is *mutation,* a change in the chemical structure of a gene. The second is *genetic recombination,* which occurs when an individual's genes are intermingled during meiosis. Mutation and genetic recombination provide variations acted upon by natural selection.

Figure 16–3. Genetic recombination produced the variation in coat color exhibited by this litter of kittens.

Reviewing the Section

1. What part do acquired traits play in Lamarck's explanation of evolution?
2. How does natural selection work?
3. How was Darwin's thinking influenced by life forms he saw on the Galapagos Islands?
4. What is the importance of Weismann's work?

Mechanisms of Evolution

The modern concept of evolution is broader than that first proposed by Darwin. Since the rediscovery of Mendel's ideas about genetics in the early 1900s, genetic principles have been added to Darwin's ideas to form the modern theory of evolution.

16.4 Species and Populations

Recall that a species is a group of similar individuals that have the capacity to produce fertile offspring. Stated simply, members of a species usually look alike and have offspring that can reproduce in nature. All redwood trees are members of the same species. All blue whales are members of the same species. You will learn more about species and the way in which living things are classified in Chapter 18.

Members of a species that live in the same area are members of a **population.** All the bluebells in a field, for example, are in the same population. All the rabbits in a forest also make up a population.

Evolution occurs when there is a change in the genetic makeup of a population. To understand how populations change, biologists study the kind and number of genes in a population. This field of study is called **population genetics.** Population geneticists study a population's **gene pool**—that is, all the alleles of all the genes in all of the individuals in a population.

16.5 The Hardy-Weinberg Principle

Consider a population of wild fruit flies living on an island. If you could list all the alleles of all the genes in every fruit fly, you would know many facts about the population's gene pool. You would know, for example, the frequency of every allele in the population. Suppose you found that the allele for normal-sized wings occurred 750 times and the allele for undeveloped wings occurred 250 times. Since the wing-size gene has only two alleles, you could determine that the frequency—the percentage of occurrence expressed as a decimal—of the normal-wing allele is 0.75 and that of the undeveloped-wing allele is 0.25. In doing so you would have determined what biologists call the **gene frequency** of the wing-size alleles. Gene frequency is a measure of the relative occurrence of alleles in a population.

In 1908 a British mathematician named Godfrey Hardy and a German physician named Wilhelm Weinberg were independently studying gene frequencies in populations. Through their

- *State* the main point of the Hardy-Weinberg principle.
- *List* the five ways in which genetic equilibrium may be disrupted.
- *Contrast* the three main types of natural selection.
- *Summarize* the effects of migration and isolation.
- *Describe* the process of genetic drift.

Figure 16–4. The purple lupines in the field shown above all belong to the same biological population.

studies Hardy and Weinberg arrived at the same conclusion, which came to be known as the **Hardy-Weinberg principle.** The Hardy-Weinberg principle states that the frequency of alleles in a population stays the same generation after generation unless it is altered by some external factor. In other words, the genetic makeup of a population will remain relatively stable unless something happens to make it change. This stability is called **genetic equilibrium,** also known as the *Hardy-Weinberg equilibrium.*

16.6 Changes in Genetic Equilibrium

A population in genetic equilibrium does not evolve. Genes determine the traits of a population. If the kinds of genes in a population never change, the traits of the population can never change, either. *For evolution to take place, something must upset the genetic equilibrium of a population.* In fact, evolution occurs when any of the following processes upset a population's genetic equilibrium: natural selection, migration, genetic drift, isolation, and mutation.

Natural Selection   Natural selection disrupts a population's genetic equilibrium by allowing fit individuals to survive and reproduce at a greater rate than unfit individuals. To see how natural selection works, reconsider the question of how giraffes acquired long necks. Imagine a population of an ancestral species of giraffe living on an African plain hundreds of thousands of years ago. Assume that variation in this species is such that neck length ranges from very short to very long. The numbers of individuals and the length of their necks are graphed in Figure 16–5. The graph of neck length versus number of individuals is shaped like a bell, and so is called a *bell curve.* Most individuals have average-length necks, but some individuals have very short necks and some have very long necks.

Assume that this population of ancestral giraffes has been at genetic equilibrium for centuries. Now what would happen if another species, such as leaf-eating deer, entered the area? These deer would be able to compete effectively with shorter-necked giraffes for leaves on the lower parts of trees. Soon the leaves would become scarce. The shorter-necked giraffes would be in danger of starvation because they are not fit to acquire food in this competitive environment. Their reproductive rates would be lower, and their genes would be slowly selected out of the gene pool. Longer-necked giraffes, however, would survive in greater numbers and thus have greater reproductive advantage. Over many years, the average neck length would change.

This shift is graphed in Figure 16–5. The new distribution is still a bell curve, but it has shifted to the right. After thousands of years of natural selection, modern giraffes exist.

Three fundamental kinds of natural selection exist. The example above is called **directional selection** because evolution has proceeded in the direction of longer necks. Figure 16–5 shows two other kinds: stabilizing and disruptive selection.

Stabilizing selection eliminates the extremes of a trait, causing a reduction of variation in species. For example, imagine a population of rabbits with varying leg lengths. In an environment with coyotes, rabbits with long legs are eliminated because they cannot crawl into small holes to escape coyotes. Rabbits with short legs are eliminated because they cannot run fast enough to evade coyotes. The result is a rabbit population with ''average-length'' legs. Variation has been reduced and the population has been stabilized.

Disruptive selection selects against the average and favors the extremes of a trait. Consider the size of acorns in a population of oak trees. Acorn sizes range from small to large. Suppose a species of acorn-eating squirrel now invades the oak forest. The squirrels do not eat small acorns, however, because they are too difficult to locate. The squirrels cannot eat large acorns because they are too large to carry. After many years, the average-sized acorns would be eliminated, but the small and large acorns would survive and germinate. The oak forest would soon have trees with two different sizes of acorns.

Figure 16–5. The graphs below illustrate the three main types of natural selection: directional selection (bottom left), stabilizing selection (top right), and disruptive selection (bottom right).

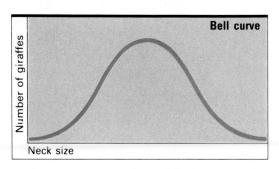

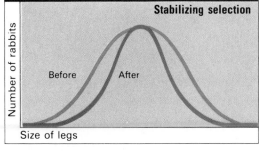

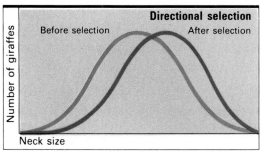

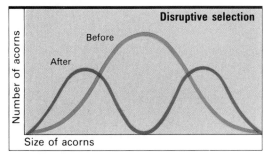

Highlight on Careers: Biology Teacher

General Description

A biology teacher is a professional who teaches life sciences in high school. Basic to the job is presenting biological facts and explaining complex subjects such as evolution.

Biology teachers bring new ideas to students and present them not only with biological facts but also with the principles of the scientific method. They must keep current to bring the latest and most exciting news to the class.

Biology teachers prepare lesson plans, lead student discussions, and assign and grade homework and tests. Teaching biology also requires ability to motivate and communicate with students.

A biology teacher has to prepare laboratories so students can gain lab experience. He or she should be able to lead field trips and arrange for guest speakers.

Career Requirements

All high school biology teachers must be certified to teach in accredited schools. To earn a certificate, one needs a bachelor's degree with a major

in biology and some courses in education. Some states require high school teachers to have a master's degree.

For Additional Information
National Association of
  Biology Teachers
11250 Roger Bacon Drive
Reston, VA 22090

Migration A population's genetic equilibrium may also be upset by **migration,** the movement of organisms into or out of a population. Organisms that leave a population take their genes out of the gene pool. Organisms that enter a population add their genes to the gene pool.

Consider a herd of caribou in northern Alaska. The herd is a population with a characteristic gene pool. Suppose another herd of caribou, one from northern Canada, migrates and becomes part of the Alaskan herd. The frequencies of genes will change because a whole new set of genes has been added from outside the population. The genetic equilibrium is thus disrupted. This is an example of *immigration,* the movement of new individuals into a population. Gene frequencies also change when individuals leave, a process called *emigration.*

Genetic Drift A process called genetic drift can also change the genetic equilibrium in small populations. **Genetic drift** is the change in gene frequency of a very small population due to chance. Consider an isolated population of 15 long-horned beetles. Suppose one of these beetles is red and all the others are black. If random mating occurs, it is mathematically probable that the one red beetle will not mate. In this case the genes for redness would not be transferred to offspring, and the gene frequency would change. Genetic drift is thus a result of the laws of probability or chance.

Isolation The equilibrium of a gene pool can also be upset by isolation—the separation of populations into groups that no longer interact. *Geographic isolation* occurs when a physical barrier separates populations. Barriers can include rivers, mountains, and canyons. Geographic isolation may change a population's gene frequency because the gene frequency of one resulting population may be different from that of the combined group. Geographic isolation often results in the two populations being unable to interbreed. This phenomenon is called **genetic isolation.**

Isolation often results in the development of a new species. For example, fossil evidence indicates that the camel originated in the western United States. Scientists believe that the camel population spread north and south, crossing land bridges into both Asia and South America. The disappearance of the northern land bridge geographically isolated the Asian and American populations. Different environmental pressures acted on these two populations. Over millions of years, the modern camel evolved in Asia, and the llama evolved in South America.

Mutation The change in genes by mutations also upsets a population's genetic equilibrium. Many mutations are harmful. An organism with a harmful mutation may be less fit than a normal organism. If the organism fails to live and reproduce, the mutant gene is removed from the gene pool. Some mutations, however, are beneficial rather than harmful. They make an individual better adapted to its environment. When a mutation is retained, the frequency of alleles in the population changes. Most mutations do not have a pronounced evolutionary effect. As a source of variation, however, mutations provide traits acted upon by natural selection.

Figure 16–6. The South American llama (top) and the Asian camel (bottom) are examples of two species that developed from a common ancestor as a result of genetic isolation.

Reviewing the Section

1. What five processes upset genetic equilibrium?
2. What are the evolutionary effects of isolation and migration?
3. How does mutation make natural selection possible?
4. What are the three types of natural selection?

- *Explain* how divergent evolution occurs.
- *Define* the term *convergent evolution.*
- *Summarize* two current explanations for the appearance of new species.

Patterns of Evolution

All of the processes that disrupt genetic equilibrium contribute to the general phenomenon of evolution. Evolution has produced many interesting similarities and differences in organisms. In some cases, organisms that are only distantly related resemble one another. In other cases, organisms that are closely related look very different.

16.7 Divergent Evolution

The process by which related organisms become less alike is called **divergent evolution.** Divergent evolution begins after **speciation**—the formation of a new species. The two new species at first are quite similar, but they may undergo divergent evolution if natural selection exerts a strong effect on one or both of the species. For example, consider the case of a group of brown bears that became geographically isolated in northern regions from the main group of bears. In time, the small group was genetically isolated and became a new species. Acted upon by natural selection, this group diverged into polar bears. These bears have many traits not found in their relatives. For example, their coats are white, and they have heads and necks modified for swimming. They diverged from their ancestors.

Divergent evolution also results in **adaptive radiation,** the process by which members of a species adapt to a variety of habitats. The finches Darwin saw on the Galapagos Islands are examples of adaptive radiation. A number of species of finches

Figure 16–7. The brown bear (left) differs from the Arctic polar bear (right) as a result of divergent evolution. The polar bear's white coat is an adaptation to hunting sea mammals on ice floes.

exist on the islands. While each species is adapted to a different habitat, the finches share enough features to show that they evolved from a common ancestor. Adaptive radiation occurs when species with suitable adaptations move into a new habitat. There they encounter less competition and have greater reproductive success.

16.8 Convergent Evolution

The process by which distantly related organisms develop similar characteristics is called **convergent evolution.** This pattern of evolution occurs when different species share the same environment and are therefore subject to the same pressures.

For example, whales and dolphins were once land mammals that adapted to an aquatic environment. As a result, their front limbs evolved into flippers for swimming. Today, the limbs of whales and dolphins resemble those of fish because they were shaped by the same environment. Yet the two groups of organisms—marine mammals and fish—are not closely related. If two organisms possess the same kinds of organs, this does not necessarily mean that they are closely related. It indicates only that they were subject to the same kinds of environmental pressures that led to the formation of similar structures.

Convergent evolution can often lead to cases of mimicry— the evolution of one organism so it comes to resemble another. Mimicry often occurs in cases where one animal is poisonous or distasteful. A nonpoisonous insect, for example, that looks like

Q/A

Q: *What is another example of adaptive radiation?*

A: Biologists believe that a few small mammals lived during the age of dinosaurs. As the dinosaurs disappeared and many new habitats became available, the mammals quickly evolved into many species, the ancestors of today's 4,500 kinds of mammals.

Figure 16–8. The similarities of dolphins (top left), salmon (top right), and whales (bottom) are the result of convergent evolution. The two species of mammals and the fish all share a marine environment.

Bottlenose dolphin

Atlantic salmon

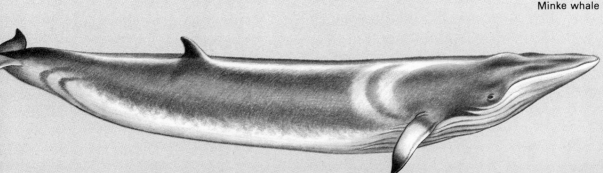
Minke whale

▶ Evolutionary Theory Today

One of the most important areas of evolutionary studies today concerns ways in which natural selection brings about the appearance of new groups of organisms. Scientists know that environmental conditions cause slow, gradual changes in species, a process called **microevolution.** However, they are not sure whether these slow, gradual changes can account for the dramatic appearance of totally distinct species, or **macroevolution.**

Traditionally, scientists searched the fossil record for clues to how major groups evolved. *Archaeopteryx,* the animal in the picture, is an example that shows how a major group, the birds, might have arisen. This organism lived 140 million years ago and

combines traits of two different animal groups. Its heavy skeleton and claws resemble that of a reptile, but its feathers indicate it is birdlike. Biologists believe that *Archaeopteryx* shows that birds evolved from reptiles.

Few fossils indicate gradual change, however. The fossil record more commonly shows that species appeared suddenly, lived for several million years, then disappeared as a new form took over.

Recently, several American scientists have suggested that new species

evolve on the fringes of a larger group's territory. If the new species adapts quickly, it might replace the old group. Some scientists think whole groups might have been rapidly extinguished by the effects of meteors and other cataclysmic phenomena.

A second theory proposes that groups are subject to natural selection just like individuals. If two animal groups exist at one time, the one with a better social organization may be better adapted to the environment and would survive and dominate.

Today scientists debate whether evolution is slow and gradual or a rapid process. The answer may be that the process differs depending on the species and the condition of the environment.

a poisonous insect will appear to predators to be poisonous, and hence will be left alone. Sometimes animals evolve to look like plants, and in doing so blend into their surroundings. These animals have a *selective advantage* in being able to avoid being eaten because they are well hidden.

Reviewing the Section

1. How does divergent evolution occur?
2. What is convergent evolution?
3. State two theories on the appearance of new species.

Investigation 16: Evolution in Bacteria

Purpose
To simulate the process of evolution in bacteria

Materials
Several sterile petri plates containing nutrient agar, sterile cotton swab, stock culture of *Escherichia coli,* container of disinfectant, forceps, disks with varying concentrations of antibiotics, incubator, safety goggles, inoculating needle, Bunsen burner

Procedure
CAUTION: Although the bacteria used in this investigation normally are not pathogenic, all bacteria may become dangerous under certain conditions. Always be careful when working with bacteria. Carefully follow your teacher's instructions for proper handling and disposal of bacteria.

1. Make a petri plate as shown in the drawing. Mark it with your initials.

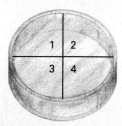

2. Moisten a sterile cotton swab in the *E. coli* bacterial culture. Inoculate the dish by gently moving the cotton swab over the agar. Place the contaminated swab in a container of disinfectant.
3. Using the forceps, place disks containing various concentrations of antibiotics on top of the inoculated agar.
4. Cover and invert the plate. Incubate at 37° for 24 hours.
5. After incubation, remove the cover and look for bacterial growth. Do not open the plate unnecessarily. *What sign indicates bacterial growth?*

6. Look for clear areas around the antibiotic disks. These areas are called zones of inhibition. *What do these clear areas indicate?*
7. Look for a colony of antibiotic-resistant bacteria growing very near one of the disks. *What is the strength of the antibiotic in this disk?*
8. Put on safety goggles. Using forceps, sterilize an inoculating needle in a Bunsen burner flame. Transfer a small amount of the resistant bacterial colony to a sterile petri plate. If you find more than one resistant colony, inoculate a separate petri plate with each one.
9. With forceps, place antibiotic disks of varying concentrations on the inoculated agar. (Do not use the concentration to which the bacteria were resistant.) Mark the plate with the name of the bacteria and your initials.
10. Incubate the plate(s) at 37° for 24 hours.
11. After incubation, observe the plate(s) for resistant bacteria. If you see resistant bacteria, repeat steps 9 through 11.
12. At the conclusion of this investigation, your teacher will explain how to dispose of the contaminated petri plates.

Analyses and Conclusions
1. At what concentration were the bacteria first resistant?
2. Did any bacteria develop a resistance after two or more generations? Explain.
3. Why might the new bacterial colonies that developed be more dangerous than the original ones?
4. How do the results of this investigation simulate evolution in bacteria?
5. State a conclusion about using antibiotics and the potential for developing new and more dangerous species of bacteria.

Chapter 16 Review

Summary

Since the beginning of the 19th century, much has been learned about how organisms evolve. Jean Baptiste de Lamarck proposed the first theory of evolution in 1809. He suggested that organisms acquire the traits they need, maintain them with use, and transmit these traits to their offspring.

In 1859 Charles Darwin presented the ideas that are the basis of current evolutionary theory. According to this theory, variations exist in species. Those members with variations suited to the environment pass their traits on to their offspring; those poorly suited are more likely to die before they can reproduce. Darwin called this process natural selection.

The Hardy-Weinberg principle, presented in 1908, states that gene frequencies do not change unless acted upon by something external. The processes that disrupt genetic equilibrium are natural selection, migration, isolation, genetic drift, and mutation. The process of evolution cannot occur without these disruptive processes.

Together, these forces of evolution operate in two major patterns. Through divergent evolution, one species becomes different from its ancestors. Through convergent evolution, distantly related organisms develop similar traits. Sometimes convergent evolution can cause unrelated species to develop similar structures. This phenomenon is called mimicry.

BioTerms

adaptation (244)
adaptive radiation (250)
convergent evolution (251)
directional selection (247)
disruptive selection (247)
divergent evolution (250)
gene frequency (245)

gene pool (245)
genetic drift (248)
genetic equilibrium (246)
genetic isolation (249)
Hardy-Weinberg principle (246)
macroevolution (252)
microevolution (252)

migration (248)
natural selection (244)
population (245)
population genetics (245)
speciation (250)
species (241)
stabilizing selection (247)

BioQuiz   (Write all answers on a separate sheet of paper.)

I. Completion

1. _____ and genetic recombination are the processes through which variations occur within a species.
2. Darwin believed that natural _____ was responsible for evolution.
3. Populations in _____ isolation are not able to interbreed.
4. Birds and bats illustrate the pattern of _____ evolution.
5. A _____ is a group of related organisms that produce fertile offspring.

II. Modified True and False

Mark each statement TRUE or FALSE. If false, change the underlined term to make the statement true.

6. Mimicry occurs when one unrelated animal looks like another.
7. Macroevolution refers to small, gradual changes in organisms.
8. Evolution takes place when gene frequencies change.
9. Mutations are one of the results of natural selection.

III. Multiple Choice

10. Adaptive radiation is a form of
    a) convergent evolution. b) divergent evolution. c) disruptive selection.
    d) stabilizing selection.
11. All squirrels in a forest form a a) genus.
    b) species. c) gene pool. d) population.
12. Both Lamarck and Darwin believed that a major force in the process of evolution is a) acquired traits. b) the environment. c) variations in species.
    d) use of organs.
13. The Hardy-Weinberg principle concerns
    a) genetic equilibrium. b) natural selection. c) speciation. d) adaptation.
14. The type of selection that leads to two species from a single species is
    a) directional. b) stabilizing.
    c) disruptive. d) natural.
15. Genetic isolation may result in
    a) acquired traits. b) mimicry.
    c) divergent evolution. d) inheritance.
16. The Galapagos finches and the specialization of their beaks illustrate the principle of a) adaptive radiation.
    b) convergent evolution. c) acquired traits. d) genetic equilibrium.

IV. Essay

17. What role does the environment play in natural selection?
18. What is geographical isolation?
19. Why is it impossible for convergent evolution to result in identical organisms?
20. How does stabilizing selection work?

Applying and Extending Concepts

1. Industrial pollution in England once resulted in directional selection in a species of moth. They changed from light bodied to dark bodied. This process has since been reversed as air pollution has been reduced. Use your school or public library to research this phenomenon and write a one-page report for class.
2. The shark has changed so little over thousands of years that it is referred to as a living fossil. Why, do you think, do sharks remain the same from generation to generation? What type of selection is acting on them?
3. According to the Hardy-Weinberg principle, all possible genotypes in a population can be represented by the formula $p^2 + 2pq + q^2 = 1$. Do library research to learn how this formula works. Then use it to answer the following question. What are the frequencies of the dominant and recessive alleles in a population of 100 cattle, of which 36 have the recessive trait? How many of the cattle are heterozygous? If natural selection eliminated all the cattle with the recessive phenotype, what would the gene frequencies be in the next generation?

Related Readings

Gorman, J. "The Tortoise or the Hare." *Discover* 1 (October 1980): 88–89. This article explains the theory of evolution through punctuated equilibrium.

Moore, R. *Evolution*. New York: Silver Burdett, 1977. An introductory text, this book explains how genetics and paleontology support evolution.

Rensberger, B. "Evolution Since Darwin." *Science 82* 3 (April 1982): 40–45. The past century's research on evolution is summarized in this article.

17

Human Evolution

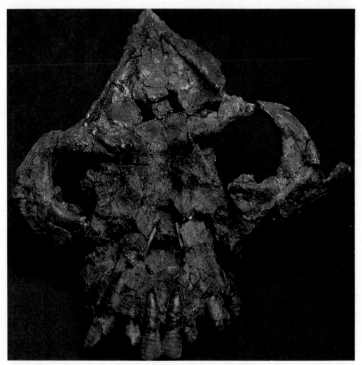

Sivapithecus skull, 8 million to 17 million years old

Introduction

One of the most fascinating areas of modern biology concerns the search for human beginnings. Little evidence exists—just partial fossils of a few hundred individuals and related clues, such as 3.75 million-year-old footprints in East Africa. From such limited records, paleontologists and anthropologists interpret the evolution of human beings. As new evidence is unearthed, the interpretations are often changed.

Human beings are among the more than 200 living species that belong to the order Primates. Scientists think that the primates began to evolve early in the Cenozoic era, which began about 70 million years ago. This process of evolution eventually led to the divergence of the ancestors of *Homo sapiens* within the last 3 million to 5 million years.

Early Human Evolution

One of the most hotly debated areas in the study of human evolution concerns the history of the hominids—that is, the earliest humanlike species. The main reason for this disagreement is the small number of hominid fossils. With no complete record of human evolution, scientists are forced to speculate about the sequence of events. Scientists do not agree on the earliest common ancestor of apes—gorillas, chimpanzees, orangutans, and gibbons—and humans. However, the fossils of several primate genera have features that make them likely candidates for the ancestors of modern apes and humans.

17.1 *Ramapithecus*

Possible candidates for the ancestor of both apes and human beings is ***Ramapithecus*** and its close relative *Sivapithecus*. These species of small, monkeylike primates lived between 17 million and 8 million years ago. *Ramapithecus* is believed to have originated in Africa and migrated to Asia and parts of Europe.

Most scientists agree that **Ramapithecus** *was not a hominid.* Scholars disagree, however, about where *Ramapithecus* fits into the human evolutionary line. Some think the Asian *Ramapithecus* evolved into the orangutan. They also argue that the African *Ramapithecus* was only distantly related to the human ancestor, which has not yet been found. Other scientists believe that *Ramapithecus* is the ancestor from which both apes and human beings evolved.

17.2 *Australopithecus afarensis*

Most scientists agree that the oldest known hominid is *Australopithecus afarensis*. **Australopithecus** is a genus of primitive hominid whose name means "southern ape." The species *A. afarensis* was discovered in Africa's Great Rift Valley in 1977 by American anthropologist Donald Johanson. The most important record of this species is a 3.6 million-year-old skeleton of an erect-walking female, which Johanson nicknamed Lucy.

Lucy's skeleton reveals that she stood about 1 m (3.3 ft.) tall and weighed less than 22.5 kg (50 lb.). Her arms were long, and her face was apelike with a large, thrusting jaw, receding forehead, and no chin. Lucy's skeleton, however, is humanlike and shows evidence that she was **bipedal**—that is, she walked on two legs. Her brain was significantly larger than that of

Section Objectives

- *State* two hypotheses for the evolutionary role of *Ramapithecus*.
- *Explain* the significance of *Australopithecus afarensis*.
- *Contrast Australopithecus africanus* and *A. robustus*.
- *Explain* why *Homo erectus* is considered the first representative of the genus *Homo*.
- *Differentiate* between ape skulls and human skulls.

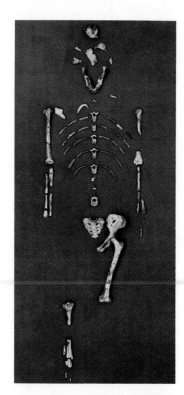

Figure 17–1. The *A. afarensis* skeleton called Lucy represents the oldest known hominid. The pelvis shape indicates a female.

Ramapithecus. However, sites where *A. afarensis* lived showed no evidence of tool use. For this reason, some scientists consider Lucy and her relatives to be more likely ancestors to apes than humans.

Johanson and his associate Timothy White propose that *A. afarensis* represents an early species of *Australopithecus*. Furthermore, they argue that *A. afarensis* is the common ancestor of two other species of *Australopithecus* and the earliest ancestor of human beings.

This interpretation is not shared by all scientists, however. Mary Leakey and her son Richard, members of a famous family of British anthropologists, contend that *A. afarensis* was part of a line of *Australopithecus* that later died out. In spite of the controversy, however, Lucy has clarified an important point about human evolution significantly: The brain size of early human beings did not increase until after they developed the ability to stand upright.

17.3 Later Species of *Australopithecus*

According to the existing fossil record, two other species of *Australopithecus* appeared thousands of years after Lucy. The first one discovered was *A. africanus*, which lived between 3 million and 2.5 million years ago. The earliest fossil record is the skull of a five-year-old child studied in 1924 by Raymond Dart, a South African anthropologist.

Based on this and other fossils, scientists have found that *A. africanus* had a rounded skull and a larger brain capacity than *A. afarensis*. In *A. africanus,* the cranial capacity, or the space available for the brain, averaged 485 cm^3 (30 cu. in.). The cranial capacity of modern human beings is 1,000 to 2,000 cm^3 (61 to 122 cu. in.).

Fossils also indicate that *A. africanus* stood about 1.5 m (5 ft.) tall and weighed about 36 to 45 kg (80 to 100 lb.). *A. africanus* also moved on two legs and had the broad, flat thumb that is common to human beings. However, the teeth of *A. africanus* resembled those of some modern apes. The back molars had thick enamel for grinding, and small front teeth were used for slicing. Evidence collected near *A. africanus* fossils has established that these early primates lived in small groups. They used simple stone tools, and their litter suggests that they ate meat.

The second species of *Australopithecus*—*A. robustus*—appeared later, about 2.2 million to 1.4 million years ago. *A. robustus* was larger than *A. africanus*. Individuals of this species weighed 69 kg (152 lb.) or more and had huge teeth set in

Figure 17–2. The skull of *A. africanus* (top), shows a combination of human and primate characteristics. The drawing (bottom) of the erect figure is based on the skull and other fossils.

heavy jaws. Some scientists believe that *A. robustus* may have evolved from *A. africanus* along a separate line from human evolution. Some suggest that *A. robustus* lived at the time of the earliest human beings and that competition between the two groups may have led to the extinction of *A. robustus*.

Some scientists think that human beings did not develop from any of the australopithecines. Louis, Mary, and Richard Leakey found fossils in East Africa of the first humanlike species to make tools. They named this species *Homo habilis*, which means "handy human." This species lived between 2.2 million and 1.5 million years ago—the same period as the australopithecines. Therefore, say the Leakeys, *H. habilis* could not have descended from the australopithecines. Other researchers, however, believe *Homo habilis* is merely an example of a different australopithecine.

Q/A

Q: *What may have caused early hominids to become bipedal?*

A: During the Miocene, forests became smaller. Some early hominids may have been driven out of the forests. Natural forces on the prairie would have favored those walking on two feet; they could see farther and use their hands.

▶ Ape or Human?

How can a paleontologist tell an ape from a human when only a skull is available? Comparing the skull of a modern human with that of an ape reveals some of the differences scientists look for.

One important difference is the location of the *foramen magnum,* the hole in the skull through which the spinal cord passes. As you see in the drawing, the foramen magnum is located near the rear of the ape's skull. In human beings, however, the foramen magnum is located at the bottom of the skull and the head sits atop the spinal column.

The angle at which the spinal cord enters the skull is one clue. The angle of

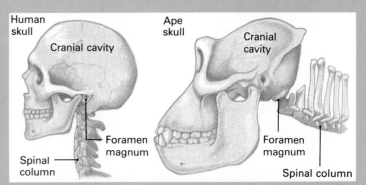

the face in relation to the spine is another. In apes, the face slopes outward from the skull, just as the spinal cord does. In humans, however, the face does not angle outwards. Instead, it is vertical and parallels the spinal column, as you see in the drawing.

A third major clue is the size of the cranial cavity, the space inside the skull

that houses the brain. As you see in the illustration, the cranial cavity of humans is about three times larger than that of apes.

Finally, humans generally have a higher, vertical forehead in place of the short, slanting one found in apes. Furthermore, the thick ridges above the eyes in apes have virtually disappeared in present-day humans.

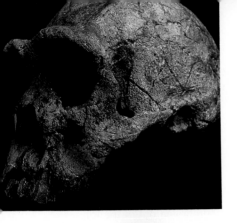

17.4 *Homo erectus*

Scientists do not agree on the number of australopithecine species or their relationship to human beings. However, most researchers do agree that the first species of the genus *Homo* appeared 1.6 million years ago and survived for 700,000 years. This species is called **Homo erectus.**

The first fossil of *Homo erectus* was found on the Pacific Ocean island of Java in 1891. Later, other fossils were unearthed in areas as diverse as Africa and China. This evidence has led some scientists to conclude that *Homo erectus* originated in Africa and later migrated elsewhere.

The skeleton of *Homo erectus* has features that are similar to those of apes and of humans. The bones of the *H. erectus* skeleton are thicker and heavier than those of modern human beings. In addition, the skull has a massive, low forehead. The chinless jaw has large, heavy teeth. However, the junction of the backbone and skull indicates that *Homo erectus* walked upright on two legs. In addition, the brain capacity of *Homo erectus* is 700 to 1,200 cm^3 (43 to 73 cu. in.), which approaches that of modern human beings.

Evidence found at the campsites of **Homo erectus** *indicates that these hominids had a brain far more advanced than that of more primitive ancestors.* For example, members of this species were tool makers. At first they used simple stone tools to prepare food and fashion other implements. Eventually, however, they made hand axes. These early human beings were also hunters, judging by the bones of bears, rhinoceroses, and elephants found at their campsites. Furthermore, they were nomads who traveled out from base camps in search of food. Campsites have been found in caves, which were located near water. Hearths in these campsites indicate the first controlled use of fire. The date for the first use of fire has been placed at nearly 1.4 million years ago.

Figure 17–3. The skull of *Homo erectus* above (top) was found in Kenya. The drawing (bottom) shows that the first human species walked more erectly than its predecessors.

Reviewing the Section

1. What relationship may exist between *Ramapithecus* and human beings?
2. Describe the theory of Johanson and White regarding *A. africanus*.
3. What were the differences between *A. africanus* and *A. robustus?*
4. Name three ways in which *Homo erectus* was more advanced than were previous human ancestors.

Modern Human Evolution

Section Objectives

- *Describe* the physical and social characteristics of the Neanderthals.
- *Compare* the Neanderthals and the Cro-Magnons.
- *Summarize* one hypothesis for the cause of the agricultural revolution.

Homo sapiens, which means "wise human being," is the species in which modern humans are classified. The earliest evidence of *H. sapiens* are fossils from Swanscombe, England, and Steinheim, Germany. Anthropologists have determined that these fossils are between 400,000 and 200,000 years old. Over the next 175,000 years, *H. sapiens* evolved into a species indistinguishable from the human beings of today.

17.5 The Neanderthals

In 1856 human bones were discovered in a limestone cave in the Neanderthal, a valley in Germany. Soon after, between 1866 and 1910, excavations in France and Belgium unearthed the remains of similar individuals. These were the **Neanderthals,** human beings that lived in Europe and Central Asia between 130,000 and 35,000 years ago, during the Ice ages.

Neanderthals stood about 1.5 m (5 ft.) tall, with erect posture. They had thick skulls, sloping foreheads, heavy brow ridges, and protruding jaws. However, their brain capacities were large, even slightly larger than those of modern human beings.

The Neanderthals used simple hand-held tools and wore animal skins. They were the first to bury their dead, perhaps indicating that their clans had rituals and ceremonies. The first evidence of violence between human beings is also found among the Neanderthals. Fossils found in Yugoslavia suggest a battle between human beings.

Some scientists think that Neanderthals are a separate species of human being, *Homo neanderthalis.* Others believe that these fossils are those of a human subspecies.

Figure 17–4. The Neanderthal skull (below left) was excavated at La Ferrassie in France. The drawing (below right) shows that Neanderthals were robust people with elongated heads, front to back.

17.6 The Cro-Magnons

Neanderthals disappeared from the fossil record about 35,000 years ago and were superseded by the **Cro-Magnons**, who belonged to the same species as modern human beings. The first fossil record of this species was found in 1868 by French railroad workers. It consisted of five skeletons at the rear of the Cro-Magnon Cave, near Les Eyzies, France.

In physical terms, the Cro-Magnons were identical to modern human beings. They had large brains; small, even teeth; and rounded skulls. They also had high foreheads and protruding chins like modern people.

The Cro-Magnon people developed sophisticated tools and weapons with flint blades. Fossil evidence also suggests that Cro-Magnons lived in bands of 30 to 100 individuals but had begun to create larger associations. In addition, the Cro-Magnons developed a complex culture that included large-scale cooperative hunting, art, and shared rituals and ceremonies. Many well-established camps have been discovered, many with cave paintings and engravings.

The Cro-Magnons were sophisticated people. They were efficient hunters who had highly developed social structures. Whether through competition with Cro-Magnons or through interbreeding, the Neanderthals disappeared after these highly capable people appeared.

Figure 17–5. Cro-Magnons produced the earliest known cave paintings. These often depicted animals of the hunt, such as the bison and goat above. Below is a drawing of how these first human beings might have looked (left), based on fossils such as the skull (right).

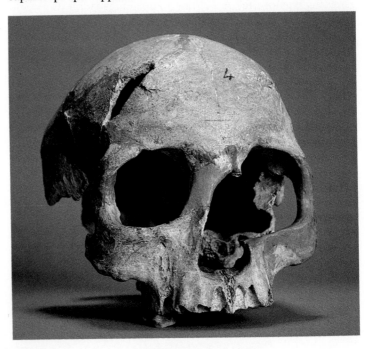

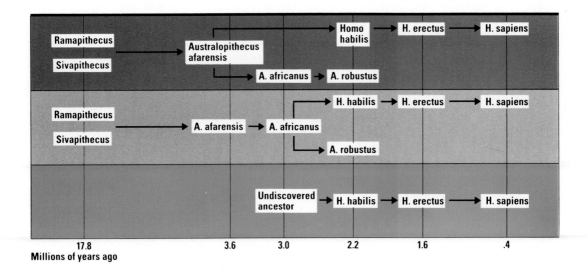

17.8   3.6   3.0   2.2   1.6   .4

Millions of years ago

17.7 The Agricultural Revolution

Rather abruptly, about 11,000 years ago, most of the world's people stopped hunting and gathering and adopted some form of agriculture. Within a few thousand years, most human societies gave up a way of life that their ancestors had followed for millions of years.

What caused such a widespread change from the hunting-gathering way of life? According to one explanation, the world climate began to grow warmer about 15,000 years ago and the glaciers made their final retreat. The grasslands were replaced by forests. The level of the sea rose, and much of the coastal plain disappeared.

Some scientists believe that these changes significantly reduced populations of mammoths, bears, and other large animals that were the major source of food for early human beings. As hunting became difficult, human groups turned to agriculture to maintain their food supplies.

The settled agricultural life caused widespread social changes. Because people no longer had to be on the move constantly hunting and gathering, they were able to set up better economic systems than they had ever had. A constant and abundant food supply supported an increasing population. As the new social systems developed, they accommodated more people than could a hunting society.

As a consequence of the agricultural revolution, the size of the human population began a steady increase that eventually became an explosion. Scientists estimate that 25,000 years ago the human population numbered 3 million. By 1650, 500

Figure 17–6. The illustration above shows three possible lines of human evolution that have been proposed by various scientists.

Q/A

Q: *Where did agriculture begin?*

A: The earliest evidence of farming has been found in the Middle East, in the modern countries of Turkey, Iraq, and Iran. Scientists think that the first crops were wheat and barley.

Spotlight on Biologists: Mary Nicol Leakey

Born: London, England, 1913

In 1959 British-born archaeologist Mary Leakey unearthed an ancient skull in the Olduvai Gorge in what is now Tanzania. The skull, which was later determined to be almost 1.75 million years old, had characteristics of both humans and apes. The Leakey discovery evoked enormous interest among scientists and led to a new surge of research on the origin of the human species.

Leakey and her archaeologist husband, Louis, named the species represented by the skull *Zinjanthropus.* Other scientists have identified it as related to, or a member of, *Australopithecus robustus.*

Mary Leakey did not attend college. She was educated by private tutors in southern France, where she first became attracted to archaeology from her exploration of prehistoric caves. Her interests blossomed after her marriage in 1942 to Louis Leakey.

Mary Leakey is also the mother of a well-known archaeologist, Richard E. Leakey. The Olduvai Gorge area studied by the Leakey family has proved to be an almost continuous record of human evolution from 1.75 million years ago to the present. Among their other finds in Kenya are pieces of a jaw and teeth about 14 million years old.

Since her husband's death in 1982, Mary Leakey has continued the family's work in Tanzania as director of the Olduvai Gorge excavations. Her own contributions to archaeology have been centered on prehistoric technology.

million people lived on the earth. Today the human species numbers nearly 4.5 billion. In terms of numbers, *Homo sapiens* has been highly successful. Will *Homo sapiens* surpass the 700,000-year span of *Homo erectus,* a primitive ancestor? Only time will tell.

Reviewing the Section

1. Describe the size and appearance of Neanderthals.
2. In what ways were Cro-Magnons more advanced than the Neanderthals?
3. How are Cro-Magnons similar to modern humans?
4. How could the agricultural revolution have caused an increase in population?

Investigation 17: Variations in Humans

Purpose
To measure a number of variations in humans

Materials
Metric ruler or tape measure

Procedure
1. Make a chart like the one shown below.

Student	Height (cm)	Length of left index finger (cm)	Length of left arm bone (cm)
1.			
2.			
3.			
4.			

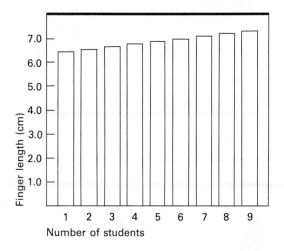

2. Working with a partner, first measure your height. Next, measure the length of the index finger on your left hand. Finally, measure the length of your left arm from the elbow to the wrist. Express all measurements in centimeters (cm) and record them on the chart.
3. Compile the data from step 2 for all members of your class.
4. List the height of each student in the class in descending order—from the tallest to the shortest.
5. Repeat step 4, making separate tallies for finger and arm measurements.
6. Find the class average for each of the three measurements.
7. Many students may be of the same height. Others may have the same finger and arm measurements. Count the number of students who have the same measurements.
8. Make three bar graphs of the class data— one graph for each of the measurements. The illustration in the next column shows how a graph might look.

Analyses and Conclusions
1. What is the average height of the members of your class? The average arm length? The average finger length?
2. All of your classmates are approximately the same age. How can you account for the differences in measurements?
3. How might the data change if you repeated this investigation after two years?

Going Further
Compile data similarly for all students in all classes taking this biology course. Find the average measurements and note the number of students having common measurements. Make three bar graphs of your data.

Chapter 17 Review

Summary

Much has been learned about human evolution since human fossils were first discovered about 100 years ago. For instance, anthropologists have found the bones of *Ramapithecus,* a primate that lived between 17 million and 8 million years ago. Some scientists think the ramapiths were the common ancestor of modern apes and human beings.

The oldest fossils of early humans found thus far are the 3.6 million-year-old remains of *Australopithecus afarensis.* Some scientists think that *A. afarensis* evolved into *A. africanus* and *A. robustus,* species that lived between 3 million and 1.5 million years ago. These scientists also believe that *A. afarensis* was the ancestor of modern human beings. Other anthropologists think human beings did not descend from australopithecines. The reason for their disagreement is *Homo habilis,* a humanlike, tool-making species that lived at the same time as the australopithecines.

Most scientists think that only two species belong to the genus *Homo,* which includes today's human beings. One is *Homo erectus,* which appeared 1.6 million years ago and survived for 700,000 years. The other is *Homo sapiens.* This species includes the Neanderthals, which lived between 130,000 and 35,000 years ago. It also includes the Cro-Magnons, sophisticated human beings who appeared about 35,000 years ago.

About 20,000 years after the appearance of the Cro-Magnons, most hunting and gathering societies turned to farming. As a result, human populations began to grow rapidly and have increased more than a thousandfold in the last 25,000 years.

BioTerms

Australopithecus **(257)**   *Homo erectus* **(260)**   *Ramapithecus* **(257)**
bipedal **(257)**   *Homo sapiens* **(261)**
Cro-Magnon **(262)**   Neanderthal **(261)**

BioQuiz   (Write all answers on a separate sheet of paper.)

I. Completion

1. The _____ is a hole in the skull through which the spinal cord passes.
2. Primates began to evolve in the _____ era.
3. The first hominids to bury their dead were the _____.
4. Hominids that are _____ walk on two legs.
5. A small, monkeylike, nonhominid primate that lived between 17 million and 8 million years ago is _____.

II. Modified True and False

Mark each statement TRUE or FALSE. If false, change the underlined term to make the statement true.

6. Cro-Magnons are classified as members of *Homo sapiens*.
7. *A. africanus* was the first hominid known to use fire.
8. Cro-Magnons were less advanced than Neanderthals.
9. The first hominid tool maker was *A. afarensis*.

III. Multiple Choice

10. The age of the earliest fossil resembling a human being is a) 17 million years. b) 3.6 million years. c) 12 million years. d) 1.5 million years.
11. The oldest known hominid is thought to be a) *Neanderthal*. b) *A. africanus*. c) *Ramapithecus*. d) *A. afarensis*.
12. Fossils show that increased brain size came a) after upright posture. b) after the use of tools. c) after the agricultural revolution. d) after the Cro-Magnons.
13. Modern human beings belong to the genus a) *Ramapithecus*. b) *Homo*. c) *Sivapithecus*. d) *Australopithecus*.
14. The hominid with the largest brain was a) *A. afarensis*. b) *H. habilis*. c) *H. erectus*. d) *A. africanus*.
15. Human beings shifted from hunting and gathering to agriculture a) 11,000 years ago. b) 35,000 years ago. c) 130,000 years ago. d) 200,000 years ago.

IV. Essay

16. According to Johanson and the Leakeys, what evolutionary role does *A. afarensis* play?
17. Why do scientists consider *H. erectus* an ancient human?
18. How are Neanderthals and Cro-Magnons different from each other?
19. What is the importance of the agricultural revolution?
20. Why do scientists disagree about the human family tree?

Applying and Extending Concepts

1. Assemble a bulletin board comparing the cultures of the Neanderthals and the Cro-Magnons, including drawings or photographs of their tools, clothing, and art. Summarize the differences between the two cultures for the class.
2. In 1912 a British paleontologist and student of early humans announced the discovery of an early human fossil in Britain. This fossil, called *Piltdown Man,* was later found to be a fake. Write a paper explaining how the hoax was detected.
3. Imagine you are an anthropologist or paleontologist excavating an ancient campsite. What might you conclude about the early people who used the site if you found the charred bones of large animals and various stone blades lying near an early human skeleton?
4. Human beings are the result of millions of years of primate evolution and have characteristics that make them fit to live in many environments. For each of the traits listed, suggest how the trait helps humans cope with their environments: eyelashes, blinking, speech, and having both eyes face forward.

Related Readings

Ingber, D. "The First Supper." *Science Digest* 90 (September 1982): 54–59. This article explains how fossil teeth and ancient garbage reveal the diet of ancient humans.

Leakey, R. E. and R. Lewin. *Origins: What New Discoveries Reveal About the Emergence of Our Species and Its Possible Future.* New York: E. P. Dutton, 1977. This beautifully illustrated book surveys human history to the development of agriculture.

Rensberger, Boyce. "Bones of Our Ancestors." *Science 84* (April 1984): 29–39. Lavish photographs of human fossils illustrate this article.

BIO*TECH*

Mapping Ancient Climates

Core-sampling device

Microfossils provide clues to the climatic conditions under which the organisms that produced them must have lived.

Traces of ancient human settlements have been found in the Sahara. How could people have survived in such a dry, barren place? The answer is that the climate of the Sahara, now hot and extremely arid, was once far more hospitable. Over time, climate changes. As a result of technological advances, the nature of these changes in past climates can be studied in detail.

The prevailing weather patterns known as climate result from complex interactions of the atmosphere with the earth's oceans, ice cover, and land masses. Scientists called *paleoclimatologists* study the climate of the earth's past. In many cases, they can provide a description of the climate of a particular place at a particular time in history, such as the ancient climate of the Sahara.

How do paleoclimatologists gather their data? Obviously they cannot measure the temperature or precipitation at a particular place thousands of years in the past. Likewise, they cannot measure how much of the earth's surface was covered by ice 5,000 or 10,000 years ago. Paleoclimatologists must rely instead on indirect sources for information.

One major source of information lies in layers of sediment buried deep beneath the floors of oceans and lakes. Within the sediment are microfossils, the fossils of microscopic organisms. These microfossils can be removed and studied through the use of a variety of technological tools. The microfossils provide clues to the climatic conditions under which the organisms that produced them must have lived.

Ocean Core

Age of core segment in millions of years

Temperature of surface

Present
5°C

1 million
10°C

5 million
15°C

15 million
20°C

25 million
25°C

For example, marine organisms called foraminifera live in waters near the surface of oceans. Different varieties of foraminifera, with different shell structures, flourish at different temperatures. The fossil shells of foraminifera record their shapes and thus indicate the temperature of the ocean water in which the creatures once lived.

Researchers examine microfossils under a scanning electron microscope to discover details of texture and structure necessary for identification. Next, they determine the age of microfossils with another technological tool: radiocarbon dating. Evidence from the scanning electron microscope suggests what climatic conditions

Scanning electron microscope

supported the organisms. The radiocarbon dating indicates at what point in the past that climate prevailed.

Using these and other techniques, paleoclimatologists have painstakingly collected information about ancient precipitation patterns and temperatures from all over the world. To interpret their data, they feed it into powerful computers that integrate the data with information about current climate.

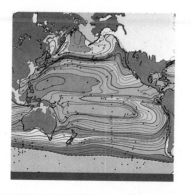

Computer-generated maps of sea surface temperature from 18,000 years ago (left) and from present (right)

CHAPTER

18

Diversity and Classification

A diversity of fossils from various geological periods

Introduction

Thirty million kinds of organisms share the earth with humans. To study such a vast number of groups of organisms, a universally accepted system of classification is necessary. Just as you arrange your clothes or record collection, so biologists throughout the world arrange all living things into groups. Your classification system allows you to find a particular article of clothing or record much more quickly than if all of your possessions were just piled in a corner of your room. Likewise, a system of biological classification allows scientists to store and retrieve information about living things quickly and efficiently. Although the system that scientists use has been developed over many centuries, biologists today continue to expand and modify it almost daily.

The History of Classification

Section Objectives

- *Define* the term *taxonomy*.
- *Compare* Aristotle's system of classification with that of Linnaeus.
- *List* the problems associated with the use of common names.
- *Explain* the system of binomial nomenclature.

The need to classify living things gave rise to **taxonomy**—the science of grouping organisms on the basis of their similarities. One of the first taxonomic systems was proposed by the Greek philosopher Aristotle around 350 B.C. He divided living things into two groups, animals and plants. Aristotle subdivided animals on the basis of habitat and behavior and plants on their size and structure. He said, for example, that herbs, shrubs, and trees are the three major divisions of the plant kingdom. Although Aristotle's system contained many errors, it was used for more than 2,000 years.

18.1 The System of Linnaeus

During the mid-1700s, biologists began to explore the world to search for previously unknown forms of life. As a result, thousands of newly discovered organisms were collected and described yearly. Although biologists used Aristotle's system of taxonomy to classify these organisms, most recognized that this system did not sufficiently explain the relationships between the organisms. For example, two plants might have the same kind of flowers and leaves. However, because one was a shrub and the other a tree, the scientists were forced to place them in different groups. In addition, the methods naturalists used to name these newly discovered organisms varied greatly. The use of many ineffective, competing, and often contradictory systems made communication between biologists very difficult.

A way out of these difficulties was provided by a Swedish botanist named Carolus Linnaeus. *Linnaeus developed a new classification system that revolutionized taxonomy.* He suggested that organisms with similar structures should be placed in the same taxonomic group and suggested that this group be called a *species*. As you will recall from Chapter 16, two organisms of the same species will produce fertile offspring if crossed. Linnaeus further suggested that similar species be grouped into a larger category called a **genus.** For example, dogs, wolves, and jackals—each a different species—are similar enough to be considered members of the same genus.

18.2 The Scientific Name

An organism is often known primarily by its **common name**— that is, the name given it by the people of an area. This practice has caused a great deal of confusion because one kind of

Figure 18–1. Many small seed-eating birds are commonly called sparrows. The house sparrow shown above, the *Passer domesticus,* is related more to weaver-finches than to other sparrows. Sometimes called the English sparrow, it is actually found worldwide.

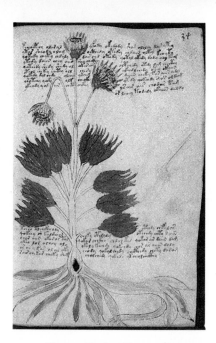

Figure 18–2. The sunflowers above are part of an herbal listing in a manuscript dating from the 1400s or the 1500s.

organism may have many names. For example, in one region the mountain lion is called a puma, in another a cougar, in others a catamount or panther. Many organisms have also been misnamed, implying misleading relationships. Prairie dogs may yelp and bark, but they are more like squirrels than dogs. Starfish, silverfish, and jellyfish differ greatly from one another—and not one of them is a fish.

To avoid such problems, early taxonomists introduced the idea that each organism be given a **scientific name**—that is, a short, standard name that is accepted by all scientists. Linnaeus suggested that the scientific name of an organism consist of its genus name followed by its specific name. This practice is known as **binomial nomenclature.** *Binomial* means using two names, and *nomenclature* is the system of naming things. Linnaeus continued the practice of using Latin, which was considered the "language of science" and was understood by all scientists of the time. The practice of writing scientific names in Latin is still followed today.

An organism's scientific name is written in a precise way. The genus name begins with a capital letter; the specific name with a lower-case letter. A scientific name is generally written in italic type or underlined. *Felis concolor,* or *F. concolor* for short, is the scientific name of the puma or mountain lion. This name is recognized and accepted by scientists throughout the world.

Many scientific names are descriptive. For example, the red maple is *Acer rubrum,* which means "red maple" in Latin. Sometimes a scientist names an organism after a scientist he or she admires. The genus *Linnea,* an herb of cold areas, was named in honor of Linnaeus. In many cases, a specific name describes where the organism lives, such as *Darlingtonia californica,* the carnivorous cobra lily of California.

In addition to giving an organism a scientific name, taxonomists may also cite the name of the person who first described the organism. For example, the scientific name of the cobra lily is often written *Darlingtonia californica* Torr. to indicate that it was given its scientific name by John Torrey, a famous American botanist.

Reviewing the Section

1. What is taxonomy?
2. On what did Linnaeus base his system of classification?
3. Why does the use of common names lead to confusion?
4. How is an organism's scientific name determined?

Modern Taxonomy

Linnaeus' system, published in 1753, remains the starting point for all modern taxonomy. Since the publication of Charles Darwin's *The Origin of Species* in 1859, the theory of evolution has influenced classification. As a result, taxonomists now base classification on evolutionary relationships.

18.3 Bases of Modern Classification

Though similarity of structure still remains the basis for grouping organisms, biologists also use other kinds of evidence in classification. *Modern taxonomists study chromosome structure, reproductive potential, biochemical similarities, and embryology to determine the relationships among organisms.*

Comparing one organism's chromosome makeup, or **karyotype,** with that of another organism helps biologists determine relationships. Taxonomists know that similarity of karyotypes usually indicates a close taxonomic relationship.

Taxonomists also study reproductive potential. Botanists, for example, perform many experimental crosses with related plants to see which will produce fertile offspring. Those that do are said to belong to the same species.

An organism's biochemical makeup also provides evidence of its relationships. Similar sequences in the amino acids of proteins from two organisms may indicate that they are closely related. For example, the horseshoe crab was given its common name and classified according to its external appearance. Examination of blood proteins, however, revealed that these so-called crabs more closely resemble spiders. As a result, taxonomists reclassified them.

Scientists also study the embryological development of *homologous structures* to determine relationships. For example, a taxonomist might note that the wing of a bat and the flipper of a whale originate from the same embryonic tissues and initially develop in a similar manner. The taxonomist might then suggest that the two animals had a common ancestor.

Figure 18–3. These images produced by electrophoresis show a comparison of blood proteins from (top to bottom) a rat, a cat, and a dog.

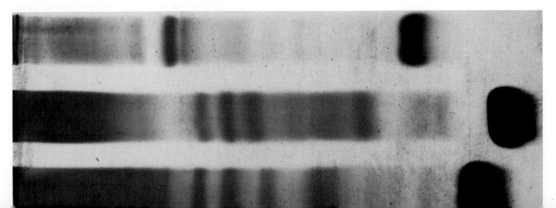

Highlight on Careers: Natural History Museum Curator

General Description

A curator in a museum of natural history is a professional scholar who specializes in one of the fields represented in the museum. The curator may conduct research on various kinds of plants, or on birds, snakes, and other animals.

Another important duty of a curator is supervising a staff of technicians or assistants. Under the curator's direction, this staff maintains and preserves the items in the collection and adds new ones to it. The curator is responsible for making certain that each item added to the collection is classified properly.

Career Requirements

Most curators have a Ph.D. degree in botany or zoology. Most often, a prospective curator first works under the supervision of an established curator, learning the preferred methods for maintaining the collection.

A few universities offer a program leading to an M.S. degree in management of museum collections. Graduates of these programs generally become employed as technicians or as administrative assistants.

For Additional Information
American Association of
   Museums
1055 Thomas Jefferson
   Street S.W.
Washington, DC 20007

18.4 Categories of Classification

Taxonomists do not use only genus and species in classifying organisms. Other classification levels also exist. When taken in order from the largest, most generalized group to the smallest, most specific group, these levels are *kingdom, phylum, class, order, family, genus,* and *species.*

Each organism has a place in this taxonomic system. For example, a tiger is a member of the kingdom Animalia, the animal kingdom. It is also a member of the phylum Chordata, a group composed mostly of animals with backbones. Because it has hair and nurses its young with milk, it is placed in the class Mammalia with animals that share these traits, such as whales and monkeys. Like all meat eaters with enlarged canine teeth, the tiger is a member of the order Carnivora. Tigers and other cats are members of the family Felidae. Tigers, lions, and Old World panthers are members of the genus *Panthera;* tigers are put in the species *tigris.* The scientific name of the tigers, therefore, is *Panthera tigris.* An African or Asian lion belongs to all

Q/A

Q: *How many species can one genus contain?*

A: There is no limit. Some genuses contain only a single species. The genus of fruit-flies, *Drosophila,* however, contains over 1,000 species.

the same higher taxonomic categories as a tiger, but is a different species, *Panthera leo*.

Members of the same species that differ in some important way—such as flower size or ear shape—are said to be members of different **varieties,** or subdivisions of a species. The variety of a species is often listed as a part of the scientific name. It is written after the species name. When two varieties are separated geographically from each other, many taxonomists prefer to call them *subspecies*.

18.5 Systems of Classification

When Linnaeus developed his new system of classification, he retained the idea of two kingdoms. All **autotrophs,** or organisms that produce their own food, were placed in the plant kingdom. All **heterotrophs,** which are organisms dependent on others for food, were placed in the animal kingdom. As scientists learned more, they realized that the two-kingdom system was inadequate. Some organisms, such as the single-celled *Euglena*, share important features with both plants and animals. It is incorrect to call *Euglena* an animal and equally incorrect to call it a plant. Some taxonomists suggested that a third kingdom, Protista, be established for organisms like *Euglena*.

The development of the light microscope enabled scientists to learn that bacteria, blue-green algae, and some other kinds of microorganisms do not have nuclei. Certainly, said some taxonomists, these organisms cannot be considered members of any of the three kingdoms. Therefore, a fourth kingdom, Monera,

Table 18–1: Comparison of Classification Systems

Number of Kingdoms				
Kingdom	Two	Three	Four	Five
Animalia	Animals, protozoa	All multicellular animals	All multicellular animals	All multicellular animals
Plantae	Plants, algae, fungi, slime molds	Plants, algae, fungi, slime molds	All multicellular plants and all fungi	All multicellular plants
Protista		Unicellular organisms and colonial protozoa	Most unicellular organisms	Most unicellular organisms
Monera			Blue-green algae, bacteria, and other organisms that lack nuclei	Blue-green algae, bacteria, and other microorganisms that lack nuclei
Fungi				All fungi

was established and organisms without nuclei were assigned there. Many taxonomists, however, were still troubled by mushrooms and molds, which have nuclei but are not plants or animals or similar to protists. A fifth kingdom, Fungi, was established to accommodate these organisms. Today some scientists favor the three-kingdom system; some prefer four kingdoms; and many favor five. Biologists continue to debate which classification system is most accurate and how best to categorize organisms. In this textbook, the five-kingdom system of classification is used. The kingdoms, their characteristics, and their major groups are shown in Table 18–2.

Table 18–2: The Five-Kingdom System of Classification

Kingdom	Characteristics	Major Groups	
Monera	Simple organisms without nuclei	Schizophyta Cyanophyta Prochlorophyta	
Protista	A varied group of organisms with nuclei; many unicellular; both autotrophic and heterotrophic forms, includes protozoa and algae	Mastigophora Sarcodina Ciliophora Sporozoa Euglenophyta	Pyrrophyta Chrysophyta Phaeophyta Rhodophyta Chlorophyta
Fungi	Multicellular heterotrophs with nuclei, absorb food through cell wall	Myxomycophyta Eumycophyta	
Plantae	Multicellular, nucleated autotrophs with photosynthesis in chloroplasts	Bryophyta Psilophyta Sphenophyta Lycophyta Pterophyta	Cycadophyta Gnetophyta Coniferophyta Anthophyta
Animalia	Multicellular, heterotrophs with nuclei	Porifera Coelenterata (Cnidaria) Platyhelminthes Nematoda Mollusca	Annelida Arthropoda Echinodermata Chordata

Some taxonomists study details of the evolution of species. These taxonomists are called *biosystematists.* Their field of study is **biosystematics.**

Biosystematists document the differences in traits between populations of the same species. For each population they may record data on size, color, shape, and other characteristics of individual members. In this way, the biosystematists develop a population profile that shows variation of traits. For example, a biosystematist studying lilies might determine variation in length of leaves, time of blooming, number of flowers, and width of seeds.

These data are then analyzed, usually with the aid of a computer. Graphs showing the range of variation of individual traits within one lily population are then compared with variations of those traits in a hundred lily populations over thousands of miles.

By interpreting the differences and similarities of populations, the biosystematists develop hypotheses regarding how the species is evolving. The systematists may find, for example, that some lily populations bloom in August, whereas others bloom only in May. This may indicate that the one species is beginning to become *genetically isolated* into two subspecies and is in the process of evolving into two species.

18.6 The Ongoing Science of Taxonomy

Since the time of Aristotle, taxonomists have continued to classify organisms. Today, however, taxonomists use computers that can analyze much more data on species than was possible in the past. With the help of computers, modern taxonomists determine relationships quickly and accurately.

Most of today's taxonomists work in museums or herbariums where they have access to large numbers of organisms collected from all over the world. Many, however, are adventurers who travel to almost uncharted regions of the world in search of undiscovered forms of life.

Reviewing the Section

1. Name five features used in modern classification.
2. What are the categories or levels of classification from the least specific group to the most specific group?
3. What are the five kingdoms in the five-kingdom system of classification?
4. How do biosystematists go about determining relationships?

Investigation 18: Classifying Organisms

Purpose
To design a classification system

Materials
Living specimens, preserved specimens, reference books such as field guides to plants and animals

Procedure
1. Your teacher will display a number of living and preserved specimens. Working with a partner, make a list of all the characteristics you can identify in each of the specimens on display.
2. First, group the specimens into two large categories—using a very broad, general characteristic such as plant or animal.
3. Subdivide each group using a more specific feature; for example, animals with a backbone versus those without a backbone.
4. Continue to subdivide each classification group, always using two opposing characteristics.
5. The illustration shows part of a dichotomous key. Using this key as an example, make a dichotomous key for the organisms you are observing.
6. When you have completed your key, pass it on to one of your classmates to see if he or she can use it to identify the organisms on display.
7. Using reference books, identify the common and scientific names of the organisms on display. Check your key to see how well you were able to classify them.

Analyses and Conclusions
1. Why is it an advantage to work in a team for this investigation?
2. What difficulties did you encounter in trying to classify the specimens on display?
3. How was the work you did in this investigation similar to that of a taxonomist? How was it different?

Going Further
Repeat this investigation outdoors in a park or a natural environment near your school, classifying the organisms naturally found in this area.

1a. skeleton made of cartilage		go to 2
b. skeleton made of bones		go to 6
2a. unpaired fins; jawless		go to 3
b. paired fins; jaws present		go to 4
3a. mainly parasitic; oral disc; undergoes metamorphosis		
b. scavenger; slitlike mouth; hermaphroditic		
4a. gills on ventral side		go to 5
b. gills on lateral side; torpedolike shape; internal fertilization		
5a. oviparous		
b. ovoviviparous		
6a. lungs present		
b. lungs not present		go to 7
7a. gills; dorsal and pectoral fins with fleshy bases supported by bones		
b. gills; fins supported by long bones		

Chapter 18 Review

Summary

Taxonomy is the science of classifying living things. Aristotle, one of the first to develop a taxonomic system, generally considered all organisms either animals or plants. This system continued until the mid-1700s, when Carolus Linnaeus developed a system based on similarities of structure between organisms.

To avoid problems of common names, Linnaeus assigned each organism a scientific name composed of its genus and species names. This two-part naming system is called binomial nomenclature.

Today's systems of classification are based on evolution, using such characteristics as structure of chromosomes, reproductive poten-

tial, biochemical makeup, and embryology. These systems are composed of a number of different levels. From most general to most specific these levels are kingdom, phylum, class, order, family, genus, and species.

The modern systems of taxonomy have been repeatedly modified to reflect advances in biological knowledge, including the discovery of new species. Many scientists today use the five-kingdom system of classification, which has the following kingdoms: Monera, Protista, Fungi, Plantae, and Animalia. Biological classification will continue to change as new information is collected.

BioTerms

autotroph (275)
binomial nomenclature (272)
biosystematics (277)
common name (271)

genus (271)
heterotroph (275)
karyotype (273)
scientific name (272)

taxonomy (271)
variety (275)

BioQuiz   (Write all answers on a separate sheet of paper.)

I. Completion

1. _____ is the system of giving an organism a two-word scientific name in Latin.
2. An organism's _____ is the structure of its chromosomes.
3. Similarities in molecules such as the _____ of proteins may indicate that two organisms are related.
4. In the five-kingdom system, most unicellular organisms with nuclei are placed in the kingdom _____.
5. The examination of homologous structures in the _____ often gives clues to organisms' relationships.

II. Modified True and False

Mark each statement TRUE or FALSE. If false, change the underlined term to make the statement true.

6. Mushrooms are placed in the kingdom Plantae in the five-kingdom system.
7. The first part of an organism's scientific name is the name of the species to which the organism belongs.
8. A phylum is a group of related classes of organisms.
9. Two varieties that are geographically separated are known as subspecies.
10. Aristotle introduced the use of binomial nomenclature.

III. Multiple Choice

11. The most general of all classification groups is a) species. b) order.
    c) variety. d) kingdom.
12. The local term for an organism is its
    a) common name. b) species.
    c) scientific name. d) genus.
13. Two organisms that look alike and produce fertile offspring are members of the same a) species. b) genus.
    c) phylum. d) family.
14. The discovery that bacteria had no nuclei led to the establishment of the
    a) Protista. b) rules of nomenclature.
    c) Monera. d) Fungi.
15. Aristotle developed a system of classification with a) two kingdoms.
    b) genus and species. c) five kingdoms.
    d) six families.

IV. Essay

16. What determines whether two individuals are the same species?
17. What characteristics distinguish biosystematics from classical taxonomy?
18. Why is Aristotle's two-kingdom approach no longer used?
19. What types of information can be communicated by a taxonomist when picking the species name of an organism?
20. What are two contributions of Linnaeus to the science of taxonomy?

Applying and Extending Concepts

1. Aristotle's division of the plant kingdom into trees, shrubs, and herbs is a classification based on structure, yet it is not followed by modern botanists. Use your school or public library to find out which structures of plants are used in the modern botanical classification.
2. Two kinds of daisies both belong to the genus Aster. Although they look exactly alike and produce fertile hybrids in the laboratory, they never cross in nature because one kind flowers only in May and the other only in October. Are they the same species? State your reasons for saying yes or no. What criteria could be used to determine the answer?
3. Use your knowledge of evolution from Chapter 16 to write a paragraph telling how one species gradually becomes two subspecies and, ultimately, two separate species.
4. Two species of oak tree, the burr oak and the white oak, are common in many parts of the United States. Some scientists argue that they are the same species because hybridization produces individuals that are fertile and show a blend of traits between the two parents. Other scientists say the two kinds of oaks look so different they must be considered two separate species. How can such a disagreement be resolved? Can it be resolved at all?

Related Readings

Conniff, R. "The Name Game." *Science 82* (June 1982): 66–67. This entertaining article offers interesting explanations of why some newly discovered organisms were given their particular genus and species names.

Margulis, L. and Schwartz, K. *Five Kingdoms.* San Francisco: W. H. Freeman, 1982. This illustrated reference book presents the phyla and divisions of organisms and describes each group.

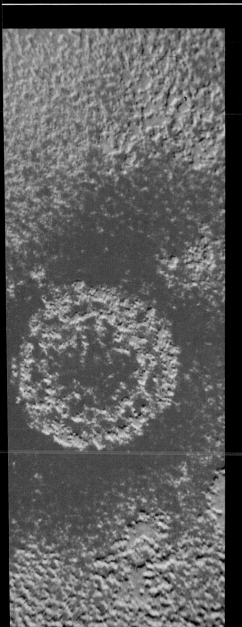

19 Viruses
Characteristics of Viruses • Viruses and Disease

20 Bacteria and Related Microorganisms
The Kingdom Monera • Bacteria • Blue-Green Algae

19

Viruses

Egyptian relief from about 1400 B.C.

Introduction

Nearly everyone suffers occasionally from a cold or from the aches and fever of the flu. People generally blame such minor illnesses as well as some potentially more serious or mysterious diseases on the presence of an unseen virus. A **virus** is a microscopic life form that reproduces only inside a living cell. Viruses are incapable of reproducing independently.

Diseases caused by viruses affected people long before scientists actually discovered viruses. An Egyptian carving from about 1400 B.C. depicts the effect of the viral disease poliomyelitis. Chinese literature describes a disease similar to smallpox as early as 900 B.C. Through an understanding of the nature of viruses, scientists have been able to prevent these and many other dreaded diseases.

Characteristics of Viruses

Section Objectives

- *Name* the characteristics of living things that viruses have and those they lack.
- *Identify* the main parts of a typical virus.
- *List* the main events that take place in the lytic cycle.
- *Explain* how viruses can transmit genetic information from one cell to another.
- *Distinguish* between a virus and a viroid.

Viruses do not easily fit into the classification systems used for all other life forms. Some biologists even question whether viruses should be considered alive at all. *Viruses exhibit some but not all of the characteristics of living things.* Viruses, like all living things, contain protein and nucleic acid. Yet, unlike cells, viruses can be solidified into crystals. When placed in a solution, they become active again. Viruses can reproduce themselves, but only within a living cell, called a **host cell.** Unlike cellular organisms, viruses do not respire, grow, or respond to stimuli.

19.1 Size of Viruses

Viruses differ greatly in size. They range in length from 0.01 micrometer (μm) to over 0.3 μm. The virus that causes influenza is of medium size, about 0.1 μm; yet over 500 of them can fit on the point of a pin.

19.2 Structure and Shape of Viruses

A typical virus consists of two parts, an inner core of nucleic acid and a protective outer coat of protein. The nucleic-acid core may consist of either DNA or RNA, the chemicals that contain coded genetic information. Unlike cells, which contain both DNA and RNA, each virus has only one type of nucleic acid. The DNA or RNA enables a virus to reproduce new viruses

▶ Viroids

Viruses are not the smallest known disease-causing agents. In 1967 biologists discovered that the potato tuber disease is caused by a tiny agent that consists entirely of a short strand of RNA. This life form requires a host cell to reproduce. Scientists gave the name **viroid** (VY royd) to this disease-causing agent.

Viroids have been identified as the cause of disease in at least seven kinds of plants, including cucumbers and tomatoes. They may also cause some diseases in animals and humans.

A viroid lacks the protective protein coat of a virus. Like a virus, a viroid reproduces itself only inside a host cell and causes disease by interfering with the normal functioning of the host cell.

Scientists still do not understand the exact process of viroid replication. Some evidence, however, suggests that reproduction takes place inside the nucleus of the host cell rather than in the cytoplasm.

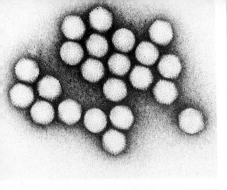
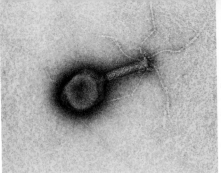

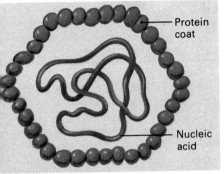
Protein coat

Nucleic acid

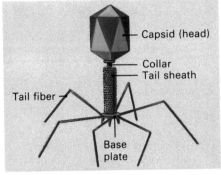
Capsid (head)

Collar
Tail sheath

Tail fiber

Base plate

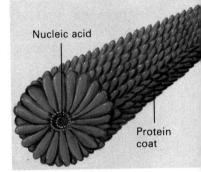
Nucleic acid

Protein coat

Figure 19–1. Polyhedral viruses (left), bacteriophages (center), and rod-shaped viruses (right) each have a protein coat that surrounds a central core of RNA or DNA.

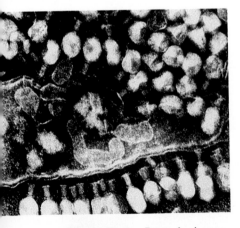

Figure 19–2. Bacteriophages attack a bacterium by first adhering to the cell wall. They then inject viral nucleic acids that alter the cell's genetic code.

exactly like itself. In cells, the DNA is double-stranded and the RNA is single-stranded. A virus, on the other hand, may have a nucleic-acid strand that is single or double, linear or circular.

The outer protein coat, called the **capsid,** makes up 95 percent of the body of the virus. Some of the proteins of the capsid are enzymes. The arrangement of the proteins in the outer coat determines the shape of a virus. Figure 19–1 shows that some viruses are polyhedral, having many sides, or facets. Others are rod-shaped viruses made up of repeating units of protein in a spiral arrangement. Viruses that invade bacteria, called **bacteriophages** (bak TIHR ee uh fayj uhz), have a polyhedral head and a hollow tail, usually with several fibers at the tip. Some viruses also have a membrane that surrounds the capsid. This membrane consists of proteins, lipids, and carbohydrates.

19.3 Reproductive Cycles of Viruses

Viruses cannot reproduce themselves unless they have invaded a host cell. Each type of virus attaches itself to specific plant, animal, or bacterial cells.

Lytic Cycle   During reproduction, many viruses kill the host cell. Such a process is called a **lytic** (LIHT ihk) **cycle.** Most knowledge of the lytic cycle comes from the study of bacteriophages, also known as *phages* (FAYJ uhz). The cycle begins when the phage comes into contact with a host cell. As Figure 19–3 (page 286) shows, the cycle has five main stages.

Highlight on Careers: Epidemiologist

General Description

Scientists who study the distribution of diseases and other health disorders in populations are called epidemiologists. These scientists also investigate the causes of widespread health problems in order to prevent or control them. For example, only a few years ago smallpox was eliminated worldwide through the work of epidemiologists.

In some ways, epidemiologists are like detectives. To understand why a disease breaks out among a group of people, epidemiologists try to answer four questions: When did the disease appear? How quickly did it spread? What are the characteristics of the place where the disease appeared? What are the characteristics of the people who developed the disease?

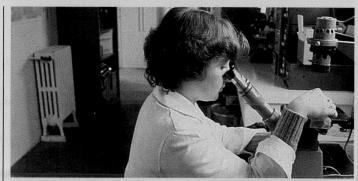

The answers to these questions usually reveal the cause of the problem. An epidemiologist, for example, many years ago discovered that people who became ill with a disease called *cholera* had all drunk water from a polluted well.

Universities and schools of public health employ most epidemiologists. Some positions for epidemiologists are also available in departments of public health and other government agencies, as well as in hospitals.

Career Requirements

Most epidemiologists have either an M.S., an M.D., or a Ph.D. Individuals with an M.S. degree usually coordinate investigations. Those with M.D. or Ph.D. degrees conduct independent research or teach. Epidemiologists must be well-organized, and conscious of details.

For Additional Information

The Association of Schools
  of Public Health
1015 15th Street, N.W.
Suite 404
Washington, DC 20005

---

1. *Adsorption.* The phage attaches itself to the cell wall. A chemical bond forms between specific molecular sites on the tail of the virus and corresponding sites on the cell wall, called **receptor sites.** The match between virus and receptor site is like that between a lock and key.
2. *Entry.* The phage releases an enzyme that breaks down the cell wall. The outer covering of the phage tail contracts, forcing the tail through the weakened cell wall. The nucleic acid of the phage passes through the hollow tail into the host cell, leaving the empty capsid outside.

Q: *Could viruses have been the first forms of life on the earth?*

A: Probably not. Since viruses require a living host for their reproduction, there must have been other living things before them. Viruses may have descended from cellular ancestors that became dependent on others, then reverted to simpler structures.

Figure 19–3. Viruses reproduce by both lytic and lysogenic cycles. In a lytic cycle, viruses invade, reproduce, and exit immediately. In a lysogenic cycle, viral genes are initially inactive.

3. *Replication.* Once viral nucleic acid enters the cell, it begins to replicate new virus parts. In a DNA virus, the viral DNA enters the host cell's nucleus and acts as a template for the formation of messenger RNA. The messenger RNA then migrates to the cytoplasm and causes the synthesis of viral proteins and viral RNA. Most RNA viruses contain an enzyme called *RNA transcriptase*. This enzyme causes the replication of viral RNA, which then acts as messenger RNA. Some viruses use an enzyme to make viral DNA from viral RNA. The viral DNA then migrates to the nucleus and directs the synthesis of new viruses.

4. *Assembly.* The viral nucleic acid and proteins are assembled into new, complete virus particles, called **virions** (VY ree ahnz).

5. *Release.* The new phages release an enzyme that weakens the cell wall. The host cell breaks open, or *lyses* (LY suhz), and releases the newly created viruses. Up to 300 new viruses can be produced in one cell. These viruses can then invade other cells.

The lytic cycle is similar for all viruses, though animal and plant viruses differ from bacteriophages in the way they enter

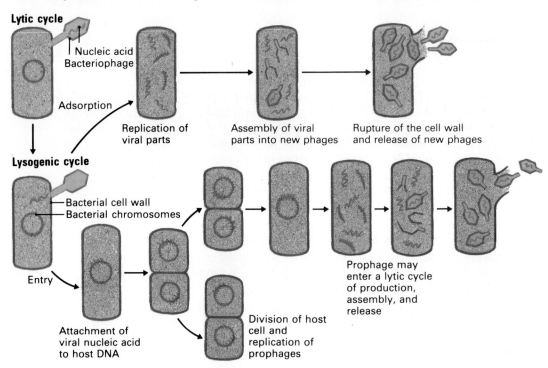

Lytic cycle

Nucleic acid
Bacteriophage

Adsorption

Replication of viral parts

Assembly of viral parts into new phages

Rupture of the cell wall and release of new phages

Lysogenic cycle

Bacterial cell wall
Bacterial chromosomes

Entry

Attachment of viral nucleic acid to host DNA

Division of host cell and replication of prophages

Prophage may enter a lytic cycle of production, assembly, and release cells. The whole animal virus passes through the cell membrane by *phagocytosis,* the same process by which large food particles enter the cell. Once inside the cell, the protein outer coat is destroyed by enzymes. Most plant viruses are injected through cell walls by insects.

Lysogenic Cycle   After entering a host cell, some phages remain inactive for many generations. Then suddenly the phages may become active and enter a lytic cycle of destruction. Scientists do not yet understand how inactive phages are activated. These phages are known as **temperate phages,** and the inactive cycle they undergo is called a **lysogenic** (ly suh JEHN ihk) **cycle.** This cycle goes through the following stages:

1. *Attachment.* The nucleic acid of the invading phage attaches itself to the DNA of the host cell. Such viral nucleic acid is called **prophage** (PROH fayj).
2. *Replication.* The prophage is replicated along with the DNA of the host cell during cell division.
3. *Activation.* The prophage enters a lytic cycle and orders the assembly of new viral parts. This process usually leads to the release of new phages and the destruction of the cell.

19.4 Transduction

Viruses have the ability to transfer genetic information from one host cell to another. This process is called **transduction** (trans DUHK shuhn). Two types of transduction have been identified. One, called *general transduction,* transfers random fragments of the host's DNA to the receiving cell. A second type, called *special transduction,* involves the transfer of specific genes from one cell to another. In each case, the gene fragments are packed into the virus just prior to lysis, and then transferred to a receiving cell during subsequent infection. ***Through transduction, a virus can alter the hereditary code of a cell.***

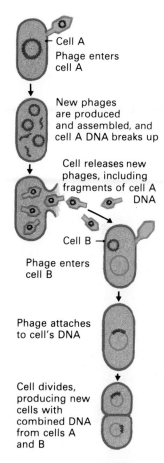

Cell A
Phage enters cell A

New phages are produced and assembled, and cell A DNA breaks up

Cell releases new phages, including fragments of cell A DNA

Cell B
Phage enters cell B

Phage attaches to cell's DNA

Cell divides, producing new cells with combined DNA from cells A and B

Figure 19–4. Transduction occurs when viruses carry genetic information from one host cell to another.

Reviewing the Section

1. Describe the parts of a typical virus.
2. How does a virus reproduce?
3. Compare a phage with a prophage.
4. What role may viruses play in spreading certain hereditary traits to other members of the species they infect?

- *Summarize* the early research that led to the discovery of viruses.
- *Explain* how some inactive viral infections recur.
- *List* three ways in which the body defends itself against viral infections.

Figure 19–5. When healthy tobacco plants (above) are attacked by the tobacco mosaic virus, their leaves become mottled yellow (inset). Tobacco mosaic is just one of many viral diseases.

Viruses and Disease

Viruses cause disease in plants and animals by destroying or altering the cells they inhabit. Disease-causing viruses, as well as certain bacteria and other microscopic parasites, are called **pathogens.**

19.5 Discovery of Viruses

A disease of tobacco plants known as tobacco mosaic first led to the identification of a virus as a pathogen. This disease causes the leaves of tobacco plants to become mottled with a yellow and green mosaic pattern. In 1892 the Russian biologist Dimitri Iwanowski squeezed the fluid from a diseased plant and passed it through a filter designed to hold back the smallest bacteria. He examined the fluid and the remains left on the filter under a light microscope. In both instances he found nothing. Yet rubbing the fluid on a healthy plant caused the plant to become diseased.

In 1898 the Dutch botanist Martinus Beijerinck repeated Iwanowski's work. Beijerinck concluded that the fluid contained an unknown factor, smaller than bacteria. He called this factor a *virus*, the Latin word for "poison."

An American biologist, Wendell Stanley, finally isolated the tobacco mosaic virus in 1935. Stanley extracted and crystallized the virus from the fluid of thousands of diseased plants. From a ton of diseased tobacco leaves Stanley produced a teaspoonful of crystals. In their crystallized form, the viruses did not appear to be alive. However, when the virus crystals were put back into solution and rubbed on a healthy plant, the plant contracted the disease.

19.6 Kinds of Viral Infections

Diseases caused by viruses range from minor infections that may go unnoticed to serious diseases, such as hepatitis and yellow fever. Some viruses cause major disturbances in a cell's growth and reproduction, resulting in tumors or cancer.

Viruses tend to attack a particular species of animal or plant and a specific type of cell within that organism. A cold virus, for example, attacks cells of the respiratory system, and a polio virus invades nerve cells. Many viruses attack more than one species, however. Rabies, for example, can be transmitted from dogs, raccoons, foxes, and bats to other mammals, including humans. Cowpox can be transmitted to humans handling diseased herds. The virus responsible for a pneumonialike disease called *psittacosis* (siht uh KOH sihs) can travel to humans from parrots and other birds infected with the disease.

Infections produced by viruses that undergo a lysogenic cycle may remain latent, or inactive, for a long period, then become active again. One example is the recurring infection caused by the *herpes simplex* virus. One type of herpes causes cold sores, and another type causes genital sores. These sores disappear during the inactive, lysogenic cycle and reappear during the virulent lytic cycle.

The ability of a virus to cause disease is called **virulence.** Several factors determine the virulence of a virus. One factor is the presence and activity of receptor sites on the surface of the cell, which enable the virus to become attached. Another factor is the speed with which the virus multiplies once it penetrates a host cell. A third factor is the response of the host organism to the invading viruses. For example, the cell may die immediately, it may divide abnormally, or it may produce defenses against the viruses.

19.7 Defenses Against Viral Infections

Many drugs used to treat other infections cannot be used on viruses. Those drugs that can kill the viruses would also harm the host cells. *The body provides its own best natural defense against viral disease.* This natural resistance to disease is called **immunity.**

The skin and the mucous membranes are the body's first line of defense. Viruses usually cannot penetrate the skin except through an existing cut or the bite of an animal such as a rabid dog. Cilia and mucus in body linings inhibit the entry of some pathogens. Nevertheless, most viruses enter the body through the nose and mouth.

Once viruses enter, the body attacks them directly. White blood cells called *phagocytes* (FAG uh syts) engulf and destroy invading viruses. A second method of defense involves the production of certain protein molecules. The foreign protein of the virus is called an **antigen** (AN tuh juhn). When the antigen enters the body, it triggers the production of **antibodies.** The antibodies are highly specific and attack only the antigen that triggered their production. Then, attaching themselves to the antigen, antibodies destroy the viruses completely or make it easier for the phagocytes to engulf and destroy the viruses.

Active and Passive Immunity Immunity to disease resulting from the production of antibodies is called **active immunity.** Antibodies can be produced in response to exposure to a

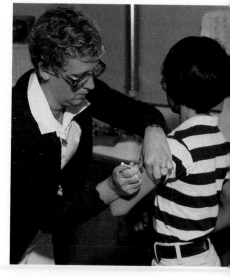

Figure 19–6. Vaccination is an important method of providing individuals with active immunity to many viral diseases.

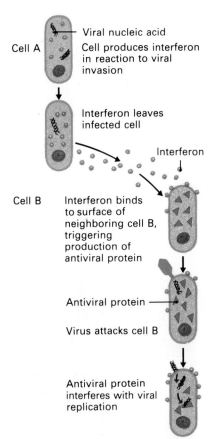

Cell A — Viral nucleic acid

Cell produces interferon in reaction to viral invasion

Interferon leaves infected cell

Interferon

Cell B — Interferon binds to surface of neighboring cell B, triggering production of antiviral protein

Antiviral protein

Virus attacks cell B

Antiviral protein interferes with viral replication

Figure 19–7. Production of interferon is triggered when a human body cell is invaded by a virus. This interferon then induces other cells to produce protective proteins. These proteins defend against subsequent viral attacks.

pathogen or in response to injection of a vaccine. A **vaccine** is a solution of weakened viruses. A vaccine serves as an antigen to stimulate the production of specific antibodies. In many cases active immunity is lifelong. One attack of measles or one measles vaccination, for example, usually provides permanent resistance against a second attack. **Passive immunity** occurs when a person receives antibodies produced in another person or in an animal that has developed immunity to the disease. One example of passive immunity is that received by an unborn child from its mother through antibodies passed in the blood. *Gamma globulin* is a blood protein often used to provide passive immunity to certain diseases. Passive immunity is always temporary.

Interferon The body's third method of defense is the production of the protein **interferon** (ihn tuhr FIHR ahn). Interferon "interferes" with viral replication. Unlike antibodies, it will affect any type of virus that invades the body. Interferon produced by cells under viral attack starts a chain of reactions in other neighboring cells. *Interferon triggers the production of an enzyme that enables a cell to recognize a virus as a foreign invader.* The enzyme prevents the reproduction of the invading viruses.

Interferon is *species specific*—that is, interferon produced in one species will work only in that species. For example, interferon produced by a mouse will not be effective in a rabbit or in humans.

The large quantities of human interferon needed for experimentation and treatment are difficult and expensive to obtain. However, recombinant DNA technology is providing a solution to this problem. If the gene containing the codes for interferon production is spliced into the DNA of a bacterium, the bacterium and its offspring will produce interferon in quantity. Human interferon produced by recombinant DNA has been used with some success in patients with some forms of cancer. Although doctors are still uncertain about the ultimate value of interferon therapy, they have been encouraged by the promising results they have obtained with their patients.

Reviewing the Section

1. Why do the symptoms of certain viral infections recur after having disappeared for a period of time?
2. Name three ways in which the body attacks invading viruses.
3. How does interferon work against viral infections?

Investigation 19: Making Models of Viruses

Purpose

To recognize the size relationships in a number of different viruses

Materials

Toothpicks, clay, plastic foam in different shapes, construction paper, pipe cleaners, wooden dowels, wire hangers, insulated electrical wire, tape, cork

Procedure

1. Look at the sketches of the five viruses on this page. The sketches show the basic structure of the viruses, but they are not drawn to scale. The actual size of each virus is shown in the table below.

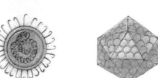

Influenza virus

Mumps virus

Polio virus

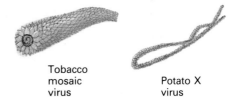

Tobacco mosaic virus

Potato X virus

2. Make a three-dimensional model of each virus. You may be as original as you wish in your choice of materials, but the five models must be to scale—that is, they must represent the actual size of each virus in relationship to every other virus.
3. Use the following procedure to determine accurate size relationships: Set the size of the smallest virus as 1. The actual diameter of the smallest virus is 0.01 μm. To get this size to 1, you must multiply by 100. Therefore, you must multiply the dimensions of every other virus by 100. You now have sizes that are easier to work with but have not changed in relationship to one another.
4. Next determine a size range. If you look at the column headed "Relative Size" in the table, you will see that the smallest size is 1 and the largest is 20. Therefore, the largest virus is 20 times larger than the smallest.

5. You must still decide on an actual size for your models. It would be convenient to make the relative size of 1 equal 5 mm. All the sizes must be multiplied by the same number. Multiplying the other sizes by 5 mm will give you the scale sizes shown in the table below.
6. You are now ready to make your models. Display your completed work in the classroom or laboratory.

Analyses and Conclusions

1. How many times larger is the mumps virus than the influenza virus?
2. Why are models such as yours useful in recognizing relative size?
3. The diameter of the *E. coli* bacterium is 1 μm; its length is 2.5 μm. What size is a model of an *E. coli* bacterium built according to the scale used in the investigation?

Going Further

- Use a microbiology textbook to determine the sizes of some other viruses. Make models of these viruses.
- Make a model of the *E. coli* bacterium mentioned in question 3 above.

Virus	Actual Size	Relative Size	Scale Size
Mumps	0.2 μm	20.0	100 mm
Potato X	0.01 × 0.5 μm	1.0 × 50	5 × 250 mm
Tobacco mosaic	0.018 × 0.3 μm	1.8 × 30	9 × 150 mm
Polio	0.028 μm	2.8	14 mm
Influenza	0.1 μm	10.0	50 mm

Chapter 19 Review

Summary

Viruses consist of a nucleic-acid core surrounded by a protein outer coat. Viruses can reproduce themselves, but only within a living host cell. The lytic cycle of viral reproduction results in the production of many new viruses and the destruction of the host cell. In the lysogenic cycle, viral nucleic acid fuses with the host DNA and is replicated during cell division. Such viruses may lie dormant for some time, then begin a lytic cycle. Scientists do not know how they are activated.

Transduction occurs when a virus carries a bit of the DNA from one host cell to another.

This process may alter the traits of the second host cell.

By interfering with the operation of the host cells, viruses can cause diseases in bacteria, plants, and animals. In humans, the body's own immune system is the best defense. Phagocytes produced by the body directly attack and destroy invading viruses. Antibodies combine with specific viral antigens, resulting in destruction of viruses. Immunity against diseases is either active or passive. Vaccines stimulate the production of antibodies. Interferon stimulates the body's defenses.

BioTerms

active immunity (**289**)
antibody (**289**)
antigen (**289**)
bacteriophage (**284**)
capsid (**284**)
host cell (**283**)
immunity (**289**)

interferon (**290**)
lysogenic cycle (**287**)
lytic cycle (**284**)
passive immunity (**290**)
pathogen (**288**)
prophage (**287**)
receptor site (**285**)

temperate phage (**287**)
transduction (**287**)
vaccine (**290**)
virion (**286**)
viroid (**283**)
virulence (**289**)
virus (**282**)

BioQuiz   (Write all answers on a separate sheet of paper.)

I. Completion

1. Viral nucleic acid that attaches itself to the DNA of the host cell is known as _____.
2. The smallest known disease-causing agent is a _____.
3. _____ and phagocytes will work against any type of invading virus.
4. An invading virus attaches itself to a _____ site prior to injecting its nucleic acid.
5. The process of transferring DNA from one host cell to another by a virus is called _____.

II. Modified True and False

Mark each statement TRUE or FALSE. If false, change the underlined term to make the statement true.

6. A vaccine is a solution of weakened viruses that can increase the body's immunity against disease.
7. Viruses that attack bacteria are known as bacteriophages.
8. The lysogenic cycle results in the rapid destruction of the host cell.
9. A temperate phage will remain inactive over a long period before becoming active and completing its life cycle.

III. Multiple Choice

10. Viruses are like cells in that they
    a) grow. b) reproduce. c) respond to
    stimuli. d) respire.
11. Antibodies are produced by the body in
    response to a) phagocytes. b) antigens.
    c) enzymes. d) capsids.
12. Scientists hope to get more interferon
    through a) chemical synthesis.
    b) animal breeding. c) recombinant
    DNA technology. d) human donations.
13. The outer coat of a virus is made of
    a) prophages. b) DNA. c) fats.
    d) protein.
14. The term *lysis* means a) to break open.
    b) to infect. c) to reproduce. d) to
    grow.

IV. Essay

15. How do phagocytes attack viruses?
16. What characteristics of living things do
    viruses show?
17. How can a virus introduce new traits
    into the cell it attacks?
18. How did Wendell Stanley first isolate a
    virus?
19. What is the difference between active
    immunity and passive immunity?
20. How does a bacteriophage enter a host
    cell?

Applying and Extending Concepts

1. Research the various folk remedies used to
   treat the common cold. Make a list of
   these remedies. Based on what you have
   learned about viruses, which remedies do
   you think are apt to be most helpful, and
   why?
2. Draw a series of diagrams showing how a
   bacteriophage penetrates the bacterial cell
   wall. Label receptor site, release of enzymes, contraction of tail, penetration,
   and passage of nucleic acid.
3. Most animal viruses pass through the cell
   membrane whole. Use your knowledge of
   cellular transport to explain how this
   might take place.
4. Select a viral disease that attacks economically important animals or plants, such as
   Newcastle disease or potato spindle tuber
   disease. Prepare a report for the class, including information on the symptoms,
   transmittal, and control of the disease.
5. Prepare a report comparing viruses,
   viroids, and prions. See the "Related
   Readings" for an article on prions. Discuss the structure of each life form and
   show how each damages host cells.

Related Readings

Eron, C. *The Virus That Ate Cannibals*. New
York: Macmillan, 1981. This collection of
essays examines six human viral diseases
and the scientific detective work that was
involved in understanding and treating
these diseases.

Grady, D. "The Mysterious Prion." *Discover*
4 (April 1983): 76–78. The article discusses the search for the prion, a hypothetical infectious agent smaller and simpler
than the virus.

Hall, Stephen S. "The Flu." *Science 83* 4
(November 1983): 56–64. This fascinating article discusses the influenza virus—
how it works, how it is transmitted, and
why it continues to infect humans year
after year.

Lenard, L. "The Battle to Wipe Out Herpes."
Science Digest 90 (November 1982): 36–
40. The article describes the herpes viruses
and their treatments.

*BIO***TECH**
Biosafety

The universal biohazard symbol appears on every shipment of dangerous substances.

A gravely ill man was admitted to the hospital, lapsed into a coma, and died within hours. Soon, hospital workers who had attended him also fell ill, and one third of them died. Finally, physicians agreed on a diagnosis: the mystery illness was Lassa fever, a rare viral disease that has no known cure.

Fortunately, the ill hospital workers had immediately been quarantined, and the disease spread no further. Lassa fever and many other rare and lethal viral diseases can spread like wildfire, killing large numbers of people and leaving survivors with permanent physical damage.

H ow do researchers study viruses like the type that causes Lassa fever without risking infection themselves? Until recently, the risk of contracting a rare disease was an accepted part of viral research. Now, however, new technological tools and techniques are helping to ensure the safety of research workers and the population at large.

At the Centers for Disease Control (CDC) in Atlanta, Georgia, researchers study viruses in a *maximum containment laboratory.* This laboratory combines many

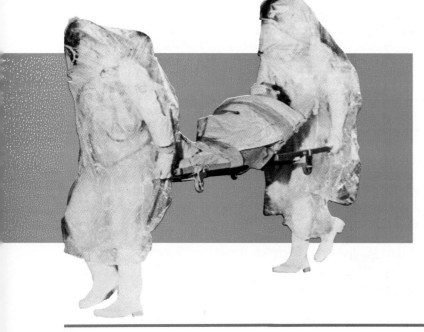

Technology has ensured the safety of researchers studying lethal viruses.

Researchers in a corporate maximum containment laboratory (left) and at CDC in Atlanta (below) wear protective suits.

technological innovations to keep viruses away from people. The laboratory itself is like a spaceship—self-contained, with its own repair workshop and air supply. Pressure detectors constantly monitor air pressure. If a system malfunctions, an alarm sounds immediately.

CDC workers enter the laboratory through an airlock. The outer door must be closed before the inner one is opened.

Lethal Marburg virus

They wear special one-piece plastic suits that are tightly sealed to prevent viruses from entering. They receive air through a central supply hose, which inflates the suit like a balloon. Should a suit develop a leak, the steady stream of air flowing out of the hole will prevent viruses from entering. In addition, each suit has its own air supply and alarm system, for use in an emergency or in case the central air supply system fails.

Anything that has been inside the laboratory is decontaminated before it is allowed to leave. CDC workers take a series of water and disinfectant

showers, and the waste liquid is itself treated to remove possible viruses. Researchers may not remove their notes and reports from the facility. A special photocopy machine makes a copy of written material. The copy is then delivered outside the laboratory.

The CDC laboratory is one of many maximum containment facilities around the world. These research centers freely exchange not only information but also samples of viruses. Researchers ship viruses in a special container that can withstand even the impact of an airplane crash.

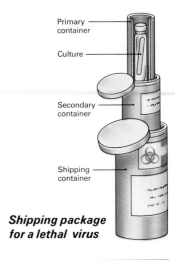

Primary container
Culture
Secondary container
Shipping container

Shipping package for a lethal virus

Bacteria and Related Microorganisms

A culture of *Escherichia coli* bacteria, ×14

Introduction

bacteria (singular, *bacterium*)

Simple, one-celled organisms called **bacteria** and related micro-organisms may have been among the earliest forms of life to appear on the earth. These one-celled organisms have evolved vastly different ways of sustaining and reproducing themselves. Some cannot live without oxygen. Others die in the presence of oxygen. These may be descendants of life forms that existed when the earth was young, before its atmosphere could sustain organisms that use oxygen.

Scientists use bacteria to study the structure and function of more complex cells. The bacterium *Escherichia coli* is one of the most studied of all living organisms. Research on *E. coli* has provided scientists with fundamental information on cellular structure and metabolism.

The Kingdom Monera

Bacteria and other members of the kingdom Monera live almost everywhere, including places in which few other organisms can survive. Some thrive in hot springs where the temperature is about 95°C (203°F.). Others have been found in Antarctica growing slowly at −7°C (22°F.). Monerans also live on and in every living thing. Human skin, even after a thorough cleansing, is home to millions of monerans.

Section Objectives

- *Name* the distinguishing characteristics of monerans.
- *List* factors used to classify monerans.

20.1 Characteristics

All monerans are prokaryotes. That is, their cells lack a true nucleus and membrane-bound organelles such as chloroplasts and mitochondria. The genetic material of monerans is a single, continuous loop of nucleic acid in direct contact with the cytoplasm. The ribosomes of monerans are smaller than those in eukaryotes.

Like plant cells, moneran cells have cell walls. However, almost all moneran cell walls contain muramic acid and sugars instead of the cellulose found in plant cell walls.

20.2 Classification

The kingdom Monera consists of three phyla. Bacteria are classified in the phylum Schizophyta (skihz AHF uh tuh). Photosynthetic blue-green algae make up the phylum Cyanophyta (sy uh NAHF uh tuh). The members of the third phylum, Prochlorophyta (proh klawr AHF uh tuh), are also photosynthetic algae but differ from the Cyanophyta mainly in the pigments used to carry out photosynthesis.

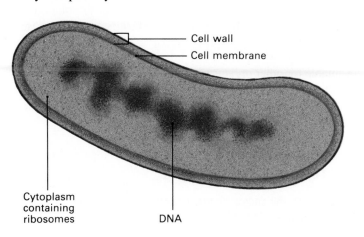

Figure 20–1. This bacterium is typical of moneran cells. A protective cell wall surrounds a selectively permeable cell membrane. The cytoplasm within the membrane contains both ribosomes and DNA.

Cell wall
Cell membrane
Cytoplasm containing ribosomes
DNA

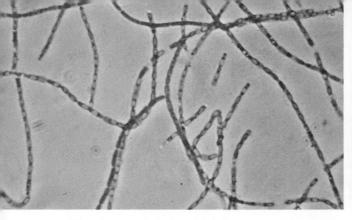
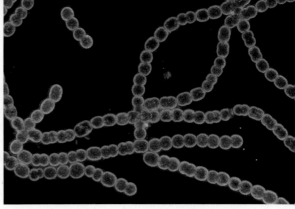

Figure 20–2. Representative organisms from two phyla of the kingdom Monera are shown above. Bacteria (left) are members of the phylum Schizophyta. Blue-green algae (right) belong to the phylum Cyanophyta.

Biologists at one time classified monerans mainly on the basis of appearance, such as size, shape, and movement. Modern tools, such as the electron microscope and chemical analyses, have enabled scientists to develop a better understanding of the biochemical characteristics of monerans.

Today biologists group the members of the phylum Schizophyta into 19 classes based on their biochemical characteristics. All of the members of this phylum are generally called bacteria. The term *eubacteria,* or true bacteria, is sometimes used to refer to 12 of the classes. Other Schizophyta organisms include **rickettsias** (rih KEHT see uhz), **mycoplasmas** (my koh PLAZ muhz), and **spirochetes** (SPY ruh keets). Rickettsias, unlike eubacteria, can live only inside other cells. Mycoplasmas are the only monerans that lack a cell wall. Spirochetes differ from other bacteria mainly in their relatively large size.

As scientists have learned more about the biochemistry of monerans, they have continued to revise their classification scheme. The blue-green algae, for example, were once classified along with other algae because both groups carry out photosynthesis with the resulting production of oxygen. As scientists learned about the prokaryotic structure of the blue-green algae, they began to classify them with the monerans. Prochlorophytes were not classified as a phylum separate from the blue-green algae until 1976. More recently some scientists have proposed that one group of monerans, the *archaebacteria,* be placed in a separate kingdom entirely. These bacteria live in environments unsuitable for other organisms, such as the bottom of swamps or water habitats seven times as salty as sea water. Some archaebacteria live in volcanic vents deep in the ocean where the pressure is great and the temperature is 306°C (581°F.).

Reviewing the Section

1. What are the major characteristics of monerans?
2. What features distinguish the three phyla of monerans?

Bacteria

Bacteria show a wide diversity of structure and function and carry out all the biological activities required for life. They play many important roles in the biosphere. Bacteria decompose and recycle the remains of dead organisms and thus return essential elements and compounds to the soil. Although some bacteria cause serious diseases, most are beneficial. Many are used to produce food and life-saving drugs.

20.3 Size and Shape

Bacteria vary widely in size, although all bacteria are microscopic. An average-sized bacterium measures about 1 μm in length. It would take about 300,000 such bacteria to cover the period at the end of this sentence. Spirochetes range in length from 5 to 500 μm. At the other extreme, rickettsias average about 0.05 μm. Mycoplasmas are the smallest free-living cells, measuring about 0.1 μm in length.

Bacteria can be characterized by their shape and by the way they group together. Bacteria have three basic shapes. Sphere-shaped bacteria are called **cocci** (KAHK sy), rod-shaped bacteria are called **bacilli** (buh SIHL eye), and corkscrew-shaped bacteria are called **spirilla** (spy RIHL uh). Cocci that form pairs are called *diplococci*, those that form clusters are called *staphylococci*, and those that form chains are called *streptococci*. Most bacilli separate after dividing. Those that stick together tend to form filaments, or threads. Spirilla do not group together but separate after division.

Section Objectives

- *Name* the parts of a bacterial cell and describe the function of each part.
- *Describe* the methods by which bacteria obtain or produce food.
- *List* some useful and harmful bacteria.
- *Name* and *explain* the major methods used to defend the body against bacterial invasion.
- *Describe* how and why Gram stains are performed.

cocci (singular, *coccus*)
bacilli (singular, *bacillus*)
spirilla (singular, *spirillum*)

Figure 20–3. The photos below show two basic shapes of bacteria. Bacilli (left) are rod-shaped. Cocci (right) are sphere-shaped.

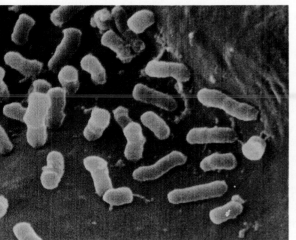

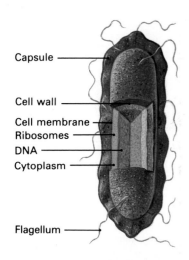

Capsule
Cell wall
Cell membrane
Ribosomes
DNA
Cytoplasm
Flagellum

Figure 20–4. Bacteria are well-protected organisms. A slimy capsule covers and protects the entire cell. The cell wall and cell membrane also protect the cytoplasm, ribosomes, and DNA.

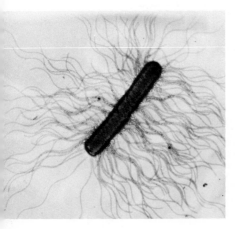

Figure 20–5. The large flagella that help this bacterium move are visible here, magnified 16,000 times.

flagella (singular, *flagellum*)

pili (singular, *pilus*)

20.4 Structure

Some bacteria have rigid cell walls and others have flexible walls. Some infectious bacteria produce a layer of slime that surrounds the outer surface of the cell wall, forming a protective **capsule.** The capsule may protect the bacteria from attack by a host's immune system. A thin cell membrane lies just inside the cell wall. Mycoplasmas, although they have no cell wall, do have a cell membrane and all the other major characteristics of bacteria.

The cytoplasm of a bacterial cell contains many ribosomes. Most of the cell's DNA forms a single, circular chromosome. In addition, some DNA may exist as smaller separate segments called **plasmids.** Like the other bacterial DNA, plasmids are circular and self-reproducing. One bacterium may have as many as a dozen different plasmids. These play a role in certain kinds of genetic transfers.

Some bacteria have long thin extensions called **flagella.** A bacterium may have a single flagellum or numerous flagella, all over its cell surface. **Pili** (PIHL ee) are extensions similar to flagella, but are shorter, thinner, and more numerous. Bacteria use pili to attach themselves to a source of food or oxygen or to another bacterium.

20.5 Movement

Not all bacterial cells are capable of movement. Those that move are called *motile,* and those that do not move are called *nonmotile.* Some motile bacteria move by gliding over a layer of slime. Other bacteria, such as the spirochetes, move by twisting or turning in a corkscrew fashion through water or other fluids. Still other bacteria propel themselves by using their flagella, which move in a rotating fashion.

20.6 Nutrition

Some bacteria are *autotrophic*—that is, they produce their own food. Others are *heterotrophic,* depending on autotrophs for food. Some autotrophic bacteria derive their energy from sunlight. Others derive their energy from the breakdown of inorganic chemicals—a process called **chemosynthesis.** Most heterotrophic bacteria have specialized enzyme systems that allow them to digest certain foods. Some species digest cellulose, while others digest only starch. Because of these various specializations, many types of bacteria can coexist in a single area with little or no competition among them for food.

Spotlight on Biologists: Lynn Margulis

Born: Chicago, 1938
Degree: Ph.D., University of California, Berkeley

Microbiologist Lynn Margulis has shaken the scientific community with her theory of how eukaryotes originated. She argues that cells with a true nucleus—and thus all multicellular organisms—evolved from one-celled prokaryotes that lived in close association with one another.

According to Margulis, cell organelles, such as mitochondria and flagella, had their origin in free-living bacteria. In her view, the first mitochondria evolved from oxygen-metabolizing bacteria that invaded other bacteria that could not metabolize oxygen. The invading bacteria helped the host bacteria survive by providing them a means to use oxygen, which is poisonous to cells unequipped to handle it. Margulis further theorizes that flagella evolved from spirochetes that similarly invaded other cells.

Obtaining conclusive evidence to support Margulis's theory would involve removing various types of organelles from cells, culturing each type separately, and then reassembling the organelles into a living cell. This effort now involves much of Margulis's time as a pro-

fessor of biology at Boston University.

Margulis's theory was highly controversial when first presented in 1966, but today her ideas are gaining acceptance by the scientific community. In 1983 she was elected to membership in the prestigious National Academy of Sciences.

Photosynthetic Bacteria

Bacteria that perform photosynthesis use different pigments than plants do. The purple sulfur bacteria and purple nonsulfur bacteria contain a chlorophyll chemically different from plant chlorophyll. Purple sulfur and purple nonsulfur bacteria also have red and yellow pigments called carotenes. Green sulfur bacteria contain a chlorophyll similar to the chlorophyll a in plants.

Another way photosynthetic bacteria differ from plants is that the bacteria do not use water and do not produce oxygen as a byproduct. Sulfur bacteria produce carbohydrates by combining carbon dioxide and hydrogen sulfide, using the energy of sunlight. Sulfur is produced as a byproduct. The reaction is:

$$CO_2 + 2H_2S \xrightarrow{\text{light, pigments}} (CH_2O) + H_2O + 2S$$

Purple nonsulfur bacteria use a variety of organic substances as raw materials instead of water.

Figure 20–6. The food preservation industry controls bacterial growth in a variety of ways. The photos above show a worker drying fish in the Philippines (top) and others curing Virginia hams with salt (bottom).

Chemosynthetic Bacteria Various types of chemosynthetic bacteria use different energy sources, including nitrogen and sulfur compounds. *Methanogens* convert CO_2 and H_2 to methane gas (CH_4), and in doing so create usable chemical energy for the cell. This process usually takes place in the mud at the bottom of swamps or marshes. The resulting methane gas gives some swamps a characteristically foul odor.

Heterotrophic Bacteria Most bacteria feed on dead organisms. Such bacteria are called **saprophytes** (SAP ruh fyts). These bacteria of decay break down organic matter and recycle it for use by other organisms. Some saprophytic bacteria live in the soil, others live on rotting bread or fruit. Still other saprophytes break down organic matter inside the intestines of humans and other animals.

20.7 Respiration

Some bacteria need oxygen to carry out respiration. Others use oxygen when it is available. Still others do not need oxygen at all, and in fact die in the presence of oxygen.

Bacteria that require oxygen to live are called **obligate aerobes.** Obligate aerobes generally live where there is an ample supply of oxygen, such as in the air or in loose soil. *Mycobacterium tuberculosis,* the bacterium that causes tuberculosis, is an obligate aerobe that can live in the lungs of human beings.

Bacteria that cannot live in the presence of oxygen are called **obligate anaerobes.** These organisms are found where there is little or no oxygen, such as deep in the soil or in mud at the bottom of lakes. Methanogens are obligate anaerobes. An obligate anaerobe called *Clostridium botulinum* causes a rare but extremely dangerous type of food poisoning known as *botulism*. If canned food is improperly prepared, botulism bacteria may begin to grow in the almost oxygen-free environment inside the can or jar.

A third type of bacteria, called **facultative anaerobes,** can grow with or without oxygen. The most common bacterium in the human digestive tract, *E. coli,* is a facultative anaerobe.

20.8 Growth

Bacterial growth usually refers to an increase in the number of bacteria, rather than to an increase in the size of an individual cell. A **colony** is a large group of bacteria, such as that grown on a nutrient plate in a laboratory. All the members of a colony are descendants of a single bacterium.

All bacteria need food and water and many need oxygen. Other factors that influence growth are temperature, sunlight, and chemicals. Limiting these growth factors helps prevent food spoilage due to bacterial action. Figure 20–6 shows how bacterial growth is controlled in food.

Some bacteria, particularly certain bacilli, survive harsh conditions such as high or low temperatures, lack of nutrients, or lack of water by forming special cells called **endospores.** The process typically begins when a colony of bacteria has begun to use up its food supply. The DNA replicates and a cell membrane forms around one strand of DNA and a bit of cytoplasm, creating a cell within a cell. A coat then develops around the smaller cell, forming the endospore. Once the endospore is developed, the rest of the cell may die. The endospore can lie dormant for hours or even years. When conditions in the environment

▶ The Gram Stain

Most bacteria are colorless. This feature makes them difficult to view under a microscope. In 1894 the Danish scientist Christian Gram developed a method to stain bacteria. The Gram stain is a differential stain—that is, a stain used to distinguish between types of bacteria.

To perform a Gram stain, a drop of bacteria culture is placed on a slide and dried. The bacterial smear is stained with *crystal violet* stain for one minute. The slide is then washed with water, and iodine is added for one minute. After another wash with water, the slide is washed in a decolorizing agent such as alcohol. If the bacteria have taken up the

crystal violet stain, the decolorizing agent will have no effect. If the bacteria have not been stained by crystal violet, then another stain must be used. A pink counterstain called *safranin* is added for 15 seconds. The slide is then washed and blotted dry. A bacteriologist then views the slide using a microscope and an oil immersion lens. Cells that have been stained violet are *Gram-positive.* Those that have taken up the pink

color of the safranin stain are *Gram-negative.*

The chemical makeup of the cell wall determines how bacteria react to the Gram stain. Since many antibiotics work by attacking the cell wall, the Gram stain also helps researchers identify the types of bacteria likely to be affected by antibiotics. Researchers have found that antibiotics are more likely to be effective against Gram-positive bacteria than Gram-negative bacteria.

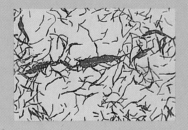

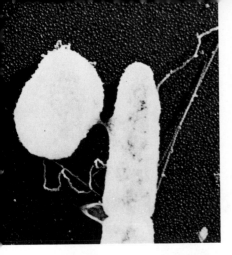

Figure 20-7. When bacteria conjugate, DNA from one bacterium passes through a long conjugation tube into another bacterium.

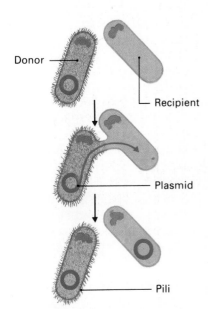

Donor

Recipient

Plasmid

Pili

Figure 20-8. During conjugation, a DNA plasmid is transferred from a donor to an acceptor bacterium.

are again favorable for growth, such as an increase in the supply of nutrients, the spore coat dissolves. The endospore then develops into a normal bacterial cell.

20.9 Reproduction and Genetic Transfer

Under ideal conditions, some bacteria can reproduce every 20 to 30 minutes. *Bacteria usually reproduce through binary fission, or splitting in two.* Except for occasional mutations, each new cell is exactly like the parent cell. Mutations, the result of errors or changes in the genetic code, account in part for the extraordinary ability of bacteria to adjust to differing conditions.

Genetic material in some cases is transferred from one bacterium to another, resulting in *genetic recombination*. When the bacterium later divides, it passes on its new genes to the daughter cells. In this way, resistance to antibiotics can be transferred from one strain, or genetic type, of bacteria to another.

In a process called **conjugation,** a donor bacterium transfers genetic material to an acceptor bacterium through direct contact. Conjugation occurs between two bacteria of the same species when one bacterium has a plasmid that the other lacks. Some of the genetic material of the donor passes to the recipient through a connective pilus. Conjugation occurs rarely and only in some species of bacteria.

As explained in Chapter 19, viruses can transfer genetic material from one bacterium to another through a process called *transduction*. Yet another process, called **transformation,** transfers genetic material from a dead strain of bacteria to a live strain. An example of transformation is provided by experiments of the British bacteriologist Frank Griffith, described in Chapter 12.

20.10 Beneficial and Harmful Bacteria

Some bacteria live independently. Others live in a close, permanent association with organisms of other species. When at least one of two organisms in close association with one another benefits from the relationship, the condition is called **symbiosis.** Three forms of symbiosis occur. In *mutualism* both organisms—the bacterium and its host—benefit. In *commensalism* one organism benefits and the other is neither helped nor harmed. In *parasitism* one organism benefits and the other is harmed. Bacteria that are **parasites,** or harmful to the host, are pathogens. These bacteria cause disease.

Under various circumstances, one type of bacteria may exist in a condition of commensalism, mutualism, or parasitism with its hosts. Consider as an example the *E. coli* that inhabits the digestive tract of humans and certain other mammals. In a commensal relationship, the *E. coli* absorbs nutrients from its host but does not help or harm the host. When producing enzymes, the *E. coli* benefits its host and is in a mutual association. If the *E. coli* enters the bloodstream, however, it may cause disease and is considered parasitic.

Beneficial Bacteria The majority of bacteria that live on and in the human body are harmless and many are even helpful. The inside of the human intestinal tract, for example, is lined with millions of bacteria. Many of these bacteria aid digestion by breaking down proteins, starches, and fats, and also by producing vitamins. Skin is covered with bacteria. Although some bacteria are harmful and cause skin irritations, most do not.

Many types of bacteria are used in food production. The genus *Lactobacillus,* for instance, is noted for its effect on milk products, producing buttermilk, yogurt, sour cream, and cheese. Lactobacilli are also used in the commercial production of sauerkraut and pickles.

The most important role of bacteria is ecological. Bacteria break down and decompose organic matter. Large, complex organic molecules are consumed and changed into simple chemicals that are then available for use by other organisms.

Nitrogen, an essential element of all proteins, makes up 78 percent of the atmosphere. However, most organisms cannot convert atmospheric nitrogen for their own use. Some bacteria, particularly the *Rhizobium,* take nitrogen from the atmosphere and incorporate it into nitrogen compounds. This process is called **nitrogen fixation.** The *Rhizobium* bacteria live in symbiosis with legumes, which are members of the pea family. The bacteria form *nodules,* or swellings, on the legume roots. The bacteria release "fixed" nitrogen into the plant's cytoplasm, and also release surplus nitrogen into the soil. Animals acquire usable nitrogen by eating the plants that have received nitrogen from the bacteria or from the soil.

Bacterial Infection Not all bacteria are harmless. Many bacteria are pathogens. These bacteria cause minor skin infections as well as such serious diseases as diphtheria, typhoid, tuberculosis, pneumonia, cholera, and leprosy. *Pasteurella pestis,* for example, is the bacterium responsible for a disease called *plague* or *black death,* which killed one out of every four people in Europe between 1348 and 1350.

Pathogens damage the body by direct attack and also by the production of *toxins,* or poisons. Diphtheria, for example, is a

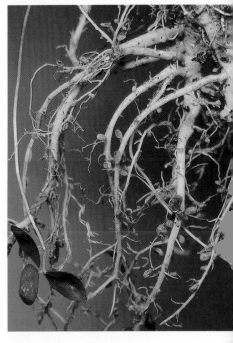

Figure 20–9. Nodules on the roots of this plant are home to millions of *Rhizobium* bacteria. The bacteria fix nitrogen and thus provide compounds that the plant needs.

A: Rocky Mountain spotted fever is a serious disease caused by rickettsias that live in the salivary glands of ticks. The organisms are transmitted to humans when a person is bitten by an infested tick. Although the disease was first discovered in the Rocky Mountains, it occurs in wooded areas throughout the United States.

disease caused by *Corynebacterium diphtheriae*. The bacteria invade the respiratory tract. The toxin produced by the bacteria enters the bloodstream. It is then absorbed by body cells and interferes with the functioning of the cells, sometimes resulting in the death of the victim.

Protection Against Bacterial Infection
The human body defends against bacterial invaders much as it does against viral infections. Attack by white blood cells and the production of antibodies are part of the body's own immune system. Another defense against bacterial infection has come through the use of antibiotics. **Antibiotics** are chemicals capable of inhibiting the growth of some bacteria.

The first antibiotic was discovered in 1929 by Alexander Fleming, a British bacteriologist. While growing a *Staphylococci* culture, he noticed that a mold had contaminated the bacterial culture. At first Fleming was annoyed at this. He felt that this experiment was ruined. Before throwing out his cultures, however, he noticed something peculiar about how the bacteria grew. Fleming noted that the bacteria were not growing in the area around the mold. Apparently the mold secreted a substance that inhibited bacterial growth. Fleming later named the substance **penicillin,** after the mold, which was *Penicillium notatum*. Penicillin acts by inhibiting the growth of bacterial cell walls. Penicillin is effective against several pathogens including *Streptococcus*. The discovery of penicillin is one of the greatest scientific advances of this century. Hundreds of thousands of lives have been saved through use of this antibiotic since methods were developed to produce it in large quantities. It is particularly effective when used for common infections.

Since Fleming's discovery, many other antibiotics have been identified. *Tetracycline* and *streptomycin* are common antibiotics that interfere with the protein synthesis of some pathogenic bacteria.

Reviewing the Section

1. Name and describe the three basic shapes of bacterial cells.
2. Name all the major structural features of a cell of bacteria.
3. How does photosynthesis in bacteria differ from photosynthesis in plants?
4. What is the ecological importance of bacteria?
5. How do antibiotics work in fighting disease?

Blue-Green Algae

Section Objectives

- *List* the major characteristics of blue-green algae.
- *Describe* how blue-green algae reproduce.
- *Explain* what causes algal bloom.
- *Compare* blue-green algae and prochlorophytes.

The blue-green algae are an extremely hardy group of organisms that live throughout the world. Some live in moist soil; others are found in the desert. Most are aquatic, living in fresh water or salt water.

20.11 Characteristics

Most of the 200 distinct species of blue-green algae form filaments, or long threads of attached cells. The others are unicellular, shaped like rods or spheres. The cell walls usually have a thick outer covering or sheath. Some, such as *Anabaena*, form single filaments. Others, such as *Nostoc*, cluster together in colonies of filaments organized in a gelatinous ball. The cells of blue-green algal filaments have interconnecting cytoplasm, and some cells may even have specialized functions. For this reason, some scientists consider them to be a simple type of multicellular organism.

Blue-green algae are photosynthetic. They contain chlorophyll *a*, the pigment found in plants, rather than the chlorophyll found in bacteria. Blue-green algae have xanthophyll and carotenes, also found in plants, plus additional accessory pigments

Figure 20–10. Blue-green algae have diverse shapes and sizes. *Trichodesmium* (top left), *Anabaena* (bottom left), and *Oscillatoria* (below) are examples of this phylum of monerans.

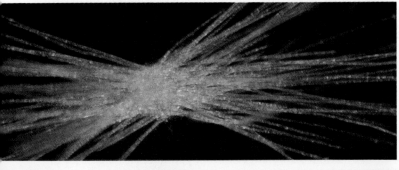

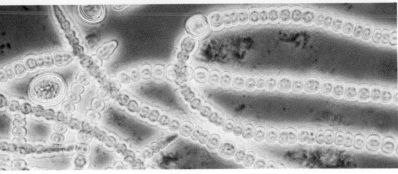

▶ Prochlorophyta, A New Phylum

In 1976 a new phylum was discovered by Roger Lewin of the Scripps Oceanographic Institute in La Jolla, California. This phylum, called Prochlorophyta, has some characteristics of blue-green algae and some characteristics of higher green algae. The prochlorophytes consist of prokaryotic organisms that are photosynthetic. These organisms almost always live in close association with certain small marine animals.

The bright green prochlorophytes contain both chlorophyll a and b, the same photosynthetic pigments used by more complex eukaryotic algae and plants. Blue-green algae contain only chlorophyll a. Prochlorophytes and blue-green algae also differ in the accessory pigments. Prochlorophytes have the standard xanthophylls and carotenes of plants. However, they lack the red and blue phycobilins possessed by blue-green algae.

News of the discovery of prochlorophytes aroused considerable interest among scientists, because the existence of these organisms helped to explain the origin of chloroplasts in eukaryotic cells. Biologists had earlier speculated that chloroplasts in plant cells came about as the result of an invasion of the cytoplasm of other cells by blue-green algae billions of years ago. This theory did not account, however, for the differences in the pigments of blue-green algae and the pigments found in plants.

The close similarity between prochlorophytes and plants suggests that the earlier hypothesis may be substantially correct but may need some revision. Some scientists now suggest that prochlorophytes, not blue-green algae, may be the living descendants of the organism that first gave rise to chloroplasts through invading other cells.

called *phycobilins* (FY koh by luhnz). The chlorophyll and other pigments are not enclosed in chloroplasts as they are in plants. Instead the pigments are located on sheets of membrane located in the cytoplasm. Like plants, and unlike photosynthetic bacteria, blue-green algae use water as a raw material of photosynthesis. They also produce oxygen as a byproduct of the process of photosynthesis.

Not all blue-green algae are blue-green in color. The presence of the accessory pigments causes these organisms to have a range of colors. Various species of blue-green algae are bright green, golden yellow, blue-black, violet, and many other colors.

Like some forms of bacteria, some blue-green algae carry out nitrogen fixation. For example, in Southeast Asia nitrogen-fixation by blue-green algae in rice paddies enables farmers to grow rice on the same land year after year without adding fertilizers.

20.12 Reproduction and Growth

Blue-green algae reproduce by binary fission. Colonies of algae also reproduce through **fragmentation,** a process in which the colony breaks into pieces and each piece forms a new organism or colony. Like bacteria, some blue-green algae can also produce resistant spores that survive in harsh conditions.

The rate at which blue-green algae grow depends in part on the chemical content of the water in which they live. Dumping phosphates and certain other chemicals into lake water can result in an uncontrolled growth of blue-green and other algae called an **algal bloom.** The water takes on the color of the algae living in it. The decay of overabundant algae reduces the amount of oxygen in the water. This process in turn causes fish to die and makes the treatment of the water for human use more difficult. For this reason, the use of phosphate in detergents has been reduced in recent years.

Q/A

Q: *How did the Red Sea get its name?*

A: The waters of the Red Sea are not red, but occasionally an algal bloom will cause the water to have a red tint. The alga responsible for the bloom is a species of blue-green alga that has a very high content of red phycobilin.

Reviewing the Section

1. Compare the pigments in blue-green algae and plants.
2. Describe the process of fragmentation.
3. What conditions can lead to an algal bloom? Why should this condition be avoided?

Investigation 20: The Gram Stain

Purpose
To distinguish between types of bacteria by means of the Gram-staining technique

Materials

Stock of culture of *Escherichia coli*, inoculating loop, Bunsen burner, three medicine droppers, two slides, forceps, Gram's differential stain (which consists of crystal violet, Gram's iodine, 95% ethyl alcohol, and safranin), beaker, paper towels, immersion oil, compound microscope with oil immersion objective, colored pencils, stock culture of *Sarcina lutea*, safety goggles.

Procedure
1. **Caution: Put on safety goggles and leave them on through step 2**. Obtain a stock culture of *E. coli*. Heat an inoculating loop in the Bunsen burner flame. Using proper bacteriological technique, place a drop of the culture on a slide. Spread the drop and allow it to dry.
2. Using forceps, quickly pass the slide through the flame of the Bunsen burner three times. Allow the slide to cool.
3. Flood the slide with crystal violet. Allow it to stand for one minute. Pour off the excess stain and rinse the slide gently in a beaker of water. Flood the slide with Gram's iodine. Allow it to stand for one minute.
4. Wash the slide with water. Add ethyl alcohol with a medicine dropper while tilting the slide to permit the alcohol to drain off. Continue to rinse the slide with alcohol until no more stain washes off.
5. Wash the slide with water.
6. Flood the slide with safranin. Allow it to stand for 15 seconds.
7. Rinse the slide with water and carefully blot it dry with a paper towel.

8. Add a drop of immersion oil to the stained bacteria portion of the slide. Place the slide on the microscope stage while the microscope is set for low power. After focusing, switch to high power. Then switch the oil immersion objective into place and observe the bacteria. The oil immersion objective should touch the oil on the slide. The objective will be very close to the slide, but the slide and objective should not touch.
9. Draw your observations. *Are the bacteria Gram-positive or Gram-negative?*
10. Repeat steps 1 through 9 using the bacteria *S. lutea.*

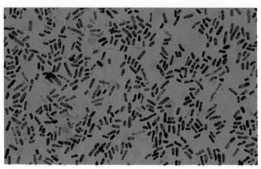

11. When you have finished using the oil immersion objective, carefully clean the oil from the objective using a solvent provided by your teacher.

Analyses and Conclusions
1. Why is the Gram stain a useful technique in classifying bacteria?
2. Explain why some bacteria are Gram-positive and some are Gram-negative.
3. Why is an oil immersion lens used to observe the stained bacteria?
4. Other stains are also used to observe bacteria. Explain why this is done.
5. How does the result of the Gram stain relate to the bacteria's reaction to antibiotics?

Chapter 20 Review

Summary

Monerans, which are all prokaryotes, are found in nearly every habitat on the earth. The three phyla of the kingdom Monera are the Schizophyta, which includes the bacteria; the Cyanophyta, also known as blue-green algae; and the Prochlorophyta, monerans that use chlorophyll for photosynthesis.

Most bacteria are heterotrophic, although some are autotrophic. Some autotrophic bacteria are photosynthetic, and others are chemosynthetic. Obligate aerobes need oxygen to grow; obligate anaerobes die in the presence of oxygen. Although some bacteria live independently, others exist a form of symbiotic relationship.

Bacteria are important ecologically because they help break down dead organisms into useful organic nutrients. Bacteria convert nitrogen into forms usable by plants. Some bacteria cause disease.

Blue-green algae are photosynthetic. They have many pigments found in eukaryotic algae and plants plus phycobilins. Their uncontrolled growth can cause an algal bloom.

BioTerms

algal bloom (309)
antibiotic (306)
bacillus (299)
bacterium (296)
capsule (300)
chemosynthesis (300)
coccus (299)
colony (302)
conjugation (304)

endospore (303)
facultative anaerobe (302)
flagella (300)
fragmentation (309)
mycoplasma (298)
nitrogen fixation (305)
obligate aerobe (302)
obligate anaerobe (302)
parasite (304)

penicillin (306)
pilus (300)
plasmids (300)
rickettsia (298)
saprophyte (302)
spirillum (299)
spirochete (298)
symbiosis (304)
transformation (304)

BioQuiz   (Write all answers on a separate sheet of paper.)

I. Completion

1. Dumping phosphates into lake water can result in ____.
2. Some bacteria form a protective ____ by producing slime.
3. ____ bacteria are light pink in color after a Gram stain.
4. ____ organisms derive nutrition from dead organisms.
5. ____ are the only monerans that contain chlorophyll *a* and *b*, which they use for photosynthesis.

II. Modified True and False

Mark each statement TRUE or FALSE. If false, change the underlined term to make the statement true.

6. Eukaryotic cells lack a membrane-bound nucleus.
7. The first antibiotic to be discovered was penicillin.
8. Pathogens exist with other organisms in a condition of mutualism.
9. Endospores allow bacteria to survive harsh environmental conditions.

III. Multiple Choice

10. All monerans have a) chloroplasts.
    b) nuclear membranes. c) ribosomes.
    d) cellulose cell walls.
11. Bacterial pili a) allow movement.
    b) move material between bacteria.
    c) control temperature extremes.
    d) protect against attack.
12. Methanogens produce energy from
    a) inorganic compounds. b) decayed
    plants. c) sunlight. d) host cells.
13. Bacteria reproduce through a) spore
    formation. b) fragmentation.
    c) transduction. d) binary fission.
14. Blue-green algae are a) saprophytic.

b) heterotrophic. c) chemosynthetic.
d) photosynthetic.

IV. Essay

15. What three forms of symbiosis occur
    among bacteria and other organisms?
16. How do motile bacteria move about?
17. How did new technology lead to
    changes in moneran classification?
18. How is the growth of bacteria controlled
    in food products?
19. Why is it possible for many different
    bacteria to coexist in a small area of
    space?
20. How do bacteria benefit humans?

Applying and Extending Concepts

1. Scientists think that the atmosphere of the
   ancient earth contained little or no free
   atmospheric oxygen gas. In that case,
   which types of bacteria probably were the
   first to appear? Explain.
2. Use your school library or public library to
   research penicillin. Write a paragraph
   about how this antibiotic kills bacterial
   cells.
3. Under ideal conditions, *E. coli* bacteria,
   common in the human digestive tract, reproduce every 20 minutes. Given these
   ideal conditions, how many bacteria could
   one *E. coli* bacterium reproduce in an
   eight-hour period?

4. The bacterium *Streptococcus mutans* lives
   on teeth and produces an acid that can
   cause tooth enamel to decay. Use this information to explain why people who eat a
   lot of food containing sugar generally
   have more cavities than those who do not.
5. Tetanus is a disease caused by an anaerobic bacterium that lives in the soil. Anyone who has stepped on a nail or received
   any other deep puncture wound must usually get an injection to prevent tetanus.
   Since bacteria could enter through any
   break in the skin, why is tetanus more
   likely to result from a puncture wound
   than from a scratch or a cut?

Related Readings

Asimov, I. *How Did We Find Out About
    Germs.* New York: Avon, 1981. This
    book traces the history of bacteriology
    from Anton Van Leeuwenhoek to the sci-
    entists of modern times.

Dixon, B. *Magnificent Microbes.* New York:
    Atheneum, 1979. The beneficial functions
    of bacteria and other microbes are ex-
    plored in this book.

Lappe, M. *Germs That Won't Die.* Garden
    City: Anchor/Doubleday, 1982. This book
    discusses the ability of disease-causing
    bacteria to develop resistance to antibiotic
    drugs.

Zinsser, H. *Rats, Lice, and History.* Boston:
    Little, 1935. This book remains a classic
    for its coverage of the cause, spread, and
    impact of epidemics.

UNIT

6 Protista and Fungi

Protozoa

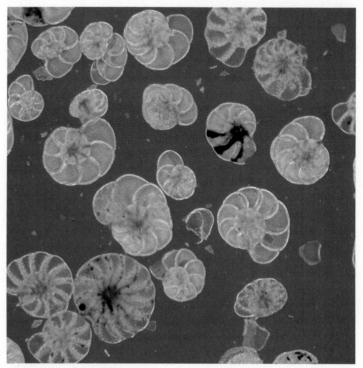

Foraminifera, marine protozoa

Introduction

Most of the plants and animals with which you are familiar are eukaryotes. These organisms have cells with a membrane-bound nucleus. Most eukaryotes are composed of millions of cells. However, one group of eukaryotes consists largely of diverse, unicellular organisms. Together these microscopic eukaryotes are called *protists* and are classified in the kingdom Protista.

The largest group of protists are the **protozoa** (proht uh ZOH uh). The word *protozoan* means "first animal." Even though they are single-celled, protozoa are complex organisms that perform all the basic functions of life. Protozoa ingest food, excrete wastes, carry out respiration, and reproduce. They also respond to stimuli from the environment.

protozoa (singular, *protozoan*)

Overview of Protozoa

Section Objectives

- *Identify* the characteristics of protozoa.
- *Summarize* the two ways in which protozoa take in food.
- *Describe* the ways in which protozoa may live in association with other organisms.
- *Name* the basis for classifying protozoa.

Like most protists, protozoa are one-celled eukaryotes. Unlike many other protists, however, protozoa are heterotrophic—that is, they obtain energy by feeding on other organisms. In general, protozoa are tiny organisms. Some protozoa grow to be 5 cm (2 in.) in diameter, but most kinds are so small they cannot be seen with the unaided eye.

The approximately 27,000 species of protozoa vary greatly in shape as well as in size. Many are jellylike blobs that continually change shape. Others are oblong or round cells that maintain a constant shape by secreting hard outer shells or with internal "skeletons."

21.1 Habitats and Ways of Life

Most protozoa are found in aquatic habitats. They flourish in freshwater lakes and streams and in saltwater environments. Marine protozoa and other microscopic heterotrophs drifting in the ocean are collectively called **zooplankton** (zoh uh PLANK tuhn). Protozoa also thrive in moist soil. The upper layers of soil may contain an average of 40,000 protozoa per gram (0.04 oz.) of soil. Some kinds of protozoa live in the body tissues and fluids of plants and animals.

Protozoa obtain nutrients directly from their environment. Oxygen, potassium, and other dissolved substances pass through the cell membrane by diffusion, a process in which molecules move from an area of greater to lesser concentration. Because the cell membrane allows certain dissolved substances to pass through, it is called *semipermeable*. Protozoa take in dissolved substances and solid bits of food. In many protozoa the flexible cell membrane surrounds and engulfs food particles that are too large to pass through it by diffusion. Protozoa also expel carbon dioxide, ammonia, and other molecular wastes by diffusion.

Protozoa live independently or in **symbiosis**—that is, in close association with another species of organism. Symbiotic relationships take one of three forms. In **parasitism,** one species is helped and the other is harmed. The species that benefits is called a *parasite*. The species that is harmed is called a *host*. Most protozoa that live in symbiosis are parasites. Other protozoa are found in two other forms of symbiosis. In *mutualism*, both species benefit from the association. In *commensalism*, one species is benefited and the other is unaffected by the symbiotic relationship.

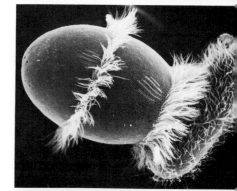
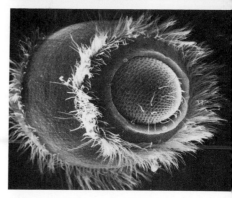

Figure 21–1. One protozoan, *Didinium,* attacks (top) and ingests (bottom) another protozoan, a paramecium. The scanning electron microscope (×1,450) captured this life-and-death struggle.

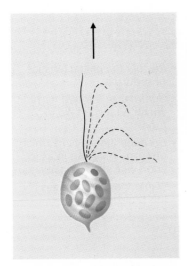

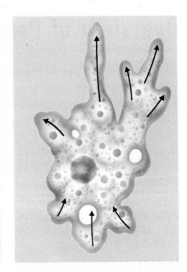

Figure 21–2. A euglenoid (left) moves by using a flagellum. A paramecium (center) uses cilia. An amoeba (right) extends pseudopodia to crawl about.

flagella (singular, *flagellum*)
cilia (singular, *cilium*)

Q/A

Q: *Do all protozoa need oxygen for respiration?*

A: No. Anaerobic protozoa will die if exposed to too much oxygen. Anaerobic protozoa commonly live in deep layers of mud or as parasites deep within animal tissues.

21.2 Classification

The majority of protozoa are *motile* organisms—that is, they use energy to move about freely. Motile protozoa are constantly in search of food. Many of them can detect and pursue prey, including other protozoa.

Nonmotile protozoa have no means of independent movement, so they cannot actively capture prey. Most nonmotile protozoa are parasites that live in a rich food supply, such as the bloodstream of an animal.

Protozoa are classified according to their method of movement. Three phyla of protozoa are motile. Protozoa in the phylum Mastigophora propel themselves by twirling or lashing one or more hairlike structures called **flagella.** Organisms in the phylum Sarcodina creep forward by the pressure of *cytoplasm* against the cell membrane. Members of the phylum Ciliophora move by the synchronized beating of a great many short flagella called **cilia** (SIHL ee uh). Sporozoa, the fourth phylum, consists entirely of nonmotile protozoa. All members of the phylum Sporozoa are parasites.

Reviewing the Section

1. What trait distinguishes protozoa from other members of the kingdom Protista?
2. How do protozoa obtain food?
3. What is parasitism?
4. What is the basis for classifying protozoa?

Kinds of Protozoa

Section Objectives

- *Contrast* the ways in which amoebas and paramecia capture and digest food.
- *Compare* the features of protozoa and *Euglena.*
- *Explain* how paramecia reproduce sexually.
- *Describe* the life cycle of a typical sporozoan.

Each of the four phyla of protozoa has a characteristic way of moving. The phyla also differ in their typical cell shape and structure, and in the various ways in which they obtain food and reproduce.

21.3 Flagellates

Protozoa in the phylum Mastigophora are called *flagellates* because they move by means of one or more flagella. About 2,500 species of protozoa are flagellates. Biologists consider flagellates the least complex protozoa because movement by flagella requires a minimal amount of cellular coordination. A few flagellates are free-living, but most are parasites. Some species of parasitic protozoa have complex life cycles involving two or more hosts.

Figure 21–3 shows flagellates in the genus *Trypanosoma.* Each of these protozoa moves by means of a single flagellum and the wavelike motion of a thin membrane that runs along one side of the cell. Trypanosomes are parasites that cause African sleeping sickness in humans and cattle.

The life cycle of trypanosomes begins not in humans or cattle, however, but in the tsetse fly. Here the trypanosomes reproduce but cause no harm to their hosts. When the tsetse fly bites humans or cattle, trypanosomes enter the bloodstream of their new host in the fly's saliva. There the trypanosomes reproduce again and secrete poisons that cause fever and the common symptom of extreme sleepiness that gives the disease its name. In some cases, trypanosomes even cause death. In general, however, they do not kill their hosts. A tsetse fly that bites an infected organism ingests trypanosomes and continues to spread the disease.

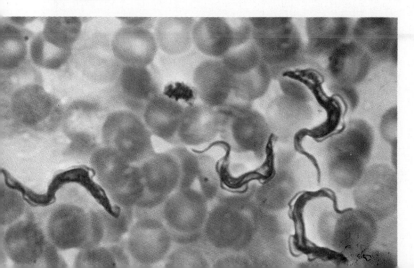

Figure 21–3. These protozoa of the genus *Trypanosoma* cause African sleeping sickness in humans and cattle.

Other flagellates are mutualistic. Members of the genus *Trichonympha* live in the digestive system of termites. Termites feed on cellulose fibers in wood. However, they lack the enzyme that can digest the tough fibers. The flagellates produce an enzyme that breaks down the fibers into a carbohydrate that both they and the termites can use. The termites benefit because they can digest their food. The protozoa benefit because they are protected from the external environment and provided with a regular food supply.

▶ *Euglena:* **Protozoa or Plantlike Protists?**

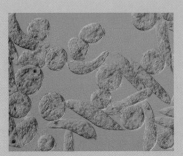

Euglena is a genus of freshwater, unicellular organisms that have features of protozoa and of plantlike protists called *algae.* Like mastigophorans, members of the genus *Euglena* move by using flagella. Like algae, they have chloroplasts and carry out photosynthesis. *Euglena* is sometimes classified as a separate phylum because of this unique combination of traits.

Normally, *Euglena* cells are autotrophic, making their own food by photosynthesis. They also have special traits that enhance their food-producing abilities. If a population of

Euglena is left in a glass by a window, a green cloud forms on the sunny side of the glass. This shows that the organisms move toward light, a trait called **positive phototropism.** A red-pigmented **eyespot** guides *Euglena* cells to areas that provide maximum opportunity for photosynthesis.

If *Euglena* cells are kept in the dark for a long period, their heterotrophic traits become dominant. They engulf and digest microorganisms and other prey just as amoebas and paramecia do.

The *Euglena* cell is surrounded by protein strips that form the **pellicle,** a layer just under the cell membrane. The pellicle's flexibility permits *Euglena* to creep through mud or water that is filled with debris. The organism stretches its front half forward and draws up its rear

half in a wormlike motion called **euglenoid movement.** A euglenoid cell may also move by means of flagellum.

Euglena reproduces asexually by dividing lengthwise in a mitotic process called *longitudinal binary fission.* Normally, a euglenoid cell divides once a day. If conditions become unfavorable, the cell forms a protective covering. Other kinds of protists—protozoa and algae alike—also respond to harsh conditions this way.

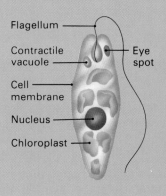

21.4 Sarcodines

About 11,500 species of protozoa belong to the phylum Sarcodina, commonly called *sarcodines*. Sarcodines move by extending parts of their cytoplasm and cell membrane to form footlike projections. These projections, called **pseudopodia** (soo duh POH dee uh), are pushed out by cytoplasm flowing in the cell. The word *pseudopodia* means "false feet."

Many sarcodines have shells. Members of the marine genus *Foraminifera* secrete hard outer shells of calcium carbonate. *Foraminifera* extend pseudopodia through spaces in their shells. Members of the genus *Radiolaria* have internal shells made of silicon. Radiolaria have pointed pseudopodia, which are coated with a sticky substance that traps food.

Other species of sarcodines have no shells. The most familiar of these belong to the genus *Amoeba* (uh MEE buh). Amoebas are soft, jellylike organisms. They live in muddy lake bottoms, the ocean floor, and on the surface of water plants.

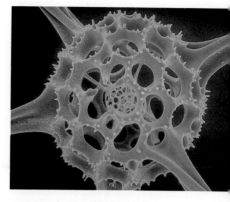

Figure 21–4. Many sarcodines secrete shells that protect their soft bodies. This radiolarian shell is made of silicon compounds.

Structure of Amoebas   Amoebas look like shapeless, irregular masses of cytoplasm, yet they are highly organized cells. Each amoeba contains at least one nucleus and other organelles. It also has a specialized organelle called the *contractile vacuole* that collects excess water and pumps it out of the cell. This regulation of water in the cell keeps the amoeba from bursting.

Most of an amoeba is filled with **endoplasm,** a thick, grainy kind of cytoplasm. Between the endoplasm and the cell membrane is a clear, thin layer of cytoplasm called **ectoplasm.** Ectoplasm acts as a lubricant by allowing the cell membrane to slide over the contents of the cell. When amoebas move, the endoplasm pushes forward to form new pseudopodia while existing pseudopodia are drawn up from behind. This characteristic creeping motion is called **amoeboid movement.** It gives amoebas their ever-changing shape.

Capturing Food   Amoebas absorb water and dissolved nutrients through their cell membrane, but they also actively seek food. Amoebas use pseudopodia to capture unicellular organisms and other bits of food in a process called *phagocytosis.* When an amoeba comes into contact with a food particle, one or more pseudopodia engulf the food. The cell membrane that surrounds the particle then pinches together and separates from the outer membrane, forming a *food vacuole.* Enzymes diffuse into the vacuole and digest the food. Undigested food is excreted by the opposite process of *exocytosis.* In exocytosis a vacuole forms around the waste, moves towards the outside of the cell,

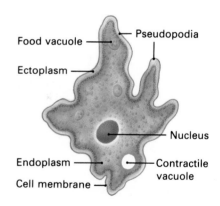

Figure 21–5. Amoebas are the most familiar sarcodines. Although irregular in shape, amoebas are highly organized cells.

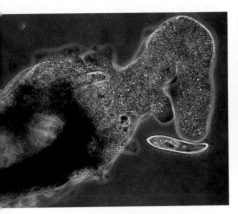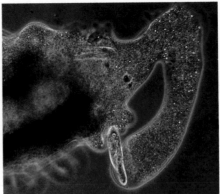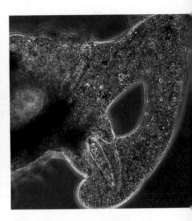

Figure 21–6. Amoebas engulf food through phagocytosis. An amoeba first encircles the food with pseudopodia (left). The pseudopodia then close around the food and draw it close to the cell (center). Later the amoeba engulfs the food and digests it (right).

fuses with the cell membrane, and expels its contents through a temporary opening. The vacuole then remains fused with the cell membrane.

Reproduction   Amoebas reproduce asexually by **binary fission,** a process in which two identical daughter cells form from one parent cell. The nucleus of the parent cell divides and the cell membrane pinches in half, dividing the cytoplasm between the two new cells. Binary fission occurs about every 24 hours unless conditions are unfavorable for survival. Then the amoeba's cell membrane thickens into a protective outer structure called a **cyst** (sihst). Within the cyst, the organism's nucleus may divide many times in a process called **multiple fission.** Many new amoebas are released from the cyst when conditions again become favorable for their survival.

Response   Like all organisms, amoebas are sensitive to their environment. In a response called **negative phototropism,** amoebas move away from light. Amoebas feed on the nutrient-rich bodies of dead organisms, called *detritus,* that collect on the dark floors of lakes and the ocean. Negative phototropism helps amoebas survive by leading them into regions where they are most likely to find food.

Amoebas show sensitivity to chemicals in their environment by moving toward those they sense as food and away from those they sense as harmful. Amoebas are also able to detect and avoid objects that block their paths.

21.5 Ciliates

The phylum Ciliophora, called *ciliates,* includes about 7,200 species. Ciliates are the most complex protozoa. Instead of one or two flagella, ciliate cells have hundreds of cilia. Each cilium looks like a fine eyelash. The cilia may cover the entire surface of the cell or they may grow only from a specific part of the cell. Cilia beat continuously and rapidly, sometimes as fast as 60 times per second. Their movements are highly synchronized.

Most ciliates are free-living and use their cilia to propel themselves through water in pursuit of prey. However, some types of ciliates become permanently attached to a rock or other surface. Organisms that remain anchored in one place for most of their lives are said to be **sessile.** Sessile ciliates are not free to move through the water and pursue prey. Instead, the beating of their cilia creates a whorl of water that sucks food particles into the cell.

Structure of *Paramecium* The genus *Paramecium* is a familiar representative of the ciliates. Paramecia live in fresh water and are easy to collect. These organisms have distinct anterior and posterior ends, as shown in Figure 21–8 on page 322. The shape of a paramecium is maintained by a sheath of protein called the pellicle that surrounds the cell membrane. Rows of beating cilia move the cell forward in a spiral path.

Like most ciliates, paramecia contain two nuclei. The smaller nucleus, called the **micronucleus,** controls reproduction. The larger nucleus, called the **macronucleus,** directs the

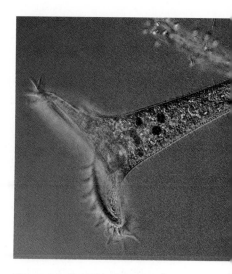

Figure 21–7. This funnel-shaped ciliate called *Stentor* collects food by creating a whirlpool with cilia located around its mouth. As organisms swim by, they are swept into the funnel.

Spotlight on Biologists: Robert Whittaker

Born: Wichita, Kansas, 1920
Died: 1980
Degree: Ph.D., University of Illinois

In 1969 ecologist Robert Whittaker presented a logical scheme for classifying organisms into five kingdoms—a system widely used by scientists today. Whittaker's studies of ecological communities gave him the broad perspective needed to work out such a classification system.

Whittaker developed statistical procedures used to study plant communities and the individuality of species. In his book *Communities and Ecosystems,* Whittaker helped demonstrate that plant species overlap in their distribution, rather than form associated groups.

Whittaker was noted for the interest he took in young scientists throughout the world. In 1984 the Ecological Society of America established in his

honor the R. H. Whittaker Travel Fellowship, which enables young ecologists from other countries to study in the United States.

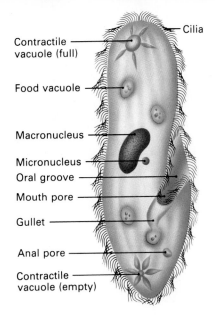

Contractile vacuole (full)

Cilia

Food vacuole

Macronucleus

Micronucleus

Oral groove

Mouth pore

Gullet

Anal pore

Contractile vacuole (empty)

Figure 21–8. A paramecium gathers food into an oral groove and gullet, digests it in food vacuoles, and excretes wastes through an anal pore. Water balance is regulated by a contractile vacuole.

Figure 21–9. The process of conjugation, illustrated below, allows the transfer of genes from one paramecium to another.

other metabolic functions in the cell. A paramecium cell has endoplasm, ectoplasm, and the organelles found in amoebas. In paramecia, however, the contractile vacuole is more specialized. Structures called **radiating canals** surround the vacuole. The radiating canals function like drainpipes by collecting excess water from a wide area around the contractile vacuole.

Capturing Food Paramecia have more complex means of capturing and digesting food than do amoebas. Figure 21–8 shows the food passageway in a paramecium. Cilia sweep food particles into a channel known as the **oral groove.** From there the food travels to an opening called the **mouth pore,** where food enters the endoplasm. The food is then stored in a chamber beneath the mouth pore called the **gullet.** At the end of the gullet, a food vacuole forms. As in amoebas, enzymes diffuse into the vacuole and digest the food. Undigested wastes are carried to the **anal pore,** where the vacuole expels the wastes through exocytosis.

Reproduction Paramecia may undergo a form of sexual reproduction called *conjugation,* a process by which two cells exchange genetic material. In some species of paramecia, one of the two donor cells dies after conjugation. This type of conjugation resembles bacterial conjugation.

Conjugation begins when two paramecia join at their oral grooves. The micronucleus in each paramecium then undergoes meiosis, producing two haploid micronuclei in each organism. One of these new micronuclei and the macronucleus of each cell disappear. Then the remaining micronuclei move to the oral groove, where they undergo mitosis. As Figure 21–9 shows, the oral groove now contains four haploid micronuclei, one matching pair from each cell.

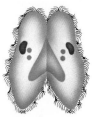  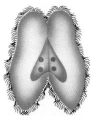 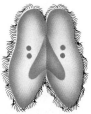 

Paramecia join at oral groove; micronuclei divide

One new micronucleus in each moves to oral groove; macronuclei disintegrate

Micronuclei divide by mitosis, leaving two matching pairs (4 micronuclei in groove)

One from each pair transfers to the other paramecium

New pairs fuse; paramecia separate; new macronuclei develop

Next, the paramecia exchange genetic material. Each cell keeps one micronucleus and donates the second one to the other cell. The new pairs of haploid micronuclei in each paramecium fuse, forming diploid micronuclei with a new combination of genetic material. As the two paramecia separate, a macronucleus containing the new combination of genetic material forms in each organism.

Like amoebas, paramecia also reproduce asexually by binary fission. However, this process is somewhat different in paramecia. The micronucleus divides by mitosis, and the resulting micronuclei move to opposite ends of the cell. A new set of organelles forms so that each daughter cell has a complete set. The genetic material in the macronucleus is divided between the daughter cells. The cytoplasm also divides, forming two separate paramecia.

Response When a paramecium bumps into an object, it stops, reverses direction, and then continues forward at a different angle. It repeats this *avoidance reaction* as often as necessary to get past the obstacle. When attacked by another organism, a paramecium shows a *defense reaction* by discharging tiny poisonous hairs called **trichocysts** (TRIHK uh sihsts), which are located underneath the pellicle. Trichocysts can drive away attacking organisms or paralyze organisms that the paramecium then consumes.

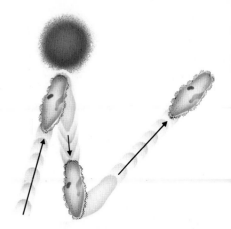

Figure 21–10. When a paramecium encounters an obstacle, such as a drop of ink, it backs off, alters its course slightly, and proceeds forward. Scientists call this behavior an *avoidance reaction.*

21.6 Sporozoans

The phylum Sporozoa contains about 6,000 species, all of which are nonmotile parasites. These protozoa do not have flagella, cilia, pseudopodia, or any structures that produce independent movement. They are carried along by currents in the blood or other body fluids of their hosts.

As parasites, sporozoans cause a wide variety of serious diseases. *Plasmodium,* shown in Figure 21–11, is the genus that causes *malaria.* **The disease cycle of Plasmodium illustrates the life cycle of a typical sporozoan.**

Like most sporozoans, *Plasmodium* has a complex life cycle that involves more than one host. This life cycle begins when a female *Anopheles* mosquito bites a person with malaria. Only female *Anopheles* mosquitoes carry malaria. The males do not feed in the adult stage. The mosquito ingests the *Plasmodium* cells in the person's blood. The cells reproduce sexually in the digestive system of the mosquito. In this process, male and female *Plasmodium* cells develop into sperm and egg cells, which fuse into zygotes. The nucleus of a zygote may divide

Q/A

Q: *How serious is malaria?*

A: Malaria has killed more people than any other infectious disease in history. Due to shortages of quinine, more people died from malaria in World War II than were killed in battle.

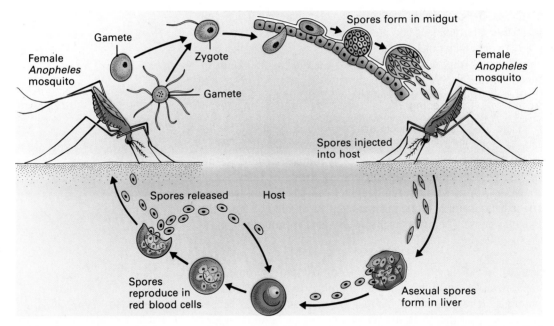

Female
Anopheles
mosquito

Gamete

Zygote

Gamete

Spores form in midgut

Female
Anopheles
mosquito

Spores injected
into host

Spores released

Host

Spores
reproduce in
red blood cells

Asexual spores
form in liver

Figure 21–11. *Plasmodium,* the cause of malaria, is carried from host to host by mosquitoes. Sexual reproduction takes place in the mosquito's gut. The formation of asexual spores takes place in the host's liver, and spores then spread to the blood.

several times, forming immature *Plasmodium* cells called *spores*. Eventually the zygote breaks apart, releasing the spores.

When a mosquito bites someone, the spores are carried in its saliva to the person's bloodstream. From there the spores travel in the blood to the liver. In the liver the spores reproduce asexually, then invade the red blood cells of the host. Every 48 to 72 hours, the spores break out of the red blood cells, destroying them and secreting poisons into the bloodstream. These poisons cause extreme weakness and intense fever. Often, they cause death. The spores may develop into male and female cells, but they do not fuse until they are taken up by a female *Anopheles* mosquito. Development then proceeds in the insect and the disease cycle begins again.

Although sporozoans are simple in structure, scientists consider them to be among the most evolutionarily advanced forms of protozoa. Sporozoans appear to have lost many complex adaptations as they became parasites. Thus, their very simplicity is considered to be an advanced adaptation.

Reviewing the Section

1. How do paramecia capture food?
2. Why could *Euglena* be called a protozoan?
3. What are two ways in which paramecia reproduce?
4. How are *Trypanosoma* and *Plasmodium* alike?

Investigation 21: Observing Protists

Purpose
To compare the structural characteristics, movement, and feeding responses of some typical protists

Materials
Stock culture of paramecia, individual stock cultures of two or three other types of protists, medicine droppers, slides, coverslips, compound microscope, yeast/congo red solution

Procedure
1. Make a chart like the one shown below. Your teacher will provide you with the names of the protists you will be observing. Put the name of each protist in column 1. As you observe each protist, fill in the rest of the chart.

4. Examine the slide under the microscope. Focus on one organism and note any feeding response. Describe how the organism feeds in the last column of your chart.
5. Using a clean dropper for each culture, repeat steps 2, 3, and 4 for each culture type provided. Complete your chart.

Analyses and Conclusions
1. What characteristic(s) did all of the protists you observed have in common?
2. In what ways were the protists different?
3. Based on the feeding mechanisms you observed, describe how the protists might obtain food in their natural environment.

Protist Name	Sketch	General Description (Size/Shape/Color)	Type of Movement	Feeding Mechanism

2. Obtain a culture of protists from your teacher. Make a wet mount of one drop of the culture. Observe the wet mount under the microscope under low power, then high power. Try to focus on one individual organism if possible. Observe the organism for several minutes. Note any distinctive characteristics. Record these on your chart. If the organism moves, describe the type of movement.
3. Place another drop of the same culture on a clean slide. Add one drop of yeast/congo red solution to the culture. Add a coverslip. Describe what happens.

Chapter 21 Review

Summary

Protozoa are unicellular, eukaryotic, heterotrophic organisms. Most of them live in water. They take in solid food by phagocytosis and absorb nutrients through their cell membrane by diffusion.

Each phylum is distinguished by its method of movement. Flagellates move by flagella. *Euglena* may be classed as a flagellate, although it has chloroplasts.

Sarcodines use footlike extensions called pseudopodia both to move and to capture food. Amoebas feed by engulfing food, and they reproduce by binary fission. Like other protozoa, sarcodines respond to light, chemicals, and food in the environment.

Ciliates move by the synchronized beating of cilia. The complex cells of paramecia contain food vacuoles, radiating canals, and an oral groove leading to a mouth pore and gullet. Paramecia reproduce by binary fission and conjugation.

Sporozoans are nonmotile, parasitic protozoa. They are represented by *Plasmodium*, the parasite that causes malaria.

BioTerms

amoeboid movement (**319**)
anal pore (**322**)
binary fission (**320**)
cilia (**316**)
cyst (**320**)
ectoplasm (**319**)
endoplasm (**319**)
euglenoid movement (**318**)
eyespot (**318**)

flagella (**316**)
gullet (**322**)
macronucleus (**321**)
micronucleus (**321**)
mouth pore (**322**)
multiple fission (**320**)
negative phototropism (**320**)
oral groove (**322**)
parasitism (**315**)

pellicle (**318**)
positive phototropism (**318**)
protozoa (**314**)
pseudopodia (**319**)
radiating canal (**322**)
sessile (**321**)
symbiosis (**315**)
trichocyst (**323**)
zooplankton (**315**)

BioQuiz   (Answer all questions on a separate sheet of paper.)

I. Completion

1. Protozoa are classified according to their method of _____ .
2. Hard-shelled protozoa such as *Foraminifera* belong to phylum _____ .
3. _____ phototropism directs amoebas away from sources of light.
4. Paramecia exchange genetic material through the reproductive process called _____ .
5. Protozoa that inhabit and harm a host are called _____ .

II. Modified True and False

Mark each statement TRUE or FALSE. If false, change the underlined term to make the statement true.

6. Amoebas take in solids by phagocytosis.
7. *Trichonympha* and termites have a mutualistic relationship.
8. Sarcodines move by flagella.
9. The sexual phase of the *Plasmodium* life cycle takes place in mosquitoes.
10. Paramecia respond to attack by discharging trichocysts.

III. Multiple Choice

11. All members of the protozoa are
    a) unicellular. b) autotrophic.
    c) parasitic. d) motile.
12. In paramecia, reproduction is controlled
    by the a) gullet. b) micronucleus.
    c) oral groove. d) macronucleus.
13. Amoebas move by extending a) cilia.
    b) pseudopodia. c) trichocysts.
    d) flagella.
14. Protozoa form cysts in response to
    a) reproduction. b) the presence of
    food. c) obstacles in their path.
    d) unfavorable environmental conditions.
15. Paramecia get rid of solid wastes

through the a) gullet. b) cilia.
c) contractile vacuole. d) anal pore.

IV. Essay

16. What role does diffusion play in the nutrition of protozoa?
17. What traits of a euglenoid cell make it autotrophic?
18. How do amoebas and *Euglena* cells respond to light? Why is it adaptive for these organisms to respond as they do?
19. How do paramecia carry out the process of sexual reproduction?
20. How do *Plasmodium* cells reproduce?

Applying and Extending Concepts

1. Many people suffer from African sleeping sickness, but not all of them die from the disease. Write a paragraph explaining what would happen to parasitic species, such as the trypanosomes, if they killed every member of the host species in which they live.
2. One of the most common diseases caused by amoebas is amoebic dysentery. Use your school or public library to research amoebic dysentery. Then write a paragraph summarizing the life cycle of *Entamoeba histolytica* and explain how the disease can be prevented.
3. Many scientists think that petroleum originated as deep, accumulated layers of dead bacteria and protozoa. Use your school or public library to research the origin of petroleum and write a brief report about how geologists use protozoa to locate oil-rich strata of rock.
4. *Radiolaria* secrete hard shells of calcium carbonate, or limestone. Amoebas, on the other hand, have no protective covering outside the cell membrane. What are the advantages of each cell structure for the survival of the organism? What are the disadvantages?

Related Readings

Jahn, T. L., E. Bovee, and F. Jahn. *How to Know the Protozoa.* 2d ed. Dubuque, Iowa: William Brown Co., 1979. A discussion of the biology of protozoa, this book contains identification guides to common species.

Engemann, J. G. and R. W. Hegner. *Invertebrate Zoology.* 3d ed. New York, New York: Macmillan Publishing Co., 1981. This book provides an overview of the invertebrates. The chapters on protozoa are very comprehensive and well-illustrated, but the discussion of other kinds of invertebrates is helpful also.

Weiner, J. ''Misfits.'' *The Sciences* 22 (May–June 1982): 5. This article explains how the reclassification of a protozoan once thought to be an ancestor of the ciliates has changed ideas about how multicellular life evolved.

Algae

Spirogyra, a filamentous green alga

Introduction

algae (singular, *alga*)

Life on Earth involves a delicate balance between organisms that produce food and organisms that consume it. Among the organisms that produce food—the autotrophs—no group is more important to maintaining this balance than the **algae** (AL jee). Algae are the autotrophic members of the kingdom Protista. They are considered plantlike protists because they contain chlorophyll and carry out photosynthesis. Like marine and freshwater protozoa, algae provide food for countless species of water-dwelling animals.

Algae also make it possible for animals to exist on land. As algae carry out photosynthesis, they release oxygen into the atmosphere. Algae are so plentiful that they produce 90 percent of the world's atmospheric oxygen.

Overview of Algae

Algae are simple, water-dwelling organisms that are more like plants than animals. ***Unlike other protists, algae are autotrophic organisms.*** Most algae float near the water's surface, where sunlight is most direct and the photosynthetic cells can produce a maximum amount of food. Algae absorb nutrients by diffusion across the cell membrane.

22.1 Characteristics

Algae vary greatly in size and shape, from microscopic, hard-shelled forms to rubbery kelps that grow as long as 70 m (230 ft.). Like those of all protists, the cells of algae are eukaryotic. Most algal cells are supported by an inner wall of cellulose. Layers of cells are held together by a jellylike substance called **pectin**.

Some algae are unicellular; others are multicellular. Many unicellular algae move by means of flagella. Kelps and many other multicellular algae are nonmotile. The bodies of multicellular algae consist of filaments that become meshed together into a solid mass. An unspecialized, multicellular body is called a **thallus.**

22.2 Habitats

Algae are common in freshwater lakes, streams, and the oceans. There they may form a layer of green scum on the water's surface or hang in strands from rocks or logs. Along with protozoa and other small organisms, algae make up **plankton** (PLANK tuhn). ***Plankton is the food source for most of the world's water-dwelling organisms.*** Billions of tiny, drifting algae are called *phytoplankton* (fy tuh PLANK tuhn).

Algae also live on land in the thin films of water found on rocks and soil particles. Some species of algae thrive in snowfields, deserts, cold springs, or the almost boiling water of hot springs. Air samples from 15,000 m (49,000 ft.) above sea level have contained algal cysts, which are much like protozoan cysts.

22.3 Classification

Algae are classified into five groups according to the pigments they contain. These five groups are golden algae, fire algae, green algae, brown algae, and red algae. The pigments give the

- *Name* three characteristics found in all algae.
- *Contrast* the features of algae and protozoa.
- *Name* the main characteristic used to classify algae.
- *Summarize* the economic importance of algae.

thallus (plural, *thalli* or *thalluses*)

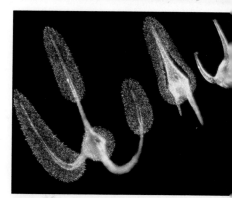

Figure 22–1. Algae range in size from microscopic unicellular dinoflagellates (top) to giant kelps (bottom).

▶ Economic Importance of Algae

Algae and products made from algae have a long history of usefulness. Dulse, a red alga, has been used as food for people and domestic animals for centuries. The alga now most widely used for food is *nori,* grown chiefly by coastal villagers in Japan. Nori is dried into sheets that are used in soups, biscuits, and as a flavoring in many foods.

Although most algae are low in protein, they contain concentrated minerals and some starch. In many countries, dried algae are ground into a powder and added to animal feed as a mineral supplement.

Algae yield valuable extracts. Agar, an extract taken from red algae, is a jellylike substance used for laboratory cultures in which bacteria are grown.

Carrageenin is an extract from red algae that is used to keep small particles in suspension in many foods. For example, carrageenin prevents chocolate from separating out in chocolate milk. Carrageenin is a common ingredient in jams and jellies, instant coffee, honey, wine, and ice cream. Another extracted algal product is mannitol, an alcohol sugar used in (sugarless) candy and gum.

algae in each phylum their characteristic color. Regardless of their color all algae contain a green pigment called chlorophyll *a*. Most also contain a second type of chlorophyll. In some algae, the dominant colors of other pigments mask the chlorophyll's green. These other pigments do more than simply change the appearance of the algae. The pigments also enhance photosynthesis by capturing photons of other colors of light and transferring them to the chlorophyll. Algae are also commonly classified by the form in which they store food and by their means of reproduction.

Reviewing the Section

1. What are three traits found in all algae?
2. What is the basic difference between algae and protozoa?
3. In what ways are algae economically important?
4. What pigment do all algae share?

Kinds of Algae

The five phyla of algae, listed in Table 22–1, are Chrysophyta (kruh SAHF uh tuh), the golden algae; Pyrrophyta (puh RAHF uh tuh), the fire algae; Chlorophyta (klaw RAHF uh tuh), the green algae; Phaeophyta (fay AHF uh tuh), the brown algae; and Rhodophyta (roh DAHF uh tuh), the red algae. Most species of algae are free-living. Some species, however, live on or in other organisms. For example, one species of alga lives on the fur of South American sloths. The green of the algae matches the color of surrounding trees, helping to conceal the tree-climbing sloth from its enemies.

22.4 Golden Algae

The phylum Chrysophyta, the golden algae, includes about 12,000 species of algae that live in fresh water or in the sea. Golden algae contain chlorophylls *a* and *c*. The characteristic golden color of the algae comes from three accessory pigments: orange *carotenes,* yellow *xanthophylls* (ZAN thuh fihlz), and brown *fucoxanthins* (FYOO koh zan thuhnz).

Table 22–1: Characteristics of Algae

Phylum	Species	Pigments	Food Storage	Reproduction
Chrysophyta (golden algae)	12,000	Chlorophylls *a* and *c*, carotenes, xanthophylls, fucoxanthins	Usually oils	Asexual by binary fission, sexual by fusion of gametes in zygote
Pyrrophyta (fire algae)	1,100	Chlorophylls *a* and *c*, xanthophyll	Starch and oils	Usually asexual by binary fission
Chlorophyta (green algae)	7,000	Chlorophylls *a* and *b*, carotene	Starch	Various sexual means; sexual by alternation of generations
Phaeophyta (brown algae)	1,500	Chlorophylls *a* and *c*, fucoxanthin	Laminarin and oils	Sexual by alternation of generations
Rhodophyta (red algae)	4,000	Chlorophylls *a* and *d*, carotenes, phycobilins	Starch	Sexual by alternation of generations

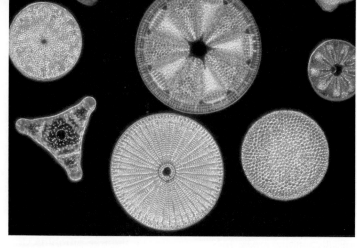

Figure 22–2. Distinctive shapes and etchings characterize the glasslike shells of each species of diatoms.

Complete shell

Shell halves separating

New shell halves added; two new shells formed

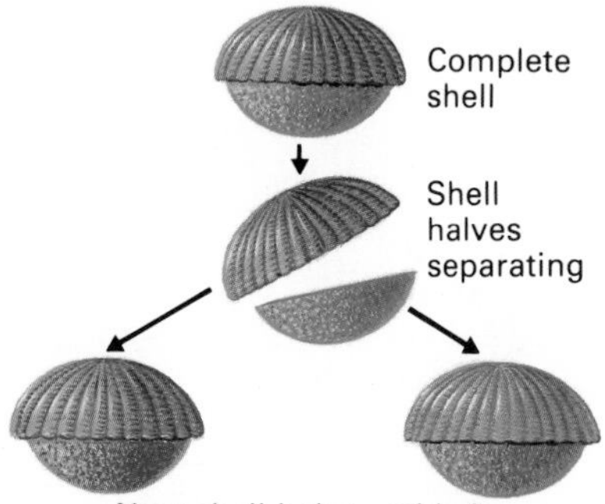

Figure 22–3. When a diatom reproduces asexually, its shell first splits into an upper and a lower half. The cell inside divides, and each new cell associates with one half of the original shell. A new half-shell then forms for each new diatom.

Most golden algae are unicellular, shelled organisms called **diatoms** (DY uh tahmz). The cell wall of a diatom is made of a hard, glasslike substance called *silica*. The two halves of the diatom's shell fit together like the sides of a petri dish. In asexual reproduction, the halves separate and each forms a new shell, as shown in Figure 22–3. Diatoms also reproduce sexually by forming gametes.

Diatoms store food in the form of oils. The oils give an unpleasant taste to fish that eat diatoms. When diatoms die, their cytoplasm and cell walls decay while the outer shell of silica remains intact. The shells accumulate on the ocean floor, creating deposits called *diatomaceous earth*. Diatomaceous earth is used as a filtering material and in abrasives.

22.5 Fire Algae

The phylum Pyrrophyta, with just 1,100 species, accounts for a large part of the sea's phytoplankton. Fire algae get their red color from chlorophylls *a* and *c* and xanthophyll. They are called "fire algae" because they sometimes look like a fire smoldering in the water. When passing ships or dolphins stir up the water, disturbed fire algae glow like a neon light. The production of light by living things is called **bioluminescence.** It also occurs in land organisms such as fireflies.

The largest group of fire algae is the **dinoflagellates,** unicellular algae with stiff, armorlike cell walls. One flagellum pulls the cell forward while the other flagellum wraps around the cell, making it spin. Dinoflagellates store food as starch or oil. They usually reproduce asexually by binary fission. Certain chemicals or changes in water temperature may cause an uncontrolled growth of algae called an **algal bloom.** Blooms of dinoflagellates, or **red tides,** may release poisons that kill thousands of fish. Red tides also make algae-eating shellfish dangerous for humans to eat.

22.6 Green Algae

Green algae make up the phylum Chlorophyta, the most diverse of all algal phyla. The 7,000 species of green algae range from microscopic single cells to multicellular organisms over 8 m (25 ft.) long. Green algae, unlike any other group of algae, contain the same three pigments found in land plants: chlorophyll *a*, chlorophyll *b*, and a type of carotene. Like many land plants, green algae store food as starch. *The similarities between plants and green algae have led some scientists to suggest that plants evolved from green algae ages ago.*

Protococcus The unicellular green algae *Protococcus* is common in damp forests, where it forms a slippery green film on moist rocks or a green dust on tree trunks. *Protococcus* reproduces asexually. The oval cell divides by binary fission, producing two genetically identical daughter cells.

Volvox The order *Volvocales* includes examples of **colonial algae,** organisms made of individual cells held together by a jellylike substance or strands of cytoplasm. Colonial algae differ from multicellular organisms because their cells do not have

Q/A

Q: *How do algae survive the low temperatures of winter?*

A: Many species form cold-resistant spores or cysts, or they break into fragments. These fragments sink to deeper levels where the water does not freeze.

Highlight on Careers: Limnologist

General Description
A limnologist is a biologist who specializes in studying freshwater life. For example, a limnologist may test a species of alga to find out how much food the organism produces.

Many limnologists study the ecosystems of lakes and streams. They help write environmental impact statements. These reports predict how a major change, such as building a dam, will affect biological conditions. The task of limnologists is to help preserve environmental quality by gauging the effects of pollutants.

Most limnologists work for universities or government agencies. However, a growing number of limnologists work for private firms.

Career Requirements
Positions as research assistants or technicians require a B.S. degree in limnology. Most limnologists also have an M.S. or Ph.D. in limnology or aquatic ecology.

For Additional Information
American Society of
    Limnology and
    Oceanography
Great Lakes Research
    Division
University of Michigan
Ann Arbor, MI 48109

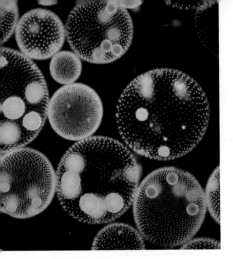

Figure 22–4. *Volvox* is a colonial green alga that rolls through the water, propelled by hundreds of flagella.

Connecting tube forms

Contents combine in one cell

Cells begin to fuse

Zygote forms

Zygospore develops

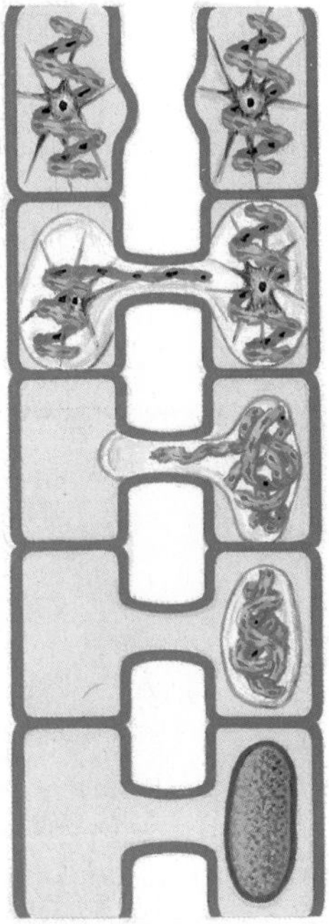

Figure 22–5. *Spirogyra* reproduces sexually through the process of conjugation illustrated above.

specialized functions. Cells in a colony can reproduce sexually more readily than single cells because mating cells are always nearby. The size of the colony protects the members from organisms such as protozoa that feed on single cells.

One of the most beautiful colonial species is *Volvox*. A *Volvox* colony is a hollow ball formed by hundreds or thousands of bright green cells. The entire colony spins slowly through the water by the synchronized beating of the cells' flagella.

Spirogyra *Spirogyra* (spy ruh JY ruh) is a multicellular green alga that grows in shallow freshwater pools. Its cells lengthen and divide without separating, forming long, slender filaments that look like transparent green ribbons. Each chloroplast contains a small protein body called a **pyrenoid** (py REE noyd), which stores starch.

Spirogyra reproduces asexually in two ways. The cells may undergo binary fission, which lengthens the filament. If the filament is broken, each fragment continues to grow on its own. This process is called **fragmentation.** Fragmentation does not harm the individual cells, and it helps disperse the algae.

Sexual reproduction in *Spirogyra* involves a process of *conjugation* that differs from conjugation in paramecia. The process begins when two neighboring *Spirogyra* filaments form connecting tubes, as shown in step 1 of Figure 22–5. The contents of one cell flow through the tube into the adjacent cell. A diploid zygote forms when the contents of the two cells join together. The wall of the receiving cell then thickens around the zygote, forming a durable **zygospore** that can survive harsh conditions. When conditions become favorable for growth, the zygospore becomes active once again. It then undergoes meiosis and develops into a new *Spirogyra* filament.

Ulva *Ulva,* the sea lettuce, is a genus of multicellular green alga whose life cycle involves two distinct forms of the organism. Although the two forms of *Ulva* look alike, they are genetically different. In one of the forms, *Ulva* cells are haploid—that is, they have (n) chromosomes. In the other form, the cells are diploid—that is, they have (2n) chromosomes. To see how these two life forms alternate, examine Figure 22–6. The haploid form of the organism is called the **gametophyte** (guh MEET uh fyt) because it produces gametes. When gametes from two *Ulva* fuse, they form a diploid zygote. All the cells that develop from the zygote are diploid. The resulting diploid form of the alga is called the **sporophyte** (SPAWR uh fyt), because its cells undergo meiosis and produce spores. Each haploid spore then develops into a haploid gametophyte.

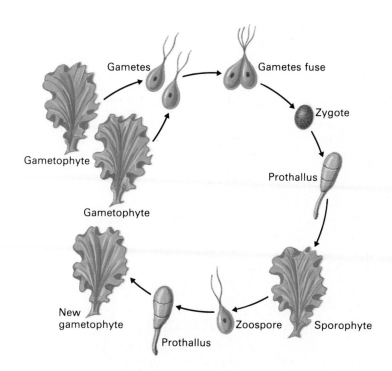

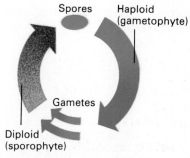

Figure 22-6. The life cycle of the green alga *Ulva* (left) involves an alternation of generations. The diagram above shows the alternation of haploid and diploid phases in this life cycle.

The alternation between sporophyte and gametophyte stages in a life cycle is called **alternation of generations.** *Many species of algae and all plants go through an alternation of generations.* This life cycle is widespread because it has great survival value. The species benefits from the recombination of parents' traits through the fusion of gametes. The species also benefits from the opportunity to reproduce by the less risky process of forming spores. In *Ulva*, the gametophyte and sporophyte forms look identical. In other algae and in plants, the two forms may look very different.

22.7 Brown Algae

The world's rocky coasts and colder oceans abound with tough seaweeds and kelps. These are the brown algae, members of the phylum Phaeophyta. About 1,500 species of brown algae have been identified, including the largest forms of algae. Brown algae commonly form extensive underwater forests, creating a relatively sheltered habitat for other organisms. Members of the phylum Phaeophyta contain chlorophylls *a* and *c* and the accessory pigment fucoxanthin. Brown algae may store food either as oil or as an unusual carbohydrate called *laminarin*. The life cycle of most species of brown algae exhibits an alternation of generations.

Q/A

Q: *How do algae in tidal areas survive exposure to the air and wind?*

A: Algae that have developed structures to retain water, such as water bladders and thick cell walls, can survive several hours of exposure during low tide. Even these algae, however, show slower rates of photosynthesis and respiration during exposure.

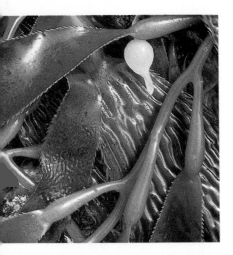

Figure 22–7. This kelp is an example of a marine brown alga.

Many brown algae have tissues that resemble the roots, stems, and leaves of plants. Algae in the genus *Laminaria,* including the most familiar kelps, are firmly anchored by a root-like structure called a **holdfast.** The long, stemlike portion of *Laminaria* ends in a broad, leaflike structure that carries on photosynthesis. Despite its outward resemblance to plants, however, *Laminaria* is a true alga. Its cells are not highly specialized. Unlike roots, the holdfast does not absorb minerals from the soil. Instead, the kelp thallus absorbs minerals directly from sea water.

Both anchored and free-floating brown algae have a leathery, highly flexible thallus that sways freely and can withstand the motion of waves. Many brown algae that grow from the sea floor have air-filled structures called **air bladders.** The buoyant air bladders help keep the photosynthetic parts of the algae near the water's surface.

22.8 Red Algae

In warm tropical oceans, algae in the phylum Rhodophyta are the predominant form of seaweed. Species of red algae have been found in most climates, however, as well as in fresh water. The 4,000 species of red algae contain chlorophylls *a* and *d,* carotenes, and red or blue accessory pigments, which are known as *phycobilins.*

The red phycobilins absorb blue light, which penetrates into water more deeply than any other color of light. This trait enables red algae to grow in deeper water than any other algae. Red algae are found as far down as 150 m (490 ft.) below sea level. Wave motion at this depth is minimal. The thallus of a typical red alga is a delicate network of filaments that fan out in the water. This fragile kind of thallus would not survive well in the more turbulent water along exposed, rocky shores where brown algae live.

Red algae store food as starch. They are usually sessile, or stationary. Most species of red algae undergo an alternation of generations.

Figure 22–8. Some red algae live in very deep water. Their red phycobilin pigments capture blue light that penetrates the oceans to 265 m (884 ft.).

Reviewing the Section

1. How does *Spirogyra* reproduce asexually?
2. What is the difference between the gametophyte and sporophyte generations in *Ulva*?
3. Why can red algae live in deeper water than brown algae?

Investigation 22: Observing Algae

Purpose
To compare several species of algae

Materials
Cultures of three different species of algae, three medicine droppers, three slides, three coverslips, stereomicroscope, compound microscope

Procedure
1. Make a chart like the one shown below. Your teacher will provide several species of algae. As you examine each kind, complete the chart.

Comparison of Species of Algae

Name of Species	Sketch	Relative Size	Distinctive Features

2. Observe one kind of alga with out using any magnification. List the characteristics and sturctures you observe, such as its color and whether it has broad leaves, long filaments, and single cells or many cells.
3. Observe your specimen using the stereomicroscope. *What additional details can you observe?*
4. Tear off a small piece of the specimen and make a wet mount of it. Examine it under the compound microscope under both low and high power. *What additional details can you see?*
5. Repeat steps, 2, 3, and 4 for each species of alga, completing the charts as you make your observations.

Analyses and Conclusions
1. What characteristics do each of the species of alga you observed have in common?

2. In what ways are they different?
3. Algae contain pigments that enable them to carry out photosynthesis. Protozoa cannot photosynthesize. Describe the differences in the culture media you would use to grow protozoa and algae.

Going Further
- Collect algae from a local pond or stream. Make wet mounts of the specimens and observe them under a microscope. Using reference books, identify each species of alga you observe. What is the relationship between the type and number of algae present in a pond and the ecological health of the pond?
- If you live near a seashore, collect a sample of marine algae. Make a wet mount and observe the specimen under a microscope. Compare the marine species of alga you observe with the freshwater species. Explain how the colors and other adaptations of marine algae help them survive in their natural environment.

Chapter 22 Review

Summary

Algae are autotrophic members of the kingdom Protista. They produce most of the world's atmospheric oxygen and provide the basis for much of its food supply. Most algae are aquatic. Asexual reproduction occurs by binary fission, fragmentation, or the production of spores. Many species also reproduce sexua¹ly.

Almost all golden algae and fire algae are unicellular organisms. Diatoms are tiny golden algae with silica shells. Dinoflagellates, the most common fire alga, move by two flagella. Dinoflagellates show bioluminescence when disturbed and cause dangerous algal blooms called red tides.

Most green, brown, and red algae are multicellular organisms. The green algae have the most diverse forms, including the colonial *Volvox* and filamentous *Spirogyra*. The life cycle of *Ulva* typifies the pattern of alternating haploid gametophyte and diploid sporophyte generations. This pattern is called alternation of generations.

Most brown algae live primarily in colder oceans. They include many seaweeds and kelps. Red algae grow primarily in tropical oceans. They grow at greater depths than other algae and typically have a delicate, fan-shaped thallus.

BioTerms

air bladder **(336)**
algae **(328)**
algal bloom **(332)**
alternation
   of generations **(335)**
bioluminescence **(332)**
colonial algae **(333)**

diatom **(332)**
dinoflagellate **(332)**
fragmentation **(334)**
gametophyte **(334)**
holdfast **(336)**
pectin **(329)**
plankton **(329)**

pyrenoid **(334)**
red tides **(332)**
sporophyte **(334)**
thallus **(329)**
zygospore **(334)**

BioQuiz   (Write all answers on a separate sheet of paper.)

I. Completion

1. A _____ is a type of biologist that specializes in the study of freshwater life forms.
2. Kelps are one kind of _____ algae.
3. Alternation of generations involves two forms; a diploid sporophyte and a haploid _____.
4. _____ are a dangerous kind of algal bloom caused by a sudden growth of dinoflagellates.
5. Algae are classified according to the kinds of _____ they contain.

II. Modified True and False

Mark each statement TRUE or FALSE. If false, change the underlined term to make the statement true.

6. A delicate, fan-shaped thallus is the typical body form of a red alga.
7. Unicellular, free-floating algae are called zooplankton.
8. A holdfast anchors brown algae.
9. The cell walls of dinoflagellates contain silica.
10. Algae grow most abundantly where conditions are dry.

III. Multiple Choice

11. All algae are a) green. b) autotrophic.
    c) unicellular. d) multicellular.
12. The characteristic color of green algae
    comes from a) chlorophyll. b) carotene.
    c) xanthophyll. d) fucoxanthin.
13. When diatoms reproduce asexually, their
    shells a) fuse. b) decay. c) break into
    bits. d) separate in half.
14. The unspecialized body of multicellular
    algae is called the a) air bladder.
    b) holdfast. c) thallus. d) filament.
15. *Spirogyra* filaments become longer when
    the cells undergo a) binary fission.
    b) fragmentation. c) conjugation.
    d) bioluminescence.

IV. Essay

16. Why is the life cycle shared by *Ulva*
    and many other kinds of algae known as
    alternation of generations?
17. In what ways is the holdfast of a kelp
    similar to the root of a plant? In what
    ways is it different?
18. How does the arrangement of flagella in
    a dinoflagellate affect the organism's
    movement?
19. How are deposits of diatomaceous earth
    formed, and how is this substance commonly used?
20. How does the process of asexual reproduction in *Ulva* differ from that in *Spirogyra*?

Applying and Extending Concepts

1. The Sargasso Sea is a unique area of the
   Atlantic Ocean named for its predominant
   life form, the brown alga *Sargassum*. Research the Sargasso Sea and write a paragraph explaining how *Sargassum* influences its environment.
2. In freshwater lakes, blooms of green algae
   can cause severe pollution problems. Use
   your school or public library to research
   desmids and other green algae that bloom
   in polluted water. Then write a brief report
   describing the impact and prevention of
   algal blooms.
3. Some scientists consider algae one way to
   help alleviate the problem of feeding the
   world's growing population. Research
   alga farming and write a report describing
   how algae are grown and how they might
   be used more extensively in the future to
   augment the production and equitable distribution of the world's available food
   energy.
4. In fireflies, flashes of light serve as signals
   to help organisms find a mate of their
   own species. Some algae, including some
   poisonous dinoflagellates, bioluminesce
   when disturbed by would-be predators.
   How might this aspect of bioluminescence
   help promote the survival of a species of
   algae?

Related Readings

Chapman, A. R. O. *Biology of Seaweeds: Levels of Organization*. Baltimore, MD: University Park Press (1979). A lively and thorough introduction to the structure and physiology of seaweeds.

Hoover, R. B. "Those Marvelous Myriad Diatoms." *National Geographic* 155 (June 1979): 871–878. As its title suggests, this beautifully illustrated article discusses the variety and importance of diatoms.

Swann, C. "Seaweed Power—Renewable Energy from Offshore Kelp Farms." *Popular Science* 157 (October 1981): 86–88. This article describes and illustrates the process of kelp farming.

BIO*TECH*

Biosynthesis

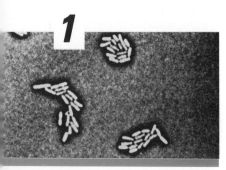

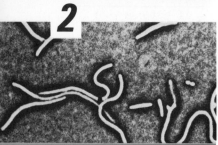

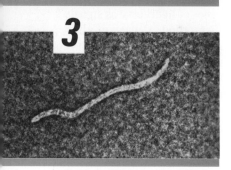

Before antibiotic is added (1), bacilli divide in lab dish. Antibiotic causes bacilli to elongate (2) and later (3) to rupture and die.

Rod-shaped *Pseudomonas* bacteria infect a deep wound. Dividing quickly, they invade healthy tissue and soon overwhelm the body's natural defenses. The antibiotic moxalactam is administered and a dramatic change takes place. Instead of dividing, the bacteria begin to elongate until they are many times their normal length. Their cell walls burst and their cellular contents leak out. A few hours later, the microorganisms are dead and the wound is healing.

Moxalactam is produced by the mold *Cephalosporium.* It is a product of **biosynthesis**—the formation of chemical compounds by the cells of living organisms. Moxalactam is one of many new antibiotics developed through technological advances in pharmaceutical research and manufacturing. These new antibiotics kill more kinds of bacteria and kill them faster, causing fewer side effects than ever before.

Cephalosporium and *Penicillium* are two of the six genera of molds from which researchers make antibiotics. *Cephalosporium* produces a group of antibiotic compounds known as cephalosporins, one of which is moxalactam. *Penicillium* produces penicillins. These two kinds of molds are used in the biosynthesis of well over 1,000 different antibiotics.

Improved antibiotics are possible because of advanced biosynthesis technology.

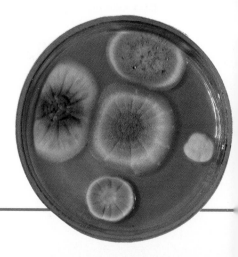

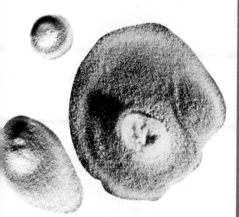

Growths of Penicillium mold are shown above. After the mold is processed, the antibiotic penicillin kills bacteria in a treated culture (right).

Today's pharmaceutical manufacturing plants produce huge quantities of antibiotics. Much of the manufacturing process is directed from computerized remote-control rooms, where temperature and nutrient composition are monitored and growth conditions can be adjusted.

Antibiotic fermentation tanks hold up to 200,000 L (53,000 gal.) of mold and medium mixture.

Molds used in the biosynthesis of antibiotics are special strains—"supermolds" developed over many years of careful research. A mold is exposed to an agent, such as a chemical or radiation, that causes mutation in some of the microorganisms. Those molds that show promise of producing antibiotics in larger amounts or of greater effectiveness are chosen for further research. More mutations are induced; the selection process is repeated again and again. This technique has been used in combination with improvements in fermentation technology to produce *Penicillium* strains that yield 10,000 times as much antibiotic as the wild strains from which they came. By altering the chemical structure of these substances, scientists can develop more powerful antibiotics.

The technology of antibiotic manufacture has kept pace with new techniques of biosynthesis.

CHAPTER

23

Fungi

Bracket fungi attached to tree bark

Introduction

Organisms in the kingdom Fungi play an important ecological role. Fungi act as decomposers, breaking down organic matter. The process of fungal decay returns valuable nutrients to the soil, where living organisms can use them for new growth. Parasitic fungi, however, can cause deadly diseases in both plants and animals.

The name *fungi* comes from the Latin word *fungus,* which means "mushroom." People have long used mushrooms and other fungi as food. More recently, people have produced antibiotics such as penicillin from a certain type of fungus. However, other types of fungi can also cause serious diseases and can destroy millions of dollars worth of corn, wheat, and other food crops.

Overview of Fungi

Fungi are nonmotile organisms that obtain food by decomposing organic matter. Fungi were once considered plants, but studies later revealed that fungi have characteristics that no plant possesses. Unlike most plants, fungi lack chloroplasts and cannot carry out photosynthesis. Neither do fungi have animal characteristics. Because of their unusual combination of traits, fungi are classified in a separate kingdom.

23.1 Characteristics

Fungi are eukaryotic organisms; most species are multicellular. The cell walls of most fungi contain a hard substance called **chitin** (KYT uhn). Chitin is found only in fungi and in the hard outer skeletons of insects.

The body of a typical fungus consists of many individual filaments called **hyphae** (HY fee). Hyphae contain cytoplasm and one or more nuclei. Hyphae secrete enzymes that digest food. The fungus then absorbs the nutrients from the food through its cell walls.

Intertwined hyphae form the body of the fungus, or **mycelium** (my SEE lee uhm). Most of a fungus lives under the *substrate,* or material in which the fungus is growing. The visible part contains the spore-producing structures and is called the **fruiting body.** *Saprophytic* fungi feed on dead matter. *Parasitic* fungi feed on living organisms.

23.2 Habitats

Fungi have adapted to almost every environment where organic material and moisture are available. They flourish in forests, grasslands, and other areas where dead wood and leaves are abundant. Some species of fungi live in deserts. Others live high atop mountains. Certain marine fungi live on the remains of dead bacteria and plankton trapped in polar icecaps. You may have seen molds—small, fuzzy growths of fungi on fruit, bread, or other foods.

Though nonmotile, fungi can reach these diverse environments by means of spores that drift in the wind. A single fungus may produce millions or even trillions of spores at a time. Many of these spores land in unsuitable environments and perish. However, many others will survive and germinate. Most kinds of fungi rely on their spores to disperse the species and to find new food sources.

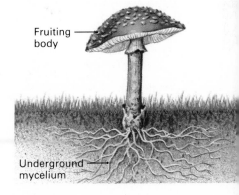

Fruiting body

Underground mycelium

Figure 23–1. A mushroom is the reproductive organ of one class of fungi. The vegetative portion, the mycelium, is an underground saprophyte that gathers food by feeding on decaying organisms.

▶ Lichens

The next time you are strolling in a woods, look for rocks with orange or green patches. The patches are **lichens** (LY kuhnz), organisms that consist of a fungus and an alga. The fungus and the alga live in a symbiotic relationship. Fungal hyphae give the lichen its internal structure and characteristic shape. Algal cells are embedded in the mycelium, as the photograph shows.

Most lichens are mutualistic. In a lichen, the fungus shields the alga from excessive sunlight and retains water that the alga

 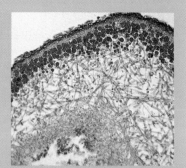

needs for photosynthesis. The alga secretes carbohydrates that the fungus absorbs as food. The alga allows the lichen to survive on bare rocks where the fungus alone cannot grow. Despite their hardiness, lichens are among the first organisms to suffer from

air pollution. Lichens absorb rainwater directly, rather than water that has soaked into the ground. As a result, lichens absorb more dissolved toxic substances than do plants that use water that has first been filtered through the ground.

Q/A

Q: *Why do mushrooms sometimes grow in circles called "fairy rings"?*

A: In popular legend, a fairy ring marks the spot where fairies danced in the night. The mushrooms actually mark the outer edge of a large underground mycelium.

23.3 Ecological and Economic Roles

Fungi help perform the important ecological function of decomposing dead organic matter. This process not only helps clear dead plants and animals from the environment but also returns nitrogen, phosphorus, and other nutrients to the soil. Fungi are also economically important. They are used directly as food or in making such foods as bread and cheese. Fungi also produce medically valuable antibiotics, such as penicillin and streptomycin. However, fungi can be extremely destructive when they attack crops. Saprophytic fungi destroy millions of dollars worth of food crops each year. Some fungi also cause certain diseases in animals, including humans.

Reviewing the Section

1. How do fungi obtain food?
2. What is a lichen?
3. How are fungi economically important?

Kinds of Fungi

Section Objectives

- *Summarize* the methods of feeding and reproduction in terrestrial molds.
- *Identify* the basic parts of a club fungus.
- *Name* three kinds of fungal plant diseases and the fungi that cause them.
- *Discuss* the traits of yeasts.
- *List* the characteristics that slime molds share with fungi and those they share with protozoa.

Fungi include two main groups of organisms. One group, the **true fungi,** account for about 81,500 species. The other group consists of 600 species called **slime molds.** Slime molds possess a mixture of traits found in fungi and in protozoa. Individual species within both groups are further classified by their means of reproduction.

23.4 Terrestrial Molds

About 600 species of terrestrial molds make up the class Zygomycetes. The fuzzy part of the mold actually consists of the specialized hyphae that produce tiny spores in structures called **sporangia.** These hyphae are called **sporangiophores.**

The common black bread mold, *Rhizopus stolonifer,* shows the development of a typical terrestrial mold. An airborne spore that lands on a piece of bread forms hyphae, which branch out over the bread's surface. These surface hyphae, called **stolons,** form short extensions that penetrate into the bread. These extensions are called **rhizoids.** Rhizoids anchor the mold to its food supply, secrete digestive enzymes, and absorb the nutrients.

Sexual reproduction occurs in terrestrial molds when contact occurs between hyphae from two genetically different molds, called **mating strains.** The mating strains are called *plus* (+) and *minus* (−) rather than male and female, because they have identical shapes and functions. The tip of each mating hypha contains nuclei that fuse and form a diploid zygote. In unfavorable conditions the zygote may become a durable zygospore. In favorable conditions the zygote undergoes meiosis and develops into a new sporangiophore. Terrestrial molds and other fungi do not go through alternation of generations.

Figure 23–2. The common black bread mold, *Rhizopus stolonifer,* has a mycelium composed of both rhizoids and stolons. When nutrients are depleted, the fungus produces spores in sporangia held up by sporangiophores.

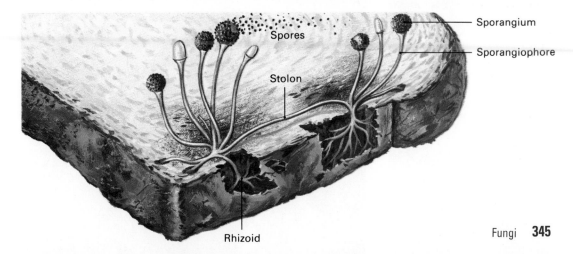

Spores

Stolon

Sporangium

Sporangiophore

Rhizoid

23.5 Water Molds

More than 500 species of water molds make up the class Oomycetes. Most are water-dwelling saprophytes. Water molds are the only fungi whose cell walls consist mainly of cellulose rather than of chitin. *Unlike all other fungi, water molds produce distinctly male and female gametes.* The differentiation of gametes into distinctly male (sperm) and female (egg) forms is called **oogamy** (oh AHG uh mee).

A water mold called *Phytophthora infestans* ruined potato crops in Ireland between 1845 and 1847. Potatoes were Ireland's main crop, and the resulting famine caused the death of more than 2 million people. Other water molds cause disease in fish, a common problem in aquariums.

23.6 Club Fungi

The class Basidiomycetes contains about 25,000 species, including mushrooms, shelf fungi, puffballs, rusts, and smuts. Basidiomycetes are called *club fungi* because they produce spores on club-shaped, microscopic structures called **basidia.** The basidia develop within the fruiting body.

Mushrooms The most common club fungi are mushrooms. What you think of as a mushroom is the spore-producing structure of an underground mycelium. The mushroom first develops as a tight mass of hyphae called a **button.** A stemlike structure known as the **stipe** pushes the button above ground. There the button opens into a **cap,** the fruiting body of the mushroom. The underside of the cap contains thin sheets of tissue, or **gills,** to which the basidia are attached. Within each basidium are two

Figure 23–3. The fruiting bodies of club fungi vary. Mushrooms (left) form spores under a cap. Bracket fungi (top right) form spores under a shelflike fruiting body. Puffballs (bottom right) have a ball-like fruiting body.

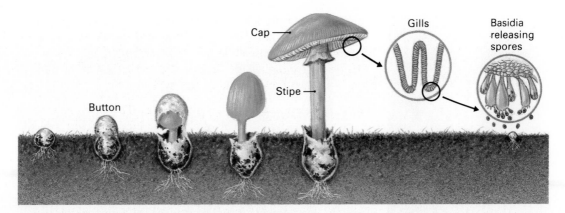

Cap — Stipe — Button Gills Basidia releasing spores

haploid nuclei that fuse to form a diploid nucleus. This nucleus undergoes meiosis, and the resulting haploid nuclei produce four **basidiospores.** Each basidiospore is capable of developing into a new mushroom.

Some kinds of mushrooms are edible, but others are poisonous. Because some poisonous mushrooms resemble harmless ones, you should never eat wild mushrooms. For example, even a small portion of the "destroying angel," *Amanita bisporigera,* can be lethal. A toxin in this mushroom damages the liver so severely that death results.

Rusts and Smuts Rusts and smuts are parasitic club fungi that cause severe damage to cereal and vegetable crops. Rusts and smuts produce spores, but not in mushroomlike fruiting bodies. The mycelia spread throughout the host plant, destroying the plant's cells while producing billions of basidiospores.

The life cycle of wheat rust involves two alternate hosts. In the spring, the fungal spores infect young wheat plants. In the summer, a second cycle of spore production infects barberry plants. The disease can be controlled by destroying all barberry bushes growing near wheat fields.

Corn smut produces large deformed growths on ears of corn. The life cycle of corn smut involves only one host. Corn smut can be eliminated by burying or burning the infected plants before the fungi produce spores.

23.7 Sac Fungi

About 30,000 species belong to the class Ascomycetes. Members of this class include the gourmet delicacies morels and truffles as well as the single-celled yeast used in making bread. Ascomycetes are called *sac fungi* because sexually produced spores form in an **ascus,** or "little sac." An ascus begins to

Figure 23–4. A mushroom starts to form when a button breaks through the soil. A stipe then lifts the cap above the ground. Later, basidia form on gills under the cap.

Q/A

Q: *Do any serious fungal diseases affect humans?*

A: Yes. One is ergotism, a disease caused by eating fungus-infected rye. The fungus causes severe abdominal pain, hallucinations, gangrene, and even death. The fungus is also a source of lysergic acid diethylamide, also known as LSD.

Figure 23–5. The fruiting bodies of this ascomycete resemble small cups. Hyphae on the outside of the cup help protect it from predators.

develop when two gametes or two mating strains fuse. Nuclei divide as the hypha grows, resulting in a row of haploid **ascospores** within the ascus. In sac fungi and club fungi, hyphae are divided by cross walls. Nuclei and cytoplasm flow through pores in these cross walls as the hyphae grow.

Yeasts   Yeasts are unusual sac fungi. They contain chitin and reproduce sexually by forming ascospores, but they are unicellular and do not form hyphae. Yeasts also reproduce asexually by budding. Many yeasts grow most rapidly in environments with a high sugar content. In bread dough, yeast cells feed on carbohydrates. As yeast cells grow, they produce carbon dioxide gas by fermentation. The process of fermentation causes bread dough to rise and creates the bubbles in beer.

Parasitic Sac Fungi   The powdery mildews are among the most destructive of the parasitic sac fungi. The mycelia of these fungi form a white powder on the leaves of apples, roses, grapes, and other economically important plants. The growth of

Highlight on Careers: Agricultural Extension Agent

General Description
Agricultural extension agents are specially trained to inform farmers on certain agricultural matters. As representatives of state and federal agencies, extension agents keep farmers informed about laws that affect the farming business. They also tell farmers about new fertilizers and other products that can increase crop yield.

A typical agent might begin the day by giving a report on weather and crop conditions over the radio. Agents often hold seminars on special topics, such as the effectiveness of new fungicides that help prevent crop damage.

Keeping up with current trends and using that information to help people produce food more efficiently are major challenges for extension agents. Farming is a rapidly changing business. Agents learn about new developments by reading scientific publications, attending classes, and talking with farmers about their work.

Career Requirements
Agents must have a B.S. degree in agricultural science. Some states require

agents to have an M.S. degree. State programs give specific training in extension work.

For Additional Information
U.S. Department of
   Agriculture
Personnel Division
Hyattsville, MD 20782

the powdery mildew destroys the tissues of the host plant. This process hinders the plant's ability to carry out photosynthesis, and further damage results.

Dutch elm disease is caused by another sac fungus. The hyphae of the fungus grow into the wood of an elm tree and clog the tissues that carry water and nutrients from the soil up to the leaves. Dutch elm disease threatens to wipe out all American elms. Chestnut blight, caused by a related fungus, poses a similar threat to chestnut trees.

23.8 Imperfect Fungi

The class Deuteromycetes includes about 25,000 species. Fungi in this class are called ''imperfect'' fungi because they do not reproduce sexually or because their sexual life cycles are not fully understood. The most familiar imperfect fungi belong to the genus *Penicillium*. These fungi are used to produce penicillin. Other imperfect fungi cause diseases including ringworm and thrush.

23.9 Slime Molds

The slime molds are difficult to classify. *In body form, slime molds resemble protozoa, but their method of reproduction is typical of a fungus.* The body of a slime mold is a brightly colored, jellylike network of cytoplasm called a **plasmodium.** The plasmodium has no cell walls but many nuclei. It creeps by amoeboid movement over the ground and dead logs. Like an amoeba, the plasmodium engulfs microscopic prey and digests the food in food vacuoles.

When its food supply runs short, the plasmodium separates into tiny mounds of cytoplasm that develop into funguslike sporangia. The spores produced in these sporangia may be scattered to new areas where food is more plentiful. Slime mold spores grow into either flagellated or amoebalike cells. A new plasmodium develops when two flagellated cells fuse or when a mass of the amoebalike cells join and form the larger organism.

Figure 23–6. A plasmodium (above) is the vegetative portion of a slime mold life cycle. Dramatic changes convert a jellylike plasmodium into numerous sporangia (below) that produce spores.

Reviewing the Section

1. How does mold grow on a piece of bread?
2. Why are basidiomycetes called club fungi?
3. What characteristics make yeasts unusual fungi?
4. At what stage of its life cycle does a slime mold most resemble a fungus?

Investigation 23: Common Molds

Purpose
To observe several kinds of household molds

Materials
Three moldy food materials, hand lens or stereomicroscope, forceps, slides, coverslips, compound microscope

Procedure
1. Make a chart like the one shown. As you examine each type of mold, fill in your observations on the chart.

Common Molds

Name of Species	Sketch	Distinctive Features

2. Obtain a sample of mold from your teacher. Examine the mold with a hand lens or a stereomicroscope. Make a sketch of the mold. Describe its appearance.
3. Using forceps, remove a small section of the mold and place it on a slide. Add a drop of water and a coverslip.
4. Examine the slide under a microscope. Focus on a low power first, then switch to high power. Describe the appearance of the mold. *How does the appearance differ when observed under the compound microscope?*
5. Repeat steps 2, 3, and 4 with other mold samples. As you make your observations, complete the chart.

Analyses and Conclusions
1. How many different kinds of molds did you observe?

2. What medium seemed to be the best for the growth of mold?
3. What is the origin of the mold growth in each of the food materials?

Going Further
Determine whether any of the molds have antibiotic properties. Prepare a petri dish with nutrient agar and inoculate it with bacteria. Place a small sample of one or more molds in the petri dish. Incubate it for 24 hours at 37°C. After incubation, examine the plate for clear areas in which bacteria are not growing.

Chapter 23 Review

Summary

Fungi are classified in a separate kingdom. The cell walls of most fungi contain chitin. Fungi feed by decomposing organic matter. Hyphae secrete digestive enzymes and absorb food through their cell walls. Many intertwined hyphae form the mycelium.

Fungi include two main groups, the true fungi and the slime molds. Slime molds have traits of fungi and of protozoa. True fungi include terrestrial molds, water molds, club fungi, sac fungi, and imperfect fungi.

Terrestrial molds reproduce asexually by forming spores and sexually when two mating strains come into contact. The water mold life cycle involves gametes that are distinctly male and female.

Club fungi produce spores in basidia, club-shaped structures. Sac fungi produce spores sexually in an ascus, or tiny sac. Imperfect fungi either do not reproduce sexually or have poorly understood sexual cycles.

BioTerms

ascospore (348)
ascus (347)
basidiospore (347)
basidium (346)
button (346)
cap (346)
chitin (343)
fruiting body (343)

fungus (343)
gill (346)
hypha (343)
lichen (344)
mating strain (345)
mycelium (343)
oogamy (346)
plasmodium (349)

rhizoid (345)
slime mold (345)
sporangiophore (345)
sporangium (345)
stipe (346)
stolon (345)
true fungus (345)

BioQuiz   (Write all answers on a separate sheet of paper.)

I. Completion

1. Fungi feed by secreting _____ and by absorbing nutrients from the digested food.
2. Fungal spores are usually formed in structures called _____.
3. The differentiation of gametes into distinctly male and female forms is called _____.
4. Dutch elm disease is caused by a parasitic _____ fungus.
5. Terrestrial molds reproduce sexually when hyphae from two _____ strains come into contact.

II. Modified True and False

Mark each statement TRUE or FALSE. If false, change the underlined term to make the statement true.

6. Unlike plants, fungi do not contain chloroplasts.
7. The sexual life cycles of many club fungi are unknown.
8. Yeasts are unicellular sac fungi that do not form hyphae.
9. Smuts and rusts cause severe damage to crops.
10. In mushrooms, basidia are attached to gills on the underside of the cap.

III. Multiple Choice

11. Fungi reproduce chiefly by forming
    a) spores. b) hyphae. c) amoebalike cells. d) rhizoids.
12. The cell walls of most fungi contain a hard substance called a) cellulose.
    b) cytoplasm. c) chitin. d) hyphae.
13. One particularly destructive group of sac fungi is called a) black bread mold.
    b) powdery mildew. c) yeast.
    d) penicillin.
14. The cap is part of a mushroom's
    a) stipe. b) basidium. c) fruiting body.
    d) basidiospores.

15. A plasmodium is one stage in the life cycle of a) slime molds. b) imperfect fungi. c) water molds. d) sac fungi.

IV. Essay

16. How do parasitic fungi injure their host organisms?
17. Why are fungi ecologically important?
18. In what way does a lichen demonstrate mutualism?
19. What kinds of fungi are economically beneficial?
20. Why are slime molds considered a link between fungi and protozoa?

Applying and Extending Concepts

1. Downy mildew of grapes is a virulent disease caused by a fungus. It almost destroyed the French wine industry in the nineteenth century. Use your school or public library to research downy mildew of grapes. Write a brief report explaining how the mildew grows, why it is so destructive, and how grapes can be protected against this disease.
2. Farmers and agricultural researchers have developed various ways to prevent fungi from spoiling fresh food in storage. Use your school or public library to research food preservation and fungicides. Then write a paragraph explaining one method of keeping fresh fruits or vegetables safe from fungal infections.
3. Mycorrhiza is the association of a fungus and the roots of a plant. Research mycorrhizal associations in citrus trees or orchids and write a report on how the plant is helped or hurt by the fungus.
4. Basidiomycetes are considered the most advanced group of fungi. How do the structure and life cycle of a typical club fungus support this idea? Compare club fungi with other classes of fungi in your answer.

Related Readings

Ahmadjian, V. "The Nature of Lichens." *Natural History* 91 (March 1982): 31–36. This article explains relationships between fungi and algae in lichens that are parasitic rather than mutualistic.

Dickenson, C., and J. Lucas. *The Encyclopedia of Mushrooms*. New York: G. P. Putnam's Sons, 1979. This text is essentially a reference book. It combines stunning photographs of fungi with explanations of their history and biology.

Lee, D. "Slime Mold: The Fungus that Walks." *National Geographic* 160 (July 1981): 130–136. This article describes the beauty and the life cycles of various kinds of slime molds. It also includes details about their reproductive cycles.

Pappagianis, D. "Dangerous Dust." *Natural History* 92 (March 1983): 36–40. "Valley fever" is a disease caused by a fungus. This article describes the symptoms of the disease and discusses possible cures.

24

Nonvascular Plants

Dew on the capsules of the moss genus *Polytrichum*

Introduction

Modern land plants probably evolved from green algae. Scientists have suggested the following theory of how this development might have occurred. About 400 million years ago, the sea was teeming with life, but the land was mostly barren rock. Algae and other marine organisms began to grow near shore because of the availability of direct sunlight and of minerals washed off the shore. As competition for resources in the sea increased, some of the algae gradually became adapted to conditions on land.

The green algae that adapted successfully were probably the ancestors of today's land plants. This chapter will examine a group of land plants with many of the adaptations that originally allowed algae to survive on land.

Origin of Land Plants

Modern land plants share certain characteristics with algae. For example, the life cycle of plants resembles that of algae. Like most kinds of algae, plants have cellulose in their cell walls. Land plants and some groups of algae store food as starch. However, only green algae contain the two types of chlorophyll, *a* and *b,* found in modern land plants. Because of these chemical similarities, scientists claim that modern land plants evolved from forms of the green algae, Chlorophyta.

- *List* the evidence that indicates land plants probably developed from green algae.
- *Describe* the adaptations that enabled plants to survive on land.
- *Distinguish* between the two major groups of land plants.

24.1 Adaptations to Life on Land

Adapting is the process by which a species gradually becomes better able to survive in a given environment. ***The specialized structures of land plants are adaptations that allowed water-dwelling algae to overcome the problems of living on land and to best use land resources.***

Most of the problems of a land habitat result from the lack of surrounding water. Algae absorb water and minerals directly by diffusion across the cell membrane. In some plants, pores evolved. Pores absorb moisture from the environment. Other plants developed specialized structures that draw water and minerals from the soil.

In the open air, plants are in danger of losing moisture because of evaporation. A protective outer coating called a **cuticle** (KYOOT ih kuhl) is an adaptation that helps prevent evaporation. The cuticle also protects the plant from the relatively wide and abrupt temperature changes encountered on land.

Unfortunately, the cuticle also prevents the exchange of oxygen and carbon dioxide with the air. Small pores evolved in the cuticle. These pores, called **stomata** (stoh MAH tah), allow the necessary exchange of gases. In some plants, *guard cells* regulate the opening and closing of stomata in response to various environmental conditions. In other plants, stomata are always open.

Surrounding water supports algal cells. Multicellular land plants form a complex carbohydrate called **lignin** (LIHG nihn). Lignin, combined with cellulose, forms an extremely tough material that supports soft plant tissues. By keeping plants exposed to the direct sunlight available on land, this support allows land plants to maximize opportunities for photosynthesis.

For aquatic plants, surrounding water allows flagellated sperm to swim to egg cells. In order to reproduce sexually, some land plants still need water for sperm to swim in. Other

stomata (singular, *stoma*)

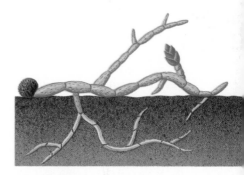

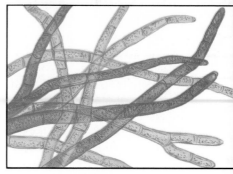

Figure 24–1. Land plants, such as mosses (top), have adaptations that enable them to live on land. Algae (bottom) live only in water.

plants evolved ways for sperm to travel through the air, in the wind, or on the bodies of insects or animals. Many plants also developed multicellular reproductive structures that protect the developing zygote and keep it from drying out.

About 400 million years ago, green algae evolved those structures necessary to survive on land—cuticles, stomata, lignified cells, and multicellular reproductive organs. Plants then colonized the land and exploited a habitat where they had no competitors.

Highlight on Careers: Park Ranger

General Description

Park rangers are specially trained employees of the National Park Service or a state, regional, or local park district. They are charged with protecting plant and animal life in the parks.

Environmental education is one of the rangers' chief duties. Rangers give lectures and slide shows to visitors. They cover such topics as the structure of nonvascular plants, the importance of forests to the environment, and the relationships among plants and animals in an ecosystem.

Park rangers also protect plants and animals from disease and from damage by visitors. Because they act as security guards in many parks, rangers are trained to use weapons. They are also trained in first aid. Park rangers work mainly outdoors.

This career has immense appeal for people who want to combine a love of the outdoors with a concern for how people can function harmoniously with nature and understand it better.

Career Requirements

A B.S. degree with a major in park management or field natural sciences and a civil service exam are usually required for many positions.

Competition is keen for ranger positions. Prospective rangers should be in excellent health, like working with plants and animals, and be skilled at dealing with all kinds of people.

For Additional Information

National Park Service
U.S. Department of the Interior
C Street at 18th Street, N.W.
Washington, DC 20240

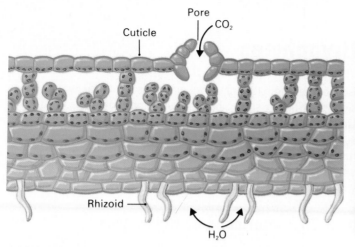

24.2 Vascular and Nonvascular Plants

Not all plants adapted to life on land in the same ways. One of the major distinctions between groups of plants is the way they transport water and nutrients throughout the plant body. The majority of land plant species have an internal system of inter-connected tubes and vessels called **vascular tissues.** These plants, grouped as **vascular plants,** will be discussed in the next chapter. Most of the plants you are familiar with—oak trees, roses, grasses, and house plants—are vascular plants and have vascular tissue. The other main group of plants, the **bryophytes** (BRY uh fyts), lack vascular tissues. These **nonvascular plants** transport water and nutrients by osmosis and diffusion, much as algae do.

The earliest fossils of vascular plants are about 400 million years old, but the earliest fossils of bryophytes are only about 350 million years old. For this reason, some scientists claim that bryophytes developed from vascular plants that gradually lost their vascular tissues. However, other scientists claim that bryophytes evolved independently. If many early bryophytes decomposed before they were fossilized, perhaps bryophytes developed earlier than the fossil record indicates.

Figure 24–2. A protective cuticle enables land plants to have leafy surfaces for photosynthesis (right). Plants such as common mosses have pores (left) for gas exchange in those areas covered with a cuticle.

Q/A

Q: *Do any bryophytes live in water?*

A: Yes. The brook moss, *Fontinalis,* grows long streamers that are supported on the surface of flowing water.

Reviewing the Section

1. In what ways are green algae and modern land plants similar?
2. Why is lignin important for the survival of modern land plants?
3. What is the main difference between vascular and nonvascular plants?

- *List* the major characteristics of bryophytes.
- *Diagram* the life cycle of mosses.
- *Name* the two ways in which liverworts reproduce asexually.
- *Distinguish* the characteristics of the liverwort sporophyte from those of the hornwort sporophyte.
- *Name* a kind of moss that is commercially important.

Figure 24–3. Mosses have small stems that cannot conduct water. Bryophyte stems (inset) lack the vascular tissues characteristic of other land plants.

Bryophytes

Because bryophytes transport materials by osmosis and diffusion, they need a large and constant supply of water to survive. Bryophytes also need water for sexual reproduction. Like algal sperm, bryophyte sperm must swim to the egg to fertilize it. For these reasons, most bryophytes grow in moist environments such as riverbeds, rain forests, and low-lying areas where water tends to collect.

24.3 Characteristics of Bryophytes

Almost all bryophytes are small plants, ranging in height from 1 to 20 cm (0.3 to 8 in.). Because gravity restricts the processes of osmosis and diffusion, nonvascular plants grow close to the ground. Bryophytes also lack the rigid tissues that vascular plants have to support vertical growth. Some bryophytes, however, grow to a large size. Most are aquatic species that live in rivers and streams. Supported by the buoyancy of water, these aquatic bryophytes can grow larger than terrestrial species.

Because they do not have vascular tissue, bryophytes do not possess true roots, leaves, or stems. What appear to be "roots" and "leaves" in bryophytes are not specialized structures like those of vascular plants, but mere elongations of the "stem." The *roots* of vascular plants anchor the plant body and absorb water from the soil. Rootlike **rhizoids** in bryophytes perform the same functions but do not channel water to other parts of the plant. The upper parts of bryophytes obtain moisture from their leaves, which absorb water through pores. Bryophyte leaves are usually only one cell thick.

The life cycles of bryophytes exhibit alternation of generations. The **gametophyte** (n) form produces gametes (n) by mitosis. In sexual reproduction, the gametes fuse. The resulting zygote grows into the **sporophyte** (2n) form, which produces spores (n) by meiosis. In asexual reproduction, these spores in turn develop into the new gametophyte (n) generation. *In all bryophytes, the gametophyte is the dominant form.* In other words, the gametophyte is the green leafy plant that makes up the major portion of the organism's life cycle.

The phylum Bryophyta is grouped into three classes. Over 9,500 species are included in the class Muscopsida, the mosses. About 6,000 species of liverworts belong to the class Hepaticopsida. The smallest class, Antherocerotopsida, has about 100 species. Members of this class are commonly called hornworts.

Figure 24–4. Mosses thrive in cool, moist environments such as this forest floor. By forming thick mats, mosses prepare the soil for colonization by other plant species.

24.4 Mosses

Mosses are small, soft plants that grow in clumps close together. They grow in a wide variety of moist, shaded habitats—on the sides of trees, in sidewalk cracks, on rocks and logs. Some mosses form a dense carpet on the floor of coniferous forests. The greatest number of mosses grow in areas of high humidity, such as the Olympic and Great Smoky mountains, the rain forests of the tropics, and in colder regions as well.

The body of a moss is composed of leaves arranged in a spiral around a central stem. Moss plants range in size from 1 or 2 cm (0.4 to 0.8 in.) to more than a meter (39 in.) long. Moss plants may stand erect or trail along the ground.

Life Cycle of Mosses   As in all bryophytes, the dominant generation in the moss life cycle is the haploid gametophyte. This form is the familiar green, leafy moss plant. The sporophyte generation of mosses, which appears as a stalk tipped with a spore-bearing capsule, does not photosynthesize. Because it is dependent on the dominant generation for nutrition, the sporophyte remains physically attached to the gametophyte throughout its life.

Figure 24–5 on page 360 illustrates the life cycle of a moss. The asexual part of the cycle begins when the sporophyte (2n) releases spores. When the environment is suitably warm and moist, a spore (n) will germinate and produce a horizontal filament called the **protonema** (proht uh NEE muh). Protonema cells contain chloroplasts and carry out photosynthesis. As the protonema grows, it periodically produces buds that develop into gametophytes.

Q/A

Q: *Are all plants that are called "moss" really mosses?*

A: No. Only members of the class Muscopsida are "true" mosses; other plants called "moss" are not. For example, Reindeer moss is a lichen; Spanish moss is a flowering vascular plant; and Irish moss is a red alga.

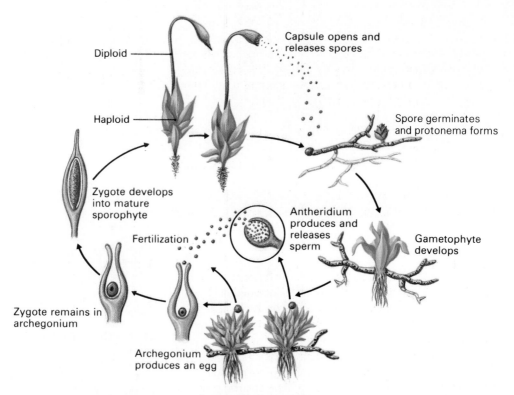

Diploid

Capsule opens and releases spores

Haploid

Spore germinates and protonema forms

Zygote develops into mature sporophyte

Fertilization

Antheridium produces and releases sperm

Gametophyte develops

Zygote remains in archegonium

Archegonium produces an egg

Figure 24–5. The moss life cycle (above) involves an alternation of generations in which the gametophyte is the dominant phase. A simplified cycle is diagramed below.

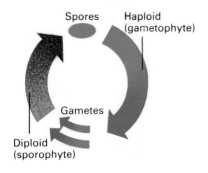

Spores
Haploid (gametophyte)

Gametes

Diploid (sporophyte)

antheridium (plural, *antheridia*)
archegonium (plural, *archegonia*)

Gametophytes produce gametes through mitosis. Sperm are produced in the male reproductive structure, called an **antheridium** (an thuh RIHD ee uhm). Each female reproductive structure, or **archegonium** (ahr kuh GOH nee uhm), contains an egg cell. In some species of moss, both antheridia and archegonia are found on the same plant, but other species have separate male and female plants.

In the gametophyte generation of mosses, fertilization can only take place in a moist environment. Sperm, which must swim to reach the egg cells, are not released from the antheridia unless the moss is covered with moisture such as heavy dew, fog, or rain.

The sporophyte generation begins with the fertilized egg (2n). As the zygote develops into a mature sporophyte, it grows up through the neck of the archegonium. The asexual phase of reproduction occurs inside the spore capsule of the sporophyte. There, layers of haploid cells undergo meiosis and form spores. When the sporophyte is fully mature, the end of the spore capsule drops off. Wind can then shake the spores from the capsule and scatter them.

Importance of Mosses Because mosses are among the first plants to grow in otherwise barren areas, they are sometimes called *pioneer* plants. As mosses become established, the growth of rhizoids splits off tiny bits of rock. This process slowly creates new soil. Rhizoids also help prevent erosion by anchoring existing soil. When the bryophytes die, their bodies add organic matter to the soil. Over time, an extremely fertile mixture called *topsoil* is formed. Many vascular plants need topsoil in which to grow.

One kind of moss, sphagnum moss, is commercially important. Sphagnum is the main component in peat moss, an organic fuel used in homes in Ireland, Canada, the Soviet Union, and other northern countries. Peat is cut from areas called *bogs* that are made up largely of the decomposing bodies of bryophytes. At one time, peat bogs were also the source of some coals that formed millions of years ago.

24.5 Liverworts

Liverworts get their name from the flat, liverlike shape of the main body of the plant. Liverworts are much smaller than mosses. The upper surface of the liverwort plant is covered with scaly flakes. On the surface of these flakes are many small pores through which gases are exchanged. Liverworts are found where conditions are warm and moist.

Liverworts have a sexual life cycle similar to that of mosses. The major difference between the two cycles is the way the plants produce archegonia and antheridia. As shown in Figure 24–6, liverworts send up stalks that bear the reproductive structures.

Liverworts also reproduce asexually. Small pieces that break off will form new plants by *fragmentation*. In some liverworts, special spores called **gemmae** (JEHM ee) are produced in small cupped structures that form on the upper part of the plant. When rain hits the cup, the gemmae splash out and begin to grow into new plants.

Figure 24–6. Liverworts (top) are bryophytes whose stems often grow pressed to the soil surface. Liverworts reproduce asexually through the production of gemmae in gemmae cups (bottom).

24.6 Hornworts

Hornworts are the smallest group of bryophytes. Only one genus, *Anthoceros* of the class Antherocerotopsida, now exists. Most hornworts grow in damp areas such as ditches, along the edges of streams, or near the shores of lakes.

Hornworts look like liverworts. The gametophyte generation is flat and circular, with lobed leaves. Hornworts take their name from the sporophyte, which resembles a long animal horn.

▶ Peat Moss

A peat bog is an unusual ecosystem. Certain features of the moss genus *Sphagnum* prevent other plants from growing there and promote the accumulation of peat.

Peat moss is formed when dead moss plants become compressed by the weight of the plants growing above them. The living moss cells, which have cytoplasm, conduct photosynthesis. The dead cells, which lack cytoplasm, absorb water. In fact, sphagnum can absorb about 20 times its own weight in water. As a result, drainage in bogs is very poor. As sphagnum grows, it draws minerals out of the water and gives off acidic wastes. Most other plants cannot grow well in the stagnant, acidic moss-filled surroundings of a peat bog.

Growing conditions are poor for saprophytes like bacteria and fungi as well. Peat accumulates quickly because dead matter decays slowly. In Denmark, "bog people" buried in bogs hundreds of years ago show almost no signs of decay. In other bogs, 10,000-year-old pollen has been discovered.

Bogs provide a number of useful substances. When peat is compressed over a long period, the water in it is replaced with carbon. This process yields coal. Peat moss itself is used in some homes as fuel, and in gardens to hold moisture and prevent erosion. Because sphagnum is mildly antiseptic, it was widely used in World War I as a surgical dressing. It is antiseptic because of its acidity.

Unlike that of liverworts, the sporophyte generation of hornworts conducts photosynthesis. Nevertheless, in hornworts as well as in liverworts, the sporophyte remains physically attached to the gametophyte. The sporangium of a hornwort splits open longitudinally to release its spores.

Reviewing the Section

1. Why are bryophytes said not to have true roots, leaves, and stems?
2. In what two ways do liverworts carry out asexual reproduction?
3. Why are mosses called pioneer plants?
4. Why is peat moss economically important?

Investigation 24: Life Cycle of a Moss Plant

Purpose
To study the life cycle of a moss plant

Materials
Sphagnum moss, 15-cm pot, humus soil, centimeter ruler, hand lens, slide, coverslip, compound microscope, triple-beam balance, 150-mL glass beaker, 100-mL graduated cylinder

Procedure
Part A
1. Carefully place a 7.5-cm square of moss in the center of a pot that has been filled with humus soil to within 2 cm of the top.
2. Observe the moss with a hand lens. Draw the different types of sporophyte and gametophyte structures you see.

Sphagnum moss with spore capsules

3. Flood the adult plant (gametophyte) with water for about one hour. This will permit fertilization to occur.
4. Following the procedure described on page 831, make a wet mount using some of the water drained from the moss. Examine the wet mount under a compound microscope. Describe what you observe.

Part B
1. Weigh out 8 g of sphagnum moss.
2. Place the moss in a 150-mL glass beaker.
3. Measure out 100 mL of water using a graduated cylinder. Pour the water into the beaker containing the moss.

4. Wait two minutes, then pour any water that was not absorbed back into the graduated cylinder.
5. Measure the volume of water not absorbed and record the figure.
6. Using the following formula to determine the absorption capacity of the moss.
   Formula:

$$\text{Absorption} = \frac{\text{Volume of water lost (in mL)}}{\text{Weight of moss (in g)}}$$

Analyses and Conclusions
1. From your observations, name ways in which the moss plant is adapted to carry on its life processes?
2. Why is sphagnum moss sometimes used as a packing material?

Going Further
- Investigate how different intensities of light affect antheridial and archegonial growth. Place one pot of moss in direct sunlight, another in shade, and a third in darkness. Observe the reproductive organs of each plant at the end of one month.
- Combine the following salts with 1 L of distilled water: 0.25 g each of KCl (potassium chloride), $MgSO_4$ (magnesium sulfate), and KH_2PO_4 (potassium dihydrogen phosphate), and a trace amount of $FeCl_3$ (ferric chloride). Mix these ingredients thoroughly. To this solution add 1 g of $Ca(NO_3)_2$ (calcium nitrate), 20 g of agar, and 20 g of glucose. Sterilize the solution and pour about 6 mm of it into a petri dish. Add spores from a ripe spore case of a moss plant. Observe growth on the artificial medium over the next several weeks.

Chapter 24 Review

Summary

Modern land plants are probably descendants of green algae that adapted to life on land. Modern plants are divided into two groups. Vascular plants have special tissues that transport water and nutrients throughout the plant. Nonvascular plants, or bryophytes, lack vascular tissue. Bryophytes transport materials by osmosis and diffusion. They are generally small and are found in moist habitats. The life cycle of bryophytes exhibits an alternation of generations in which the haploid gametophyte is the dominant form. Bryophytes do not have true roots, leaves, or stems.

Bryophytes are divided into three classes: mosses, liverworts, and hornworts. The sexual phase of the moss life cycle occurs when gametes fuse. The zygote develops into a sporophyte, which in turn produces haploid spores by meiosis. When the spores germinate, they give rise to the protonema. The new gametophyte generation develops from the protonema. The gametophyte produces archegonia and antheridia. Egg and sperm are formed. When they unite, a new zygote is established.

The gametophytes of liverworts and hornworts are flat-bodied. Both of these groups have a life cycle similar to that of mosses. Liverworts also reproduce asexually by fragmentation and by the production of gemmae usually formed in special gemmae cups.

BioTerms

antheridium (**360**)
archegonium (**360**)
bryophyte (**357**)
cuticle (**355**)
gametophyte (**358**)
gemmae (**361**)
lignin (**355**)
nonvascular plant (**357**)
protonema (**359**)
rhizoid (**358**)
sporophyte (**358**)
stomata (**355**)
vascular plant (**357**)
vascular tissue (**357**)

BioQuiz   (Write all answers on a separate sheet of paper.)

I. Completion

1. _____ moss is the main component of peat bogs.
2. Nonvascular plants, such as mosses and liverworts, transport water and nutrients by diffusion and _____ .
3. The establishment of mosses in a barren area helps create _____ .
4. The waxy, protective _____ prevents land plants from losing too much water through evaporation.
5. The most important advantage of life on land for plants is the availability of direct _____ .

II. Modified True and False

Mark each statement TRUE or FALSE. If false, change the underlined term to make the statement true.

6. The gametophyte is the dominant generation in all bryophytes.
7. Hornworts have the fewest species of the bryophytes.
8. A sporophyte results from the germination of a moss spore.
9. The egg cell of a moss plant is fertilized in the antheridium.
10. Fragmentation is one method of asexual reproduction in liverworts.

III. Multiple Choice

11. Stomata allow land plants to carry out
    a) reproduction. b) osmosis. c) gas exchange. d) mitosis.
12. Land plants can grow upright because they contain a) lignin. b) a cuticle. c) minerals. d) spores.
13. The process of meiosis in moss produces
    a) rhizoids. b) roots. c) spores. d) gametes.
14. Mosses have rootlike structures called
    a) gemmae. b) rhizoids. c) protonemas. d) stomata.
15. In mosses, sperm are produced in the
    a) protonema. b) archegonium. c) antheridium. d) gemmae.

IV. Essay

16. Why do scientists think that plants such as mosses and vascular plants evolved from green algae?
17. What are the advantages of multicellular reproductive structures over the unicellular type found in algae?
18. Why are mosses and other bryophytes generally small?
19. What role does dew play in the reproduction of mosses?
20. Did bryophytes evolve from vascular plants? Explain your answer by citing evidence from bryophyte structure.

Applying and Extending Concepts

1. Use your school or public library to research the process of asexual reproduction in liverworts that produce gemmae. Then draw a diagram showing each step of this process. Include captions and labels in your diagram. The labels should name each part of the liverwort and indicate whether they are haploid (n) or diploid (2n). The captions should explain what happens during each stage of the asexual reproductive process.
2. The spores of some mosses can remain dormant in conditions that are unfavorable for survival. Write a paragraph explaining how this characteristic of moss spores might be beneficial to mosses in their role as pioneer plants.
3. The adaptation of algae to life on land entailed various structural changes. Explain how conditions in a terrestrial environment made the presence of stomata an effective adaptation.
4. Bryophytes produce gametes through mitosis rather than meiosis, the method of gamete formation in animals. Explain why bryophyte gametes are nevertheless haploid, just as animal gametes are.

Related Readings

Conrad, H. S. and P. L. Redfearn, Jr. *How to Know the Mosses and Liverworts*. 2nd. ed. Dubuque, Iowa: William C. Brown, 1979. This book is a useful and well-organized field guide for the identification of common bryophytes.

Knauss, F. "The Varied World of Lichens and Mosses." *The Conservationist* 35 (July/ August 1980): 32–34. This article describes the structures of mosses and lichens. It also comments on the adaptations that enable these bryophytes to survive in unfavorable conditions.

Massey, C. "Peat Moss." *Flower and Garden* 22 (September 1978): 30–32. This article describes peat mosses and bogs and discusses the various uses of sphagnum mosses and methods of harvesting them.

Vascular Plants

Deciduous angiosperm trees in fall coloration

Introduction

All the grasses, trees, ferns, shrubs, and wildflowers that cover the earth are vascular plants. More than 250,000 species have been identified, 15 times the number of existing species of bryophytes. Vascular plants display a tremendous variety of sizes, structures, and methods of reproduction because they have adapted to almost every kind of climate. Their complex adaptations also give them advantages over other kinds of plants. These advantages allow vascular plants to flourish where bryophytes cannot.

The plants in turn influence their environment. Wherever vascular plants flourish, they create habitats and provide food for various species of insects as well as for many other kinds of animals.

Development of Vascular Plants

Section Objectives

- *List* the basic plant structures that vascular and nonvascular plants have in common.
- *Describe* the vascular system of plants.
- *Explain* how vascular plants show adaptations to life on land.
- *Name* the dominant generation in the life cycle of vascular plants.

Like nonvascular plants, vascular plants developed from water-dwelling green algae approximately 400 million years ago. The **tracheophytes** (TRAY kee uh fyts), or vascular plants, share some basic adaptations to life on land with the bryophytes. Both groups have a waxy outer *cuticle* that retains water, and stomata in the cuticle to allow an exchange of gases. Multicellular reproductive structures that protect delicate zygotes evolved in both vascular and nonvascular plants.

Unlike bryophytes, tracheophytes have an internal network of tubes known as the **vascular system.** The tubes carry water, nutrients, and the products of photosynthesis throughout the plant. The vascular system can transport fluids over long distances, from roots buried deep in the soil to treetops perhaps hundreds of meters above the ground. The cell walls of tracheophytes contain *lignin,* a substance that helps support the plant body.

25.1 The Vascular System

The body of a vascular plant is made up of three types of structures. The *roots* absorb moisture and nutrients from the soil and anchor the plant. *Leaves* have chloroplasts and produce food by photosynthesis. The *stem* contains vascular tissues that transport substances between the roots and leaves and support the plant body. Because they contain vascular tissue, tracheophytes are said to have true roots, stems, and leaves.

The vascular system includes two distinct kinds of vascular tissues. The **xylem** (ZY luhm) transports water and minerals absorbed by the roots up to those parts of the plant that are above the ground. The **phloem** (FLOH ehm) carries sugar and other soluble organic materials created in photosynthesis from the leaves to the rest of the plant.

25.2 Reproduction in Vascular Plants

The life cycles of vascular plants are significantly different from those of nonvascular plants. *In vascular plants, the sporophyte is the dominant generation.* The sporophyte is physically larger, shows more complex development, and produces more varied types of cells than the gametophyte.

Tracheophytes are traditionally divided into two groups, seedless plants and seed plants. Seedless plants developed first and still have traits that show their watery origin. Most seedless plants require water for sexual reproduction.

Figure 25–1. Vascular plants have tissues that conduct water and nutrients. The inset shows a cross section of the plant's stem with xylem in red and phloem in blue.

Figure 25–2. About 300 million years ago, vascular plants dominated the earth. Giant club mosses and giant horsetails were common in the great hot swamps of the Carboniferous period.

Seed plants developed an important adaptation; they reproduce sexually by forming seeds. The details of the life cycles of specific vascular plants will be covered later in this chapter.

25.3 Adaptations in Vascular Plants

Certain adaptations in tracheophytes gave them important evolutionary advantages over nonvascular plants. Consider for example the different ways in which bryophytes and vascular plants cope with an environment where the top soil layer is permanently frozen. To survive in this environment, a plant must have mechanisms to overcome the lack of available water. Nonvascular plants cannot easily tolerate these conditions. Bryophytes need an abundant supply of unfrozen water to transport materials by osmosis and diffusion and to reproduce sexually. However, tracheophytes are able to obtain the water they need from roots growing beneath the frozen surface. The sperm of vascular plants, protected in pollen, can reach egg cells through the air. Seeds from seed plants can remain dormant until conditions are favorable for their germination. These adaptations enabled tracheophytes to survive in unfavorable conditions and contributed to the worldwide proliferation of vascular plants.

Q/A

Q: *What kind of tree lives the longest?*

A: Cone-bearing trees. The oldest living tree on Earth, a bristlecone pine in California, is 4,900 years old.

Reviewing the Section

1. What traits do vascular and nonvascular plants share?
2. What functions do xylem and phloem perform?
3. Which is the dominant generation in vascular plants?
4. How did the development of the vascular system help plants survive on land?

Seedless Vascular Plants

The seedless vascular plants include living representatives from four phyla: Psilophyta (sy LAHF uh tuh), the whisk ferns; Sphenophyta (sfee NAHF uh tuh), the horsetails; Lycophyta (ly KAHF uh tuh), the club mosses and their relatives; and Pterophyta (tehr AHF uh tuh), the ferns. Seedless vascular plants reproduce sexually by means of flagellated sperm that need water to swim to the egg cells. The sporophyte generation produces haploid spores that develop into small, independent gametophytes.

25.4 Whisk Ferns

The few surviving species of whisk ferns are the simplest vascular plants living today. Whisk ferns are unique among vascular plants because they lack roots and leaves. Most of the whisk fern body consists of an unspecialized, branching stem that contains vascular tissues. The green portion of the stem above the ground carries out photosynthesis. The underground stem, called a **rhizome,** anchors the plant and produces rhizoids that absorb water and nutrients from the soil. Whisk ferns grow from 10 to 40 cm (4 to 16 in.) tall.

25.5 Horsetails

The 15 living species of horsetails all belong to the genus *Equisetum*, Latin for "horsetail." Perhaps their ribbed stems and whorls of leaves suggested the name. Horsetails are the only seedless vascular plants that have hollow stems. *Silicon*, a sandlike substance, gives the stem a coarse texture. The ancestors of horsetails were treelike swamp plants that lived over 300 million years ago. Horsetails today live primarily in warm, moist environments.

Section Objectives

- *Name* the features that distinguish whisk ferns and horsetails from all other seedless vascular plants.
- *Describe* the life cycle of club mosses.
- *Summarize* how the remains of seedless vascular plants were converted into coal.
- *List* the parts of a fern.
- *Explain* the differences between sexual and asexual reproduction in ferns.

Q/A

Q: *Why are horsetails also called "scouring rushes"?*

A: Pioneers used the plants' tough stems to scour dirty pots and pans.

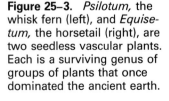

Figure 25–3. *Psilotum,* the whisk fern (left), and *Equisetum,* the horsetail (right), are two seedless vascular plants. Each is a surviving genus of groups of plants that once dominated the ancient earth.

25.6 Club Mosses and Their Relatives

Club mosses and their relatives date back to the Devonian period, 345–395 million years ago. About 1,000 species survive today. Of the five genuses of club mosses, the two most important are *Lycopodium* and *Selaginella*. Both of these genuses have many representatives in both temperate and tropical areas.

Lycopodium   There are about 200 species of club mosses in the genus *Lycopodium*. Most club mosses live in shady, moist places like the floor of a forest. *Lycopodium* is called a club moss because it produces spores on narrow, clublike cones. These cones have many small leaves called **sporophylls.** At the base of each sporophyll is a *sporangium*, in which spores are formed by meiosis. In club mosses, the sporophyte is clearly the dominant generation. The gametophyte is so small it is difficult to see. Its only function is to produce male and female reproductive structures, *antheridia* and *archegonia*, which are often present on the same plant. Some gametophytes lie dormant underground for almost 10 years.

Figure 25–4. The club moss, *Lycopodium,* produces spores in distinct reproductive structures that resemble cones. The leaves of these cones are called sporophylls.

▶ Coal: A Fossil Fuel

Most coal was formed from the remains of plants that lived on the earth long ago. During the Carboniferous period, which occurred about 250 million to 350 million years ago, much of the earth was covered by dense forests and swamps of giant horsetails, club mosses, and ferns. As these plants died, they formed thick layers of partly decomposed organic material called *peat.* In time, the peat was buried under minerals and sand. The weight of these overlying layers put tremendous pressure on the peat, causing it to transform into a form of coal called *lignite.*

As more and more organic matter accumulated, the lignite was further compressed. It was gradually transformed into *subbituminous coal, bituminous coal*, and finally *anthracite,* the hardest of all types of coal. Geologists estimate that it takes a layer of compressed plant material 2.1 m (7 ft.) thick to make 0.4 m (1 ft.) of anthracite.

The plant species that were formed into coal are now almost all extinct.

Their descendants are generally small plants of little economic importance.

Coal formation is still occurring. Under proper conditions, plants growing in today's swamps may one day be transformed into coal. However, the process of coal formation is very slow and coal reserves in the United States and in the world are being rapidly depleted. Some large deposits of anthracite are still found in Pennsylvania and Virginia, but coal and other fossil fuels such as petroleum will soon be used up.

Selaginella *Selaginella* grows in tropical regions and desert environments. Over 700 species have been identified. These plants are characteristically small. Their trailing stems branch frequently and produce many small leaves.

Selaginella produces two kinds of spores. The smaller spores are called *microspores*. They develop into male gametophytes that produce only antheridia. The larger spores, called *megaspores,* develop into female gametophytes that produce only archegonia.

25.7 Ferns

Approximately 12,000 species of ferns have been identified, more than any other group of seedless vascular plants. Most ferns prefer moist, fertile soil and live in the tropics, but ferns have adapted to almost every climate. Certain types of ferns are even found in very cold areas north of the Arctic Circle or high atop mountains.

Ferns have a wide range of sizes as well. Some are very small plants, but others grow as tall as trees.

Figure 25–5. This species of *Selaginella* lies flat on the ground. *Selaginella* differs from most seedless vascular plants in producing male and female spores and male and female gametophytes.

Figure 25–6. Ferns range in size from small species (far left) to large tree ferns (left).

Physical Structure Some ferns are delicate plants scarcely 3 mm (0.04 in.) tall. In contrast, huge tree ferns can reach 28 m (93 ft.). Few plant phyla show such wide variation.

Generally ferns are supported by underground rhizomes that produce roots. Each fern leaf, called a **frond,** has two parts. The *stipe* is the stemlike structure that attaches the leaf to the rhizome. The *blade* is the broad, green part of the leaf that carries on photosynthesis. Fronds spread out over a wide area. In this way they catch the dim light that reaches the forest floor.

Spotlight on Biologists: Harlan Banks

Born: Cambridge, Massachusetts, 1913

Degree: Ph.D., Cornell University

What is the origin of land plants and how did they first evolve? Paleobotanist Harlan Banks uses fossil plants collected from rocks in the Catskill Mountains of New York to answer these questions.

Four hundred million years ago, central New York state was a shallow inland sea. Then, towards the end of the Silurian period, 405 million years ago, the sea bottom rose up, creating land. The succeeding period, the Devonian, is when plants first began to grow on land. Today central New York state is particularly rich in early land-plant fossils.

The shoreline of the inland sea that once covered central New York is only 242 km (130 mi.) east of

Cornell University, where Banks taught for 31 years. Now a professor emeritus, he has compared the New York fossils with similar collections made in Poland, Wales, Canada, and Libya. From these studies, Banks and his colleagues are reconstructing what the earliest plants looked like and how they grew and reproduced.

The information collected by Banks and his colleagues may also help determine whether early plants had a single origin or whether they evolved independently in a number of places. Placed in an evolutionary timetable, Banks's study of fossil plants may shed light on other questions, such as how continents were formed.

Preparing plant fossils for study is tedious, time-consuming work. Rock fossils must often be cut and

then ground into hundreds of separate layers, each thin enough to be viewed under a microscope. Banks and his colleagues reconstructed the structure of Devonian plants by following this and other painstaking procedures.

Banks has received many scientific honors during his lifetime. Banks was enormously popular as a teacher, and during his long career at Cornell also received many student-nominated awards for excellence in teaching.

Life Cycle of Ferns The fern life cycle is typical of seedless vascular plants. Gametes from the gametophyte generation fuse to form zygotes. The sporophyte generation grows from the zygote. As the sporophyte matures, it forms a tightly curled leaf called a *fiddlehead*. When the fiddlehead is exposed to sunlight, it gradually opens into a new frond. Haploid fern spores develop in sporangia on the underside of fronds. Sporangia usually occur in clusters called **sori**. Mature spores are released from the sporangia and dispersed by the wind.

sori (singular, *sorus*)

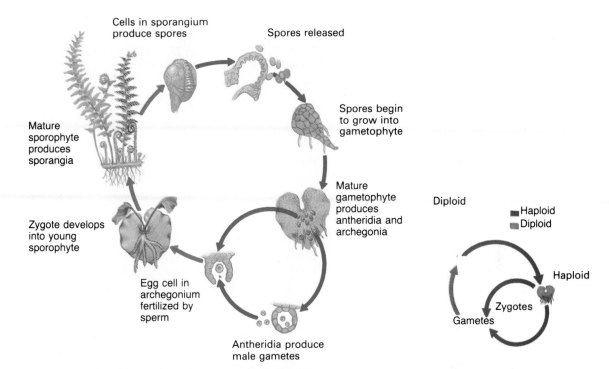

Cells in sporangium produce spores

Spores released

Spores begin to grow into gametophyte

Mature gametophyte produces antheridia and archegonia

Mature sporophyte produces sporangia

Diploid

Haploid
Diploid

Haploid

Zygote develops into young sporophyte

Zygotes

Gametes

Egg cell in archegonium fertilized by sperm

Antheridia produce male gametes

Fern spores develop into the gametophyte generation, called the **prothallus.** The prothallus is a green organism, often heart-shaped, that attaches itself to soil, rocks, or tree bark by rhizoids. The typical prothallus is only 1 cm (0.4 in.) wide. Archegonia and antheridia develop on the underside of the prothallus, where a film of rain or dew collects and enables sperm to swim to the archegonia. The fertilized egg cell produces a diploid zygote that in time develops into a new fern sporophyte.

Ferns can be cloned by cutting a section of rhizome from a mature fern and replanting it in soil. Cloned ferns are genetically identical to the parent plant.

Figure 25–7. The life cycle of ferns (above) involves an alternation of generations in which the sporophyte is the dominant phase. The inset shows a simplified diagram of this life cycle.

Reviewing the Section

1. What unique characteristics distinguish whisk ferns from other vascular plants?
2. Why is water necessary for sexual reproduction in club mosses?
3. What stages of fern reproduction involve the prothallus?

- *Identify* the parts of a seed.
- *Name* three ways in which seeds increased the adaptability of vascular plants.
- *List* the similarities and differences between gymnosperms and angiosperms.
- *Distinguish* between monocots and dicots.
- *Summarize* the economic and ecological importance of angiosperms.

Seed Plants

Seed plants reproduce chiefly by forming seeds. Every seed contains a plant **embryo,** or partially developed plant that is capable of growing into a mature plant. A seed also contains one or two embryonic leaves called **cotyledons** (kaht uhl EED uhnz). Cotyledons may become the young plant's first leaves or may be used as a food supply as the plant grows. A hard covering called the **seed coat** encases the embryo and cotyledons and protects them from physical injury and drought.

The development of seeds greatly increased the ability of tracheophytes to survive in unfavorable environments. Seeds protect plant embryos from harsh conditions. As a result, embryos can lie dormant for years and still produce healthy plants when conditions allow. Some seed-bearing structures, including stickers, burrs, and thistledown, travel long distances on animals or in the wind. In this way seeds substitute for mobility, allowing vascular plant species to spread to new areas.

Two groups of seed plants developed from early vascular plants. **Gymnosperms** (JIHM nuh spuhrmz) produce their seeds in cones and generally keep their leaves throughout the year. **Angiosperms** (AN jee uh spuhrmz) produce flowers, bear their seeds in fruit, and in general lose their leaves annually. Like all tracheophytes, the dominant generation in gymnosperms and angiosperms is the sporophyte. The roots, stems, and leaves of the sporophyte make up the plant's **vegetative body** and carry out the processes of photosynthesis and normal growth.

25.8 Gymnosperms

The seeds of most gymnosperms develop uncovered on *cone scales.* Four groups of gymnosperms have living representatives. Most of the 550 species of gymnosperms are conifers. Cycads (SY kadz) have about 100 species, and the order Gnetophyta (NEHT uh fy tuh) has approximately 70. Ginkgoes, the rarest of gymnosperms, have only one species.

Conifers *Conifer* means "cone-bearer." Pines, spruces, firs, and other conifers are characterized by their stiff cones and needlelike leaves. Coniferous forests were once common in temperate zones. Now, conifers are mostly found in the north temperate zone and other arid regions with sandy soil, cold winters, and moderate rainfall.

Conifers can thrive in harsh conditions because of special adaptations. Their needles are sheathed in a hard, waxy cuticle

Figure 25–8. Conifers like this pine have stiff cones and needlelike leaves.

and have little exposed surface area. As a result, needles retain moisture through hot, dry summers and the coldest of winters. Needles are shed and replaced throughout the year, rather than being lost every autumn and replaced every spring. Conifers send roots out over a wide area rather than deep into the soil. This shallow root system holds the tree stable even where soil is scarce.

Sexual reproduction in most conifers involves separate male and female cones that grow on the same tree. The male *pollen cones* produce microspores that develop into pollen grains, each one an immature male gametophyte. Within the female *seed cones*, megaspores develop into female gametophytes that contain egg cells. When the egg cells mature, the female cones secrete a sticky sap that traps pollen drifting in the wind. As the sap dries, it draws the pollen toward the egg cells. The pollen grains then produce mature male gametes, or sperm, which fertilize the egg cells. The resulting diploid zygote develops into a conifer embryo.

Conifers do not need water to carry out sexual reproduction because their male gametes are dispersed by the wind. Conifers can therefore reproduce in areas where nonvascular plants and seedless vascular plants cannot.

Other Gymnosperms Forests of cycads were once common. Now, these gymnosperms live mainly in the tropics. Cycads

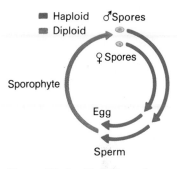

Figure 25–9. The life cycle of a conifer (below) involves an alternation of generations. Male cones produce pollen that travels to the female cone. A simplified life cycle diagram is shown above.

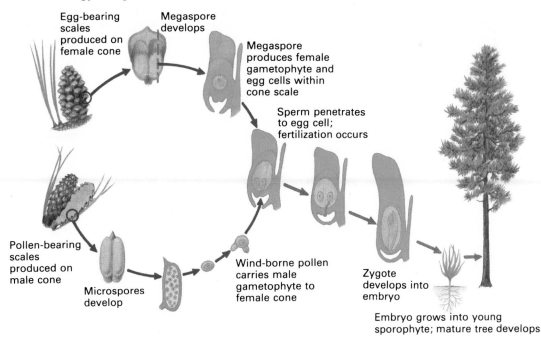

Figure 25–10. Gymnosperms are a diverse group of seed plants. *Welwitschia* (left) grows only in the deserts of southern Africa. Cycads (center) are palmlike plants from the tropics. Ginkgo trees (right) are native to China.

Q/A

Q: *Why are female ginkgoes seldom planted as ornamental trees?*

A: The seeds they produce give off a foul odor when mature.

resemble palms but are unrelated to them. The cycad reproductive cycle is similar to that of conifers, but most cycad trees are either male or female.

The order Gnetales includes trees and woody vines that have traits of gymnosperms, but some species have reproductive structures like those found in angiosperms. One species produces an edible plumlike fruit. Another, *Welwitschia*, grows only in the deserts of southern Africa.

The ginkgo is the last species of a once widespread family of trees. They are called ''living fossils'' because virtually no ginkgoes are now known to live in the wild. Ginkgoes have unique fan-shaped leaves. Pollen is produced in *catkins,* or small conelike structures that dangle from the tree branches. Only male ginkgoes produce catkins. After fertilization, female trees produce fleshy seeds that look like pale berries.

25.9 Angiosperms

Angiosperms are flowering plants. They produce enclosed seeds, as opposed to the uncovered seeds of the gymnosperms. To botanists, the history of angiosperms is still a mystery, because the flowering plants appeared so suddenly in the fossil record about 280 million years ago. Without a doubt, however, the angiosperms have been extremely successful. Of the more than 250,000 species of vascular plants, about 235,000 are angiosperms. They include most green plants. Oaks, birches, vegetables, and grasses are all angiosperms.

Physical Structure Angiosperms are classified according to the number of cotyledons in their seeds. Plants with one cotyledon are called **monocots;** those with two are called **dicots.** Monocots include about 89,000 species; dicots have about

170,000. Monocots and dicots can also be identified by the characteristics illustrated in Figure 25–11. For example, the flower petals of monocots usually occur in threes or multiples of three, while the petals of dicots usually occur in fours, fives, or multiples of four or five. Monocots usually have leaves with parallel veins. Dicots usually have netlike veins.

Angiosperms are also commonly classified by the characteristics of their stem tissues. In **woody plants,** the xylem and phloem produce cumulative layers of new plant tissue that increase the width of the stem and make it strong and hard. This cumulative growth is called **secondary growth.** Woody plants often live for many years and tend to produce relatively few seeds. Most woody plants are dicots, such as maple, walnut, and chestnut trees.

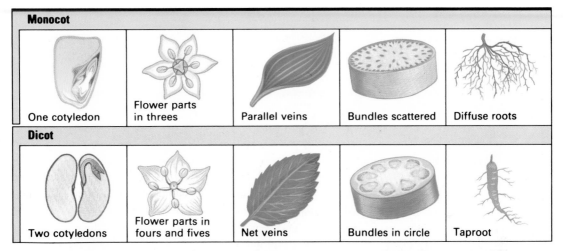

Almost all monocots are **herbaceous plants.** Their stems are usually green and lack secondary growth. Herbaceous plants typically have shorter lives and produce more seeds than woody plants. Corn, tomatoes, and orchids are examples of herbaceous plants.

Figure 25–11. The differences between the two groups of angiosperms, dicots and monocots, are shown in the diagrams. Each group represents a distinct line of angiosperm evolution.

25.10 Flowers and Fruits in Reproduction

All angiosperms produce reproductive structures called *flowers*. In most species, the flower contains both male and female reproductive organs. After the male sperm fertilize the female eggs, the flower petals usually die and the remaining flower structures develop into a *fruit*. Fruits protect seeds and help disperse them in various ways. When an animal eats fruit, for example, it may scatter the seeds or deposit them unharmed in a new area.

Table 25-1: Summary of Characteristics of Nonvascular and Vascular Plants

Common Name	Vascular System	Structure	Life Cycle and Reproduction	Habitats
Mosses, Liverworts Hornworts	None	Relatively simple; no true roots, stems, or leaves	Require water for sexual reproduction; gametophyte dominant	Moist areas
Club mosses Horsetails Whisk ferns	Relatively simple	True roots, stems, and leaves	Require water for sexual reproduction; sporophyte dominant	Areas with at least periodic moisture
Ferns	Relatively simple	True roots, stems, and leaves	Require water for sexual reproduction; sporophyte dominant; asexual reproduction from rhizome	Areas with at least periodic moisture
Conifers	Complex	True roots, stems, and leaves	Do not require water for sexual reproduction; naked seeds; gametophyte reduced to a few reproductive cells	Wide range of land environments
Flowering Plants	Complex	True roots, stems, and leaves	Do not require water for sexual reproduction; enclosed seeds; gametophyte reduced to a few reproductive cells	Almost all land environments

The development of seeds gave the animal world a new high-energy food source. For example, seeds provide excellent food for mammals, who need a great deal of energy to maintain their body heat. For this reason, biologists link the rise of mammals to the development of angiosperms and other types of seed plants.

Prehistoric people gathered fleshy fruits and seeds for food. About 11,000 years ago, people began to cultivate wild grains and to herd livestock on grassy pastures. People have depended on angiosperms ever since for food, lumber, fibers, clothing, and medicines. The most important grains are corn, rice, wheat, and sorghum. All of these are the seeds of grasses—a highly evolved group of angiosperms.

Reviewing the Section

1. What are the main parts of a seed?
2. How do seeds increase a plant's ability to survive?
3. How have conifers adapted to cold, dry climates?
4. What is secondary growth?

Investigation 25: The Effects of Overcrowding on Plants

Purpose
To investigate the effects of crowding on plant growth

Materials
Potting soil, three 15-cm pots, centimeter ruler, twenty lima bean seeds, calipers, scissors, triple-beam balance

Procedure
1. Place an equal amount of potting soil in each of the three pots. The soil should be no higher than 2.5 cm from the top of the pot.
2. Plant three seeds in pot 1, seven seeds in pot 2, and ten seeds in pot 3.
3. Allow the plants to grow to a height of about 10 cm.
4. Cut the plants off at soil level. Obtain the following data for each plant: diameter of stem, number of leaves, width of largest leaf at widest point.
5. Dry the plants overnight and weigh them the following day. Record the data.

Analyses and Conclusions
1. Make a bar graph of the weight data using the illustration on this page as a guide.

Bar Graph—Average Weight

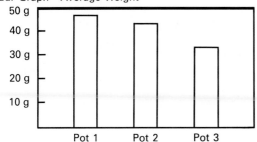

2. Make a table displaying the data on stem diameter, number of leaves, and width of largest leaf for each plant. Under what conditions did the plants do best?

3. Name the constants in this experiment (that is, name the factors contributing to plant growth that did not vary from plant to plant).
4. Why is it likely that some of the seeds in pot 3 did not germinate?
5. When planting a garden, why is it important to thin out the plants after they have started to emerge from the soil?

Going Further
- Using the steps given under Procedure, investigate the effects of overcrowding on different types of seeds such as corn, radish, and lettuce.
- Investigate whether the effects of overcrowding can be partly eliminated by using different kinds of soil such as humus, loam, or sand.

Chapter 25 Review

Summary

Tracheophytes and bryophytes both adapted to life on land. The vascular plants, however, developed specialized vascular tissues as well as roots, stems, and leaves. These structures enabled vascular plants to grow taller, disperse their reproductive cells more widely, and withstand harsher climates than nonvascular plants can.

Seedless tracheophytes require water for sexual reproduction. Whisk ferns, horsetails, club mosses, and ferns are living relatives of the early seedless vascular plants. Club mosses and ferns have similar life cycles, involving sporangia and spores that develop into the gametophyte generation. Ferns, the most varied seedless tracheophytes, also commonly reproduce asexually.

Seed plants include gymnosperms and angiosperms, the flowering plants. Both produce seeds that consist of an embryo plant, cotyledon(s), and a seed coat. Conifers have adapted well to life in cold, arid regions. They do not require water for reproduction. Angiosperms are the dominant plants on Earth. They are classified as monocots or dicots. Angiosperms produce flowers and fruits that provide much of the food essential for animal life.

BioTerms

angiosperm (**374**)
cotyledon (**374**)
dicot (**376**)
embryo (**374**)
frond (**371**)
gymnosperm (**374**)
herbaceous plant (**377**)

monocot (**376**)
phloem (**367**)
prothallus (**373**)
rhizome (**369**)
secondary growth (**377**)
seed coat (**374**)
sori (**372**)

sporophyll (**370**)
tracheophyte (**367**)
vascular system (**367**)
vegetative body (**374**)
woody plant (**377**)
xylem (**367**)

BioQuiz   (Write all answers on a separate sheet of paper.)

I. Completion

1. The most widespread tracheophytes are the _____ .
2. In the sexual reproductive cycle of vascular plants, pollen grains are part of the _____ generation.
3. _____ are fern leaves, which usually spread out and capture light on the forest floor.
4. The leaves of _____ usually have parallel veins.
5. Food reserves in seeds may be stored in the _____ .

II. Modified True and False

Mark each statement TRUE or FALSE. If false, change the underlined term to make the statement true.

6. The sporangia of ferns are attached to the <u>prothallus</u>.
7. The <u>embryo</u> of a seed plant is an immature plant.
8. Most <u>gymnosperms</u> produce their seeds in female cones.
9. Male <u>cycad</u> trees produce pollen in drooping catkins.
10. Most <u>dicots</u> show secondary growth.

III. Multiple Choice

11. The only vascular plants with hollow stems are the a) ferns. b) conifers. c) horsetails. d) whisk ferns.
12. Embryonic leaves are called a) fronds. b) cotyledons. c) fiddleheads. d) sori.
13. In ferns, the gametophyte generation is called the a) prothallus. b) microspore. c) sporangium. d) spore.
14. The hardest coal known is called a) peat. b) carboniferous. c) organic. d) anthracite.
15. The vascular system in tracheophytes consists of xylem and a) lignin. b) phloem. c) cuticle. d) gametes.

IV. Essay

16. In what ways are angiosperms, the flowering plants, important to humans and other animals?
17. What is the advantage of a shallow root system to a conifer?
18. Why do seedless vascular plants such as horsetails grow primarily in warm, moist habitats?
19. What are three basic differences between monocots and dicots?
20. What is the basic difference between the two major groups of seed plants, gymnosperms and angiosperms?

Applying and Extending Concepts

1. Florists sometimes dye carnations bright colors by putting dye in the flowers' water. What parts of the flowers' internal structure make this method effective?
2. Herbaceous plants are more sensitive to a lack of soil moisture than are woody plants. What characteristics of herbaceous plants might account for their sensitivity? Write directions for a class experiment that will test your answer, noting especially the kinds of gametophytes.
3. What aspects of the fern life cycle are similar to aspects of the life cycle of typical nonvascular plants? Make a list of their similarities and differences.
4. Pine forests tend to have much less undergrowth than hardwood forests do. Research coniferous forests. Then write a paragraph about the traits that could cause this phenomenon.
5. Grasses are highly specialized angiosperms. It is easy to overlook their flowers because they are generally small and pale. What might the size and color of these flowers indicate about the way grasses are pollinated?

Related Readings

Frankel, E. *Ferns: A Natural History*. Brattleboro, VT: The Stephen Greene Press, 1981. This lively book gives information about the history and characteristics of ferns. It also includes simple instructions for growing ferns and building a fern greenhouse. Care of fern plants is also emphasized.

Mickel, J. T. *How to Know the Ferns and Fern Allies*. Dubuque, IA: Wm. C. Brown, 1979. This book is a good introduction to the structure and life history of ferns, whisk ferns, club mosses, and horsetails. It is also useful as a field guide for identifying the plants.

Mulcahy, D. L. "Rise of the Angiosperms," *Natural History* 90 (September 1981): 30. This article gives an overview of angiosperm development and speculates about how and why the angiosperms appeared on Earth. It also discusses the question of angiosperm origin.

A grove of banyan trees

Introduction

Like a well-designed building, the body of a vascular plant is sturdy and functions efficiently. Unlike a building, however, a plant constantly grows. The plant's foundation is its root system. Roots anchor the plant in the soil. They also absorb water and minerals the plant needs to carry out photosynthesis and other cell functions.

Stems support the plant's softer tissues, such as leaves and flowers. Stems also contain the vascular tissues that transport water, minerals, and food between the roots and leaves. Leaves produce food, which the plant may convert to energy right away or may store for later use. Together, roots, stems, and leaves are the organs that carry out the major functions of plant life. These organs are often adapted for life in unusual habitats.

Kinds of Plant Tissue

- *State* the function of meristems.
- *Name* three kinds of tissue found in plants.
- *Contrast* vessel members and sieve-tube members in terms of their structures and functions.

One of the chief differences between plants and animals is the location of cell division in the organism. Certain kinds of cells anywhere in an animal's body may undergo division. In plants, cells divide only in specific areas called **meristems** (MEHR uh stehmz). The meristems located near the tips of roots and stems are called *apical meristems.* ***The apical meristems produce the most rapid growth.***

As the apical meristem deposits new cells behind it, the root or stem grows longer. The lengthening of roots and stems is called **primary growth.** Young plants show primary growth as they send up green shoots. Mature plants also show primary growth in the lengthening of their roots and stems.

The cells produced by meristems become specialized and carry out particular plant functions. **Differentiation** is the name for this specialization. ***Plant cells differentiate into three basic kinds of tissue: the epidermis, vascular tissue, and ground tissue.***

26.1 Epidermis

The outermost layer of plant cells develops into protective tissue called the **epidermis.** The epidermis helps retain moisture in several ways. It secretes a waxy layer of *cutin,* which not only slows evaporation from the plant's surface but also protects the plant against the invasion of parasites. Cells in the epidermis may develop hairlike structures that trap water vapor next to the plant surface.

26.2 Vascular Tissue

Vascular tissue consists primarily of tubelike xylem and phloem. These tissues transport water, dissolved minerals, and food throughout the plant.

Two kinds of conducting cells develop in the xylem. **Tracheids** (TRAY kee ihdz) are long, tapered cells. Water and minerals pass from one tracheid to another through small pits in the cell end walls. Tracheids are the main form of xylem in gymnosperms. The xylem in angiosperms consists mostly of **vessel members,** which are short, open tubes that are more efficient conducting tubes than tracheids. The cytoplasm in both tracheids and vessel members dissolves at maturity, and the cells die. The remaining parts of the cell are hollow tubes through which water and minerals pass easily.

Apical meristem

Root meristems

Figure 26–1. A seedling grows when cells in its meristems divide and enlarge. Cell divisions in the apical meristem establish the stem. Divisions in the root meristem form the root, which gathers water and nutrients.

Figure 26–2. A cross section of angiosperm wood (left, ×500) shows that it is composed of many small, thick-walled tracheids and a few large, thin-walled vessels. Vessels are composed of vessel members, which are connected to each other by perforated, overlapping cell walls (right, ×1200).

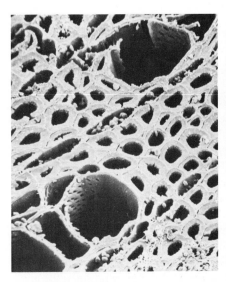 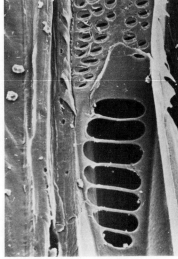

Phloem cells are called **sieve-tube members** because their end walls look and act something like sieves. Food passes freely from one cell to another through small holes in the end walls. *Unlike xylem cells, sieve-tube members are living cells.* Although they are living, sieve-tube members have no nuclei. A *companion cell* is attached to each sieve-tube member and regulates its metabolism.

26.3 Ground Tissue

Ground tissue is relatively unspecialized tissue that cushions and protects vascular tissues from physical injury. It usually consists of two concentric regions. The ground tissue in the center of roots and stems is generally called **pith.** Pith contains soft, spongy cells called **parenchyma** (puh REHN kih muh). Pith is surrounded by an outer layer of more rigid cells that make up the **cortex.** Generally, ground tissue stores food and water and supports the vascular tissue. The specific makeup and function of ground tissue varies in each organ of the plant.

Reviewing the Section

1. What areas of the plant produce primary growth?
2. What are the three basic kinds of tissue in plants?
3. Where are the two main regions of ground tissue located in a plant?
4. Describe how a plant might be harmed if its epidermis were torn.

Roots

Through variations in their roots, plants have adapted to many different kinds of soil. Some roots split rocks as they grow through cracks in the rock. Others push through dense clay, or anchor plants in shifting sand. In any environment, roots carry out three basic functions. *Roots anchor the plant, absorb water and minerals from the soil, and store food produced in the leaves and stem.*

- *List* the three main functions of roots.
- *Label* the three basic layers of tissue in a diagram of a root.
- *Contrast* primary growth and secondary growth in roots.
- *Give* examples of two kinds of root adaptations and explain the survival value of each.

26.4 Root Systems

The first root produced by a young plant is called the **primary root.** The primary root develops into one of two types of root systems. In many dicots, such as carrots and oak trees, the primary root matures into a large, thick root, called a **taproot,** that reaches deep into the soil. As a taproot grows, it produces smaller roots called *secondary roots.* In most monocots, such as grasses, the primary root shrivels and dies as the plant matures. It is replaced by secondary roots that grow from the base of the stem. These roots grow out over a wide area, forming a **fibrous root** system. A fibrous root system often extends farther underground than the visible parts of the plant extend above ground.

26.5 Root Structure

Both taproot systems and fibrous root systems consist of individual roots with a similar internal structure. The roots of most angiosperms and gymnosperms contain three concentric layers of tissue: the epidermis, the cortex, and the vascular cylinder.

Epidermis The root epidermis forms a protective outer layer around internal root tissues. The epidermis also produces tiny outgrowths called **root hairs,** which absorb water and minerals from the soil.

Water enters the root hairs by osmosis. Recall that osmosis is the movement of water across a membrane into a region of lower water concentration. Because root hairs contain higher concentrations of solutes than the soil does, water moves into the hairs. Tiny pores in the cuticles of the root hairs allow water and minerals to pass into the cell. However, these pores prevent the escape of the larger molecules of sugar and starch that are stored in root cells.

Root hairs greatly increase the surface area of the root system and so increase the plant's capacity to absorb water.

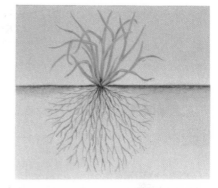

Figure 26–3. Most monocots (top) have fibrous root systems. Many dicots (bottom) have a taproot that develops directly from the primary root.

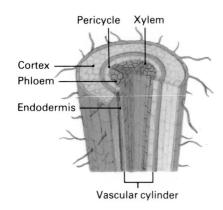

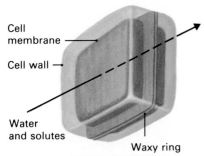

Figure 26–4. A root (top) is composed of distinct tissues and cells. A cell of the endodermis (bottom), for example, has a waxy ring that prevents water from flowing around it.

Root hairs are very numerous. One rye plant, for example, may have 14 billion root hairs. These hairs obtain minerals and other nutrients from throughout the soil.

Cortex   In a young plant, the cortex forms the bulk of root tissue. The root cortex primarily consists of loosely packed parenchyma cells. Water moves easily around these cells. The innermost layer of cells in the root cortex, however, is tightly packed. This single layer of cells is called the **endodermis.** Water must flow *through* the endodermis cells. Each endodermis cell has a waxy ring that prevents water from flowing around it. Scientists believe that the endodermis functions as a kind of "checkpoint" by regulating the intake of minerals and other substances into the plant body. A substance must be taken in by the endodermis to reach the central root tissues.

Vascular Cylinder   The endodermis forms a sheath around the tissues in the center of the root. These central tissues, which form the **vascular cylinder,** consist of xylem, phloem, and special layers of meristematic cells. One of the layers of meristematic cells, called the **pericycle,** forms the outer layer of the vascular cylinder. The pericycle produces secondary roots. The other layer of meristematic cells develops between the xylem and phloem. This layer, called the **vascular cambium,** produces new xylem and phloem cells.

26.6 Root Growth

Roots grow in both length and width. The lengthwise growth occurs at the tip, or apical meristem. The cells in the apical meristem divide and then elongate. The new cells may grow up to 10 times their original length. Either during or after elongation, the root cells differentiate and form the epidermis, cortex, and vascular cylinder.

The sequence of cell development in the growing root tip is visible in three regions of growth, which are shown in Figure 26–5. New cells result from divisions of cells in the region of the apical meristem. The new cells gradually lengthen in the region of elongation. Farther up is the region of differentiation, where all the various tissues of the root mature. Lignification of cells takes place in this region.

The elongation of cells pushes the apical meristem through the soil. The fragile meristem would be torn and crushed if it were not protected by an outer layer of cells called the **root cap.** The cells in the root cap are continuously scraped away and replaced by new cells formed in the apical meristem.

As the root matures, the meristematic cells in the vascular cambium and pericycle become active. Cell divisions in the vascular cambium produce new xylem and phloem, resulting in *secondary growth*, which increases the plant's diameter rather than its length or height. **In both roots and stems, all secondary growth is produced by the vascular cambium.** Roots with secondary growth can become very large. Often these woody roots are used to help prop up the tree by forming a strong, stable base upon which a solid, tall trunk can form.

Secondary roots that grow out from a mature root are produced by cell divisions in the pericycle. A new root develops a root cap and vascular tissue while it grows through the cortex of the mature root. By the time the secondary root reaches the soil, it is fully protected against abrasion and ready to absorb water and minerals for the plant.

Older roots with many layers of secondary growth often develop a tough outer tissue called *bark*. Bark replaces the epidermis and much of the cortex. It consists of phloem and *cork* cells, which are produced by a meristem called the **cork cambium.** As you will see later in this chapter, the cork cambium is especially important in stems.

26.7 Adaptations

Roots that arise from unusual places on the plant are called *adventitious* (ad vuhn TIHSH uhs) roots. The stems of corn and some other monocots develop adventitious roots that help brace and support the plant. Tropical orchids grow on tree trunks, with their roots exposed to the air. These *aerial* roots absorb water and minerals directly from the air.

Mangroves have especially unusual root systems. These large trees grow in or near warm coastal waters in the tropics. They often grow in soil that is muddy and that contains little oxygen. Mangrove stems produce adventitious roots that hold

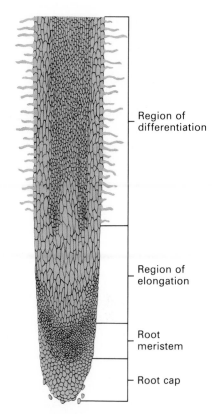

Region of differentiation

Region of elongation

Root meristem

Root cap

Figure 26–5. A longitudinal section of a root shows that it is composed of three growth regions: the meristem, the region of elongation, and the region of differentiation. A root cap protects the meristematic tissue.

Figure 26–6. Many roots are modified in ways that help plants survive. The prop roots of corn (left) support the narrow, slender base of the stem. Adventitious roots of a mangrove (right) stabilize the plant in the shifting soil and changing tides.

Roots, Stems, and Leaves   **387**

General Description

Tree surgeons care for trees by removing harmful or unattractive growth and by treating the plants' wounds. They are specially trained to recognize a variety of diseases in tree trunks, roots, branches, and leaves. If a tree has been attacked by insect pests, a tree surgeon may spray or dust the tree with a pesticide. He or she may also recommend fertilizers that will help the tree grow vigorously.

Clients often ask tree surgeons to cut off dead or decaying branches. Tree surgeons sometimes prune trees or shrubs into special shapes. They are also asked to remove branches that interfere with power lines or with a neighbor's property.

Most tree surgeons are employed by private companies that provide service to private homeowners and to city institutions, such as schools. Tree surgeons help preserve plants. They also help to enhance the environment by caring for plants.

Career Requirements

Some tree surgeons enter the field after earning a B.S. degree in urban forestry or horticulture. Others get their training on the job. Anyone considering a career as a tree surgeon should like the outdoors, have no fear of heights, and be willing to do the difficult physical labor demanded of a tree care professional.

For Additional Information
National Arborist
  Association
3537 Stratford Road
Wantagh, NY 11793

the plant steady in shifting tides. Mangrove roots also grow up into the air. From the air, the exposed root tips absorb the oxygen that the mangrove needs for cellular respiration. Mangrove seedlings germinate while still on the tree. The primary root is pointed. When the seed falls, the root sticks into the mud.

Reviewing the Section

1. What are the three main layers of root tissue?
2. Where is the vascular cylinder located?
3. Which layer of root cells produces secondary growth?
4. What root function would be greatly affected if the root lost all of its root hairs?
5. Why might a fibrous root system survive better in shallow, rocky soil than a taproot system?

Stems

A plant's roots need food from the leaves to carry out their functions. At the same time, the leaves need water and minerals absorbed by the roots to carry out photosynthesis. Stems contain the tissues that make this exchange of materials possible. *Stems transport food, water, and minerals between the roots and leaves, and support plant growth above the ground.* The thin stem of a four-leaf clover and the mammoth stem of a redwood tree both perform these tasks.

26.8 Structure of Stems

Like young root cells, stem cells differentiate into the epidermis, vascular tissues, and ground tissues. The strands of xylem and phloem that begin in the root continue up through the stem. In the young stem, the vascular tissues are arranged in groupings called **vascular bundles.** The pattern of vascular bundles in a monocot differs from the pattern in a dicot, as Figure 26–7 shows. In monocots, bundles are scattered through the ground tissues. In dicots, bundles are arranged in a ring that separates the inner core of pith from the outer ring of cortex cells.

Herbaceous Stems Most herbaceous monocots and dicots have soft, fleshy stems that produce little or no secondary growth. For this reason, the structure of a young herbaceous stem does not change significantly as the plant matures, though some of the stem tissues may become slightly lignified.

Figure 26–7. In monocots (left), the vascular bundles are distributed throughout the stem. In dicots (right), the bundles are arranged in a ring and form a boundary between the cortex and pith.

Monocot flower and stem

Dicot flower and stem

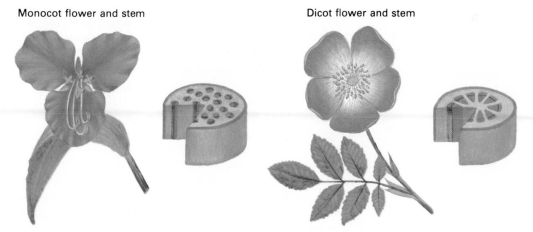

Figure 26–8. A woody stem develops from divisions in the vascular cambium (top). Xylem forms on the inside; phloem, on the outside (bottom left). In a woody stem (bottom right), xylem is wood and phloem is bark.

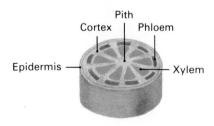

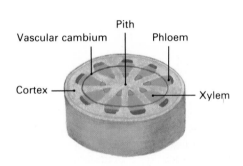

Q: *Why does sap rise in the spring?*

A: Sap (dissolved sugars) flows down to be stored in the roots and stem in the fall. In the spring, the tree transports sap back up to nourish the growth of young leaves and branches.

Herbaceous stems are supported by the water that fills their cells. The water in each cell presses against the cell wall. This creates a force called *turgor pressure*, which makes the cell rigid. How might the loss of water from its cells affect the stem of a herbaceous plant?

Woody Stems Many dicots and most gymnosperms produce woody stems. As a woody plant matures and produces secondary growth, the structure of its stem changes. A young dicot stem contains vascular bundles arranged in a ring. As the stem matures, the vascular cambium produces xylem toward the center of the stem and phloem toward the outside. Each year the plant adds another ring of xylem. This increasingly wider column of wood supports the plant's vertical growth.

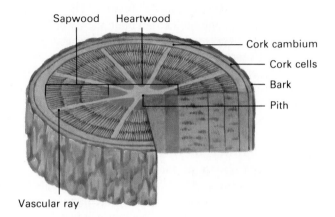

In time, the conducting xylem cells, called *sapwood*, surround a core of older xylem cells that stop conducting water. This nonfunctioning xylem, called *heartwood*, becomes plugged with substances that make the wood hard and dry. Heartwood forms the bulk of a mature woody stem.

At the same time, the increasing diameter of the stem splits the epidermis and triggers cell divisions in the cork cambium. The new cork cells produced by this cell division gradually replace the cortex. Together with phloem, cork forms a layer of bark around the stem. Bark protects stems from physical injury and insects. The hollow, air-filled cork cells insulate the stem from extremes in temperature. They also contain a waxy substance that helps retain water in the stem. These traits are important to the survival of trees and other woody plants that must endure harsh winters. Bark is not airtight, however. Tiny openings in the bark, called *lenticels,* permit air to pass through the

Leaves

Young, growing stems often carry out a small amount of photosynthesis. In most plants, however, food production takes place mainly in the leaves. The broad part of the leaf, called the **blade,** contains most of the plant's photosynthetic cells. The **petiole** (PEHT ee ohl), or leaf stalk, supports the blade. The petiole is attached to the stem at the **leaf base.** A leaf that has a single, undivided blade is called a **simple leaf.** If the blade is divided into several separate parts that are attached to an extension of the petiole, the leaf is called a **compound leaf.** These and other traits, such as leaf arrangement, used to classify leaves are illustrated in Figure 26–10.

Section Objectives

- *State* the chief function that leaves perform.
- *List* some leaf adaptations found in carnivorous plants.
- *Describe* the internal structure of a typical leaf.
- *Summarize* how water and food are transported through plants.
- *Explain* the survival value of abscission.

26.11 Structure of Leaves

Leaves use sunlight, water, and carbon dioxide to carry out photosynthesis. They also transport the food they produce to the rest of the plant in a process called **translocation.** In addition, leaves exchange gases and water vapor with the atmosphere. The structure of the leaf is well suited to performing these and other functions.

Epidermis The epidermis in most leaves is a single, transparent layer of cells. Sunlight passes directly through these cells to the photosynthetic cells below. The epidermis secretes a layer of *cutin*, which slows evaporation from the leaf blade.

Figure 26–10. A leaf has a blade, a petiole, and a leaf base (right). Monocot leaves have parallel veins; dicot leaves have net veins and can be simple or compound (left). Leaf arrangement is alternate, opposite, or whorled.

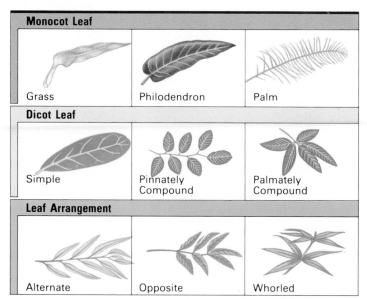

Monocot Leaf: Grass, Philodendron, Palm
Dicot Leaf: Simple, Pinnately Compound, Palmately Compound
Leaf Arrangement: Alternate, Opposite, Whorled

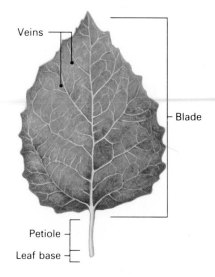

Veins, Blade, Petiole, Leaf base

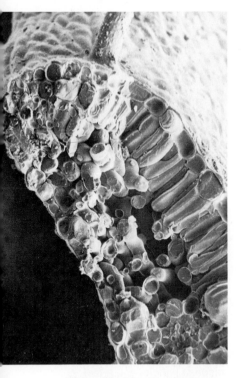

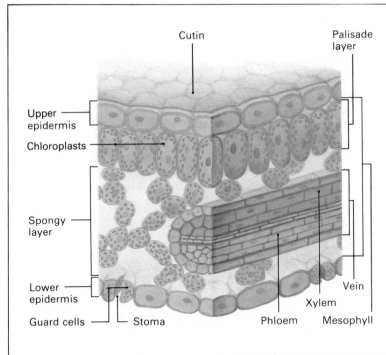

Figure 26–11. The two types of mesophyll cells (×500) as they actually appear (above) are clearly shown in the drawing (right) to lie between two epidermal layers.

Figure 26–12. Stomata occur primarily on the lower epidermis of leaves.

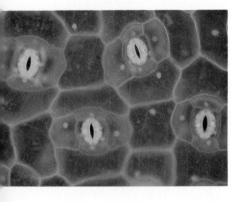

Carbon dioxide, oxygen, and water vapor enter and exit the leaf through openings in the epidermis called *stomata.* Each stoma is flanked by two kidney-shaped **guard cells.** When the plant contains excess water, *turgor pressure* increases in the guard cells. The swelling opens the stoma, allowing water vapor to escape from inside the leaf. When the plant needs to retain water, the guard cells lose turgor pressure and close. Most guard cells are located on the underside of the leaf, where the surface is shaded and somewhat protected from dust that might clog the stomata.

Mesophyll The epidermis encloses the middle portion of the leaf, called the **mesophyll.** The parenchyma cells in the mesophyll contain chlorophyll and other pigments. Cells in the upper layer of the mesophyll are arranged in close-fitting columns that expose a maximum number of chloroplasts to the sun. This layer is called the *palisade layer.* The layer below, called the *spongy layer,* consists of loosely bunched, irregularly shaped cells surrounded by air spaces. Gases and water vapor accumulate in these spaces.

The mesophyll also contains vascular bundles, which are called *veins* in leaves. The veins transport water and minerals into the leaf and carry out food.

26.12 Leaves and Water Loss

Ninety percent of the water that enters the roots is lost as water vapor, most of it through open stomata in the leaves. The process by which plants lose water is called **transpiration.** How can plants survive losing so much of their moisture? The answer is that the water lost through the top of the plant is continuously replenished from the roots below.

Several forces are involved in the movement of water in plants. Osmosis continuously pulls water into the roots, building up internal *root pressure* that pushes water into the stem. In addition, xylem cells form long, narrow tubes, and the water in these tubes is attracted to the xylem cell walls. This attraction tends to pull the water up the walls. The force of attraction that causes water to move up narrow tubes is called *capillary action*. However, the major force behind the movement of water in plants is the attraction of water molecules for one another. The upward movement of one molecule tugs on the molecules below. Each water molecule that evaporates from a leaf exerts a pull all the way down to the roots. The loss of water from the leaves thus helps maintain water flow throughout the plant.

Plants benefit from transpiration in many ways. It prevents the accumulation of excess water in the plant body. The evaporation of water from the leaves cools the plant, much as evaporation cools animals. Transpiration also creates a force that results in the steady supply of fresh water to all the plant organs.

26.13 Adaptations

Plants have developed several adaptations to climates in which there are cold winters. During cold winters, less water is available for absorption by a plant's roots, partly because the soil may freeze.

One of the most valuable adaptations in plants that endure cold winters is **abscission,** the shedding of leaves. Abscission reduces water loss due to transpiration. Seasonal changes trigger abscission. As the air cools and the days grow shorter, a row of cells called the *abscission layer* develops in the leaf base. These cells cause the vascular bundles and connecting tissue between the leaf base and the stem to degenerate. The leaf then drops off. Trees that shed their leaves every year, such as maple and oak trees, are said to be *deciduous*.

Conifers, which also grow in climates with cold winters, have continual leaf cover. Conifer leaves have a thick epidermis and a minimal surface area, which together make the leaves less vulnerable to evaporation.

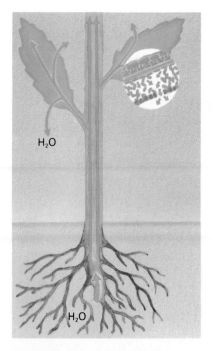

H_2O

H_2O

Figure 26–13. This illustration shows that water is pulled upwards through the xylem during transpiration. The inset shows that water vapor is lost through stomata on the lower surface of leaves.

Q/A

Q: *Is the poinsettia flower bright red?*

A: No. The red petal-like structures are modified leaves. The small, yellow structures in the center of the red leaves are the plant's flowers.

▶ Carnivorous Plants

Among the most remarkable of all leaf modifications are those that occur in carnivorous plants. The leaves of carnivorous plants are specialized to capture insects and other small invertebrates. Because the ingested animals contain nitrogen, the carnivorous plants can live in nitrogen-poor soil where most other plants cannot survive.

Pitcher plants catch animals in leaves that are modified into tubes. The tube is lined with downward-pointing hairs and filled with water. An insect is lured to the tube by sweet-smelling nectar. Once inside the tube, the insect cannot crawl back out against the slanting hairs. It eventually falls into the water and is digested by enzymes.

The sundew has another type of insect trap. The leaf of the sundew is covered with epidermal hairs that secrete a sticky substance. An unsuspecting insect that wanders onto the leaf will become stuck. The leaf then folds over like a closing fist, and enzymes digest the insect.

The Venus's flytrap captures its prey in a different way. The blade of the leaf is modified into two halves, which have special trigger hairs. If an insect trips these hairs, the entire leaf closes rapidly on the insect.

Aquatic plants and desert plants show highly specialized adaptations. Some aquatic plants have narrow leaves that ride easily over waves. Others are full of holes that allow water to flow through the leaf. In contrast, cactus leaves are hard, non-photosynthetic spines. The spines transpire almost no water.

Reviewing the Section

1. What functions do guard cells perform?
2. How is food transported out of the leaves?
3. How does osmosis help plants transport water?

Investigation 26: Comparing Root Systems

Purpose
To investigate how the roots of different plants anchor the plants to the soil

Materials
Three lima bean seeds, six 7 cm pots, three corn seeds, potting soil, centimeter ruler, adhesive tape, cord, hand-held spring scale, paper, triple-beam balance

Procedure
1. Place one lima bean seed in each of three pots. Cover with soil and water.
2. Place one corn seed in each of three pots. Cover with soil and water. *Why are you growing more than one sample of each seed type?*
3. Allow the plants to grow to a height of about 10 cm.
4. Place a piece of adhesive tape around each of the stems at soil level as shown in the illustration.

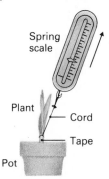

Spring scale

Plant

Cord

Tape

Pot

5. Attach a cord to the tape.
6. Attach the other end of the cord to the spring scale.
7. Carefully pull the scale upward and away from the plant.
8. As the plant is uprooted, note the calibration on the scale. Record your results. Repeat the process for each plant.
9. Allow the plants to dry overnight. Then carefully tap the root system of each plant several times against a piece of paper to remove as much as possible of the soil still adhering to the roots. Weigh the loose soil from each plant. Record your results.

Analyses and Conclusions
1. Make a bar graph of the data gathered as the plants were uprooted.
2. Based on the bar graph and the data gathered in step 9, what can you conclude about the relative effectiveness of the two root systems in anchoring themselves to the soil?
3. Why were these plants grown in fairly loose soil?

Going Further
- Investigate different types of seeds, such as radish versus lettuce.
- Plant the same type of seed in different types of soil. Which soil type is the most effective medium for growth and development?

Chapter 26 Review

Summary

Plants consist of three kinds of tissue: epidermal, vascular, and ground. All new plant cells are produced in specific areas called meristems. Cells produced in the root's apical meristem differentiate into the epidermis, cortex, and vascular cylinder. Roots may develop as taproots or fibrous roots. All roots anchor the plant, absorb water and minerals through root hairs, and store food.

Stems support plant growth above the ground and transport water and food between the roots and leaves. Herbaceous stems are supported by turgor pressure. Woody stems are supported by the wood produced by secondary growth.

Most of the plant's photosynthetic cells are contained in the leaf blade. Leaves exchange gases and release water by the opening and closing of their stomata. Water movement in plants results from root pressure, capillary action, and the pull created by transpiration.

BioTerms

abscission (395)
blade (393)
bud (392)
compound leaf (393)
cork cambium (387)
cortex (384)
differentiation (383)
endodermis (386)
epidermis (383)
fibrous root (385)
ground tissue (384)

guard cell (394)
leaf base (393)
meristem (383)
mesophyll (394)
parenchyma (384)
pericycle (386)
petiole (393)
pith (384)
primary growth (383)
primary root (385)
root cap (386)

root hair (385)
sieve-tube member (384)
simple leaf (393)
taproot (385)
tracheid (383)
translocation (393)
transpiration (395)
vascular bundle (389)
vascular cambium (386)
vascular cylinder (386)
vessel member (383)

BioQuiz   (Write all answers on a separate sheet of paper.)

I. Completion

1. The stems of woody plants protect their meristems during winter by enclosing them in _____ .

2. Food produced by photosynthesis conducted in the leaves moves through the plant by the process known as _____ .

3. The _____ added each year to the stems of woody plants support the plant's vertical growth.

II. Modified True and False

Mark each statement TRUE or FALSE. If false, change the underlined term to make the statement true.

4. A <u>compound</u> leaf has a single, undivided <u>blade</u>.

5. The lengthening of roots and stems is called <u>primary</u> growth.

6. Oak trees have a <u>fibrous root</u> system that extends deeply into the surrounding soil.

III. Multiple Choice

7. Secondary roots are produced by cell divisions in the a) root cap. b) pericycle. c) tracheids. d) root hairs.
8. Stomata open and close due to changes in a) turgor pressure. b) internodes. c) capillary action. d) translocation.
9. The cork cells that help protect woody stems are produced by the cork a) xylem. b) phloem. c) nodes. d) cambium.
10. The stems of monocots are braced by a) adventitious roots. b) root caps. c) taproots. d) aerial roots.
11. All new plant cells are produced by a) meristems. b) ground tissue. c) cutin. d) vascular tissue.
12. Companion cells in phloem regulate the metabolism of a) tracheids. b) vessel members. c) sieve-tube members. d) xylem cells.
13. As it pushes through the soil, the root meristem is protected by the a) vascular cambium. b) root cap. c) endodermis. d) pericycle.
14. Photosynthesis takes place in the leaf a) epidermis. b) veins. c) cutin. d) mesophyll.
15. An adaptation to cold winters involving the shedding of leaves is called a) abscission. b) translocation. c) transpiration. d) secondary growth.

IV. Essay

16. What are the functions of vessel members and sieve-tube members?
17. How are carnivorous plants adapted to living in nitrogen-poor soil?
18. How is water transported from a plant's roots to its leaves?
19. How do the tissues in a woody stem change as the stem matures?
20. What do growth rings show?

Applying and Extending Concepts

1. Furniture makers use heartwood rather than sapwood to make furniture. What characteristics of heartwood and sapwood might explain this preference? Use a cross-sectional diagram of a tree trunk to illustrate your answer.
2. If you drove a nail into the trunk of a young tree, the nail would remain at that height regardless of how tall the tree grew over the years. Write a paragraph explaining which characteristics of plant growth could account for this phenomenon.
3. Pines have needlelike leaves and sunken stomata similar to those in cacti. Why might both kinds of plants benefit from these traits, given that pines live in regions of considerable rainfall while cacti live in deserts?
4. Bonsai plants are usually dwarfed by trimming away some of their roots. Considering the fact that food is produced in the leaves, write a brief report explaining why trimming the roots produces dwarf plants.

Related Readings

Rosen, E. "Trees in Winter Wind and Ice." *Science Digest* 87 (March 1980): 49–51. This article explains the methods trees use to survive cold weather. It explains how meristematic areas are protected.

Zimmerman, M. H. "Piping Water to the Treetops." *Natural History* 91 (July 1982): 6–13. This article describes the hydraulic system of plants. It explains how water gets to the top of tall trees.

Reproduction in Flowering Plants

Orchids

Introduction

Sexual reproduction in flowering plants involves structures of amazing complexity and beauty. Each kind of flower shows adaptations to a particular environment and often to the animals that share that environment. Sexual reproduction makes such adaptations possible. The combination of male and female gametes introduces genetic variations that can produce valuable new traits.

Sexual reproduction has some disadvantages. The fragile gametes may be destroyed by animals or bad weather. Many flowering plants have an alternate means of survival—asexual reproduction. Although offspring produced asexually do not have new traits, they preserve existing ones and may quickly disperse the species to more favorable environments.

Sexual Reproduction

Section Objectives

- *State* the function of flowers in the life of a plant.
- *Name* three essential flower parts and three nonessential flower parts.
- *Summarize* the processes of pollination and fertilization in a typical flowering plant.
- *Explain* the survival value of double fertilization.
- *Describe* some adaptations of flowers that help to ensure pollination.

Most people think of flowers only as beautiful decorations. For insects and some other animals, however, flowers produce important kinds of food. *In the plants that produce them, flowers function in sexual reproduction.* The brilliant colors of flowers, as well as their varied scents and shapes, all help plants reproduce sexually.

27.1 Structure of the Flower

The various flower parts evolved from leaves that bore reproductive structures along their edges, as shown in Figure 27–1. Over time, the leaves curled inward and provided greater protection for the developing gametes. These leaves gradually became closed structures. Today flowers consist of several kinds of highly modified leaves, each of which is classified as either essential or nonessential.

Essential Flower Parts   The parts of the flower that produce the gametes and carry out sexual reproduction are called the *essential flower parts*. These include the male parts, called **stamens,** and the female parts, called **pistils.**

Most flowers have three, four, or five stamens. The thin, stemlike portion of a stamen is called the **filament. Pollen** is produced at the tip of the filament, generally in an oblong structure called the **anther.**

Figure 27–1.   A flower is the result of a series of evolutionary changes by which an entire fertile branch became a compressed reproductive organ. The ovary and stamens evolved from the top two whorls of leaves. The petals and sepals arose from the two lowest whorls.

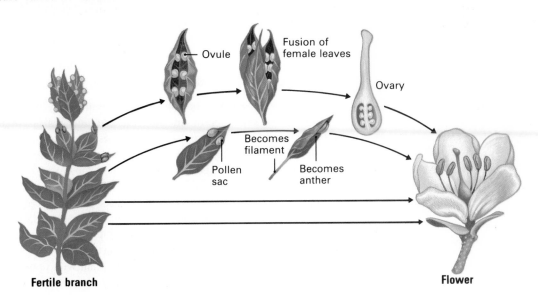

Fertile branch — Ovule — Fusion of female leaves — Ovary — Pollen sac — Becomes filament — Becomes anther — Flower

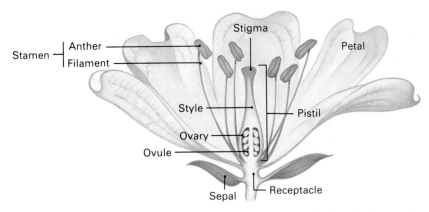

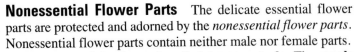

Figure 27–2. A perfect, complete flower is composed of both essential and nonessential parts. The pistil and the stamens are essential parts. The sepals and petals are nonessential parts.

Most flowers contain only a single pistil. The pistil contains three parts. The swollen base of the pistil is called the **ovary.** Within the ovary, one or more **ovules** produce the egg cells. The slender middle part of the pistil is called the **style.** At the tip of the style is the **stigma.** The stigma produces a sticky substance to which pollen grains become glued.

Nonessential Flower Parts   The delicate essential flower parts are protected and adorned by the *nonessential flower parts.* Nonessential flower parts contain neither male nor female parts.

The base of the flower is called the **receptacle.** The **sepals** grow out from the receptacle and enclose the flower bud before it blooms. Sepals look much like tiny leaves and in many species are green. The sepals collectively form a structure called the *calyx,* which protects the ovary.

Petals grow between the sepals and the essential flower parts. They protect the pistil and stamens, and are often fragrant and brightly colored. Together, the petals form the *corolla.*

27.2 Kinds of Flowers

Flowers differ in the number and kinds of parts they possess. **Complete flowers** contain all the essential and nonessential parts. Roses, violets, and mustard blossoms, for example, are complete flowers. **Incomplete flowers** lack one or more of the essential or nonessential parts. The flowers of most grasses lack developed petals and sepals, and so are incomplete.

The flower in Figure 27–2 is a **perfect flower** because it contains both stamens and a pistil. In many other species, however, the male and female structures develop on separate flowers. Flowers that contain the reproductive structures of only one sex are called **imperfect flowers.** Corn is a plant with imperfect flowers. The tassel of corn is made of many male flowers. The

Figure 27–3. The male flowers of corn are located on the tassels. The female flowers are located on the ear. Each corn flower is thus an incomplete flower.

ear of corn is made of many female flowers. A corn plant is also an example of a plant that has male and female flowers on the same individual. Other plants, like spinach, have male and female flowers on separate plants.

27.3 Formation of Gametes

Stamens and pistils are part of the diploid, sporophyte generation of the plant's life cycle. The male gametes formed in the anthers and the female gametes formed in the ovules are parts of the plant's haploid, gametophyte generation. Both male and female parts are needed to complete the life cycle.

Pollen Grain Formation Most anthers have four pollen sacs, each of which contains hundreds of cells called *microspore mother cells*. Each of these cells undergoes meiosis and produces four haploid *microspores*. The nucleus of each microspore then divides by mitosis, but the new cell wall forms internally. The result is a *pollen grain*, which is made up of two cells of unequal size located inside a hard, outer wall. The larger internal cell is called the **tube cell.** The smaller cell is called the **generative cell.** The pollen grain is the male gametophyte.

Egg Cell Formation The development of the female gametophyte occurs in the ovule, located inside the ovary. A *megaspore mother cell* in each ovule undergoes meiosis and forms four haploid *megaspores*. Three of the four megaspores die, leaving one megaspore in the ovule.

The surviving megaspore enlarges to several times its original size, filling up most of the ovule. The nucleus of the

Q/A

Q: *Why do certain kinds of pollen make people sneeze?*

A: Spiny pollen grains, such as ragweed pollen, irritate the sensitive tissues of the eyes and nasal passages. The body responds to this irritation by producing a sneeze.

Figure 27–4. The outer surfaces of the pollen grains of different plant species vary. The SEM photos below show phlox (left, ×1700), common ragweed (center, ×4000), and northern white cedar (right, ×4200).

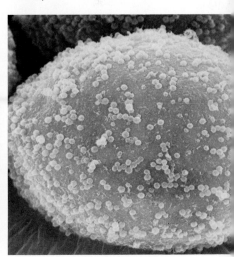

Figure 27–5. The life cycle of flowering plants is shown below. In the male part of the cycle, microspores form pollen grains. In the female part, a megaspore forms an embryo sac. Pollen grains are dispersed, land on the stigma, and form sperm that fertilize the egg.

megaspore then undergoes mitotic divisions and produces eight haploid nuclei. As Figure 27–5 shows, three of these nuclei migrate to one end of the cell, three migrate to the other end, and two migrate to the center. The two central nuclei are called **polar nuclei.** The two polar nuclei join together to form a single, diploid nucleus.

Cell walls form around each of the three nuclei at each end of the cell and around the polar nuclei in the center. The result is a large structure which, in most species, contains eight nuclei enclosed in seven cells. This structure, called the **embryo sac,** is the female gametophyte. The cell located nearest the ovule opening is the egg cell.

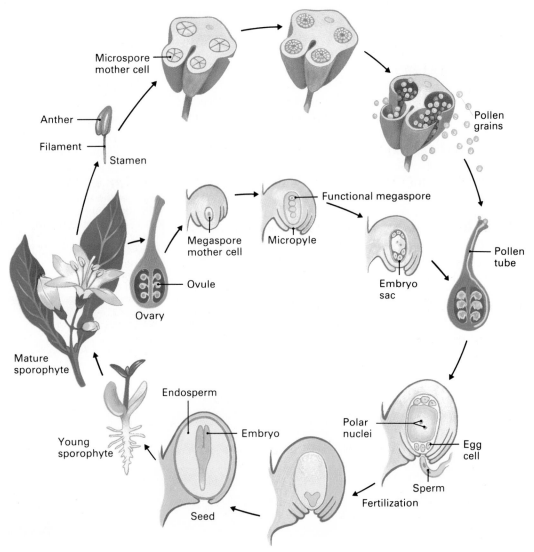

27.4 Pollination and Fertilization

For sexual reproduction to occur, pollen must first reach the stigma. The process of transferring ripe pollen from the anther to the stigma is called **pollination.** *Self-pollination* occurs when pollen falls from an anther onto the stigma of a flower on the same plant. A flower is *cross-pollinated* when its stigma traps pollen from another plant. Because cross-pollination mixes gametes from two different plants, it tends to produce offspring with more varied genes.

Pollination   Animals, wind, and water all transport pollen from flower to flower. *The nonessential flower parts are modified to aid the specific type of pollination a plant undergoes.* In flowers that are pollinated by animals, for example, the stem and receptacle hold the flower out where its colors and scent are most obvious. Insects and birds attracted by the petals feed on pollen or the sweet liquid called *nectar* that some flowers produce. As it feeds, an animal is dusted with pollen. This pollen is then transferred to a stigma when the animal goes to another flower.

In contrast, flowers pollinated by wind-blown pollen grains are not showy. The nonessential parts are not modified to attract animals. Similarly, plants with pollen that floats on water do not have attractive petals or scents.

Fertilization   The landing of a pollen grain on a stigma starts the process of *fertilization,* the union of male and female gametes. Sugars and enzymes in the stigma cause the tube cell inside the pollen grain to grow. The tube cell breaks out of the pollen grain and forms a tube, called a **pollen tube,** that grows through the stigma. Meanwhile the generative cell divides mitotically into two sperm cells, the male gametes. These sperm cells move down through the pollen tube towards the ovule. When they reach the embryo sac, one sperm cell enters and fertilizes the egg cell, forming a diploid *zygote.* The other sperm cell joins with the now fused polar nuclei, forming a triploid (3n) nucleus that divides to form a special nutritive tissue called **endosperm.** The white part of a piece of popcorn is an example of endosperm.

Flowering plants are the only organisms that undergo two kinds of fertilization, one that forms the zygote and one that forms the endosperm. This process is called **double fertilization.** Double fertilization has great survival value for the plant because each new generation carries its own temporary source of nutrition.

Figure 27–6.   Wind transports the pollen released by cockfoot grass. The pollen then lands on other grass flowers and completes the process of pollination.

▶ Flowers and Pollination

Throughout evolution modifications in flower parts occurred as a species' requirements for pollination changed. Some flowers evolved deep-set, cuplike structures that hold nectar. Simultaneously, butterflies evolved long tongues ideally suited to sucking nectar from deep within the flowers.

Insects, especially bees, are the most important animal pollinators. Flowers attract a distant bee with red, yellow, or purple petals. Close by, the bee is attracted by fragrant oils the petals produce. Plants such as violets and pansies provide landing platforms in the form of a broad, sturdy petal.

Color patterns called *nectar guides* lead the bee to the place the nectar is found inside the flower.

One of the most remarkable of all flower modifications is that of the orchid genus *Orchis.* In order to attract pollinators, the flower has evolved to resemble the female of one certain species of wasp.

Male wasps, while attempting to mate with what they believe to be females of the species, inadvertently pick up pollen from the orchid. Then the wasps carry this pollen to the next orchid on which they land.

While the endosperm divides and grows, the zygote itself develops to form the *embryo*. At the same time, the outer layer of the ovule loses moisture and develops a hard *seed coat*. The seed coat surrounds and protects the embryo and its nutritive endosperm.

Reviewing the Section

1. Why are petals considered nonessential flower parts?
2. What kind of cell forms the embryo sac? Where is the egg cell located?
3. How do male gametes produced in one flower reach the female gametes in another flower?
4. Why is double fertilization so important to the plant?
5. What special adaptations of flowers ensure pollination?

Fruits and Seeds

Once an egg has been fertilized, the ovule undergoes changes to become a seed. All of the seeds are located inside the ovary. As the seeds ripen, so does the ovary. The ripened ovary is called a **fruit.** A fruit, therefore, always contains seeds. Tomatoes, squash, cucumbers, pumpkins, and eggplants are all fruits because they all contain seeds. Whole peanuts are also fruits; the edible portions are seeds.

27.5 Fruit Formation

Fruit formation begins when the ovary begins to swell and ripen. It soon changes color and may become either fleshy or dry. Many fleshy fruits become sweet as sugars translocated from the leaves accumulate in the fruit. Peaches, apples, and berries are examples of such fruits. Nuts, burrs, and the winged fruit of maple trees are examples of dry fruits. The paperlike "wings" of the maple protect the seed from being eaten and help disperse the seed in the wind.

All fruits, both fleshy and dry, can be classified into one of three groups. Beans, peaches, tomatoes, and other fruits that form from a single ovary are called *simple fruits*. *Aggregate fruits* form from flowers that have many pistils on the same flower. Blackberries, raspberries, and strawberries are aggregate fruits. Pineapples and figs are *multiple fruits*. A multiple fruit consists of numerous single fruits that have grown so close together that they form a single structure.

27.6 Seed Dispersal

Seeds are dispersed with or without their surrounding fruits. Both seeds and fruits are modified to ensure successful dispersal of the young plant. ***Because flowering plants are nonmotile, they rely on the dispersal of their seeds to prevent overcrowding as well as to carry new plant generations to more favorable environments.***

Dispersal by Animals Sweet, fleshy fruits attract animals of all kinds. Birds eat fruit and transport the seeds in their digestive tracts. The seeds are eliminated, unharmed, in the feces some distance from the parent plant. Squirrels bury dry fruits, such as acorns, before the onset of winter. Some of the seeds are never retrieved and later grow into new plants. Some dry fruits have natural hooks or claws that catch on animals' fur or people's

Section Objectives

- *Describe* how a typical fruit forms.
- *Name* the three groups of fruit and explain the basic differences between them.
- *Explain* the importance of seed dispersal to the survival of plant species.
- *Describe* the germination of a typical dicot seed.

Figure 27–7. Three kinds of fruits are shown above. A cherry (top) is a simple fruit. A strawberry (middle) is an aggregate fruit formed from many ovaries on a single flower. A fig (bottom) is a multiple fruit.

Figure 27–8. Plants disperse fruits and seeds in many ways. A dandelion fruit (top) floats on the air. Winged maple fruits (middle) twirl like the blades of a helicopter as they blow in the wind. Peas (bottom) are dispersed when animals searching for food open the pod.

clothing. You may have transported small fruits yourself while hiking through a forest or prairie.

Dispersal by Wind and Water Most fruits and seeds dispersed by wind have special adaptations for traveling by air. The wind carries winged maple seeds far from the parent tree. Tiny dandelion fruits have feathery plumes that act like parachutes. The thousands of tiny seeds produced by an orchid plant have no special structures, but they are so tiny and light that the wind can blow them for miles.

Seeds dispersed by water can float. The best-known fruits dispersed by water are those of the coconut palm. Ocean currents carry these fruits from one island to another, often over great distances. Air trapped inside the coconut keeps it afloat, and a waxy coating prevents salt water from entering and rotting the seed.

Self-Dispersal Many dry fruits disperse their seeds by propelling them away from the parent plant. The dwarf mistletoe fruit absorbs water until the pressure within the tissues finally bursts the fruit open, throwing the seeds as far as 14.5 m (16 yd.). Seed pods of the Scotch gorse, an evergreen shrub, become warped and dry in the summer. On a particularly hot day, the pods will suddenly explode with a force that scatters the seeds in all directions.

The seeds of some grasses produce long bristles. As the bristles coil and uncoil in response to air moisture, the seeds creep along the ground. Similar structures in a species of wall ivy help bury the ivy seeds in small crevices in rocks or bricks.

27.7 Seed Germination

After a seed is dispersed from the parent plant, the enclosed embryo does not begin to grow immediately. The seed usually undergoes *dormancy,* a period of rest in which metabolic activity is low. For the seed to **germinate,** or resume its growth, the seed coat must first undergo modifications that allow water and oxygen to penetrate the seed. In some cases the modifications are stimulated by internal chemical changes. In other cases environmental influences, such as animals or changes in weather, cause the modifications. After water and oxygen enter the seed, the embryo swells, grows, and cracks through the seed coat. The young plant is a *seedling.*

The embryo of flowering plants possesses all the basic plant organs in embryonic form. The embryonic leaves are called *cotyledons.* The embryonic root of a plant is called the

radicle. The embryonic stem consists of two parts. The part above the radicle and below the cotyledons is called the **hypocotyl.** The part of the stem above the cotyledons is called the **epicotyl.**

Dicot and monocot embryos differ in structure. Dicots have two cotyledons; monocots have one. In monocots the long, thin epicotyl is called a *plumule*. The plumule is surrounded by a protective sheath called a *coleoptile*.

Dicot and monocot seedlings develop in different ways. In dicots, such as the bean shown in Figure 27–9, the endosperm transfers its food reserves to the cotyledons, which become thick. As the seed germinates, the growing hypocotyl pushes the thick cotyledons above the ground. At the same time, the radicle grows downward and becomes the *primary root*. The first leaves form from the apical meristem located atop the epicotyl.

Q/A

Q: *What are the seed and fruit in a corn plant?*

A: The corn kernel contains both the fruit and seed. The ovary wall fuses with the seed and a single structure develops.

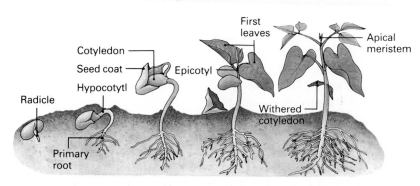

They begin photosynthesis and produce food for the seedling's further growth.

In many monocots the endosperm retains the seed's food reserves. In most monocots the cotyledon remains below the ground as the plumule grows upward and forms leaves. At the same time, the radicle grows to become the primary root, which immediately begins to form lateral roots. In both monocots and dicots, the cotyledons shrink or fall off the stem as the young plant grows.

Figure 27–9. When a dicot seed such as a bean (left) germinates, the expanding hypocotyl lifts the cotyledons out of the ground. Later, leaves form at the apical meristem and the cotyledons wither. In a monocot such as corn (below), the cotyledon remains underground as the plumule pierces the soil.

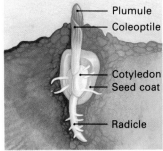

Reviewing the Section

1. What is the difference between a simple and an aggregate fruit?
2. How does a fruit change as it ripens?
3. Name three methods by which seeds are dispersed with or without their fruit.
4. What organs are formed by the radicle and epicotyl?

- *Explain* the advantages plants obtain from vegetative propagation.
- *Name* four types of modified stems that are used in vegetative propagation.
- *Describe* two kinds of artificial propagation.

Asexual Reproduction

Many species of flowering plants produce new plants without the aid of sex cells. Asexual reproduction is common in strawberries, potatoes, irises, spider plants, and grasses. *Any plant produced asexually has the same genes as its parent plant.* Since identical offspring are often desirable, plant cultivators use artificial propagation to ensure that offspring have certain traits.

27.8 Natural Propagation

Producing new individuals from roots, stems, or leaves of existing plants is called **vegetative propagation.** Vegetative propagation occurs naturally in one of six kinds of plant structures: runners, rhizomes, tubers, bulbs, food-storing roots, and leaves.

Runners, also called stolons, are modified stems produced by low-growing plants such as strawberries. Lateral stems grow along the top of the ground and send adventitious roots into the ground at nodes. Once a root is anchored in the soil, the node produces leaves. The runner then breaks off from the parent plant and forms a new plant.

Rhizomes are long, modified stems that grow under the soil. Like runners, rhizomes produce new plants at nodes along the stem. Lawn grasses and irises reproduce by rhizomes. *Tubers* are also modified storage stems. They are shorter and thicker than rhizomes. Potatoes and yams are familiar tubers. Each potato "eye" is a bud capable of producing a new plant. *Corms* are similar to tubers but are smaller and usually round.

Bulbs are slightly different organs of propagation. A bulb, such as an onion, consists of a short stem surrounded by layers

Figure 27–10. Three forms of vegetative propagation are shown below. A strawberry plant (left) sends out runners. Grasses (center) produce underground stems called rhizomes. A potato (right), a tuber, can sprout new plants.

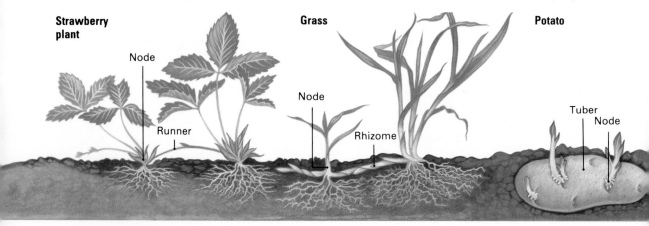

Strawberry plant

Node

Runner

Grass

Node

Rhizome

Potato

Tuber
Node

of modified leaves. The leaves protect the stem and produce food. The food nourishes the young plant that grows up from the buds of the stem.

Some plants propagate with *food-storing roots*. Carrots and beets are food-storing roots capable of producing a new stem and leaves.

An unusual plant called *Kalanchoe* reproduces asexually by growing tiny plants along the edges of its leaves. These plants drop off the ''mother'' plant and root in the soil. Because it produces so many offspring, *Kalanchoe* is called the ''maternity plant.''

27.9 Artificial Propagation

A few kinds of economically important plants are reproduced vegetatively. Commercial plant growers, however, use methods of artificial propagation, collectively called *cloning*. These methods are generally faster and easier to control than natural methods.

Cuttings   *Cuttings* are pieces of stem that are cut from the parent plant and kept in water, moist soil or sand, or some other medium. Adventitious roots develop at the base of the stem. When the roots are sufficiently developed, the stem is planted. Stem cuttings are used to propagate many garden plants. Cuttings can also be made from leaves. In this method, used with such plants as African violets, cut leaves are placed in water or soil until roots and stems form.

Grafting   *Grafting* is a method used to propagate fruit trees, roses, and grapes. Buds or sections of a stem, called *scions* (SY uhnz), are cut from the top of one plant and attached to another plant, called a *stock,* that is already rooted in soil. The scion produces flowers, fruits, and seeds. Because it provides the root system, the stock furnishes the plant with nourishment as well as support.

Tissue Culture and Layering   Some commercial crops are often propagated by *tissue culture*. Pieces of *pith*, the center portion of stems, are removed from the plant and placed in flasks. The flasks contain a growth medium. Whole plants develop from the pith tissue. When plants develop, they are removed from the flasks and planted. Factors such as nutrient supply and invasions by disease-causing organisms can be better controlled in the laboratory than if the seedlings were grown in the field.

Figure 27–11. Tiny plants form along the edge of leaves of the genus *Kalanchoe.* These plants later drop off the leaf, root in the ground, and form new individuals.

Q/A

Q: *Why do dandelions grow back if you cut them off at ground level?*

A: Dandelion roots produce adventitious buds that are capable of growing new stems. Also, they have roots that contract and pull the apical meristem to safety under the soil.

General Description

Ornamental horticulturists grow plants to beautify homes, office buildings, parks, botanical gardens, and other public places. They often develop new varieties of plants to highlight features with aesthetic appeal, such as slender leaves or brightly colored flowers.

When choosing plants for a large project, a horticulturist may consult with clients and travel to the planned site. Many horticulturists work in nurseries or greenhouses or for landscape firms. A growing number are employed by interior landscape design services, which select and care for plants in large office buildings, restaurants, and shopping malls. These firms hire young people to water and fertilize plants, check for pests, and deliver plants to customers.

A horticulturist combines knowledge of plant biology with a talent for design. People who enjoy growing plants and beautifying the environment will find horticulture satisfying.

Career Requirements

Many community colleges offer two-year programs leading to an associate's degree in horticulture. More advanced horticultural work requires knowledge of plant reproduction and genetics. A B.S. degree in horticulture or in a related field is needed. Research or teaching at the college level requires an M.S. or a Ph.D. in horticulture or botany.

For Additional Information
American Society for
    Horticultural Science
701 North Saint Asaph
    Street
Alexandria, VA 22314

Layering is another type of artificial vegetative propagation. A branch of the plant is folded down and covered with soil. Roots are produced at nodes and new plants form. The new growth is then cut off and planted. Commercial crops such as blackberries are produced by layering.

Reviewing the Section

1. Why do plants reproduced asexually look identical to the parent plant?
2. Why do runners produce new plants at their growth nodes?
3. When growing potatoes from pieces of a tuber, why is it essential that each piece have an "eye"?
4. What is the difference between a graft and a cutting?
5. Why is it advantageous to tissue culture plants?

Investigation 27: Examining a Complete Flower

Purpose
To identify the parts of a complete flower and to state the functions of each part

Materials

A complete flower, hand lens or stereomicroscope, scissors, forceps, scalpel or single-edged razor blade, glue or tape, sheet of plain white paper, medicine dropper, slide, coverslip, compound microscope

Procedure
1. Carefully examine the flower that your teacher has provided. Using Figure 27–2 on page 402 as a guide, identify the petals, sepals, stamens, and pistil. *What other parts are obvious on your specimen?*
2. Using a hand lens or stereomicroscope, examine the flower more closely. *What additional details can you see?*
3. As you remove each of the flower parts, arrange them on a sheet of plain white paper. Using scissors, forceps, or both, first remove the sepals. *What is the name given to all of the sepals collectively?*
4. Carefully remove the petals. If the petals are fused, try to remove them in one piece. *What is the name given to all of the petals collectively? Once the sepals and petals are removed, what additional structure(s) can you see?*
5. Carefully remove the stamens. Count the number of stamens. *Based on the number of sepals, petals, and stamens, in which group is the flower classified?*
6. Carefully examine the remaining flower part, which should be the pistil. Describe the stigma.
7. **CAUTION: Use extreme care when working with a scalpel or a razor blade.** Cut the pistil in half above the ovary. Then make a longitudinal cut through the ovary itself. Examine the cut ovary under either a hand lens or stereomicroscope. *What structure(s) can*

you identify inside the ovary? What is the function of each of the structures you have observed?

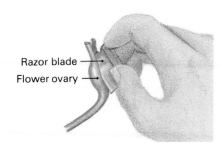

Razor blade —
Flower ovary —

8. Glue or tape all of the flower parts on the paper. Label each part.
9. Place a small drop of water on a clean slide. Shake or tap an anther while holding it above the drop of water so that pollen falls into the water. Add a coverslip. Examine the slide under a compound microscope, first using low power then switching to high power. Make a sketch of what you see. Draw the sketch above one of the anthers that you have placed on the white paper.

Analyses and Conclusions
1. Name the function of each of the following flower parts: filament, anther, stigma, ovary, calyx.
2. What characteristic identifies a flower as complete or incomplete? Perfect or imperfect?
3. What evidence suggests that the flower you have dissected would probably be pollinated by an insect?

Going Further
• Repeat this investigation using other types of flowers. Identify all flower parts and identify each flower as complete or incomplete and perfect or imperfect.
• Use reference books to help you identify the parts of a composite flower, such as a daisy.

Chapter 27 Review

Summary

Flowers consist of highly modified leaves that are specialized to carry out sexual reproduction. Essential flower parts include the pollen-producing stamen and the pistil, which contains the ovary. The nonessential flower parts protect and adorn the reproductive structures and aid in pollen dispersal.

Pollination is the process of transferring ripe pollen from the anther to the stigma. Each pollen grain contains a generative cell and a tube cell. The generative cell produces two sperm cells. In double fertilization, one sperm cell fuses with the egg cell while the other fuses with the polar nuclei. This forms a diploid zygote and endosperm, which nourishes the developing embryo.

A fruit develops from the ovary of a flower. Fleshy fruits are attractive to animals, which eat them and then disperse their seeds. Dry fruits have structures that help carry their seeds in wind and water currents.

Plants produced asexually have the same genes as the parent plant. Plant cultivators use methods of artificial propagation such as cutting and grafting.

BioTerms

anther (**401**)
complete flower (**402**)
double fertilization (**405**)
embryo sac (**404**)
endosperm (**405**)
epicotyl (**409**)
filament (**401**)
fruit (**407**)
generative cell (**403**)
germinate (**408**)

hypocotyl (**409**)
imperfect flower (**402**)
incomplete flower (**402**)
ovary (**402**)
ovule (**402**)
perfect flower (**402**)
petal (**402**)
pistil (**401**)
polar nuclei (**404**)
pollen (**401**)

pollen tube (**405**)
pollination (**405**)
radicle (**409**)
receptacle (**402**)
sepal (**402**)
stamen (**401**)
stigma (**402**)
style (**402**)
tube cell (**403**)
vegetative propagation (**410**)

BioQuiz   (Write all answers on a separate sheet of paper.)

I. Completion

1. The male gametophyte of flowering plants is the _____ .
2. The endosperm provides _____ for the developing embryo.
3. Runners are modified stems that grow along the ground and produce new plants at _____ .
4. Flowers function solely to carry out _____ .

II. Modified True and False

Mark each statement TRUE or FALSE. If false, change the underlined term to make the statement true.

5. In pollination, bees deposit pollen on the stigma.
6. Nonessential flower parts are directly involved with sexual reproduction.
7. Before reaching the egg, the generative cell divides into a pollen grain.

III. Multiple Choice

8. Only flowering plants undergo the process of a) fertilization. b) asexual reproduction. c) sexual reproduction. d) double fertilization.
9. The process by which pollen is transferred to the stigma from the anther is called   a) asexual reproduction. b) double fertilization. c) germination. d) pollination.
10. The female gametophyte in flowering plants is the a) embryo sac. b) megaspore mother cell. c) microspore mother cell. d) tube cell.
11. In grafting, a scion is attached to a a) pistil. b) anther. c) stock. d) radicle.
12. Fruit develops from a flower's a) filament. b) embryo sac. c) sepals. d) ovary.
13. The structure of the seedling that grows to become the stem is the a) radicle. b) hypocotyl. c) epicotyl. d) cotyledon.
14. A flower that has both stamens and a pistil is a) perfect. b) imperfect. c) female. d) male.

IV. Essay

15. What is the difference between self-pollination and cross-pollination?
16. What role do a flower's petals play in the reproductive process of the plant?
17. How does the embryo of a monocot differ from the embryo of a dicot?
18. How do fleshy fruits help disperse their seeds?
19. What is the difference between a tuber and a bulb?
20. Why do plant cultivators use artificial propagation rather than sexual propagation to produce fruit trees and other special crops?

Applying and Extending Concepts

1. Groups of flowers that are in clusters, such as sunflowers, are called *inflorescences*. What advantage might a plant have in producing an inflorescence instead of a single flower?
2. Research bees in your local school library. Then write a brief report about the special structures bees possess that trap pollen. What characteristics of bees' vision help explain why they are attracted to brightly colored flowers?
3. Seeds ordinarily reach maturity before the fruit is fully ripe. How might the lag in fruit development increase the plant's chances for survival?
4. Why is it impossible to reproduce hybrid roses from seed?

Related Readings

Cook, R. E. "Reproduction by Duplication." *Natural History* 89 (March 1980): 88–93. This article discusses the advantages plants gain from reproducing themselves sexually.

Grossman, M. L. "Ours Was a World Without Flowers 'Until Just Recently.'" *Smithsonian* 9 (February 1979): 119–130. This article explains how animals and flowering plants adapted to each other and went through parallel stages of evolution in relation to one another.

Knutson, R. M. "Flowers That Make Heat While the Sun Shines." *Natural History* 90 (October 1981): 75–81. This article describes how plants that bloom at low temperatures act as solar collectors to attract insects.

Growth and Response in Plants

A tendril of wild grape clinging to a twig for support

Introduction

A small seedling becomes a large plant by adding and expanding cells. All the time a plant is growing, it responds to variations in the amount of light, water, and nutrients in its environment. Special internal substances regulate how and when the plant uses energy and food for growth. These internal substances determine how fast the plant grows and when a seed will germinate. They also affect the differentiation of plant cells into roots, stems, leaves, and flowers.

Plants respond in specific ways to environmental stimuli. For example, stems usually grow toward light, and roots grow down in response to gravity. Responses to light, gravity, moisture, and other stimuli are important adaptations that help plants survive in changing environments.

Factors Affecting Plant Growth

Plant growth is best understood as the result of interactions between the plant and its environment. In your study of plants thus far, you have focused on individual plant tissues and functions. In this section, you will see how the development of the entire organism is influenced by various factors, both external and internal.

28.1 Influences from the Environment

Light, moisture, and temperature are some of the important external factors that influence plant growth. Every kind of plant has certain requirements for light, moisture, and temperature. Changes in these external factors trigger some of the basic phases in the plant's life cycle, such as growth, flowering, and movement.

Light Light supplies the energy that plants use to carry out photosynthesis. Without light, plants cannot produce the glucose that they need to grow. The duration of light also affects plant growth. There are more hours of daylight in the summer than there are in the winter. Strawberry plants and apple trees are examples of plants that monitor the length of day. They contain a pigment called *phytochrome* (FYT oh krohm). As light gives way to darkness, and darkness to light, phytochrome changes from one chemical form to another. By detecting the type and amount of phytochrome present, plants determine

Figure 28–1. These two flowers bloom in response to changing day length. The squash flower (left) is a short-day plant. The blackberry flower (right) is a long-day plant.

the length of darkness and light each day. The response that plants show to changing light and dark periods is called **photo-periodism** (foht oh PIHR ee uhd ihz uhm).

When photoperiodic plants are in the dark for specific amounts of time, they begin to flower. This amount of darkness is called the **critical dark period.** Plants are classified according to their critical dark periods. A **short-day plant** begins to produce flowers during the short days of spring or fall, when the nights are the same length as or longer than the plant's critical dark period. A **long-day plant** produces flowers during the long days of summer, when the nights are shorter than the plant's critical dark period. Plants that do not flower in response to the duration of light are called **day-neutral plants.**

One of the effects of photoperiodism is that plants bear fruit at different times. Strawberry and blueberry plants are short-day plants that produce ripe fruit by early summer. Squash plants, as well as pear and apple trees, are long-day plants that produce fruit in the fall. Day-neutral plants such as carrots and tomatoes can produce fruit throughout the entire growing season.

Moisture   Plants need water for photosynthesis and other cellular functions. Water carries dissolved minerals throughout the plant. Water, which is stored in the enlarged central vacuole, makes up most of the volume of plant cells and gives the cells their internal pressure.

The amount of water a plant needs for proper growth varies greatly. Some plants, called *xerophytes* (ZIHR uh fyts), are adapted to dry conditions. Plants of the southern African genus *Xerophyllum* can survive with as little as 5 percent of their body weight as water. Some plants thrive in very wet conditions. These plants, such as water lilies and duckweeds, are *hydrophytes*. Most plants, however, maintain an amount of water in their cells intermediate between that of xerophytes and hydrophytes. These plants are called *mesophytes*. Most house plants are mesophytes.

Some plants have special mechanisms for getting rid of excess water. Plants in humid areas get rid of excess water by a process called *guttation,* in which the plant secretes droplets of water from its leaves.

Figure 28–2. Drops of water extruded at the edges of this leaf are the plant's way of expelling excess water.

Temperature   Temperature is a crucial factor in plant growth. Most plants grow best in temperatures between 10°C and 38°C (50°F. and 100°F.). Temperatures colder than 10°C slow down chemical reactions in the cells. Temperatures above 38°C begin to destroy proteins, enzymes, and other important molecules in most plants.

In temperate regions, as autumn days grow shorter and the nights colder, plant activity slows down dramatically. Woody plants stop producing secondary growth, and most plants stop producing new leaves. The plants gradually enter **dormancy,** a condition in which little growth occurs and metabolism proceeds at a very low rate. Dormant plants survive through the winter by using only the minimal amount of energy and food needed to keep their cells alive.

Many kinds of seeds are adapted to cold regions. In addition to becoming dormant, these seeds must go through a period of chilling called **vernalization** (vuhr nuhl eye ZAY shuhn) before they will germinate. If a seed germinated in autumn, the seedling would be killed by frost. Vernalization keeps the seed inside its protective coat until spring, when the seedling has a much better chance for survival. Seeds that need vernalization can be germinated by storing them first in a freezer. After a month or two of freezing, the seeds can be thawed and germinated.

28.2 Internal Factors

Because plants are sensitive to light, moisture, and temperature, they can respond to major environmental changes, such as winter or periods of drought. Most internal changes in plants, including normal cell growth, involve tiny amounts of powerful chemicals called **hormones.** Hormones are internally produced chemicals that regulate the functions of tissues and organs in an organism.

Hormones are produced in one part of a plant and transported to another part, where they cause specific changes in cellular activity. Hormones may stimulate or inhibit cell growth and development. The three major groups of plant hormones are *auxins* (AWK sihnz), *gibberellins* (jihb uh REHL ihnz), and *cytokinins* (sy tuh KY nihnz).

Auxins The term *auxin* comes from the Greek word meaning "to increase." **Auxins** are hormones that tend to stimulate the elongation of cells. Auxins produce their effects by breaking chemical bonds in the cell wall. As a result, the cell becomes more flexible and grows longer as it absorbs water and synthesizes new proteins.

The amount of auxin necessary to cause the elongation and differentiation of cells varies for each plant organ. In growing plant stems, relatively high concentrations of the auxin *indoleacetic acid (IAA)* produce growth at the apical meristem. The IAA in the stem tip promotes growth in the terminal bud while inhibiting the growth of lateral buds. This characteristic pattern

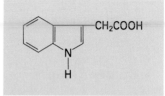

Figure 28–3. Indoleacetic acid (IAA) is an auxin that consists of a two-ring indole portion and an acetic acid group.

of growth is called *apical dominance* because growth at the tip controls growth elsewhere on the stem. The tall, pyramidal shape of spruce and fir trees is a result of the dominance of the central trunk over the lateral branches. The lower branches are large because they are far away from the apical meristem.

Auxins also inhibit the process of *abscission*—that is, the loss of plant parts such as leaves, flowers, and fruit. Fruit trees are often sprayed with auxin to prevent the fruit from dropping, so that it will accumulate as much sugar from the tree as possible. Artificially produced auxins are widely used commercially. Horticulturists use small amounts of auxins to stimulate root growth in cuttings. Auxins are also used to prevent stored potatoes from sprouting.

▶ Discovery of Auxin

A Dutch plant physiologist named Frits W. Went conducted a series of four experiments with oat seedlings in 1926. In these experiments, Went identified auxin and showed how auxins regulate plant growth.

In the first experiment, Went removed the *coleoptile,* the sheath that covers the shoot tip. He observed that the shoot then stopped growing. When the coleoptile was replaced, normal growth resumed. This showed that a substance produced in the coleoptile influenced plant growth.

In the second experiment, a block of agar was placed between the coleoptile and the rest of the shoot. The shoot continued to grow. Based on this

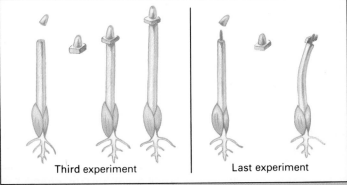

Third experiment   Last experiment

observation, Went concluded that the substance produced in the coleoptile was transported to the rest of the shoot.

In the third test, the coleoptile was placed cut side down on a block of agar and left there for one to four hours. When the agar block was placed on the cut shoot, the shoot grew. Went concluded that a particular substance, not the coleoptile itself, was regulating the growth.

Went began the last experiment, illustrated here, just as he did the third experiment, with the coleoptile removed and placed on a block of agar. A small piece of the agar was then placed on one side of the cut shoot. The shoot bent as it grew more rapidly on the side where the agar was placed. Went concluded that a chemical from the coleoptile caused the seedling to bend. He named the chemical *auxin.*

Gibberellins

Gibberellins Hormones called **gibberellins** stimulate rapid growth. Gibberellins were discovered in 1935 by Japanese researchers who were studying "foolish seedling disease" in rice. The diseased seedlings grew so tall in such a short period of time that they were unable to support themselves. The rice seedlings were infected with a fungus called *Gibberella fujikuroi*. This fungus produced the chemical that the researchers named *gibberellin*.

Gibberellin and its acidic form, gibberellic acid, begin the process of converting a seed's endosperm into sugars and amino acids that the seed can use to grow and develop. In many mature plants, gibberellins cause the stem to elongate suddenly just before the plant flowers. This process, called *bolting*, produces a long stem that holds the flower up to pollinators and the wind. Most plants can be made to bolt artificially by applying gibberellin to the stem.

Cytokinins The group of hormones called **cytokinins** stimulate cell division in plants. These hormones are mostly concentrated in endosperm and young fruits. Cytokinins cause plant cells to divide many times without differentiating or specializing. When a balance of auxins and cytokinins work in combination, however, the hormones stimulate normal growth in the plant.

Other Hormones In addition to auxins, gibberellins, and cytokinins, plants produce three other kinds of hormones that affect the development of fruits and seeds. *Abscissic acid* causes dormancy in seeds and buds by inhibiting cellular activity. *Maleic hydrazine* works with abscissic acid to maintain dormancy. *Ethylene* is an unusual hormone because it exists as a gas. Ethylene ripens fruit by stimulating color change, softening fruit cell walls, and stimulating the conversion of starches and acids into sugar. Fruit is often picked while it is still green and then treated with ethylene on its way to market.

Figure 28–4. Gibberellic acid causes normal lettuce plants (top) to bolt. Bolting causes the stems to elongate and the leaves to spread apart (bottom).

Reviewing the Section

1. What role does phytochrome play in photoperiodism?
2. What is the difference between a short-day plant and a long-day plant?
3. How does dormancy help plants survive the winter?
4. What kind of growth do gibberellins stimulate?
5. In what way does vernalization benefit seeds?
6. Where is auxin produced?

- *Distinguish* between tropisms and nastic movements.
- *Explain* the survival value of phototropism.
- *Describe* how geotropism affects the growth of roots and stems.
- *List* three kinds of nastic movements.

Plant Movements

Photoperiodism, dormancy, and vernalization are all adaptations that enable plants to survive winter. Plants have other mechanisms that help them survive. These mechanisms allow plants to move in response to changes in their environment. New generations move through seed dispersal and vegetative propagation. Mature plants also move—stems twine around posts, flowers open and close, and leaves turn toward the sun. The stimuli for these limited but important movements may come from the environment or from the plant itself.

28.3 Tropisms

A plant may respond to a stimulus by growing toward or away from the stimulus. Such a movement is called a **tropism.** *Tropisms occur when one part of a plant organ grows faster than other parts, causing the organ to bend.* In a *positive tropism*, the plant moves toward the stimulus. In a *negative tropism*, the plant moves away from the stimulus.

Phototropism The response of a plant to the direction of its light source is called **phototropism.** The above-ground organs of a growing plant tend to bend toward light. You may be familiar with this type of phototropism in house plants. The sun shines directly on the side of a plant next to a window, while the other side is partially shaded. This uneven stimulus causes auxins in the plant's stem to move to the shady side of the stem. There they stimulate cell growth and so cause the dark side of the stem to expand more rapidly than the sunny side. As Figure 28–5 shows, this uneven growth causes the stem to bend toward the light. This positive phototropism exposes the plant's photosynthetic cells to as much light as possible.

Unlike stems and leaves, roots show negative phototropism by growing away from light. This response helps direct roots toward water and minerals deep in the soil.

Geotropism Roots are also directed down into the soil by a response to gravity called **geotropism.** Roots show *positive geotropism*—they grow toward the earth. Stems and leaves show *negative geotropism*—they grow upward against gravity. Scientists are not certain what causes geotropism, though the response appears to be stimulated in part by the movement of particles called *statoliths*. These tiny pellets of starch are located in the root cap. When a plant is placed on its side, the statoliths

Figure 28–5. The movement of a plant's stems and leaves towards a source of light is called positive phototropism.

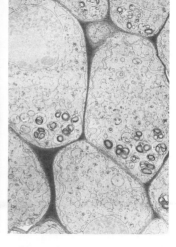

settle to the lower side of the cap. The root then curves downward and the stem curves upward, perhaps in response to the redistribution of hormones. As the root returns to a vertical position, the statoliths settle back in the cells of the root cap. Figure 28–6 shows an electron photomicrograph of statoliths.

Other Tropisms   Two other kinds of tropic movements important to plant survival are **thigmotropism** and **hydrotropism.** Thigmotropism is a curving response to contact with a solid object. Grape and ivy plants have thigmotropic stems that coil tightly around posts, trees, rocks, or other supporting objects. Hydrotropism is the movement of a plant's roots toward water. Willow trees and other plants that need a great deal of water show positive hydrotropism.

28.4 Nastic Movements

A seedling bends toward light, but not all plant responses are related to the location of the stimulus. For example, morning glory flowers open and close in response to daylight and darkness. The direction of the petals' movement is independent of the direction of the light. Plant movements unrelated to the direction of an external stimulus are called **nastic movements.** Nastic movements occur rapidly and do not necessarily involve the growth of cells.

The most common kinds of nastic movements involve changes in the internal pressure of cells. For example, turgor pressure in the base of clover leaves decreases after sunset, causing the leaves to droop and fold. Such movements in leaves and flowers help plants conserve water, heat, and energy by exposing less surface area to the air. When the sun rises, the turgor pressure increases, and the leaves return to their upright, daytime positions.

Figure 28–6. The stem and leaves of an impatiens plant placed on its side for 16 hours exhibit negative geotropism (left), as statoliths resettle in the root cap. The electron photomicrograph (right) shows statoliths in a single root cell of a maize plant, magnified 5,000 times.

Q/A

Q: *How does a Venus' flytrap catch insects?*

A: The trap of a Venus' flytrap is a modified leaf. Hairs on the upper surface of the plant are stimulated by insects that walk on the leaf. This stimulus sets up an electrical signal in the plant that causes the chemistry of the cell membranes to change. As a result, the two halves of the leaf close rapidly over the insect. The insect is then digested by enzymes.

General Description

A horticultural technician, or greenhouse worker, is a worker assigned to handle the daily tasks necessary to grow and ship plants commercially. Horticultural technicians generally work in a production nursery where ornamental plants are grown.

The horticultural technician performs a variety of tasks, working with a botanist, ornamental horticulturist, or plant propagator. Most technicians routinely water plants, fertilize them, and inspect them for diseases. They also prepare plants for shipping and keep records of what was done to the plants.

Some horticultural technicians become salespersons, working either in the wholesale or retail market. Other technicians work in a greenhouse for a few years to gain experience in growing plants, then begin their own nurseries or landscaping businesses.

Career Requirements

No formal requirements must be met for a career as a horticultural technician. Most workers, however, have a high school diploma and some experience with plants. Technicians develop their skills through on-the-job training in nurseries. Classes in botany or horticulture can

also be very helpful. Success on the job, however, depends on a liking for plants as well as knowledge of them.

For Additional Information
American Society for
    Horticultural Science
101 North Asaph Street
Alexandria, VA 22314

Nastic movements are especially rapid in touch-sensitive plants such as *Mimosa pudica*. The touch of an insect or even a gust of wind causes the leaves to release a chemical that suddenly lessens the turgor pressure in special motor cells. The leaves close almost instantly. The stamens of many flowers also show nastic movements. When touched by an insect, the stamens of these plants lose turgor pressure and droop over the insect, depositing pollen on the insect.

Reviewing the Section

1. What changes cause a stem to bend toward light?
2. How do plants benefit from phototropism?
3. How does positive geotropism help a plant survive?
4. What is the major difference between a tropism and a nastic movement?

Investigation 28: Growth Rates in Plants

Purpose
To compare the rates of growth among different kinds of plants

Materials
Seeds from four kinds of plants, such as bean, corn, sunflower, and oat; four clay pots; potting soil; centimeter ruler

Procedure
1. Plant three seeds of each kind of plant. Make sure that each pot contains only one kind of seed. Allow the seeds to grow, making sure that each pot receives the same amount of water and light. *Why should the seeds selected come from plants that grow under similar environmental conditions?*
2. Using a centimeter ruler, measure the height of each plant once a week for four weeks. Prepare a chart similar to the one shown below and record your measurements. *Why are the measurements taken weekly instead of daily?*

Plant Growth

Type of Plant	Height of Plant			
	Week 1	Week 2	Week 3	Week 4

Analyses and Conclusions
1. Which kind of plant grew most rapidly? Which plant grew most slowly?
2. Explain the similarities and differences in the growth rates of the different kinds of plants.

Going Further
- Collect additional data on growth and development by measuring the width of the stems of each kind of plant.

- Place some plant seeds on top of a moist blotter or moist filter paper in a petri dish. Allow the seeds to germinate. When the roots are about 2–3 cm long, mark every 2 mm of the roots with an ink pen. Allow the embryonic plants to continue growing in the petri dish. Compare the rates of growth in different parts of the root.

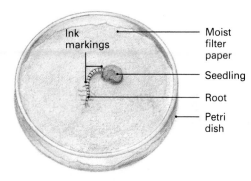

Chapter 28 Review

Summary

Growth and response in plants result from interactions between plants and their environment. All plants need a certain amount of light and water to grow. The length of daily exposure to light also triggers flowering in some plants. This response is called photoperiodism. Each photoperiodic species is either a short-day or a long-day plant. Day-neutral plants can produce flowers throughout the growing season. Dormancy and vernalization are responses to temperature that enable plants to survive winter in temperate regions.

Hormones are internally produced substances that alter the internal chemistry of plants. Auxins stimulate cell elongation while inhibiting the growth of buds and roots. Gibberellins stimulate rapid growth, and cytokinins stimulate cell division. Other hormones affect the development of fruits and seeds. A hormone may have different effects in different organs.

Plants also respond to external stimuli by moving. Tropisms such as phototropism result from growth toward or away from a stimulus. Nastic movements result chiefly from changes in turgor pressure and are independent of the direction of the stimulus. Typical nastic movements involve the opening and closing of leaves or flowers in response to daylight and darkness.

BioTerms

auxin (**419**)
critical dark period (**418**)
cytokinin (**421**)
day-neutral plant (**418**)
dormancy (**419**)
geotropism (**422**)

gibberellin (**421**)
hormone (**419**)
hydrotropism (**423**)
long-day plant (**418**)
nastic movement (**423**)
photoperiodism (**418**)

phototropism (**422**)
short-day plant (**418**)
thigmotropism (**423**)
tropism (**422**)
vernalization (**419**)

BioQuiz   (Write all answers on a separate sheet of paper.)

I. Completion

1. A long-day plant will flower when the daily period of darkness is _____ than its critical dark period.
2. Plants in temperate regions undergo a period of _____ when their metabolic rate slows down dramatically.
3. In seeds, _____ begin the process of converting endosperm into sugars and amino acids.
4. Many kinds of nastic movements are caused by changes in the _____ of plant cells.

II. Modified True and False

Mark each statement TRUE or FALSE. If false, change the underlined term to make the statement true.

5. In phototropism, auxin produced in the apical meristem moves to the light side of the stem.
6. Cytokinins cause stems to elongate suddenly just before flowering.
7. All tropisms result from the unequal growth of plant cells.
8. Hormones called auxins tend to stimulate the elongation of cells.

III. Multiple Choice

9. The hormone that causes fruit to ripen is
   a) ethylene. b) phytochrome.
   c) abscissic acid. d) maleic hydrazine.
10. Plants conserve food and energy during
    winter by entering a) a critical dark
    period. b) photoperiodism.
    c) guttation. d) dormancy.
11. If seeds did not undergo vernalization in
    temperate climates, many would be
    killed by a) heat. b) moisture. c) cold.
    d) light.
12. Apical dominance is caused by the hor-
    mone a) indoleacetic acid. b) ethylene.
    c) maleic hydrazine. d) gibberellin.
13. Photoperiodism affects the production of
    a) roots. b) stems. c) leaves.
    d) flowers.
14. The response in which roots curve
    toward water is called a) geotropism.
    b) phototropism. c) hydrotropism.
    d) thigmotropism.
15. Leaves show one kind of nastic
    movement when they a) open and close.
    b) fall from trees. c) turn yellow.
    d) grow toward sunlight.

IV. Essay

16. Why do apple growers spray the fruit
    trees with auxin as the apples mature?
17. What kinds of functions do nastic move-
    ments perform in plants?
18. What kind of growth do cytokinins tend
    to produce when they act alone on plant
    cells?
19. How does photoperiodism help flowering
    plants to survive?
20. How do tropisms affect root growth?

Applying and Extending Concepts

1. The leaves of trees and ivy often grow so
   that they barely overlap. How might phototropism account for this pattern of
   growth? Does this pattern have survival
   value for the plant? Explain your answer.
2. Design an experiment to test the effects of
   applying various concentrations of the
   hormone IAA to plant stems. Then conduct the experiment and write a paragraph
   summarizing the results.
3. The flowering of photoperiodic plants can
   be controlled artificially. If a red light
   flashes once in the middle of every night,
   for example, a short-day plant will never
   flower. Use your school or public library
   to research photoperiodism and write a
   report explaining why such an interruption
   prevents flowering. Include the role of
   phytochrome in your answer, noting the
   differences between its chemical forms.

Related Readings

Billings, W. D. ''Plants in High Places.'' *Nat-ural History* 90 (October 1981): 82–88. This article describes the special adapta-tions typical of plants that thrive in moun-tainous regions.

Cook, R. E. ''Long-Lived Seeds.'' *Natural History* 88 (February 1979): 54–61. This article describes the process of germinat-ing seeds that have been dormant for many centuries.

Ray, T. S., Jr. ''Slow-Motion World of Plant 'Behavior' Visible in Rain Forest.'' *Smith-sonian* 9 (March 1979): 121–130. This article describes how tropical vines reach sunlight.

Wolkomir, R. ''It's a Jungle Out There.'' *Na-tional Wildlife* 21 (August–September 1983): 19–21. This article explains plants' chemical defense mechanisms.

BIO*TECH*

Growing Plants in Space

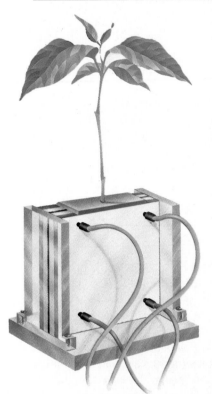

During the twenty-first century, people may spend months and even years in space. Orbiting space stations and bases on moons and planets must be designed with balanced, self-sustaining life-support systems. Technology must provide a means to supply air, water, food, and warmth, and to dispose of or recycle wastes. Green plants could serve several functions as part of these complex technological systems.

Plant roots (left) are contained between plates for hydroponic growing. A scientist (below) researches tissue culture propagation.

An adequate food supply for extended missions would take up a large amount of storage space and the additional mass would require extra fuel. Plants may help solve the problem of food supply. People could obtain some of their food by raising food crops in space. Such a supplement would doubtless prove a welcome change from prepackaged, freeze-dried fare.

Plants could also serve to recycle wastes, taking up carbon dioxide exhaled by human inhabitants and releasing oxygen back into the air. Human wastes could be used to fertilize plants.

Growing plants in space requires new propagation techniques.

Experimental methods of plant propagation include tissue culture (top right) and synthetic plant seed germinating (bottom right).

Finally, plants could provide a psychological benefit by surrounding people with living reminders of Earth. They could create a pleasant, natural oasis in the high-technology environment of a space station or lunar base.

How will plants grow in the confined area of a space station? Scientists are experimenting with growing plants in a water solution using a technology called *hydroponics.*

The solution contains a special combination of elements essential for plant growth.

For space stations that do not have gravity, scientists are designing special devices that put plant roots in contact with the hydroponic solution without letting the solution escape into the surrounding air. One system currently being researched by NASA scientists gently compresses the roots inside a semi-permeable membrane held in place by two specially designed plates. The hydroponic solution is circulated on one side of the membrane. As the roots grow on the other side of the membrane, they draw the solution through the membrane.

To propagate plants in space, two methods might be used. Seeds, treated before launch to prevent the growth and transfer of unwanted microorganisms, could be germinated. Plants could also be propagated through *tissue culture.* In this technique, a piece of plant is cultured on a medium of salts, sugars, vitamins, and hormones. The tissue can be used to clone many identical plants.

CHAPTER

29

Applied Plant Biology

Outline

Plants for Food
Cereal Grains
Legumes
Root Crops
Improvements in Food
    Crops
Foods for the Future

Other Uses of Plants
Forestry
Fibers
Medicines and Seasonings
New Uses for Desert Plants

Irrigated farmland in the Coachella Valley of California

Introduction

Scientists now know a great deal about how plants grow, respond, and reproduce. The field of applied plant biology applies this knowledge in developing new varieties of plants for many uses. New species of crop plants have multiplied the amount of food that can be produced per acre. Genetic research has led to crops that can withstand extreme cold or scorching heat and crops that can resist disease.

The efforts of applied plant biology are directed toward providing adequate food and plant products for a rapidly growing world population. To meet the rising demand throughout the world for food and manufactured goods, scientists are seeking ways to make agriculture more efficient and to expand the areas where agriculture is possible.

Plants for Food

Out of the approximately 275,000 species of plants, scarcely 100 are cultivated as major food crops. People began to cultivate crops from wild plants 10,000 years ago. They favored plants that produced highly nutritious food in forms that could be stored over long periods. Today, farmers and agricultural scientists continue to refine many of the same crops. The major food crops fall into three main categories: cereal grains, legumes, and root crops.

29.1 Cereal Grains

Cereal grains are the small, one-seeded fruits of grasses. Wheat, rice, corn, barley, sorghum, oats, millet, and rye are all important cereal grains. Because grains are a dry form of fruit, they can be kept for months or years without spoiling if stored under the proper conditions.

Cereal grains form the basis of the human diet in many countries. The *endosperm* in each grain consists of carbohydrates and protein. Endosperm is plant material that supplies food for the plant embryo. As food for humans, endosperm is the main component of flour. The embryo and seed coat contain vitamins as well as fibers that help move food through the digestive system.

Most cereal grains are grown in temperate environments with fertile soil and moderate rainfall. Wheat grows best in the sunny grasslands of the central plains in the United States and Canada and in similar areas of the Soviet Union. Corn, the only cereal grain that originated in the Americas, grows chiefly in

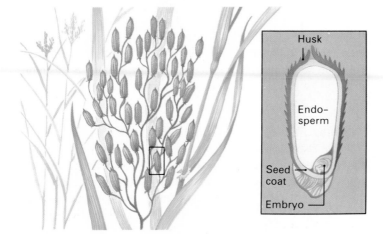

Figure 29–1. Rice is the principal food in many parts of Asia. The rice grain is actually a fruit.

areas that receive more than 20 cm (8 in.) of rain during the growing season. Rice, which requires even more rainfall, originated in the warm, humid tropics. Many tropical varieties of rice thrive in fields flooded by heavy rainfall.

29.2 Legumes

Plants that produce seeds in pods are called **legumes.** Peanuts, soybeans, peas, and beans are important legumes in the human diet. Alfalfa and clover are legumes that are grown primarily to feed animals.

Most legumes grow in the same environment in which cereal crops grow. Some legumes, including peanuts, also grow in the tropical areas of Africa and Asia. *Legumes have nutritional value comparable to cereal grains.* Peanuts, the most nutritious of all legumes, have more protein, minerals, and vitamins per gram than beef liver. Soybean oil is used widely to make margarine and cooking oil. Protein from soybeans is often added to animal feed.

29.3 Root Crops

Edible roots and underground stems are called **root crops.** Potatoes, sweet potatoes, and sugar beets are root crops with a high carbohydrate content. Cassava, a large, whitish root, is an important food in the tropics. *Root crops contain less protein and fewer vitamins than legumes and grains.* Potatoes, however, are a good source of vitamin C.

29.4 Improvements in Food Crops

As growing populations use up more and more farm land on which to live, farmers and agricultural scientists strive to make the remaining land produce food more efficiently. Efforts to increase the amount and quality of food production have focused on developing new varieties of crops, making plants more resistant to diseases and pests, and improving the nutritional value of crops.

Increased Yield Many food crops use up the available nutrients in the soil in just a few growing seasons. Farmers maximize the productivity of their crops by using fertilizers or by rotating crops to enrich the soil. **Crop rotation** is the alternating cultivation of two or more crops in the same field. After a season of growing corn, for example, a farmer may plant the field in soybeans. Soybeans and other legumes attract a particular type of

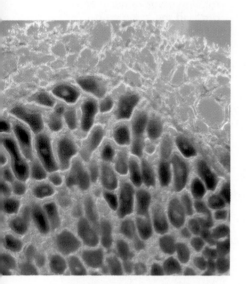

Figure 29–2. This photograph shows nitrogen-fixing bacteria within a nodule on the root of a legume.

Q/A

Q: *Are potatoes native to Ireland or the United States?*

A: Neither. Potatoes originated in the valleys of the Andes Mountains in South America. Spanish explorers introduced them to Europe in the 1500s.

bacteria to their roots. These bacteria then convert the nitrogen in the atmosphere to a form of nitrogen that the plant can absorb. This process is called *nitrogen fixation*. Some of the nitrogen compounds remain in the soil and provide nutrients for the next crop of grain. Farmers in dry areas use **irrigation,** the artificial watering of crops, to ensure large crop yields.

The most dramatic improvements in crop yield have come from the development of new plant varieties. Scientists have used *cross-breeding* to produce high-yield crop species. In cross-breeding, two different varieties of plant are used as parents in order to combine the most desirable traits of each in the offspring. *Genetic engineering,* the process of transferring genes from one species to another, is also being used to create improved crops. Scientists are attempting to transfer the genes that make nitrogen fixation possible from legumes to grains. If these grains could be made to add nitrogen to the soil, farmers could use less fertilizer.

▶ The Green Revolution

One of the most comprehensive efforts to introduce high-yield crops into poor agricultural regions is the **Green Revolution.** The Green Revolution began in Mexico and India in the 1950s. Intensive plant-breeding programs produced varieties of "miracle" wheat and rice that could yield two to five times as much grain per acre as ordinary varieties.

The miracle varieties grow best with heavy fertilization that produces large, heavy heads of seed. In ordinary wheat or rice, these heads would topple the long, slender stems before harvest, wasting the grain. The

miracle varieties, however, have genes of dwarf varieties that grow short, strong stalks that can support the extra weight.

Although these crops produced surplus food, they presented problems for poor farmers. The grains required expensive fertilizers, irrigation, and pesticides. Many poor farmers could not afford to buy the materials needed to cultivate the grains properly, nor the expensive machinery needed to apply them. Some lost their crops to pests or disease. Others, unable to compete with big, mechanized farms, had to sell their land. The Green Revolution is still going on, but it is no longer viewed as entirely positive.

Increased Resistance Both cross-breeding and genetic engineering have produced new crop species that resist drought, heat, and cold as well as certain diseases and parasites. Winter wheat, the result of years of breeding experiments, can survive cold Canadian winters.

Improved Nutritional Value Although food crops are rich in carbohydrates, most are relatively poor sources of protein. Grains, legumes, and root crops provide *incomplete protein* because none of them alone provides all of the amino acids needed by human beings. Twenty amino acids make up the building blocks of protein. The eight amino acids that humans cannot produce themselves are called *essential amino acids*. To improve the protein content of crops, genetic engineers are attempting to transfer genes for essential amino acids to corn and other grains. This research could prove especially important in countries where many people depend on a plant-based diet.

29.5 Foods for the Future

In addition to improving existing crops, scientists are seeking new sources of food. One promising discovery is *grain amaranth,* once a staple food of the Aztec empire. Amaranth outranks other common grains in protein content and contains an important amino acid that most grains lack. When eaten in combination with grains such as corn that are deficient in amino acids, the protein in both grains can be more completely used by the human body. Some species of amaranth are also grown for their outer leaves, rich in essential nutrients.

Plant researchers are also looking for ways to enable some genetically engineered plant varieties to pass on their new, improved traits in their seeds. One solution is artificial seeds. In this technique plant embryos are produced from stems or leaves. The embryos are then encased in artificial seed coats. These seeds produce plants of unusually high quality. Technologists are also experimenting with new crop varieties that will grow in the desert with the help of irrigation and special fertilizers.

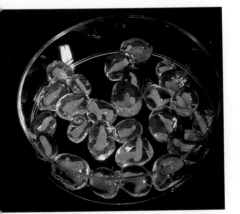

Figure 29–3. Artificial seeds like these are produced by encasing plant embryos in artificial seed coats.

Reviewing the Section

1. What are cereal grains? How do they differ from legumes?
2. What two new technological advances have most improved crop yield?
3. How does crop rotation improve soil for crops?

Other Uses of Plants

Plants have many uses in addition to providing food. Trees provide fuel, lumber, and raw materials for manufactured products such as paper. Various parts of other plants are used as fibers, medicines, and food seasonings.

29.6 Forestry

Forestry is the business of cultivating trees to provide fuel, lumber, and wood products. Half of the wood cut from the world's forests each year is used as fuel. Wood has always been an important building material. Thousands of products are made from wood, including paper goods of all kinds. Waste liquids from paper pulp mills are used in cleaning compounds, insecticides, cosmetics, and medicines. Wood chips and sawdust are used to make particle board, soil mulches, and soil conditioners. Pine extracts yield turpentine and rosin, which are used in paints and varnishes.

A crop of trees may need 20 to 50 years to mature. ***To ensure a steady production of wood over many years, forests must be carefully managed.*** Forestry companies may use chemicals to prevent disease and fertilizers to stimulate growth. Most companies choose a limited number of areas with mature trees for cutting each year, replacing the cleared areas with seedlings. Enough mature trees are left standing to protect the animals that use them as shelter and as sources for food. The newly planted seedlings prevent the erosion of soil as well as the loss of nutrients the soil contains.

Section Objectives

- *List* three important wood products.
- *Explain* how forests are managed to produce a steady supply of wood.
- *Describe* the source of two kinds of natural fibers.
- *Name* two kinds of plants that are used in medicines and two that are used as herbs.
- *Discuss* the value of developing desert plants as substitutes for other natural resources.

Q/A

Q: *Is paper money all paper?*

A: No. Paper money printed in the United States also contains flax fibers for added strength and flexibility.

Figure 29–4. As an important part of forest management, foresters take core samples to monitor the health of trees.

Figure 29–5. The stem of the flax plant is the source of a strong, durable fiber that is used to make linen cloth.

Figure 29–6. The leaves of the foxglove contain digitalis, a powerful heart stimulant.

29.7 Fibers

Cotton is the world's most important plant fiber. Cotton thread is spun from the strong, fine fibers attached to cotton seeds. The stems of flax plants yield a soft, durable fiber called *linen* which is used to make cloth. Although synthetic fibers now make up more than 30 percent of the world's fibers, natural fibers are still prized for their strength and resilience.

29.8 Medicines and Seasonings

Long before the age of modern medicine, people used plants to cure illness and soothe discomfort. Some medicinal plants are still in use. Dried foxglove leaves yield a substance called *digitalis,* which is used to treat heart disease. Extracts from the opium poppy are used in the powerful pain relievers *morphine* and *codeine*. The bark of the cinchona tree contains *quinine,* a drug used to treat malaria.

Food seasonings may come from various parts of plants. The seasonings rosemary, chives, parsley, and basil are leaves. Dill, pepper, anise, nutmeg, and mustard are derived from seeds. Capers and cloves are flowers, horseradish is a root, and coriander, a fruit. Parts of stems may also provide seasonings. Cinnamon comes from the bark of certain trees, and ginger, from a rhizome.

29.9 New Uses for Desert Plants

Many substances produced by desert plants can be used as substitutes for scarce natural resources. Often these plants are easier to obtain than the scarce natural resource they replace. For example, the sperm whale, now an endangered species, once supplied oil used in cosmetics, lubricating oils, and floor wax. Scientists have found, however, that oil from the desert shrub

Spotlight on Biologists: Norman Borlaug

Born: Cresco, Iowa, 1914
Degree: Ph.D., University of Minnesota

In 1970 plant pathologist Norman Borlaug became the first agricultural scientist to win the Nobel Peace Prize. He was honored for developing new varieties of wheat that increased world crop production and helped reduce the threat of famine. Borlaug's achievements gave developing nations the means to grow high-yield crops in tropical climates where wheat had not previously been grown successfully.

Borlaug developed the new grain types in association with a group of American scientists. They sought to control crop losses from weeds, disease, and insects. By

crossing hardy strains of wheat with varieties from Mexico, Borlaug produced new strains that thrived in many environments. Widespread use of these new strains helped lead to the Green Revolution.

Since his original experiments in Mexico, Borlaug has gone on to advise the governments of India, Pakistan, Turkey, Afghanistan, Tunisia, and Morocco on grain production. He has been honored by many of these nations as well as by the University of Minnesota and the American Society of Agronomy.

Borlaug is the type of scientist who would prefer to be out in the wheat fields than in an office writing administrative reports. He sees his achievements as merely a partial

solution to the problem of worldwide food shortages. Borlaug urges governments to develop ongoing research programs focused on keeping crops free of disease and of pesticide-resistant insects. In addition, Borlaug strongly supports a major commitment of funds by local authorities to purchase modern machinery, fertilizers, and pesticides.

jojoba (hoh HOH buh) has the same properties as sperm whale oil. Guayule (gwah YOO lee), another desert plant, produces a substitute for natural rubber. Discovering how desert plants survive their harsh conditions may some day make it possible to transfer those traits to other plants. Then desert land could be used for farming without extensive irrigation.

Reviewing the Section

1. What are three important products made from wood?
2. What are three medicines derived from plants?
3. In what new ways are desert plants useful?

Investigation 29: Hydroponics

Purpose
To compare the growth of plants in water with their growth in soil

Materials
Bean seeds, two small clay pots, potting soil, glass wool wick, sand, beaker, commercial water-soluble fertilizer, centimeter ruler

Procedure
1. Plant several bean seeds in a pot filled with potting soil. Extend a glass wool wick through the hole in the bottom of a second pot as shown in the illustration. Then fill the

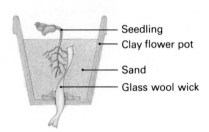

Seedling
Clay flower pot
Sand
Glass wool wick

second pot with sand, and plant several bean seeds in it. Water the seeds in both containers and allow them to germinate.
2. Remove all but one seedling from each pot so that you are left with two plants as close as possible in height. Place the seedling from the sand-filled pot in a beaker that contains fertilizer dissolved in water, as illustrated. *Why should the plant not be placed directly in the water?*

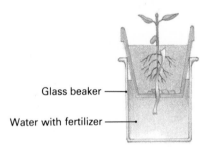

Glass beaker
Water with fertilizer

3. Observe the plants once a week for four weeks. Measure and record the growth of each plant in centimeters.

Analyses and Conclusions
1. Which plant grew faster—the one raised in soil or the one raised in sand and water? Describe any differences between the two in the development of various plant structures, such as stems and leaves.
2. Explain why the soil or the water provided a better medium for plant growth.

Going Further
- Vary the hydroponic solution by adding more nitrogen, nitrates, or ammonium compounds to the water. Describe the effect on the plant's rate of growth. Vary the pH of the solution and describe its effect on the rate of growth.
- Use Pfeffer's solution to grow bean seeds. Test the effect of various mineral deficiencies on plant growth by varying the solution as indicated below. Photograph your results.

Pfeffer's Solution

$Ca(NO_3)_2$ (Calcium nitrate), 4 gm
KNO_3 (Potassium nitrate), 1 gm
$MgSO_4 \cdot 7H_2O$ (Magnesium sulfate), 1 gm
KH_2PO_4 (Potassium dihydrogen phosphate), 1 gm
KCl (Potassium chloride), 0.5 gm
$FeCl_3$ (Ferric chloride), trace
H_2O (Distilled water), 3 to 7 liters

Chapter 29 Review

Summary

Only about 100 species of plants are cultivated as food crops. Cereal grains, the small, one-seeded fruits of grasses, include wheat, rice, and corn. Grains form the basis of people's diet in much of the world. Legumes are seeds that are produced in pods. Peanuts, soybeans, and all other legumes permit nitrogen-fixing bacteria to convert nitrogen in the atmosphere to nitrogen compounds useful to plants. Rotating grain and legume crops thus helps replace lost nutrients by adding nitrogen compounds to the soil. Root crops, including potatoes and cassava, are generally less nutritious than grains and legumes.

Food crops have been greatly improved over the last century. By cross-breeding plants with desirable traits or transferring genes by genetic engineering, scientists have developed new varieties of crops with higher resistance, increased yield, and better nutritional value. Many foods lack essential amino acids. As a result, their proteins are incompletely utilized by humans. Amaranth is a grain high in both proteins and amino acids and is a promising new food source. Future needs for food will require further improvements in technology and the development of new land for farming.

Trees are cultivated to provide fuel, lumber, and thousands of manufactured products. Forests must be carefully managed to ensure continued production. Other plants are grown for fibers, medicinal effects, and flavors. Desert plants are being cultivated as substitute sources for whale oil, natural rubber, and other scarce resources.

BioTerms

cereal grain (431)
crop rotation (432)
forestry (435)

Green Revolution (433)
irrigation (433)
legume (432)

root crop (432)

BioQuiz   (Write all answers on a separate sheet of paper.)

I. Completion

1. Genetic engineering is the process of transferring _____ between species.
2. _____ plants such as the jojoba are being developed as substitutes for several scarce natural resources.
3. Plants that produce seeds in pods are called _____.
4. _____ is an herbal medicine that is used in the treatment of heart disease.
5. _____ are the small, one-seeded fruits of grasses.
6. _____ produces a substitute for rubber.

II. Modified True and False

Mark each statement TRUE or FALSE. If false, change the underlined term to make the statement true.

7. Legumes are said to provide incomplete protein because they do not contain all essential amino acids.
8. Replanting harvested forest areas helps prevent soil erosion.
9. The endosperm in cereal grains contains protein and minerals.
10. Winter wheat is an example of a crop bred for increased nutritional value.

III. Multiple Choice

11. Animal feed is often prepared with protein added from a) alfalfa. b) peanuts. c) wheat. d) soybeans.
12. Cereal grains can be stored for long periods of time because they are a) small. b) dry. c) high in carbohydrates. d) one-seeded fruits.
13. The world's most important plant fiber is a) linen. b) jute. c) hemp. d) cotton.
14. Legumes are useful in crop rotation because they help supply a) weed killer. b) nitrogen compounds. c) irrigation. d) seeds.
15. The "miracle" grains of the Green Revolution were developed for their a) resistance to pests. b) long stems. c) high yield. d) flavor.

IV. Essay

16. What was the major contribution provided to poor agricultural regions by the Green Revolution?
17. How might genetic engineering be used to improve the nutritional value of grains or legumes?
18. What advantage do the products of plants such as the jojoba have over the scarce resources that they replace?
19. What might happen to a forest if all its mature trees were cut for timber at the same time? How would the cutting affect forest animal life?
20. Why are root crops generally considered less nutritious when consumed than grains and legumes?

Applying and Extending Concepts

1. Plants may have other uses besides foods, fibers, and medicines. For example, landscapers and interior decorators use plants to add beauty to indoor and outdoor areas. Design a setting in which plants are used as decoration. Draw a diagram of it, including instructions on the climate and care the plants would need.

2. What advantage would a grain plant gain if genetic engineers could introduce genes for nitrogen fixation?

3. Much of the world's unused land is desert; yet few countries have developed extensive desert agriculture. What special problems are posed by desert cultivation? Discuss difficulties plants must overcome to survive in the desert.

4. Some scientists fear that the popularity of high-yield "miracle" crops could result in genetic uniformity and the loss of many species of crop plants. Why might this be a dangerous trend?

Related Readings

Asimov, I. "Passing of Agriculture? Your Food by Chemistry!" *Science Digest* 87 (May 1980): 22–25. This article suggests how the use of plants in space stations might introduce an entirely new approach to agriculture.

"Foods for the Future" *Science 82* (January–February 1982): 70–76. This article discusses the dramatic results of new agricultural technologies, particularly those that

are used to produce new varieties of crop plants.

Vogelmann, H. W. "Catastrophe on Camels Hump." *Natural History* 91 (November 1982): 8–14. Acid rain threatens to destroy forests in northeastern America. This article describes the effect that acid rain has on trees and the impact that the loss of forests could have on the world's climate and economy.

30

Sponges and Coelenterates

Coral and sponges on a reef

Introduction

One major difference between animals and plants is the ability that animals have to move. Unlike plants, animals cannot internally produce their own food. Instead, most animals rely on their ability to move about to obtain their food. However, the corals and sponges in the photograph above do not move. These simple animals are **sessile**—that is, they attach themselves to an object and remain in that one place all their lives. Sponges feed on microorganisms filtered out of the water currents that flow through their bodies. Corals capture prey that swim within their reach.

In this chapter you will learn about characteristics of animals. You will also find out why sponges and coelenterates such as corals are considered the simplest animals.

Introduction to Animals

An elephant, a hummingbird, and a jellyfish differ greatly in their size, shape, and behavior. Nevertheless, they are all animals. In distinguishing between types of organisms, the characteristics animals have in common far outweigh the differences among them. However, differences in body structure are useful in classifying animals. *The arrangement of body parts is related to how a particular animal species meets the challenges of living, which include gathering food, protecting itself, and reproducing.*

30.1 Symmetry and Body Plans

The arrangement of an animal's body parts around a center point or line determines its **symmetry.** Most animals are symmetrical in some way. Only a few animals can be described as *asymmetrical,* or having an arrangement of body parts that cannot be divided into corresponding sections. Sponges are asymmetrical. They grow in varied and irregular shapes.

The general form of an animal frequently provides a clue to its way of life. An organism with a round form—with no front or back, no right or left side—shows **spherical symmetry.** This form is well suited to certain protozoa that roll and float in water. Such organisms face all directions at once.

No animal exhibits complete spherical symmetry. Animals whose body parts are arranged around a central point, like spokes around the hub of a wheel, exhibit **radial symmetry.** Their sensory organs are located around the wheel's circumference. Jellyfish and sea anemones show radial symmetry. Such animals do not move efficiently. Either they are sessile, or they float in the water or crawl along the bottom of the sea.

Section Objectives

- *List* three types of body plans that are characteristic of animals.
- *State* the feature that distinguishes vertebrates from invertebrates.

Figure 30–1. The large purple tube sponge below has an asymmetrical, or irregular, body plan. The body parts of the blue sea anemone (left) are organized around a central point. What is this kind of body plan called?

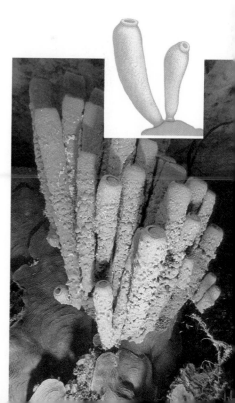

Figure 30–2. The body plan of the swallowtail butterfly displays bilateral symmetry. For example, the markings on the right wing are a mirror image of the markings on the left wing.

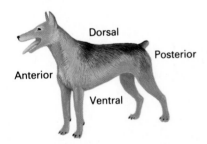

Figure 30–3. This diagram shows the location of the dorsal and ventral surfaces and the anterior and posterior ends of a dog, a bilaterally symmetrical animal.

Most animal species have **bilateral symmetry**—that is, one-half of the body is a mirror image of the other half. A butterfly is a good example of bilateral symmetry. If a line were drawn lengthwise through the butterfly's body—the longitudinal axis—the right half would appear to be the exact opposite of the left half.

Most bilaterally symmetrical animals also have upper and lower sides, and front and hind ends. The upper side is called the **dorsal** surface, and the lower side the **ventral** surface. The front is the **anterior** end, and the hind the **posterior** end. For example, the back of a dog is the dorsal surface, and the stomach is the ventral surface. The head is the anterior end, and the tail is the posterior end.

Animals that have a definite anterior end and move head first generally exhibit **cephalization** (sehf uh lih ZAY shuhn). Cephalization is the adaptation that allows the neural and sensory organs to become concentrated in the anterior end in animals. When such animals move, their sensory organs go first, providing information about the environment that lies ahead of the animals.

30.2 Vertebrates and Invertebrates

Scientists make a major distinction between **vertebrates,** animals with a backbone or a spine, and **invertebrates,** animals without a backbone. Humans and other mammals are vertebrates, as are fish, amphibians, reptiles, and birds. Vertebrates are included in the phylum Chordata. Although they are the most widely recognized and familiar of all animals, vertebrates make up only about 3 percent of the more than 1 million species of animals. Vertebrates are categorized by bilateral symmetry and cephalization.

Invertebrates make up 97 percent of the animal kingdom. Invertebrates include sponges, jellyfish, starfish, worms, mollusks, insects, and crabs. Some species of invertebrates have radial symmetry, and some bilateral. Some, but not all, species are characterized by cephalization.

Reviewing the Section

1. Name the type of symmetry shown by each of the following: a jellyfish, an eagle, and a sponge.
2. What distinguishes vertebrates from invertebrates?
3. How does the arrangement of an animal's body parts affect its movements?

Sponges

Sponges are so unlike other animals that they are sometimes put in their own subkingdom. *Sponges have the simplest body organization of any animal.* Their cells are not organized into tissues and organs. Thus they have no head, no mouth, and no digestive, circulatory, or nervous system. Early naturalists classified sponges as plants, or as plant-animals, mainly because of their branchlike forms and their inability to move around. Sponges were not classified as animals until the mid-1800s, when scientists first began to closely study sponges and their method of feeding.

Most sponges live in shallow seas, though some have been found at depths of 8,500 m (27,800 ft.). A few species live in fresh water. In size, sponges range from less than 1 cm (0.4 in.) to more than 2 m (6.6 ft.). In color, they range from white and gray to brilliant shades of red, yellow, green, purple, and black. Sponges have many shapes—they may look like balls, discs, vases, goblets, branching shrubs, or small trees. The approximately 5,000 species of sponges make up the phylum Porifera (puh RIHF uhr uh). *Porifera* means "pore body." Sponges have been described as looking like sacks full of holes.

30.3 Characteristics of Sponges

Once a sponge attaches itself to a rock, shell, or other submerged object, it does not move. *Sponges feed by filtering food and nutrients out of the water.* The body of a sponge consists of two layers of cells, with a jellylike layer between them. The outer layer is called the **ectoderm,** and the inner layer the **endoderm.** Water is drawn in through the **incurrent pores** in the ectoderm and leaves through a larger opening at the end of the body, called the **osculum** (AHS kyuh luhm). The water carries food and dissolved oxygen. Cells that digest the food, called **collar cells** or *choanocytes* (koh AN uh syts), are located in the endoderm. Each collar cell has a flagellum and many hairlike filaments. The movement of the flagella sets up a current of water through the filaments. The filaments catch and remove bacteria, unicellular algae, and other microorganisms from the water. This food is drawn into the collar cells and digested. Amoebalike cells called **amoebocytes** (uh MEE buh syts) carry the nutrients to the cells of the endoderm and take away waste matter.

Without some kind of skeletal structure, sponges would collapse under their own weight. The composition of their skeletal

Section Objectives

- *List* the major characteristics of sponges.
- *Describe* how sponges feed.
- *Name* three ways in which sponges reproduce.
- *List* some common commercial uses of sponges.

Figure 30–4. Each of the large openings in the sponges above is an osculum, through which water leaves the body of a sponge.

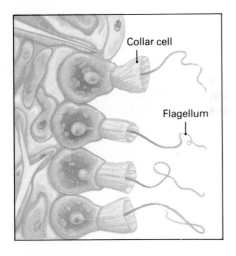

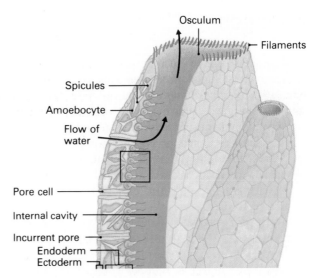

Q/A

Q: *How much water does an average sponge circulate during a day?*

A: One type of sponge 10 cm (4 in.) tall and 1 cm (0.4 in.) in diameter pumps about 22.5 L (23 qt.) of water through its body in a day.

structure is the basis of classification of sponges. Some sponges have skeletons made of **spicules** (SPIHK yoolz), tough interlocking spikes of either calcium or silicon. Other sponges have a framework made of a flexible protein called **spongin.** Bath sponges are composed of spongin, which makes them firm but soft. Some types of sponges have a mixture of both spicules and spongin.

Sponges may grow individually or in colonies. Some colonies are so dense that it is difficult to distinguish one sponge from its neighbor.

Sponges have a remarkable ability to **regenerate**—that is, to grow new parts to replace those that are lost. *A sponge not only can regenerate parts of its body, but can also regenerate the entire body from fragments.* A sponge can also reform after being separated into single cells by being pushed through a fine silk cloth. The cells will move around, form clumps, and then larger groups. Within a few days the fragments will reform into several new sponges. If sponges of two different species are pushed through a screen and mixed together, the fragments will regroup into sponges of the original two species.

30.4 Reproduction of Sponges

Sponges reproduce both sexually and asexually. Most species are **hermaphrodites** (huhr MAF ruh dyts)—that is, an individual sponge can produce both eggs and sperm, though at different times. Sperm cells produced by one sponge are carried to another sponge by water currents. Once inside the sponge, sperm cells are captured by carrier cells, which are modified collar

▶ Commercial Uses of Sponges

Sponges have been used since ancient times as bath aids, cleaning tools, and even as drinking vessels. When dried, the skeletons of certain species of sponges are soft and elastic yet still capable of absorbing large amounts of water.

The most valuable commercial sponge is the silk cup sponge found off the coasts of Egypt and Greece. Other commercial varieties live off the west coast of Florida, the Florida Keys, and the West Indies.

Sponges are harvested in shallow waters from glass-bottomed boats. As one person maneuvers the boat, a second person pulls up the sponges, using a long pole with a hook at the end. In deeper waters divers go down and collect the sponges.

Once collected, the sponges are left to dry until all the flesh decays and falls off. The sponge skeletons then are shipped to market.

In recent years synthetic sponges have largely replaced natural sponges for household use. Natural sponges are still preferred by artists and craftworkers and by hospital surgical teams.

cells. The carrier cells take the sperm to the egg. The fertilized egg develops into a **larva** (LAHR vuh), an immature form. Sponge larvae have flagella and swim about. They soon attach themselves to objects in the water and develop into adult sponges.

Fragmentation is a common method of asexual reproduction among sponges. Even a small branch that breaks off from the parent sponge can develop into a full-grown animal. Sponges also reproduce asexually by producing **gemmules** (JEHM yoolz)— special food-filled balls of amoebocytes surrounded by protective coats. Generally only freshwater sponges produce gemmules. The gemmules can survive harsh conditions, such as extreme cold or periods of dry weather. When conditions are favorable for growth, sponge cells emerge through an opening in the gemmules and grow into new sponges.

larva (plural, *larvae*)

Reviewing the Section

1. Where do sponges live?
2. What keeps sponges from collapsing under their own weight?
3. How do sponges feed?
4. During what part of their life cycle are sponges able to move about? How do they move?
5. Name two commercial uses of sponges.

Coelenterates

- *Name* four common coe-
  lenterates.
- *List* the major characteristics common to all coelenterates.
- *Distinguish* between a
  polyp and a medusa.
- *Describe* how hydras feed.
- *Name* the major organisms
  that inhabit a coral-reef
  community.

Coelenterates (sih LEHN tuh rayts) are baglike animals with long, flexible appendages called **tentacles.** Most coelenterates live in sea water. One group, the hydras, lives in fresh water. The 9,000 species of coelenterates also include jellyfish, sea anemones, and corals. Although corals are one of the simpler forms of animal life, they have had profound effects on the geography of the earth's surface. Millions of tiny coral skeletons massed together over centuries have formed entire islands and offshore reefs. These islands and reefs are found primarily in the South Pacific Ocean, in the Caribbean Sea, and along the Florida coast.

30.5 Characteristics of Coelenterates

Coelenterates get their name from their *coelenteron,* or "hollow gut." The coelenteron is a digestive cavity with only one opening. Coelenterates also have special stinging cells called **cnidocytes** (NYD uh syts). For this reason, they are also known as *cnidarians* (ny DAIR ee uhnz).

Coelenterates generally exhibit radial symmetry. They have two body plans: vase-shaped and bell-shaped. Hydras and some other coelenterates develop only a vase-shaped body, called a **polyp.** Jellyfish and some other coelenterates go through a polyp stage but spend most of their lives as a bell-shaped **medusa** (muh DOO suh).

Coelenterates live singly or in colonies. Individual animals within a coelenterate colony may have specialized functions. The Portuguese man-of-war is an example of a colonial coelenterate. Some of the individual animals specialize in reproduction or feeding, others in gathering food.

The bodies of both polyps and medusae consist of two layers of cells, the endoderm and the ectoderm, separated by a jellylike substance called **mesoglea** (mehz uh GLEE uh). The mesoglea of a polyp is thin. In a medusa, however, the mesoglea often makes up the major part of the body substance.

The tentacles of most coelenterates circle the mouth of the animal, and the cnidocytes are in the tentacles. Inside each cnidocyte is a coiled stinger, called a **nematocyst** (neh MAT uh sihst). When discharged, the stinger can paralyze and lasso small prey. Once discharged, the cnidocyte cannot be used again, and a new one grows to take its place. The tentacles draw the food up to the mouth. Enzymes inside the digestive cavity break down the food, and the food particles are absorbed by the

medusae (singular, *medusa*)

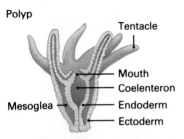

Medusa
— Mesoglea
— Endoderm
— Coelenteron
— Ectoderm

Mouth
Tentacle

Polyp
Tentacle

Mouth
— Coelenteron
Mesoglea — Endoderm
— Ectoderm

Figure 30–6. Coelenterates are either vase-shaped polyps (bottom) or bell-shaped medusae (top).

cells that line the cavity. Undigested waste products are expelled through the animal's mouth. Individual cells carry out respiration directly, taking in oxygen from the surrounding water by diffusion.

30.6 Hydras

Hydras and related animals make up the class Hydrozoa (hy druh ZOH uh). Hydras are the most extensively studied coelenterates, but they are not typical of hydrozoans in some ways. For example, hydras have only the polyp form, while most hydrozoans go through a medusa stage as well.

Hydras are only about 1 cm (0.4 in.) long. Most hydras are orange, brown, white, or gray in color, but some are green due to algae that grow in the cells of their endoderm. Hydras live in freshwater streams and ponds. They attach themselves to leaves and other debris in the water by means of a flattened adhesive base called the *basal disc*. Hydras move by floating, gliding, or somersaulting.

Although primitive animals, hydras show a great degree of specialization in their nematocysts. Four distinct types have been identified. One nematocyst anchors the tentacles when the animal moves, and another repels animals other than prey. A third holds the prey by winding around the animal, while a fourth nematocyst stings the prey, paralyzing it.

Like other coelenterates, hydras have no brain or central nervous system. A network of nerves, the *nerve net*, permits some coordination of responses and some simple movements. Hydras also have sensory cells that respond to chemical and mechanical stimuli. The animals have little control in their responses to stimuli. If touched with a probe on one part of the body, for example, the whole body will contract.

When a hydra catches a shrimp or a water flea in its tentacles, a feeding response begins. The tentacles move the prey towards the mouth, the mouth opens in response to a chemical given off by the prey, and the prey is pushed in whole. Enzymes are released that digest the food, and the undigested remains are expelled through the mouth.

Hydras reproduce asexually by forming small buds on the outside of their bodies. These buds grow, and within two or three days they fall off and begin life as independent animals. Hydras also reproduce sexually, usually in autumn. When the water temperature drops, individual hydras begin to develop either egg-producing **ovaries** or sperm-producing **testes.** The sperm swims to the egg, which remains attached to the body of the hydra. After fertilization has occurred, the egg begins to

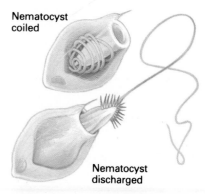

Figure 30–7. This diagram shows the nematocyst, or stinger, found in the tentacles of coelenterates. The nematocyst is coiled inside the tentacle (top) until released to kill small organisms (bottom).

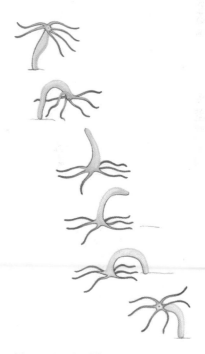

Figure 30–8. The nerve net of the hydra enables its muscles to produce a somersaulting motion, one way in which hydras travel.

Figure 30–9. The Portuguese man-of-war is a jellyfish made up of many animals attached to a large float. Polyps with stinging nematocysts make up the purple tentacles.

Figure 30–10. This diagram shows the sexual and asexual phases of the Aurelia jellyfish life cycle.

divide and falls off the adult female. The young hydra emerges in the spring.

Hydras live individually and independently. Most other hydrozoans, however, live in colonies. Most hydrozoans also have both polyp and medusa forms at different times in their life cycle. Certain species, for example, spend most of their lives as colonial polyps. A single polyp multiplies by budding, forming a colony. Within the colony, the polyps become specialized. Some are feeding polyps, and some are reproductive polyps. Tiny medusae bud off the reproductive polyps. These medusae produce eggs or sperm. Thus, colonial polyps exhibit a division of function between feeding and reproductive forms. Other colonial coelenterates also show specialization of individuals.

30.7 Jellyfish

From a distance, jellyfish resemble inflated plastic bags. Observed more closely, jellyfish can be seen swimming by rhythmically contracting and relaxing their "bells." Jellyfish belong to the class Scyphozoa (sy fuh ZOH uh), which means "cup animals." The tentacles of jellyfish with their stingers may reach up to 70 m (230 ft.) in length. Their central discs may range from 4 cm (1.5 in.) to a meter (3.3 ft.) in diameter, though one observer has recorded a central disc 3.6 m (12 ft.) in diameter.

In the life cycle of a jellyfish, the medusa reproduces sexually, and the polyp reproduces asexually. Figure 30–10 shows the life cycle of one species of jellyfish. The male medusa releases sperm cells through its mouth, and the female releases eggs. The eggs lodge in pockets on the tentacles of the female

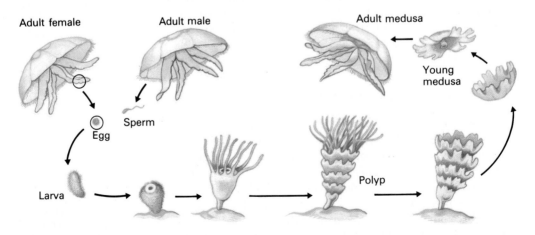

Adult female Adult male Adult medusa

Young medusa

Sperm

Egg

Larva

Polyp

and are fertilized. The eggs grow and develop into small, free-swimming larvae.

The larvae swim away and attach themselves to the sea floor. There they develop into a polyp stage that resembles the hydra. The jellyfish polyp grows and eventually produces buds. This development is the asexual phase of its reproductive cycle. The buds grow and eventually begin to form medusae, which, as they build up, resemble a stack of plates. These medusae move off one by one and begin the cycle again. The polyp may repeat the process the following year.

30.8 Sea Anemones and Corals

The sea anemones (uh NEHM uh neez) and corals belong to the class Anthozoa (an thuh ZOH uh). *Anthozoa* means "flower animals." These coelenterates are beautifully colored and have varied forms.

Sea anemones are marine polyps that inhabit coastal areas. A sea anemone has a basal disc like that of the hydra, by which the anemone attaches itself to rocks or other objects. Sea anemones are solitary. They feed on fish and crabs that swim within

Figure 30–11. Sea anemones look like flowers but are animals that use stinging cells in their tentacles to kill smaller organisms for food.

▶ Coral Reef Communities

A coral reef is a crowded community. In spite of the vast size of the ocean, corals and sponges sometimes end up competing for the same underwater space. Other plants, animals, and algae benefit from one another's presence.

Coral reefs grow best in tropical waters, where the average temperature is 27°C (80°F). Corals can live in much colder waters. However, the warm waters and sunlight are necessary to ensure the presence of photosynthesizing algae that live in a *symbiotic,* or mutually helpful, relationship with coral. Some algae live inside coral tis-

sues. They use the wastes from the coral and provide oxygen for the coral. The algae also speed up the production of calcium deposits by the coral. These deposits, which make up the coral's skeleton, form the reef.

The algae are a link in a vast food web in and around the coral branches. The corals and small fish eat the tiny marine organisms called *plankton,* and big fish eat the small fish.

Corals have their enemies within the coral-reef community. The most deadly is the crown-of-thorns starfish, which consumes soft corals. Some borer sponges compete

with corals for space. Humans are also a threat. They take away corals as souvenirs and pollute the waters. Drastic changes in one group of plants and animals can upset the precarious balance of undersea life and cause all to suffer.

Q/A

Q: *Does a coral reef grow faster at the bottom or top?*

A: Due partly to water temperature differences, the top of a reef grows about 1,000 times faster than corals in the waters deep below.

reach of their tentacles. Sea anemones digest their food in much the same way as hydras do.

Corals resemble sea anemones but have skeletons and live in colonies. Soft coral species have internal skeletons. These species include the so-called precious coral, which is used to make jewelry. The stony coral has an external skeleton that is almost pure limestone. The skeletons of dead corals accumulate and form the reefs often seen rising above the surface in tropical waters.

Reviewing the Section

1. Name four characteristics of coelenterates.
2. Compare the manner of movement in hydras, jellyfish, and sea anemones.
3. How do algae and coral benefit each other?

Investigation 30: Competition Between Two Species of Hydra

Purpose
To describe and explain how two different species of hydra survive in the same environment

Materials
Four culture dishes, one culture of brown hydra, one culture of green hydra, brine shrimp eggs, spring water, medicine dropper, stereomicroscope

Procedure
1. Several days before the shipments of hydra arrive, prepare a culture stock consisting of spring water and brine shrimp eggs in each of the four culture dishes. The eggs should have hatched by the time the hydras arrive.
2. Inoculate two of the culture dishes with one medicine dropperful each of brown and green hydra.
3. Keep the rest of the two species apart by maintaining separate cultures in the remaining two culture dishes. These can be used in further experiments.
4. Place one mixed hydra culture in a well-lighted area and the second in a closet or desk drawer.
5. Observe the cultures the next day and continue to do so each day for a week. Record your observations in a table like the one shown. Put an X in the proper column to indicate which species predominates each day. Put Xs in both columns if neither predominates.

Predominant species

	Brown hydra	Green hydra
Day 1		
Day 2		
Day 3		

Analyses and Conclusions
1. Which of the two species was more successful? Under what conditions? Explain why.
2. Compare your results with those of other students.
3. What would be the effect on each culture if you were to reverse the conditions, placing the culture from the dark environment in the light and the one from the light environment in the dark?

Going Further
- Add a small amount of baking soda to one of the mixed hydra cultures. What is released into the water? What effect does this have on the pH of the water and the growth of each species?
- Inoculate mixed cultures with a medicine dropperful of *Chlamydomonas,* a unicellular biflagellate green algae. How does the introduction of this new organism affect the growth of the two species?
- Place both mixed cultures first under ordinary light, then in turn under green light, red light, and blue light. Note the effect on the growth of each species and explain the results.

Chapter 30 Review

Summary

Animals differ in their symmetry, or the arrangement of their body parts around a center point or line. Most animals exhibit one of these forms of symmetry: spherical, radial, or bilateral. Some simple animals, such as sponges, have no symmetry. Another major division among animals is between vertebrates, which have a backbone, and invertebrates, which do not. All vertebrates are bilaterally symmetrical. Mammals, fish, amphibians, reptiles, and birds are vertebrates.

The simplest type of invertebrate is the sponge. Sponges are sessile, chiefly marine animals that feed by filtering nutrients from water that circulates through their bodies. They reproduce sexually and asexually.

Coelenterates include hydras, jellyfish, sea anemones, and corals. Coelenterates feed by grasping their prey with tentacles and stinging the prey with nematocysts. Many coelenterates have a polyp form, which reproduces asexually, and a medusa form, which reproduces sexually. Coelenterate colonies exhibit specialization in the functions of individual animals. One coelenterate, the stony coral, forms coral reefs from its skeletal deposits.

BioTerms

amoebocyte (**445**)
anterior (**444**)
bilateral symmetry (**444**)
cephalization (**444**)
cnidocyte (**448**)
coelenterate (**448**)
collar cell (**445**)
dorsal (**444**)
ectoderm (**445**)
endoderm (**445**)
gemmule (**447**)

hermaphrodite (**446**)
incurrent pore (**445**)
invertebrate (**444**)
larva (**447**)
medusa (**448**)
mesoglea (**448**)
nematocyst (**448**)
osculum (**445**)
ovary (**449**)
polyp (**448**)
posterior (**444**)

radial symmetry (**443**)
regenerate (**446**)
sessile (**442**)
spherical symmetry (**443**)
spicule (**446**)
spongin (**446**)
symmetry (**443**)
tentacle (**448**)
testis (**449**)
ventral (**444**)
vertebrate (**444**)

BioQuiz   (Write all answers on a separate sheet of paper.)

I. Completion

1. _____ move by floating or by somersaulting.
2. _____ carry nutrients to cells in the endoderm of a sponge's body.
3. Vertebrates show _____ symmetry.
4. Jellyfish polyps develop from _____ .
5. Tough spikes called _____ form the skeleton of many sponges.

II. Modified True and False

Mark each statement TRUE or FALSE. If false, change the underlined term to make the statement true.

6. The hydra is a freshwater coelenterate.
7. Water enters the central cavity of a sponge through the osculum.
8. A sea anemone shows spherical symmetry.

III. Multiple Choice

9. Animals without a backbone are
   a) vertebrates. b) invertebrates.
   c) hermaphrodites. d) sessile.
10. Budding is an example of a) sexual
    reproduction. b) cephalization.
    c) asexual reproduction. d) bilateral
    symmetry.
11. The tail of an animal is a) ventral.
    b) posterior. c) dorsal. d) anterior.
12. Water moves through a sponge's body
    because of special cells called
    a) amoebocytes. b) spicules. c) collar
    cells. d) osculum.
13. A jellyfish moves by contracting its
    a) tentacles. b) nematocysts. c) bell.
    d) coelenteron.
14. In a coelenterate with both a polyp and
    a medusa stage, the medusa reproduces
    a) asexually. b) sexually. c) by
    fragmenting. d) by regenerating.
15. Reefs are formed from the calcium
    deposits of a) sponges. b) sea fans.
    c) hydras. d) corals.

IV. Essay

16. What is a benefit of cephalization?
17. What advantage does a hydra have over
    a sponge in obtaining food?
18. Why do corals grow best in tropical
    waters?
19. How do the individuals in some
    coelenterate colonies show specialization
    of function?
20. How does water enter and leave a
    sponge?

Applying and Extending Concepts

1. Use your school or community library to
   research how precious coral differs from
   other varieties of coral. Report to your
   class how precious coral is harvested and
   polished. If possible, accompany the report with a display of color photos showing some of the interesting shapes in
   which coral forms.
2. The life cycle of some animals resembles
   the alternation of generations exhibited in
   the life cycle of plants. Compare and contrast the life cycles of a jellyfish and a
   fern. How are they the same? How are
   they different?
3. Sexual reproduction in hydras usually occurs in the autumn. A scientist might wish
   to establish whether it is the change in
   temperature or the diminished light in autumn that prompts their sexual development. Design an experiment that would
   show which of these factors is responsible. Be sure to control for other factors as
   well.

Related Readings

Dean, L. "Sponging." *Oceans* 16 (March–
April 1983): 34–39. This article recounts
the history of sponge fishing off the Florida Keys and discusses the biology of valuable sponges.

Gibson, M. E. "The Plight of *Allopora*." *Sea
Frontiers* 27 (July–August 1981): 211–
218. The article discusses the *Allopora*
hydrocorals. The survival of these organisms has become seriously threatened due
to overcollection.

Walls, J. G., ed. *Encyclopedia of Marine Invertebrates*. Neptune, N.J.: TFH, 1982.
Chapters 2 and 3 of this text deal with
sponges and coelenterates.

Worms

Tube worms on the ocean floor

Introduction

When you hear the word *worm,* you probably think of the common earthworm that comes to the surface of the ground after a heavy rain. However, worms show an incredible diversity of size, appearance, and way of life. Of the approximately 36,000 species of worms, many are so small they can be seen only with a microscope, but giant tube worms grow to be more than 1.5 m (5 ft.) long. Giant earthworms that live in the Transvaal area of Africa can be over 6.0 m (20 ft.) long.

Some worms live in the soil, some in salt water, and some in fresh water. Many are parasites that live in the bodies of plants and animals, including humans. In this chapter you will read about three major phyla of worms—flatworms, roundworms, and segmented worms.

Flatworms

Flatworms make up the phylum Platyhelminthes (plat ee HEHL mihn theez). The phylum name is derived from two Greek words: *platy,* which means flat, and *helminthes,* which means worm. The 13,000 species of flatworms belong to three major classes. Members of the class Turbellaria are called *free-living* because they live independently of other animals. Members of the class Trematoda, the flukes, and the class Cestoda, the tapeworms, are parasites. They derive nutrition from living hosts.

Flatworms have a digestive cavity with only one opening for receiving food and expelling wastes. However, flatworms lack circulatory and respiratory systems. The flat shape of their bodies allows cells to obtain oxygen and food molecules through diffusion. Flatworms range in length from a fraction of a millimeter to over 15 m (17 yd.).

31.1 Planarians

The most common flatworm is the planarian (pluh NEHR ee uhn), which belongs to the class Turbellaria. Planarians live in both aquatic and terrestrial environments. Some kinds grow to be 30 cm (1 ft.) long, and many are brightly colored. The freshwater planarians that live in North America are gray, brown, or black and average 3 to 15 mm (0.1 to 0.6 in.) in length.

Planarians have a spade-shaped head and a body covered with cilia. The mouth is located in the middle of the ventral side. Planarians move by laying down a trail of slime from mucus-producing cells and using their cilia to propel themselves along the trail.

Nervous Control in Planarians A planarian will move away from light and toward food. These responses are made possible by the planarian's simple nervous system. A planarian has two light-sensitive areas on its head called *eye spots* that contain pigments. When light strikes the pigments, they give off electrical impulses that are picked up by special photoreceptor cells. The head area also has cells sensitive to chemicals and to touch. Messages from these receptors travel to a concentration of nerve cells in the worm's head known as the **ganglion** (GAN glee uhn). From the ganglion two main nerve cords run the length of the planarian's body. Messages from the receptors cause nerves to stimulate action by the planarian's muscles. In this way, the planarian can respond to changes in light or chemical surroundings.

Section Objectives

- *List* the major characteristics of flatworms.
- *Describe* the feeding behavior of planarians.
- *Diagram* the life cycle of the liver fluke.
- *Describe* how tapeworms reproduce.
- *Explain* how the body plan of flatworms differs from that of the two other major worm phyla.

Q/A

Q: *Are planarians cross-eyed?*

A: No. Planarians do not have image-forming eyes. The eyespots that give them the appearance of being cross-eyed do not have lenses. However, the eyespots are sensitive to light intensity and direction.

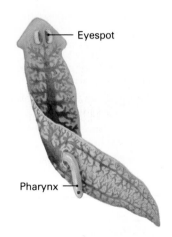

- Eyespot

- Pharynx

Figure 31–1. This diagram shows the major structures of a planarian.

Digestion and Elimination in Planarians Planarians eat dead animals or slow-moving organisms, including smaller planarians. The worm feeds by extending a muscular organ called a **pharynx** out of its mouth. Enzymes released through the pharynx partially digest the food, which is then sucked into the digestive cavity. Nutrients travel through branches of the digestive cavity to all parts of the body. Solid wastes are removed through the pharynx.

A network of fine tubules extending the entire length of the body of the animal allows for the removal of excess water. Some of the tubes end in cells called **flame cells.** Cilia in these cells beat back and forth, much like the flickering of a candle flame. The beating of the cilia moves water and liquid wastes along the tubules to surface pores where the wastes are expelled.

Reproduction in Planarians Planarians are hermaphrodites—that is, each animal has both male and female reproductive structures. Sexual reproduction occurs by a mutual exchange of sperm. Each worm sheds fertilized eggs in capsules containing up to 10 eggs. The capsules attach to objects in water until the eggs hatch.

Planarians may reproduce asexually by horizontal fission. The head fragment then grows a new tail, and the tail fragment, a new head. Because of this capacity for regeneration, a planarian that is accidentally or deliberately cut will grow into two new organisms.

31.2 Flukes

Flukes belong to the class Trematoda. Flukes are parasites and are a serious health hazard in many areas of the world. Many kinds of flukes cause serious and even fatal diseases. Suckers or hooks on the anterior end of the fluke attach the parasite to its host. Flukes usually live on the fluids of a host, such as blood and mucus. Many flukes may live for years off a single host.

Mature flukes do not have cilia. Instead their skin is covered with a thick protective coating called a **cuticle** (KYOOT ih kuhl), which prevents the host from digesting them.

Some flukes are *ectoparasites*—that is, they live on the outside of their host's body. Most flukes are *endoparasites* and live inside the body of their host. Flukes have a complex life cycle often involving two or more hosts. For example, the liver fluke spends most of its life in the digestive tract of a sheep or other vertebrate. There the fluke lays thousands of eggs, many of which are carried out of the sheep's body in its **feces,** or solid wastes.

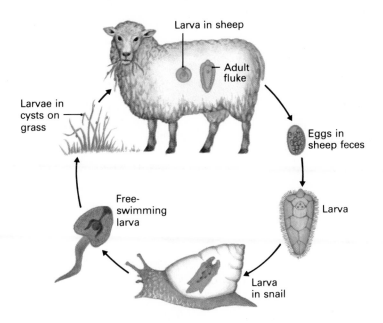

Larva in sheep

Adult fluke

Larvae in cysts on grass

Free-swimming larva

Larva in snail

Eggs in sheep feces

Larva

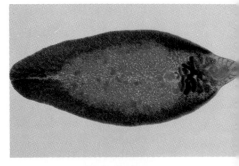

Figure 31–2. The adult liver fluke (above) lives in the digestive system of sheep and other vertebrates. It releases eggs that develop into adults through a complex life cycle involving two hosts (left).

If the eggs fall into water, immature flukes called larvae develop. The microscopic larvae must next enter the body of a certain species of snail. Fluke larvae die if they do not find the right type of snail soon enough. Once inside the host snail, the larvae remain only long enough to reproduce asexually. The many new parasites that result then leave the snail.

After swimming about, a larva attaches itself to a blade of grass near the water. It then forms a thick-walled structure around itself, called a **cyst.** The fluke remains inactive inside the cyst until it is taken into the digestive system of a sheep grazing on the grass. Inside the body of the sheep, the cyst dissolves and the larva develops into an adult fluke.

31.3 Tapeworms

Tapeworms, another group of common parasitic flatworms, belong to the class Cestoda. *Tapeworms live in the intestines of vertebrates where they feed by absorbing food that has already been digested by their host.*

The head of a tapeworm is called the **scolex** (SKOH lehks). Hooks and four suckers on the scolex attach the tapeworm to the intestinal wall of the host. Tapeworms have no mouth or digestive system of their own.

Most tapeworms measure about 1 meter (3 ft.) in length, but some grow much longer. The tapeworm's long ribbonlike

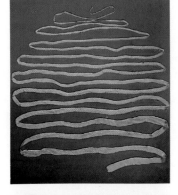

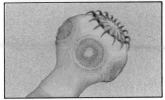

Figure 31–3. The tapeworm (top) clings to the intestinal wall of its host by means of hooks and suckers on its head or scolex (bottom).

▶ The Body Plans of Worms

Sponges and coelenterates have two embryonic layers of cells—an ectoderm and an endoderm. Worms and other bilaterally symmetrical animals have a third layer in between, called the **mesoderm.** These three layers of cells are called the **germ layers.**

All worms have three germ layers, but various types of worms differ greatly in their general body plan. A major difference is the presence or absence of a **coelom** (SEE luhm), a body cavity in the mesoderm layer.

Of the three major types of worms, flatworms have the simplest structure. A

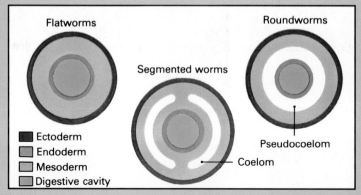

Ectoderm
Endoderm
Mesoderm
Digestive cavity

flatworm's tissue layers are packed together without any cavity other than its digestive cavity. Flatworms and other animals without a coelom are called *acoelomatic* (ay see luh MAT ihk).

Segmented worms have a true coelom. In a true

coelom, the stomach, heart, and other internal organs within it are suspended by double layers of membrane.

Roundworms have a false body cavity called a **pseudocoelom** (SOO doh SEE luhm). The organs within a pseudocoelom are

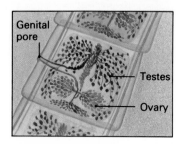

Genital pore

Testes

Ovary

Figure 31–4. The dark areas on a tapeworm's segments, or proglottids, are its reproductive structures.

body is divided into many segments called **proglottids** (proh GLAHT ihdz). Tapeworms grow by adding proglottids at the anterior end.

Each proglottid is a hermaphroditic reproductive unit in which the male structures develop first. As the proglottid matures, the male organs decline and female organs develop. When two or more tapeworms are present in a host, they may cross-fertilize. Tapeworms also self-fertilize by folding in upon themselves. In this case sperm is usually transferred from a forward, sperm-producing segment to a more mature proglottid farther back. One tapeworm proglottid may contain as many as 100,000 eggs.

A person may ingest beef tapeworms by eating raw or undercooked beef containing tapeworm cysts. The undeveloped worm inside the cyst is called a **bladderworm.** The digestive juices of the host dissolve the cysts, freeing the young worms. The immature tapeworms attach themselves to the wall of the intestine and begin to add proglottids. Proglottids containing not suspended by membrane layers, and the intestine lacks a lining.

To understand how the germ layers and coelom form, you must first know about the earliest stages in the development of an animal. Development begins when the sperm penetrates the egg and fertilizes it. The fertilized egg, called a **zygote,** divides into two halves in a process called **cleavage.** Repeated cleavage forms a ball of cells, the **blastula** (BLAS choo luh). This sphere has a fluid-filled cavity in the center. Because the cells grow very little as they divide, the blastula is about the same size as the original egg cell.

Next the blastula begins to fold in upon itself. As the cells of the blastula move inward, they form a cup-shaped arrangement, the **gastrula** (GAS troo luh). This process, called *gastrulation,* can be compared to what happens when you press the surface of an underinflated balloon. The surface moves inward and forms a dent.

The illustration below shows that in worms the mesoderm forms between the ectoderm and the endoderm. The coelom results from the splitting of the solid mesoderm. Each of the germ layers gives rise to different organ structures in the fully developed animal.

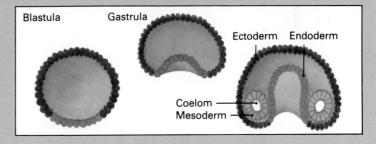

Blastula   Gastrula   Ectoderm   Endoderm   Coelom   Mesoderm

fertilized eggs break off and pass out of the host in the feces. Cattle eat grass contaminated by feces and take in the eggs, which hatch in the intestines. The worms then travel through the blood to the muscles. There the bladderworms form cysts in the muscles, allowing the cycle to begin again. Other kinds of human tapeworms use fish and small crustaceans as intermediate hosts. Tapeworms of all kinds compete with the body for nutrition, and the host is often fatigued. In serious infections, death may occur.

Reviewing the Section

1. What are the major characteristics of flatworms?
2. How does a planarian use its pharynx to feed?
3. What are the two hosts in the life cycle of a liver fluke?
4. How do tapeworms self-fertilize?
5. What is a coelom?

Roundworms make up the phylum Nematoda, so they are also called *nematodes*. Like flukes and tapeworms, most roundworms are parasites. Almost all species of plants and animals are affected by one of the 12,000 species of roundworms.

One shovelful of garden soil may contain over 1 million nematodes. Roundworms feed on plants by sucking the juices from them. Growers of fruit trees, strawberries, vegetables, and cotton suffer annual financial losses due to roundworms. Humans are hosts to about 50 species of roundworms. In fact, more than a third of the world's population suffers from diseases caused by roundworms. These diseases are most common in areas of Africa, Asia, and South America where sanitation is poor. Pinworms, hookworms, and intestinal roundworms are common human parasites.

Nematodes have tubular bodies covered by a tough cuticle and tapered at both ends. A fluid-filled *pseudocoelom* provides a structure against which the worm's longitudinal muscles can contract. The absence of circular muscles gives roundworms their characteristic thrashing or whipping motion in water.

The roundworm's digestive system consists of a tube called the *alimentary canal,* which is open at both ends. The anterior opening is the mouth. The posterior opening is the **anus,** through which solid wastes leave the body. Liquid wastes are collected by a system of tubes and are expelled through an *excretory pore* in the worm's posterior end.

Most roundworm species have separate male and female sexes. The females, which produce thousands of eggs, are usually larger than the males. The male guinea worm, for example, is about 2.5 cm (1 in.) long, while the female is 600 to 1,200 cm (240 to 480 in.) long. In the male reproductive system, sperm passes from the testis into the **cloaca** (kloh AY kuh), a common chamber into which digestive, reproductive, and excretory systems empty. During mating the sperm leaves the cloaca and enters the female's reproductive opening. Sperm is then stored in the female's body and used to fertilize eggs when they mature. Young roundworms develop in the body of the female or, in some species, outside of the female's body.

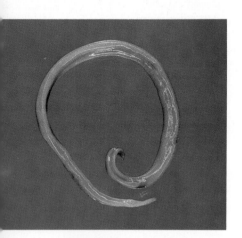

Figure 31–5. Ascaris, or intestinal roundworm, is thought to infect more than 650 million people worldwide.

31.4 *Ascaris*

One of the largest nematodes that lives in humans is the *Ascaris* (AS kuh rihs), or intestinal roundworm. These worms grow to 30 cm (12 in.) in length. *Ascaris* also lives in the intestines of pigs and horses. Usually an intestinal roundworm does not cause serious health problems, but when many worms are present they may block the intestine and cause the death of the host.

The life cycle of *Ascaris* is typical of many roundworm species. The cycle begins when a human or other host eats vegetables grown in soil containing the eggs. The eggs hatch in the intestine of the host. The larvae bore through the intestinal wall, enter the bloodstream, and are carried to the lungs. The larvae are then coughed up into the mouth, swallowed, and returned to the small intestine. There they develop into mature adults and reproduce. The fertilized eggs leave the host's body in the feces and may be picked up by other hosts. *Ascaris* eggs are well protected by tough shells that allow them to survive outside a host for as long as five years.

31.5 Hookworms

The hookworm gets its name from its way of hooking onto the small intestine of its host. The worm feeds by sucking blood. This process can greatly reduce the number of red blood cells in the host's bloodstream, causing severe anemia. Hookworms can also damage the host's small intestine.

The hookworm is a serious problem in warm, moist areas throughout the world, including some parts of the United States. The problem is most acute in areas where people walk barefoot in contaminated soil. Hookworm larvae develop in the soil and enter a human host through microscopic cracks in the soles of the feet. Like *Ascaris*, hookworms travel through the circulatory system to the lungs. There they are coughed up and swallowed, thus entering the digestive system where they complete their life cycle.

31.6 Trichina

The trichina (trih KY nuh) worm lives throughout the world and ranges from 1.5 to 4 mm (0.06 to .15 in.) in length. The trichina worm causes a serious disease called *trichinosis* (trihk uh NOH sihs) that is transmitted primarily through eating raw or undercooked pork.

The trichina life cycle begins with sexual reproduction in a host mammal such as a pig. The female trichina hatches the fertilized eggs in her body. The larvae are then deposited in the lining of the intestine of the host. The larvae work their way into the bloodstream where they travel to all parts of the body. The worms eventually lodge in muscle tissue where they develop cysts and cause cramps.

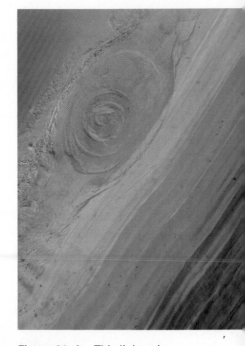

Figure 31–6. This light microscope image, enlarged 150 times, shows a trichina encysted in the muscle tissues of a pig.

Spotlight on Biologists: Libbie Henrietta Hyman

Born: Des Moines, Iowa, 1888
Died: 1969
Degree: Ph.D., University of Chicago

American zoologist Libbie Hyman gained worldwide notoriety before her death as a researcher of invertebrates. During her career she also wrote many textbooks and reference works on invertebrate and vertebrate zoology that are still used in classrooms.

The author of more than 145 articles for professional journals, Hyman is most famous for *The Invertebrates,* a 10-volume work on invertebrate zoology. Hyman wrote and illustrated 6 volumes of this

study before her death, and the text was finished by scientists at Oregon State University. The collection remains one of the most comprehensive of its kind in English.

Hyman first became fascinated with invertebrates while studying at the University of Chicago. After graduation she continued to work at the university for 16 years as a research assistant, pursuing her longtime special interest in planarians.

In 1937 Hyman joined the American Museum of Natural History in New York City as a research associate. There she pursued her study of such invertebrates as jellyfishes,

corals, and microscopic organisms.

Hyman was the first woman to receive several distinguished science awards. These included the Daniel Giraud Medal from the National Academy of Science and gold medals from the Linnean Society of London and the American Museum of Natural History.

The life cycle of the trichina continues if humans or animals eat muscle tissue containing the cysts. A single gram of raw pork meat may have as many as 3,000 trichina cysts. Cooking pork thoroughly can prevent the transmission of these cysts. The heat of cooking destroys trichina worms by destroying the cysts in the tissue. For this reason, you should always cook pork thoroughly and avoid tasting the meat while it is cooking. Trichinosis is uncommon in the United States because farmers cook meat scraps before feeding them to hogs.

Reviewing the Section

1. What are two traits all roundworms share?
2. What are the stages of the life cycle of *Ascaris?*
3. Why is it important to cook pork thoroughly?

Segmented Worms

Section Objectives

- *List* the major characteristics of segmented worms.
- *Identify* the main structures of an earthworm.
- *Explain* why earthworms can live only in moist soil.
- *Compare* earthworms and leeches.
- *Describe* how giant tubeworms that lack mouths and digestive systems obtain nourishment.

Worms that make up the phylum Annelida have bodies that are divided into a series of segments, which are often visible as rings on the outside of the body. These worms are also called *annelids,* which means "little rings." The 9,000 species of annelids include earthworms, leeches, and a variety of marine worms.

Annelids are the most complex of all worms. Like flatworms and roundworms, annelids have three tissue layers and a bilaterally symmetrical body plan. ***Annelids differ from the other worms in having a true coelom, giving them a "tube-within-a-tube" body plan.*** Two sets of muscles, circular and longitudinal, allow annelids to move more efficiently than other worms. Annelids also have more complex circulatory, respiratory, and nervous systems than other worms.

31.7 Earthworms

Earthworms, which belong to the class Oligochaeta (ahl uh goh KEET uh), live in soils all over the world. Earthworms vary in size from a few centimeters to 3.3 m (11 ft.) long. A common North American species has a dark dorsal surface and light ventral surface. Most of the worm's 100 to 150 segments are identical, except for the pointed anterior and posterior ends. Between segments 35 and 37 lies a swelling called the **clitellum,** which plays an important role in reproduction.

Earthworms have pairs of bristles, called **setae** (SEET ee), on each body segment except the first and last. An earthworm moves by anchoring the setae on its posterior segments and then contracting the circular muscles in front of the anchored segments. These contractions extend the body forward.

setae (singular, *seta*)

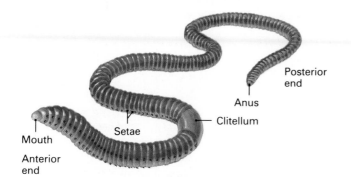

Figure 31–7. This diagram illustrates the external structures of the earthworm.

Posterior end

Anus

Mouth

Setae

Clitellum

Anterior end

Figure 31–8. Setae are clearly visible in this closeup view of a section of an earthworm.

Figure 31–8. Setae are clearly visible in this closeup view of a section of an earthworm.

Figure 31–9. The specialized organs of the earthworm's nervous, digestive, circulatory, and reproductive systems occupy the segments at its anterior end.

Digestion in Earthworms An earthworm feeds by taking in soil with its muscular pharynx. The soil moves down a tube called the **esophagus** and then enters a storage chamber called the **crop.** From the crop, the soil moves to another chamber called the **gizzard.** Here, the grinding together of soil particles swallowed by the earthworm crushes pieces of organic matter. The food next moves into the intestine, which extends to the posterior end of the worm. Folds in the wall of the intestine increase the surface area where absorption of digested food takes place. Solid wastes pass out of the body through the anus.

Earthworms are very valuable to gardeners and farmers. As the worms eat their way through the soil, they break up soil clumps and add nutrients. Earthworms break down organic material faster than normal bacterial decomposition does. In fact, an earthworm can produce its weight in fertile soil every 24 hours.

Circulation in Earthworms Unlike flatworms and roundworms, earthworms have a *closed circulatory system,* as shown in Figure 31–9. In a closed circulatory system, the blood circulates through a series of vessels. Dorsal and ventral blood vessels run the length of the worm's body. The blood absorbs molecules and carries them through the dorsal vessel to five pairs of muscular pumping tubes, or "hearts." These "hearts" pump blood into the main ventral blood vessel. Smaller blood vessels carry the blood to all parts of the body.

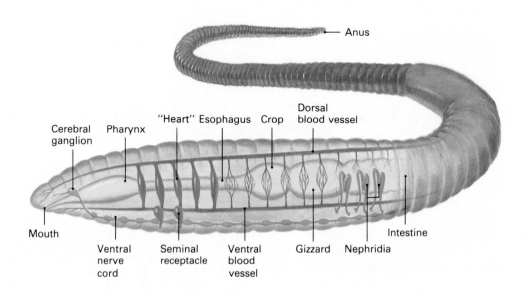

Respiration and Excretion in Earthworms Earthworms take in oxygen and give off carbon dioxide by diffusion through the skin. Because diffusion can occur only across a moist membrane, an earthworm must remain in an environment that is neither too wet nor too dry. Earthworms can drown in soil that is saturated with water because there is not enough oxygen. During dry periods, earthworms dig deep into the soil in search of moisture.

Earthworms eliminate liquid wastes through ciliated tubes called **nephridia** (neh FRIHD ee uh). The beating of cilia draws fluid from the coelom into a funnel-shaped opening in the nephridia. As the fluid passes through the tubules, needed water is reabsorbed by tiny blood vessels. Waste materials then pass out of the body through a pore in the skin. Each body segment has a pair of nephridia.

Nervous Control in Earthworms An earthworm can respond rapidly to changes in its environment because of a concentration of nerve cells called a *cerebral ganglion,* near the worm's anterior end. The cerebral ganglion is connected to the rest of the body by a ventral nerve cord that extends the entire length of the animal. A ganglion connects each body segment to the ventral nerve cord.

The earthworm has no external eyes or ears, but receptors in the skin enable the worm to react to light, sound, and chemicals. Earthworms are active mainly at night and will move away from bright light. However, the light-sensitive cells of earthworms do not respond to red light. For this reason, earthworms to be used as fishing bait may be most easily dug up at night if red light is used for illumination.

Reproduction in Earthworms Like planarians, earthworms are hermaphrodites. The female structures are located towards the anterior portion of the earthworm, and the male structures towards the posterior. Fertilization occurs when two worms exchange sperm. A mucous secretion from the clitellum holds the two earthworms together while they mate. The sperm each worm receives is stored in a *seminal receptacle* until just before the eggs are laid.

Two or three days after mating, the earthworm produces an external mucous case formed of sticky secretions from the clitellum. Muscular contractions push the case along the body. Mature eggs and sperm held in the seminal receptacle enter the case as it passes over the body of the worm. The case then seals, forming a coat that protects the fertilized eggs until they hatch.

Q/A

Q: *Why does "the early bird catch the worm"?*

A: Worms come to the surface of the soil at night when there is no significant light. The "early bird" happens to catch the worms before they burrow back into the soil.

Figure 31–10. After mating, earthworms produce a mucous case that is moved along the body to collect sperm and eggs. The case slips off the anterior end of the earthworm and seals to protect the developing eggs.

Polychaetes (PAHL ih keets) are a class of segmented worms that live mainly in salt water. Unlike earthworms, polychaetes have tentacles, antennae, and specialized mouthparts. Each polychaete segment has a pair of appendages called **parapodia** (pa ruh POHD ee uh) that assist in movement.

Some polychaetes are mobile and feed on small animals. Others live in elaborate tubes constructed from sand or mud on the ocean floor.

Giant tube worms that grow to be 1.5 to 3 m (5 to 10 ft.) in length were discovered in 1977. These strange worms, named *Riftia pachyptila* Jones, have no eyes, mouths, or digestive tracts. The worms belong to their own phylum, the Pogonophora (poh guh NAHF uh ruh).

The giant tube worms live in an area called the Galapagos Rift. Here water is heated by volcanic rocks, which create a warm "oasis" in the near-freezing waters.

The Galapagos Rift is a highly unusual environment. It is deep beneath the surface of the ocean where sunlight does not penetrate. Because photosynthesis cannot occur, organisms in this environment need another way of producing energy. The *Riftia* worms get energy from bacteria that live inside their bodies. The bacteria produce energy through chemical reactions involving sulfur compounds that the worms absorb from sea water.

31.8 Leeches

Another common segmented worm is the leech, which belongs to the class Hirudinea (hihr yuh DIHN ee uh). Most leeches live in fresh water, although some may be found in salt water or in moist soil. Unlike other kinds of segmented worms, leeches do not have setae.

Leeches feed on the blood of other animals using suckers found on both ends of their flat, tapered bodies. A leech uses its posterior sucker to attach to its host and its anterior sucker to extract blood. Inside the anterior sucker is a mouth with three sharp teeth, which make a small incision in the host. The leech secretes a substance that prevents the host's blood from clotting.

Reviewing the Section

1. How does the body plan of earthworms differ from that of flatworms and roundworms?
2. Why do earthworms come to the surface when it rains?
3. Compare the physical characteristics of earthworms and leeches.

Investigation 31: Ecological Significance of Earthworms

Purpose
To compare the growth of plants in an eco-system that contains earthworms with one that does not

Materials
Sphagnum moss; potting soil; two 15-cm pots; six small house plants such as coleus, bryophyllum, or spider plant; earthworms; gauze or wire mesh

Procedure
1. Prepare a mixture of one-third sphagnum moss and two-thirds potting soil.
2. Plant three plants in each pot.
3. To one pot add three to six medium-sized earthworms. Place gauze or wire mesh over the pot as shown below to prevent the earthworms from climbing out.

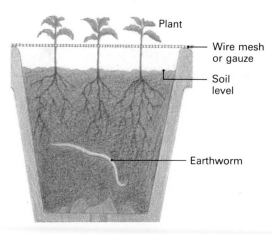

Plant

Wire mesh or gauze

Soil level

Earthworm

4. Do not add earthworms to the second pot.
5. Both pots should receive adequate water and light. The temperature should be kept relatively stable. *Why is it important that water, light, and temperature be similar for both pots?*
6. Observe the general growth patterns of each set of plants each day for two weeks. Record your observations in your notebook.

Analyses and Conclusions
1. Explain differences in the growth patterns of the two sets of plants.
2. How would overwatering affect the plants? How would it affect the earthworms?
3. Why is the type of plant selected important in this experiment?

Going Further
- Use the same procedure suggested in this experiment but change the type of pot. (Use a plastic pot if you first used a clay pot, or vice versa.)
- Use three pots and vary the type of soil in each. Use potting soil, humus, and sphagnum moss. Place earthworms and similar plants in each pot and observe the various growth rates. Prepare another group of three pots exactly like the first set but do not add earthworms. These pots will serve as controls.
- Investigate the effect of earthworms on the pH of the soil. To do this, first test the pH of the three types of soil named above. Then place earthworms in each pot and test daily for changes in pH. How would pH changes affect the growth of the plants?

Chapter 31 Review

Summary

The three main worm phyla discussed in this chapter are all bilaterally symmetrical and have three tissue layers. However, they differ significantly in their body plans. Acoelomatic flatworms have a digestive cavity with one opening. Roundworms, or nematodes, have a pseudocoelom and a tubular digestive system that is open at both ends. Segmented worms have a true coelom and efficient digestive, circulatory, and respiratory systems.

Flatworms include planarians, which are free-living, and flukes and tapeworms, which are parasites. The life cycle of the liver fluke is typical of many parasitic worms. The adult develops and reproduces inside a host. The eggs hatch into larvae outside the host's body. They must be picked up by another host for the cycle to continue. After reproducing asexually, a larva leaves the host and forms a cyst, which may be picked up by a new host.

Common parasitic roundworms that infest humans and other mammals include the *Ascaris* worm and the hookworm. Transmission of parasites to human hosts can largely be prevented by improved sanitation and thorough cooking of certain foods.

Segmented worms, or annelids, include the common earthworms and blood-sucking leeches. A wide variety of segmented worms also live in fresh water and salt water.

BioTerms

anus (462)
bladderworm (460)
blastula (461)
cleavage (461)
clitellum (465)
cloaca (462)
coelom (460)
crop (466)
cuticle (458)

cyst (459)
esophagus (466)
feces (458)
flame cell (458)
ganglion (457)
gastrula (461)
germ layer (460)
gizzard (466)
mesoderm (460)

nephridia (467)
parapodia (468)
pharynx (458)
proglottid (460)
pseudocoelom (460)
scolex (459)
setae (465)
zygote (461)

BioQuiz   (Write all answers on a separate sheet of paper.)

I. Completion

1. An earthworm crushes particles of food in its _____ .
2. _____ can be contracted from eating undercooked pork.
3. The secondary host in the life cycle of a liver fluke is a _____ .
4. _____ enable planarians to expel liquid wastes.

II. Modified True and False

Mark each statement TRUE or FALSE. If false, change the underlined term to make the statement true.

5. Leeches have suckers at both their anterior and posterior ends.
6. Roundworms have two body openings.
7. Tapeworms use their posterior suckers to suck blood.

III. Multiple Choice

8. Planarians use their eyespots to a) see.
   b) navigate. c) respond to light.
   d) respond to chemicals.
9. Gastrulation occurs when a) sperm
   fertilizes the egg. b) the mesoderm
   splits in two. c) the egg divides
   repeatedly. d) the blastula cells fold
   inward.
10. Gastrulation occurs during a) blood
    circulation. b) digestion. c) respiration.
    d) embryonic development.
11. The type of symmetry shown by worms
    is a) circular. b) bilateral. c) radial.
    d) amorphous.
12. Cilia help planarians a) move.
    b) reproduce. c) communicate.
    d) digest food.
13. Earthworms' nephridia a) eliminate
    liquid wastes. b) allow movement.
    c) diffuse oxygen. d) transmit nerve
    impulses.

IV. Essay

14. How do giant *Riftia* worms get energy?
15. What role does the clitellum play in the
    reproduction of earthworms?
16. Why does the tapeworm not need an
    extensive digestive system?
17. What are the effects of ingesting trichina
    worms?
18. Through which parts of the body does
    the hookworm travel?
19. How are the body plans of flatworms,
    roundworms, and segmented worms the
    same and how are they different?
20. Why must earthworms live in a moist
    environment?

Applying and Extending Concepts

1. A tapeworm in a forest might inhabit as
   hosts a wolf and a moose. Draw and label
   a diagram showing this tapeworm's life
   cycle.
2. Some religious groups enforce strict dietary laws among their members. Jews and
   Muslims, for example, do not as a rule eat
   pork. What might this suggest about the
   among members of those groups?
3. As a species evolves over time, it becomes
   better adapted. In other words, the species
   develops bodily structures and systems
   that allow it to thrive in the environments
   it inhabits. A tapeworm proglottid develops male structures first and female structures later. Why is this characteristic of
   proglottids adaptive?
4. At one time, "leeching" was a common
   medical practice. Write a brief report explaining when leeches were used by doctors, why they were used, and if they are
   used any more.
5. Design an experiment to test whether the
   presence of earthworms speeds up the
   decay of organic matter in soil.
6. People who are infested by *Ascaris* worms
   frequently develop pneumonia. Suggest
   an explanation for this fact.

Related Readings

Blonston, G. "To Build a Worm." *Science 84*
5 (March 1984): 63–70. This article discusses the significance of recent studies of
the process of embryonic development in
the nematode.

Gould, J. "The Importance of Trifles." *Natural History* 91 (April 1982): 16. This
discussion of Darwin's 1881 book on
earthworms emphasizes his process of historical reasoning.

32

Mollusks and Echinoderms

A starfish opening a clam

Introduction

This chapter deals with two phyla of animals that are usually found in a marine environment. The clam in the photo above belongs to the phylum Mollusca. Oysters, snails, slugs, scallops, and squid are also mollusks. One member of the phylum Mollusca, the octopus, is probably the most intelligent of all invertebrates. Mollusks vary greatly in size and habitat, but are linked together as a phylum by common characteristics in their body plans.

The starfish shown above belongs to the phylum Echinodermata, which includes sea urchins, sand dollars, and sea cucumbers. Echinoderms are marine animals that have never adapted to life on land or in fresh water. They are among the few invertebrates with an internal skeleton.

Mollusks

Section Objectives

- *Explain* how mollusks are classified.
- *Describe* how clams feed.
- *Compare* respiration in land snails and sea snails.
- *Identify* features of the octopus that suit it to a predatory life.

The term *mollusk* means "soft." It describes the bodies of organisms in the phylum Mollusca. Besides being soft, however, the 47,000 species of mollusks differ greatly. They range in size from tiny snails 5 mm (0.2 in.) long to giant squid 20 m (66 ft.) long. Mollusks live on ocean bottoms and in fresh water, as well as on land.

Many species of mollusks are protected by one or more shells. Imprints in rock made by the shells of prehistoric mollusks remain millions of years after the animals themselves died. As a result, the fossil record of mollusks is particularly clear. The fossils indicate that some classes of mollusks existed 500 million years ago. For centuries people have used the bodies of mollusks as food, and their shells as currency, utensils, and ornaments.

32.1 Characteristics of Mollusks

Despite their diversity, all mollusks share certain characteristics. Their body plan is basically bilaterally symmetrical, and they have a true coelom. *Mollusks have three distinct body parts: the head-foot, the visceral mass, and the mantle.* The **head-foot** area contains the mouth and the sensory organs, as well as the foot and other motor organs. Clams and some other mollusks use the foot for burrowing into the sand. In squid and related species, the head-foot forms a circle of armlike tentacles. Some mollusks have a toothed organ called a **radula** (RAJ oo luh). A mollusk uses its radula to tear or scrape loose bits of plant or animal matter and to transfer the food to the digestive tract. The **visceral mass** contains the digestive, excretory, and reproductive organs of the mollusk. The **mantle** is a thin membrane that surrounds the visceral mass. The mantle secretes the shell. Between the mantle and the visceral mass is an area called the *mantle cavity*. The respiratory organs are located in the mantle cavity.

Many mollusks pass through a similar larval stage of development. The mollusk larva, called a **trochophore** (TRAHK uh fawr), is pear-shaped with a band of cilia around its middle. Because a trochophore swims about freely, it helps disperse the species. This opportunity is especially important for mollusks, which are encumbered with heavy shells as adults. Some marine annelids also pass through a trochophore larval stage. This similarity has led scientists to conclude that annelids and mollusks are closely related groups.

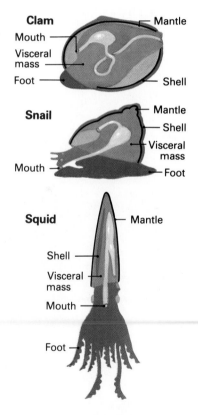

Figure 32-1. The diagram above shows how the three body parts found in all mollusks differ among the three major groups of mollusks.

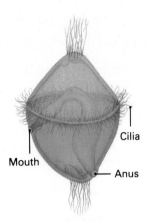

Figure 32–2. The larval stage of a mollusk, called a trochophore, is shaped like a top. It has cilia that help it move and also help guide food to its mouth.

Figure 32–3. The siphons are clearly visible in this photograph of Florida coquinas, colorful clams that measure less than 2.5 cm (1 in.) in length.

Mollusks are classified according to the kind of shell they have. Of the seven classes in the phylum Mollusca, the three major classes are two-shelled mollusks, such as clams, scallops, and oysters; one-shelled mollusks, such as snails; and head-footed mollusks, such as octopuses, squids, and cuttlefish.

32.2 Mollusks with Two Shells

Clams, oysters, and scallops belong to a class of mollusks that have two shells hinged together. These shells are called *valves.* The animals themselves are referred to as **bivalves.** A characteristic of bivalves is the shape of their muscular foot. For this reason, the class is called *Pelecypoda* (peh luh SIHP uh duh), which means "hatchet-foot."

Clams Most clams live in salt water, buried in the sand or mud at the sea bottom. Some clams, however, live in fresh water. Clams range in size from the tiny *Condylocardia* 0.1 mm (0.004 in.) across to South Pacific giants 1.2 m (4 ft.) across.

A clam's shell is usually off-white and consists of three layers: a tough horny outer layer, a smooth shiny inner layer called the *pearly layer,* and a middle *prismatic layer* made up of calcium carbonate crystals. The two shells are held together by ligaments. Two *adductor muscles* open and close the shell.

Like all bivalves, clams are entirely encased in their shells. As a result, they have no real head and no radula. Sense organs are poorly developed, though sensory cells along the edge of the mantle do respond to light and touch. Two long pairs of nerve cords connect three sets of nerve cells. In other organisms, these nerve cells are located in the head, but in clams they are found above the mouth, in the digestive system, and in the foot.

Clams obtain both food and oxygen from the water that flows through their bodies. Clams usually remain buried in the sand with their two valves slightly open and two siphons, or tubes, projecting into the water. Water enters the clam through the **incurrent siphon.** Cilia move the water across respiratory organs, called **gills,** in the mantle cavity. Gills have a large surface area and an abundant supply of blood to allow for the exchange of gases. Water drawn into the clam by the incurrent siphon has more oxygen than the blood supply of the gills has. For this reason oxygen diffuses from the water to the blood, and carbon dioxide diffuses from the blood to the water. The clam then expels the water through the **excurrent siphon.**

Clams are *filter-feeders.* They live on microscopic organic matter carried in the water that flows through their siphons. Mucus on the gills traps the food matter, and cilia push the

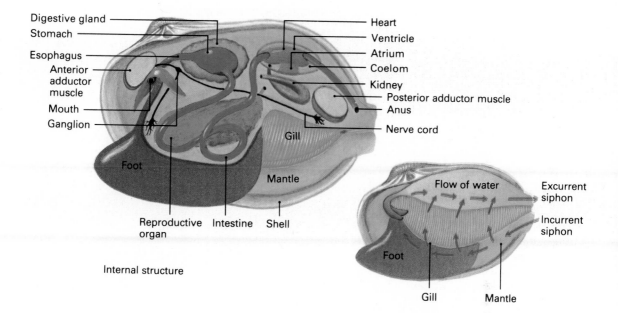

Internal structure

Digestive gland
Stomach
Esophagus
Anterior adductor muscle
Mouth
Ganglion
Foot
Reproductive organ
Intestine
Shell
Mantle
Gill

Heart
Ventricle
Atrium
Coelom
Kidney
Posterior adductor muscle
Anus
Nerve cord

Flow of water
Excurrent siphon
Incurrent siphon
Foot
Gill
Mantle

food-laden mucus on to the clam's mouth. From there the food passes into the stomach. Undigested food particles leave the clam through the anus.

Clams have an *open circulatory system.* This means that the blood flows through large open spaces, or *sinuses,* rather than through a system of blood vessels. A three-chambered heart pumps the blood through the clam.

Most clam species have separate sexes. The sperm and eggs are shed into the water, where fertilization takes place. The fertilized egg becomes a trochophore larva that settles on the bottom and develops into an adult clam.

Other Bivalves Scallops live in all oceans, mainly in shallow waters but also in ocean depths. Scallops range in diameter from 2.5 cm (1 in.) to 15 cm (6 in.). They have a fan-shaped shell that may be smooth or sculptured. Scallops may be purple, red, orange, yellow, or white. Scallops have a single large adductor muscle. When their valves are open, tentacles hang like a curtain between the shells. The scallop's eyes are located along the edge of the mantle. One scallop may have as many as 100 eyes. Although the eyes cannot focus, they can distinguish between light and dark and can sense passing shadows. Scallops propel themselves by opening and closing their valves and expelling water in bursts.

Unlike scallops, oysters cannot move about. Early in its life, an oyster permanently attaches its flat lower shell to a hard

Figure 32–4. All the clam's organs are contained within its protective shell. Water is drawn across the gills (inset), which extract oxygen from the water.

Figure 32–5. Scallops have two rows of sightless blue eyes located at the edge of the mantle. Note the fringe of tentacles also located at the mantle's edge.

Highlight on Careers: Aquaculturist

General Description

The job of an aquaculturist is to cultivate food animals that live in water, such as fish or mollusks. Most aquaculturists in the United States raise trout, catfish, oysters, clams, and crayfish. Most fish and mollusk farms are small businesses owned and operated by a single family. Some large companies raise their own clams in hatcheries and also process the clams before marketing them.

Aquaculturists generally raise clams in seawater beds leased from a state or township. Workers place young "seed" clams on protected rafts. When the clams are larger, workers transfer them to the ocean bottom and cover the clams with netting to protect them from predators.

The aquaculture industry in the United States today is small. It is bound to grow, however, as Americans attempt to reduce the vast amount of seafood imported into this country.

Career Requirements

There are no special requirements for a career as an aquaculturist. Most people enter the field by acquiring on-the-job experience at a fish farm or clam bed, then going into business for themselves.

Professional or technical jobs in aquaculture research generally require a bachelor's degree in biology. Courses in fish and

mollusk culture are available at community, junior, or four-year colleges in many parts of the United States.

For Additional Information
U.S. Aquaculture
 Federation
Box 276
Lacey Spring, VA 22833

---

surface. The outer shell is rough in texture, while the inner surface of the shell is smooth and white. If an irritant such as a grain of sand enters an oyster shell, the oyster protects itself by covering the foreign matter with several layers of calcium carbonate. This process forms a pearl.

32.3 Mollusks with One Shell

The largest class of mollusks is the *Gastropoda* (gas TRAHP uh duh), a name that means "belly-footed." The 37,500 or more species include snails and slugs that live in water and on land. Most gastropods are **univalves**—that is, they have only one shell. The snail's protective shell is usually coiled, which allows the long pointed body of the snail to enclose itself in a compact form. Slugs have no outer shell. Some slugs have an internal shell, which is actually an external shell that has been reduced and covered over by the mantle.

A characteristic of gastropods is **torsion,** or a twisting of the body, which occurs during larval development. Before torsion, the body plan of a snail is bilaterally symmetrical. One half of the body grows faster than the other, causing the anus to curve forward while the head and foot remain in place. The organs on one side of the visceral mass twist over to the other side. The mantle cavity, which originally faced backwards, now faces forward. This arrangement of body parts allows the snail to draw its head into the shell, then plug the hole with its foot.

The nervous system of gastropods is more developed than that of bivalves. In gastropods, six pairs of ganglia are interconnected with nerve cords. Gastropods can detect light and shadows by means of eyes located on tentacles that extend from their heads. Like bivalves, gastropods have an open circulatory system. Unlike bivalves, most gastropods reproduce by fertilizing eggs internally.

Figure 32–6. Land snails emerge from their shells only when the air is moist and warm. The eyes, located at the tips of the two tentacles, are pulled back into the tentacles when touched.

Snails Most snails are less than 2.5 cm (1 in.) long. The Australian sea snail, however, sometimes grows a shell more than 60 cm (2 ft.) in length. On land the largest is the giant African snail, which grows a shell 20 cm (8 in.) in length.

Snails that live in water breathe through gills. They have one pair of tentacles. *Land snails breathe through a network of blood vessels in the mantle cavity.* To allow for the exchange of gases by diffusion, the blood tissues in the mantle cavity must be kept moist. Consequently, snails are most active at night or early morning when the air is moist. In dry weather, snails seal themselves inside their shells with a mucus plug in order to retain moisture. Some snails have a flat plate on the side of their foot called an **operculum** (oh PUHR kyoo luhm), which can be used like a trap door to close off the shell from the outside. Land snails have two pairs of tentacles.

Snails move by contracting their foot in a wavelike motion from back to front. They glide over a trail of mucus laid down by the front of the foot.

Most land snails feed on plants. They scrape off bits of plant matter using the radula. Snails help break down decayed matter, but too many snails in a garden can cause serious damage to plants.

Slugs Slugs can survive without shells because they live in moist environments. Like land snails, slugs that live on land respire through blood vessels in the mantle cavity. Most sea slugs breathe through gills. However, some sea slugs, called

Q/A

Q: *How fast is a "snail's pace"?*

A: Many snails move at a speed of less than 8 cm (3 in.) per minute. This means that if a snail did not stop to rest or eat it could travel 4.8 m (0.003 miles) per hour. A really speedy snail, such as *Haliotis*, however, might manage almost 48 m (0.03 miles) per hour.

Figure 32-7. Like other nudibranchs, this golden dirona lives in shallow sea water and feeds on sea anemones.

Q/A

Q: *Does jet propulsion allow the octopus and the squid to move only backwards?*

A: No. The siphon with which the octopus and the squid eject water is flexible, and can be turned in any direction. This enables the animals to move quickly in any direction.

nudibranchs (NOO duh branks), lack gills, shells, and mantle cavities. These slugs have decorative plumes on their backs that may function as respiratory organs.

32.4 Head-Foot Mollusks

The class Cephalopoda (sehf uh LAHP uh duh) includes the most evolutionarily advanced of all mollusks—the squid, octopus, cuttlefish, and nautilus. *Cephalopod* means "head-foot." All members of the class have a large well-developed head and a foot divided into many armlike tentacles. Only the nautilus has an outer shell. In the squid and the cuttlefish, the shell is reduced in size and overgrown by the mantle. The octopus has no shell at all.

Cephalopods have a closed circulatory system and a well-developed nervous system with many ganglia and a complex brain. The central mouth has jaws and a radula. It is surrounded by tentacles—up to 8 in the octopus, 10 in the squid, and 94 in the nautilus. Suckers on the tentacles help the animals to grasp their prey and to move along the ocean bottom.

All cephalopods are marine animals and live at all depths. Cephalopods are **predators**—that is, they kill and eat other animals, such as fishes, crabs, and bivalves.

Octopuses The octopus is highly specialized for its predatory way of life. Because the octopus is not enclosed by an outer shell, it has great freedom to move about in search of prey. The octopus moves rapidly by jet propulsion. By forcibly contracting the muscles of its mantle cavity, the octopus squirts out a jet of water through its siphon and speeds off in the opposite direction. The octopus also has special senses that help it locate prey. Its large eyes form images, but the octopus does not have stereoscopic vision. The suckers on its tentacles, more sensitive than human fingertips, contain special receptors that respond to chemicals in the water. The octopus uses its tentacles to reach into crevices for prey it cannot see.

The sexes are separate in the octopus, as they are in all cephalopods. The male octopus uses one of its tentacles, specialized for this function, to transfer sperm from its mantle cavity to the mantle cavity of the female. Later the female lays a mass of fertilized eggs encased in a gelatinous cover. The female **broods** the eggs—that is, she guards and cleans the eggs until they hatch.

Scientists have studied the behavior of the octopus extensively. By using a simple system of rewards and punishments, experimenters have easily conditioned these animals to pick up

Figure 32–8. With its tentacles extended, this octopus is propelling itself forward by jet propulsion.

certain objects and to ignore others. An unusual aspect of the behavior of the octopus is its coloration. Octopuses have sacs of pink, blue, or purple pigment just below the surface of the skin. An octopus pales before a predator, making itself seem larger than it is. Excited by the sight of a crab, however, an octopus becomes dappled with color. Scientists think this behavior may be a form of communication among octopuses.

Other Cephalopods   Squid range in size from about 1.5 cm (0.6 in.) to the giant 20-meter (66 ft.) species that weighs 3,360 kg (8,800 lbs.). Squid do not use their tentacles for crawling. They move by jetting out streams of water through siphons and using two finlike extensions of the mantle cavity for steering. Squid have been studied much less than octopuses due to the difficulty of maintaining them in captivity.

   The nautilus lives in the outermost chamber of a many-chambered coiled shell. A tube that runs from its visceral mass secretes a gas into all but the outermost chamber. By adjusting the amount of gas in the chambers, the nautilus can control the depth at which it floats. The cuttlefish can adjust its buoyancy in a similar fashion by controlling the amount of gas in its porous inner shell.

Figure 32–9.   Individual chambers are clearly visible in this cross-section of a chambered nautilus.

Reviewing the Section

1. How do cephalopods differ from other mollusks?
2. How do clams feed?
3. Name three features of the octopus that enable it to lead a predatory way of life.

Echinoderms

- *List* the major characteristics of echinoderms.
- *Describe* the water-vascular system of the starfish.
- *Name* the features of the sea cucumber that are typical of echinoderms.

The 6,000 living species of echinoderms (ih KY nuh durmz) include starfish, sea urchins, sand dollars, and sea cucumbers. Echinoderms live only in marine habitats. They are found in all oceans, both along coastal regions and at considerable depths.

The term *echinoderm* means "spiny-skinned" and refers to the spines of calcium found on these invertebrates. The spines are projections from an interior skeleton, or **endoskeleton,** that is covered by a thin layer of skin.

Echinoderms are considered the most advanced form of invertebrates and are thus classified closest to the vertebrates. In their embryonic development, echinoderms more closely resemble vertebrates than they do most other invertebrates.

Echinoderms begin life as bilaterally symmetrical creatures, but the left side of the embryo grows more than the right side. This results in an adult body plan that is radially symmetrical. A mature echinoderm has five or more separate extensions, called arms, around a central disc.

32.5 Starfish

Starfish range from 1 to 65 cm (0.4 to 25 in.) in diameter. Starfish, also known as sea stars, are in many ways typical echinoderms. Most species have five hollow arms, which extend from a central disc, but some species have up to 50 arms. The lower side of the arms are covered with rows of **tube feet,** which are hollow cylinders tipped with suckers. Each arm contains a pair of digestive glands.

The underside of the central disc is called the *oral surface*. The mouth of the starfish is located in the center of the oral surface, and the stomach lies just above the mouth. Starfish feed on mollusks, such as clams. The starfish clasps the clam and begins to pull it apart. When the starfish's arms get tired, it

Figure 32–10. Three examples of echinoderms are (left to right) the starfish, the sea urchin, and the sand dollar.

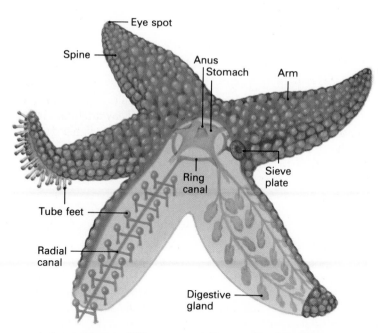

Labels in figure: Eye spot, Spine, Anus, Stomach, Arm, Ring canal, Sieve plate, Tube feet, Radial canal, Digestive gland

Figure 32–11. A distinctive feature of the starfish's internal anatomy is its water vascular system, which moves water through a ring canal into extensions in the arms of the starfish.

switches and uses different arms. Eventually the adductor muscles of the clam become tired, and the clam opens. The starfish then turns its stomach inside-out and slips it into the mantle cavity of the clam. Enzymes dissolve the clam's body inside its own shell, and the starfish sucks up the liquid. Starfish also feed on coral polyps and on other echinoderms.

The starfish has no brain, but coordinates its movement by means of a nerve ring that surrounds the mouth and extends down each arm. Light-sensitive eyespots are found at the end of each arm. The starfish breathes through skin gills that are protected by spines.

Starfish and other echinoderms have a unique method of locomotion. Their **water-vascular system** uses water pressure to create a ''skeleton'' for muscles to contract against. Water in the system flows through a *ring canal* and into *radial canals* that extend from each arm to rows of hollow tube feet. One end of each tube foot is a sucker, and the other end is a muscular sac. When the starfish contracts its sac muscles, water is forced into a tube foot. The sucker of the extended foot attaches to a surface. The arm muscles of the tube foot then contract, forcing water back into the sac at the base of the tube foot. This pulls the animal forward.

Most starfish reproduce sexually, depositing the sperm and eggs into the water. Sexes are usually separate, but a few species are hermaphroditic. The starfish may release as many as 2,500,000 eggs at one time. The female of some species brood their eggs and the young larvae.

Q/A

Q: *Are any echinoderms edible?*

A: Yes. The sea cucumber is dried and used in Chinese cooking. The sex organs of sea urchins are considered a delicacy in many parts of the world, eaten either raw or cooked.

▶ The Unique Sea Cucumber

The sea cucumber scarcely resembles other echinoderms. The animal has a soft cylindrical body that is dark in color. It is named for its warty skin, which resembles a cucumber. The sea cucumber's endoskeleton is reduced to microscopic spicules. Five rows of tube feet extend from the mouth at one end to the anus at the other. The animal uses short tentacle-like feet around the mouth for feeding and for burrowing.

The sea cucumber has an unusual defense mechanism. It can expel sticky threads from its anus that entrap its enemy. When irritated, some species can even eject their internal organs and then regenerate new ones.

The pearlfish has a peculiar relationship with the sea cucumber. This fish takes shelter inside the body of the sea cucumber. The fish leaves the sea cucumber at night to feed. When it returns, it first

pokes its head into the anus of the sea cucumber. Then it quickly turns so that it can be drawn tail first into the body of the sea cucumber.

Some starfish also reproduce asexually, by splitting in half and growing new parts. ***The starfish has a remarkable ability to regenerate new body parts.*** Most species can regenerate parts from a fragment that remains attached to a portion of the central disc. One species can grow a new body from a single arm.

32.6 Other Echinoderms

Not all echinoderms have five arms, but they all show some radial characteristics. Sand dollars and sea urchins have no arms but do have five rows of tube feet. Both animals have strong jaws and the spines characteristic of echinoderms.

The brittle star takes its name from its five easily broken arms. The brittle star will shed an arm to escape from a predator. The animal moves by bending its flexible arms in a snakelike movement.

Reviewing the Section

1. What structural features characterize echinoderms?
2. Why are sea cucumbers classified as echinoderms?
3. How does a starfish move?

Investigation 32: Starfish Behavior

Purpose
To investigate the competitive behavior of starfish

Materials
A 5- or 10-gallon aquarium, aquarium salt, gravel, aquatic plants, air pump, light source, three starfish, small- to medium-sized clams

Procedure
1. Set up an aquarium several days before receiving the shipment of starfish. *Why are the plants added to the aquarium?*
2. On the day of the experiment, test the temperature and salinity of the water. *What is the significance of temperature and salinity to this experiment?*
3. Place three starfish in the aquarium along with several clams to serve as their food source.

Analyses and Conclusions
1. Observe and record the manner in which the clams attempt to avoid the starfish and how the starfish pursue their prey.
2. What evidence of competition exists among the starfish? How is it resolved?

3. Why must the water be allowed to age several days before the starfish are placed in the aquarium?

Going Further
- A starfish has incredible regenerative abilities. A small portion of the organism can regenerate an entire specimen. Upon request, suppliers will provide specimens that have lost an arm so that you can observe the regenerative process.
- Place clams of different sizes in the aquarium. Note which are selected by the starfish. Explain this behavior.
- To investigate interspecies competition among echinoderms, place different species of starfish or brittlestars in the same aquarium. Research each species prior to your investigation so that you are familiar with their normal life patterns.

Chapter 32 Review

Summary

Mollusks are divided into three classes. These classes are bivalves, or pelecypods; univalves, or gastropods; and head-footed mollusks, or cephalopods. Bilateral symmetry has undergone extreme modification in the body plans of snails, squids, and related species. Mollusks have three body parts: the head-foot; the visceral mass; and the mantle, which secretes the shell. Most mollusks except land snails and slugs breathe by using gills.

Clams and other bivalves are filter-feeders. They take both food and oxygen out of the water that circulates through their bodies.

Gastropods include snails and slugs. Snails retreat into a coiled shell to protect themselves and to conserve moisture. Land snails breathe by using blood vessels in their mantle cavities. Slugs have no shell or only an internal shell.

Cephalopods are predators with well-developed nervous systems. The absence or reduction of a shell aids the great mobility of the octopus and the squid.

Unlike mollusks, echinoderms are never found on land or in fresh water. The phylum includes starfish, sea urchins, sand dollars, and sea cucumbers. Echinoderms have an endoskeleton and a radially symmetrical body. They move by using a water-vascular system in which water pressure is controlled by muscular action. Echinoderms are considered the most advanced form of invertebrates.

BioTerms

bivalve (**474**)
brood (**478**)
endoskeleton (**480**)
excurrent siphon (**474**)
gill (**474**)
head-foot (**473**)

incurrent siphon (**474**)
mantle (**473**)
operculum (**477**)
predator (**478**)
radula (**473**)
torsion (**477**)

trochophore (**473**)
tube foot (**480**)
univalve (**476**)
visceral mass (**473**)
water-vascular system (**481**)

BioQuiz (Answer all questions on a separate sheet of paper.)

I. Completion

1. The trochophore is the _____ form of many mollusks.
2. A _____ coordinates the movements of the starfish.
3. The body of a sand dollar shows _____ symmetry.
4. Starfish attach themselves to surfaces by means of _____.
5. Clams feed by trapping organic matter with _____ as water passes through their gills.

II. Modified True and False

Mark each statement TRUE or FALSE. If false, change the underlined term to make the statement true.

6. The scallop has image-forming eyes.
7. The blood of snails flows through sinuses.
8. Cuttlefish have an internal skeleton.
9. Scallops move by burrowing into the mud with a muscular foot.
10. The starfish can eject its internal organs and grow new ones.

III. Multiple Choice

11. All mollusks except bivalves use their radula for, a) defense. b) reproduction. c) feeding. d) respiration.
12. Squids move by a) jet propulsion. b) somersaulting. c) waving their tentacles. d) crawling on tentacles.
13. Slugs survive without shells because of a) a lack of enemies. b) powerful radula. c) a moist environment. d) well-developed gills.
14. Starfish use a water-vascular system a) to feed. b) to move. c) to breathe. d) to communicate.
15. The nautilus adjusts the gas in its shell chambers as a means of a) defense. b) movement. c) excretion. d) respiration.

IV. Essay

16. Why do biologists think that annelids and mollusks are closely related?
17. In what way do oysters benefit from forming pearls?
18. How is the bilaterally symmetrical body plan of a snail altered during its development?
19. How does a starfish feed?
20. How do land snails function without gills?

Applying and Extending Concepts

1. Starfish are especially abundant around coral reefs. What do you suppose is the reason for this?
2. Use your school or public library to research ''left-handed'' and ''right-handed'' snail shells. If possible, use actual snail shells to demonstrate to the class how to tell which shells are which.
3. Use your school or public library to find ways in which humans have used mollusks and echinoderms, other than as food and jewelry.
4. Oyster farmers once made a practice of taking all the starfish predators out of their oyster beds and chopping them into pieces. The farmers then threw the starfish pieces back into the water. Why do you think this practice was abandoned?
5. How do the feeding habits of clams and other bivalves increase the risk of contamination to seagulls and humans who eat bivalves from polluted waters?
6. At one time slugs were found only in areas such as rain-drenched mountainsides, but now slugs have migrated into lowland gardens. What must the gardens provide to make it possible for slugs to survive there?

Related Readings

Bowser, H. ''Indestructible!'' *Science Digest* 89 (May 1981): 46–51. This article describes how starfish are able to grow new arms.

''Evolution at a Snail's Pace.'' *Science News* 120 (November 7, 1981): 292. This article shows how fossils of snails provide an unbroken record of the evolution of snail species.

Laycock, G. ''Vanishing Naiads.'' *Audubon Magazine* 85 (January 1983): 26–28. This article explains why these once-common mollusks are now rare and describes how their shells are used to detect pollution.

Ware, P. D. ''Nautilus: Have Shell, Will Float.'' *Natural History* 91 (October 1982): 64–69. This article describes the structure and life processes of the nautilus.

BIO*TECH*
Probing the Ocean Depths

Invertebrates have been found almost everywhere on Earth, from the upper layers of the atmosphere to the deepest reaches of the ocean floor. Biologists who specialize in marine invertebrates have been studying invertebrate inhabitants of relatively shallow parts of the oceans for many years. However, technological advances in marine research have enabled them to begin to learn about life in the ocean depths.

Technological tools can overcome the crushing pressure of thousands of meters of water, the inky darkness of the ocean floor, and the uncertainties of undersea currents and ocean-floor terrain. A recent innovation called *Bathysnap* is a remote-control camera attached to a special module. When it is released, Bathysnap's ballast carries it

to the sea floor, where the device takes time-lapse photographs. When an acoustic signal is transmitted from its mother ship, Bathysnap releases its ballast. Floats then carry it to the surface, where researchers recover the device and the exposed film.

Marine researchers can study life in the ocean depths using new technological tools.

Bathysnap (below) photographs invertebrate life along the ocean floor (below left).

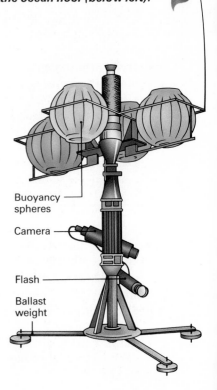

Buoyancy spheres

Camera

Flash

Ballast weight

Alvin (left) and Mantis (above) allow scientists to travel to the ocean floor to conduct research.

Probably the best-known name in deep-sea exploration is *Alvin.* This tiny ocean-research vessel can transport scientists to depths of more than 4,300 m (13,000 ft.). *Alvin* has a thick hull made of a titanium alloy that surrounds a personnel compartment that has its own life-support system. Battery power permits the vessel to move slowly through the underwater environment. While its mechanical arm collects samples of sea life from the ocean floor, cameras and other technological equipment on board collect data about *Alvin's* surroundings.

While Bathysnap takes pictures and *Alvin* carries researchers to the ocean floor, some biologists have begun an even more daring exploration aided by startling new technology. They are exploring the depths in specially developed diving suits called *body submarines.* The Wasp is one such diving suit. It enables researchers to descend 670 m (2000 ft.) below the ocean's surface.

The Wasp has arms and claws that can be used to collect specimens. The diver uses foot pedals and other control devices within the suit to manipulate the arms and claws. The Wasp and similar diving suits are opening a new realm of observation that few biologists have ever seen.

"Wasp"

Overview of Arthropods

A green lynx spider devours its insect prey

Introduction

Arthropods are one of Earth's most successful groups of life forms in terms of number of species and adaptability. Animals such as spiders, lobsters, crabs, millipedes, centipedes, and insects—all arthropods—make this the largest phylum. More than 85 percent of all known animal species are arthropods.

The first arthropods appeared on Earth over 550 million years ago. Most of the early species have become extinct and are known only from their fossils. Scorpions evolved about 400 million years ago in forms that have been retained to the present day. The arthropods of today thrive in every livable habitat. One species of crab lives 4,000 m (13,000 ft.) below sea level. One type of jumping spider lives 6,700 m (22,000 ft.) above sea level on Mount Everest.

The Phylum Arthropoda

Section Objectives

About 1 million species of arthropods have been identified. Scientists estimate, however, that as many as 10 million species of arthropods may live on Earth. The great success of arthropods is due in part to their body structure. Their large population is also due to their extreme diversity in form, feeding habits, and patterns of behavior.

- *Name* three major characteristics of arthropods.
- *State* the advantages and disadvantages of the exoskeleton.
- *Name* the five major classes of arthropods and give an example of each.

33.1 Characteristics

The phylum name *Arthropoda* (ahr THRAHP uh duh) means "jointed leg." Legs or other movable extensions of the body are called **appendages.** *Arthropods are characterized by having jointed appendages, a segmented body, and an outer skeleton.*

Most of the jointed appendages of arthropods function for movement. Some, however, are specially modified for other purposes, such as grabbing prey or injecting poison.

The number of body segments of arthropods varies. Insects, for example, have three body segments; spiders have two. The arthropod's body is covered by a skeleton that is external to the body. This type of skeleton is called an **exoskeleton.** The exoskeleton is secreted by the outer layer of cells and is made of protein and a carbohydrate called **chitin** (KYT ihn). The exoskeleton is composed of three layers. The waxy outer layer is waterproof; it keeps out water while retaining body fluids. The hard middle layer protects the organism. The inner layer, flexible at the joints, allows the animal to move freely.

Though an exoskeleton provides a flexible armorlike protection, it has two major disadvantages. First, the exoskeleton does not grow with the animal, but must be shed and replaced periodically. This process is called **molting.** During a molt the animal faces two dangers—attack from predators and drying out before the new skeleton forms. Many arthropods hide in cool places during molting.

A second disadvantage of an exoskeleton is its weight. On a large land animal, an exoskeleton would be very heavy. Most arthropods, therefore, are small and move with ease. Arthropods move by contracting muscles that are attached to the inside of the exoskeleton. In spite of the hard exoskeleton, arthropods are capable of a wide variety of movements.

Arthropods have well-developed circulatory and nervous systems. The arthropod's heart is a long dorsal tube. The circulatory system is open—that is, blood bathes the animal's cells and tissues and is not always contained in vessels. The nervous

Figure 33–1. A cricket sheds its old skeleton in a process called molting. A special layer of cells secretes a new exoskeleton.

▶ Horseshoe Crabs

The horseshoe crab is one of the strangest of all arthropods, and a true "living fossil." This animal is not a crab, however. It is more closely related to the scorpions, the first modern arachnids to evolve.

Horseshoe crabs inhabit shallow coastal waters off the eastern coast of North America. They live 3.5 to 6 m (12 to 20 ft.) below sea level and forage for small animals in the sand and mud of the ocean floor.

The head and chest of this odd animal are fused into one. Both this body part and the abdomen are covered by a horseshoe-shaped shell. Spines emerge from the back part of the shell. A large, mobile spine projecting at the end of the body is used for protection.

Limulus is the most common of the three genuses of horseshoe crab. The *Limulus* horseshoe crab has existed on Earth for

over 200 million years. Fossils that old, which are identical to the present-day *Limulus* in every detail, have been found.

system consists of two chains of nerves that run along the ventral side of the organism. A ganglion located above the esophagus serves as the brain.

33.2 Classification

The five major classes of the phylum Arthropoda are Insecta, Arachnida (uh RAK nihd uh), Crustacea (kruhs TAY shee uh), Diplopoda (dih PLAHP uh duh), and Chilopoda (ky LAHP uh duh). The largest of these classes is Insecta, the insects. Because insects are such a large and important group, they are discussed separately in Chapter 34.

The class Arachnida consists of spiders, scorpions, ticks, and mites. Class Crustacea is made up of crayfish, lobsters, crabs, and related organisms. Millipedes form the class Diplopoda, and centipedes, Chilopoda. Millipedes and centipedes are often collectively called **myriapods** (MIHR ee uh pahdz).

Reviewing the Section

1. What are three characteristics of arthropods?
2. What are the five major classes of arthropods?
3. Why are horseshoe crabs called "living fossils"?

Arachnids

Section Objectives

- *Name* the major body parts of arachnids.
- *Describe* how spiders hunt and feed.
- *Describe* how spiders make and use web silk.
- *Name* the distinguishing characteristics of scorpions, ticks, and mites.

The 57,000 species of arachnids include spiders, scorpions, ticks, and mites. These organisms are found in most land areas of the world. Arachnids range in size from mites 0.1 mm (0.004 in.) long to a species of African scorpion 18 cm (7 in.) long.

Arachnids typically have two body segments and eight legs. The head and the *thorax*, the chest area, are fused to form the **cephalothorax** (sehf uh luh THAWR aks). The second segment, the **abdomen,** contains most of the organs. The arachnids' eight legs are attached to the cephalothorax.

Arachnids also have two other pairs of appendages attached to the cephalothorax. The **chelicerae** (kih LIHS uh ree) are claw-like fangs with poison glands at their base. The chelicerae are used to inject poison into prey. The other pair of appendages, called **pedipalps** (PEHD uh palps), aid in chewing and also serve as sense organs. Arachnids may have up to 12 **simple eyes.** These eyes can detect light but most are not structured to form images. The jumping spider has very fine eyesight and detects prey at 10 to 20 cm.

33.3 Spiders

Spiders, the most familiar arachnids, are among the world's most fascinating animals. Although some people shy away from spiders, few of the 30,000 species are harmful to humans. Spiders actually contribute to human welfare by eating insect pests.

The largest known land spider is the ''bird-eating'' spider from South America that is 8.75 cm (3.5 in.) long and has a leg span of 25 cm (10 in.). The smallest type of spider lives in Samoa and measures only 0.4 cm (0.02 in.) long.

Respiration and Senses Oxygen is carried through the spider's body in two unusual ways. One way is through tubes called **tracheae** (TRAY kee ee), which carry air directly to the cells. The tubes receive air through slits in the exoskeleton called **spiracles** (SPIHR uh kulz). The opening and closing of the spiracles regulates air flow.

In the second method of respiration, blood circulates through a structure called the **book lung,** so named because its sheets of tissue hang down like the pages of a book. The book lung is located near the front of the abdomen and absorbs oxygen through the spiracles.

Some spiders have either tracheae or book lungs, but most species have both. These structures are unique to arthropods.

Figure 33–2. Spiders are the best known arachnids. They have four pairs of legs, unlike insects which have only three pairs.

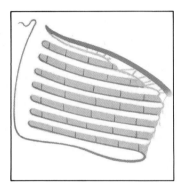

Figure 33–3. The diagram on the right shows the major body parts of the spider. The detail (above) is of the spider's book lungs.

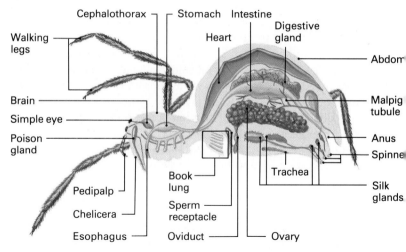

Cephalothorax — Stomach   Intestine

Walking legs

Heart

Digestive gland

Abdom

Brain

Simple eye

Poison gland

Pedipalp

Chelicera

Esophagus

Book lung

Sperm receptacle

Oviduct

Ovary

Trachea

Malpig tubule

Anus

Spinne

Silk glands

However, the structures serve the same purpose as gills and other respiratory structures—they provide a large surface area over which oxygen can diffuse into cells.

Most spiders have eight eyes arranged in two rows at the top of the cephalothorax. Spiders also have hairs called **sensory setae** all over their bodies, but especially on the legs. The setae detect pressure and movement.

Feeding and Reproduction   All spiders are skillful hunters. Some catch prey by chasing and catching it or jumping on it. Many others trap prey with "trap doors" built into the ground or catch it in webs.

Webs are formed when short, fingerlike organs called **spinnerets** release a substance formed in *silk glands*. This substance is a strong, elastic protein that hardens on contact with air. Once the prey lands in a web, a spider captures it by wrapping the prey with silk.

Not all spiders make webs for trapping prey. Web silk is also used to build nurseries for the young, to line nests, and to hold sperm or eggs during reproduction. Baby spiders may even leave the nest by letting out a tiny strand of silk that acts like a parachute. The wind blows the baby spiders away by the hundreds. This form of travel is called *ballooning*.

Spiders feed by injecting enzymes into their prey to dissolve the body substances. The spiders then suck out and swallow the liquified remains. **Malpighian** (mal PIHG ee uhn) **tubules** found near the base of the abdomen form the excretory system of the spider. The Malpighian tubules remove excess nitrogen, a by-product of metabolism, from the blood. Spiders also have waste-removing organs on the first and third pairs of legs.

Like most arthropods, spiders are either male or female. A male spider approaches a female with caution. Since spiders are habitually solitary animals, a female might mistake a male suitor for potential prey. For this reason many species engage in complex courtship rituals, such as tapping on the web or stroking the female.

Having captured the attention of the female, the male puts sperm on his pedipalps and places it into her genital opening on the underside of the abdomen. Eggs, laid in special webs or cocoons, hatch in about two weeks.

33.4 Scorpions

Scorpions have two distinctive features. Their greatly enlarged pedipalps are held in a forward position, and they have an abdomen that ends in a tapered stinger. Scorpions are most common in tropical areas and deserts. The 800 known species range in length from 1.3 to 17.6 cm (0.5 to 7 in.).

Scorpions hide under rocks and in crevices by day and are active mainly at night. They feed on insects and spiders. A scorpion catches and holds its prey in its pedipalps until the stinging abdomen can curl over the top of its body and inject poison into the prey.

33.5 Ticks and Mites

Ticks and mites are tiny animals, usually less than 1 mm (0.04 in.) long. Scientists have identified over 30,000 species, but they estimate that there may be as many as 500,000. *Ticks and mites differ from other arachnids in that the cephalothorax and abdomen are fused to form a single body part.*

Although many ticks and mites are free-living, most are parasites. Ticks attach themselves to an animal's skin and suck blood. Many ticks carry disease-causing organisms that are transmitted to other animals through the tick's bite. Most human beings have mites living in their hair follicles. The mites live off bacteria and are so small that they go unnoticed.

Figure 33–4. A scorpion uses its huge pedipalps like pincers to capture and hold prey. It then injects the prey with poison from its long stinger.

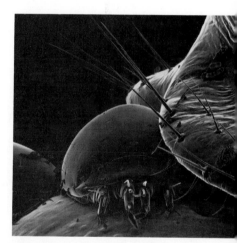

Figure 33–5. This scanning electron micrograph shows a mite that lives on a termite magnified 1,750 times.

Reviewing the Section

1. What characteristics distinguish arachnids from other arthropods?
2. How do spiders feed?
3. How does a scorpion inject poison into its prey?
4. How do scorpions and ticks differ from spiders?

- *State* how crustaceans differ from arachnids.
- *List* the functions of the crayfish's appendages.
- *Describe* the compound eye and its advantages over the simple eye.
- *Summarize* the process of molting.

Crustaceans

The 25,000 species that make up the class Crustacea include crayfish, lobsters, crabs, shrimps, pill bugs, sow bugs, water fleas, and barnacles. Most crustaceans are small—one type of water flea measures only 0.25 mm (0.01 in.) long. The Japanese spider crab, on the other hand, has limbs that extend 3.6 m (12 ft.) or more. Almost all crustaceans live in the sea, although a few species live in fresh water or on land.

Crustaceans have chewing jaws. The jaws, called **mandibles,** are absent in arachnids. The crayfish, sometimes called a crawdad, is representative of a group of crustaceans that also includes lobsters, crabs, and shrimps.

33.6 Crayfish

Crayfish are freshwater animals that live at the bottoms of lakes and streams. Crayfish and related saltwater crustaceans are important sources of food in many areas.

As Figure 33–7 shows, the body of a crayfish is divided into two main parts—a cephalothorax and an abdomen. The cephalothorax consists of thirteen segments and is covered by a protective shield of exoskeleton called a **carapace.** The abdomen is made up of six segments.

Each crayfish segment has a pair of appendages. The first segment has two small, antenna-like structures called **antennules** that maintain balance and are sensitive to taste and touch. Behind the antennules are another pair of sense organs called **antennae,** which may be as long as the body of the crayfish. Behind the antennae lie the mandibles, which chew and crush food. Behind the mandibles are two pairs of **maxillae.** The first pair holds food. The other pair, which look like flat shovels, scoop water over the gills. These are the *gill bailers.*

The first pair of appendages on the thorax part of the cephalothorax are the **maxillipeds,** or jaw feet. The maxillipeds taste and hold food while it is being chewed. The next pair of appendages, the **chelipeds** (KEE luh pehdz), are large claws. They grab food and also protect the crayfish against enemies. Each of the last four segments of the cephalothorax has a pair of **walking legs** that are used for locomotion.

Five pairs of **swimmerets** located on the first five segments of the abdomen are used in swimming. The next pair of appendages has become fused and is used as a flipper, called a **uropod** (YOOR uh pahd). The **telson** is the middle part of the tail. By flipping the telson forward, the crayfish can move backwards.

Figure 33–6. Bottom-dwelling crayfish are most active at night or early morning. They feed on other freshwater animals, such as snails, tadpoles, and small insects.

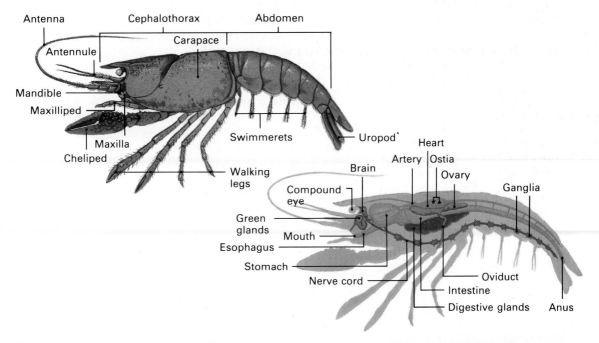

Figure 33–7. This diagram shows the external and internal structures of a crayfish.

Respiration The crayfish removes oxygen from the water by means of its gills. The featherlike gills contain a high concentration of blood vessels. The gills are attached to the walking legs, between the thorax and the carapace. As the animal walks, water flows over the gills. When the crayfish is not moving, the gill bailers move water over the gills.

Circulation Crayfish, like spiders, have an open circulatory system. Blood enters the heart through three pairs of pores called **ostia.** Valves seal off the ostia, the heart contracts, and blood is forced into seven large arteries. These arteries then discharge blood into the spaces surrounding the organs. The blood drains out of these spaces and collects in a cavity called the *sternal sinus*. From there, blood travels through other vessels to the gills. Blood returning from the gills enters the *pericardial sinus* and then returns to the heart through the ostia.

Digestion and Excretion Crayfish eat living and dead plants, worms, larvae, and tadpoles. Food moves from the mouth down the esophagus to the stomach, where it is ground by teeth of chitin. Undigested wastes move through the intestine and leave the body through the anus. Excretory organs called **green glands** are located at the base of the antennae. The green glands remove liquid wastes from the blood.

Q/A

Q: *Can crustaceans climb trees?*

A: Not most crustaceans. But *Bingus latro,* also known as the "robber crab," can. This crab lives on South Pacific islands and climbs coconut palms in search of food. It uses its pincers to cut the nut from the tree and to open the husk.

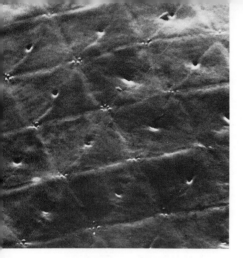

Figure 33–8. The scanning electron micrograph magnifies a part of the compound eye of a lobster 550 times. The eye is made up of over 2,500 lenses.

Nervous System   The nervous system of crayfish consists of a dorsal brain formed from a pair of ganglia. Branches from the brain run to the eyes, antennules, and antennae. Other nerves encircle the esophagus and connect with a ventral nerve cord that runs the length of the body. Each of the crayfish's segments has a pair of ganglia.

The crayfish has two eyes located on movable stalks on the front of the body. Each eye has over 2,500 lenses. Eyes that have more than one lens are called **compound eyes.** Compound eyes form images by combining the sensations from multiple lenses. Such eyes respond rapidly to light and readily detect motion. Arthropods are the only animals with compound eyes.

Reproduction and Growth   Crayfish mate in the spring or fall. The male deposits sperm into the female, who stores it. About two weeks later, the female produces 200 to 300 eggs. They are then fertilized by the stored sperm and hatch in six weeks. The fertilized eggs are attached to the last three pairs of swimmerets of the female until they hatch.

Crayfish molt twice a year. Molting begins when the outer layer of body cells digests the inner layer of the exoskeleton, weakening it. The crayfish swells by absorbing water and taking in excess air. This causes the exoskeleton to crack, and the crayfish backs out. The outer body cell layer secretes salts that harden the new, developing exoskeleton.

33.7 Other Crustaceans

Lobsters, crabs, and shrimp closely resemble crayfish. Lobsters are much larger than crayfish. Crabs have a short, wide body. The abdomen is curled under the thorax, and the carapace forms a single, large shell over all the body segments. Shrimp typically have semitransparent bodies and long, whiplike antennae.

Pill bugs and sow bugs, called *isopods,* are common in gardens. Their exoskeletons are segmented, allowing them to roll into a tight ball when threatened. Water fleas are tiny organisms often seen twisting and turning in a drop of pond water.

Figure 33–9. The Sally Lightfoot crab lives along the rocky shores of the Galapagos Islands off the coast of Ecuador.

Reviewing the Section

1. What distinguishes crustaceans from arachnids?
2. Describe a compound eye and state its advantages over a simple eye.
3. What is the function of maxillipeds and mandibles?
4. What features characterize crabs?

Myriapods

About 11,000 myriapod species have been identified. The name *myriapod* means "many legs." Both centipedes and millipedes have numerous legs with which they scurry over the ground. All myriapods have a head and numerous body segments. The head and thorax are not fused as in arachnids and some crustaceans. Like crustaceans, myriapods have mandibles.

Myriapods respire by means of tracheae. The openings to these tracheae do not close, which can result in loss of body fluids. In addition, the body of myriapods lacks a waxy outer layer that retains moisture. For these reasons, myriapods must live in moist environments.

33.8 Centipedes

Although the name *centipede* means "100 legs," not all centipedes have that many. Some, like the common house centipede, have only about 15 pairs of legs. On the other hand, *Himantarum gabrelis* of southern Europe has 177 pairs of legs. Some centipedes are as long as 33 cm (13 in.). Others, however, measure as little as 0.47 cm (0.18 in.) in length.

Centipedes have flattened bodies and one pair of legs per body segment. Every segment of the body has a pair of legs except the first segment, which has poisonous claws. These claws inject poison to kill prey, mainly insects and worms. Most centipedes have simple eyes, one pair of antennae, one pair of mandibles, and two pairs of maxillae. Liquid wastes are excreted through Malpighian tubules.

Centipedes live in such moist places as forest floors or tropical jungles. During mating centipedes undergo an elaborate courtship ritual that involves mutual caressing of the antennae. In most species the young have fewer segments than the adults and pass through several growth stages.

33.9 Millipedes

Like centipedes, millipedes have a pair of simple eyes, a pair of antennae, mandibles, and maxillae. ***Millipedes differ from centipedes by having a rounded body and two pairs of legs per body segment.*** The word *millipede* means "1,000 legs." The most legs ever recorded for a millipede, however, is 710. The largest millipede measures about 27.5 cm (11 in.) in length.

Millipedes feed mainly on decayed plants. These myriapods live in the tropics or in temperate areas under cool and wet

Figure 33–10. A centipede (top) has one pair of legs per body segment; a millipede (bottom) has two pairs per segment.

Highlight on Careers: Marine Biologist

General Description

A marine biologist is a scientist who studies sea life. Most marine biologists specialize in either plants or animals, and many select a particular species to study. They record its life history and food cycles. Often they collect samples from the ocean to study in the laboratory.

Marine biologists may also study the organism's environment. They may, for example, measure the temperature and chemical content of the water.

Some jobs in marine biology at colleges and universities combine teaching with research. An increasing number of jobs are becoming available in private industry. For example, marine biologists are needed by pharmaceutical companies that are developing drugs from sea life. Undoubtedly, more opportunities will exist in the future for marine biologists who want to help make use of marine resources for food.

Career Requirements

Entry-level positions in industry are available to those with a B.S. degree in biology, zoology, or botany. A Ph.D. degree is required to teach or to initiate research projects.

The marine biologist does not need to engage in deep-sea diving. Some biologists enjoy diving,

however, and take diving courses in order to enhance their research.

For Additional Information
American Society of
   Limnology and
   Oceanography
Great Lakes Research
   Division
University of Michigan
Ann Arbor, MI 48109

---

conditions. Millipedes avoid bright sun, and many dig burrows in which they escape high temperatures and dry conditions.

When attacked, millipedes may defend themselves in two ways. First, they roll into a ball and do not open up until the predator goes away. If this method does not work, many millipedes excrete a foul-smelling and often toxic substance from their *stink glands*.

Reviewing the Section

1. Compare the bodies of millipedes and centipedes.
2. How do the feeding habits of millipedes and centipedes differ?
3. How does a millipede defend itself when attacked?
4. Compare myriapod and arachnid body structures.

Investigation 33: Land-Dwelling Crustaceans

Purpose
To study the responses of the common sowbug to pH changes

Materials
A small terrarium, a culture of sowbugs, potato or carrot, pencil, ruler, paper towels, medicine dropper, vinegar, baking soda solution, salt solution, sugar solution

Procedure
1. Place the sowbugs in a terrarium containing a rich mulch base. Keep the base moist with water. Occasionally place a slice of potato or carrot on top of the mulch.
2. With a ruler and a pencil, divide a piece of paper towel into four sections.
3. In the center of the towel, draw a circle approximately 8 cm in diameter.
4. On the first section, place several medicine dropperfuls of vinegar, being sure to keep the solution inside the section dividing line. Repeat the process using baking soda solution, salt solution, and sugar solution on the remaining three sections.
5. Carefully place the sowbugs in the center of the circle and observe their response to each of the four solutions.

Analyses and Conclusions
1. Do all of the organisms respond in the same manner? Explain.
2. Provide an explanation for their movement toward or away from portions of the paper.
3. Compare your results with those of your classmates. What are some conditions that may differ from yours in the investigations set up by your classmates? (Hint: Consider factors such as lighting.)

Going Further
- Another version of this experiment can be conducted by varying the concentration of each solution; for example, pure vinegar, and 75%, 50%, and 25% vinegar. Try to determine the concentration that provides the greatest attraction; the greatest repulsion.
- Place a live earthworm on a paper towel prepared in the same manner as for the sowbugs. Is the earthworm's behavior comparable to that of the sowbug? If so, why?
- Using different insecticides, spray a small amount of each on a different area of a paper towel. Do all the insecticides repel the sowbugs equally?

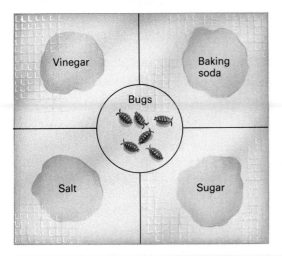

Chapter 33 Review

Summary

Arthropods are invertebrates that have jointed appendages, segmented bodies, and exoskeletons. They shed and regenerate their exoskeleton by molting. The five principal classes of arthropods are Insecta, Arachnida, Crustacea, Chilopoda, and Diplopoda.

Arachnids include spiders, scorpions, ticks, and mites. Arachnids typically have two body segments, a cephalothorax and an abdomen. Spiders have four pairs of legs. They also have two clawlike chelicerae and two chewing pedipalps. Many spiders spin webs to catch prey. Scorpions have an elongated abdomen

that ends in a stinger. Ticks and mites are small and have fused body parts.

Crustaceans are mostly saltwater arthropods and have mandibles for chewing. Each body segment of the freshwater crayfish has a pair of appendages, each of which serves a special function.

Centipedes and millipedes are myriapods. They inhabit moist, cool places. Centipedes are characterized by one pair of legs per body segment, and millipedes by two. Centipedes are meat eaters; millipedes eat mostly decayed plants.

BioTerms

abdomen (**491**)
antenna (**494**)
antennule (**494**)
appendage (**489**)
book lung (**491**)
carapace (**494**)
cephalothorax (**491**)
chelicera (**491**)
cheliped (**494**)
chitin (**489**)

compound eye (**496**)
exoskeleton (**489**)
green gland (**495**)
Malpighian tubule (**492**)
mandible (**494**)
maxilla (**494**)
maxilliped (**494**)
molting (**489**)
myriapod (**490**)
ostia (**495**)

pedipalp (**491**)
sensory seta (**492**)
simple eye (**491**)
spinneret (**492**)
spiracle (**491**)
swimmeret (**494**)
telson (**494**)
trachea (**491**)
uropod (**494**)
walking leg (**494**)

BioQuiz   (Write all answers on a separate sheet of paper.)

I. Completion

1. Millipedes have _____ pairs of legs per body segment.
2. _____ remove wastes from the spider's blood.
3. An arthropod moves by contracting muscles that are attached to the inside of its _____.
4. The _____ covers the cephalothorax in crustaceans.

II. Modified True and False

Mark each statement TRUE or FALSE. If false, change the underlined term to make the statement true.

5. The telson helps crayfish move backwards.
6. Centipedes and millipedes have compound eyes.
7. Spiders spin webs of protein from structures called silk glands.

III. Multiple Choice

8. Air enters the tracheae of a spider through its a) mouth. b) nose. c) spiracles. d) antennae.
9. The crayfish holds food and moves water over the gills with its a) maxillae. b) telson. c) antennules. d) pedipalps.
10. Scorpions grasp prey with their a) uropods. b) pedipalps. c) thorax. d) telson.
11. Crustaceans differ from arachnids in that they have a) a cephalothorax. b) mandibles. c) an abdomen. d) appendages.
12. Crayfish use green glands for a) photosynthesis. b) courtship. c) camouflage. d) excretion.
13. Chelicerae function in a) feeding. b) respiration c) excreting wastes. d) maintaining balance.
14. Mites and ticks are characterized by the fusion of their abdomen and a) cephalothorax. b) segments. c) pedipalps. d) tail section.
15. The middle layer of the exoskeleton functions for a) water retention. b) flexibility. c) movement. d) protection.

IV. Essay

16. What is the function of each layer of the exoskeleton?
17. What are the major advantages and disadvantages of the exoskeleton?
18. For what purposes do spiders spin silk?
19. How do tracheae differ from lungs and gills in oxygen distribution?
20. How does molting occur in the crayfish?

Applying and Extending Concepts

1. Keep a record for a week of all spiders or evidences of spiders in your home or neighborhood. Research and identify if possible the species observed. If you fail to observe any spiders, write a report suggesting why that was so.
2. Krill are tiny marine crustaceans that are an important source of food for some of the largest animals on Earth. Use your school or public library to prepare a short report on krill. Explain how a reduction in their population would affect other animal life.
3. Write and illustrate a report on the various hunting methods of spiders. Include the wolf spider, the trap door spider, and the bird-eating spider.
4. What advantage might a compound eye have over a simple eye for animals such as crayfish?

Related Readings

Bond, Constance. "The Swift Spider That Is Nature's Smallest 'Angler'." *Smithsonian* II (July 1980): 78–83. The article profiles spiders that capture aquatic insects and fish.

McWhinnie, Mary A. and Charlean J. Danys. "The High Importance of the Lowly Krill." *Natural History* 89 (March 1980): 66–70. The article discusses the importance of krill in the Antarctic food chain and the implications of human harvesting of this tiny, protein-rich crustacean.

Tuthill, Jo Ellen. "Spider Webbing." *Science 82* 3 (November 1982): 100–102. This article describes how the many different webs are constructed.

Insects

Honeybee pollinating a plant while extracting nectar

Introduction

Sit in the countryside—or even in a city garden—on a warm summer day and you can observe the variety and abundance of insects and the important roles insects play in the world around them. A honeybee flies from blossom to blossom, pollinating plants while extracting nectar that will be converted to honey in the hive. A grasshopper, more destructive in its feeding habits, chews on a cabbage leaf. A ladybug crawls along a tomato leaf, ridding the plant of destructive aphids.

Bees, grasshoppers, ladybugs, and aphids are but a few of the world's insects. In fact, there are more than 750,000 different species of insects. In their diversity—the wide variety of ways in which they solve the problems of survival—insects are the most successful of all groups of animals.

Overview of Insects

Section Objectives

Almost three-quarters of all animal species on Earth are insects. Although insects are not found in salt water, they live in almost every freshwater and land habitat. The great diversity of insects extends also to size. Insects range from the fairy fly, only 0.2 mm (0.008 in.) long, to the African Goliath beetle, over 10 cm (4 in.) long. Insects are the only invertebrates that can fly.

The evolutionary success of insects is due in part to the characteristic structure they share with other arthropods—exoskeletons, segmented bodies, and jointed appendages. A key to their success is that most insects are small and have adapted to specific habits and needs. In many cases this allows multiple species of insects to exist in a small area without competing with one another for scarce resources. Their ability to fly has also added to their evolutionary success.

Insects are both harmful and beneficial to human society. Only about 1 percent of insect species are destructive to crops and property. Nevertheless, this small group causes several billion dollars of damage each year in the United States alone. Harmful insects include household pests, such as termites; crop and livestock pests, such as boll weevils; and hosts of disease-causing organisms, such as mosquitoes infected with parasitic protozoa.

Many insects, on the other hand, are beneficial to human society. Insects pollinate fruit trees, flowers, and many field crops. Bees produce honey and beeswax, silkworms form cocoons from which silk is spun, and lac insects provide the raw material for commercial shellac. Some kinds of insects are natural enemies of destructive insects. For example, the larvae of certain wasps feed on caterpillars that destroy plants.

34.1 Characteristics

All insects share certain characteristics. ***Unlike other arthropods, insects have three distinct body parts and three pairs of legs.*** The three body parts of an insect are a head, a thorax, and an abdomen.

On the head are antennae, compound eyes, and mouthparts. An insect's mouthparts are modified for different methods of feeding. Butterflies have mouthparts shaped like coiled tubes, which they uncoil and use to suck nectar from deep within flowers. A praying mantis tears its food apart with mandibles. Flies have no mandibles and therefore lap up food with a tonguelike lower lip.

Section Objectives

- *State* reasons for the evolutionary success of insects.
- *Discuss* the importance of insects to human society.
- *Distinguish* insects from other arthropods.
- *Identify* three feeding methods of insects and the modified mouthpart used.
- *Name* eight insect orders and give an example of each order.

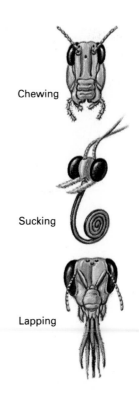

Chewing

Sucking

Lapping

Figure 34–1. Modified mouthparts of insects allow different feeding methods. A grasshopper (top) chews its food; a butterfly (center) sucks nectar from flowers; a fly (bottom) laps up food.

Born: Birmingham, Alabama, 1919
Degree: Ph.D., Harvard University

The work of American entomologist Edward Wilson on the social behavior of ants and other insects has had far-reaching implications on current thinking in many fields. Wilson, a professor at Harvard University, is one of the foremost advocates of sociobiology, the study of the biological basis of social behavior. The extent to which sociobiology can be applied to human behavior has aroused considerable controversy in scientific circles.

Wilson has been fasci-nated by ants since he was a schoolboy and has discovered some 100 ant species. This interest has also led him to study the chemicals used by social insects and some other animals to communicate through taste and smell. Wilson was one of the first to suggest using the knowledge of these chemicals to control insect pests.

Wilson is also an authority on island biogeography, a field of study that focuses on plant and animal life on islands, both natural and habitat islands. Habitat islands are communities that are surrounded by widely different habitats. Wilson has determined that the more

isolated a habitat island the fewer the number of species and individuals on that island. Wilson was awarded the Tyler Prize for Environmental Achievement in 1984 for his work on island biogeography.

Three pairs of legs and any wings the insect has are attached to the thorax. The legs are usually adapted for jumping, walking, or running.

The abdomen may have as many as eleven segments. The abdomen houses the heart, the respiratory and excretory organs, and the spiracles through which the insect breathes. In respiration, tracheae carry air to body cells. All insects have an open circulatory system, through which blood flows into large open spaces surrounding the organs.

34.2 Classification

Scientists who study insects are called **entomologists.** Entomologists have grouped insects into more than 30 orders. Classification is based primarily on the number and type of wings the insect has. Eight important orders are shown in Table 34–1. Others are listed on pages 827–828.

Table 34-1: Insect Orders and Characteristics

Order		Examples	Characteristics
Orthoptera ("straight-winged")		Grasshoppers, crickets	Two pairs of straight wings; legs modified for jumping
Isoptera ("equal-winged")		Termites	Two pairs of similar wings that are shed after mating
Hemiptera ("half-winged")		True bugs, squash bugs	Two pairs of membranous wings overlap to form V over abdomen
Homoptera ("uniformly winged")		Cicadas, aphids, scale insects	One or two pairs of membranous wings roof over the body when at rest (some species are wingless); mouthparts adapted for piercing-sucking; feed on plants
Diptera ("two-winged")		Flies, gnats, mosquitoes	One pair of membranous wings; mouthparts adapted for lapping or piercing-sucking
Lepidoptera ("scale-winged")		Butterflies, moths	Two pairs of wings; body and wings covered with scales
Coleoptera ("shield-winged")		Beetles (includes fireflies, ladybugs)	Two pairs of wings; hard forewings join to form a straight line down the back
Hymenoptera ("membrane-winged")		Ants, bees, wasps	Two pairs of membranous wings; chewing mouthparts; social insect

Reviewing the Section

1. Why can many insect species live in one area?
2. How are insects beneficial to humans?
3. How do insects differ from other arthropods?
4. How does a butterfly's mouthpart aid feeding?
5. To what order of insects do beetles belong? What are their distinguishing characteristics?

Grasshoppers

- *Identify* the major external parts of the grasshopper and state the function of each part.
- *State* how grasshoppers respire.
- *Describe* the reproductive process of grasshoppers.

Grasshoppers are familiar insects found in almost all parts of the world. Because grasshoppers share many features with other species, they are often studied as representative insects.

Grasshoppers belong to the order Orthoptera. Most are large insects, averaging 7.5 cm (3 in.) in length. Common North American species range in color from green to olive or brown, making them hard to see amidst the grasses and plants on which they feed. Some orthopteran species are known as locusts. Locusts often move in large masses called **swarms** that migrate from place to place, destroying crops along their way.

34.3 External Structure

Figure 34–2. The external structure of the grasshopper shows the three-part division characteristic of all insects. Less typical are the strong back legs that propel the grasshopper into the air for long distances.

Like all insects, a grasshopper has three main body parts—the head, thorax, and abdomen. The grasshopper's mouthparts are modified for chewing. A grasshopper feeds by holding a blade of grass in its upper lip, called the **labrum.** The mandibles, or primary jaws, chew the food. The maxillae, or secondary jaws, help hold and tear the food. The lower lip, called the **labium,** is located behind the maxillae. The function of the labium is to press food against the jaws.

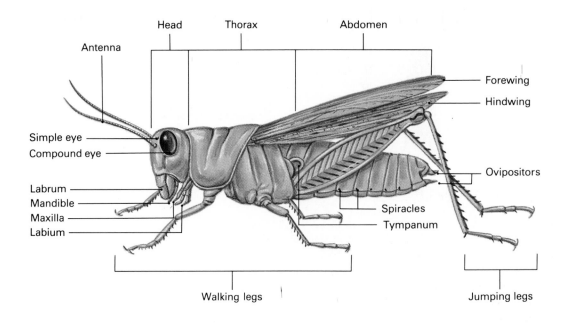

The grasshopper has two large compound eyes and three simple eyes. Between the compound eyes are a pair of antennae, which have highly sensitive receptors for smell.

The grasshopper's thorax is divided into three segments. A protective shield covers the first segment, the **prothorax.** The *forewings* are attached to the second segment, the **mesothorax.** The *hindwings* are attached to the third segment, the **metathorax.** Both pairs of wings are used in flight. The narrow, stiff forewings cover and protect the transparent, membranous hindwings. Hindwings fold up when they are not in use.

Each segment of the thorax has a pair of legs attached. The first and second pairs are modified for walking, and the hindlegs, for jumping. Each leg has five segments. The *coxa* attaches the leg to the thorax. The short *trochanter* connects the coxa to the *femur*, which houses the large leg muscles. The *tibia* is the lower part of the leg, which ends in a segmented, grasping foot called the *tarsus*.

The grasshopper's abdomen consists of 10 segments. The eardrums, thin flexible membranes called **tympanums**, are located near the front of the abdomen. In some species the tympanums are found at the posterior end of the thorax.

Figure 34–3. The grasshopper's mouthparts are derived from ancestral legs. The segmented character of legs is most clearly evident in the maxillae and labium, which hold and tear food.

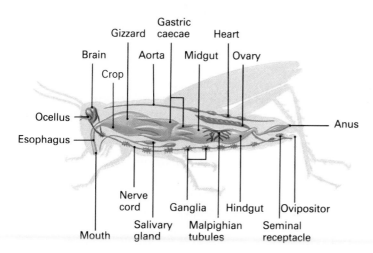

Brain · Ocellus · Esophagus · Crop · Gizzard · Aorta · Gastric caecae · Midgut · Heart · Ovary · Anus · Mouth · Nerve cord · Salivary gland · Ganglia · Malpighian tubules · Hindgut · Seminal receptacle · Ovipositor

Figure 34–4. This diagram of the internal structure of a grasshopper shows its respiratory, digestive, and nervous systems. Activity increases the flow of blood through the insect's circulatory system.

34.4 Internal Structure

An insect's organ systems differ in some significant ways from the systems of other animals. As you read about the internal structure of a grasshopper, locate its major organs in Figure 34–4. Pairs of spiracles are found on the second and third segments of the thorax and on the first eight segments of the grasshopper's abdomen. As the flexible abdomen expands, air enters the spiracles and moves into the tracheae. The tracheae then

Q: *Why is an insect's blood usually colorless?*

A: An insect's blood carries nutrients to the cells and removes wastes but has little to do with the transport of oxygen. The red blood of mammals contains a red pigment, *hemoglobin,* that carries oxygen to the cells.

transport air throughout the body, allowing the oxygen to diffuse directly into the tissues. When the grasshopper's abdomen contracts, the waste gases are forced out of the insect's body through the spiracles.

Food passes from the mouth to the crop. There the food is stored and moistened with saliva from **salivary glands.** Moistened food moves to the gizzard, where it is ground by teeth of chitin. Food then enters the insect's stomach, called the **midgut.** It is digested by enzymes secreted by pockets of the stomach called **gastric caecae** (SEE kee). Undigested food enters the intestine, called the **hindgut,** and leaves through the anus.

The grasshopper's heart is found in the dorsal part of the abdomen. The heart pumps blood through a large vessel, called an **aorta,** into the body cavity. There the blood bathes the organs, then reenters the heart through large pores called *ostia.*

Wastes are removed from blood by Malpighian tubules attached to the hindgut. These organs send nitrogenous wastes into the hindgut. From there, the wastes leave the body through the anus. Grasshoppers have an important adaptation that helps conserve the insect's bodily fluids; nitrogenous wastes are excreted as nearly dry uric acid crystals.

Like all arthropods, the grasshopper has a brain that consists of two ganglia located in the head. Two nerve cords run from the brain down the ventral side of the body. These cords form a pair of ganglia in every segment. Nerve branches run from the ganglia to all parts of the body.

Because grasshoppers have ganglia in every segment, a grasshopper can continue functioning for some time if its head is cut off. Headless grasshoppers and other insects can move about and even mate.

Grasshoppers are either male or female. The reproductive organs of both sexes are located in the upper abdomen. During the mating season in early summer or fall, the male deposits sperm into a storage pouch in the female, called a *seminal receptacle.* When eggs are produced, they are fertilized by the stored sperm. The female then uses a pair of pointed organs called **ovipositors** to dig a hole in the ground and to deposit the eggs. The eggs hatch the next spring.

Reviewing the Section

1. Name the structures attached to each thorax part of a grasshopper.
2. What is the advantage of excreting uric acid crystals?
3. How does a grasshopper deposit its fertilized eggs?

Insect Development

Only a few wingless insects, such as silverfish, emerge from the egg as miniature versions of their parents. ***Most young insects differ in appearance, movement, and feeding habits from the adult insects they become.*** The profound series of molts and changes that transform the immature form into the adult insect is termed a **metamorphosis.** Metamorphosis is governed by the action of *hormones,* or chemicals secreted by the insect's body.

34.5 Incomplete Metamorphosis

Some insects, such as grasshoppers and lice, go through a gradual transition from egg to adult called *incomplete metamorphosis*. In this type of development, the insect that emerges from the egg is called a *nymph*. The nymph is a smaller version of the adult insect of its species, similar in structure but without wings or mature reproductive organs.

34.6 Complete Metamorphosis

Most insects, including butterflies, beetles, ants, bees, and flies, go through a *complete metamorphosis*. This kind of development involves four stages: egg, larva, pupa, and adult.

Silverfish

Butterfly

Egg

Larva

Pupa

Adult

Figure 34–5. Insects differ in their development. The grasshopper exhibits incomplete metamorphosis. The butterfly, as most insects, exhibits complete metamorphosis. Others such as silverfish do not change in form after hatching.

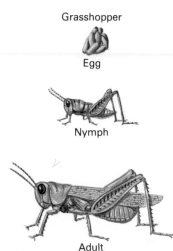

Grasshopper

Egg

Nymph

Adult

Adult

▶ Hormones and Metamorphosis

What controls the dramatic changes that transform a caterpillar into a butterfly? These changes result from the interaction of three hormones: brain hormone; molting hormone, also called *ecdysone;* and juvenile hormone.

Eating and growing are the major activities of a caterpillar. As the caterpillar grows, cells in the brain periodically secrete brain hormone into the blood. This hormone triggers the release of ecdysone from a gland in the thorax. The ecdysone in turn causes the caterpillar to molt, or shed its outer covering to accommodate its larger size. A caterpillar may molt many times as it progresses through the larval stage.

Metamorphosis is actually a molt that produces not a larger larval form but an adult insect. Ecdysone controls this molt also, but its action is regulated by the presence of juvenile hormone. As long as juvenile hormone is present in sufficient quantity, ecdysone does not trigger metamorphosis. The molts caused by ecdysone will result in a larger caterpillar, not in a butterfly.

The insect reduces its production of juvenile hormone as it nears the end of the larval stage of development. When the amount of juvenile hormone is sufficiently reduced, ecdysone triggers the start of the pupal stage. The pupa forms a protective covering, the **chrysalis,** from which it will emerge as a butterfly.

Q/A

Q: *Do any insects ever develop from unfertilized eggs?*

A: Yes. Aphids and gall wasps are two insects that can produce young without the benefit of fertilization by a process called *parthenogenesis.* In this process, the aphid produces a great number of eggs that develop even though they have not been fertilized.

When the egg of an insect hatches, an immature form called a *larva* emerges. The larva looks nothing like the adult it will become; it has a wormlike, segmented body. Needless to say, larvae, like nymphs, cannot yet fly. Some larvae feed on entirely different kinds of food than the adults do. The fact that adults and their young do not compete with each other for food is a key advantage of complete metamorphosis.

When the larval stage is complete, the insect enters an inactive stage called the **pupa.** Many kinds of pupae form a protective covering around themselves called a **cocoon.** For a period of several months, the insect lies still and does not feed while larval tissues and organs are replaced with new tissues and organs. At the end of the pupal stage, a fully formed, sexually mature adult insect emerges.

Reviewing the Section

1. What is metamorphosis?
2. What are the developmental stages of a grasshopper?
3. How is a nymph like a larva? How is it different?
4. What causes a caterpillar to begin spinning a chrysalis?

Insect Behavior

All insects, whether they live singly or in groups, need to communicate and to defend themselves. Different kinds of insects accomplish these tasks in different ways. Part of the job of an entomologist, then, is to find explanations for the individual and collective behavior of insects.

34.7 Social Insects

Social insects live in communities and engage in a *division of labor*. In other words, individuals perform different tasks necessary for the survival of the group. Termites, ants, and some bees and wasps are social insects. Insect societies are probably the most complex of all nonhuman societies.

A termite colony may have over a million individuals. The most numerous kind are the workers, which collect food and build and repair the nest. Soldiers defend the colony against attack. They have large heads and powerful jaws. The only duty of the queen and king is to produce young.

All ant colonies have one or more queens, reproductive males, and female workers. Army ants that live above ground form an unusual society. When the food supply dwindles, the entire colony swarms out in search of a new home, chewing down everything in its way.

The most widely studied insect society is that of the honeybee. One of the 30,000 to 40,000 bees is the queen. A few are males, or drones. The vast majority are female workers. The queen is the only reproductive female in the hive. She flies from the hive only once. During this flight she collects enough sperm from drones of nearby hives to last throughout her five- to seven-year life of laying eggs. The sperm are stored in a special organ and released one at a time. In the springtime the queen lays unfertilized eggs, which develop into drones.

The only function of a drone is to inseminate the queen. In the fall, when the food supply dwindles, the workers usually drive the drones out of the hive or sting them to death.

The workers develop from fertilized eggs deposited in the waxy cells of the hive. Adult workers feed the developing bees until the larvae pupate and emerge as new adults. The first job of the worker is to feed the queen, drones, and larvae. After about a week of this work, she begins to excrete wax, which is used to build and repair the hive. In the final stage of her life, the worker helps forage for food. Her entire life cycle lasts about six weeks.

Q/A

Q: *Are any bees solitary?*

A: Yes. Over 85 percent of all bees are solitary; they do not live in hives. Many crops, like alfalfa, are pollinated entirely by solitary bees.

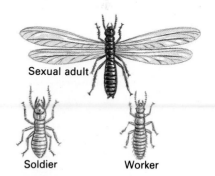

Sexual adult

Soldier   Worker

Fig. 34–6. Termite classes can be identified by their body structures. The king and queen shed their wings after mating. Workers and soldiers are wingless.

Fig. 34–7. A scout bee indicates the location of food by waggling her abdomen while running forward, then circling. The orientation of the run to the vertical hive symbolizes the direction to the food source. The number of turns indicates the distance from the hive.

Fig. 34–8. The yellow jacket wasp (left) stings its foes. Similar color patterns protect its mimic, the stingless syrphid fly (right).

34.8 Communication

Both social and solitary insects need to communicate about food sources, predators, and possible mates. *Insects communicate through chemicals, visual signals, and motions.*

The most common means of communication is chemical. **Pheromones** are chemicals that influence the behavior of other insects. For example, the queen bee, on her flight from the hive, secretes a pheromone that attracts drones. Ants secrete pheromones that mark a trail from the nest to a food source. Only other members of the ant's own colony recognize the trail.

Fireflies communicate during the mating season by flashing a light. The female of each species emits a special series of flashes that males of her species recognize. Other insects communicate by tapping, rubbing, or stroking each other. Sometimes these are part of an elaborate courtship ritual.

Perhaps the most complex of all forms of insect communication is the dance of the honeybee. When a forager bee finds a source of nectar, she returns to the hive and does a "waggle dance." The dance lets other bees know where the food is.

Does all of this mean that insects "talk" to each other, using chemicals and light and signals instead of words? Research has shown that insects do not communicate as people do. For example, when an ant dies, it excretes a certain chemical that tells other ants to carry it out of the nest. If a live ant is painted with this chemical, its fellow ants will drag it out of the nest as it struggles. The ant will be dragged out of the nest again and again even though it is obviously not dead.

34.9 Defense

Almost all insects will flee when threatened. Many insects, however, have more specialized means of defense. Roaches and stink-bugs, for example, secrete foul-smelling chemicals that deter aggressors. Bees, wasps, and some ants have poisonous stings that can kill smaller predators and cause pain for larger ones. The larvae of some insects have hairs filled with poison. If a predator eats one of these larvae, it may suffer a toxic reaction. Insects that defend themselves by unpleasant or dangerous chemicals gain two advantages. On one hand, they often deter a predator from eating them. On the other, predators learn not to bother them in the first place.

Other insects gain protection by **mimicry,** or similarity of appearance. In one kind of mimicry, insects with similar defense mechanisms look alike, and predators learn to avoid them all. Bees and wasps mimic each other in this way. In another

▶ Insect Control

Because some insects can be destructive, the growth of their population must be controlled. One method of control is to use chemicals called **insecticides** that kill insects. Insecticides are effective, but they have an important drawback. These deadly chemicals tend to accumulate in the environment where they may harm not only pests but also useful insects and other animals.

For this reason, scientists have developed *biological controls,* or ways of controlling insect populations without insecticides. These include such methods as controlling the entry of insect pests into an area, developing plants that are resistant to insect pests, and using the natural enemies of these pests.

Biological controls are often effective. For example, ladybugs were imported from Australia in 1888 to control the cottony scale insect that was devastating the California citrus groves. Pheromones are often used to lure insects into a trap. A small amount will attract insects from several kilometers away.

Recently scientists have sterilized male insects as a way of eliminating insect pests. The screwworm fly, for example, causes great financial losses to ranchers by laying its eggs in the open sores of livestock. The maggots that hatch then feed on the wounds, enlarging them.

Scientists raise screwworm flies and subject the cocoons to radiation. The sterile flies that emerge are then "dumped" into an area to mate with the local flies. The resulting eggs do not hatch, and in time the fertile flies die out.

kind of mimicry, insects with no defenses of their own mimic the appearance of stinging or bad-tasting insects. As a result, predators avoid the mimic as well as the insect with the unpleasant taste or sting. For example, syrphid flies look like bees but do not sting.

Another kind of defense based on appearance is **camouflage,** or the ability to blend into surroundings. Many kinds of insects and animals have distinctive color markings that make them difficult to see. Predators have trouble locating prey that looks like its background. For this reason an insect is more likely to survive and produce offspring if it is camouflaged than if it is not.

Reviewing the Section

1. What is the role of the queen bee?
2. Give three examples of ways insects communicate.
3. How are pheromones used to control pest insects?
4. How does mimicry help mimic insects survive?

Investigation 34: Natural Pesticide Control

Purpose
To determine the effectiveness of caffeine as a pesticide

Materials
Sleep suppressant (caffeine) tablets, *Drosophila melanogaster* (fruit fly) culture, four culture vials, instant coffee, tea, nonether anesthesia, stereomicroscope

Procedure
1. Add a solution made of caffeine tablets to a vial containing a fruit fly culture. To two additional vials, add solutions of instant coffee and tea. Prepare the solutions according to directions given by your teacher. Different groups will use different concentrations of the three forms of caffeine. *What is the advantage of varying the concentration?*
2. Add six male and six female fruit flies to each vial.
3. Maintain 12 fruit flies in a culture vial to which no stimulant has been added. These will serve as the control group.
4. After one week anesthetize the flies in your culture vials. Follow your teacher's instructions for anesthetizing.
5. Remove the flies from the vial and count adults and larvae under the stereomicroscope.
6. Make a chart like the one shown.
7. Repeat steps 4 and 5 after one additional week and again after two additional weeks. Record your data on the chart.

Analyses and Conclusions
1. Compare the four vials for the effects of caffeine. Distinguish between the effects on the adult flies and on their larvae.
2. Have one student make a chart like the one shown, using data from the entire class. Indicate the concentration of each caffeine product and its effect on adult flies. What concentrations of caffeine appear to be most effective as a pesticide? Why?
3. What advantages does a caffeine-type pesticide have over a commercial product?

Going Further
- Repeat the experiment using other insects, such as grasshoppers.
- Investigate other naturally occurring substances, such as powdered herbs or spices, for their effectiveness as a pesticide.

Solution	At 1 week		At 2 weeks		At 3 weeks	
	No. of adults	No. of larvae	No. of adults	No. of larvae	No. of adults	No. of larvae
None 1 tablet 2 tablets 3 tablets 4 tablets						
1 t coffee 2 t coffee 3 t coffee 4 t coffee						
1 t tea 2 t tea 3 t tea 4 t tea						

Chapter 34 Review

Summary

There are over 750,000 different species of insects. They differ from other arthropods in having three pairs of legs and three main body parts—a head, a thorax, and an abdomen. Insects are the only arthropods that fly.

The grasshopper, a representative insect, has legs modified for jumping and two pairs of wings. It uses spiracles and tracheae to breathe. The female uses an ovipositor to deposit her eggs.

Most insects go through complete metamorphosis—a four-stage development from egg to larva to pupa to adult. The immature insects do not resemble the adults. Species that go through incomplete metamorphosis emerge from the egg as juveniles that resemble the adult.

Termites, ants, and some bees are social insects that live in large groups and engage in a division of labor. Honeybee hives consist of workers, drones, and a queen.

Many kinds of insects communicate with each other using flashes of light, motions, or chemicals called pheromones. Honeybees do a complex "waggle dance." Insects defend themselves by stinging or poisoning aggressors, mimicking other insects, and camouflaging themselves.

BioTerms

aorta (**508**)
camouflage (**513**)
chrysalis (**510**)
cocoon (**510**)
entomologist (**504**)
gastric caeca (**508**)
hindgut (**508**)
insecticide (**513**)

labium (**506**)
labrum (**506**)
mesothorax (**507**)
metamorphosis (**509**)
metathorax (**507**)
midgut (**508**)
mimicry (**512**)
ovipositor (**508**)

pheromone (**512**)
prothorax (**507**)
pupa (**510**)
salivary glands (**508**)
swarm (**506**)
tympanum (**507**)

BioQuiz   (Write all answers on a separate sheet of paper.)

I. Completion

1. The _____ carries blood from an insect's heart to its body cavities.
2. The _____ is a wingless, immature version of the adult grasshopper.
3. Insects differ from other arthropods in having _____ legs.
4. A protective shield covers a grasshopper's _____ .
5. Worker bees let other bees know where food is located by _____ .

II. Modified True and False

Mark each statement TRUE or FALSE. If false, change the underlined term to make the statement true.

6. An insect is inactive while in the larval stage.
7. Grasshoppers breathe by using spiracles.
8. Pheromones control metamorphosis.
9. Food is digested in the hindgut.
10. The female grasshopper digs a hole and lays eggs with an ovipositor.

III. Multiple Choice

11. Insects differ from other arthropods in having a) a cephalothorax.
    b) mandibles. c) jointed appendages.
    d) three body parts.
12. An insect's tympanum allows it to
    a) breathe. b) hear. c) deposit sperm.
    d) attract females.
13. Food is ground up by chitinous teeth in the insect's a) gizzard. b) crop.
    c) gastric caecae. d) hindgut.
14. When an insect enters the pupal stage, it spins a cocoon to a) keep warm.
    b) protect itself. c) provide food.
    d) attract a mate.
15. In bee societies the drones are the
    a) stingers. b) workers. c) fighters.
    d) reproducers.

IV. Essay

16. Why are bees said to engage in a division of labor?
17. How do chemical control and biological control of insects differ?
18. Is insect communication like human communication? Explain.
19. Is it beneficial for a bee to have a mimic? Why or why not?
20. How can the idea of natural selection help explain camouflage?

Applying and Extending Concepts

1. Prepare a report to the class on the sounds insects make. Explain how flies and bees hum, crickets chirp, cicadas sing, and June beetles buzz.
2. Many beekeepers have threatened to close their hives if the "killer bees" arrive in the United States, because the beekeepers do not wish to deal with aggressive bees. In addition to raising the cost of honey and beeswax, what other effect do you think the closing of the hives might have?
3. The flashing of fireflies is a complex form of communication. Its primary function is to allow the insects to find mates of their own species. Some female fireflies, however, have developed other uses for their flashing. Research the flashing patterns of *Photuris versicolor* and write a report explaining how females of this species use their flashing other than to attract mates. If time permits, present your findings to the class.
4. Worker bees protect their hive by stinging any intruding animals. This behavior is a clear example of altruism—or self-sacrificing behavior—because the act of stinging kills the bee. E. O. Wilson and other scientists working in the field of sociobiology have proposed an explanation of altruistic behavior based on modern genetics. Use your school or public library to prepare a report on how sociobiologists explain altruistic behavior.

Related Readings

Bland, R. G. and H. E. Jaques. *How to Know the Insects*. Dubuque, Iowa: Wm. C. Brown, 1978. This book is a complete guide to identifying insects and the setting up of a collection.

"Of Sting and Fang." *Science 85* 6 (January/February 1985) 64–69. The dramatic photos and text in this article tell the story of how a female wasp battles a tarantula in order to perpetuate her species.

Gould, J. L. "Do Honeybees Know What They Are Doing?" *Natural History* 88 (June 1979): 66–75. The article discusses how bees find and remember food sources.

Fishes

Pacific angel fish

Introduction

The Pacific angel fish shown above is a member of the phylum Chordata and of the subphylum Vertebrata. Like other vertebrates, including birds and mammals, fishes have backbones. However, they differ from birds and mammals in significant ways. Among these is the inability of fishes to maintain a constant body temperature.

Both in number of species and in individuals, fishes total more than all other vertebrates combined. Fishes are divided into three main classes: jawless fishes, jawed fishes that have skeletons of cartilage, and jawed fishes that have bony skeletons. The differences among these three classes are as great as those that separate the amphibians, reptiles, birds, and mammals from one another.

Chordates and Vertebrates

The 43,000 chordate species are the most recent products of evolution and the most complex of all animals. The chordates are amazingly diverse. They have adapted to life on land, in water, and in the air. They range in size from the tiny hummingbird to the giant blue whale, which is the largest animal ever to have lived on earth.

The vertebrates, animals with backbones, make up the largest subphylum within the phylum Chordata. The majority of the animals most familiar to you—sparrows and elephants, goldfish and human beings—are vertebrates.

35.1 Characteristics of Chordates

Despite their variety, all chordates share certain traits. *At some time during their life cycles, all chordates possess four distinctive structures: a notochord, a nerve cord, gill slits, and a tail.* The **notochord** is a long, firm rod that extends along the back of the animal's body. In vertebrates the notochord is present in the embryo and is later replaced by a backbone. The hollow, tubular **nerve cord** runs the dorsal length of the animal, just above the notochord. In most chordates the anterior end of this nerve cord becomes a brain. Paired openings in the throat region are called **gill slits.** In fish the gill slits serve a respiratory function. In more complex vertebrates, the gill slits have become modified for other uses. Finally, most adult chordates possess **tails,** which are blocks of muscle tissue surrounding the posterior end of the animal's skeleton.

35.2 Classification of Chordates

In addition to the vertebrates, the phylum Chordata includes two other minor subphyla. Members of the subphylum Urochordata (YOOR uh kawr DAHD uh) are called *tunicates,* or *sea squirts.* Except for their gill slits, tunicates show little resemblance to other chordates. These cylinder-shaped animals attach themselves to underwater rocks, reefs, and ocean floors. The third subphylum, the Cephalochordata (SEF uh loh kawr DAHD uh), includes the lancelet, a thin, fishlike animal that lives in warm, shallow ocean waters. As Figure 35–1 shows, the lancelet retains all four chordate characteristics throughout its life. Members of the subphyla Urochordata and Cephalochordata are known as the *lower chordates.* The more complex species of the subphylum Vertebrata are called the *higher chordates.*

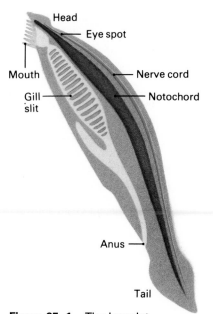

Figure 35–1. The lancelet exhibits all four chordate characteristics: notochord, nerve cord, gill slits, and tail.

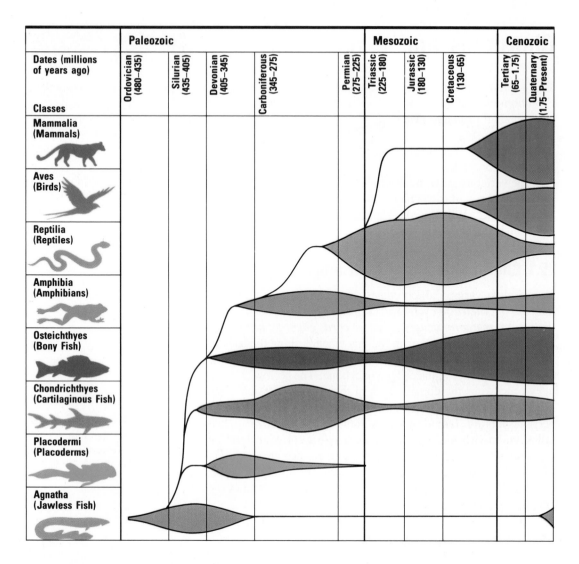

Classes / Dates (millions of years ago)	Paleozoic					Mesozoic				Cenozoic	
	Ordovician (480–435)	Silurian (435–405)	Devonian (405–345)	Carboniferous (345–275)	Permian (275–225)	Triassic (225–180)	Jurassic (180–130)	Cretaceous (130–65)	Tertiary (65–1.75)	Quaternary (1.75–Present)	

Mammalia (Mammals)

Aves (Birds)

Reptilia (Reptiles)

Amphibia (Amphibians)

Osteichthyes (Bony Fish)

Chondrichthyes (Cartilaginous Fish)

Placodermi (Placoderms)

Agnatha (Jawless Fish)

Figure 35–2. A comparison of the changing populations of vertebrate classes over the last 500 million years shows only one extinct class. Placoderms died out by the end of the Paleozoic era.

35.3 Characteristics of Vertebrates

A strong, flexible backbone and complex body systems have enabled the vertebrates to inhabit many environments. Three of the seven living vertebrate classes—the jawless fishes, cartilaginous fishes, and bony fishes—live entirely in water. The four other living vertebrate classes are amphibians, reptiles, birds, and mammals. Amphibians are adapted to life both on land and in the water, while reptiles and mammals are primarily land dwellers. All but a few birds, such as penguins, can fly.

All vertebrates share a number of physical characteristics that set them apart from the other chordates and all invertebrates. Vertebrates are bilaterally symmetrical with two pairs of

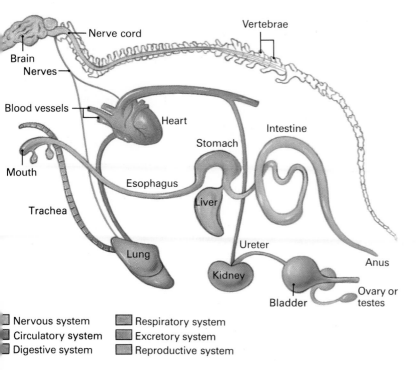

Nervous system
Circulatory system
Digestive system
Respiratory system
Excretory system
Reproductive system

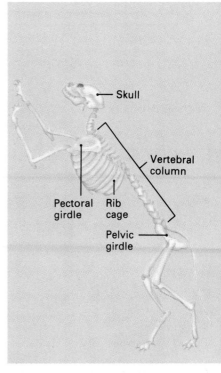

appendages such as limbs, fins, or wings. They exhibit cephalization—that is, their main sense organs are located in their heads. All vertebrates also have a closed circulatory system and a **coelom,** which is a large central body cavity that contains their vital organs. Finally, all vertebrates have an internal support system called an **endoskeleton.** The endoskeleton is made of bone or cartilage or a combination of both.

Figure 35–3 shows the skeletal system and most other major systems of a representative vertebrate, the domestic cat. A major part of the endoskeleton, and the feature that distinguishes vertebrates from all other animals, is the **vertebral column,** or backbone. The backbone is composed of bony parts called **vertebrae** cushioned by cartilage. Together, the backbone, the bones of the skull, and the rib cage make up the **axial** (AK see uhl) **skeleton.**

The endoskeleton of most vertebrates also includes two structures called *girdles*. The girdles are wide, flattened surfaces that connect the limbs—arms, legs, fins, wings, or flippers—to the axial skeleton. The **pectoral** (PEHK tuhr uhl) **girdle** is located toward the top or front of the animal. The **pelvic girdle** is found near the bottom or back of the animal. The girdles and their associated limbs make up the **appendicular** (ap uhn DIHK yuh luhr) **skeleton.**

Q/A

Q: *Do humans ever have tails?*

A: Yes. Tails form in human embryos and reach their greatest length during the second month of embryonic life. Afterwards, the tails usually disappear. In a few cases, however, infants are born with short tails. These are usually surgically removed soon after birth.

Table 35–1: Vertebrate Systems

System	Description	Function
Skeletal	Endoskeleton of bone and/or cartilage	Provides support and protection
Muscular	Contractile tissue attached to bone or cartilage; some are part of internal organ walls	Together with the skeleton, enables animals to move; protects some organs
Integumentary	Body coverings of skin, hair, scales, or feathers	Provides support and protection; also involved in excretion, respiration, and perception
Digestive	Tube extending from mouth to anus, and associated organs	Prepares food for use by the animal's cells; removes solid wastes from the body
Respiratory	Gills or lungs and associated structures	Exchanges gases between the animal and its environment
Circulatory	Closed system of blood vessels, with two-, three-, or four-chambered heart	Carries blood from the heart to the rest of the body
Excretory	Pair of kidneys and associated tubes; skin, lungs, and gills also may be involved	Removes cellular wastes from the body
Nervous	Spinal cord, brain, nerves, and sense organs	Monitors the environment; controls and coordinates many body functions
Reproductive	Male or female reproductive organs	Produces and carries eggs or sperm; in some vertebrates, allows for internal fertilization and development of offspring
Endocrine	Glands	Secretes chemicals that regulate body growth, reproduction, and development

35.4 Vertebrate Systems

Both invertebrates and vertebrates have body tissues that are organized into organs that perform specific body functions. Look at Figure 35–3 on page 521. You can see that the mouth, esophagus, and stomach, for example, form one integrated system, the digestive system. On the whole, the organs of vertebrates are more highly developed than those of invertebrates and form 10 complex systems, shown in Table 35–1.

Reviewing the Section

1. What four structures do chordates have in common?
2. Compare the appendicular and the axial skeletons.
3. What is the function of the excretory system?

Jawless Fishes

Section Objectives

- *Name* the major characteristics of the class Agnatha.
- *Describe* the feeding habits of lampreys and hagfishes.
- *Summarize* the life cycle of the lamprey.

Lampreys and hagfishes are the only existing members of the class Agnatha (AG nuh thuh). *Agnatha* means "without jaws." The 60 living species of agnathans all feed by latching onto their prey with suckerlike, jawless mouths. These fishes are the only parasitic vertebrates.

Agnathans have smooth, cylinder-like bodies with flexible skeletons formed of cartilage. They have a notochord at all stages of their life cycle. Like all fishes, agnathans have a heart with two chambers. The *ventricle* pumps blood to the body, and the *atrium* receives the blood as it returns from the body.

35.5 Lampreys

Lampreys live in cool, fresh and coastal waters. Their long, thin bodies measure 40 to 80 cm (16 to 32 in.) in length. Lampreys have one or two fins that extend along their dorsal surface and a tail fin. They have seven circular gill slits on each side of their bodies. **Gills** are the respiratory organs of fish. Lampreys have a single nostril on the top of the head and well-developed eyes. The lamprey's mouth, a round sucking organ called an **oral disc,** has sharp, rasping teeth.

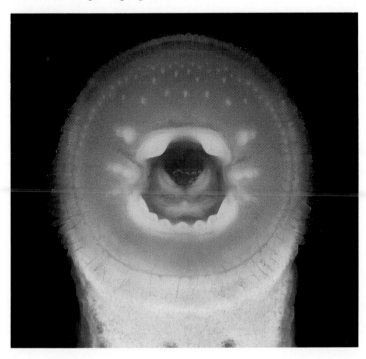

Figure 35–4. A toothed oral disc is used during feeding by a lamprey (left). This fish, which is jawless, attaches the oral disc to other fish and sucks out the blood.

Q/A

Q: *Do lampreys always kill the fish they feed on?*

A: Not always. Sometimes fish are caught that have circular scars caused by lampreys feeding on them.

Figure 35–5. The jawless hagfish often attacks fish caught in fishing nets. The hagfish bores inside its prey and eats the inside of the fish, leaving the skin intact.

Most species of lampreys are parasites. The lamprey uses its oral disc to fasten itself onto a fish. It then scrapes a hole in the side of the fish with its teeth and sucks out the blood and tissues. This usually results in the death of the host fish. Because of their feeding methods, lampreys can cause severe economic loss to commercial fishing industries. In the 1950s one species of sea lamprey threatened the entire Great Lakes fishing industry. The lamprey was controlled by chemicals that killed newly hatched larvae.

Lampreys reproduce by laying eggs in nests hollowed out in the gravelly bottom of freshwater streams or lakes. Eggs are fertilized externally and hatch into wormlike larvae that burrow into the gravel. The larval lamprey remains burrowed with its head protruding for three years, feeding on microorganisms. It then changes into an adult, a process called *metamorphosis*. The adults of most species swim into the ocean to feed and return to fresh water to reproduce. Some species remain in fresh water all their lives.

35.6 Hagfishes

Hagfishes are bottom dwellers that live only in cold ocean water. They have 5 to 15 pairs of gill openings. Hagfishes have mucus-secreting glands all over their bodies and so are sometimes known as *slime eels*. Hagfishes have poorly developed eyes that are covered with skin. They have a slitlike, toothed mouth but lack the lamprey's oral disc.

Hagfishes locate their food by scent. They are mainly scavengers that feed on dead or dying fish and marine invertebrates. Hagfishes feed much like lampreys, drilling a hole and sucking the blood and insides from the animal. When not feeding, hagfishes live burrowed in the ocean floor with just the tips of their heads protruding.

Hagfishes are *hermaphrodites*—that is, a single fish has both male and female sex organs. *A hagfish may produce sperm one season and eggs the next.* Eggs are fertilized externally. The young of the hagfish hatch from the egg as miniature versions of the adult.

Reviewing the Section

1. What are two major characteristics of agnathans?
2. Why is the presence of the lamprey in an area a threat to the local fishing industry?
3. Compare the lamprey and hagfish life cycles.

Cartilaginous Fishes

Section Objectives

Sharks, rays, and skates are members of the class Chondricthyes (kahn DRIHK thee eez). *Chondrichthyes* means "cartilage fishes"; all members of this class have skeletons made of cartilage. Unlike the jawless agnathans, sharks and related fishes have hinged jaws lined with rows of teeth that are continuously replaced when worn or lost. Their skin is covered with small, pointed teeth, giving the skin the texture of rough sandpaper.

Sharks, skates, and rays have five to seven pairs of gills and a two-chambered heart. A spiral membrane or valve extends through the intestine. The valve delays the passage of food through the intestine, and furthers digestion. The fishes have separate sexes, and fertilization of eggs is internal.

Fossil remains indicate that fishes similar to modern chondrichthyes existed over 100 million years ago. Approximately 625 living species have been identified. Most of these species live in salt water.

- *List* the major characteristics of cartilaginous fishes.
- *Name* three features that contribute to the shark's success as a predator.
- *Distinguish* skates and rays from sharks.

35.7 Sharks

Sharks are carnivorous, or meat-eating, fish with torpedo-like bodies that are well-adapted to a predatory life. Sharks live in every ocean and are particularly abundant in warm seas. The whale shark is the largest of all fishes. It grows to about 18 m (60 ft.) long and weighs about 14 metric tons (15.4 tons).

Figure 35–6. The sand tiger shark inhabits warm, shallow Atlantic waters. Small fish called *remoras* (lower left) travel with the shark and feed on its leftovers. Remoras attach themselves to the shark's body by means of adhesive discs on the top of their heads.

A: No. Of the 250 known species of sharks, only 27 species have been involved in attacks on humans. One expert estimates that the chances of being attacked by a shark are equal to the chances of being struck by lightning—extremely rare.

Figure 35–7. This illustration shows the lateral line system of the shark.

Structure Sharks are powerful swimmers that can move rapidly when excited. Scientists estimate that blue sharks can swim 69 km (43 mi.) per hour for short bursts. Their rapid speed is made possible by strong muscles and streamlined bodies that are stabilized by two pairs of fins. The **pectoral fins** grow on the sides just behind the head. The **pelvic fins** grow farther back on the underside of the body. Two **dorsal fins** extend along the back, one behind the other. Sharks also have a vertical **caudal fin,** which is an expansion of the tail. Because sharks are heavier than water, they will sink if they do not continually move forward.

Cone-shaped toothlike structures called **placoid scales** cover the shark's tough leathery skin. Each scale consists of an enamel-like outer layer and a tooth, called a denticle, that projects from the center. Modified scales form the two rows of backward pointing teeth in the animal's jaws. The backward slant of the teeth makes it possible for the shark to hold food more securely in its jaws.

Senses Sharks have keenly developed senses. As much as two-thirds of the shark's brain is given over to the sense of smell. Sharks can smell the odor of blood as far away as 0.4 km (0.25 mi.), in concentrations as small as one part of blood in 1 million parts of water. The eyesight of sharks is not as well developed as their sense of smell, but they can see moving objects as far away as 15 m (50 ft.).

Sharks also detect movement and locate objects through a system of fluid-filled canals called a **lateral line system.** This system extends down the head and along the side of the shark. Openings to the body's surface occur at intervals.

Vibrations transmitted through the ocean water jostle the fluid in the lateral line system. The movement of the fluid stimulates special receptor cells in the canals that send nerve impulses to the brain. Combined with the animal's other sense organs, the lateral line system makes sharks keen monitors of their environment.

Reproduction Unlike most fishes, the eggs of sharks are fertilized inside the female's body. The male grasps the female with his teeth and inserts the sperm into her with two organs called *claspers*. In most species the eggs develop inside the mother, which later gives birth to live "pups." A few species lay eggs, which may take as long as 15 months to hatch. The eggs of the whale shark are among the largest in the animal kingdom. They may measure up to 30 cm (12 in.) in length and 14 cm (5.5 in.) in width.

35.8 Rays and Skates

Unlike sharks, rays and skates have flat, broad bodies well suited for life on the bottom of the ocean. Their pectoral fins are greatly enlarged and flap like wings when the fish are swimming. The gill openings are on the underside of the head. Water for breathing enters the body through openings called **spiracles** on the top of the head. The spiracles move water to the gills.

Rays are typically less than 1 m (3.3 ft.) long. The giant manta ray measures as much as 7 m (23 ft.) in breadth. Most of the 350 species of rays are harmless to humans. Most rays eat smaller fishes and invertebrates. Some have unique methods of defense and attack. The sting ray has a long, slender tail with sawlike spines that can inject poison. Electric rays have powerful organs that emit an electric charge, stunning their prey.

Skates are similar to rays, though most are somewhat smaller. Their sandlike coloration enables them to blend in with the ocean bottom. When a source of food passes overhead, a skate leaves its hiding place, swims above the other animal, and settles down upon it, pinning the victim between itself and the ocean floor.

Figure 35–8. The manta ray on the left and the skate on the right exhibit flattened bodies suited for living on the ocean floor.

Reviewing the Section

1. What do sharks and skates have in common with lampreys and hagfishes? How do the two classes differ?
2. What features of sharks help them swim rapidly?
3. How does a shark's lateral line system help it to locate food?
4. What physical features of rays make them well suited for life as bottom dwellers?

- *Distinguish* bony fishes from the other main classes of fishes.
- *Label* the main skeletal structures in a diagram of a typical bony fish.
- *Describe* how fish respire.
- *Explain* what is meant by *spawning*.
- *List* three patterns of fish migration and give an example of each.

Bony Fishes

The vast majority of the world's fishes belong to the class Osteichthyes (ahs tee IHK thee eez). *Osteichthyes* means "bony fishes." As the name indicates, these fishes have skeletons made of bone instead of cartilage. Like sharks, skates, and rays, however, bony fishes also have jaws and scaly skin.

35.9 Characteristics

Bony fishes number about 20,000 species, or 95 percent of all fish species. The species differ greatly in appearance and behavior. Some are as thin as pipes. Others have protective coloring and shapes that make them barely distinguishable from the rocks in which they hide. Some species are sluggish and move as little as possible. Others move rapidly in pursuit of prey. Some species live solitary lives, while other species live and travel together in large groups, called *schools*.

Moving together in schools protects the individual members from predators. A school that normally spreads out over several hundred meters may, in the presence of danger, condense to form a sphere only a meter (3.3 ft.) in diameter. When this happens, a predator may mistake the school for a single large, powerful fish. Or the predator, unable to concentrate on a single individual, may not be able to catch any fish at all. About 20 percent of fish species travel together in schools.

Figure 35–9. Bony fishes vary greatly in their protective adaptations to underwater life. The pipefish (left) is hard to see among aquatic grasses, while the leaf fish (center) resembles a leaf floating on the water. Hussars (right) protect themselves by traveling in large schools.

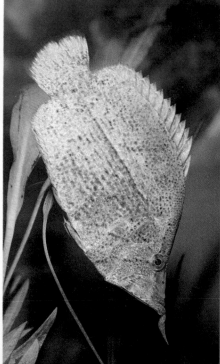

Scientists divide bony fishes into three major groups: lobe-finned fishes, lungfishes, and ray-finned fishes. The first group, the **lobe-finned fishes,** have dorsal and pectoral fins with large fleshy bases supported by leglike bones. Only one living species is known, the *coelacanth* (SEE luh kanth). Lobe-finned fishes were considered extinct until 1938 when a coelacanth was caught off the coast of South Africa.

Lungfishes use lungs as well as gills for breathing. *Lungs* are the internal respiratory organs used by all air-breathing land animals. The three surviving genuses of lungfishes differ in the amount of time they can survive out of water using only their lungs. The Australian lungfish can survive in oxygen-poor water by coming to the surface to gulp air, but it cannot live out of water. The South American and African lungfishes bury themselves in mud when the streams in which they live dry up. They breathe through their lungs for periods of up to two years until rains come and the streams fill up again.

Ray-finned fishes have fins that are supported by a number of long bones called *rays*. Most fishes, including perch, bass, and all the familiar fishes of the world, belong in this category. Ray-finned fishes evolved about 400 million years ago. Scientists speculate that they arose in freshwater habitats and some later migrated to the oceans. They soon replaced cartilaginous fishes as the dominant kind of fish on Earth.

35.10 The Trout: A Typical Bony Fish

The trout is a bony fish that lives in northern lakes and rivers of the United States or returns from the sea to breed in northern streams. One species, the brook trout, grows to about 46 cm

Figure 35–10. This illustration shows the three major groups of bony fishes: the coelacanth, the only living lobe-finned fish (top left); the lungfish (bottom left); and the ray-finned fish (right), here represented by a perch.

Q/A

Q: *How do fish sleep?*

A: Fish have no eyelids, so they cannot close their eyes to sleep. Some rest on the bottom; others doze in mid-water. Some fishes, such as catfishes and some eels, feed at night and sleep during the day.

(18 in.) long and has light markings on a dark background. The trout's external and internal anatomy is in many ways representative of all bony fishes.

External Structure As Figure 35–11 shows, the trout has an elongated body. Its mouth, nostrils, and eyes are located on its head. The trout has two sets of gills, one on either side of the body. A protective flap of tissue called an **operculum** covers the gills.

The trout's fins stabilize and maneuver the fish and propel it foward. The forwardmost pair are the pectoral fins. Midway along the sides of its body are the paired pelvic fins. The pectoral and pelvic fins help the fish to steer and brake. Two dorsal fins extend from the trout's back. An **anal fin** extends from its ventral surface, near its anal opening. The trout's deeply forked caudal fin forms part of the tail and adds force to the fish's swimming movements.

A trout swims by moving its body from side to side while swinging its tail in the opposite direction. The paired fins help change course. The caudal fin provides the power; by pushing against the water, it forces the fish forward.

Figure 35–11. The external and internal anatomy of the trout is shown in the illustration below.

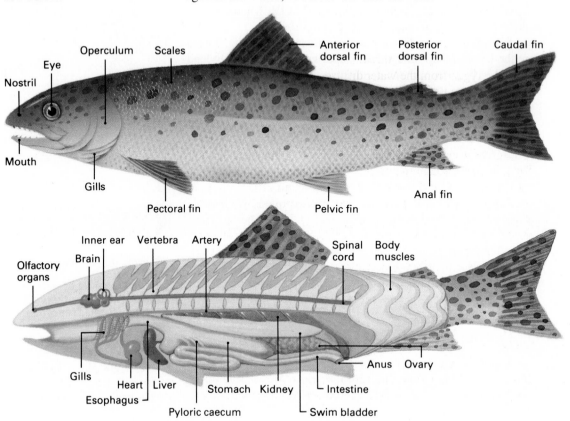

Most fishes have skin covered by thin overlapping segments called **scales.** The trout's scales are thin, bony, and rounded at the edges. Fish are born with a certain number of scales. Though the scales enlarge throughout the life of the fish, new ones are never grown.

Figure 35–12. The trout, like most bony fishes, has thin, flexible scales that are arranged in overlapping rows.

Skeletal and Digestive Systems The trout's skeleton is composed almost entirely of bone. The anterior end of the vertebral column is the skull, which covers and protects the brain. Ribs project from the backbone. The spinal cord runs parallel to the backbone and nerves branch from it to various parts of the body. Trout feed on insects and fish eggs. The food moves from the throat cavity down the *esophagus,* a short tube that leads to the stomach. There the food is stored and digestion begins. Near the stomach is the *liver,* a large organ that secretes *bile,* a substance that breaks down the fats in the food. Most of the fish's digestion takes place in special intestinal pouches called *pyloric caeca.* Undigested material leaves through its anus.

Respiratory and Circulatory Systems Like most fishes, trout obtain oxygen by means of gills. Trout have four gills, two each in *gill chambers* on either side of the head. Each gill consists of a bony *gill arch* fringed with thin-walled tissues called *gill filaments.* The gill filaments contain many small blood vessels. As the mouth and throat force water over the gills, dissolved oxygen from the water diffuses through the thin walls of the blood vessels and enters the blood.

Trout also have an organ unique to bony fishes—a gas-filled **swim bladder** in the coelom that acts as a float. Gases pass into and out of the swim bladder from the blood. As the bladder fills up, the fish becomes more buoyant and rises in the water. As the bladder deflates, the fish becomes less buoyant and sinks. Glands regulate the gas content in the swim bladder, enabling the fish to remain at a specific depth in the water with little effort.

A two-chambered heart pumps blood through a series of vessels to all parts of the body. *Arteries* carry blood away from the heart; *veins* carry blood back to the heart. The small vessels that form the connecting network between arteries and veins are called *capillaries.* The exchange of nutrients and waste products takes place in the capillaries.

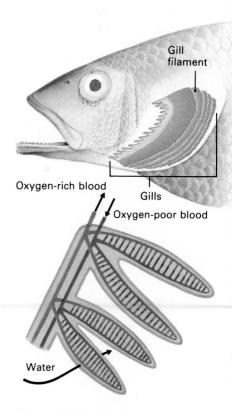

Figure 35–13. The illustration above shows how trout obtain oxygen by means of gills.

Excretory System In fishes, as in all vertebrates, the *kidney* is the main excretory organ that removes wastes from the blood. In trout the kidney also plays an important role in maintaining the proper *osmotic balance* of water and salts in the body.

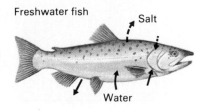

Freshwater fish

Salt

Water

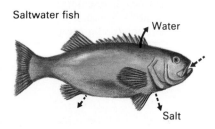

Saltwater fish

Water

Salt

Figure 35–14. To maintain a proper osmotic balance, freshwater fishes (top) must continually take up salt from the water and expel water. Saltwater fishes (bottom) do the reverse. They have cells that excrete salt and they expel only small amounts of water.

Q/A

Q: *Do walking catfish really walk?*

A: No. The walking catfish makes its way from one pond to another using its side fins and tail to help it crawl over the ground.

The concentration of salt in the bodies of freshwater trout is higher than in the water around them. Because water moves from areas of lower salt concentration to areas of higher salt concentration, trout and other freshwater fishes tend to gain water, lowering the salt concentration in the fish's cells. To maintain a proper osmotic balance, freshwater fishes have special salt-absorbing glands in their gills that take up salt from the water and transport it to cells. Also, large amounts of water are expelled as urine. The result is a system that is osmotically balanced.

Fishes that live in the ocean have the opposite problem. The concentration of salt in their bodies is lower than that in the water. As a result, marine fishes constantly lose water to the sea. To protect against dehydration, ocean fishes take in a lot of water. They drink almost continuously. Along with the water, the fishes also take in much more salt than their bodies can use. Saltwater fishes have salt-secreting cells in their gills and kidneys to rid themselves of extra salt and thus maintain an osmotic balance. Furthermore, they excrete a concentrated urine that has little water.

Fishes that move back and forth between saltwater and freshwater environments can adjust their bodies for each environment. Trout, salmon, and sturgeon are fishes that have well-developed kidneys for life in fresh water and salt-secreting cells for adjustment to life in salt water.

Nervous System The trout's brain coordinates the information received from its sense organs. A fish's keenest sense is smell. The **olfactory organs,** which monitor smell, are located above and on each side of the mouth. Water enters the olfactory organs by way of the nostrils. Chemicals in the water stimulate certain nerve cells to send an electrical message to the brain.

Fish have no external ears or eardrums; yet they hear. Sound vibrations arrive through the water and are transmitted through the fish's skin and bones to an *inner ear*. The inner ear consists of a series of tubes on each side of the head that are lined with nerve cells. When stimulated, the nerve cells send impulses to the brain. Like sharks, bony fishes monitor vibrations in the water by means of a lateral line system. The sensory cells of the system can detect very low frequency vibrations in the water.

The trout's eyes can see both to the left and right at the same time—a definite advantage for an animal that has no neck to turn the head from side to side. Most fish have poor vision, however. They can probably see objects no farther than 0.5 m (1.6 ft.) away.

► Migration of Fish

Some fish travel vast distances in search of food or spawning grounds. Movements like these that follow a regular pattern year after year are called **migrations**. Scientists have discovered the travel routes of many migratory species by "tagging" fish in one spot and recording where the tagged fish were later caught.

Migratory fish can be classified according to their travels. Migrants such as herring, cod, and white tuna live and travel only in the ocean. These fish are called *oceanodromous.* White tuna, for example, winter near the Azores and the Canary Islands off Africa, spawn in the spring, and then migrate to the waters off Iceland to spend the summer. Food is plentiful there during the summer months.

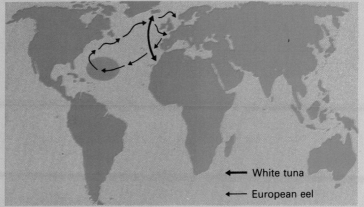

← White tuna

← European eel

Anadromous fish live in the ocean and migrate to fresh water to spawn. The Pacific salmon spends its adult life in the ocean. When it is ready to spawn, it returns to the same stream in which it was born. Even those anadromous fish hatched in fish hatcheries and later placed in the ocean return to the same stream in which their ancestors were born.

Catadromous fish live in fresh water and return to the sea to breed. Eels are catadromous fish. The adult European eel lives in lakes and streams in Europe. It migrates 5,000 km (3,100 mi.) to a specific spot in the Sargasso Sea in the western Atlantic to spawn. The larval eels drift with the Gulf Stream. Several years are required for the larval eels to float back to the European continent. Once there, the young eels migrate up rivers by the millions.

Reproductive System   Like most bony fishes, trout fertilize their eggs externally. The females **spawn**—that is, they shed eggs from their bodies into nests that have been hollowed out on the floor of the river or lake bed. The males then release **milt,** which is a fluid containing sperm, over the eggs. In many species, this process takes place in special locations called *spawning grounds*. Some types of fish travel very great distances to return to the same spawning ground where they themselves were hatched.

Spawning is an inefficient method of reproduction. Only a small percentage of the eggs are fertilized and actually develop.

Spotlight on Biologists: Eugenie Clark

Born: New York City, 1922
Degree: Ph.D., New York University

When American scientist Eugenie Clark earned her Ph.D. in 1951, she may have been the only woman marine biologist. The challenge to prove herself in a male-dominated profession and her avid interest in diving and observing fish prodded her into becoming one of the world's leading experts in fish behavior.

Clark is especially noted for her fearless underwater study of poisonous tropical fish and the behavior of sharks. She has observed firsthand many of the nearly 350 species of sharks worldwide to study their feeding, learning, and reproductive behavior.

During trips to the Yucatan and Japan, Clark observed "sleeping" sharks. She determined that their sluggish behavior was a result of the high oxygen and low salt content of the water in which they lived.

Clark described her work with sharks in a highly successful popular book, *Lady with a Spear.* The book resulted in an offer

from the Vanderbilt family to set up a marine research laboratory for her. Clark helped found and direct what is now the Mote Marine Laboratory near Sarasota, Florida. Currently she teaches marine biology at the University of Maryland.

Even when the eggs develop, the baby fishes are in constant danger of being eaten by predators. Some fish species compensate for this inefficiency by producing a large number of eggs. A female cod, for example, produces over 9 million eggs during a spawning season. Although most of these are destroyed or eaten, a few survive and develop into new cod.

A few fish species are *live-bearers*. In live-bearers, the eggs are fertilized internally, and the young fish develop within the mother's body. Guppies, mollies, and swordtails are examples of live-bearing fish.

Reviewing the Section

1. Explain how trout, lampreys, and sharks differ from one another.
2. Name the fins of a trout and give their location.
3. How do a trout's gills function to get oxygen?
4. How do saltwater fish maintain an osmotic balance?
5. Describe the migration pattern of the white tuna.

Investigation 35: Learned Behavior in Siamese Fighting Fish

Purpose
To investigate how fish can learn to modify their behavior

Materials
Two-gallon aquarium tank, section of glass or clear hard plastic, Siamese fighting fish (alternative: angel fish), guppies (five to nine days old), tubifex worms

Procedure
1. In a two-gallon aquarium tank, place gravel, plants, and water. Partition the tank with the piece of glass or clear plastic, as shown below.

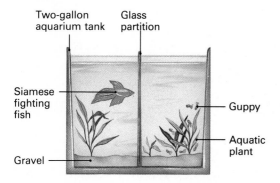

Two-gallon aquarium tank   Glass partition

Siamese fighting fish

Guppy

Aquatic plant

Gravel

2. Place a number of guppies on one side; place a single fighting fish on the other side. (If angel fish are used, you may place more than one on the side opposite the guppies.) *Why should only one Siamese fighting fish be placed in the tank?*
3. Observe the behavior of the predator fish for several minutes as it attempts to eat the guppies. *What changes do you see in the behavior of the fish?* Record your observations.
4. Provide several tubifex worms as a food source for the Siamese fighting fish. Observe the feeding. *What advantage does a live food source like tubifex have over a dry powdered product?*

Analyses and Conclusions
1. Make a chart like the one shown. Complete your chart by describing and comparing the general behavior of each type of fish.

Fish	Behavior		
	Feeding	Swimming	Other
Guppy			
Siamese fighting fish			

2. After how long did you notice behavioral changes?
3. Account for the change in the fishes' behavior.

Going Further
- Carefully remove the partition from the tank. Observe, record, and analyze the fishes' behavior.
- Repeat the same experiment (with partition) using different species of fish. Compare your new observations with the original ones. Do all fish learn at the same rate?

Chapter 35 Review

Summary

Phylum Chordata includes as its most advanced and familiar class the vertebrates, or animals with backbones. The seven living classes are: jawless fishes, cartilaginous fishes, bony fishes, amphibians, reptiles, birds, and mammals.

Jawless fishes, such as lampreys and hagfishes, feed by attaching themselves to other fish and sucking out blood and tissues. Cartilaginous fishes include sharks, skates, and rays. Most sharks are predators. They are powerful swimmers with a keen sense of smell.

Rays and skates have flattened bodies and are mostly bottom dwellers.

Although bony fishes are divided into three groups—lobe-finned fishes, lungfishes, and ray-finned fishes—all common species are ray-finned. Bony fishes, such as trout, contain typical vertebrate organs, but have a special swim bladder that regulates their buoyancy in water. Fish obtain oxygen through gills. Fertilization usually takes place externally after the female spawns.

BioTerms

anal fin **(530)**
appendicular skeleton **(521)**
axial skeleton **(521)**
caudal fin **(526)**
coelom **(521)**
dorsal fin **(526)**
endoskeleton **(521)**
gill **(523)**
gill slits **(519)**
lateral line system **(526)**
lobe-finned fish **(529)**

lungfish **(529)**
migration **(533)**
milt **(533)**
nerve cord **(519)**
notochord **(519)**
olfactory organ **(532)**
operculum **(530)**
oral disc **(523)**
pectoral fin **(526)**
pectoral girdle **(521)**
pelvic fin **(526)**

pelvic girdle **(521)**
placoid scale **(526)**
ray-finned fish **(529)**
scale **(531)**
spawn **(533)**
spiracle **(527)**
swim bladder **(531)**
tail **(519)**
vertebra **(521)**
vertebral column **(521)**

BioQuiz   (Write all answers on a separate sheet of paper.)

I. Completion

1. The manta ray takes in water for breathing through openings on the head called _____.
2. The _____ system produces eggs and sperm.
3. The gill filaments are attached to the bony _____.
4. Rays and skates swim by flapping their _____.

II. Modified True and False

Mark each statement TRUE or FALSE. If false, change the underlined term to make the statement true.

5. The two chambers of a fish's heart are the <u>artery</u> and the ventricle.
6. <u>Hagfishes</u> emerge from the egg as larvae.
7. <u>Male</u> trout fertilize eggs by releasing milt over them.

III. Multiple Choice

8. The vital organs of vertebrates are found in the a) pyloric caeca. b) coelom. c) oral disc. d) operculum.

9. The hind legs of vertebrates are attached to the a) pectoral girdle. b) vertebral column. c) pelvic girdle. d) coelom.

10. Freshwater fish maintain osmotic balance by excreting a) salt. b) water. c) oxygen. d) blood.

11. Vertebrates differ from other chordates in having a) a notochord. b) a dorsal nerve cord. c) gill slits. d) a vertebral column.

12. The function of the operculum is best described as a) protection. b) excretion. c) reproduction. d) digestion.

13. Lampreys feed by a) stealing food from others. b) filtering out plankton. c) eating live fish. d) swallowing prey whole.

14. Sharks use their claspers for a) communication. b) fertilization. c) feeding. d) respiration.

15. Rays have bodies modified for a) hunting. b) bottom dwelling. c) burrowing in mud. d) rapid swimming.

IV. Essay

16. How do trout swim?

17. What is the function of the pectoral and pelvic girdles in vertebrates?

18. What does a trout do to regulate the depth at which it floats?

19. How do the habitat and feeding habits of the immature lamprey differ from those of the mature fish?

20. What takes place at a spawning ground?

Applying and Extending Concepts

1. Make a list of all the kinds of vertebrates you see in a single day. Write down the name of each animal and the class to which it belongs. For each animal, list one characteristic that is an adaptation to living on land or in water.

2. Use your library to research data for a chart on sharks. Include basking sharks, hammerheads, nurse sharks, white sharks, and whale sharks. Make columns giving their scientific name, habitat, and feeding habits. You may also wish to extend the chart to include their lengths and whether they have ever been involved in attacks on people.

3. Gills enable fish to withdraw dissolved oxygen from the water. Why are most fish unable to live outside the water even when oxygen is readily available?

Related Readings

Clark, E. ''Sharks: Magnificent and Misunderstood.'' *National Geographic* 160 (August 1981): 138–186. The author describes her experiences researching sharks and offers a number of fascinating observations on their everyday behavior.

Levine, J. S. ''Life Imitates Art: Living Canvases of Marine Animals.'' *Smithsonian* 13 (March 1983): 126–135. This article describes how pigmentation is used for defense and for courtship.

Romer, A. S. and T. S. Parsons. *The Vertebrate Body, Shorter Version*. Philadelphia: W. B. Saunders Company, 1978. This classic textbook includes excellent drawings showing the evolution of vertebrate body structures from early chordates through the various classes of vertebrates.

36

Amphibians

Red-eyed tree frog, *Agalychnis callidryas*

Introduction

About 370 million years ago the earth was a much different place than it is today. The land alternated between periods of extreme droughts and periods of flooding. The first land plants had just begun to appear. These were small, leafless organisms less than 0.6 m (2 ft.) tall. Fish dominated the oceans, but no vertebrates yet lived on land.

In the oceans, animals competed fiercely for food and space. Some animals sought refuge on land. Unable to breathe air or to move easily on soil, they stayed only briefly, then retreated back to the water. Some lobe-finned fishes were better suited for life on shore. They had lungs as well as gills and limblike fins that enabled them to crawl from pond to pond. Their descendants became the first amphibians.

From Water to Land

Section Objectives

The word *amphibian* means "double life." It is thus a fitting name for organisms such as frogs, toads, and salamanders that during their life cycle live in two worlds—the world of water and the world of dry land.

- *State* three problems animals faced in making the transition from water to land.
- *Name* four characteristics of amphibians that are adaptations to living on land.
- *List* the three orders of amphibians.

36.1 Movement to Land

The movement to land was a significant biological event that opened up new habitats and new sources of food for animals. Land also offers more oxygen to breathe, for air has 20 percent more available oxygen than water. Furthermore, land provides more shelter than water for breeding and for raising young.

The transition from water to land was not an easy one, however, because terrestrial environments are more harsh than aquatic ones. One major problem of living on land is that body structures dry out more quickly in air than in water. The external gills of fish, for example, are constantly bathed when in water. In air, the gills dry out and cannot function. Therefore, an aquatic species could not move onto land until it was equipped with respiratory organs that would not dry out in air. Similarly, the skin of aquatic animals is constantly moistened by water. Thus the skin also had to undergo modifications before a species could colonize land.

Another problem of life on land is that air is not as dense as water and so provides less support against gravity. For this reason animals living on land need strong skeletons.

A third obstacle to survival on land is the continuous fluctuation of temperature there. Animals that live in large seas and lakes are buffered by water's fairly constant temperature. Success on land requires adaptations for surviving wide variations in temperature.

Q/A

Q: *Do all amphibians have either lungs or gills?*

A: No. Salamanders of the family Plethodontidae have neither lungs nor gills. They respire almost entirely through the skin, which contains many capillary networks. Salamanders supplement this "breathing" by mouth breathing.

36.2 Characteristics of Amphibians

Many amphibian characteristics are adaptations for living on land. One example is the amphibian's mode of respiration. Most adult amphibians have internal lungs rather than external gills. These lungs are contained in the chest cavity and are constantly moistened by water condensed from air and by body fluids. Amphibians have simple saclike lungs that are not as efficient as the lungs of other land vertebrates. Amphibians supplement lung breathing by diffusing oxygen directly through their moist skin and the lining of the mouth.

Figure 36–1. Many frogs have powerful hind legs that permit huge leaps (top). Some toads can bury themselves in mud to conserve body moisture (bottom).

Amphibian skin is kept moist in several ways. Frogs, for example, have **mucus glands** on the skin. These glands secrete a slimy substance that protects the skin from drying out. Toads have relatively dry skin. They avoid drying out by confining most of their activity to night time and to wet areas and by burying themselves in wet soil.

Amphibians have skeletons strong enough to support their weight. Most have four limbs, which are specialized for various functions. Frogs and toads, for example, have strong back legs for jumping on land. Amphibian feet are clawless. In many species, the feet are webbed for swimming.

Like fishes, amphibians are cold-blooded—that is, they have no way of maintaining a constant internal body temperature. Amphibians withstand temperature fluctuations by avoiding temperature extremes. Amphibians remain active as long as temperatures are favorable for movement. They may cease activity for long periods if conditions become too hot or too cold.

Amphibians represent only a partial adaptation to land. Most species return to water to lay their eggs. The young of many species pass through a larval stage in water before beginning their life on land.

36.3 Classification of Amphibians

Biologists classify the 2,500 living species of amphibians into three orders. Tailless amphibians, such as frogs and toads, make up the order Anura (uh NYUR uh). *Anura* means "without a tail." Salamanders and other amphibians with legs and tails are placed in the order Urodela (yur uh DEE luh). *Urodela* means "visible tail." Legless amphibians called *caecilians* (sih SIHL yuhns) make up the order Apoda (A puh duh), a name that means "without legs."

Amphibians range in size from a Cuban frog only 1.2 cm (0.5 in.) long to a giant Asian salamander 160 cm (63 in.) long. Amphibians live on every continent except Antarctica, but they are most abundant in the tropics.

Reviewing the Section

1. What advantages and disadvantages did life on land offer for the first amphibians?
2. What adaptation allowed amphibians to withstand the greater stress of gravity on land?
3. Name the three orders of amphibians and give an example of each order.

Frogs

The most successful amphibians are frogs. Most of the 2,000 known species live in the moist tropics, but frogs also are found in all temperate areas. Because frogs are both familiar and easy to obtain, they are often used in biology classes to introduce vertebrate anatomy.

36.4 External Structure

Frogs are short, broad amphibians with long limbs and bulging eyes. Their heads and trunks are fused so they have no visible necks. Adult frogs do not have tails.

The smooth, moist skin of frogs contains both mucus glands and poison glands. The poison glands secrete substances to protect against predators that would find such chemicals distasteful or harmful. Most frogs produce poison that is only mildly irritating to humans, and does not deter snakes, the frog's major predator. One South American species, however, produces a poison that is the most powerful venom known. Most frogs that produce powerful venoms have highly colored skin that warns potential predators of danger. In contrast, the majority of frogs have green, gray, or brown skin, which allows them to blend into their surroundings.

Frogs have large powerful back legs that enable them to jump and leap. Frogs have been known to cover more than 5 m (16 ft.) in a single leap. The front legs, which absorb the shock on landing, are much shorter. Most frogs have webbed feet. Tree frogs, however, have grasping feet or suckerlike pads specially adapted to life in trees.

A frog's eyes are large, round, and protruding. In most species, the eyes protrude above the top of the skull. This adaptation allows frogs to remain submerged in water while watching for predators. Frogs can also pull their eyes into their sockets. The retracted eyes exert pressure on the top of the throat and hold food tightly in the mouth. Each eye is covered by a special third eyelid called a **nictitating** (NIHK tuh tay ting) **membrane** that keeps the eyeball moist.

Two nostrils located at the front and top of the head open into the mouth cavity. Frogs hold their nostrils out of the water to breathe while submerged.

Frogs do not have external ears. Instead, they receive sounds through two **tympanic membranes,** which are circular structures located posterior to each eye. Sound and balance are sensed in the internal ear, which lies directly under the tympanic

Figure 36–2. A frog's nostrils and protruding eyes are located at the top of the head. This arrangement permits the frog to see and breathe while submerged in water.

Q: *Does croaking differ among various kinds of frogs?*

A: Yes. The croaking of frogs is *species specific.* Each species of frog or toad has specific characteristics of tone, pitch, and pulse. A female of a species can recognize the male of her species by listening to the call.

membrane. The ear is connected to the mouth by a channel called the **Eustachian** (yoo STAY shuhn) **tube.** By allowing air to enter the inner ear, the Eustachian tube helps to maintain equal air pressure on both sides of the tympanic membrane.

Frogs communicate by making a distinctive croaking sound. To make this sound, a frog forces air from its lungs across bands of tissue called *vocal cords.* The vibrations of the vocal cords give off sound. Male frogs have more developed vocal cords than females, and use the cords to call to the females during the spring mating season. Some species have a large vocal pouch in the throat or on each side of the head that collects the air after it passes over the vocal cords. When air flows in, these pouches balloon outward. Frogs with vocal pouches make louder sounds than frogs that have no vocal pouches.

36.5 Digestive System

Frogs, like other adult amphibians, are carnivorous. They eat insects, worms, and other animals small enough to be swallowed whole. A frog uses its sticky tongue to snare insects. The

Figure 36–3. The illustrations below show the major parts of the frog's skeletal system (left) and digestive system (right).

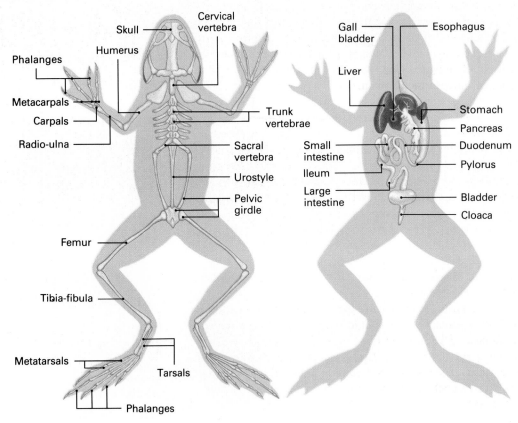

frog's tongue attaches to the front of the mouth and flips outward to catch its prey. Small teeth that line the upper jaw and two **vomerine teeth** that project from the roof of the mouth aid in holding prey. In the mouth, food is moistened by saliva produced in *salivary glands*. The food then enters the esophagus, the tube that leads to the stomach. Enzymes secreted by glands in the stomach break down food further.

Food leaves the stomach through an opening called the *pylorus* and enters the **duodenum,** the first part of the small intestine. The **small intestine** is a long tube in which most of the absorption of food chemicals occurs. A thin, tough membrane called the **mesentery** holds the intestine in place. Food next passes to the second part of the small intestine, the **ileum.** From there it travels to the **large intestine,** where much of the water in the food is absorbed. Any remaining waste then enters the **cloaca,** a cavity that leads to the outside. The cloaca also receives materials from the kidneys, the urinary bladder, and the sex organs. Waste materials pass out of the body through the cloacal opening.

Two organs, the pancreas and the liver, aid digestion by secreting enzymes. The **pancreas** lies anterior and slightly dorsal to the stomach and secretes digestive enzymes into the stomach. It also secretes **insulin,** a substance that controls the level of sugar in the blood. The liver fills most of the body cavity. It produces bile and also stores the carbohydrate *glycogen,* a by-product of glucose metabolism.

36.6 Skeletal System

Figure 36–3 shows that a frog's short trunk has a backbone consisting of nine vertebrae. The *cervical vertebra,* located at the anterior end of the backbone, allows the head to move. Seven *trunk vertebrae* connect the cervical vertebra to the *sacral vertebra,* which supports the hind limbs. Attached to the sacral vertebra is the *urostyle,* a long, undivided portion of the backbone that forms a ridge along the frog's back. The urostyle consists of several fused vertebrae.

A fundamental feature of amphibians is the presence of limbs. Limbs are adaptations that made it possible for a frog to move around on land. Forelimbs are attached to the main skeletal frame by the pectoral girdle; hind limbs are attached by the pelvic girdle. The skeleton of each forelimb consists of a long *humerus,* a *radio-ulna,* and the bones of the wrist and hand. The long bone in the hind limb is called the *femur.* It is attached to a *tibia-fibula,* which in turn is connected to the bones of the ankle and the foot.

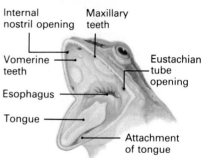

Internal nostril opening

Maxillary teeth

Vomerine teeth

Eustachian tube opening

Esophagus

Tongue

Attachment of tongue

Figure 36–4. A frog catches its prey with a quick flip of its sticky tongue (top). The tongue and other parts of the frog mouth are shown in the illustration (bottom).

36.7 Respiratory System

A frog takes oxygen into its body in three ways. A frog breathes through its lungs, its mouth, and its skin.

To force air into its lungs, a frog lowers the bottom of its mouth while keeping its mouth shut. This procedure creates a temporary vacuum in the mouth. As a result, air from the outside rushes in through the nostrils to fill the vacuum. When the frog raises its mouth and closes its nostrils, air is forced on into the lungs. Often a frog requires a number of gulps of the mouth to fill its lungs.

Frogs diffuse some oxygen into the blood through the membranes of the mouth, which are rich in blood vessels. This process is called **mouth breathing.**

Highlight on Careers: Wildlife Biologist

General Description

A wildlife biologist works to protect natural habitats and to maintain reproducing populations of animals. Many wildlife biologists work to save wilderness habitats. Some also restore waterways and parklands in or near cities.

Wildlife biologists specialize in a variety of areas. Some concentrate on preserving endangered species, such as the Wyoming toad and the gray wolf. These wildlife biologists research the causes of the species' decline and the type of habitat needed to sustain the species. Through the efforts of wildlife biologists, a number of animals in this country have been saved from extinction.

Some wildlife biologists do field research to determine the populations of various animals. Others hold wildlife management positions, or work as game wardens helping to enforce hunting and fishing regulations.

Most wildlife biologists work for state and federal agencies, such as the U.S. Fish and Wildlife Service and the U.S. Forest Service. Some hold positions with county or city park and recreation agencies, or with private companies that have holdings of large tracts of land.

Career Requirements

Individuals interested in wildlife management and research need an M.S. or Ph.D. with a specialty in an

endangered species or one class of animals. Persons with a B.A. in biology may qualify for jobs with local governments and conservation organizations or as wildlife technicians assisting a biologist doing field research. A few jobs as game wardens are open to high school graduates.

For Additional Information

Department of the Interior Fish and Wildlife Service 18th and C Street, N.W. Washington, DC 20240

Frogs also respire by absorbing dissolved oxygen directly through the skin. Dissolved oxygen diffuses through the frog's thin skin into the blood capillaries underneath. This trait allows frogs to remain under water for long periods. Most carbon dioxide also leaves the frog's body by diffusion through the skin.

36.8 Circulatory System

The frog, like all amphibians, has a three-chambered heart. A membrane separates the atrium into two chambers. Blood returning from the body enters the *right atrium*. Blood returning from the lungs enters the *left atrium*. Both of these chambers contract, and blood is pumped into the *ventricle*. Oxygen-poor blood from the body and oxygen-rich blood from the lungs thus mix in the ventricle. When the ventricle contracts, blood is forced out of the heart. Some goes to the lungs and some goes to the rest of the body.

Amphibians differ from birds and mammals, which have four-chambered hearts. Four-chambered hearts have a *left ventricle* and a *right ventricle*. Blood returning from the body enters the right atrium, is pumped to the right ventricle, and then to the lungs. It returns to the left atrium, is pumped to the left ventricle, and then to the body. There is thus a separation between the **systemic circulation**—that is, blood moving to and from the body—and the **pulmonary circulation,** blood moving to and from the lungs. The separation of blood flow to body and to lungs is called *double circulation*.

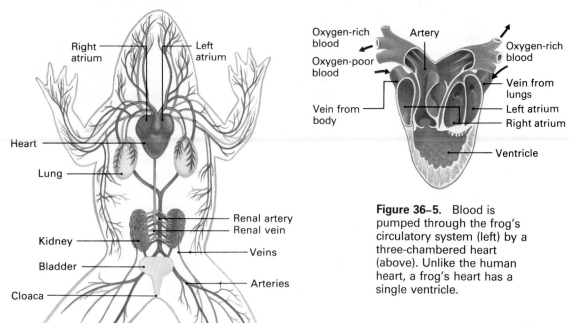

Figure 36–5. Blood is pumped through the frog's circulatory system (left) by a three-chambered heart (above). Unlike the human heart, a frog's heart has a single ventricle.

Frogs, like all amphibians, have no way of regulating their internal body temperatures. As the temperature of the external environment falls, their own body temperature also falls. This is why amphibians are termed "cold-blooded."

To survive extreme cold, frogs and other amphibians often bury themselves in mud or huddle together inside logs. As their body temperature falls, their respiration and level of activity slow down. What little energy they require comes from stored fat and glycogen in their bodies. Such a period of severely reduced activity in winter is called **hibernation.** As the weather warms, the frog's body gradually returns to normal, with no ill effects.

Frogs also go through a period of inactivity during the summer, called **estivation.** To avoid the extremes of hot temperature, frogs remain immersed in water. During a drought, when water is scarce, they burrow themselves in mud. Under such conditions frogs breathe almost entirely through their skin.

Other animals also go through similar periods of hibernation or estivation. Lungfishes survive seasonal droughts by burying themselves in the mud at the bottom of dried-out ponds. Honeybees cluster together to hibernate and to protect the queen bee with their accumulated body warmth.

36.9 Nervous System

Adult frogs lack the lateral line system used by fishes to monitor their environment. As a result adult frogs depend more on their senses of smell, vision, and hearing. *The frog's brain is the central organ of its nervous system and processes information from the various sense organs.*

The frog's brain resembles a swollen portion of the spinal cord. The front parts of the brain are the *olfactory lobes*, which receive and interpret smell impulses. Behind these lobes is the **cerebrum,** the control center of the brain. Two *optic lobes*, located behind the cerebral hemispheres, receive and interpret impulses from the eyes. The **cerebellum** controls movement and muscular coordination. Behind the cerebellum is the **medulla oblongata,** where many of the more automatic responses of the nervous system are carried out.

Ten pairs of cranial nerves connect the brain and the sensory organs. These nerves carry information from the sense organs to the brain, which interprets the information and selects the correct response.

The medulla oblongata connects with the spinal cord. Nerves that carry impulses to and from the body are attached to the spinal cord.

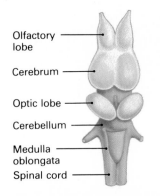

Olfactory lobe
Cerebrum
Optic lobe
Cerebellum
Medulla oblongata
Spinal cord

Figure 36–6. A frog's brain looks like a mere widening of its spinal cord. However, it functions as a true brain, processing information from the entire nervous system.

36.10 Excretory System

Two kidneys are the major excretory organs of the frog. The kidneys lie along the lower side of the frog's back directly under the skin. Blood flows to a kidney through the **renal artery,** which branches into many smaller vessels and capillaries as it enters the kidney. In the kidney, wastes and excess water are removed from the blood. The wastes then leave the body from the cloaca. A frog takes in a great deal of water in a day through its skin and must remove the excess in urine. Each day a frog typically excretes approximately 25 percent of its body weight in urine.

The skin also serves an excretory function. In addition to carbon dioxide, excess salts and minerals leave the frog's body by diffusion through the skin.

36.11 Reproductive System

Male and female frogs are difficult to distinguish because their sex organs are internal. During the mating season, however, the thumbs of many male frogs enlarge and become black.

A female frog's two egg-producing organs, called **ovaries,** are located near the kidneys. Thousands of eggs produced by the ovaries are released directly into the body cavity. Cilia sweep the eggs into coiled tubes called *oviducts.* There the eggs are coated with a jellylike covering and stored in two *ovisacs* until they are released for external fertilization. The eggs leave the body through the cloaca.

Male frogs produce sperm in two **testes** located on the front portion of the kidney. The sperm travel through narrow tubes into the kidneys and from there into storage sacs called *seminal vesicles.* Sperm are released from the seminal vesicles through the cloaca during mating.

Both testes and ovaries have **fat bodies** on their external surface. These bodies store fat that supplies nutrition for developing eggs and sperm.

36.12 Life Cycle

During the spring mating season, a male clasps a female from behind and presses his large black thumbs into her back. This process, called **amplexus,** stimulates the female to expel her eggs. The male then spreads sperm on the newly released eggs. The sperm are able to penetrate the jellylike coating. This covering is distasteful to potential predators and also helps regulate the temperature of the fertilized eggs.

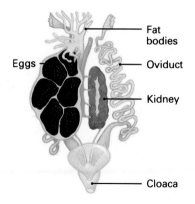

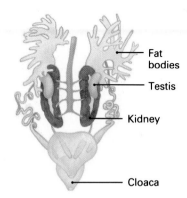

Figure 36–7. The illustrations above show the major organs of the excretory and reproductive systems of the female frog (top) and the male frog (bottom).

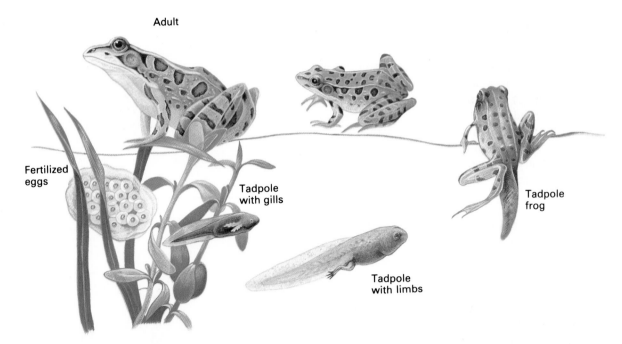

Adult

Fertilized eggs

Tadpole with gills

Tadpole with limbs

Tadpole frog

Figure 36–8. The major stages in the life cycle of a frog are shown in the illustration above.

The eggs hatch 2 to 30 days later, depending on the species and the water temperature. The eggs hatch into **tadpoles,** the larval form of frogs. *The aquatic tadpole goes through a metamorphosis, or change, as it develops into the adult amphibious frog.* Follow the life cycle of a frog in Figure 36–8.

The young tadpole begins life with a short tail, lidless eyes, and adhesive discs with which it attaches to objects. The tadpole breathes through gills and feeds on vegetation in the water. Gradually the tadpole develops limbs and its tail begins to disappear. Lungs replace gills, and the young frog leaves the water for the land. The process of metamorphosis from tadpole to frog takes a year or less for the common leopard frog and up to three years for the bullfrog.

Reviewing the Section

1. Where are a frog's ears located?
2. How does a frog break down its food to digest it?
3. How does a frog take air into its lungs?
4. Describe the route of blood through the chambers of a frog's heart.
5. How does a frog keep from freezing in winter?
6. Compare the structure and feeding habits of the newly hatched tadpole with those of the adult frog.

Other Amphibians

Toads, salamanders, and caecilians are amphibians that have developed various adaptations for living on land. Desiccation, or drying out, is a major threat for each animal.

36.13 Toads

Like frogs, toads belong to the order Anura. Toads resemble frogs in many ways, and the name *toad* is often applied to any species of Anura that is specially adapted to a dry environment. True toads belong to the family Bufonidae.

Toads tend to have shorter legs than frogs, squat bodies, and thick skin covered with bumps. The common American toad ranges from 2 to 25 cm. (0.8 to 10 in.) in length. The red or brown color of the skin allows these toads to blend in with the ground.

Many toads live in deserts and arid areas, but others live in moist habitats. They move about at night and bury themselves in soil during the day to avoid heat. Toads usually return to water only to reproduce. The female toad lays eggs in stringy masses, in contrast to the egg clusters of frogs. The tadpoles of toads are typically black.

36.14 Salamanders

Salamanders are amphibians with tails. Scientists believe that, of all modern amphibians, salamanders most closely resemble ancestral amphibians.

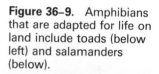

Section Objectives

- *Name* a feature that distinguishes toads from frogs.
- *List* the major characteristics of salamanders.
- *Describe* the habitats of caecilians.

Figure 36–9. Amphibians that are adapted for life on land include toads (below left) and salamanders (below).

Figure 36–10. The mud puppy never loses its larval characteristics, even when it becomes mature enough to reproduce.

Figure 36–11. Like all caecilians, this Japanese caecilian has tiny scales embedded in its skin.

Most of the 300 species of salamanders live in cooler areas of the world, but some species are common in the tropics. Salamanders tend to live near water in moist places, such as under stones or rotten logs. Some species that live mainly in water are commonly called *newts*.

Most salamanders measure about 10 to 15 cm (about 4 to 6 in.) long. Their front and hind limbs are approximately the same size and are attached at right angles to the body, resulting in a slow, labored walk. Salamanders eat almost any small animal, especially worms or small arthropods.

Salamander eggs are fertilized internally. The female uses her cloaca to pick up a packet of sperm, or *spermatophore*, that the male has deposited on a leaf or stick. Aquatic species deposit the fertilized eggs in the water, and land species place them on the ground or on logs. The newly hatched young resemble adult salamanders much more closely than tadpoles resemble frogs and toads. All young salamanders have gills, but most develop lungs as adults.

Some species of salamanders never lose their larval characteristics even though they achieve sexual maturity and begin to reproduce. This type of development is called **neoteny.** The common North American mud puppies and the congo eel are examples of amphibians that reproduce while still aquatic and using gills to breathe.

36.15 Caecilians

Caecilians have long slender bodies and no limbs. In fact, they resemble worms both in appearance and in habit. They burrow into the ground of tropical forests and live on worms and small invertebrates they find in the ground. Most adult caecilians either are blind or have very small eyes. Special sensory tentacles are located between their snout and their eyes. Caecilians have many vertebrae and typically measure about 30 cm (1 ft.) long. Some species, however, grow to 130 cm (4.3 ft.). Caecilians are rarely seen, and less is known about their behavior than about the other, more common, amphibians.

Reviewing the Section

1. How can toads be distinguished from frogs?
2. Describe the preferred habitats of land salamanders.
3. How do the sense organs of caecilians differ from those of other amphibians?
4. How do mud puppies breathe without lungs?

Investigation 36: Tadpole Development

Purpose

To investigate the effect of temperature on the metamorphosis of tadpoles

Materials

Supply of tadpoles, four five-gallon aquaria, aquarium heaters, dry fish food

Procedure

1. Fill each tank with pond water or aged tap water (water that has stood overnight). *Why is it preferable to use pond water rather than tap water?* Place 10–15 tadpoles in each of the four tanks. *Why is it necessary to place the same number of tadpoles in each of the tanks?*
2. Maintain one tank at room temperature, the second tank at 24°C (75°F.) and the third tank at 29°C (85°F.). Place the fourth tank outside. (Note: Omit the fourth tank if the investigation is performed in winter.)
3. Place some aquatic plants and other aquarium materials in each tank. *What purpose is served by including aquatic plants in the tanks?*

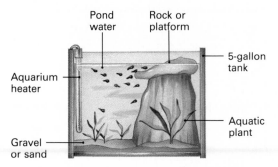

Pond water · Rock or platform · 5-gallon tank · Aquarium heater · Aquatic plant · Gravel or sand

4. Feed the tadpoles a small amount of dry fish food.
5. Observe the tadpoles' behavior and development.

Analyses and Conclusions

1. Based on your written observation of the tadpoles' behavior and developmental stages, make a chart like the one shown to compare development in each tank.

	Observations on Tadpole Development
Tank 1 Room temperature	
Tank 2 24°C	
Tank 3 29°C	
Tank 4 Outside temperature	

2. How do differences in temperature affect the metamorphosis of the tadpoles? What is the optimal temperature for the most rapid growth and development?

Going Further

- Maintain the same temperature in each of the four tanks, but vary the number of tadpoles in each as follows: 10-20-30-40. How does the population density affect the rate of tadpole development?
- A number of hormonal derivatives have been shown to affect the development of a frog. Test this finding by adding thyroid extract to the aquarium water in varying concentrations. After careful observation, determine its effect.

Chapter 36 Review

Summary

Amphibians were the first group of vertebrates to move between water and land. They adapted to the increased dryness of the land environment, to its greater gravitational stress, and to greater fluctuations in temperature.

The frog's systems show its adaptations to living on land. Frogs have a supporting skeleton and powerful legs for movement on land as well as in water. Adult frogs breathe mainly through lungs but also breathe through their mouth membranes and through their skin. A circulatory system pumped by a three-chambered heart moves the blood to and from the lungs. Frogs return to water to reproduce. The larvae, called tadpoles, live in water and breathe through gills before undergoing a metamorphosis to become adults.

Toads resemble frogs but are even more suited to life on land. Amphibians with tails include salamanders and newts. Caecilians are legless, blind amphibians that burrow like worms.

BioTerms

amplexus (547)
cerebellum (546)
cerebrum (546)
cloaca (543)
duodenum (543)
estivation (546)
Eustachian tube (542)
fat body (547)
hibernation (546)
ileum (543)

insulin (543)
large intestine (543)
medulla oblongata (546)
mesentery (543)
mouth breathing (544)
mucus gland (540)
neoteny (550)
nictitating membrane (541)
ovary (547)
pancreas (543)

pulmonary circulation (545)
renal artery (547)
small intestine (543)
systemic circulation (545)
tadpole (548)
testis (547)
tympanic membrane (541)
vomerine teeth (543)

BioQuiz   (Write all answers on a separate sheet of paper.)

I. Completion

1. Solid waste matter passes out of the frog's body through its _____ .
2. Blood returning from the lungs to the heart enters that organ through the _____ .
3. Most adult amphibians breathe mainly through their _____ .
4. A number of fused vertebrae form the frog's _____ .
5. Modern _____ most closely resemble ancestral amphibians.

II. Modified True and False

Mark each statement TRUE or FALSE. If false, change the underlined term to make the statement true.

6. A frog's tongue is attached to the back of its mouth.
7. In the frog, insulin is secreted in the liver.
8. The muscular activity of a frog is coordinated by its brain cerebellum.
9. Most species of amphibians lay their eggs on land.

III. Multiple Choice

10. Blood from the pulmonary and systemic circulations mixes in an amphibian's
    a) lungs. b) right atrium. c) ventricle. d) renal artery.
11. A frog's body rids itself of carbon dioxide through its a) nostrils.
    b) kidney. c) mouth. d) skin.
12. Amphibians meet the challenge of temperature fluctuations by a) having cold blood. b) modifying behavior. c) having mucus glands. d) undergoing metamorphosis.
13. A frog gulps air to a) mouth breathe. b) make a croaking sound. c) call to female frogs. d) fill its lungs.
14. Amphibians adapted to the gravitational stress on land by developing a) strong skeletons. b) a three-chambered heart. c) lungs. d) thin skins.

IV. Essay

15. How do frogs survive periods of extreme cold?
16. How do frogs and toads protect themselves against drying out on land?
17. How can a frog see and breathe while under water?
18. How do the limbs of frogs and salamanders differ?
19. What do large black thumbs on a frog indicate, and for what purpose are they used?
20. How are the eggs of a salamander fertilized?

Applying and Extending Concepts

1. The appearance of toads after a rain was once explained by folk superstitions that claimed the toads rained from the sky. Use your knowledge of the habits of toads to give a more reasonable explanation for this observation.
2. Caecilians have solved the problem of moving from water to the dryness of land by spending all their lives underground. Write a paragraph explaining how some caecilian features are adaptations to underground life.
3. The largest toad, *Bufo marinus,* is more than 23 cm (9.3 in.) long. The smallest toad, the *Bufo taitanus,* is only 2.3 cm (0.93 in.) long. Extremely large or small size can often be understood as adaptations that help an animal to defend itself. How might you relate this theory to the size differential between the largest and smallest toads?
4. The axolotl (AK suh laht uhl) is an American salamander that retains larval characteristics unless the water in which it lives begins to dry up. This salamander then loses its gills and develops adult lungs. What does this type of development indicate to you about the adaptive value of neoteny?

Related Readings

Bellemy, D. "As Earth's Jigsaw Puzzle Forms, the Dry Land Pioneers Emerge." *Science Digest* 87 (January 1980): 60–65. The author describes the evolution of the earliest amphibians and reports on recent New Zealand finds.

Cornejo, D. "Night of the Spadefoot Toad." *Science 82* 3 (September 1982): 62–66. This article gives an account of the breeding habits of desert toads who mate nocturnally at a precise time each year.

37

Reptiles

Marine iguanas in the Galapagos

Introduction

Snakes, lizards, and other members of the class Reptilia have long fascinated humans. Giant Galapagos tortoises, Komoda dragons, and Gila monsters all seem to be a link to prehistoric times. Modern reptiles are, in fact, the descendants of the first totally land-dwelling vertebrates. Modern reptiles are also distant relatives of the dinosaurs, the largest animals ever to live on land.

About 6,000 species of reptiles still live today. They are found in all areas of the world. Because they are cold-blooded, however, they are more common in the warm tropics and deserts than in the cooler areas. Most reptiles are carnivorous and some—such as crocodiles, snakes, and snapping turtles—are fierce predators.

History of Reptiles

Amphibians, despite their success on land, never live far from water. They need moisture to prevent their thin skin from drying out, and they must return to water to reproduce. Reptiles, by developing adaptations that met these two needs, became the first successful vertebrates to become truly independent of water. As a result, reptiles were able to move into new habitats on land where there was little competition for food and territory.

37.1 The Success of Reptiles

Unlike amphibians, reptiles do not need to return to water to reproduce. In reptiles, fertilization is internal, so water is not needed to transport the sperm to the egg. In addition the reptiles have a "land egg"—an egg that can protect a developing embryo on dry land. The development of such an egg allowed reptiles to reproduce in even the driest environment.

The reptile's egg, called an **amniotic egg,** has a fluid-filled sac enclosed by a protective, porous shell. The shell may be flexible like leather, or rigid and reinforced with calcium carbonate. Oxygen and carbon dioxide pass freely through the shell.

Inside the shell are four specialized membranes: the amnion, the yolk sac, the allantois, and the chorion. The **amnion** surrounds the embryo and forms a chamber filled with a saline fluid. The embryo thus floats in its own tiny sea, which also provides support for the embryo's tissues. The **yolk sac** is

Section Objectives

- *List* two characteristics that contribute to reptiles' success on land.
- *Name* the four membranes of the amniotic egg and state the function of each.
- *Describe* examples of four types of dinosaurs.
- *State* the difference between endothermic and ectothermic animals.

Figure 37–1. The amniotic egg (below, left) is believed to have first appeared among reptiles. Young reptile hatchlings (below, right) were able to develop on land because of the protection provided by the egg's fluid-filled amnion.

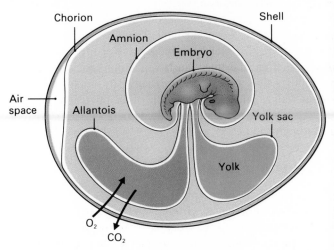

attached to the abdomen of the embryo. It forms a container for the large amount of stored food called the **yolk.** The **allantois** (uh LAN tuh wihs) makes up a sac that stores the nitrogen wastes produced by the growing cells. The allantois is rich with blood vessels. The outermost membrane, the **chorion** (KOHR ree ahn), lines the eggshell. Together the allantois and chorion control the passage of gases in and out of the egg, supply the embryo with oxygen, and remove carbon dioxide.

Because the developing reptiles are protected within the eggshell, their survival rate is much higher than that of amphibians. For this reason, most reptiles do not produce as many eggs as amphibians.

An additional reason reptiles do not need as much moisture as amphibians to survive is that the skin of reptiles is thick and contains **keratin,** a protein that makes skin hard and waterproof. The skin of reptiles is also covered with horny, overlapping scales or plates that are also made of keratin.

37.2 The Age of Reptiles

The fossil record shows that the first reptiles appeared on Earth about 310 million years ago. These reptiles, called *cotylosaurs* (KAHD uh luh sahrs), were small lizardlike animals that fed on insects. As the earth underwent profound climatic changes during the Permian and Triassic periods (275 to 180 million years ago) and the land dried out, cotylosaurs evolved into a number of new kinds of reptiles. These descendants of the cotylosaurs became the dominant animals on land during the Mesozoic era, which lasted from about 225 million to 65 million years ago. For this reason, the Mesozoic era is called the Age of Reptiles.

During the Jurassic and Cretaceous periods of the Mesozoic era, reptiles were masters of the land. They lived in every possible habitat and varied greatly in their ways of life. Some were *herbivores* that ate the thick, lush vegetation that covered the earth. Reptiles that were *carnivores* ate fish, amphibians, and insects, as well as other reptiles. Some of the prehistoric reptiles looked like modern species of lizards. Others swam like fish or glided through the air.

37.3 Prehistoric Reptiles

The most fascinating of the prehistoric reptiles were the dinosaurs. The dinosaurs ranged in size from animals as small as a chicken to the largest creatures that have ever lived on land. The largest dinosaurs were herbivores, such as the brontosaurus (brahn tuh SAHR uhs) and the diplodocus (dih PLAHD uh kuhs).

Figure 37–2. The cotylosaur was the ancestor of most modern reptiles. Only the crocodile descended from another ancestor.

▶ Dinosaurs: Cold-Blooded or Warm-Blooded?

Were dinosaurs cold-blooded like modern reptiles or warm-blooded like birds and mammals? Scientists long assumed that dinosaurs were cold-blooded like other reptiles. During the 1970s, however, new research persuaded some scientists to reconsider the question.

Warm-blooded, or *endothermic,* animals have the ability to maintain a constant internal temperature. Cold-blooded, or *ectothermic,* animals lack this ability, and their body temperature is always close to that of their surroundings. For this reason, snakes and other modern reptiles must avoid extreme heat or cold. A snake, for example, may bask in the sun to warm itself during the coolest part of the day, then crawl into the shade to avoid the midday heat.

The possibility that dinosaurs were warm-blooded was first suggested by studies of fossil dinosaur bones. Researchers found that the internal structure of these bones showed evidence of small blood vessels, a characteristic of warm-blooded animals.

The idea that at least some dinosaurs were warm-blooded gained further support from studies of the fossil remains of dinosaur communities. These communities had a high ratio of prey to predators, a pattern characteristic of warm-blooded animals. Warm-blooded animals must eat much more than cold-blooded animals in order to supply the large quantity of energy these animals use in order to maintain a constant body temperature.

These dinosaurs had immense bodies with long necks and tails. They could have easily peered over a four-story building. The diplodocus, for example, is estimated to have been more than 24 meters (80 ft.) long and to have weighed over 50,000 kg (110,000 lb.). The diplodocus had a small head with tiny teeth and weak jaws. These reptiles probably spent much of their time near shallow lakes and swamps where they fed on water plants. The water may have helped support their huge bulk.

Perhaps the fiercest of the predatory dinosaurs was the tyrannosaurus rex (tih ran uh SAHR uhs REHX). This carnivorous beast walked on its hind legs and stood 6 m (20 ft.) tall. It weighed more than 8,000 kg (17,600 lb.). The tyrannosaurus rex had a huge head with powerful jaws studded with sharp, daggerlike teeth 15 cm (6 in.) long.

Some dinosaurs were ''walking tanks'' covered with heavy, protective armor. Certain dinosaurs, such as the triceratops (try SEHR uh tahps), had huge, protective spikes on its head and armor surrounding the back of the neck. Other dinosaurs, such as the stegosaurus (stehg uh SAWR uhs), had a defensive armor that consisted of two pairs of sharp spikes on its tail and two rows of large bony plates along its spine. Some scientists think

Q/A

Q: *How long did it take a diplodocus to attain its giant size?*

A: If the diplodocus were warm-blooded, it might have grown to its full length of 24 m (80 ft.) in 50 years. A cold-blooded diplodocus might have had to live 200 years before it became fully grown.

Figure 37–3. Dinosaurs were the dominant life form of the Jurassic period. A scene such as this was common about 150 million years ago.

that the bony plates were used by the animal to regulate its body temperature.

Some ancient reptiles, such as the pterodactyl (tehr uh DAC tuhl), had the ability to fly. The fourth digit of the pterodactyl's hand was extremely long. A thin flap of skin from the animal's side was attached to this long digit to form a gliding wing. Pterodactyls were about the size of turkeys but had wing spans of more than 8 m (26 ft.).

Some prehistoric reptiles, called ichthyosaurs (IHK thee uh sawrs) looked like porpoises. These reptiles lived in the ocean and were powerful swimmers. Plesiosaurs (PLEE see uh sawrs) were also marine. They had extremely long necks, and their limbs were modified into flippers.

Dinosaurs were the dominant animals on Earth for about 160 million years, then died out. The modern vertebrates most closely related to dinosaurs are birds and crocodiles. All other modern reptiles evolved from the early cotylosaurs and their descendants. A separate group of early reptiles gave rise to mammals.

Reviewing the Section

1. What are the four embryonic membranes in an amniotic egg? What is the function of each?
2. Why did reptiles colonize the land so quickly once the amniotic egg was developed?
3. What were the largest animals ever to have lived on the land?
4. How do endothermic and ectothermic animals differ from each other?

Modern Reptiles

Though reptiles no longer rule the land, they still play a major ecological role in most of the world's biological communities. This is especially true in the tropics, deserts, and warm grasslands. Many reptiles are virtually the same as their ancient ancestors. Other groups continue to evolve to meet the demands of changing environments.

37.4 Characteristics

Like their ancient ancestors, modern reptiles have a waterproof skin and produce amniotic eggs. All reptiles also have strong, bony skeletons and well-developed lungs. Most reptiles have two pairs of limbs that enable them to walk on land. The limbs of some, but not all, reptiles are positioned more vertically under the trunk than the limbs of amphibians. The limbs of reptiles are also larger, stronger, and support more weight than those of amphibians. Reptilian feet have toes with claws that are used for digging or for climbing on trees and rocks.

The nervous system of reptiles is similar to that of amphibians. The reptile brain is small in relation to the body. For example, a crocodile 2.5 m (8 ft.) long has a brain the size of a walnut. Despite the small brain, reptiles have shown complex behavioral patterns, including elaborate courtship rituals.

The excretory system of reptiles is modified to minimize water loss. Water is absorbed into the body through the intestine, bladder, pairs of kidneys, and cloaca. Aquatic reptiles, which do not need to conserve water, excrete most of their nitrogenous wastes as ammonia dissolved in water. Terrestrial reptiles, which must retain water, secrete nitrogenous wastes in a drier form, as uric acid solids.

Like amphibians, most reptiles have a three-chambered heart. However, crocodiles and their relatives have a heart with four chambers, a structure more efficient at separating oxygenated and deoxygenated blood.

Section Objectives

- *List* four characteristics common to all reptiles.
- *Name* the four orders of reptiles and give distinguishing features of each.
- *Distinguish* between alligators and crocodiles.
- *Describe* the two different methods by which snakes kill their prey.
- *State* three methods by which snakes move.

Figure 37–4. Scaly skin and toes with claws are typical features of reptiles. The reptile shown here is an iguana.

37.5 Classification

Of the 16 orders of reptiles that once flourished, only 4 now remain:

- *Rhyncocephalia* (rihn koh suh FAY lee uh): This order consists of a single species, the *Sphenodon punctatus*, commonly called by its native New Zealand name of tuatara (too uh TAH ruh).
- *Chelonia* (kih LOH nee uh): Turtles and tortoises make up this order. The bodies of these reptiles are enclosed within two bony shells.
- *Crocodilia* (krahk uh DIHL ee uh): These are the crocodiles, alligators, and related species adapted for life in shallow water.
- *Squamata* (skwuh MAYD uh): This order consists of lizards and snakes, reptiles with long slender bodies.

37.6 Tuataras

The tuatara represents the last surviving species of a group of reptiles that appeared on Earth more than 225 million years ago. The tuataras living today have changed hardly at all from their prehistoric ancestors. Adults are about 60 cm (24 in.) long. The tuatara has strong legs and a long tail. A scaly crest runs down its neck and back. Like some lizards the tuatara has a third **parietal** (puh RY uh tuhl) **eye** located on the top of the head. The parietal eye has both a lens and a light-sensitive surface, but no muscles, and is not used for seeing. Some scientists believe it may be used for temperature control. Tuataras live only in the Cook Strait and on North Island of New Zealand. They are an endangered species and are protected by the local government.

37.7 Turtles and Tortoises

Turtles and tortoises are another group of reptiles that have changed little through time. Some of the 150 species live only on land, while others live mainly in water. The term *tortoise* is commonly used to refer to land species and *turtle* to most aquatic species. Some turtles and tortoises are enormous. The giant land tortoise of the Galapagos Islands grows to 1.5 m (5 ft.) in length and may weigh up to 255 kg (560 lb.). Tortoises often have long life spans. Some Galapagos tortoises live as long as 150 years. Many Galapagos tortoises were on the verge of extinction at the beginning of this century, but they are being reestablished by careful breeding programs.

Figure 37–5. The tuatara of New Zealand looks almost exactly like its ancestor of 200 million years ago. For this reason, the tuatara is often called a "living fossil."

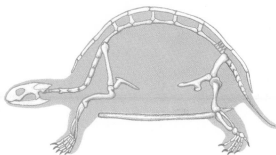

The shell is the distinguishing characteristic of turtles and tortoises. The shell is a two-part case made of modified horny scales. The top half of the shell is the **carapace** and the bottom half is the **plastron.** The shell is covered by plates called **scutes** (skyootz). Turtles and tortoises use their shells for protection, and pull their heads and limbs into their shells when threatened. Unlike other reptiles, turtles and tortoises have no teeth. Their jaws form a horny beak that crushes their food. Turtles and some tortoises are *omnivorous*. They eat both plants and animals.

Figure 37–6. Locked inside its shell, the turtle does not fit most people's picture of a reptile. The illustration above shows the hidden features of its skeleton.

37.8 Alligators and Crocodiles

The order Crocodilia includes about 250 species of alligators, crocodiles, and their smaller relatives, caimans and gavials. Crocodilians are active water-dwelling carnivores. They have broad, heavy bodies and are powerful swimmers. Their strong tails propel them through the water. The crocodilian's eyes and nostrils rise above the rest of its head. This adaptation allows a crocodilian to remain submerged and still see and breathe. A special valve between its mouth and nose passage keeps water from entering the breathing passage even when the animal opens its mouth to feed under water. Many crocodilians lie in wait for prey with only the tips of their nostrils above the water. When their prey comes within range, they use their powerful jaws and sharp teeth to capture it.

As Figure 37–7 shows, crocodilians differ from one another mainly in the structure of their head and teeth. The alligator has a broad head with a rounded snout. The crocodile has a triangular head with a pointed snout. In alligators, the teeth of the lower jaw fit inside the teeth of the upper jaw. In crocodiles, most of the upper and lower teeth mesh together.

Alligator

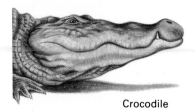

Crocodile

Figure 37–7. Crocodiles (bottom) have two protruding teeth and narrower snouts than alligators (top). Caimans and gavials have snouts that are even narrower.

Figure 37–8. Geckos have toe pads containing many tiny hooks. These hooks enable the animal to climb surfaces as smooth as a glass window.

37.9 Lizards and Snakes

Members of the order Squamata are the most successful of the modern reptiles. Together they comprise almost 4,500 species. The two major kinds of squamates are lizards and snakes.

Lizards   The most obvious difference between lizards and snakes is that most lizards have legs. Some wormlike lizards, however, have no limbs at all. Lizards live in diverse habitats. Some, like most skinks, live on the ground and walk on small, weak legs. Others, like geckos, are nocturnal and may live in trees. Geckos have claws modified into pads for clinging to vertical surfaces. Many lizards run on their hind limbs. "Flying" lizards live in trees and glide from branch to branch.

Some lizards eat only plants; others eat only insects or small animals. Most lizards, however, are omnivores.

Many lizards have the capacity to regenerate lost limbs or tails. This capacity for regeneration protects the lizards by allowing them to escape from a predator by losing an appendage.

Highlight on Careers: Zookeeper

General Description
A zookeeper is a worker in a zoo or aquarium who feeds the animals, gives them fresh water, cleans their cages, and transfers them to other cages or tanks. The zookeeper also observes the animals closely for signs of possible illness or injury. Some zookeepers prepare the animals' food, bathe and groom the animals, and answer visitors' questions.

In a small zoo, zookeepers may take care of several different kinds of animals. In large zoos, however, workers tend to specialize. The reptile keeper, for example, takes care of the snakes, lizards, crocodiles, and other reptiles. A reptile keeper's special equipment includes a hood and gloves that are worn when handling venomous snakes.

Career Requirements
Each zoo sets its own rules for hiring, and each provides on-the-job training. Most require graduation from high school. A few colleges and universities offer courses in zookeeping and animal behavior. An ability to handle animals is a prime prerequisite for being a zookeeper.

For Additional Information
American Association of
   Zoological Parks and
   Aquariums
Oglebay Park
Wheeling, WV 26003

Chameleons are tree-dwelling lizards native to Africa and Madagascar. They are capable of dramatic changes of color in response to excitement or to changes in light or temperature. These color changes often act to camouflage the chameleons, allowing them to blend in with their surroundings. Chameleons can become green, beige, or brown, and can also develop spots. Chameleons have large eyes that rotate independently of each other. Keen vision and quick reflexes enable a chameleon to catch insects with a flick of its sticky tongue.

The largest of all lizards is the Komodo dragon, so called because it comes from the Indonesian island of Komodo. These lizards may grow to be almost 3 m (10 ft.) long. Komodo dragons have extremely strong limbs and powerful claws. When they run, their limbs lift their bodies completely off the ground, and they may attain speeds of 16 km (10 mi.) per hour. Komodo dragons are highly successful carnivores. They catch and eat animals as large as small deer or wild pigs. Because of their large claws and aggressive manner, Komodo dragons have few natural enemies.

Only two species of lizards, the Gila monster of the southwestern United States and the beaded lizard of Mexico, are venomous, that is, able to inject venom, or poison, through a bite. The venom glands of the Gila monster form a row inside the lower lip at the base of grooved teeth. When it bites another animal, a Gila monster shakes its head vigorously from side to side. This action causes the venom to flow out of the mouth, through the grooves, and into the prey. The venom of a Gila monster can kill a small animal. The venom is not usually deadly to humans but can make a person extremely ill.

Iguanas are large, often colorful lizards that live in tropical areas. Many have tough, leathery skin and a large, distinctive crest on their backs. This group includes the only aquatic lizard, the marine iguana of the Galapagos Islands. This large black lizard is adapted to life in the ocean, where it eats green algae exclusively. The marine iguana maintains its salt balance by secreting salt through special glands in the nose.

Snakes Snakes are reptiles without limbs that slither over the ground or live in trees. Most features peculiar to snakes are adaptations to their limbless state. Their bodies have become extremely long and muscular. They have a large number of ribs and vertebrae that make their bodies both strong and flexible. In some species, one lung is missing or reduced in size to make room in the body cavity for the swallowing of large prey.

Both lizards and snakes periodically shed their outer layer of skin—a process called **molting.** Most snakes rid themselves

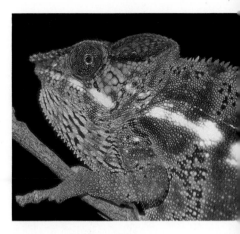

Figure 37–9. The chameleon's ability to change color is controlled by hormones that act on various pigments in the skin.

Q/A

Q: *What are the longest and the heaviest living reptiles?*

A: One reticulated python, a snake from southeast Asia, measured 10.8 m (32.7 ft.) long. The heaviest reptiles are male saltwater crocodiles that weigh an average of 418 kg (1,100 lb.).

Figure 37–10. A snake molts by literally crawling out of its skin. Active snakes molt frequently, some as many as six or more times a year.

Figure 37–11. A boa constrictor (left) subdues its prey by squeezing it to death. Flexible jaws permit some snakes, like the parrot snake (right), to open its mouth extremely wide, and feed on large prey.

of their outgrown skin in one piece. The skin around the mouth and head loosens first. Then the snake proceeds to crawl out of its old skin, turning the discarded skin inside out as it moves.

Snakes have highly developed senses of taste and smell. When a snake darts its slender, forked tongue in and out of its mouth, it is "tasting" the air. As the tongue darts out, it picks up scent particles in the environment. The tongue returns into the mouth, where it presses up against two small hollows in the roof of the mouth. These hollows, lined with nerves sensitive to chemicals, are called **Jacobson's organ.** A snake uses its Jacobson's organ to track its prey. A male snake uses its tongue and Jacobson's organ to follow the scent trail of a female snake.

Snakes have relatively poor senses of vision and hearing. They have eyes on each side of the head that are covered by transparent scales. Unlike most other reptiles, snakes do not have movable eyelids. Snakes have no external ears or eardrums and can hear only a few sounds.

Most snakes are **oviparous** (oh VIHP uhr uhs)—that is, they lay eggs that hatch outside the mother's body. Other snakes are **ovoviviparous.** The fertilized eggs of these snakes develop and hatch inside the mother, who then gives birth to live young.

All snakes are predators. Many snakes simply seize an animal and swallow it live. Some snakes, called *constrictors*, kill their prey by coiling their bodies around their victim. As the snake begins to slowly tighten its muscles, the pressure of constriction stops the animal from breathing and also stops the heart and circulation of blood. After the prey is overpowered, the snake swallows it whole.

The remarkable structure of the jaws allows the snake to swallow prey four to five times larger in diameter than its own body. The lower jaw is made up of two bones held together at the chin by elastic tissues. This allows each side of the jaw to be

moved separately. The upper and lower jaw are similarly attached. This arrangement allows the mouth to open to an enormous size. By moving one side of the jaw forward and then the other, a snake "walks" its food down its throat. After swallowing, the snake digests the prey. Digestion is slow. A boa constrictor may take five days to digest a rat.

Some species of snakes kill their prey with venom. These snakes include cobras, sea snakes, and mambas. Nineteen species of venomous snakes live in the United States. They include coral snakes, water moccasins, and several varieties of rattlesnakes.

▶ Locomotion in Snakes

Not all snakes crawl in the same way. In fact, snakes have three different ways to move on land: lateral undulating, straight crawling, and sidewinding.

The most rapid way of moving on land is by lateral undulations. The body of the snake forms waves that begin at the head and travel backwards toward the tail. These waves, or loops, are pressed against the ground, creating a thrust that pushes the snake forward. The entire body of the snake follows the same track. Each succeeding loop pushes against the same point, such as a pebble or a blade of grass.

Snakes may also use a slower, straight crawling movement. Straight crawling takes advantage of the large overlapping scales located on the snake's belly. To move, the snake stretches forward with part of its body. It then lowers the stretched part to the ground. The belly scales make contact with the ground and anchor the snake as it pulls the back part of its body forward. Some species of snakes can even climb trees by gripping the bark with their belly scales.

Sidewinding allows a snake to move over loose sand. The snake's body forms a double S-curve. The head, the center of the body, and the tail are raised. The snake's body now makes contact with the ground at only two or three points. Using the contact points for support, the snake throws the loops of its body forward and to the side. This produces a rolling, sideways form of travel.

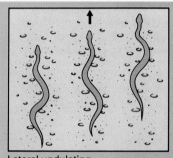

Lateral undulating

Straight crawling

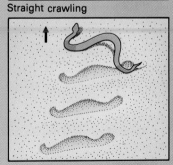

Sidewinding

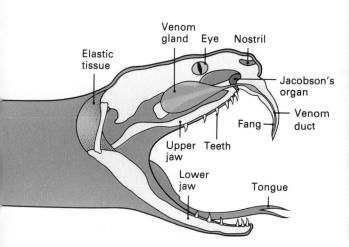

Figure 37–12. Venomous snakes, like the copperhead (above right), kill their prey with fangs, which are teeth modified for injecting venom into the victim. The illustration above shows a snake's head and a flexible jaw.

In venomous snakes some of the teeth are modified into sharp **fangs** through which venom flows. Fangs are connected to special venom glands that contain venom. A venomous snake strikes at its prey with its mouth open. When the needle-sharp fangs enter the prey, venom flows into the victim. The snake then releases the victim and waits for the venom to take effect.

Venom contains poisonous proteins or *toxins*. There are two main types of toxins, each affecting the prey's body in a different way. **Hemotoxins** destroy the red blood cells and break down the walls of the blood vessels, causing heavy internal bleeding. **Neurotoxins** attack the nervous system, paralyzing the nerve cells. Cobras and their relatives have venom that is mostly neurotoxic, whereas rattlesnake's venom is mainly hemotoxic.

Some venomous snakes, called *pit vipers,* have a special type of sensory organ. Pit vipers have small heat-sensitive pits in front of their eyes. These structures can detect changes in temperature as slight as 0.003°C (0.005°F). Using these pits, pit vipers can locate warm-blooded animals. Rattlesnakes, the most common venomous snake in the United States, are pit vipers. Water moccasins are also pit vipers.

Reviewing the Section

1. How do the limbs of reptiles differ from those of amphibians?
2. What are the four orders of reptiles?
3. How is an alligator distinguished from a crocodile?
4. By what two methods do snakes kill prey?
5. What is the advantage of sidewinding?

Investigation 37: Reptilian Behavior

Purpose
To observe some structural and behavioral characteristics of a typical reptile

Materials
Live lizard, thread, masking tape, construction paper, dead flies, straight pins, aquarium

Procedure
1. Make a noose on one end of a piece of thread. Carefully place the noose over the lizard's head and body, and slip the noose down over the body until it comes to rest just above the back legs. Tighten the noose just enough to keep it in place.

Green anole

2. Secure the other end of the thread with masking tape, so that the thread acts as a leash and defines an area in which the lizard can move about freely.
3. Observe the lizard for a few minutes. Draw the lizard. Include the following parts: scales, eyes, teeth, tympanic membrane, toes, and tail.
4. The lizard will not hurt you. Let the lizard bite your finger or a piece of paper and observe the teeth.
5. Put a piece of colored paper under the lizard. Record the amount of time it takes the lizard to change color.

6. Place a fly on the end of a straight pin. Move it around in front of the lizard's head and observe the lizard's reaction. Try different insects and different movements. Record your results.
7. Observe the fan of skin around the throat of the lizard. It is called the dewlap. *What is the purpose of the dewlap?*
8. Observe the skin of the lizard. *Explain how it differs from the skin of a frog. How is this an advancement over the frog?*

Analyses and Conclusions
1. Why are the lizard's toes considered an advancement over those of amphibians?
2. What is the function of the color change in the lizard?

Going Further
Determine the sex of your lizard. Males usually have larger dewlaps and are more colorful. Also, some will have a crest on the back of their neck. Place two females in the same area and observe their behavior. Place two males in the same area and observe their behavior. Which pair seems to get along better? Which pair seems to be more aggressive? Record your observations of both pairs of lizards.

Chapter 37 Review

Summary

Reptiles have two adaptations that give them a distinct advantage over amphibians in coping with life on land. Reptiles have an amniotic egg that allows them to lay eggs on dry land, and hard, waterproof skin. The amniotic egg consists of four embryonic membranes: the amnion, the allantois, the chorion, and the yolk sac.

Reptiles originated about 280 million years ago. The first reptiles were cotylosaurs, small lizardlike animals that fed on insects. The dinosaurs are the most notable of the ancient reptiles. Some herbivorous dinosaurs weighed more than 50,000 kg (110,000 lb.). Others were carnivores. The most closely related descendants of the dinosaurs are birds and crocodiles. Some ancient reptiles could fly, and others could swim.

Modern reptiles have two pairs of limbs that are positioned vertically under the trunk. Reptilian feet have clawed toes. Modern reptiles are classified into four orders: tuataras, large, lizardlike animals from New Zealand; turtles and tortoises, which are characterized by two shells that encase their bodies; alligators, crocodiles, and related carnivores adapted for life in shallow water; and lizards and snakes, which are the largest order. Most lizards are small, fast omnivores. Snakes are limbless and kill their prey either by constriction or by injecting victims with hemotoxins or neurotoxins.

BioTerms

allantois **(556)**

amnion **(555)**

amniotic egg **(555)**

carapace **(561)**

chorion **(556)**

fang **(566)**

hemotoxin **(566)**

Jacobson's organ **(564)**

keratin **(556)**

molting **(563)**

neurotoxin **(566)**

oviparous **(564)**

ovoviviparous **(564)**

parietal eye **(560)**

plastron **(561)**

scute **(561)**

yolk **(556)**

yolk sac **(555)**

BioQuiz   (Write all answers on a separate sheet of paper.)

I. Completion

1. The _____ bathes the embryo in an amniotic egg.
2. _____ and their relatives are the only reptiles which have a heart with four chambers.
3. The senses of taste and _____ are well developed in snakes.
4. The _____ are extinct reptiles that were able to fly.
5. The venom of rattlesnakes is an example of a _____ .

II. Modified True and False

Mark each statement TRUE or FALSE. If false, change the underlined term to make the statement true.

6. Caimans and gavials are members of the order Crocodilia.
7. Reptiles dominated land during the Mesozoic era.
8. Snakes only have one rib to allow for swallowing bulky food.
9. The pit of pit vipers functions to detect light.

III. Multiple Choice

10. The shell that covers the turtle's back is the a) carapace. b) plastron. c) scute. d) chorion.
11. Nutrition for the developing embryo in an amniotic egg is provided by a) amniotic fluid. b) embryonic membranes. c) Jacobson's organ. d) the yolk.
12. Ectothermic animals are those that are a) warm-blooded. b) cold-blooded. c) hibernating. d) migratory.
13. The earliest reptiles were a) dinosaurs. b) pterodactyls. c) cotylosaurs. d) ichthyosaurs.
14. The skin of all reptiles has a) mucus glands. b) scutes. c) keratin. d) plastron.

IV. Essay

15. What are the three ways that snakes move?
16. How do neurotoxins and hemotoxins differ in their mode of action?
17. What two characteristics are mainly responsible for the success of reptiles?
18. How do snakes produce young?
19. When did dinosaurs dominate the earth and when did they die out?
20. What special qualities of the amniotic egg allow it to survive on land?

Applying and Extending Concepts

1. What advantage does a snake have by being limbless? What are the disadvantages of having no limbs?
2. One of the most intriguing and baffling of all scientific questions is why the dinosaurs became extinct by the end of the Cretaceous period. Use your school or public library to research this question and list at least two theories that attempt to explain this phenomenon.
3. Although the senses of sight and smell in snakes are strong, snakes have no outer ears for hearing. Discuss the advantage of earlessness in snakes and research how snakes monitor sounds. How would you explain a snake weaving back and forth to the music of a snake charmer?
4. The poisonous coral snakes in the United States are strikingly colored with red, yellow, and black rings. Some harmless snakes, such as king snakes, are colored in a similar manner. What is the advantage to a poisonous snake in being brightly colored? What is the advantage to a nonpoisonous snake in displaying color mimicry?

Related Readings

Lynch, W. "Great Balls of Snakes." *Natural History*. 92 (April 1983) 64–68. Mass mating of red-sided garter snakes is discussed.

"Masters of the Tongue Flick." *Natural History* 91 (September 1982): 58. This article tells how lizards use their tongues in activities such as food seeking and courtship.

Sattler, H. R. *Dinosaurs of North America*. New York: Lothrop, Lee and Shepard, 1981. This comprehensive volume presents thorough information on all dinosaurs found in North America and includes sketches as well as beautiful full-page illustrations of dinosaurs in their natural habitats.

"Saved by a Tail." *Science 83* 4 (April 1983): 6. This article tells how the wiggling of a lizard's recently lost tail can help the lizard escape predators.

Arctic tern

Introduction

Summer has come to an end in Alaska and other regions close to the North Pole. The days are getting shorter and colder. It is time for the Arctic terns, birds that look like small seagulls, to leave the Arctic for their winter feeding grounds. Large flocks head south, flying through rain, fog, and storms. Eventually they arrive at their destination—the icy wastes of Antarctica. Within a few months, the terns have flown almost from pole to pole, a distance of 18,000 km (11,000 mi.).

The yearly journey of the Arctic tern is incredible. But it is not the only extraordinary story found among the feathered vertebrates. Some, like the ostrich, do not fly. Many search for food underwater, the loon to depths of 49 m (160 ft.). A few birds mate for life, and all care for their young.

Origin and Characteristics

All birds belong to the class Aves. Within this single class are tiny hummingbirds that weigh less than 10 gm (0.35 oz.) and ostriches that stand almost 3 m (10 ft.) tall and weigh more than 125 kg (275 lb.). The extinct elephant birds of Australia were even larger, possibly weighing as much as 450 kg (990 lb.).

Birds live on the high slopes of the Himalayas and on Antarctic icecaps. They flourish in tropical jungles and on city rooftops. Some birds live in deserts while others spend much of their lives soaring over the ocean, returning to land only to breed. You may well ask, how are birds related to other vertebrates? Where did birds come from, and what kinds of animals were their ancestors?

Section Objectives

- *Describe* the ancestors of modern birds.
- *Tell* how birds differ from other vertebrates.
- *List* the four categories of birds.
- *Discuss* how feet and beaks are modified in different kinds of birds.
- *Explain* why the use of pesticides may bring about the extinction of some bird species.

38.1 Origin of Birds

Birds arose from reptiles, possibly from small tree-climbing dinosaurs that leaped from branch to branch. These dinosaurs did not have feathers. Some scientists have suggested that feathers evolved as a method of insulation for endothermic reptiles. Another theory suggests that feathered wings were an adaptation for hunting, used to trap small land animals. Whether or not feathers originally functioned for flight, insulation, or hunting, a new kind of animal evolved that had feathers but also retained some reptile characteristics. This animal was the first bird.

The earliest known bird, called *Archaeopteryx* (ahr kee AHP tuhr ihx), lived about 140 million years ago. Now extinct, *Archaeopteryx* was about the size of a modern crow. It resembled reptiles in having bony teeth set into jaw sockets, a large bony tail, and claws that were located on each wing. It was, however, covered with feathers and had a skull like a bird. For these reasons, *Archaeopteryx* is considered a link between reptiles and birds.

Other prehistoric birds, such as the aquatic diving bird *Hesperornis* and the large predator *Diatryma*, came after *Archaeopteryx* and had fewer reptilian features. By the beginning of the Cretaceous period, about 130 million years ago, the class Aves had been established on Earth.

38.2 Characteristics of Birds

The most obvious feature of birds is that they have wings and can fly. Flight is not the feature that makes birds unique, however. Birds share this ability with bats and insects.

Figure 38–1. *Archaeopteryx* may have looked like a flying reptile, but feathers showed it to be a true bird. Fossil remains indicate that *Archaeopteryx* had short wings with claws at the end and was a clumsy flier.

Q: *What is the most abundant species of bird on Earth?*

A: The African red-billed quelea is the most common bird. Scientists estimate there are over 10 billion of these small, seed-eating birds.

Feathers distinguish birds from other classes of vertebrates. Some other features of birds are listed below.

- The body is divided into a head, neck, trunk, and tail.
- Bones are lightweight and filled with air.
- Air sacs are found throughout the body.
- Birds lack teeth; jaws are covered by a horny beak.
- Front limbs are modified into wings; hind limbs are adapted for perching, hopping, swimming, or other similar functions.
- Body temperature remains constant.
- The heart has four chambers.
- Reproduction involves the production of an amniotic egg that most species incubate in a nest.

▶ Endangered Birds

The California condor is a magnificent bird. It has a wingspan of 3 m (9.9 ft.) or more and can soar for long periods without moving its wings. Only about 30 of these birds are living today, however. They are an endangered species and may soon join the nearly 100 species of birds that have become extinct in the last 300 years.

Although extinction normally occurs in nature, human interference has speeded up the process. For example, in the 1800s billions of passenger pigeons lived in the United States. These birds were ruthlessly hunted. Within a very short time, the population dropped so low that the species could not recover. The last passenger pigeon, a bird named Martha, died in 1914 in the Cincinnati Zoo.

In most cases, bird species become extinct because humans cause changes in the environment, destroy food, or introduce new predators. One of the most disruptive factors has been the use of pesticides such as DDT, which do not break down quickly. They remain in the environment and often become concentrated in the bodies of insects, fish, and other organisms. As birds eat these animals, pesticides build up in their bodies and interfere with calcium metabolism. As a result, eggshells are fragile, and many young birds die before they can hatch. In the United States, pere-

grine falcons such as the one shown above and brown pelicans almost disappeared before the harmful effects of DDT were understood. The use of DDT in the United States was banned in 1972. It is, however, still used in many countries around the world.

38.3 Classification of Birds

The 8,600 species of birds are placed into 27 orders. Birds are classified on the basis of a number of characteristics, including body structure and behavior. The four most common types of birds are shown in Table 38–1.

Table 38–1: Four Common Bird Types

Type	Common Examples	Types of Feet	Types of Beaks
Flightless Birds	Penguins, rheas, ostriches	Adapted for running	(Beaks vary)
Water Birds	Ducks, swans, geese	Webbed	Broad and flat for filtering Long and pointed for fishing
Perching Birds	Sparrows, robins, other songbirds	Toes cling to branches	Short, thick, strong (seed eaters) Long and slender for probing (insect eaters)
Birds of Prey	Hawks, eagles, owls	Sharp, curving claws	Tearing beaks

Each group of birds is adapted to a specific way of life. Legs, feet, and beaks differ, for example, according to the ecological role of the group. Perching birds that live in trees have curved toes that are able to cling to branches. Predators capture their prey with sharp, curving claws called **talons.**

Reviewing the Section

1. What characteristic distinguishes birds from other vertebrates?
2. What are the four main categories of birds?
3. How are legs and feet modified to perform specific tasks?
4. How does DDT interfere with reproduction among birds?

- *Name* the five parts of a contour feather.
- *Explain* how a bird's skeleton and muscles are modified for flight.
- *Identify* the five organ systems of a bird's body.
- *Describe* how an egg is fertilized and provided with an egg white and shell.
- *Distinguish* between precocial and altricial young.

Anatomy of a Bird

A bird is adapted for flight. Feathers, bones, muscles, and all the internal organs must be strong enough and light enough to allow the bird to move easily through the air.

38.4 Feathers

Feathers are outgrowths of the bird's skin that facilitate flight and also provide insulation. Feathers evolved from the scales of birds' reptile ancestors.

Feathers grow from small sacs in the skin called **follicles.** They first appear as small, dark **pin feathers** that have a scaly covering. As they grow, feathers burst through the scaly covering and expand into mature feathers.

Mature feathers are strong and light because of their unique structure. Feathers have a central cylinder, or **rachis** (RAY kihs), running their length. The part of the rachis that extends out of the skin is solid, while that beneath the skin is hollow. This hollow section is known as the **quill.** Branching off the rachis are many small projections called **barbs.** Barbs make up the soft, flexible part of the feather. Barbs are linked together by small lateral projections called **barbules.** Several barbs hooked together by barbules form a **vane.** When the barbs in a vane become unhooked by wind or by striking an object, the bird rehooks them with its bill.

Birds have four types of feathers: contour feathers, down feathers, filoplumes, and bristles. **Contour feathers** are smooth, sleek feathers that cover the head, body, and wings. Contour feathers protect and streamline the bird. The contour feathers of the wing and tail are especially large and are called *quill feathers.* **Down feathers** are soft, fluffy feathers found under contour feathers. Down feathers insulate birds. Newly hatched birds have only down feathers. **Filoplumes** are short, thin feathers that look like hair. They are also called pin feathers. **Bristles,** also short, hairlike feathers, are located near the bird's nostril. Bristles filter dust.

Many birds have a special gland, called a **preen gland,** at the base of the tail. The preen gland secretes oil. In a process called *preening,* the bird rubs its beak against the preen gland, picks up oil, and spreads the oil over its feathers. Preening serves to waterproof the feathers. The oil repels water that would otherwise seep in and leave the feathers waterlogged.

Birds, like lizards and snakes, go through a process of molting. When feathers become old or damaged, they are shed and

Figure 38–2. This Stellar's jay has its feathers ruffled to increase their insulating effect.

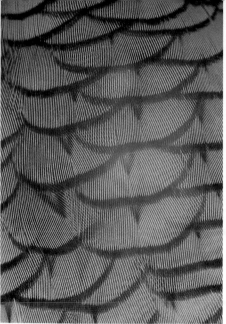

replaced by new feathers. In most birds, the feathers are not shed all at once. In birds like geese or ducks, however, the loss of even a few flight feathers makes them unable to fly. Therefore, these birds lose all of their feathers at the same time. They avoid predators during molts by hiding in safe places.

Many birds have colored feathers. Sometimes coloration helps the bird avoid predators. Birds that live in trees, for example, are often camouflaged by their green or yellow feathers. Color is also important when birds mate. In many species the males are brightly colored to warn away other males and to increase their appeal to females. Females are often duller than males because they must be concealed from predators when they sit on nests and incubate their young.

Figure 38–3. Contour feathers (left), down feathers (center), and filoplumes (right) are three major types of feathers.

38.5 Skeleton and Muscles

The prominent parts of the bird's skeleton and muscles are modified for flight. Except for the neck, the entire skeleton is rigid. This makes flight mechanically efficient because the skeleton does not wobble during flight. Individual bones are thin, strong, and filled with air pockets that make them exceptionally light.

The most noticeable part of the skeleton is the large breast bone, or **sternum.** The bottom section of the sternum forms a high, narrow ridge called a **keel.** The powerful breast muscles that move the wings are attached to the keel. The pectoralis (pehk tuh RAL uhs) muscle lowers the wing and the supracoracoideus (soo pruh kawr uh KOY dee uhs) raises it. Slow-motion photography reveals that birds fly by moving their wings in a complex, figure-eight motion.

Q/A

Q: *How many feathers does a bird have?*

A: Whistling swans often have over 25,000 feathers. On the other hand, ruby-throated hummingbirds may have as few as 950.

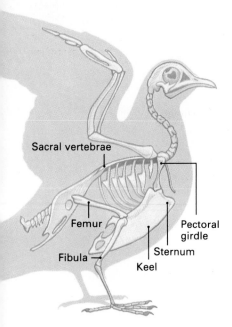

Figure 38–4. The lightweight skeleton of a bird is an adaptation for efficient flight.

Sacral vertebrae

Femur

Fibula

Keel

Sternum

Pectoral girdle

The bird's powerful wing muscles enable some species of birds to fly very rapidly. Many birds fly at 50 to 80 km (30 to 50 mi.) per hour. Racing pigeons can maintain speeds of 60 km (37 mi.) per hour over long distances. Hummingbirds can hover as well as fly. To stay in one place in the air, they must flap their wings more than 200 times each second.

A group of vertebrae, called *sacral vertebrae,* are located at the lower end of the spine. Attached to these vertebrae is the pectoral girdle. Legs are attached to the pectoral girdle. The *femur* (FEE muhr) is the upper leg bone; the lower leg bone is the *fibula* (FIHB yoo luh). Large muscles in the lower leg move the four toes. The ankle and foot are fused into a single structure. The outside of a bird's leg is covered by **leg scales,** keratinized coverings similar to those found in reptiles.

38.6 Digestive System

Birds must eat large quantities of food to generate the energy needed for both flight and temperature regulation. Birds, therefore, spend much of their time searching for and eating food. Different species of birds have different diets. A seagull may eat everything from fish it has caught to garbage it finds in a dump. Eagles and hawks eat the flesh of fish and small mammals. Other birds may eat only seeds, and some drink only the nectar of flowers.

The digestive system of a bird processes a large amount of food at a time. Food is taken in through the mouth, travels down the esophagus, and enters a large crop, which is used primarily for food storage. Upon leaving the crop, food moves into a two-chambered stomach. The first chamber is the thick-walled *proventriculus* (proh vehn TRIHK yoo luhs), where food is mixed with digestive enzymes. The second chamber of the stomach is the gizzard. The gizzard functions as a substitute for teeth; it grinds up the food. A bird's gizzard has a hard, horny lining. The duodenum, the coiled intestine, and the other digestive organs of the bird are similar in structure and function to those in other vertebrates.

38.7 Respiratory System

A bird's high body temperature and level of activity require a large amount of oxygen. In addition, birds use air to help buoy them up in flight. Air is drawn into the body through the mouth or the two nostrils in the beak. The air then passes through the *trachea* (TRAY kee uh) into a pair of tubes called *bronchi.* Each bronchus leads to one of the lungs.

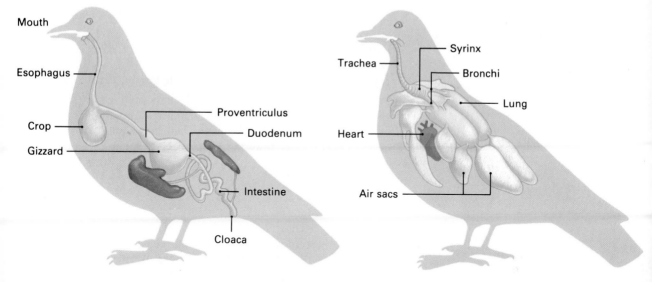

Mouth

Esophagus

Crop

Gizzard

Proventriculus

Duodenum

Intestine

Cloaca

Syrinx

Trachea

Bronchi

Lung

Heart

Air sacs

Much of the inhaled air travels into **air sacs**—cavities that make the bird lighter when filled with air. Air sacs extend into the body cavity and even into the toes. Because of the air sacs, a bird's respiratory system may take up as much as 20 percent of the body volume. The main function of the air sacs is to force air across the lung surface. Gas exchange occurs on the surface of the lungs, not in the air sacs.

Some of the air in the body is used to make sounds. Birds do not have vocal cords. They produce their calls instead by means of their **syrinx** (SIHR ihnks), or ''song box.'' The syrinx is located at the bottom of the trachea. It contains a pair of vibrating membranes. As air passes out of the lungs, the membranes vibrate and produce sound. Muscles attached to the membranes change the pitch of the sound.

Figure 38–5. A bird's digestive system (left) processes the large amounts of food required for energy. The respiratory system (right) must deliver large quantities of oxygen for converting this food to energy.

38.8 Circulatory System

Birds have four-chambered hearts. The right and left sides of the heart are completely separate. Each side of the heart is divided into two chambers, an upper atrium and a lower ventricle. The right side of the heart receives oxygen-poor blood from all parts of the body and pumps it to the lungs. After the blood is oxygenated in the lungs, it travels to the left side of the heart. From there it is pumped out to the body through the aortic arch. A bird's heart beats very rapidly, although the rate varies greatly from species to species. A mourning dove's heart rate ranges between 135 and 570 beats a minute. Because a hummingbird must deliver large amounts of oxygen to the muscles of its rapidly beating wings, its heart may beat as much as 1,000 times a minute.

38.9 Excretory System

The excretory system of a bird is modified to conserve water. Almost all the water is removed from urinary products as they pass from the kidney to the cloaca. The primary form of nitrogenous wastes is a white solid compound called *uric acid*. Wastes from the intestine and from the kidneys are excreted together.

38.10 Nervous System

The nervous system of a bird is highly developed. The brain has three well-developed areas: the cerebellum, the midbrain tectum, and the cerebrum. The cerebellum controls flying and walking by coordinating movement. The **midbrain tectum** is the center of vision. The **optic lobes,** the place where visual impulses are interpreted, are located in this area. The cerebrum is the control center for complex patterns of behavior such as eating, mating, nest building, and care of the young. The cerebrum is also the site of a bird's learning ability and serves as the control center for its voluntary muscles.

Birds have keen senses of hearing and balance but poor senses of smell and taste. They depend more on their vision than on any other sense. Birds have extremely large eyes that detect slight movements. Although the eyes are unable to rotate, a bird still has a wide range of vision because its neck is flexible due to a large number of vertebrae. Some birds, such as owls, can rotate their necks almost 270°, three-quarters of a circle. Owls and other birds that are active at night see well in the dark. They can see by starlight what humans see by moonlight.

38.11 Reproductive System

Male birds produce sperm in testes. The mature sperm pass from the testes into a long tube, called the *vas deferens*. The lower part of the vas deferens is enlarged to form a seminal vesicle where the sperm are stored until mating occurs.

The female of most bird species has only a single ovary, located on the left side. Eggs produced in the ovary are released into the body cavity. After mating occurs and sperm fertilize the eggs, the eggs enter the **oviduct.** The oviduct is a long tube that leads to the cloaca. As eggs travel through the oviduct they are covered by **albumin,** a nutritious protein that makes up the egg

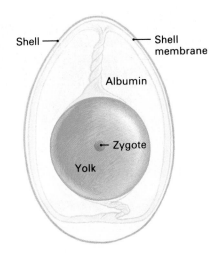

Shell — — Shell membrane

Albumin

← Zygote

Yolk

Figure 38–6. The embryo is the only living part of the egg. The yolk and albumin provide food. Other parts provide protection, waste disposal, or the exchange of oxygen and carbon dioxide.

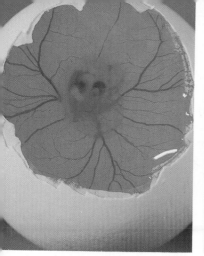
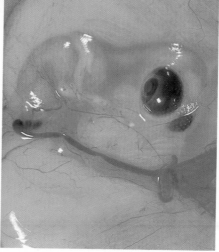

white. Later, the eggs are covered by shells made primarily of calcium carbonate, also called lime. Eggs with shells leave the body through the cloaca. Eggs that are not fertilized also receive albumin and a shell. Nearly all of the eggs bought in stores are unfertilized.

The structure of the fertilized egg is shown in Figure 38–6 on page 578. You can see that the structure is similar to that of a reptile's egg but that the yolk of a bird's egg is larger.

Figure 38–7 shows the development of a bird embryo. When embryo development is complete, hatching begins. The young bird slowly pecks its way out of the shell using a special **egg tooth.** This sharp structure located on the tip of the beak is lost a few days after hatching.

The newly hatched young of birds such as chickens and ducks are well developed. Their eyes are open and they are covered with a full coat of down feathers. Within a few hours after birth these young birds leave the nest and follow their parents. Such young are called **precocial.** Songbirds and most sea birds have **altricial** young—that is, the newly hatched birds are blind, naked, and helpless. Altricial young must remain in the nest for a time because they require constant feeding and care.

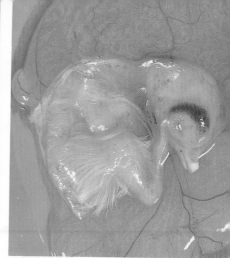

Figure 38–7. These photographs show a chick embryo at 5 days (left), 8 days, when eyes are visible (center), 14 days, when limbs and feathers can be seen (top right), and a chick emerging on the 21st day (bottom).

Reviewing the Section

1. What are the four kinds of feathers?
2. How are a bird's bones modified for flight?
3. What are the three main regions of the brain of a bird and what is the function of each?
4. What are two functions of air sacs?
5. How do birds sing?
6. How do precocial and altricial young differ?

Behavior of Birds

Reproduction in birds is accompanied by elaborate patterns of courtship, mating, nest building, incubation, and care of young. Complex behavior is also shown by the yearly movement of bird populations. Unlike the Arctic tern discussed earlier in this chapter, most birds migrate away from mating and nesting areas to warm wintering grounds.

38.12 Courtship

Males and females of most bird species form **pair bonds**—that is, they stay with one another throughout the reproductive season. They are attracted to each other through a series of courtship displays. During these displays, one of the birds—usually the male—engages in elaborate rituals designed to attract a mate. Most males attract females with song. Some, such as peacocks and lyrebirds, spread out their feathers in a beautiful, colorful display. The male bower bird builds a nest of twigs and lines it with any blue object he can find. Male grebes and herons perform acrobatics in the water to attract females.

These courtship rituals help the female recognize a mature male of her species that is capable of reproducing. In response to male courtship, a female usually performs a series of movements indicating her desire to mate.

38.13 Nesting and Care of Young

During or after courtship, birds build nests. Most bird nests are bowl-shaped structures made of twigs and leaves. Swallows and other birds build nests of mud on the sides of cliffs. Woodpeckers lay their eggs in trees, in holes lined with wood chips. Killdeer and other shore birds make shallow nests of pebbles. Some birds, such as monk parakeets, build large community nests in which each female has her own space. A few species make no nest at all. Falcons, for example, simply lay their eggs on bare ground.

The number of eggs laid and their size varies. Hummingbird eggs are no larger than a pea. Ostrich eggs may weigh 2 kg (4.5 lb.) or more. Most eggs are protectively colored.

A group of eggs in a nest is called a **clutch.** Most females remain with the clutch because bird eggs cannot develop unless the eggs are **incubated,** or kept warm. The female sits carefully on the clutch and warms the eggs with her body. In many species both parents share the job of incubation. The incubation

Figure 38–8. The male peacock displays his colors to a female during the courtship ritual. The female's drab color helps protect her from predators when brooding eggs or guarding her young.

Spotlight on Biologists: George Archibald

Born: New Glasgow, N.S., Canada, 1947

Degree: Ph.D., Cornell University

American wildlife biologist George Archibald has devoted his career to the study of cranes. His research has contributed much to the knowledge of these graceful wading birds. His breeding program at the International Crane Foundation at Baraboo, Wisconsin, has also been a factor in preserving several rare species of cranes.

Of the 15 existing species of cranes, all but 7 are now on the endangered

list. Archibald has developed many creative techniques to breed cranes in captivity. He once lived with a female whooping crane for 7 weeks, 15 hours a day, and performed courtship dances with her to help her produce an egg that would be fertilized by artificial insemination.

time ranges from a few weeks to several months. Chicken eggs hatch 21 days after they are laid. The eggs of the albatross must be incubated for three months.

38.14 Migration and Navigation

Many bird species mate, build nests, and raise their young in one part of the world and then move to another to avoid winter. A common sign of winter's approach is the sight of birds overhead, flying south. Similarly, the return of birds to the north signals the coming of spring. This regular, seasonal movement is known as **migration.** More than two-thirds of the bird species in the northern United States make an annual round trip between their winter feeding grounds in the south and their northern breeding grounds. These trips are often long and difficult. Among migrating birds, the Antarctic Adelie penguin follows an unusual pattern of behavior. These penguins travel north (away from the Antarctic) during the winter by floating on ice-

▶ Bird Songs

Beautiful bird songs have intrigued both poets and scientists. Recently, scientists have applied modern techniques to understand why birds sing. Scientists record bird calls in the wild and later make *sonograms* of their songs. A sonogram is a graph that plots the frequency of a sound over a period of time. By looking at sonograms, scientists can visualize songs and thus analyze them more carefully. The sonograms shown are from three closely related species of thrushes. Thrushes are often hard to distinguish visually but, as you can see, their songs are very different. By studying sonograms, scientists discovered that many bird species have regional dialects—that is, birds have accents just like humans.

Many times songs are used to mark and defend territories. For example, when a male wood thrush moves into an uninhabited area, it sings to warn other males to stay away. If another male wood thrush does not heed the song and enters the territory, it will be attacked. But other kinds of thrushes are safe from attack because their songs identify them as different species.

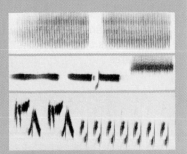

Like thrushes, many other birds use songs for species recognition. Male birds, for instance, call to attract females during the mating season. Newly hatched birds sing special "begging" songs when they get hungry. By analyzing sonograms, scientists hope to learn more about songs—the language of birds.

bergs. In spring they swim south back to the Antarctic. The penguins then cross several kilometers of land to their breeding ground by walking and sliding across the ice on their stomachs.

Year after year birds migrate over the same paths, called *migration routes*. Birds find their way partly by instinct and partly by learned behavior. Birds apparently use a number of different clues to guide their flights. They may direct themselves by landmarks or they may use wind currents, the sun, and the stars. Pigeons and certain other bird species can find their way by using the earth's magnetic field.

Reviewing the Section

1. Why do males engage in courtship displays?
2. Describe three types of bird's nests.
3. How do birds navigate their migration routes?
4. What is the biological significance of bird songs?

Investigation 38: What Is a Bird?

Purpose
To become familiar with the unique characteristics of birds that distinguish them from other vertebrates

Materials
Photographs of birds; flight, contour, and down feathers or slides of feathers; compound microscope; hand lens; chicken bones; hacksaw

Procedure
1. Make a chart like the one shown.

Feature	Habitat	Beak	Legs	Feet
Bird	Where does the bird nest and hunt? wetlands desert/prairie trees cliff tops buildings	How is the bird's beak adapted to what it eats? spear shaped chisel shaped seed cracker hooked/strainer serrated/probe	How are the bird's legs adapted to where it lives? long-short short stout-short slender positioned for walking positioned for swimming	How are the bird's feet adapted to its way of life? long toes-wading webbed-swimming stout toes with nails-walking sharp talons-grasping

2. Using the photographs of birds provided by your teacher, complete the chart. As you examine each feature, ask yourself how each characteristic enables that specific type of bird to survive in its habitat.
3. Examine each of the three types of feathers with a microscope or hand lens. Draw each type of feather and label the following parts: vane, barbs, rachis, hooklets, and barbules. *What is the function of each type of feather?*
4. Study the clean bones of a chicken. Note how lightweight the bones are. Using a hacksaw, carefully saw one of the bones in half. Use a hand lens to study the cross section of the bone. *How are the bones of birds adapted for strength and flight?*

Analyses and Conclusions
1. Describe adaptations that allow birds to do the following: live in wetlands, live in forests, live in prairies, eat fish, eat seeds, and eat insects.
2. Explain the function of the three types of feathers with respect to: streamlining the body, insulation, courtship and display, and aiding in flight.

Chapter 38 Review

Summary

Birds evolved from reptiles about 140 million years ago. The first known bird, the *Archaeopteryx,* had feathers like a bird but retained a few reptilian characteristics. Birds were a well-established class by the beginning of the Cretaceous period.

Feathers distinguish birds from all other vertebrates. Birds also have bones with air pockets, horny beaks, endothermic temperature regulation, and air sacs. Four common bird types are: flightless, water, perching, and predatory.

Birds have four kinds of feathers: contour, down, filoplumes, and bristles. Feathers protect and insulate birds and enable them to fly. The bird's skeleton, muscles, and organ systems are all modified for flight.

Birds have complex behavior patterns during reproduction. Males engage in courtship displays that attract females. Birds build nests and incubate eggs in them. Many species continue to care for their young after the eggs have hatched. In winter, many species of birds migrate to warmer environments.

BioTerms

air sac **(577)**
albumin **(578)**
altricial **(579)**
barb **(574)**
barbule **(574)**
bristle **(574)**
clutch **(580)**
contour feather **(574)**
down feather **(574)**
egg tooth **(579)**

filoplume **(574)**
follicle **(574)**
incubate **(580)**
keel **(575)**
leg scale **(576)**
midbrain tectum **(578)**
migration **(581)**
optic lobes **(578)**
oviduct **(578)**
pair bond **(580)**

pin feather **(574)**
precocial **(579)**
preen gland **(574)**
quill **(574)**
rachis **(574)**
sternum **(575)**
syrinx **(577)**
talon **(573)**
vane **(574)**

BioQuiz   (Write all answers on a separate sheet of paper.)

I. Completion

1. Feathers may have evolved for purposes of flight, hunting, or _____.
2. _____ feathers are the first to emerge from the follicle.
3. The midbrain tectum is responsible for _____.
4. Females _____ eggs to ensure that the young develop.
5. _____ birds have feet that are modified into webs for swimming.

II. Modified True and False

Mark each statement TRUE or FALSE. If false, change the underlined term to make the statement true.

6. Altricial birds are nearly helpless when they hatch.
7. Birds migrate over a different migration route each year.
8. Birds fly by moving their wings in a figure-eight pattern.
9. Talons help predators catch prey.

III. Multiple Choice

10. Feathers evolved from a) reptile scales.
    b) hair. c) amphibian glands. d) quills.
11. Albumin provides the bird embryo with
    a) support. b) nutrition. c) waste
    removal. d) protection.
12. Many barbs hook together to form a
    a) barbule. b) vane. c) rachis.
    d) filoplume.
13. The use of pesticides is a major threat to
    many bird species because pesticides
    a) destroy food sources. b) poison
    water supplies. c) weaken eggshells.
    d) kill birds.
14. Birds are able to make sound by passing
    air over the a) vocal cords. b) keel.
    c) syrinx. d) filoplume.

IV. Essay

15. On which senses do birds most rely to
    monitor their environment?
16. What are the four kinds of feathers, and
    what is the function of each?
17. What is the adaptive value of bird
    songs?
18. Why is *Archaeopteryx* considered a link
    between reptiles and birds?
19. What are the reasons for having court-
    ship behavior before mating?
20. How is the skeleton modified for flight?

Applying and Extending Concepts

1. The yellow-billed cuckoo, a common bird
   of the southeastern United States, is
   brown on top and white underneath. What
   do you suppose is the advantage of this
   particular color pattern to a tree-dwelling
   bird?
2. When Canada geese migrate, they do so in
   flocks that sometimes contain hundreds of
   thousands of individuals. Write a paragraph explaining the adaptive advantage
   of this flocking behavior.
3. The structure of a bird's wing and that of
   an airplane are similar. Describe the similarities and differences, and tell how each
   type of wing is designed to use air pressure as a means of creating the lift necessary to fly.
4. The bar-headed goose is an Asian bird that
   flies higher than any other bird. It attains
   altitudes of 7,600 m (25,000 ft.) as it flies
   over the Himalayas. Based on what you
   know about bird anatomy, suggest what
   adaptations you would expect to find in
   these birds that enables them to fly so high.

Related Readings

Gillard, T. *Living Birds of the World*. New
York: Doubleday, 1958. This book is an
illustrated comprehensive survey of the
characteristics, habits, and habitats of all
the known families of birds.

Graham, F. ''For Migrants, No Winter
Home?'' *Audubon* 82 (November 1982):
14. This article discusses how the destruction of tropical habitats affects the survival
of migratory birds.

Griffin, D. *Bird Migration*. Garden City, NY:
Anchor Books, 1964. This comprehensive
discussion of the anatomy and physiology
of bird migration explains the environmental clues by which birds navigate.

''Myriad Meaning of Birdsong.'' *Science Digest* (November/December 1980): 30.
Some of the ways birds use their songs to
communicate within and between species
are described in this article.

BIO*TECH*
Biotelemetry

Biologist attaches transmitter to a pheasant.

An all-terrain vehicle moves slowly through a snowy landscape in western New York State. An antenna atop the vehicle receives a signal; a biologist inside marks a map. The biologist is a member of a research team that has been tracking several dozen ring-necked pheasant hens for the past year. Each hen is equipped with a 27-g (0.950 oz.) radio transmitter powered by a tiny battery. When the biologists learn what kind of ground cover the hens prefer, they can recommend habitat management procedures to increase pheasant populations.

The pheasant researchers use the newest technological advances in **biotelemetry**—the measurement of living things at a distance. Biotelemetry techniques have been used for many years to study the movement and behavior of animals in the wild. Now innovations in techniques and instrumentation make biotelemetry an increasingly valuable tool for biologists in many fields. A biotelemetry system has two parts: a radio transmitter carried on or in the animal being studied, and a receiver to pick up transmitted information.

Biotelemetry is being used more and more frequently by marine mammalogists to follow the activities of whales and dolphins. Satellite monitoring has even re-

Biotelemetry aids in the study of animal movement and behavior in the wild.

Divers attach transmitter to manatee (right) and release manatee (far right) for monitoring.

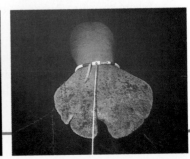

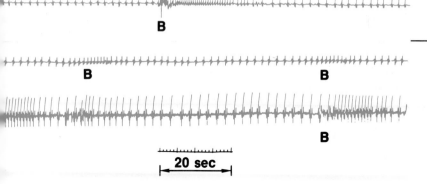

B

B   B

B

20 sec

Heartbeat records of two dolphins (top and center) and whale (bottom) provide comparisons of physiological functions.

vealed the movements of a humpbacked whale—its location and speed, the water temperature, and the number and duration of its dives.

Over the past few years, the technology of biotelemetry has been applied increasingly in monitoring animals' physiological functions. Heartbeat and breathing rates, body temperature, brain waves, blood pressure—all can be recorded using implanted transmitters. This data gives biologists new insights into the physiology of animals under natural conditions. For example, researchers have discovered that when an alligator is frightened and dives under water, its heartbeat rate drops.

Biologists have long thought that the same is true for a dolphin's heartbeat rate during a dive. However, biotelemetry monitoring shows that the heartbeat rate remains high throughout the dolphin's dive.

Now researchers have developed an extremely tiny transmitter— 0.5 g (0.002 oz.)—that could be implanted in a creature as small as a mouse. The transmitter signal is strong enough to be monitored 15 m (50 ft.) away. The inventor is looking for a way to convert body heat or some other biological source of energy into electric energy to power the device. In this way, the transmitter could be used to record an animal's physiological functions continuously over its entire lifetime.

Alligator with transmitter

Mammals

Impalas at a South African watering hole

Introduction

A cheetah races across an African plain chasing a swift, sleek gazelle. Nearby, an elephant browses on leafy bushes, while mice hide in the grass beneath. At dusk, the skies fill with flapping bats. A group of people on a photo safari linger to watch giraffes and zebras gathered at a water hole. The giraffes and zebras stampede away as a lion approaches the water hole. All of these remarkably different animals, including the humans, are closely related. They are all mammals.

The 4,500 species of mammals live throughout the world. Some burrow in desert sand, some swing through tropical trees, some swim in arctic waters. Mammals can live in such different environments because their flexible body plan has allowed the various species to undergo many special adaptations.

Characteristics of Mammals

The class Mammalia consists of vertebrates whose young are nourished by their mother's milk. Mammals include all the most familiar domesticated animals, such as dogs, cats, horses, and cows. Mammals also include human beings, as well as the many animals that still live in the wild. Mammals have been the dominant life form on Earth for the past 65 million years. This period of time, the Cenozoic era, is called the Age of Mammals.

Fossil evidence indicates that mammals evolved more than 180 million years ago from a now-extinct order of reptiles called *therapsids* (thuh RAP suhdz). Fossil therapsid tooth and jaw fragments show a mixture of reptile and mammalian characteristics. Mammalian features are jaws that do not become unhinged, and a lower jaw formed of only two bones. Although many of the first species of mammals died out, some developed into forms that are known today. Small, insect-eating mammals coexisted with dinosaurs and later survived the conditions that led to the extinction of the giant reptiles. Most of the mammals known today had come into existence in some recognizable form by the start of the Jurassic period, 180 million years ago.

Mammals have several characteristics not found in other vertebrates:

- *The nursing of young:* Mammals have modified sweat glands called **mammary glands** that secrete the milk used to feed the young after they are born. These glands give the order its name.
- *Body hair:* Only mammals have protective coverings of body hair. The hair acts as insulation and also protects the body from injury.
- *Live birth:* Most mammals are **viviparous**—that is, the young are born alive after developing inside the mother.
- *Extended parental care:* All mammalian young go through a prolonged period of development. They remain with their parents while learning to take care of themselves.
- *A large, well-developed brain:* More than any other single organ, the brain has enabled mammals to survive change and to develop complex patterns of behavior.
- *An outer ear:* Mammals are the only animals that have an outer ear, which collects sound and transmits it to the middle and inner ear.
- *Separate chest and abdominal cavities:* In mammals, a dome-shaped muscle called the **diaphragm** divides the coelom into two parts.

Section Objectives

- *List* the major characteristics that distinguish mammals from other vertebrates.
- *Name* three adaptations that help mammals maintain a high metabolic rate.
- *Distinguish* between a mammal's territory and its home range.
- *Give* three reasons why mammals migrate.
- *State* how bats locate objects they cannot see.

Figure 39–1. The woolly mammoth, a prehistoric animal, was hunted by early humans. A close relative of today's elephant, the mammoth died out about 10,000 years ago.

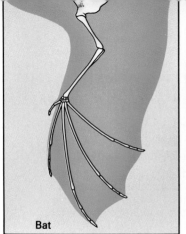

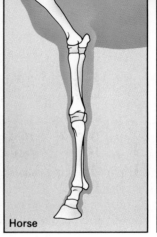
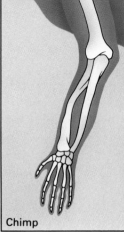

Bat

Seal

Horse

Chimp

Figure 39–2. The hands and forelimbs of mammals are adapted to various uses. Skeletal modifications form the framework for a bat's wing, a seal's paddlelike flipper, a horse's sturdy foreleg, and a chimpanzee's flexible hand.

39.1 Movement of Mammals

Mammals have made great advances over the reptiles in the way they move. A mammal's legs are directly underneath its body. Thus, little energy is required to hold the body off the ground. The elbow of the forelimb is turned backward, and the knee of the hindlimb bends forward. Also, the limbs of mammals are longer and more slender than those of reptiles. These changes have increased the speed at which some mammals can travel on land. The cheetah, for example, can reach incredible bursts of speed. It can run up to 113 km (70 mi.) per hour for short distances.

39.2 Respiratory and Circulatory Systems

Like birds, mammals are warm-blooded. To keep their body temperature at a constantly high level, mammals need to produce a great deal of heat. A high metabolic rate, in turn, requires plenty of food for energy. Unlike reptiles, mammals can remain active even when environmental temperatures are low. An arctic fox, for example, can run about and look for food in temperatures as low as −50°C (−58°F).

Mammals have efficient respiratory and circulatory systems to maintain their high metabolic rate. Mammals, like birds, have four-chambered hearts. The four-chambered heart allows oxygen-rich blood to travel from the lungs to the heart without mixing with the oxygen-poor blood returning to the heart from the rest of the body.

The relaxation and contraction of the diaphragm changes the volume of the lungs. The changed volume results in a change of air pressure inside the lungs, which in turn moves air in and out of the lungs. In mammals, the breathing and food passages are separate, allowing the animal to breathe and eat food at the same time. The air and food passages cross in a

Have you ever listened to the high, faint squeak of a bat as it flies through the sky on a summer night? This squeak is part of the bat's extraordinary ability to locate objects even in total darkness. Bats "see" by **echolocation**—that is, they locate objects by the reflection of sound waves. The process works much like the sonar of a ship.

Bats emit extremely high-pitched sounds through the mouth and nose. The sound waves travel in front of the bat and make contact with an object. Part of the wave is then reflected back toward the bat, where it is received by the bat's unusually large ears. The reflected sound, the echo, permits the bat to interpret the object's size, direction, and its distance from the bat.

Echolocation is so accurate that bats can avoid thin wires in complete darkness. Most bats use echolocation to hunt for food and to avoid obstacles. Sometimes the pitch of sounds emitted by the bats is so high that humans cannot hear it.

Not all bats echolocate, however. Fruit-eating bats have large eyes that allow them to see well even in dim light.

Bats are not the only animals that "see" by echolocation. Shrews and other nearly blind insectivores have echolocation mechanisms. Scientists studying captive porpoises have discovered that these mammals can use echolocation to locate tiny objects, detect the difference between different kinds of metals, determine the sizes of objects, and detect thin wires. Other toothed whales also echolocate, sending sounds inaudible to the human ear into the water.

common area at the back of the mouth. When mammals swallow, a thin flap of cartilage, the *epiglottis*, moves downward and prevents food from moving into the respiratory passages.

39.3 The Mammalian Nervous System

In mammals, the constant high internal temperature permits the development of an exceptionally sensitive and complex nervous system. In proportion to its body size, a mammal's nervous system is far larger than that of any other vertebrate. In addition, a large proportion of a mammal's brain correlates and integrates incoming information. In mammals, the cerebrum is the dominant as well as the largest part of the brain. The brains of fish, amphibians, and reptiles are organized differently. In mammals, a part of the cerebrum, the *cerebral cortex*, has expanded to form a layer covering most of the forebrain. The cerebral cortex has become the major coordinating center of the brain. *The exceptional learning ability and memory of mammals are attributed to the well-developed cerebral cortex.*

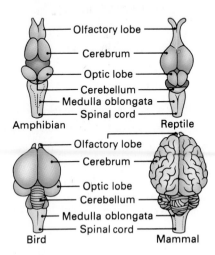

Figure 39–3. The cerebrum controls intelligence. The more complex the vertebrate, the larger its cerebrum.

Figure 39–4. Young lions cannot survive without their mother's help and protection. When the mother hunts prey, the young watch and learn by her example.

39.4 Reproduction and Parental Care

All mammals reproduce through internal fertilization. The male releases the sperm inside the body of the female where it fertilizes one or more eggs. After the eggs are fertilized, they are nurtured within the mother's body for a time.

Male mammals can mate at any time after they reach maturity. In most species, however, a female will not mate except during her fertile period, called **estrus.** Some species have estrus periods several times a year. In other species, estrus occurs once a year, timed so that the young's birth takes place during the season most favorable for survival.

One of the most important and distinctive features of mammalian reproduction is that the young have a childhood, a distinct developmental period. When a young mammal is born, it is unable to survive in the world on its own. A young mammal needs a certain amount of parental care while it is learning how to obtain its own food and to defend itself against predators. The care and protection given to the mammalian young greatly increase their chances to survive. Thus mammals bear relatively small numbers of young, compared with vertebrates such as frogs and fishes.

39.5 Behavior

Many mammals live together in social groups for purposes of hunting or defense. A zebra herd consists of one adult male and several females and their young, while a wolf pack is made up of several males and females and their young. Some hunting animals live solitary lives. Adult tigers, for example, come together only to mate.

Other patterns of behavior also differ among species. Certain mammals exhibit territoriality. Others migrate or respond to cold weather by hibernation.

Territoriality   Many mammals exhibit **territoriality**—that is, they claim a particular area as their own and defend the area against other animals of the same species. Fur seals establish a territory only during the breeding season. At this time, one male tries to prevent other males from entering his area. At the same time, he tries to move as many female seals as possible into his area. Other species establish territories as a means of protecting a food source. Many mammals mark their territories by leaving scents or body wastes.

Not all mammals defend territories, but most have a **home range**—that is, an area over which they travel during normal

Figure 39–5. A pronghorn buck uses scent glands in his mouth to mark a plant. In this way he lets other pronghorns know that this is a boundary of his territory.

Highlight on Careers: Animal Behaviorist

General Description

An animal behaviorist is a scientist who studies how animals behave and why they behave as they do. Animal behaviorists look at all possible influences on an animal's actions, such as heredity, physiology, environment, food, weather, and the behavior of other animals.

Some scientists are specifically interested in how animals learn. They may concentrate on highly intelligent mammals, such as monkeys, chimpanzees, or porpoises.

A few scientists study animals in the wild, tagging them to keep track of specific individuals. Usually animal behaviorists study animals in captivity.

Some animal behaviorists hope their research will have applications to the study of human behavior. These scientists study

such topics as the effects of crowding, of separating infants from their mothers, or of chemical substances on animal behavior.

Some animal behaviorists teach in colleges or universities. Zoological training is also useful in seeking a job as a zookeeper, game warden, or ranger in state or national wildlife and park services.

Career Requirements

Most animal behaviorists earn a B.S. degree in general biology or zoology.

They then specialize in animal behavior for the M.S. or Ph.D. degree. An advanced degree is required to teach in a college or university or to initiate research projects.

Some jobs as game wardens or rangers at zoos or in parks are available to candidates with a B.S. degree. Competition for these positions is keen, however, and the limited number of jobs are often obtained by those with advanced degrees.

The individual considering a career in animal behavior studies should have a love of animals. Acute powers of observation and patience are also essential personal qualities.

For Additional Information

Animal Behavior Society
College of Sciences
Clemson University
Clemson, SC 29631

---

activities. A meadow mouse may have a home range of 270 m^2 (333 sq. yd.). A larger grazing animal, such as a whitetail deer, may have a home range of 1 to 5 km^2 (0.5 to 2 sq. mi.).

Migration Many mammals move regularly over large distances in search of more food, better environmental conditions, or more suitable places to bear their young. Insect-eating bats, for example, migrate south each fall as the insect population decreases in their summer area. Caribou regularly move from summer mountain pastures to protected valley pastures for the

Figure 39-6. Many mammals, such as these caribou, migrate in search of better environmental conditions.

winter. Gray whales travel each autumn from arctic waters to waters off the coast of Mexico. There the young of the herd can be born in warmer waters. They return north in the spring.

Hibernation   Migration is not the only way mammals solve problems of temperature extremes. Some species of bats, insect eaters, and rodents go into hibernation during winter months. Their metabolic rate decreases. Body temperature falls and heartbeat and breathing rates slow. In such a state, the animals need little energy and can live off stored body fat. When the outside temperature increases, they slowly wake up and resume normal activities. The body temperature of some bats approaches the temperature of the surrounding air during the day and rises when the bats become active at night, all year long.

Bears, skunks, and some other mammals spend the coldest part of the winter in a den asleep in a state of *dormancy*. Unlike true hibernators, their body temperatures decrease only slightly and they may move in and out of sleep. A few mammals, mainly rodents, avoid the extremes of summer heat or drought by moving into a state similar to hibernation called *estivation*.

Q/A

Q: *Which mammals have most successfully adapted to living with humans?*

A: Rats and mice. Most large cities, for example, probably have as many rodents as people.

Reviewing the Section

1. Name four features unique to mammals and two features common only to mammals and birds.
2. Why do mammals have a higher metabolic rate than reptiles?
3. Compare the ways caribou, bats, and bears meet the problem of surviving cold winter temperatures.
4. How do bats use sounds to locate objects?

Classification of Mammals

Section Objectives

- *Name* the three major types of mammalian teeth and give the function of each.
- *Distinguish* between the three major types of development among unborn mammals.
- *List* the major orders of mammals and name a representative animal from each order.
- *State* why Australian mammals differ remarkably from most mammals found in other parts of the world.

Biologists classify the 4,500 species of mammals alive today into 18 main orders on the basis of structural differences and development of the unborn. Members of these orders live throughout the world.

Some structural differences between orders of mammals can be linked to feeding habits. The specialized teeth of mammals, for example, perform different tasks. The *incisors* at the front of the mouth are used for cutting. Behind the incisors are the *canines,* used for gripping, tearing, and stabbing. At the rear of the mouth are heavy flat *molars,* which are teeth that grind food. As Figure 39–7 shows, the number and shape of these teeth vary among mammal species.

Mammals differ in the way their young develop before birth. ***Mammals can be divided into three groups: egg-laying mammals, pouched mammals, and placental mammals.***

Monotremes are mammals that lay eggs. Young monotremes develop within a protective shelled egg much as reptiles do. Pouched mammals are known as **marsupials.** Young marsupials develop within the mother's body for only a short time. After birth, they complete their development inside a special pouch, or *marsupium,* located on the mother's abdomen. Most modern mammals are **placental mammals.** In placental mammals, the young remain within the mother's body until they are able to maintain life independently. A special organ called the **placenta** connects the unborn young to the mother.

39.6 Monotremes

Only two types of monotremes are alive today, and they live in only a few isolated regions of the world. The duck-billed platypus lives only in Australia. The spiny anteater lives in parts of Australia and New Guinea.

Monotremes are considered mammals because they have hair and produce milk for their young. In other ways, the monotremes are fundamentally different from all other mammals. Monotremes show a curious mixture of traits. They have a cloaca, as do the lower vertebrates. Their limbs are attached to the sides of the body like the limbs of reptiles, and their feet have claws. The platypus has webbed feet and a flat tail that it uses for swimming. Unlike most other mammals, monotremes do not have true teeth. The platypus uses its flat bill to probe in the mud for worms, snails, and shellfish. The spiny anteater uses its beak to probe into anthills for food.

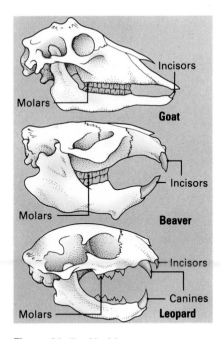

Figure 39–7. Herbivores, such as goats and beavers, need large incisors for grazing or gnawing. In leopards and other carnivores, the canine teeth predominate.

Figure 39–8. Opossums (above) are the only North American marsupial. The spiny anteater (above right) is an example of a monotreme.

The major reptilelike feature of the monotremes is that they lay eggs. Monotremes produce two to three eggs at a time. After fertilization the eggs are kept inside the mother's body and are nourished for a short period before they are laid in the nest. The female incubates her eggs by curling her body around them. The spiny anteater has a special *brood pouch* on her lower abdomen in which she keeps the eggs until they hatch.

After the young monotremes hatch, they are fed with milk secreted by their mother. The milk is produced by more than 100 specialized sweat glands on the lower abdomen. Unlike other mammals, young monotremes do not suckle. Instead, they lap up the milk as it oozes onto the mother's belly.

39.7 Marsupials

Kangaroos, koalas, and almost all other marsupials live in Australia. A few species of marsupials live in South America. North America has only one marsupial, the opossum. *The one major difference between marsupials and other mammals is that young marsupials complete their development inside their mother's pouch.*

The fertilized egg develops into an **embryo,** which remains in the female marsupial's body for only a short time. In some species, internal development lasts only a matter of days. At birth, young marsupials are still in an incomplete stage of development. They are blind, helpless, and extremely small. Newborn opossums, for example, are so tiny that more than a dozen could fit in a teaspoon. A kangaroo is only about 2 cm (0.75 in.) in length at birth.

Once inside the pouch, the young marsupial fastens its mouth onto a nipple. The nipple swells, and the young becomes firmly attached and begins to suckle. Immature marsupials are fed and protected inside the pouch until they are able to feed and care for themselves. The hind legs of the newborn marsupial are

Q/A

Q: *How do baby kangaroos first get into their mother's pouch?*

A: The babies have to crawl into the pouch themselves. Their mother does not help them, although she licks her body hair to create a path from the birth opening to the stomach pouch.

Without doubt, Australia and its nearby islands are home to the most unusual mammals on Earth. They include pig-footed bandicoots, red-bellied pademelons, pretty-faced wallabies, numbats, wallaroos, bilbies, honey possums, and wahl-wahls. These and other unfamiliar animals make the Australian biological region one of the most interesting.

Most of Australia's mammals are marsupials or monotremes. These animals became isolated about 70 million years ago when Australia split off and drifted away from the continents of Antarctica and South America. At the time Australia separated from the larger land area, no placental mammals lived in the Australian region. Freed from competition from the more intelligent placental mammals, the monotremes and especially the marsupials moved into every available habitat.

The marsupials of Australia have evolved to resemble different species of placental mammals in other areas of the world. The marsupial mouse, for example, resembles a placental mouse. Likewise, the Tasmanian devil is similar to a wolverine or badger. Australian possums, such as the cuscus and the sugar glider, live and look like monkeys. Similarly, each ecological role that is filled in other areas of the world by a placental mammal is occupied in Australia by a marsupial. Red kangaroos, for instance, occupy the role of grazers, just as antelopes do in the plains of the western United States. These similarities of shape and habitat between marsupials and placentals are examples of **convergent evolution,** the process by which two distantly related species come to resemble one another physically or ecologically.

Though protected for millions of years by their isolation, marsupials and monotremes are today threatened with extinction as more and more placentals are imported into Australia. Unless carefully protected, these fascinating mammals could disappear from the earth.

poorly developed; actually, they are little more than embryonic buds. The front legs, however, are more fully developed and tipped with claws. A young marsupial uses its front limbs to pull itself up the mother's abdomen and into her pouch.

39.8 Placental Mammals

More than 95 percent of all mammals are placentals. Early in the development of a placental mammal, the embryo becomes implanted in the wall of the mother's reproductive organ, called the **uterus.** Then the placenta forms, connecting the young mammal directly to its mother's uterus. The fluid-filled sac, the amnion, surrounds the embryo and supports it during development. Blood vessels from the amnion connect to the placenta as

Figure 39–9. Like all mammals, a newborn foal emerges from its mother still surrounded by the amnion.

Figure 39–10. The thin skin of a bat's wing exposes the long hand and forearm bones that aid in flight. Bat wings vary in shape. The speediest bats have long and narrow wings.

well. The circulatory systems of the mother and the embryo are not directly connected. Nutrients and oxygen from the mother's blood passes across the tissues of the placenta into the blood of the developing embryo. Waste materials pass from the embryo to the mother's blood.

Because the developing young mammals get their nourishment directly from the mother through the blood, development is not limited by the fixed amount of food found in an egg. The longer period of time in which to develop permits the formation of a complex brain and nervous system. The period of time during which the young mammal develops within the uterus is known as **gestation,** or **pregnancy.** The length of gestation differs among mammals.

Placental mammals differ in size, shape, diet, and the way they move. Each order of placentals shows adaptations for a particular way of life. Some are adapted for walking; others are adapted for running, leaping, swimming, or flying. Some mammals differ so greatly in appearance that it is hard to believe they are closely related.

Insect-Eating Mammals The oldest group of placental mammals are small, highly active animals called *insectivores*. As their name indicates, these animals eat mainly insects. Their diets vary, however, and may also include snakes, fruit, birds, and other insectivores. *Biologists think that insectivores are the ancestral stock that gave rise to the other placental mammals, even to the enormous elephants and whales.*

Shrews, hedgehogs, and moles are typical members of the order Insectivora. Compared with other mammals, insectivores have small brains. They also have enormous appetites. Shrews, for example, daily eat more than two times their body weight. With their extremely high metabolic rates, insectivores would starve to death within a short time if they were deprived of food. Most insectivores live in burrows or trees and are active only at night.

Flying Mammals Bats are the only mammals that can truly fly. Without their large, folding wings, bats resemble insectivores in both habits and appearance. The bat's wing is made of a flexible flap of skin stretched over extremely long arm and hand bones. The wing is supported by the bones of the last four fingers, which are exceptionally long and thin. The thumb is usually not attached to the wing and has a curved claw used for clinging or grasping.

Bats generally fly and live in groups. They are active only at night. During the day bats sleep hanging upside down in caves, hollow trees, or even barns and attics. Most bats eat insects they catch while in flight. Some bats eat nectar, and others catch fish or frogs with their clawed hind feet.

Hoofed Mammals Among land mammals, those with hoofs are called **ungulates.** Sheep, cattle, deer, pigs, camels, and other ungulates with an even number of toes make up the order Artiodactyla. Horses, tapirs, rhinoceroses, and other ungulates with an odd number of toes belong to the order Perissodactyla.

Ungulates walk on tiptoe. The weight of the animal is not supported on the entire foot, but only on one or more of its toes. The toes that touch the ground are broad, and the claw is enlarged, forming a hard, protective *hoof.* Ungulates' main means of defense is their ability to run.

Hoofed mammals generally live together in herds. The young are well developed at birth and can move with their herd within a day or two after they are born.

Hoofed mammals are herbivores that feed on grasses or leaves. All ungulates except the pig and the hippopotamus have a four-chambered stomach. The first chamber is the *rumen,* which contains bacteria and other microorganisms that digest cellulose. Animals with this chamber are called **ruminants.**

Trunk-Nosed Mammals In prehistoric times, huge herds of mammoths and mastodons roamed through Europe, Asia, and North America. Today only two living representatives of the order Proboscidea remain, the African elephant and the Asian elephant. These herbivores are noted for their enormous size and their long, grasping trunk. The trunk is really an elongated nose and upper lip. An elephant uses its trunk to take up water to drink, to spray water over its body, and to collect food and place it in its mouth. The long ivory tusks of some elephants are highly modified upper incisors.

Carnivorous Mammals Carnivores hunt other animals for food. This order includes land predators, such as tigers, lions, and wolves, as well as marine mammals, such as seals and walruses. Their long, sharp canine teeth are specialized for capturing prey and tearing flesh. Carnivores are intelligent and have keen senses of smell, vision, and hearing. On their cheeks are whiskers, called *vibrissae,* that are sensitive to touch.

Scientists generally divide carnivores into three subgroups: the cat family, the dog family, and the seal or sea lion family. The powerful limbs of land carnivores enable them to leap onto prey from trees or chase their prey across the ground. Their feet have thick pads that absorb the shock of landing or

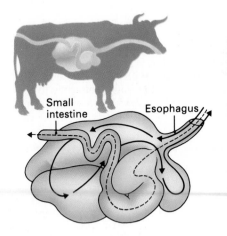

Figure 39–11. Food swallowed by ruminants passes through two of four stomach chambers, then returns to the mouth and is rechewed as cud. On its second swallowing, the food passes through all four stomach chambers.

running. Most members of the cat family have sharp retractable claws for capturing their prey. Members of the dog family have lean, muscular bodies and slender legs. Some members of the dog family are not strictly carnivores. Raccoons and bears, for example, are **omnivores;** they also eat plant materials.

Seals, sea lions, and walruses have streamlined bodies and a thick layer of insulating fat called *blubber*. Seals, sea lions, and walruses use their hind limbs as paddles to propel themselves through the water. The canine teeth of walruses have become long, heavy tusks. Members of the seal family eat a wide variety of foods, including mollusks, fish, and birds. Though adapted to a life spent mainly in the water, marine carnivores mate, bear young, and rest on land.

Whales and Related Aquatic Mammals

Whales, porpoises, and dolphins are other mammals that live their whole life in the sea. These mammals are called **cetaceans** (sih TAY shuhns); they belong to the order Cetacea. Cetaceans are probably descendants of land mammals that returned to the sea about 50 million years ago and adapted to an aquatic life. They have many adaptations that facilitate a life in water. Like fish, cetaceans have long, streamlined bodies. Underneath their skin is a thick layer of blubber. These mammals have completely lost their hindlimbs. The toes of their forelimbs have fused to form flat, paddlelike flippers used for steering and balance. Although cetaceans live in water, they must come to the surface regularly to breathe. The cloud of water vapor that whales exhale is called a *blow* or *spout*. Cetaceans are extremely intelligent animals that communicate by making sounds.

Scientists divide whales into two major groups—those that have teeth and those that do not. Toothed whales include dolphins, porpoises, and sperm whales. These whales use their

Figure 39–12. Baleen whales (top) filter plankton from sea water with thin plates that hang from the upper jaw. Toothed whales (bottom) usually have teeth only in the lower jaw. The teeth are used to catch prey.

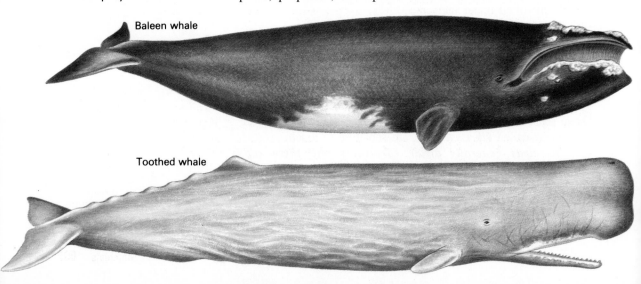

Baleen whale

Toothed whale

Table 39-1: Major Orders of Mammals

	Order	Examples	Characteristics
Egg-Laying Mammals			
	Monotremata ("one opening")	Platypus, spiny anteater	Mothers have no nipples; no teeth
Pouched Mammals			
	Marsupialia ("pouched")	Kangaroo, koala, opossum	Young poorly developed at birth; remain attached to nipple inside mother's pouch until fully developed
Placental Mammals			
	Insectivora ("insect-eating")	Shrew, hedgehog, mole	Insect eaters; high metabolic rate
	Chiroptera ("hand-winged")	Bat	Only true flying mammals; nocturnal; sharp teeth; large ears; insect or nectar eaters
	Artiodactyla ("even-number toed")	Sheep, cattle, deer, pig, camel	Ungulates with even number of toes; most have four-chambered stomach; herbivores
	Perissodactyla ("odd-number toed")	Horse, tapir, rhinoceros	Ungulates with odd number of toes; most have four-chambered stomach; herbivores
	Proboscidea ("trunk-nosed")	Elephant	Trunk-nosed mammals; large size, massive legs; herbivores; enlarged upper incisors form tusks
	Carnivora ("flesh eater")	Cat, dog, bear, seal	Meat eaters or omnivores; predators; most have sharp incisor teeth, claws, powerful limbs
	Cetacea ("whales")	Whale, dolphin, porpoise	Aquatic mammals; streamlined bodies; paddlelike forelimbs, no hindlimbs; mouths contain teeth or baleen
	Sirenia ("mermaidlike")	Dugong, manatee	Aquatic mammals; paddlelike forelimbs, no hindlimbs, flat tails; herbivores
	Rodentia ("gnawers")	Rat, mouse, squirrel, porcupine, beaver	Small gnawing mammals; one pair of upper incisors
	Lagomorpha ("hare-shaped")	Rabbit, hare	Gnawing mammals; two pairs of upper incisors; long hind legs for leaping
	Edentata ("toothless")	Armadillo, anteater, tree sloth	Toothless or with molars only; insect eaters; powerful forelimbs
	Primates ("first")	Lemur, monkey, ape, human	Most species tree dwellers; opposable thumbs; most with frontal eyes; capable of standing erect

Figure 39–13. Manatees belong to the order Sirenia. They are found in the Caribbean Sea and in rivers and coastal waters of West Africa and of North and South America.

teeth to catch prey such as seals, birds, squid, fish, and porpoises. They swallow their prey whole. Whales without teeth are called *baleen whales* because they have hundreds of thin plates called *baleen* in their mouth. Whales use the baleen to filter out plankton from the water. Baleen whales include right whales and the blue whale, which is the largest animal that has ever lived. Blue whales grow up to 30 m (100 ft.) in length and weigh more than 91 metric tons (100 tons).

Sea Cows Sea cows are aquatic mammals unrelated to whales. Today only two species are known, the dugong and the manatee. Dugongs live in warm coastal waters, and manatees live both in tropical coastal waters and in rivers. Sea cows are herbivores that feed on algae or aquatic plants. The animal has a torpedo-shaped body, 3 to 4 m (10 to 13 ft.) in length. Like whales, sea cows have lost all traces of hindlimbs and have short forelimbs and flat horizontal tails.

Gnawing Mammals Rodents are distinguished from other mammals by their teeth, which are highly specialized for gnawing. All rodents have two pairs of large, curving incisors that grow constantly. The incisors have hard enamel only on the front surface. As the rodent gnaws, the back surface of the tooth wears away faster than the front, thus keeping a chisel-sharp edge on these teeth.

Rodents number more than 3,000 species, which makes them the largest order of mammals. ***Three factors account for the rodents' remarkable worldwide distribution: their intelligence, their small size, and their rapid rate of reproduction.***

Most rodents—including mice, hamsters, guinea pigs, rats, and squirrels—are small animals. Larger rodents include prairie dogs, porcupines, and beavers. The largest rodent is the South American capybara, which grows up to 1.2 m (4 ft.) long.

Rodentlike Mammals Rabbits and hares resemble rodents in many ways, but are not closely related to rodents. Rabbits and hares belong to the order Lagomorpha and are called *lagomorphs* (LAG uh mawrfs). Both rabbits and hares have long hind legs that are specialized for leaping and hopping. Like rodents, rabbits and hares have teeth adapted for gnawing. Lagomorphs, unlike rodents, have an additional pair of teeth posterior to their upper incisors.

Both rodents and lagomorphs have a special intestinal pouch called the **cecum** that contains cellulose-digesting microorganisms. Like the rumen of ungulates, the cecum is an adaptation to a diet that consists mainly of grains and tough grasses.

Toothless Mammals Armadillos, anteaters, and tree sloths make up the order Edentata (ee dehn TAHD uh). *Edentata* means "toothless," but only the anteaters completely lack teeth. The other species have molars only. Most edentates have specialized features that are adaptations for an insect diet. The anteater, for example, uses its powerful clawed forefeet to rip open termite and ant nests. Then it inserts its long, sticky tongue into the nest to capture the insects. The largest of the edentates, the giant anteaters, weigh as much as 25 kg (55 lb.) and measure over 2 m (6 ft.) in length. The giant anteater lives only on the ground, although other anteater species live in trees. Armadillos have a unique protective shield formed of small bony plates.

Primates Lemurs, monkeys, apes, and human beings belong to the order Primates. Most primates are tree dwellers, and many of the features characteristic of primates were originally adaptations for life in the trees. Sensitive grasping hands with opposable thumbs and feet with big toes are aids in climbing and swinging through trees. Most primates have a flattened face with both eyes directed forward. The eyes of primates can focus and can discern color. Because the fields of vision slightly overlap, their eyes can also perceive depth. Generally, young primates are cared for by their parents for a longer period after birth than most other mammals.

The outstanding feature of primates is their highly developed brain. *Primates are distinguished from other mammals mainly by their active life, their curiosity, and their exceptional ability to learn.*

Of the apes, only gibbons and orangutans live in trees. Gorillas spend their days on the ground but sleep in trees at night. Apes can stand upright and walk for short distances on their hind legs. More often, they lean forward and balance on the knuckles of their hands when they stand or move. Except for the orangutan, apes live in highly developed social groups. They communicate with each other through a large number of sounds.

Figure 39–14. Monkeys use their tails and all four limbs to get from place to place. Gibbons swing with long arms. Gorillas usually walk half erect. Only humans walk fully erect at all times.

Reviewing the Section

1. Compare the teeth of herbivores and carnivores.
2. Compare the nursing habits of the young platypus and the young kangaroo.
3. How do rodents and lagomorphs differ? How are they the same?
4. How does convergent evolution help explain the similarities between the Tasmanian devil and the badger?

Investigation 39: What Is a Mammal?

Purpose
To become familiar with the unique characteristics of mammals

Materials
Mounted mammal specimens or photographs of mammals, reference books

Procedure
1. Make a chart like the one shown. Fill in the chart as you complete this investigation.

Characteristic	Vertebrate				
	Fishes	Amphibians	Reptiles	Birds	Mammals
Body Covering					
Heart (number of chambers)					
Endothermic Ectothermic					
Milk Glands to Suckle Young					
Brain Size					

2. Observe several of the mammal specimens. *What type of body covering does a mammal have? How does this compare with the body covering of the other vertebrate groups?*
3. Use your textbook to discover the characteristic of the heart of the five types of vertebrates. *How is the heart of a mammal different from other vertebrate hearts?*
4. The terms *endothermic* and *ectothermic* are often used when talking about vertebrate groups. Describe each of the vertebrate groups as either endothermic or ectothermic.
5. All female mammals possess special glands that produce milk to nourish the young. *How does this characteristic of mammals compare to feeding methods in the other vertebrate groups?*
6. Use page 591 of your textbook to determine the size ratio of the brain in the five vertebrate groups. *Which group has the largest? How might the size of a mammal's brain be related to the ability to cope rapidly with environmental conditions?*

Analyses and Conclusions
Using the complete chart, summarize the characteristics of the mammals that distinguish them from the other vertebrate groups.

Chapter 39 Review

Summary

Mammals are the only vertebrates that nurse their young. The young remain with the parent through childhood. Mammals are also unique in having large well-developed brains, body hair, outer ears, and a diaphragm.

Mammals vary in their social behavior. Some live together in complex social groups, while others hunt independently. Many animals defend their hunting or breeding territories against others of the same species. Some mammals hibernate or go into a state of dormancy during periods of extreme cold; others migrate to more favorable areas.

Mammals are grouped into 18 major orders on the basis of embryonic development,

tooth structure, feeding habits, and habitats. Young monotremes hatch from eggs. Young marsupials spend only a short time inside their mother's body. They complete their development after birth inside their mother's pouch. Most mammals, however, are placentals and remain inside the mother until they can live on their own.

The major placental orders include insect eaters, bats, hoofed mammals, elephants, carnivores, whales and related aquatic mammals, sea cows, rodents, rabbits and hares, toothless mammals, and primates. The primates, or tree dwellers, include not only monkeys and apes but also human beings.

BioTerms

cecum (602)

cetacean (600)

convergent evolution (597)

diaphragm (589)

echolocation (591)

embryo (596)

estrus (592)

gestation (598)

home range (592)

mammary gland (589)

marsupial (595)

monotreme (595)

omnivore (600)

placenta (595)

placental mammals (595)

pregnancy (598)

ruminant (599)

territoriality (592)

ungulate (599)

uterus (597)

viviparous (589)

BioQuiz   (Write all answers on a separate sheet of paper.)

I. Completion

1. A well-developed _____ accounts for the high intelligence of mammals compared to other animals.
2. Mammals evolved from a now extinct order of _____ .
3. _____ is a development period unique to mammals.
4. Most ungulates have a _____ to break down cellulose.
5. Bats can locate objects even in total darkness by using _____ .

II. Modified True and False

Mark each statement TRUE or FALSE. If false, change the underlined term to make the statement true.

6. Many mammals mark the boundaries of their territory with urine.
7. In cold weather bears go into a state of hibernation.
8. Artiodactyls have odd-numbered toes.
9. Gestation periods vary among placental mammals.
10. Whales belong to the order Cetacea.

III. Multiple Choice

11. The diaphragm in mammals separates the a) food and air passages. b) placenta and uterus. c) chest and abdominal cavities. d) two halves of the heart ventricle.
12. During estrus most mammals a) hibernate. b) migrate. c) become extinct. d) reproduce.
13. The spiny anteater uses its brood pouch to a) incubate its eggs. b) store food. c) carry the newborn. d) digest food.
14. A rabbit's cecum helps to a) digest cellulose. b) keep incisors sharp. c) protect against enemies. d) maintain body temperatures.
15. Hands with opposable thumbs enable primates to a) nurse their young. b) communicate. c) fight. d) swing through trees.

IV. Essay

16. What adaptations in the limbs of mammals allow them to move faster than reptiles?
17. Why do most marsupial and monotreme species live only in Australia?
18. How do baleen whales differ from toothed whales?
19. What do tigers, wolves, and seals have in common?
20. How do you account for the large population of rodents?

Applying and Extending Concepts

1. Some scholars have suggested that early sailors' tales of encountering mermaids may have originated with sightings of sea cows. Use your high school or public library to find out what about these animals might explain how the mermaid legend came about.
2. Mammals have many ways of communicating. Observe a pet dog, cat, or guinea pig for a week. Record the specific ways the mammal communicates its needs to human beings or to other animals in the household. Listen and look for special sounds and behavior that indicate your pet wants food, exercise, or attention.
3. Research the special features or circumstances that have enabled the blue whale to maintain itself as the largest animal on Earth.
4. Spider monkeys have eyes that can focus on a single point and discern color. Their eyes also enable spider monkeys to view objects in three dimensions. Explain how these adaptations are useful ones for tree dwellers.
5. When hunting seals, a polar bear covers its black nose with its white paw. Explain this hunting strategy in terms of evolutionary adaptation or learned behavior.

Related Readings

Ellis, R., *The Book of Whales*. New York: Alfred A. Knopf, 1980. Illustrations by the author enliven a highly readable text that covers the courtship, communication, and intelligence of whales. The book also includes a comprehensive bibliography arranged by species.

Prisico, J., "Insights into Intelligence: New Horizons of Language Open When This Gorilla Speaks." *Science Digest* 85 (February 1979): 32–35. The article describes how work done with a gorilla has provided important insights into mammal intelligence and language.

40 Overview of Human Biology

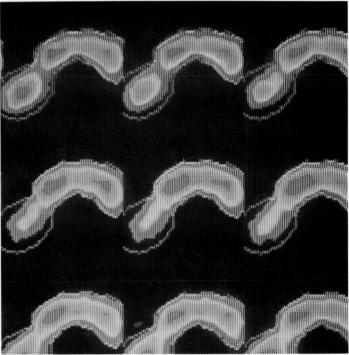

Heart function display produced by gamma scan technology

Introduction

Scientists believe that humans evolved from a primate that lived about 2 million years ago. The humans who have developed since that time have a number of unique characteristics. The most important of these features is the brain. The human brain is the most complex biological structure known to exist in any organism.

Over the last 700 years, researchers have learned a great deal about how the human body works. They have identified the dozens of types of cells that form tissues and organs and learned how groups of organs work together as systems to sustain life. In recent years scientists have even found ways to improve and prolong life by using transplanted and artificial organs to replace diseased or injured body parts.

Human Characteristics

Scientists classify humans as members of the kingdom Animalia, phylum Chordata, class Mammalia and order Primates. The animals we resemble most closely are the monkeys, apes, and more than 200 other types of the order Primates. *The traits that make us distinctly human are mostly refinements of traits found in other primates.*

40.1 Physical Characteristics

We share many physical characteristics with other primates, because humans and other primates developed from a common ancestor. This ancestor, now extinct, lived an *arboreal,* or tree-dwelling, existence. The evolution of some primate traits into human traits came about much later.

Primate Traits   One important characteristic of all primates is a complex and highly developed brain. Compared to other animals, primates have brains that are larger in relation to their overall body size.

Primates also have sophisticated eyes that distinguish minute details—an adaptation to the ancient dim forests. The keen vision of primates is due in part to the position of the eyes at the front of the face. This position produces *stereoscopic vision,* or the ability to perceive objects in three dimensions. Special eye cells called *cones* also contribute to primates' keen vision. These cells distinguish color and enable the eye to see sharp images.

A third primate characteristic is a hand with five digits. These digits include an **opposable thumb**—that is, a thumb that can be positioned opposite the fingers to grasp branches and objects.

Long arms with flexible shoulder and wrist joints are another feature of primates. Two bones in the forearm enable primates to rotate their hands a full semicircle; shoulder joints enable them to move their arms in many directions. Together, these structures and the grasping hand permit primates to swing from branch to branch. Some primates are able to maintain an upright sitting or standing posture during certain activities such as feeding.

Primates also share the same four types of teeth—*incisors* and *canines* for tearing, and broad *premolars* and *molars* for grinding and chewing. Together these teeth enable primates to eat both plants and other animals.

Section Objectives

- *Name* four traits humans share with other primates.
- *List* the traits that characterize humans.
- *Identify* the body parts that permit upright posture in humans.
- *Summarize* the major behavioral characteristics of humans.

Figure 40–1.   This orangutan in Milwaukee County Zoo uses its opposable thumb to hold the apple it is eating. This and its upright posture link it to humans.

Human Traits

Human Traits   The earliest humans possessed so many ape-like characteristics that scientists sometimes have difficulty telling whether fossil bones are those of a primitive ape or a human. As evolution continued, however, humans developed the distinctive traits that characterize them as a species.

The most important human feature is a brain larger than that of any other primate. Chimpanzees, for example, have a brain capacity of about 500 cm³ (30 cu. in.). Humans, however, have an average brain capacity of about 1,400 cm³ (85.5 cu. in.). The expansion of the human brain resulted in the vertical forehead typical of humans.

The ability to stand and walk upright under all conditions is another distinctly human trait made possible by several specially adapted structures. The *pelvis*, the girdle of bone that includes the hip bone, is wide and slightly curved. This permits it to

Figure 40–2.   A comparison of human and ape skeletal features shows the traits that distinguish humans as a species. These include (from left) an S-shaped spine, arch-shaped jaw, wide pelvis, large hand with well-opposed thumb, and a flat foot with nonopposable large toe.

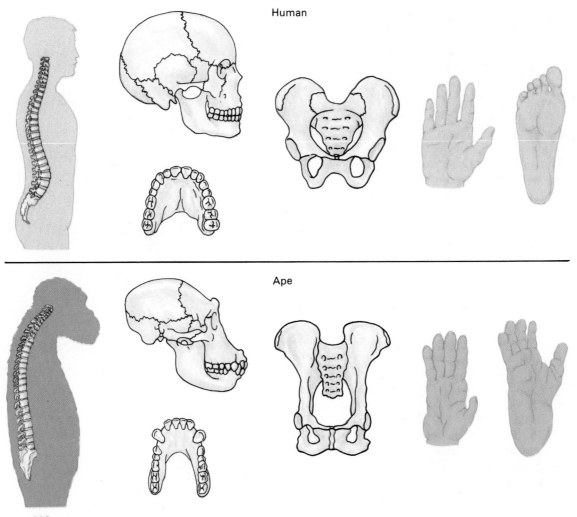

Human

Ape

support the upper part of the body. The broad rear of the pelvis provides a large area for anchoring the walking muscles. The S-shaped spine rising from the pelvis provides support and balance. The head sits erect at the top of the spine. Even the human foot is designed for standing and walking upright. Basically flat, it contains an arch for support. The large toe is not opposable, but lies parallel to the other toes. In this way the large toe is adapted for walking instead of grasping. *More than any other characteristic, upright posture with the erect head creates the distinctly "human" appearance.* This posture, with the eyes at a high level, enables humans to see distant objects.

Human teeth and jaws are also distinctive in size and shape. The canine teeth of monkeys, apes, and other primates are long and sharp. These canines are useful for tearing food. Human beings have smaller, more even teeth than other primates. Human canines are only slightly longer than the incisors and are used to hold food as well as to tear it. The premolars and molars, the back teeth that are specialized for chewing and grinding, are broader than they are in other primates. The human jaw is shaped like an arch, while the jaw of other primates has a rectangular shape.

40.2 Behavioral Characteristics

Although the physical characteristics of all primates are somewhat similar, behavioral characteristics vary greatly between humans and other primates. The reason for this difference is the enlarged human brain. The brain enables humans to process and remember a great deal of information. These mental abilities also enabled humans to develop a system of symbols that make spoken and written language possible. The use of language, in turn, allows people to share their information and ideas. Using this sophisticated brain, humans have been able to create and use tools. With the ability to speak to one another and use tools, humans have altered their social organization from a simple agrarian structure to complex societies that depend greatly upon scientific technology.

Reviewing the Section

1. What are five traits that characterize all primates?
2. Name five traits that distinguish humans from other primates.
3. Name four body structures that enable humans to stand upright.

Q/A

Q: *Are humans distinctly different from early mammals?*

A: No. Many human characteristics existed in primitive mammals. Among these are five separate digits on the hands and feet, three-segment fingers, two bones in the forearm, and shoulder sockets that move in many planes. While primates retained these characteristics, other evolving mammals lost them.

Organization of the Body

- *Describe* the plan of the human body.
- *List* the different types of epithelial tissue and tell where in the body each is found.
- *Distinguish* among connective, muscle, and nervous tissue.
- *Identify* the major systems of the body and the functions they perform.

The human body is organized in much the same way as the bodies of other *vertebrates,* or animals with a spinal cord. In overall structure the human body is bilaterally symmetrical, which means the body has two sides that, in most ways, are mirror images of each other. The organs of the human body are formed of specialized cells and are organized into complex systems that perform specific functions.

40.3 Plan of the Body

The human body is divided into four major parts—the head, neck, trunk, and limbs. The body is built around a jointed bony skeleton covered with layers of muscle and skin. Inside the trunk of the body is a cavity called the **coelom** (SEE luhm). The coelom is divided into two smaller cavities by the **diaphragm** (DY uh fram), a dome-shaped sheet of muscle. The **thoracic** (thaw RAS ihk) **cavity** lies above the diaphragm and contains the heart, lungs, and esophagus. The **abdominal cavity** lies below the diaphragm and contains the organs of digestion, reproduction, and excretion. The **cranial cavity** is inside the skull and contains the brain.

40.4 Tissues of the Body

The organs of the body are formed from four types of tissue: epithelial (ehp uh THEE lee uhl), connective, muscle, and nervous. Most types of tissue have several forms that perform different functions.

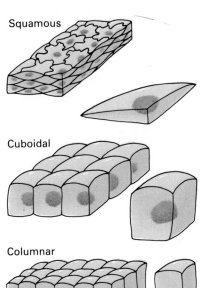

Squamous

Cuboidal

Columnar

Epithelial Tissue Tissue composed of one or more layers of cells protects all internal and external body surfaces. Such tissue is called **epithelial tissue.** *Squamous epithelium* is composed of flat, irregularly shaped cells. Squamous cells form the top layers of the skin, the protective covering of the heart and lungs, and the lining of blood vessels. *Cuboidal epithelium* is made up of cells that are basically cube shaped. They are found in many glands and in the ducts of some organs, such as the kidney, as well as in the middle ear and the brain. *Columnar epithelium* is composed of cells that are long, narrow, and tightly packed. They line much of the digestive system and the upper respiratory tract. Many columnar epithelial cells have tiny hairlike extensions called **cilia** (SIHL ee uh). The wavelike motion of cilia helps move substances along these surfaces.

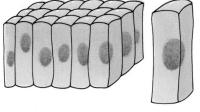

Figure 40–3. Three types of epithelial tissue cover the inner and outer surfaces of the body. This tissue, classified according to the shape of its cells, is squamous, cuboidal, or columnar.

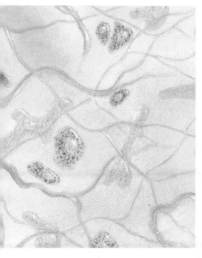

Figure 40–4. Connective tissue consists of cells embedded in a matrix. Loose connective tissue (left) has a semifluid matrix. The cells of fat tissue (right) contain large droplets of fat.

Connective Tissue   The most widely distributed tissue in the human body is **connective tissue.** It joins, supports, and protects the other types of tissue. Connective tissue is composed of relatively few cells embedded in a thick, nonliving material called the **matrix** (MAY trihks). The matrix contains many tiny, living fibers.

Four kinds of connective tissue are found in the human body. *Dense connective tissue* makes up cartilage and bone. Cartilage is a flexible but tough material consisting of small clusters of cells embedded in the matrix. Bone consists of cells in a matrix that contains hard crystals. *Loose connective tissue* is found under the skin and around nerves, blood vessels, the heart, and the lungs. Its matrix is semifluid. *Liquid connective tissue* forms blood and lymph, a clear fluid that comes from blood. The matrix in blood is a liquid called plasma. *Fat tissue* is composed of cells in which large droplets of fat are stored. This fat can be used for energy when needed.

Muscle Tissue   Specialized cells with the ability to contract and thereby produce movement make up **muscle tissue.** Muscle tissue is classified into three types. *Skeletal muscles* are attached to bones and move the skeleton. *Smooth muscles* are found in the walls of many internal organs, such as the digestive organs. *Cardiac muscle* is found only in the heart.

Nervous Tissue   Cells that can transmit messages throughout the body make up **nervous tissue.** These cells are found in the brain, spinal cord, nerves, and sensory organs. Nervous tissue provides information about the environment. It also controls many body functions.

Q: *Does a body organ ever consist of more than one type of tissue?*

A: Yes. All organs contain the four different types of body tissue.

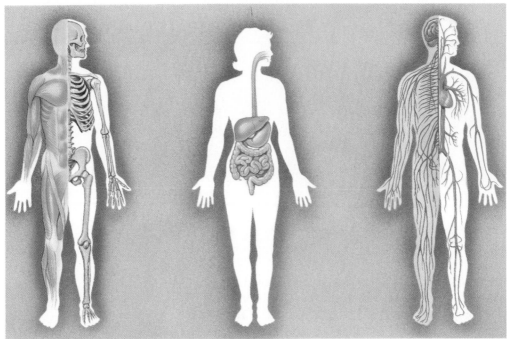

Muscular and
skeletal systems

Digestive system

Nervous and
circulatory systems

Figure 40–5. Each body system consists of a group of organs that work together.

40.5 Systems of the Body

Tissues are organized into larger units called organs. Organs that work together to perform a particular function form a *system*. All body systems are interrelated and operate in unison.

- The **skeletal system** moves, supports, and protects the body. Blood cells are manufactured inside bones, and calcium and phosphorus are stored in bone tissue.
- The **muscular system** works with bones to make the body move. Muscles also protect some of the body's organs.
- The **digestive system** includes the tube running from the mouth through the trunk and several accessory organs. In this system, food is broken down into essential nutrients, nutrients are absorbed, and solid wastes are eliminated.
- The **circulatory system** transports nutrients, gases, and chemicals to all parts of the body. It also collects waste products from cells. Blood is circulated through blood vessels by the pumping action of the heart. The **lymphatic system,** part of the circulatory system, collects fluid from tissue and returns it to the blood. Both systems also help fight disease.
- The **respiratory system** takes oxygen into the body and eliminates carbon dioxide and water.

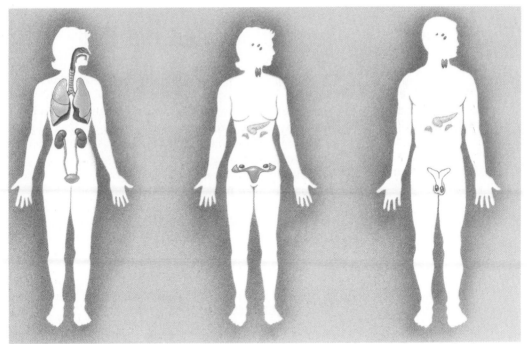

Respiratory and
excretory systems

Endocrine and female
reproductive systems

Endocrine and male
reproductive systems

- The **excretory system** removes cellular wastes from the blood. It also maintains the body's fluid and chemical balance. Wastes leave the body through the **urinary system,** a part of the excretory system.
- The **nervous system** monitors the outside environment and controls and coordinates body activities.
- The **integumentary system** forms the body's outer protective layer. It consists of the skin, hair, and nails.
- The **endocrine system** helps control body functions through chemicals called *hormones*. Hormones regulate functions such as growth and maturation.
- The **reproductive system** provides a means of producing offspring in order to maintain the species.

Reviewing the Section

1. What are the three types of epithelial tissue? Give an example of where each is located.
2. What four tissue groups make up the human body? What is the function of each?
3. What are the functions of the skeletal system?
4. Name three other systems of the body and give a major function of each.

Technology and the Body

- *Distinguish* between transplants and prostheses.
- *List* some organs that are commonly transplanted.
- *Name* several products of biomedical engineering.
- *Describe* how computers help paralyzed muscles move.

The human body is often compared with a complex machine. However, there is one major difference between the two. When a machine breaks down, it can be shut off until repairs are made. New parts can be ordered to replace worn-out ones. A human body cannot be shut off when repairs are needed, and new parts cannot simply be ordered.

Science, however, is finding ways to treat human disorders and replace some body parts. One solution may be an organ transplant—the replacement of a body part with an identical part from another person. Another solution may be replacement with an artificial part, or **prosthesis** (prahs THEE sihs). The design and development of artificial body parts is called **biomedical engineering**.

40.6 Organ Transplants

The first kidney transplant, accomplished in 1954, was a major milestone in transplant surgery. Since then about 64,000 patients have received kidney transplants. Other body parts that can be transplanted include blood, heart, lungs, cornea, liver, skin, and bone. Scientists are also studying ways to transplant the small intestine and brain tissue.

Until 1978 many transplants failed because the recipients' bodies rejected the new organs. Rejection occurred because the body recognized a transplanted organ as a foreign substance and attacked, or rejected, the organ as it would attack invading viruses or bacteria. To prevent rejection, doctors administered drugs that suppressed all the body's natural defenses. However, these drugs left the organ recipient susceptible to infections of all types. Today transplant recipients are given *cyclosporine,* an antibiotic drug that suppresses only the defenses against a transplanted organ. Since it was introduced in 1978, cyclosporine has doubled the number of transplanted organs that survive for at least a year.

40.7 Artificial Replacement Parts

Since the early 1970s, biomedical engineers have developed an amazing array of artificial parts—limbs, joints, bones, teeth, blood, hearts, and even skin. Often these prostheses involve innovative uses of modern materials and electronic equipment. For example, silicone is used in artificial skin and plastics are used in artificial joints. Researchers are also designing limbs

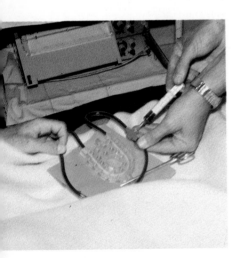

Figure 40–6. Artificial organs can now take the place of some diseased or malfunctioning organs in the human body. Here surgeons implant an artificial pancreas.

▶ Computers That Move Muscles

Spinal injuries have caused about 400,000 Americans to become paralyzed. The injuries are frequently a result of automobile or sports accidents.

In most such accidents, the brain and limbs are not damaged. The problem is that the connection between these body parts has been broken because of a broken neck or back. Muscles that move limbs get their commands from the central nervous system. Generally the commands travel by way of nerves in the neck and spine. When the nerves are severed, paralysis results.

Because paralysis victims are inactive, their muscles

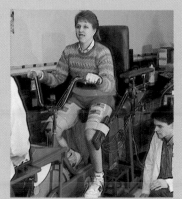

begin to deteriorate. The process of muscle deterioration leads to other problems, such as diseases of the heart and circulatory system and weakness of the bones.

Computers may soon end some of these problems. In certain experiments, researchers have enabled paralysis victims

to move their legs. The researchers strap the patient's feet to the pedals of a stationary bicycle, then use a computer to produce electrical impulses that in turn trigger movement in the paralyzed muscles. This procedure allows some paralysis victims to pedal the bicycle at a rate of more than 19.2 km (12 mi.) per hour. Researchers have also used computers to help paralyzed people walk. A small portable computer provides the impulses to the muscles.

Computers may soon be used with a pedal-operated wheelchair and a special tricycle. With this equipment, paralysis victims can move around outdoors.

equipped with high-powered batteries and microprocessors, tiny devices that receive and channel electrical signals.

The chief aim of biomedical engineers is to design prostheses that behave like normal human parts. Some prostheses come close to achieving this goal. The Utah Arm, for example, is an artificial limb equipped with microprocessors. When attached to a person who has lost an arm, the electronic equipment picks up nerve impulses generated by the wearer's muscles. Then the microprocessors translate the impulses into movements almost identical to those of a natural human arm.

Research is also under way on artificial organs that are part transplant and part prosthesis. One example is an artificial replacement for the pancreas, an important organ of digestion. Part of the artificial pancreas consists of pancreatic cells from rats that produce essential digestive juices. These cells line a system of artificial tubes in a frame of metal and plastic.

Spotlight on Biologists: Robert K. Jarvik

Born: Midland, Michigan, 1946

Degree: M.D., University of Utah

American physician Robert Jarvik developed the first permanent artificial heart implanted in a human. Although the recipient, Barney Clark, lived only four months after the 1982 implant, Jarvik's invention greatly advanced the practice of cardiac surgery.

Jarvik began designing artificial hearts in 1971 when he first joined the University of Utah artificial organ program. He drew on a facility for inventing and sculpting that he developed while still a teenager. Jarvik holds a patent on a surgical stapler that he designed when he was 17.

The Jarvik-7 heart implanted in Clark consisted of polyurethane ventricles with tubes running outside the body to controls and tanks of compressed air. Jarvik has since devised a smaller pump that attaches to a battery and microcomputer. The computer fastens to the patient's belt, affording the patient some mobility.

Jarvik is president of Symbion, a Salt Lake City company established in 1976 by Jarvik and colleagues at the University of Utah. Symbion manufactures and sells artificial organ devices developed

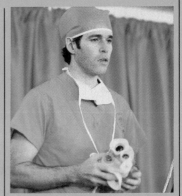

at the university.

For his work on the artificial heart, Jarvik received in 1983 the first John W. Hyatt Award for "service to mankind" given by the American Society of Plastics Engineers. In addition, he was named the 1983 Inventor of the Year by the National Inventors Hall of Fame.

Many researchers believe that transplants work better than prostheses for organs that chemically control the activities of body cells. Many also believe that transplants may always be preferred for certain organs, including the kidneys, liver, and heart. At present, however, artificial parts are superior to transplants for organs such as the arm in which strength and mechanical accuracy are most important.

Reviewing the Section

1. What is the difference between a transplant and a prosthesis?
2. Name five organs that can be transplanted.
3. List four prostheses designed by biomedical engineers.
4. How are computers used to help paralyzed people move?

Investigation 40: Looking at Human Cheek Cells

Purpose
To observe typical human cells

Materials
Slide, toothpick, coverslip, compound microscope, medicine dropper, methylene blue, paper towel

Procedure
1. Put a small drop of water in the center of a glass slide.
2. With the wide end of a flat toothpick, gently scrape the inside of your cheek. You will not see any material on the toothpick.

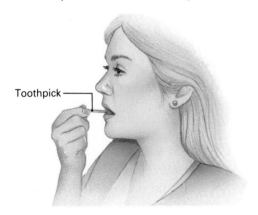

Toothpick

3. Put the same end of the toothpick into the drop of water on the slide. Mix carefully. The water should appear slightly cloudy. Add a coverslip.

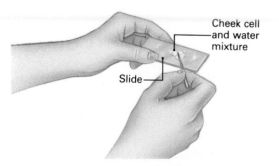

Cheek cell and water mixture

Slide

4. Place the slide on the stage of the microscope and focus under low power. Focus on an individual cell. Draw and label the cell.
5. Switch to high power. Draw and label the cell as it appears under high power.
6. Remove the slide from the microscope stage. Place one drop of methylene blue at the edge of the coverslip. Place a small piece of paper towel at the opposite edge of the coverslip. The stain will move under the coverslip. Use a paper towel to absorb any excess stain.
7. Return the slide to the microscope and locate an isolated cell under low power. Draw your observations. Switch to high power. Draw your observations and label cell parts. *How does the appearance of the stained cell differ from that of the unstained cell?*

Analyses and Conclusions
1. Why is the cheek cell considered a typical animal cell?
2. What is the advantage of staining the cells?
3. What parts of the cell absorbed the most stain? Explain.
4. Are all the cells the same? Why?
5. Compare the characteristics of this typical animal cell with those of a typical plant cell.

Going Further
You may perform Investigation 40–2 in Laboratory Investigations to observe other types of human cells.

Chapter 40 Review

Summary

Humans share many characteristics with other primates. These characteristics include an opposable thumb, color and stereoscopic vision, and four different types of teeth. However, human beings are distinctive, especially in having an enlarged, complex brain and upright posture.

Humans are the only animals with a spoken and written language. They also have great skill in making and using tools. Human social patterns differ significantly from those of lower animals because of these behavioral characteristics.

The human body consists of epithelial, connective, muscle, and nervous tissue. These tissues compose the following interrelated systems: skeletal, muscular, digestive, circulatory, lymphatic, respiratory, excretory, urinary, nervous, endocrine, reproductive, and integumentary.

A damaged or diseased body part can frequently be replaced by an organ transplant. The drug cyclosporine has helped overcome the rejection of transplanted organs. Certain damaged body parts may also be replaced by artificial parts.

BioTerms

abdominal cavity (612)
biomedical
   engineering (616)
cilia (612)
circulatory system (614)
coelom (612)
connective tissue (613)
cranial cavity (612)
diaphragm (612)

digestive system (614)
endocrine system (615)
epithelial tissue (612)
excretory system (615)
integumentary system (615)
lymphatic system (614)
matrix (613)
muscle tissue (613)
muscular system (614)

nervous system (615)
nervous tissue (613)
opposable thumb (609)
prosthesis (616)
reproductive system (615)
respiratory system (614)
skeletal system (614)
thoracic cavity (612)
urinary system (615)

BioQuiz   (Write all answers on a separate sheet of paper.)

I. Completion

1. Cells of connective tissue are embedded in a ＿＿＿.
2. All primates have an ＿＿＿ thumb adapted for grasping.
3. The body rids itself of cellular wastes through the ＿＿＿ system.
4. The ＿＿＿ system regulates body activities by producing chemicals.
5. The ＿＿＿ system makes possible the production of offspring.

II. Modified True and False

Mark each statement TRUE or FALSE. If false, change the underlined term to make the statement true.

6. Primate characteristics are chiefly an adaptation to arboreal existence.
7. The Utah Arm is a transplant.
8. Connective tissue is designed to receive and transmit messages.
9. The endocrine system takes in oxygen and expels carbon dioxide.

III. Multiple Choice

10. Stereoscopic vision allows primates to see a) dimly lit objects. b) colors. c) sharp images. d) three dimensions.
11. Epithelial tissue in the two layers of skin is a) squamous. b) cuboidal. c) columnar. d) adipose.
12. Substances are transported through the body by the a) nervous system. b) excretory system. c) circulatory system. d) respiratory system.
13. The coelom is found inside the a) head. b) trunk. c) limbs. d) neck.
14. Severing the nerves to body limbs causes a) prosthesis. b) paralysis. c) rejection. d) impulses.
15. Blood exists in a liquid matrix of a) lymph. b) fat droplets. c) plasma. d) liquid connective tissue.

IV. Essay

16. What behavioral characteristics distinguish humans from other primates?
17. What are the structural differences among the three types of epithelial tissue?
18. How do the functions of connective, muscle, and nervous tissue differ?
19. Why does the body reject a transplanted organ?
20. What are the organic and the artificial parts of the artificial pancreas?

Applying and Extending Concepts

1. Make a time line that shows the important advances in the techniques of organ transplant. Your school or public library should be able to provide the information you need.
2. Monkeys, which are among the more highly developed primates, are frequently used instead of humans in both medical and behavioral experiments. Many scientists, however, believe that these experiments cannot provide useful results. These scientists argue that the physical and mental differences between monkeys and humans are so great that the results of experiments on monkeys cannot be applied to humans. Write a report on the pros and cons of this argument. In what ways are monkeys valid substitutes for humans in experiments? In what ways are they not valid substitutes?
3. Cells of the skin and liver reproduce themselves throughout a person's lifetime. However, no cell division takes place in muscles or in the nervous system. Assume that a person has suffered skin and nerve injuries. Explain what difference might exist in the healing processes of these two types of tissue.

Related Readings

Hanken, James, and Brian K. Hall. "Evolution of the Skeleton." *Natural History* 92 (April 1983): 28. In this article, the authors explain how skeletons are analyzed to determine types of body tissues in early humans and how tissues have changed through evolution.

"The New Era of Transplants." *Newsweek* (August 29, 1983): 38–44. This informative article surveys the history of transplants and discusses the recent advances in the field.

Miller, Jonathan. *The Body in Question*. New York: Random House, 1978. Based on a successful 13-part television series, this well-illustrated book tells the fascinating story of how human beings found out about their own bodies.

41

Support, Movement, and Protection

Gymnast on balance beam

Introduction

Human appearance, in general, is determined by three systems of the body. These systems make up the bony framework, the muscular bulk, and the outer surface covering of the body. What an individual looks like specifically, however, depends upon the size of the bones, the condition of the muscles, and the texture and color of the skin and hair. As a person grows older, his or her appearance changes as modifications take place in the skeleton, muscles, and skin.

These three systems provide more than appearance. Bones and muscles support the body, protect vital internal organs, and allow for movement. The skin prevents harmful organisms from entering the body, and also protects internal tissues by covering the body.

The Skeletal System

The human skeleton is a remarkable structure. Its materials are strong and light. Bone is as strong as cast iron but several times lighter and considerably more flexible. The skeleton's design is simple and efficient. Many bones are hollow cylinders, a shape that provides the greatest strength while using the least amount of material.

- *List* the functions of the skeleton.
- *Summarize* the process of bone development.
- *Describe* the structure of a long bone.
- *Name* the main types of joints and give an example of each.

41.1 Functions of the Skeleton

The skeleton serves several vital functions. *Along with muscles, the skeleton makes possible a wide range of movements. It supports the body and protects internal organs.* Bones store calcium and phosphate, which are taken up by the blood when needed. Also, tissue called **marrow** inside some bones produces red and white blood cells.

41.2 Structure of the Skeleton

The adult human skeleton is an *endoskeleton,* or internal skeleton, consisting of about 206 bones as well as connective tissues called *cartilage* and *ligaments.* The skeleton has two main divisions—the *axial skeleton* and the *appendicular skeleton.*

The **axial skeleton** forms the body's central framework of support and protection. It consists of 80 bones in the skull, face, vertebral column, and rib cage. The skull protects the brain. In the adult 26 irregularly-shaped bones called *vertebrae* make up the vertebral, or spinal, column, which holds the body upright and protects the spinal cord. As Figure 41–1 on page 624 shows, the vertebral column has five regions: cervical, thoracic, lumbar, sacral, and coccyx. The rib cage consists of 12 sets of ribs and the *sternum,* or breastbone. These bones protect the heart, lungs, and other organs in the thoracic cavity. Each of the ribs is attached to the vertebral column. Seven pairs of ribs, called *true ribs,* are also attached to the sternum by cartilage. The other five pairs do not attach to the sternum and are therefore called *false ribs.*

The **appendicular skeleton** consists of 126 bones in the *pectoral girdle,* the *pelvic girdle,* and the arms and legs. The pectoral girdle—the bones of the shoulder area—provides support for the arms and allows them a wide range of movement. Muscles attach the pectoral girdle to the axial skeleton. The pelvic girdle—bones of the hip area—attaches directly to the lower part of the vertebral column.

Q/A

Q: *What is a slipped disc?*

A: Discs of cartilage serve as cushions between vertebrae. With age they tend to deteriorate. Extra pressure on the vertebrae may then force a disc to break. This condition, which can cause pain and even paralysis, is called a herniated, or ruptured, disc.

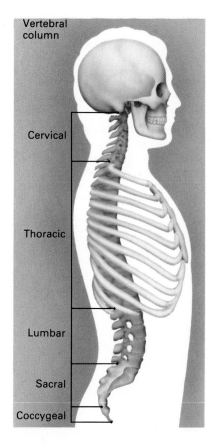

Vertebral column

Cervical

Thoracic

Lumbar

Sacral

Coccygeal

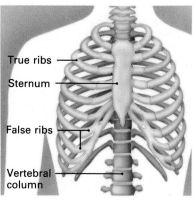

True ribs

Sternum

False ribs

Vertebral column

Figure 41–1. The vertebral column (top) and the rib cage (bottom) are two major sub-units of the axial skeleton. The entire skeleton (right) consists of 206 bones.

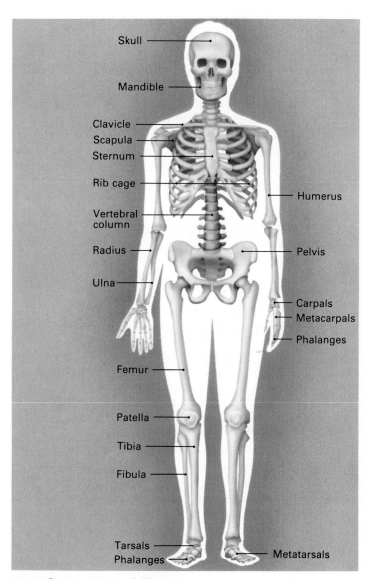

Skull

Mandible

Clavicle

Scapula

Sternum

Rib cage

Vertebral column

Radius

Ulna

Humerus

Pelvis

Carpals

Metacarpals

Phalanges

Femur

Patella

Tibia

Fibula

Tarsals

Phalanges

Metatarsals

41.3 Structure of Bones

Bones are classified according to their shape. A bone's shape is closely related to its function. For example, *long bones* in the arms and legs support weight and are involved in movements such as walking and lifting. *Flat bones,* such as the sternum and skull, have a large surface area that protects the underlying organs. The *short bones* of the wrists and ankles allow great flexibility and precise movements.

Although bones vary greatly in shape, they all have a similar structure. Bone consists of living and nonliving materials.

The living cells that make up the bone are called **osteocytes** (AHS tee uh sytz). Osteocytes are embedded in a network of tough protein fibers called *collagen*. The nonliving part of bone, the mineral portion, consists mainly of compounds containing calcium and phosphorus that surround the osteocytes and make bones hard. A protective fibrous membrane, the **periosteum** (pehr ih AHS tee uhm), covers all bones and helps connect them to muscles. Its rich blood supply nourishes the bone.

Figure 41–2 shows the internal structure of a typical long bone—the femur, or thigh bone. The middle portion, called the *shaft,* is composed of a central cavity surrounded by hard bony material. This hard material is *compact bone*. Small channels, known as **Haversian** (huh VUR shuhn) **canals,** run through this compact bone. Haversian canals contain blood vessels that nourish the osteocytes. The central cavity in long bones is filled with *yellow marrow,* which stores fat. The shaft is separated from the end of the bone by an *epiphyseal* (ehp uh FIHZ ee uhl) *line,* which marks the area where growth formerly took place.

In flat bones and at the ends of long bones, the hard material is very thin. Under this thin, hard material is *spongy bone,* which consists of loosely packed cells supplied with blood. In certain parts of the skeleton the spongy bone contains *red marrow.* It is in the red marrow that red blood cells and white blood cells are manufactured.

41.4 Development of Bones

During early embryonic development, the skeleton consists of only cartilage and layers of membrane. During the second month of development, the cartilage starts to be replaced by bone through a process called **ossification** (ahs uh fuh KAY shuhn). During ossification bone cells replace cartilage cells, and calcium compounds from the blood are deposited around the cells. Some cartilage, however, never ossifies. For example, the tip of the nose and the outer portion of the ear contain cartilage throughout life.

Portions of the skull ossify after birth. At birth the large, flat skull bones cover the brain, but they do not meet. The spaces between the bones, called *fontanels,* are covered by a tough membrane. This membrane ossifies over a two-year period after a child is born.

A person grows as bones lengthen. In long bones, growth takes place at both ends of the bone in regions called **epiphyseal plates.** An epiphyseal plate is a layer of cartilage that contains cells that undergo mitosis. Divisions of these cells increase the amount of cartilage, and thus the length of the bone increases.

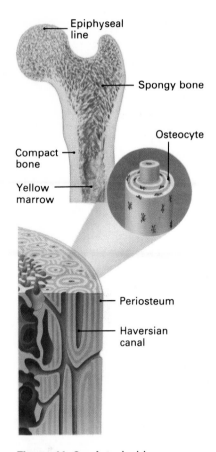

Figure 41–2. A typical long bone (top) has a shaft of compact bone surrounding a center of yellow marrow. Haversian canals like the one shown fill the compact bone.

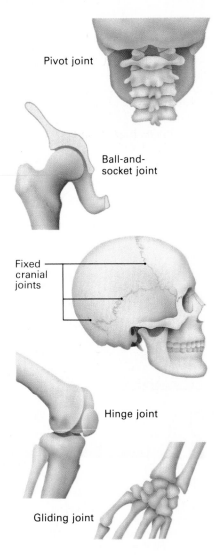

Pivot joint

Ball-and-socket joint

Fixed cranial joints

Hinge joint

Gliding joint

Figure 41–3. Each type of movable joint permits a different kind of movement. The fixed joints in the skull do not allow any motion.

At the same time, calcium is deposited around the cartilage cells in the portion of the plate closest to the shaft. Bone cells then develop in the portion of the plate closest to the shaft. As the bone cells ossify, the bone gets even longer. The plate continues to produce new cartilage cells until growth stops, at which time the plate itself is replaced by bone. Although bones grow in width throughout life, growth in length is largely complete by the time a person is 25 years old, if not sooner.

41.5 Joints

Because bones do not bend, movement can occur only where bones meet. The point where two or more bones meet is called a **joint.** Joints are of two kinds: movable and immovable. A joint that permits movement is a *movable joint*. Some movable joints, such as the shoulder joint, allow full movement. Others, such as the knee, are only partially movable. *Immovable joints* exist in bones that are fused together, as in the skull.

The body has four major types of movable joints. *Hinge joints,* such as knees and knuckles, allow forward and backward movement. *Pivot joints,* such as where the skull joins the vertebral column, permit a rotating movement. *Ball-and-socket joints,* such as the hip, allow the widest possible movement. *Gliding joints* in the wrist and ankle allow sliding movement.

Cartilage and a special lubricant called **synovial** (sih NOH vee uhl) **fluid** keep joints moving smoothly. Bones are held together at a movable joint by **ligaments,** which are strong bands of connective tissue.

Athletes and other active people frequently dislocate or sprain joints. Stretching or tearing ligaments causes a *sprain*. Joints may also become swollen and painful in a condition called *arthritis*. The most painful and crippling type is *rheumatoid arthritis,* in which cartilage becomes inflamed and enlarged. Eventually it is replaced by bone, which fuses and prevents movement. In *osteoarthritis,* which is common among elderly people, cartilage wears away, and the bones rub together.

Reviewing the Section

1. What are the functions of the skeleton?
2. How does bone change during embryonic development?
3. List three bone types and give an example of each.
4. What are the main types of joints?
5. In what two major ways does the skull differ from most of the other bones in the body?

The Muscular System

Bones would be virtually useless if there were no muscles. However, only some muscles move bones. Others assist in circulating blood and in digesting food. The body has more than 600 muscles, accounting for about 40 percent of the body weight of a healthy person.

- *Distinguish* among the three types of muscle tissue.
- *List* the main steps in a muscle contraction.
- *Describe* how skeletal muscles cause movement.

41.6 Functions of Muscles

A **muscle** is an organ made up of many muscle cells. Muscles attached to bones cause movement at joints. Some muscles are always working in the body whether a person is conscious of this effort or not. For example, the heart beats and the eyelids open and close. Though movement is the chief function of muscles, they also protect some internal organs. Additionally, sitting and standing require some muscles to be active.

41.7 Types of Muscles

Muscle tissue is made of special cells that have the ability to contract and relax. ***Three types of muscle tissue make up the muscular system: skeletal, smooth, and cardiac.*** Each differs in structure and task.

Skeletal Muscle   Muscles that move bones are called **skeletal muscles.** They attach to bones either directly or by means of strong bands of nonelastic connective tissue called **tendons.** Because skeletal muscles are generally under a person's conscious control, they are also called *voluntary muscles.* However, they sometimes move without conscious control, such as when responding to danger.

Muscle cells are called **muscle fibers.** Skeletal muscle fibers have a long tapering shape. Each fiber contains many nuclei and 1,000 to 2,000 full-length protein threads called **myofibrils** (my oh FY bruhlz). Tiny units called **sarcomeres** (SAHR koh mihrz) can be seen forming bands across the myofibrils. These units lie in single file in a way that gives myofibrils a striped, or *striated*, appearance. For this reason skeletal muscle is also called *striated muscle.*

Smooth Muscle   Smooth muscle is made up of spindle-shaped cells with one nucleus each. Most smooth muscles protect organs of the digestive, respiratory, and circulatory systems. Smooth muscles are not under conscious control, so they

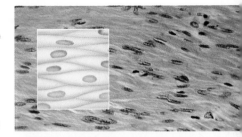

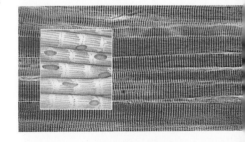

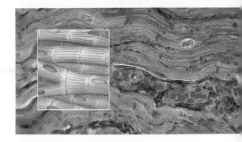

Figure 41–4. Distinctive shapes distinguish the cells of smooth muscle (top), skeletal muscle (center), and cardiac muscle (bottom).

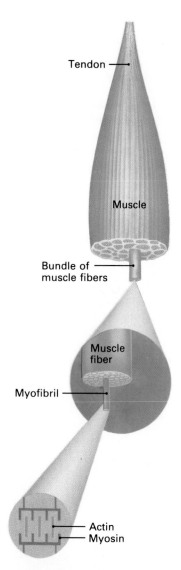

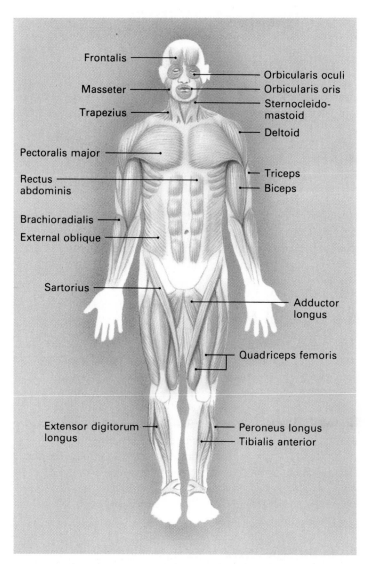

Figure 41–5. Each skeletal muscle (top) consists of individual fibers (center). These fibers, in turn, are composed of myofibrils containing sarcomeres (bottom). The skeletal muscle system is shown to the right.

are called *involuntary muscles*. They do not respond as quickly as voluntary muscles, but they do not tire as easily. These muscles lack the striations of skeletal muscles.

Cardiac Muscle **Cardiac muscle** is involuntary, striated muscle that is found only in the heart. The tightly packed cells of cardiac muscle have one or two nuclei each. Unlike other types of muscle, cardiac muscle does not receive impulses from the nervous system. Instead, the heart has its own regulator, a tiny block of special muscle fibers called the *sinoatrial node* that cause the muscle cells to contract.

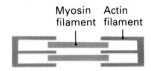

41.8 How Muscles Contract and Relax

How muscles contract and relax has been the subject of many scientific studies. Most of these studies have been done on skeletal muscles. Contraction of skeletal muscles takes place within a sarcomere. As Figure 41–6 shows, two different types of protein filaments are involved: actin and myosin. Thin *actin filaments* are twisted into double strands and are attached at the ends of a sarcomere. Thick *myosin filaments* lie in the middle of the sarcomere. Myosin filaments are also twisted into a rope-like structure that has two globular ''heads,'' one on each end of the compound myosin filament. When a muscle is relaxed, the actin and myosin filaments overlap slightly.

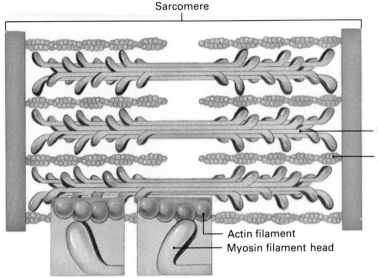

Figure 41–6. A muscle contracts as actin and myosin filaments slide past one another (top). Each individual sarcomere must shorten (left) in order to contract the muscle fiber.

Contraction begins when a nerve impulse is transmitted to a muscle cell by a *motor nerve fiber*. This nerve cell releases a substance called a *transmitter* that causes several chemical changes in the muscle cell. First, the muscle cell membranes release calcium ions into the cytoplasm of the cell. These ions then attach themselves to molecules on the actin filaments, causing the shape of actin filaments to change. At the same time, an ATP molecule binds to the head of a myosin filament to form a myosin-ATP complex. This binding changes the shape of the myosin filament.

The actin filaments then slide past the myosin filaments. This movement shortens the sarcomere and causes the muscle to contract. Exactly how this happens is not clear. According to the most widely accepted theory, the change in shape of the two

Q/A

Q: *What is rigor mortis?*

A: Soon after death a person's skeletal muscles contract and do not relax. The body becomes rigid—a state called rigor mortis.

Highlight on Careers: Athletic Trainer

General Description

An athletic trainer is a person specially trained to give first aid to athletes who are injured in sports activities. After an athlete has been seriously injured, a trainer will work with the athlete until he or she is fully recovered.

Trainers are generally hired by a team, a school, or a school district. However, an orthopedic surgeon who specializes in sports medicine may employ a trainer. A trainer working for a physician generally deals with athletes from many different sports and different teams.

Baseball, basketball, football, ice hockey, and other sports need trainers. In all cases, the trainer is on hand to give first aid on the field or on the court, and treatment before or after a game. If athletes need other treatment, the trainer refers them to physicians.

Trainers may be licensed physical therapists, physicians' assistants, nurses with masters' degrees, or former members of the military medical corps.

Career Requirements

Students interested in becoming athletic trainers should obtain a license in an allied health profession, such as physical therapy, physical assistantship, or nursing. They need sound knowledge of anatomy, physiology, chemistry, and physics.

Certification in this field is optional but desirable because of keen job competition. The American Athletic Trainers Association offers certification

exams. To become certified, a candidate must pass oral, written, and practical exams in first-aid procedures. Candidates must also know other basics of sports medicine.

For Additional Information
American Athletic Trainers
   Association and
   Certification Board
660 West Duarte Road
Arcadia, CA 91006

filaments enables the heads of the myosin filament to form attachments, or cross-bridges, with the actin filaments. After forming one cross-bridge, each myosin head appears to swivel and begin to form new cross-bridges with the actin filament. The attachment of each myosin head to an actin filament is made, broken, and re-formed five times per second. The energy for the pulling, releasing, and reattaching comes from the ATP. The result of these sequential attachments is that the actin filament is pulled by the myosin head so it slides along the myosin filament. The ends of the sarcomere are thus drawn together, causing the sarcomere to shorten. Many sarcomeres contracting at the same time cause a muscle to contract.

When an impulse to a muscle cell ends, the calcium ions leave the cytoplasm. The filament then returns to its original shape and position. The muscle is then relaxed.

Interference with the biochemistry of muscles and muscle action can lead to both simple and severe health problems. A *cramp* occurs when a muscle cannot relax. In a disorder called *myasthenia gravis,* the muscles have insufficient transmitter. They do not contract properly, and the victim has problems with speaking, eating, and other voluntary movements. In an inherited disorder called *muscular dystrophy,* muscle tissue degenerates and is replaced by fatty tissue. The *Duchenne* type is the most severe and accounts for 90 percent of all cases. No cure for muscular dystrophy has yet been discovered.

41.9 How Muscles Cause Movement

When a skeletal muscle contracts, it creates a pulling action that results in movement. ***Muscles can only pull; they cannot push.*** For this reason, muscles work in opposing pairs. A muscle pair is termed *antagonistic* if, for example, the contraction of one muscle bends a joint and the contraction of the other straightens the joint. A muscle that bends a joint is called a **flexor,** and a muscle that straightens a joint is called an **extensor.**

Most skeletal muscles are attached to two bones. During contraction one bone serves as an anchor. The point at which the muscle is attached to the anchoring bone is the **origin.** The point at which the muscle is attached to the moving bone is the **insertion.** Between these two points is a joint. Contraction of a muscle thus causes movement at the joint.

Two muscles of the upper arm—the *biceps* and the *triceps*—illustrate how antagonistic muscles produce movement. The biceps has its origin at the shoulder and its insertion on the *radius,* a bone of the forearm. When the biceps contracts, the forearm is drawn toward the front of the shoulder. If no antagonistic muscle opposed the biceps, the arm would remain bent. However, the *triceps* on the back of the upper arm has its origin on the *humerus* of the upper arm and its insertion on the radius. When the triceps contracts, it straightens the arm.

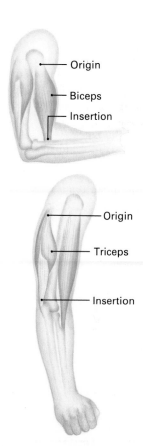

Figure 41–7. The biceps and the triceps are an antagonistic pair of muscles. Which muscle must contract to bend the arm?

Reviewing the Section

1. What are the three types of muscle tissue and where is each type found in the body?
2. What role does calcium play in muscle contraction?
3. How do skeletal muscles cause movement at joints?

- *List* several functions of the skin.
- *Explain* the function of melanin.
- *State* the functions of sebaceous glands and sweat glands.
- *Describe* the effects of sunburn on the skin.

The Integumentary System

Bones, muscles, and body organs are covered by the largest single organ, the skin. The skin is also called the **integument.** Along with the hair and nails, it makes up the *integumentary system.*

41.10 Functions of the Integumentary System

The skin performs many functions for the body, the most important of which is protection. The unbroken skin prevents harmful organisms from entering the body. It also cushions the body against physical injury. The skin is a sense organ containing receptors for touch, heat, cold, pressure, and pain. It is also an organ of elimination because it rids the body of certain waste materials through sweat. In addition, sweating cools the body and so helps control body temperature. Blood vessels near the skin's surface also allow heat to escape. When exposed to direct sunlight, components of the skin produce vitamin D. In addition, skin acts as a waterproof covering that keeps fluids inside the body.

41.11 Structure of the Integumentary System

Skin consists of all four types of body tissue: nervous, muscle, connective, and epithelial. As a result, it is elastic, flexible, and responsive. Its thickness depends upon its function. For example, an extremely thin layer of skin covers the eardrums, which must be sensitive to sound waves. In contrast, thick skin covers the soles of the feet.

Layers of Skin   The skin, shown in Figure 41–8, consists of two layers. The thin outer layer is called the **epidermis.** The thick inner layer is called the **dermis.**

The epidermis itself has two layers. The top one is actually about 20 layers of dead, scalelike, flattened cells. These cells die quickly because they are cut off from their food supply. They contain a protein called *keratin,* which makes them waterproof. The body loses several thousand of these cells each day, and new cells are produced by mitosis in the lower epidermal layer. As the surface cells disappear, those in the lower layer become the outer surface. It takes about 27 days for all of the outer skin cells to be replaced. In addition to these skin-generating cells, the lower layer has cells that contain melanin, the pigment that makes skin dark. Every person has approximately

Q/A

Q: *What forms fingerprints?*

A: The dermis attaches to the bottom of the epidermis, forming unique interlocking ridges and indentations called whorls, loops, double loops, and arches. The palms, soles of the feet, and toes also have distinctive identifying prints.

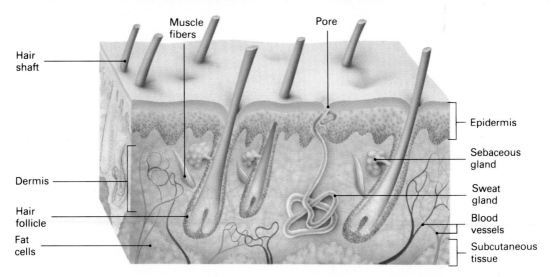

Hair shaft

Muscle fibers

Pore

Epidermis

Sebaceous gland

Dermis

Sweat gland

Hair follicle

Blood vessels

Fat cells

Subcutaneous tissue

the same number of these cells. Therefore, skin color differences result from variations in the amount of pigment produced by these cells.

The dermis is composed mainly of connective tissue, which gives the skin its strength and elasticity. Blood vessels, nerves, hair roots, and oil and sweat glands are all located in the dermis.

Figure 41–8. Although the skin is barely 4 mm (0.16 in.) thick, it is the largest organ of the body.

Subcutaneous Layer

Subcutaneous Layer A protective layer of loose fatty tissue and dense connective tissue called the *subcutaneous layer* attaches the dermis to the bones and muscles. Although not technically part of the skin, the subcutaneous layer, like skin, helps protect the body against injury and heat loss.

Glands *Sebaceous glands*, or oil glands, secrete *sebum*, an oil that reaches the skin's surface through the places where hair emerges from the skin. The oil prevents hair and skin from drying out and helps waterproof the skin. A condition called *acne* commonly occurs during adolescence. Acne occurs when oil mixes with dead cells and plugs up pores in the skin, causing blackheads. In addition, inflammation of oil glands causes pimples. Acne may be related to hormonal changes that take place during adolescence.

More than 2.5 million *sweat glands* exist in the dermis. Most consist of a tiny duct that opens to the skin's surface and rids the body of excess water and certain wastes. The evaporation of sweat also acts to cool the body when it becomes overheated.

Hair and Nails Hair is present on the skin over the entire body, except on the soles of the feet, the palms of the hands, and the lips. Hair is manufactured in **hair follicles,** which are small folds of epidermis that extend into the dermis. Tiny blood

Many people spend hours in brilliant sun getting a suntan, which they believe makes them look healthy and young. Excessive sun bathing can be harmful, however.

Suntan is actually the body's attempt to protect skin cells from an overdose of ultraviolet rays. When skin is exposed to the sun's rays, a brown skin pigment called *melanin* absorbs the harmful ultraviolet rays. Exposure to these rays stimulates melanin production. The result is a darker brown skin color, or suntan.

Damage occurs because tanning without burning is almost impossible. Melanin cannot always absorb all the ultraviolet rays. Too much ultraviolet light can kill or shock some cells and prevent mitosis in others. It can damage enzymes, cell membranes, and blood vessels in the skin. The damaged blood vessels cause the redness and chills that accompany sunburn. In severe cases of sunburn, materials from the dead skin cells enter the blood and can cause sun poisoning. The ultraviolet rays can even damage the genetic material in skin cells and cause cancer.

Although too much sun can be harmful, moderate exposure can be helpful. Sunlight stimulates the production of vitamin D and may help acne victims by removing excess skin oil. But exposure must be gradual. Several days are needed for melanin to reach the skin's surface where it provides protection. Therefore, a sunbather's goal should be to get enough sunlight to stimulate melanin production but not enough to cause redness.

A way to prevent absorption of too many harmful rays is to use sunscreens. These products have numerical ratings from 2 to 15. A rating of 2 means a person can be in the sun twice as long with the lotion as without it. A sunscreen with a rating of 15 provides maximum protection.

vessels at the base of the follicle nourish the hair root. A group of actively dividing cells near the base produce new hair. The hair *shaft*, which extends above the skin's surface is composed of dead epidermis.

Nails are mainly dead cells composed of keratin that protect the tips of fingers and toes. At the base of the hard *nail plate* is a whitish, semicircular area called the *lunula*. Cell division takes place in the root of the nailbed.

Reviewing the Section

1. How is the outer layer of epidermis replaced?
2. What causes the redness that accompanies sunburn?
3. How does the skin protect the body?
4. What is the function of sebaceous and sweat glands?

Investigation 41: The Sense of Touch

Purpose
To discover how far apart touch receptors are on different parts of the body

Materials
Cardboard, 15 cm × 25 cm; scissors; 10 straight pins; metric ruler

Procedure
1. Make a table like the one shown.

Analyses and Conclusions
1. Which area of the skin has the greatest number of pressure receptors?
2. How is the number of pressure receptors related to the functions of the hand?
3. Rank in order the various parts of the body according to their sensitivity.
4. Why would some areas of the body be more sensitive to touch than others?

Distance of pins	Fingertip	Back of hand	Palm of hand	Lips	Forearm	Back of neck
2 mm						
5 mm						
1 cm						
2 cm						
3 cm						

2. Using scissors, cut five pieces of cardboard into 3 cm × 5 cm rectangles.
3. Insert two pins 2 mm apart into a piece of cardboard. Insert 2 pins 5 mm apart into another piece of cardboard. Repeat for the remaining rectangles of cardboard, placing two pins at distances of 1 cm, 2 cm, and 3 cm respectively.
4. Blindfold your lab partner. Using the five pieces of cardboard, in any order, gently touch the heads of the pins to your partner's lips. If your partner says that he or she feels only one point, record a minus (−) in the table. If he or she feels two points, record a plus (+) in the table. Repeat this process with each of the five pieces of cardboard. Then repeat this procedure to gather information about the other parts of the body listed in the table.
5. When you are finished, trade places with your partner and repeat the procedure for steps 2, 3, and 4.

Going Further
Map the pressure receptors on your face or other part of the anatomy listed in the table.

Chapter 41 Review

Summary

Three systems of the body—the skeletal, muscular, and integumentary—provide support, movement, and protection.

Bones develop and grow through a process called ossification in which bone replaces cartilage and membranous tissue. Several types of bones make up the skeleton, including long bones, flat bones, and short bones. Movement occurs between bones at joints. Some joints are movable and others do not move. Bones are held together at joints by ligaments.

The body has three types of muscle tissue: skeletal, smooth, and cardiac. Contraction of a skeletal muscle occurs when a nerve impulse causes chemical changes in muscle cells. The reactions enable actin filaments to slide over myosin filaments of the myofibril within the cell. This movement causes the muscle to contract. When the impulse stops, the muscle relaxes.

Skin covers all body surfaces. It consists of the epidermis and the dermis. New cells are made in the epidermis. Skin pigment, melanin, is also in the epidermis. Glands, nerves, blood vessels, and hair follicles are in the dermis. A subcutaneous layer under the dermis attaches the skin to the tissues below.

BioTerms

appendicular skeleton (**623**)
axial skeleton (**623**)
cardiac muscle (**628**)
dermis (**632**)
epidermis (**632**)
epiphyseal plate (**625**)
extensor (**631**)
flexor (**631**)
hair follicle (**633**)

Haversian canal (**625**)
insertion (**631**)
integument (**632**)
joint (**626**)
ligament (**626**)
marrow (**623**)
muscle (**627**)
muscle fiber (**627**)
myofibril (**627**)

origin (**631**)
ossification (**625**)
osteocyte (**625**)
periosteum (**625**)
sarcomere (**627**)
skeletal muscle (**627**)
smooth muscle (**627**)
synovial fluid (**626**)
tendon (**627**)

BioQuiz   (Write all answers on a separate sheet of paper.)

I. Sentence Completion

1. Bones are held together at joints by _____.

2. _____ are small channels containing blood vessels, which run through compact bone.

3. Myofibrils are composed of _____ and _____ filaments.

4. The point at which a muscle is attached to a moving bone is the _____.

II. Modified True and False

Mark each statement TRUE or FALSE. If false, change the underlined term to make the statement true.

5. A muscle that bends a joint is called an extensor.

6. Bones are covered by a protective membrane called the periosteum.

7. The dermis is a layer of fatty tissue under the skin.

III. Multiple Choice

8. Mature bone cells are called
   a) cartilage. b) epidermis.
   c) myofibrils. d) osteocytes.
9. Another name for striated muscle is
   a) smooth muscle. b) skeletal muscle.
   c) involuntary muscle. d) tendon.
10. The femur is an example of a) long
    bone. b) short bone. c) flat bone.
    d) membranous bone.
11. Energy for muscle contraction comes
    from a) osteocytes. b) myosin.
    c) ATP. d) actin.
12. New skin cells are produced by the
    a) subcutaneous layer. b) epidermis.
    c) myofibrils. d) osteocytes.
13. Muscles that work in opposition to one
    another are called a) skeletal.
    b) oppositional. c) antagonistic.
    d) Haversian.

IV. Essay

14. What is the internal structure of a long
    bone?
15. What bones make up the axial skeleton
    and the appendicular skeleton?
16. Why does skeletal muscle appear to be
    striped?
17. What is the difference between the
    origin and the insertion of a muscle?
18. How is the structure of a hair follicle
    related to its function?
19. How does a skeletal muscle contract to
    cause movement?
20. What are the four kinds of movable
    joints in the body?

Applying and Extending Concepts

1. Since the skull is designed to protect the brain, why is it necessary for some athletes to wear helmets?
2. Do research and write a report about exercise and its effects on the human body. Include one or more of the following topics: resistance, aerobic, or anaerobic exercises; conditioning; and muscle building.
3. Muscle fibers will contract completely if they receive an impulse, or they will not contract at all. This is called the *all-or-none* law of muscle contraction. Given this fact, explain how the force of a muscle contraction is controlled so a pencil is not lifted with the same force as a 20-kg (50-lb.) weight?
4. When winter weather is very severe, deer may starve to death. Conservation officers often look in the marrow cavity of a deer's leg bones to determine whether starvation was the cause of death. Why would this be a valid investigation?

Related Readings

"Acne: New Approaches to an Old Problem." *Consumer Reports* 46 (August 1981): 472–477. This article discusses the causes and treatment of acne and evaluates nonprescription medications.

Allen, O. E., and the editors of Time-Life Books, *Building Sound Bones and Muscles*. Alexandria, VA: Time-Life, 1981. The functions and disorders of muscles and bones are described.

Goldstein, N., and R. B. Stone, *The Skin You Live In*. New York: A & W, 1979. This book discusses skin problems, their prevention, and treatment.

Hart, F. D., *Overcoming Arthritis*. New York: Arco, 1981. Many arthritic conditions and several disorders not directly involving joints are discussed.

42 Nutrition and Digestion

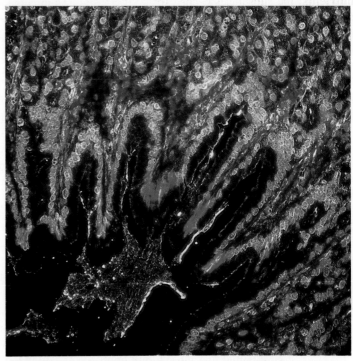

Mucous lining of human intestine stained by Azan method

Introduction

The human body is a marvelous, complex machine. Like many machines, it needs fuel in order to operate. Food is that fuel. However, food is very different from all other types of fuel. In addition to providing energy, it also supplies the materials required for the growth, maintenance, and repair of the machine—the body.

Actually, energy and building materials exist in food only in potential form. The body cannot directly use food in either its raw or cooked state. Your body cannot take the milk you drink and run it through your blood vessels. Similarly, your body cannot take a steak and attach it to your biceps to give you a stronger muscle. Instead, your body must change the food you eat into usable form.

Food and Nutrition

Section Objectives

When you sit down to a meal, you immediately notice the appearance, aroma, and taste of the foods. The most important thing about the food, however, is whether it provides the proper **nutrients,** the substances needed for body growth and maintenance. *Your daily diet should include the proper amounts of carbohydrates, fats, proteins, vitamins, minerals, and water.* All the nutrients required for the functioning of the body can be obtained by drinking water and eating the proper amount of food from four basic food groups. These groups are the milk group, the meat group, the fruit and vegetable group, and the grain and grain products group. Figure 42–1 shows some of the foods in each of these groups.

- *Describe* the role of proteins, carbohydrates, fats, vitamins, and minerals.
- *Name* three ways in which water is important to the body.
- *Define* the term *Calorie.*
- *List* the problems associated with high levels of salt, sugars, and saturated fats in the diet.

42.1 Proteins

Proteins are the major building blocks of body tissue. The body requires proteins for growth and tissue repair. Proteins called **enzymes** act as catalysts in chemical reactions in the body. The body can even use proteins to supply energy if its supplies of carbohydrates and fats have been used up.

Proteins consist of long chains of molecules called *amino acids.* The body requires 20 kinds of amino acids, which are divided into two groups based on dietary requirements. The *nonessential amino acids* are those that the body can synthesize from other amino acids. The *essential amino acids* must be obtained directly from food and so are required in the diet. There are 12 nonessential and 8 essential amino acids.

Figure 42–1. The four basic food groups are (left to right) meat, milk, fruits and vegetables, and bread and cereals.

► The Three S's and You

"Watch out for the three S's" could very well be the most important diet guideline for most Americans. It means "Be careful of *sugars, salt,* and *saturated fats."*

On the average, Americans get about 40 percent of their Calories from fat, over a third of this from such saturated fats as butter and meat fat. A high level of saturated fat in the diet has been linked directly to excessive blood *cholesterol,* a fatty organic compound. Together, saturated fat and cholesterol are considered a major

cause of heart and blood vessel diseases.

Forty-five percent of the Calories come from carbohydrates. Refined and processed sugars account for more than a third of the carbohydrate Calories. These Calories provide no additional nutritional value—no vitamins, minerals, or fiber.

For a healthful diet, a person should get about 58 percent of his or her Calories from carbohydrates and fewer from fats. The types of fats and carbohydrates a person eats are also important. Satu-

rated fats should be no more than 10 percent of the Calories. Complex carbohydrates, such as whole grains, fruits, and vegetables, should make up 48 percent of the Calories, and refined sugar, no more than 10 percent.

Salt abounds in Americans' food. Studies show that a person may need as little as a quarter gram of salt a day; yet Americans average 6 to 18 grams. A salty diet may contribute to high blood pressure, or hypertension, which in turn contributes to heart attacks and strokes.

Q/A

Q: *Why are whole grain flours more healthful than bleached flour?*

A: Whole grains contain not only starch but also protein, vitamins, and fiber that come from the seed coat, or husk, of the grains. During refining of bleached flour, the husk is removed.

Complete proteins contain the 8 essential amino acids. From these, the body can synthesize the 12 nonessential amino acids. Meat, eggs, and dairy products contain complete proteins. *Incomplete proteins* do not contain all the essential amino acids. Most plant proteins are incomplete. However, a diet that combines plant proteins can provide all the essential amino acids.

42.2 Carbohydrates

Carbohydrates supply most of the body's energy needs. The simplest form of carbohydrate is a *monosaccharide,* or a simple sugar molecule. Glucose, fructose, dextrose, and galactose are monosaccharides. They exist in such foods as fruits, honey, syrups, artichokes, and onions. *Disaccharides,* or double sugars, consist of two monosaccharides. Disaccharides include sucrose, or cane sugar; and lactose, which is found in milk.

Complex carbohydrates called *polysaccharides* consist of many simple sugar molecules. Starches and cellulose are two important polysaccharides. Starches are found in cereal grains

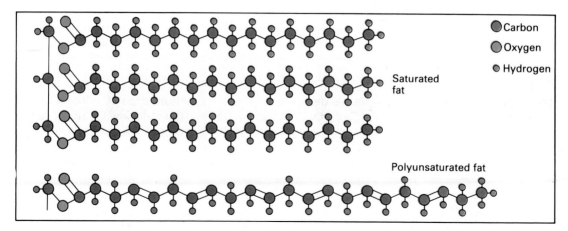

Saturated
fat

Polyunsaturated fat

Figure 42–2. A molecule of saturated fat (top) contains as many hydrogen atoms as it can hold. A molecule of unsaturated fat (bottom) has fewer hydrogen atoms because some of its carbon atoms are double-bonded.

and such vegetables as potatoes, beans, and corn. Cellulose is present in all plant tissues but cannot be digested by human beings. However, cellulose provides the body with **fiber,** which aids digestion by stimulating the muscles of the digestive tract.

42.3 Fats

Fats are highly concentrated sources of energy. They provide twice as much energy per gram as do carbohydrates or protein. The body stores fats for use when carbohydrates are not available for energy. Fats also form part of cell membranes and organelles. Thus, they are a structural material of the body.

Fats exist in two forms—saturated and unsaturated. Both forms consist of hydrogen, carbon, and oxygen atoms. In *saturated fat,* all the carbon atoms are joined by single bonds, and so the molecules contain the maximum number of hydrogen atoms. It is saturated with hydrogen. An *unsaturated fat* has at least one double bond between carbon atoms. For every double bond, two hydrogen atoms are missing. A molecule with two or more double bonds is called *polyunsaturated.* Animal fats are saturated fats. Butter, lard, and other saturated fats are solids at room temperature. Plant oils are unsaturated fats. Corn oil, olive oil, sunflower oil, and other unsaturated fats are typically liquid at room temperature.

42.4 Vitamins

Vitamins are organic substances that act as *coenzymes*—that is, they assist enzymes during chemical reactions. Although vitamins do not provide energy, they are necessary for normal growth and body activity. Vitamins are classified as either *fat-soluble* or *water-soluble.* Fat-soluble vitamins—vitamins A, D,

Table 42-1: Vitamins and Minerals

Vitamin or Mineral Sources	Use by the Body	Deficiency Symptoms
A Fish liver oils, liver, eggs, butter, yellow and green vegetables, fruits	Healthy skin, eyes, bones, teeth, urinary tract, and epithelial tissue	Night blindness, skin and mucous membrane disorders, kidney stones
B₁ (thiamine) Organ meats (liver, brain, kidney, heart), whole grains, most vegetables	Proper functioning of heart, nervous system, and digestive tract; energy release from food; growth	Beriberi (nervous system disorder), cardiovascular disorders, indigestion, fatigue
B₂ (riboflavin) Liver, poultry, milk, eggs, cheese, fish, green vegetables, whole grains	Metabolism of proteins, carbohydrates, and fats; tissue repair; healthy skin	Dim vision, premature aging, poor growth, sore mouth and tongue
Niacin Meat, whole grains, potatoes, leafy vegetables, yeast	Growth, healthy nervous and digestive systems, carbohydrate metabolism	Pellagra; nervous, digestive, and skin disorders
B₁₂ (cobalamin) Liver and other meats, eggs, cheese, yogurt, milk	Red blood cell production, healthy nervous system	Pernicious anemia
C (ascorbic acid) Citrus and other fruits, leafy vegetables, tomatoes, potatoes	Healthy blood vessels, bones, teeth, cartilage; resistance to infection; healing of wounds	Scurvy, easy bruising, bleeding gums, swollen tongue and joints
D Liver, fish oils, eggs, milk, sunlight	Growth, healthy bones and teeth, metabolism of calcium and phosphorus	Rickets, poor teeth and bones
E Whole grains, leafy vegetables, milk, butter, vegetable oils	Healthy cell membranes, possibly for reproductive functions	Red cell rupture, muscular dystrophy, sterility (in lab animals)
K Leafy vegetables, soybeans; made by intestinal bacteria	Normal blood clotting, proper liver functioning	Hemorrhages
Calcium Milk, cheese, whole grains, meat, leafy vegetables, peas and other legumes	Muscular and nervous system functioning, bone and tooth development, blood clotting, cell membrane permeability	Soft bones, poor teeth, failure of blood to clot
Iodine Seafoods, iodized salt	Cellular respiration (control of body functions)	Goiter
Iron Liver, red meat, egg yolk, whole grains, prunes, nuts	Healthy red blood cells	Anemia
Magnesium Milk, whole grains, legumes, nuts, meat	Healthy bones and teeth, carbohydrate and protein metabolism	Improper nerve and muscle functioning
Phosphorus Milk, whole grains, meats, nuts, legumes	Tooth and bone development, ATP production, nucleic acids	Poor teeth and bones
Potassium Whole grains, fruits, legumes, meat	Nerve function, cell activities	Improper nerve and muscle functioning
Sodium Seafood, table salt	Water balance, proper nerve and muscle functioning	Muscle and nerve disorders, dehydration

E, and K—are stored in the body's fatty tissue. Water-soluble vitamins—all the B vitamins and vitamin C—can be dissolved in water but cannot be stored in the body. They must be obtained directly from food.

Excessive amounts of vitamins A and D may cause disorders of the bones and other body tissues. Vitamin deficiencies cause many types of disorders. A well-balanced diet provides the proper amounts of all the necessary vitamins daily.

42.5 Minerals

Minerals are inorganic substances that form an important part of living tissue. Like vitamins, minerals do not supply energy but help regulate body functions. Teeth and bones require calcium and phosphorus. Iron is the central atom in the oxygen-carrying molecules of the blood. Magnesium, calcium, and zinc help regulate nerve and muscle function. Table 42–1 lists vitamins and minerals, their uses, and the results of deficiencies.

42.6 Water

About two-thirds of the body's weight is water, most of it in the cytoplasm of cells. Blood plasma, tissue fluids, and body cavities contain the remainder. Water is required for many body functions. Most chemical reactions in the body can take place only in a water solution. Water carries nutrients to the blood plasma and into body cells. Water also forms the major part of urine and sweat, which help rid the body of wastes.

42.7 Calories

The energy value of food is commonly measured in **Calories.** One Calorie is the amount of heat energy needed to raise the temperature of one kilogram of water 1°C. The energy potential of food and the daily energy requirements of individuals are stated in Calories. A teenage boy needs about 3,000 Calories daily, and a teenage girl, about 2,000.

Q/A

Q: *Are Calories related in any way to the nutritional value of food?*

A: No. Calories are merely a measure of heat energy available from food. Nutritional value pertains to the nutrients in food.

Reviewing the Section

1. Which would be a more serious problem for the body, a fat-poor or a protein-poor diet? Why?
2. How do vitamins contribute to a healthy body?
3. How does water help the body function?
4. Why should people limit saturated fat in the diet?

The Digestive System

- *Identify* the main organs in a diagram of the digestive system.
- *Distinguish* between mechanical digestion and chemical digestion.
- *Summarize* the digestive processes that take place in the stomach.
- *Describe* the process of digestion in the small intestine.

Food can be used by the body only after it has been broken down into small molecules. The process by which food is changed into a form the body can use is **digestion.** Digestion takes place in a continuous tubelike passageway that extends from the mouth to the *anus.* This passageway is known as the **alimentary canal,** or **digestive tract.** The alimentary canal and other organs associated with digestion make up the digestive system.

The digestive system serves two major functions. The first, of course, is digestion—the breaking down of food into molecules the body can use. The nutrient molecules must then get to the cells where they are needed. Therefore, the second major function of the digestive system is **absorption.** Absorption is the movement of nutrient molecules into blood vessels or other vessels. The blood carries these nutrients to the cells, which use the nutrients for energy, growth, and repair.

Digestion takes two forms—mechanical and chemical. Mechanical digestion is the physical tearing and grinding of food into smaller pieces. Mechanical digestion thus increases the amount of surface area of food exposed to the action of digestive enzymes. These enzymes help bring about the second form of digestion. *Chemical digestion changes food particles into molecules the body can use.*

42.8 The Mouth

Mechanical and chemical digestion both start in the mouth, or *oral cavity.* Food is bitten, cut, and torn by the *incisors,* the sharp teeth at the front of the mouth, and the teeth next to them, the *canines.* Strong muscles of the jaws and tongue move the food into position for chewing. Food is then crushed and ground by the broad, flat surfaces of the premolars and the molars at the rear of the mouth.

While in the mouth, food is moistened by **saliva,** a mixture of *mucus* and a digestive enzyme called *ptyalin* (TY uh lihn), or *salivary amylase.* Saliva is produced by three pairs of **salivary glands.** The largest of these are the *parotid glands* located in the cheek region. The *sublingual glands* are in the floor of the mouth under the tip of the tongue, and the *submaxillary glands* are along the lower jaw.

Saliva lubricates food so that it moves smoothly through the digestive tract. Saliva may also kill some bacteria in the mouth. Ptyalin starts the breakdown of starches to glucose. However, because food remains in the mouth for such a short time, ptyalin

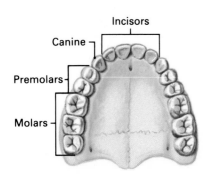
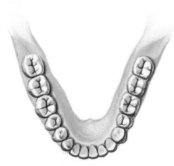

Figure 42–3. This diagram shows the location of canines, incisors, premolars, and molars in the upper and lower jaws.

Incisors

Canine

Premolars

Molars

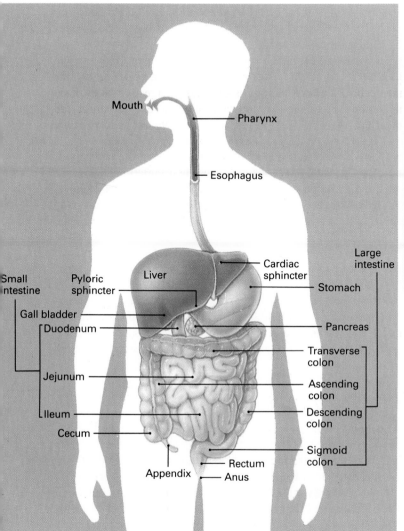

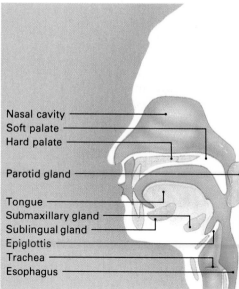

Mouth

Pharynx

Esophagus

Large intestine

Cardiac sphincter

Stomach

Liver

Small intestine

Pyloric sphincter

Gall bladder

Duodenum

Pancreas

Transverse colon

Jejunum

Ascending colon

Ileum

Descending colon

Cecum

Sigmoid colon

Appendix

Rectum

Anus

Nasal cavity

Soft palate

Hard palate

Parotid gland

Tongue

Submaxillary gland

Sublingual gland

Epiglottis

Trachea

Esophagus

Figure 42–4. After food passes through mouth and pharynx (above), involuntary muscles move it through the rest of the digestive system (left).

acts on less than 5 percent of the starches. The food and saliva eventually form a moist, soft ball called a **bolus.**

When you swallow, your tongue presses against the **hard palate,** the bony plate in the roof of the mouth. The pressure forces the bolus to muscle tissue called the **soft palate.**

42.9 The Pharynx and Esophagus

From the soft palate area, the bolus moves into the **pharynx,** a common passageway for food and air. As the bolus is forced to the back of the mouth, the soft palate moves up and closes off the nasal cavities. At the same time, a flap of tissue called the

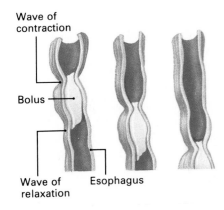

Wave of
contraction

Bolus

Wave of   Esophagus
relaxation

Figure 42–5. A continuous wave of muscle contractions called *peristalsis* moves the bolus through the esophagus.

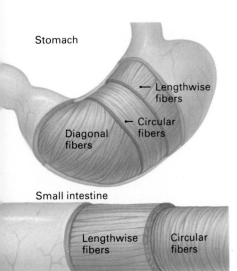

Stomach

— Lengthwise
fibers

— Circular
fibers

Diagonal
fibers

Small intestine

Lengthwise   Circular
fibers   fibers

Serosa

Figure 42–6. The stomach (top) has three layers of muscle that together produce a grinding motion. Only two layers, circular and lengthwise muscle, are found in the small intestine (bottom).

epiglottis (ehp uh GLAHT ihs) seals off the *trachea,* or windpipe. The **larynx,** or voice box, at the top of the trachea moves up against the bottom of the epiglottis. Food can then pass quickly through the pharynx, across the trachea, and into the **esophagus** (ih SAHF uh guhs), the muscular tube leading to the stomach. If a person attempts to breathe while swallowing, food gets into the trachea. An automatic coughing reflex then helps clear the trachea.

Until food enters the pharynx, voluntary muscles control the process of mechanical digestion. Once a swallow has started, however, it cannot be stopped because involuntary muscles take over in the pharynx. A strong contraction of a muscle around the pharynx propels food into the esophagus. This contraction starts a wavelike motion called **peristalsis** (pehr uh STAWL sihs), which moves food along. Peristalsis results from the action of two layers of involuntary muscles that form the walls of most of the digestive tract. One layer of muscles wraps around the tract, and the second layer runs along its length. While the circular muscles squeeze, the parallel ones relax. As the parallel muscles contract, the encircling ones relax. The squeezing and contracting action of the two sets of muscles always occurs above the bolus or liquid in the digestive tract, thus pushing it through the tract.

42.10 The Stomach

At the end of the esophagus is a muscular valve called the **cardiac sphincter** (SFINK tuhr). A sphincter is a muscle that controls a circular opening in the body. This valve prevents food from reentering the esophagus. Food passes through the valve into the **stomach,** a J-shaped, baglike organ with a capacity of 2 to 4 L (2.1 to 4.2 qt.). Both mechanical and chemical digestion continue in the stomach. In addition to the two layers of involuntary muscle, the stomach has a third, diagonal layer of muscle. Through the action of these three muscle layers, the stomach can actually grind food.

Chemical digestion of protein begins in the stomach. The stomach contains about 35 million glands that produce mucus and gastric secretions. The chief gastric secretions produced by the stomach are hydrochloric acid and an enzyme called *pepsin.* Pepsin is active only in a highly acidic environment. This enzyme splits protein into smaller groups of amino acids called *polypeptides.* Hydrochloric acid also dissolves minerals and kills bacteria. There is a muscular valve that controls the passage of food out of the stomach. This valve is known as the **pyloric sphincter.**

Much of the early study of digestion was done by literally looking into a person's stomach. In 1822 an American Indian named Alexis St. Martin was severely wounded. A shotgun blast tore open the abdominal wall and part of his stomach.

St. Martin went to Dr. William Beaumont, an army surgeon in northern Michigan. Beaumont examined the wound and

thought that St. Martin would die from it, so he merely packed the wound with cotton and waited.

To Beaumont's surprise, St. Martin did not die. Instead, the stomach healed with a permanent opening to the outside about the size of a quarter. This opening provided Beaumont with a unique opportunity to study digestive processes by looking right into St. Martin's stomach.

Beaumont removed samples of gastric juice from St. Martin's stomach and tested them on various foods. He noted that some foods were changed chemically by the juice and others were not. He placed foods directly into the open stomach and determined the amount of time needed to digest each one. He even studied the effects of emotions on gastric secretions.

What prevents the stomach from digesting itself? Mucus-secreting cells line the surface of the stomach. The mucus helps protect the stomach lining from hydrochloric acid. About 500,000 cells of the stomach lining are shed every minute. As a result of this process, the cell layer replaces itself every three days. Occasionally too little mucus or too much acid exists in the stomach. An open, painful sore called an *ulcer* may then form in the stomach lining. Ulcers may sometimes bleed severely. They can also eat completely through the stomach wall, leading to much more serious conditions.

42.11 The Small Intestine

Food leaves the stomach as a semifluid mass called **chyme.** Chyme enters the **small intestine,** a tube about 7 m (23.1 ft.) long and 3 cm (1.2 in.) in diameter.

The small intestine has three sections. The uppermost section, the *duodenum* is about 25 cm (1.2 ft.) long. The next section, the *jejunum*, is about 4 m (13.1 ft.) long. The last 2.5 m (8.2 ft.) form the *ileum*. *The majority of chemical digestion and absorption takes place in the small intestine.*

The Pancreas and the Liver The **pancreas** and the **liver** secrete digestive juices into the small intestine and so play important roles in digestion. The pancreas, located behind the

Q/A

Q: *Can a person whose stomach has been removed digest food?*

A: Yes. The stomach is the site of protein digestion and a holding area for large quantities of food. Without a stomach, a person would have to eat frequent small meals, avoid animal protein, and chew food thoroughly.

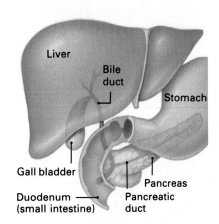

Figure 42–7. The liver secretes bile, which enters the small intestine through the gall bladder. Bile emulsifies fat, which is then digested by an enzyme produced by the pancreas.

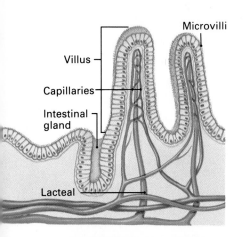

Figure 42–8. Nutrients from digested food in the small intestine pass from the villi to the microvilli, where they enter blood capillaries.

stomach, has many small lobes that secrete enzymes and sodium bicarbonate. The sodium bicarbonate neutralizes the acidity of the chyme leaving the stomach. The chief enzymes secreted by the pancreas are *pancreatic amylase, pancreatic lipase, trypsin,* and *chymotrypsin.* Pancreatic amylase continues the chemical digestion of starch that began in the mouth. It converts starches into maltose. Pancreatic lipase breaks down fats into their component molecules, fatty acids and glycerol. Trypsin and chymotrypsin break down the proteins by splitting them into smaller chains of amino acids called peptides.

The liver is the largest internal organ, weighing about 1.5 kg (3.3 lb.). It produces *bile,* which is a salt solution, not an enzyme. Bile *emulsifies* fat—that is, it breaks down large fat globules into tiny droplets. This process greatly increases the surface area of fat particles. Lipase can then act on the fat more effectively. This process is referred to as the detergent effect of bile, because detergent does the same thing to fat. Bile generally enters the duodenum from the **gall bladder,** a small sac where bile is stored.

Digestion in the Small Intestine Most chemical digestion occurs in the duodenum. A heavy layer of mucus protects the first few centimeters of the duodenum from the acidic chyme released by the stomach. If the mucus protection is not sufficient, the high acid level can cause duodenal ulcers, which are even more common than stomach ulcers.

Enzymes produced by the small intestine include *peptidases, maltase, lactase, sucrase,* and *intestinal lipase.* Various peptidases break down peptides into amino acids. Maltase, lactase, and sucrase convert disaccharides into monosaccharides. Intestinal lipase, like pancreatic lipase, splits fats into fatty acids and glycerol.

Absorption in the Small Intestine Amino acids, monosaccharides, fatty acids, glycerol, water, and minerals are all absorbed in the small intestine. *Absorption occurs quickly in the small intestine because of its lining.* The mucous lining consists of folds covered with millions of tiny projections called **villi** (VIHL eye). Each villus has a *brush border* composed of approximately 600 **microvilli,** which are extensions of the epithelial tissue covering the villi. Intestinal enzymes are not released into the cavity of the small intestine. Instead, the enzymes remain in the brush border where they act upon the food molecules. The molecules are then absorbed into the bloodstream through the microvilli walls. The lining folds, the villi, and the microvilli together increase the surface area of the small

Highlight on Careers: Dietitian

General Description

A dietitian provides information on proper nutrition. Dietitians know the chemical makeup of foods and the effects of foods on the human body. For example, they know what nutrients—how much protein or what vitamins—are in foods.

Clinical dietitians work in hospitals or nursing homes, where they plan diets as part of the patients' treatment. They often work with physicians, nurses, and other professionals in planning special diets to meet patients' individual needs.

Some dietitians teach nutrition in colleges. Others are administrative dietitians who work for commercial food service companies. They give advice on the safe handling and preparation of food.

Community dietitians work with groups or individuals, usually through local government or voluntary health agencies.

Increasing numbers of dietitians are going into private practice. They accept patients on an individual basis and work with them in planning diets designed for their special needs. With the growing awareness of nutrition and fitness in society today, dietitians have been expanding their role on the health care team.

Career Requirements

A dietitian earns a B.S. degree in dietetics, nutrition, or food science. Many dietitians also complete a 6- to 12-month internship after getting their degrees. The American Dietetic Association (ADA) accredits

the internship programs, which include clinical experience. The ADA offers a registered dietitian (RD) certificate to those who meet educational requirements and pass an examination.

For Additional Information
American Dietetic Association
430 North Michigan Avenue
Chicago, IL 60611

---

intestine 600 times. The result is an absorptive area equal in size to the area of a tennis court.

Absorption occurs through the processes of diffusion and active transport. Each villus contains tiny blood vessels called *capillaries*, through which monosaccharide and amino acid molecules enter the bloodstream. The blood carries these nutrients to the body's tissues. The liver converts excess glucose into glycogen, a form of starch, and stores it as a future energy source. The cells of the villi resynthesize fatty acids and glycerol into fats. The villi contain tiny vessels called **lacteals,** which absorb the fats. These fats eventually pass from the lacteals into the bloodstream.

42.12 The Large Intestine

Minerals, water, and undigested foods enter the last part of the digestive tract, called the **large intestine.** The large intestine is also known as the **colon.** *Absorption of water, minerals, and vitamins is completed in the large intestine.*

Approximately 9 L (9.5 qt.) of water pass through the digestive tract in one day. Some of the water is transported by osmosis into capillaries lining the walls of the large intestine. The solid waste material that remains is called **feces.** It stays in the body until it is eliminated through the the anus. The last part of the large intestine, the **rectum,** controls the elimination of feces.

A type of bacteria called *Escherichia coli,* or *E. coli,* lives in the large intestine. These organisms feed on materials that the human body cannot digest. *E. coli* produce some amino acids and vitamin K.

Among the disorders resulting from digestive problems are diarrhea and constipation. *Diarrhea* is a condition in which the feces do not remain in the large intestine long enough for the water to be absorbed. *Constipation* is the opposite condition. Constipation results when the feces remain in the colon too long. As a result, too much water is absorbed. Diarrhea may be caused by bacteria or viruses, emotional stress, or eating certain foods. Prolonged diarrhea can result in dehydration and even death. Constipation may be caused by insufficient fiber and water in the diet. It results in hard, dry feces that can make defecation, or elimination of feces, painful.

Near the beginning of the large intestine is a small, finger-like projection called the *appendix.* The appendix is a blind, saclike structure. The appendix has generally been considered a useless structure in humans. However, some scientists now believe it helps produce antibodies. Sometimes the appendix becomes infected and must be immediately removed. This painful condition is called *appendicitis.*

Reviewing the Section

1. What is the difference between mechanical and chemical digestion?
2. What are the functions of the enzymes in the mouth and stomach?
3. What role does the pancreas play in digestion?
4. How does the absorption of food molecules occur?
5. What functions does the large intestine perform?

Investigation 42: Testing a Potato for Water and Salts

Purpose
To learn how to test a food for the presence of water and certain mineral salts

Materials

Safety goggles, lab apron, scalpel, ruler, potato, five test tubes, matches or starter, Bunsen burner, test tube holder, test tube rack, sodium chloride, balance, 100-mL beaker, stirring rod, silver nitrate, distilled water, magnesium sulfate, barium chloride

Procedure
1. **CAUTION: Put on safety goggles and a lab apron and leave them on throughout the entire experiment.** Using a scalpel carefully cut a 1 cm cube of potato and place the potato cube in a test tube.
2. Light a Bunsen burner following the procedures on page 831.
3. Using a test tube holder, heat the potato over the Bunsen burner pointing the open end of the tube away from you. *Is there evidence that water is being driven off?*
4. Continue heating the potato until it burns and forms ash.
5. Set the test tube and potato ash aside in the test tube rack and allow them to cool.
6. Weigh 1 g of sodium chloride on a balance and place the sodium chloride in a 100-mL beaker.
7. Add 50 mL of distilled water to the beaker and stir the mixture well.
8. Pour 5 mL of this salt solution into a test tube.
9. Add four drops of silver nitrate solution. The precipitate that forms indicates the presence of chloride ions. *What is the color of the precipitate?*
10. Pour 5 mL of distilled water over the potato ash in the cooled test tube and shake the tube to mix thoroughly.
11. Pour half of the resultant solution into another test tube and set this test tube aside.

12. Using the procedure in step 9, test one tube of ash solution for the presence of chloride ions. Make a chart like the one shown and enter the result, if any.
13. To a clean test tube add 5 mL of magnesium sulfate solution.
14. Add three drops of barium chloride to the test tube. The precipitate that forms indicates the presence of sulfate ions. *What is the color of the precipitate?*
15. Using the procedure you learned in step 14, test the remaining half of the potato ash solution for the presence of sulfate ions. Enter the result, if any, in your chart.

Testing Food for Mineral Content

	Test Solution	Potato Ash Solution
Result of adding silver nitrate	Precipitate forms	
Result of adding barium chloride	Precipitate forms	

Analyses and Conclusions
1. How can you be certain that the liquid driven from the potato when heated was really water?
2. Does potato contain sulfate or chloride ions?

Going Further
- Test some other common foods, such as a sweet potato, for the presence of water and chloride or sulfate ions.
- Conduct research to determine what tests are available for detecting other mineral salts. Then test the potato for the presence of these other substances.

Chapter 42 Review

Summary

The nutrients in food provide the body with energy and with the materials needed for the growth, maintenance, and repair of body tissue. Carbohydrates and fats provide energy. Proteins form the building blocks of cells. Vitamins and minerals help regulate body processes. Water helps break down compounds, transports nutrients, and helps remove wastes.

Digestion is a mechanical and chemical process in which food is broken down into materials the body can use. Carbohydrates are broken down into monosaccharides. Fats are reduced to fatty acids and glycerol. Proteins are changed to amino acids.

Chemical digestion of carbohydrates starts in the mouth. Protein digestion starts in the stomach. Fat digestion begins in the small intestine. All chemical digestion is completed in the small intestine. Absorption of monosaccharides, amino acids, and fats occurs in the small intestine. Water, minerals, and vitamins are absorbed in the large intestine.

BioTerms

absorption (**644**)	fat (**641**)	peristalsis (**646**)
alimentary canal (**644**)	feces (**650**)	pharynx (**645**)
bolus (**645**)	fiber (**641**)	protein (**639**)
Calorie (**643**)	gall bladder (**648**)	pyloric sphincter (**646**)
carbohydrate (**640**)	hard palate (**645**)	rectum (**650**)
cardiac sphincter (**646**)	lacteal (**649**)	saliva (**644**)
chyme (**647**)	large intestine (**650**)	salivary glands (**644**)
colon (**650**)	larynx (**646**)	small intestine (**647**)
digestion (**644**)	liver (**647**)	soft palate (**645**)
digestive tract (**644**)	microvillus (**648**)	stomach (**646**)
enzyme (**639**)	mineral (**643**)	villus (**648**)
epiglottis (**646**)	nutrient (**639**)	vitamin (**641**)
esophagus (**646**)	pancreas (**647**)	

BioQuiz   (Write all answers on a separate sheet of paper.)

I. Completion

1. Calories are a measure of the _____ contained in food.
2. The villi of the small intestine aid in _____.
3. In the stomach proteins are digested by the enzyme _____.
4. The processes by which absorption occurs are _____ and _____.

II. Modified True and False

Mark each statement TRUE or FALSE. If false, change the underlined term to make the statement true.

5. <u>Peristalsis</u> is the main process for moving food in the digestive tract.
6. Bile, produced by the <u>liver</u>, helps in the digestion of <u>fats</u>.
7. Most absorption occurs in the <u>stomach</u>.

III. Multiple Choice

8. The enzyme ptyalin in saliva begins the digestion of a) carbohydrates. b) fats. c) proteins. d) minerals.
9. The passage of food from the stomach into the small intestine is controlled by the a) cardiac sphincter. b) esophagus. c) lacteal. d) pyloric sphincter.
10. A diet low in fiber may cause a) diarrhea. b) vomiting. c) constipation. d) appendicitis.
11. Heart and blood vessel diseases may result from a diet high in a) protein. b) polyunsaturated fat. c) carbohydrates. d) saturated fat.
12. Chemical digestion of protein starts in the a) small intestine. b) mouth. c) stomach. d) large intestine.
13. The body does not store a) fat-soluble vitamins. b) water-soluble vitamins. c) fats. d) vitamin A.
14. Most of the body's energy needs are supplied by a) proteins. b) fats. c) carbohydrates. d) vitamins.
15. Cellulose provides the body with a) starch. b) minerals. c) fiber. d) enzymes.

IV. Essay

16. How are mechanical and chemical digestion different?
17. What digestive processes occur in the mouth?
18. Which enzymes are secreted by the pancreas and what do they do?
19. How might the differences between fat-soluble and water-soluble vitamins affect the diet?
20. What is the difference between complete and incomplete proteins?

Applying and Extending Concepts

1. Record what you eat in one day. Decide whether this menu represents a balanced diet. If not, plan how it may be changed to make it more nutritious.
2. Assume you have just eaten a hamburger on a bun with lettuce and melted cheese. Trace the pathway of the food through the digestive system. Explain what happens to it, where it happens, the enzymes involved, and the source of these enzymes.

Present this information in the form of a chart.
3. You may have heard of carbohydrate loading as a technique to help athletes reach peak efficiency. Research carbohydrate loading and write a report on your findings. Speak with a coach or health club operator about the benefits and potential problems of this practice.

Related Readings

Brody, Jane E. *Jane Brody's Nutrition Book*. New York: Norton, 1981. The personal-health columnist of *The New York Times* offers a guide to a lifetime of better eating and weight control.

Mylander, Maureen, *The Great American Stomach Book*. New York: Ticknor & Fields, 1982. This book describes the digestive process and some of the problems associated with it.

Yudkin, John. *This Nutrition Business*. New New York: Norton, 1981. The personal-nutritionist discusses the fundamentals of nutrition and the most commonly asked questions about the subject.

Circulation

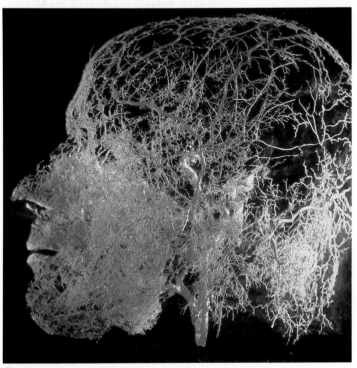

Circulatory system of the human head

Introduction

Substances within a single cell can be transported by movement of cytoplasm. Diffusion and active transport carry food, chemicals, gases, and waste products into and out of the cells of simple organisms. Large multicellular animals, however, require a more elaborate system for the transport of nutrients, oxygen, and wastes throughout the body.

Pickup and delivery within the human body are handled by an effective transport system. This system consists of a pump, carriers, and thousands of kilometers of tubes that run throughout the body. The system is called the *circulatory system*. The circulatory system not only carries nutrients, oxygen, and bodily wastes but also provides a natural defense mechanism against disease.

The Blood

Blood is the chief carrier of the body's transport system. It carries nutrients and oxygen to body cells and transports carbon dioxide and other waste products away from the cells. Blood also combats disease and helps maintain body temperature.

43.1 Composition of Blood

A human adult has about 5 L (5.3 qt.) of blood, which makes up about 9 percent of the body's weight. Blood is liquid connective tissue. It consists of a liquid called plasma and three kinds of blood cells: red blood cells, white blood cells, and platelets. Approximately 55 percent of blood volume is plasma. About 44 percent is red blood cells. The remaining 1 percent is white blood cells and platelets.

Plasma The straw-colored, nonliving part of blood, called **plasma,** has many functions. For example, it carries nutrients such as amino acids and glucose molecules absorbed in the small intestine to body cells. Plasma also takes waste products away from the cells and delivers these wastes to the kidneys and sweat glands so they can be safely removed from the body.

Plasma is more than 90 percent water. The remainder consists of minerals and thousands of other compounds, including many proteins. These proteins assist in blood clotting, help maintain the body's water balance, and influence the exchange of materials between the circulatory system and the body cells. Also present in plasma are nitrogenous waste products and respiratory gases. Some plasma, with fewer proteins, seeps through blood vessel walls. It fills spaces between body tissues and bathes every body cell. This fluid is known as *tissue fluid*.

- *Name* four functions of blood.
- *Describe* the functions of plasma, red blood cells, white blood cells, and platelets.
- *Summarize* the process of blood clotting.
- *List* the four primary blood types.
- *Explain* the cause of Rh disease.

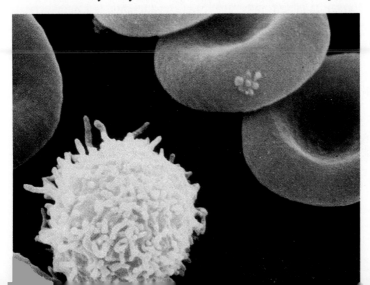

Figure 43–1. This photomicrograph shows human red blood cells and a white blood cell, ×5,200.

Q: *How does carbon monoxide poisoning affect the body?*

A: Carbon monoxide (CO) is an odorless gas in automobile exhaust and tobacco smoke. Hemoglobin combines with carbon monoxide more readily than with oxygen but does not release it as easily. When hemoglobin is carrying carbon monoxide, it cannot combine with oxygen needed by the body. Death by asphyxiation may result.

Figure 43–2. This photograph of two activated human platelets was taken by a scanning electron microscope, magnified 15,300 times.

Red Blood Cells The blood cells that transport respiratory gases are called the **red blood cells.** Red blood cells, also called *erythrocytes* or *red corpuscles,* carry oxygen from the lungs to body cells. They also transport carbon dioxide from the cells to the lungs. A red blood cell has a nucleus when it is formed in red bone marrow. However, the nucleus and other organelles disappear as the red blood cell matures. Each cell becomes a disc-shaped sac with a thick rim and thin center. Almost the entire cell fills with **hemoglobin** (HEE muh gloh bihn), an iron-containing protein molecule that is bright red when combined with oxygen. One molecule of hemoglobin carries four molecules of oxygen. Hemoglobin is therefore an effective oxygen carrier.

Red blood cells are so small that hundreds of them would be needed to encircle one strand of hair. The human body has about 25 trillion red blood cells. They are produced at the rate of over 100 million per hour and have a life span of about 120 days. The dead cells are dismantled, and the hemoglobin is stored to be reused in new red blood cells.

White Blood Cells The **white blood cells,** also known as *leukocytes* or *white corpuscles,* are the body's main defense against viruses, bacteria, and other foreign organisms. In fighting invaders, white cells pass through blood vessel walls and into tissue fluid. They move like amoebas, attracted to the site of an infection by chemicals. The chemicals may be products of blood clotting. They may also come from bacteria, other leukocytes, or from degeneration of infected tissue. The white blood cells engulf and digest the invading organisms by a process known as *phagocytosis.*

Several kinds of white blood cells are found in blood. Most white blood cells are manufactured and stored in red bone marrow until they are needed by the body. They are colorless, irregularly shaped cells with nuclei. Although white blood cells are larger than red cells, they are considerably less numerous—about 1 white cell for every 750 red cells. The normal life span of white blood cells is about three days unless they are fighting infection. In that case, they may live only a few hours.

Platelets Cell fragments called **platelets,** or *thrombocytes,* aid in blood clotting. Within five seconds after an injury occurs, the process of clotting, or **coagulation** (koh ag yoo LAY shuhn) begins. Platelets begin to stick to the rough surfaces created by damaged tissue, such as the tissue around a cut or a broken blood vessel. Some platelets break and release chemicals that cause nearby blood vessels to constrict, thus reducing bleeding. They also release an enzyme called *thromboplastin,* which triggers a process involving proteins in the plasma including *prothrombin* and *fibrinogen*. In the presence of calcium, thromboplastin causes prothrombin to change into *thrombin*. Thrombin is an enzyme that promotes the conversion of fibrinogen into *fibrin*. Fibrin forms strong, elastic protein threads into a mesh that traps blood cells and platelets around the edges of the injury. The result is a blood clot. Within minutes the clot begins to shrink, pulling together the injured ends of skin and forming a *scab*.

Like red blood cells, platelets lack nuclei and are formed in red bone marrow. Platelets are about one-third the size of red cells and number about 1 to every 20 red cells. Their life span is about 7 to 11 days.

43.2 Blood Types

Occasionally an injury or a disorder is so serious that a person must receive blood from another person. *A blood transfer, or transfusion, can succeed only if blood of the recipient and donor match.* Among the factors that must be considered in matching blood is **blood type.** Blood type is determined by the presence of an antigen on red blood cells. An **antigen** is any molecule that stimulates an organism to produce antibodies. An **antibody** is a protein that attacks, or neutralizes, the antigen that triggered its production. Microorganisms, such as bacteria and viruses, are antigenic. The antigens that result in blood types, however, are inherited. The most familiar blood-typing system is the ABO system. Under this system, the primary blood types are A, B, AB, and O. Type A blood has antigen A, and type B has antigen B. Type AB has both antigen A and antigen B, while type O has neither of these antigens.

Types A, B, and O also contain antibodies. Type A blood contains anti-B antibodies. Type B blood contains anti-A antibodies. Type O has both anti-A and anti-B antibodies. Type AB

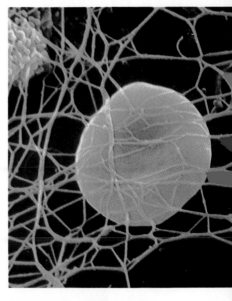

Figure 43–3. This scanning electron microscope image shows a red blood cell enmeshed in fibrin, magnified 7,700 times.

Figure 43–4. If a transfusion of type B blood were given to a person with type O blood, antibodies in the recipient's blood would attack B antigens in the donor's blood. The result would be clotting.

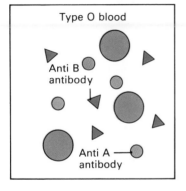

Type O blood

Anti B antibody

Anti A antibody

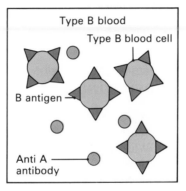

Type B blood

Type B blood cell

B antigen

Anti A antibody

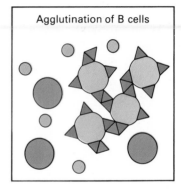

Agglutination of B cells

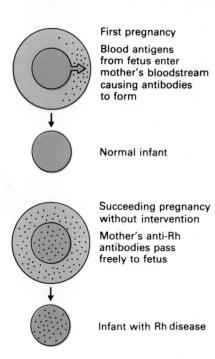

First pregnancy

Blood antigens from fetus enter mother's bloodstream causing antibodies to form

Normal infant

Succeeding pregnancy without intervention

Mother's anti-Rh antibodies pass freely to fetus

Infant with Rh disease

Succeeding pregnancy with treatment

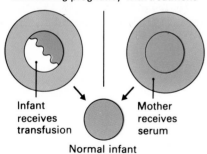

Infant receives transfusion

Mother receives serum

Normal infant

Figure 43–5. This diagram summarizes the facts regarding Rh disease.

has neither of the antibodies. If two blood types are mixed during transfusion, antibodies may cause *agglutination,* or clumping, of red cells. Agglutination results, for example, if type A blood is mixed with type B blood. In this case, the anti-B antibodies in the type A blood will attack the antigens in the type B blood.

When a patient needs blood, doctors must first determine what the patient's blood type is. Modern medical practice, however, requires that more than just blood type be analyzed. Other factors in donated blood must also be compatible for a transfusion to be successful.

43.3 Rh Factor

Another type of antigen is the **Rh factor,** which is present in about 85 percent of all people in the United States. These people are said to be Rh-positive (Rh^+). People whose blood does not contain the Rh factor are Rh-negative (Rh^-). The Rh factor can cause a problem to children of an Rh^- woman. If the father is Rh^+, the child could have Rh^+ blood. If some of the Rh^+ blood antigens from the unborn child enter the mother's bloodstream, her body produces anti-Rh antibodies. During any succeeding pregnancy, the mother's anti-Rh antibodies may pass into the child's bloodstream. If the unborn child is Rh^+, the antibodies can cause clumping and destruction of the child's red blood cells, a condition known as *erythroblastosis fetalis,* or *Rh disease.* The Rh factor problem can be a critical one. The result may be anemia, brain damage, or even death.

Two procedures are used to overcome the problem. The Rh^- mother may be given a serum containing anti-Rh antibodies within 72 hours after the birth of her first Rh^+ baby. The serum destroys the child's Rh^+ blood antigens that have entered her system before her body can develop anti-Rh antibodies. The second procedure treats the child. If the unborn child of a later pregnancy has already developed Rh disease, a blood transfusion can be given to the unborn child to remove the antibodies from its blood.

Reviewing the Section

1. What are four functions of blood?
2. What are the chief components of plasma?
3. What is the function of red blood cells?
4. How do white blood cells protect the body?
5. What causes Rh disease?

Circulation Through the Body

Section Objectives

- *Trace* the path of a drop of blood through the heart.
- *Describe* how the heartbeat is controlled.
- *Name* the three major types of blood vessels.
- *Explain* what high blood pressure is and how it can affect the body.
- *Distinguish* among the various subsystems of the circulatory system.

Blood could not meet the body's needs if it did not flow. The circulatory system, therefore, includes a pump that forces blood to move and tubes through which it flows smoothly.

43.4 The Heart

The **heart** is a muscular organ that pumps blood to all parts of the body. When a person is resting, the heart pumps about 5 L (5.3 qt.) of blood each minute. When a person is exercising strenuously, however, the heart may have to pump up to seven times that amount.

Structure of the Heart   The heart is a fist-sized organ composed chiefly of cardiac muscle, nervous tissue, and connective tissue. It lies between the lungs and behind the breastbone. An average adult human heart weighs about 350–450 g (0.5–1 lb.). A tough protective sac called the **pericardium** (pehr uh KAHR dee uhm) surrounds the heart. The pericardium secretes a slippery liquid that acts as a lubricant, allowing the heart to move smoothly within the sac.

Figure 43–6. The heart is divided into a left and a right pump by a septum that runs down the middle of the heart. The two pumps beat as one.

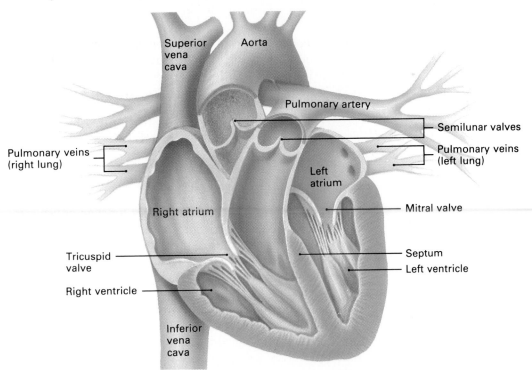

Superior vena cava

Aorta

Pulmonary artery

Semilunar valves

Pulmonary veins (right lung)

Pulmonary veins (left lung)

Left atrium

Right atrium

Mitral valve

Tricuspid valve

Septum

Left ventricle

Right ventricle

Inferior vena cava

Figure 43–7. In the circulatory system, major arteries and veins are connected to smaller and smaller blood vessels. The smallest are microscopic in size and are called capillaries.

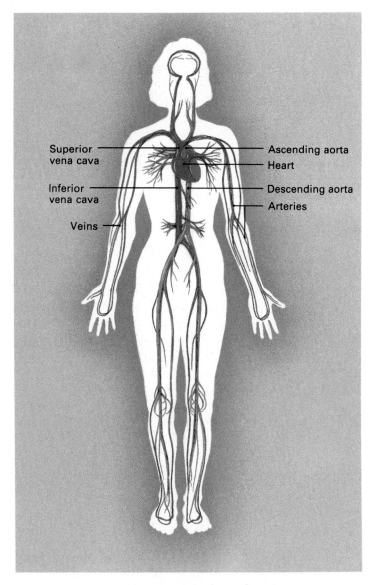

Superior vena cava
Inferior vena cava
Veins
Ascending aorta
Heart
Descending aorta
Arteries

The right and left sides of the heart function as two completely separate pumps. An interior wall called the *septum* separates the two sides of the heart. Each side has an upper section called the **atrium** (AY tree uhm) and a lower section called the **ventricle** (VEHN trih kuhl).

As you can see in Figure 43–6 on page 659, the atrium and ventricle on each side are separated by a one-way valve. The valve on the right side is called the *tricuspid valve*. The valve on the left side is the *bicuspid*, or *mitral*, *valve*. Another set of one-way valves, called the *semilunar valves*, separate the ventricles from the large blood vessels into which blood is pumped out of the heart. All the valves prevent blood from flowing backward.

Circulation Through the Heart

As shown in Figure 43–8, blood enters the right atrium through two large veins. A **vein** is a blood vessel that carries blood to the heart. The **superior vena cava** (VEE nuh KAY vuh) brings blood from the upper regions of the body; the **inferior vena cava** brings blood from the lower body. The blood entering the heart through these veins is dark red because it is *deoxygenated*—that is, without oxygen.

About 70 percent of the blood in the right atrium flows directly into the right ventricle. The remaining blood is forced into the ventricle by a mild contraction of the atrium. When the right ventricle contracts, the tricuspid valve closes, and blood is forced into the **pulmonary artery.** An **artery** is a blood vessel that carries blood away from the heart. The semilunar valve closes. The blood travels from the pulmonary artery into its two branches, one to each lung. In the lungs, the exchange of carbon dioxide from the deoxygenated blood and oxygen from freshly inhaled air takes place. The blood, now bright red and saturated with oxygen, enters the left atrium via the **pulmonary veins.** The path of blood from heart to lungs and back is called the *pulmonary circulation.*

The path of blood through the left side of the heart is similar to that through the right side. The contraction of the left ventricle is very powerful because it must force blood to the farthest regions of the body. Blood rushes from the left ventricle into the **aorta,** the largest artery. From the aorta, blood flows to all parts of the body through a system of increasingly smaller arteries.

Figure 43–8. Blood enters the two atria of the heart simultaneously from two different areas (left). It is then pumped to the two ventricles (center), before leaving the heart (right).

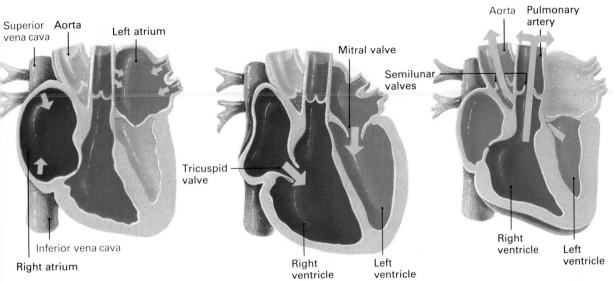

Superior vena cava · Aorta · Left atrium · Mitral valve · Semilunar valves · Aorta · Pulmonary artery · Tricuspid valve · Inferior vena cava · Right atrium · Right ventricle · Left ventricle · Right ventricle · Left ventricle

The Silent Killer

A major cause of death in this country is a "silent killer." This condition has no symptoms and no cure, but it can be treated successfully. Sufferers may be unaware they have this condition until complications such as stroke, heart attack, and kidney failure occur. Its victims are both young and old. This silent disease is *hypertension,* or *high blood pressure.*

Blood pressure is the amount of force the blood exerts against artery walls. Two numbers are used to register blood pressure. The first number, called the *systolic pressure,* tells how much pressure is exerted when the heart contracts and blood spurts through the arteries. The second number is the *diastolic pressure,* which tells how much pressure is exerted while the heart relaxes. If the blood pressure is 120 over 80 (120/80), the blood is pushing against the artery walls with a pressure of 120 as the heart contracts and 80 as the heart rests. These figures are based upon how high the pressure in the arteries can raise a column of mercury similar to that in a *sphygmomanometer,* the instrument used to measure blood pressure.

Normally, blood pressure fluctuates. It is affected by such things as exercise and anxiety. In people suffering from hypertension, however, the pressure never drops to a safe level. A pressure that remains at 140/90 or more is considered high blood pressure.

High blood pressure can have many dangerous effects. Each time the heart contracts, it pushes against the resistance exerted by artery walls. This resistance registers as the systolic pressure. The higher the resistance, the harder the heart must work. This increased pumping effort can result in heart failure.

As the heart begins to deteriorate, other problems develop. An increasing lack of oxygen to muscle tissue causes muscle fatigue and weakening. Lack of oxygen to the brain affects thinking. High pressure against the delicate walls of small blood vessels may cause them to rupture. A ruptured blood vessel in the brain causes a stroke, the effects of which can range from mild paralysis to death. High blood pressure may also affect the body's ability to get rid of waste products.

High blood pressure can be discovered through a medical checkup and can be treated or controlled through diet and exercise. An estimated 24 million Americans suffer from hypertension, and most do not know they have it.

The Heartbeat The heart is really two separate pumps that operate simultaneously at about 70 contractions—heartbeats—per minute. Blood flows into both atria at the same time, and the atria contract together. Similarly, the ventricles contract together. A ventricular contraction is called *systole* (SIHS tuh lee). Relaxation is called *diastole* (dy AS tuh lee).

The activity within the heart causes the heartbeat, a sound usually described as "lubb dup." The "lubb" sound is related to the closing of the tricuspid and mitral valves. The shorter and higher pitched "dup" comes very shortly thereafter and is related to the closing of the semilunar valves. Certain types of heart disorders can be detected through irregularity in one or both sounds.

What causes the heart to beat regularly without any conscious control? The heart has its own automatic pacemaker. It is a small region of muscle called the **sinoatrial** (sy noh AY tree uhl), or **SA, node** in the back wall of the right atrium. The SA node triggers each heartbeat with an impulse that causes the atria to contract. Within a tenth of a second, the impulse reaches the **atrioventricular** (ay tree oh vehn TRIHK yuh luhr), or **AV, node** at the base of the right atrium. Within milliseconds, the AV node triggers an impulse that causes the ventricles to contract. In a disorder called *fibrillation,* contractions become irregular and rapid. These uncoordinated contractions affect the ventricles, and therefore the pumping of blood to the body.

43.5 Blood Vessels

Blood is carried to all parts of the body through 112,000 km (70,000 mi.) of blood vessels. Different types of vessels vary in size and structure.

Arteries have especially elastic, muscular walls. As Figure 43–10 shows, these walls consist of three layers of tissue. Arteries branch into smaller and smaller arteries until they become tiny vessels called *arterioles*. Arterioles continue to decrease in diameter until they branch into **capillaries**—vessels so narrow that red blood cells must pass through them in single file.

Capillaries are the smallest and most numerous blood vessels in the body. Every body cell is within 0.13 mm (0.005 in.) of one or more capillaries. *Although other blood vessels transport nutrients and waste products, the actual exchange of these products between blood cells and body cells takes place through capillary walls.* Capillary walls are only one cell thick. As a result, diffusion of nutrient molecules, waste products, and gases can take place quickly. Capillary walls also allow plasma to filter out of the blood to become tissue fluid.

Deoxygenated blood travels from capillaries into small veins called *venules*. Veins increase in size as they approach the superior vena cava and inferior vena cava. Like artery walls, vein walls consist of three layers of tissue. However, the middle layer is less muscular than that of arteries.

Blood in veins generally must flow against the force of gravity—for example, from the feet to the heart. Two features prevent blood from flowing backward, away from the heart. The first is location. Many veins run through skeletal muscles.

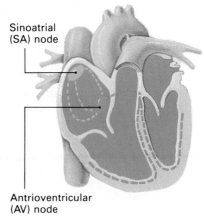

Sinoatrial
(SA) node

Antrioventricular
(AV) node

Figure 43–9. The sinoatrial node triggers a beat in the two atria. The atrioventricular node transmits the beat to the ventricles.

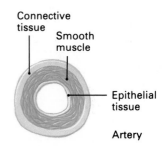

Connective
tissue

Smooth
muscle

Epithelial
tissue

Artery

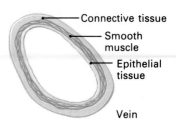

Connective tissue

Smooth
muscle

Epithelial
tissue

Vein

Figure 43–10. An artery (top) relies on pumping to move blood and so needs a thick muscle layer. A vein (bottom) relies less on pumping and needs less muscle tissue.

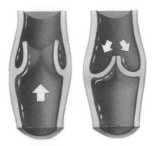

Figure 43–11. Valves in veins open under the pressure of blood flow. When pressure is relaxed, they close in a manner that prevents backflow.

As the muscles contract, the veins are squeezed and blood is pushed along. The second is a series of valves that keeps the blood from moving backward.

43.6 Circulatory Subsystems

Within the circulatory system are several subsystems. The pathway of blood from the heart to the lungs and back to the heart is called the *pulmonary circulation*. The pathway of blood from the heart to other parts of the body and back to the heart is called the *systemic circulation*.

Systemic circulation also has subsystems. *Coronary circulation*, for example, supplies the heart itself with blood. The *left* and *right coronary arteries* branch off the aorta and provide the heart continuously with oxygen and nutrients. The blood returns to the right atrium by way of a large vein called the *coronary sinus*.

Heart tissue must be nourished continuously. When something prevents blood from reaching the cardiac muscle, the lack of oxygen causes the muscle cells to die. This condition, known as a heart attack, is the leading cause of death in the United States. A heart attack may result from a blood clot that blocks a blood vessel or from a gradual buildup of cholesterol, fibrin, and other cellular material inside the blood vessels. This buildup, called *atherosclerosis*, narrows the openings inside blood vessels.

Another part of systemic circulation is *renal circulation*, which carries blood to and from the kidneys. The *left* and *right renal arteries* branch from the aorta and enter the kidneys. Nitrogenous waste products filter out of the bloodstream into renal

Figure 43–12. Subdivisions of the circulatory system transport blood to and from the lungs, heart, liver, intestines, and kidneys.

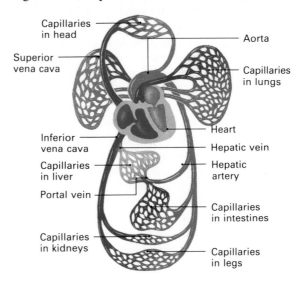

Capillaries in head

Aorta

Superior vena cava

Capillaries in lungs

Inferior vena cava

Heart

Hepatic vein

Capillaries in liver

Hepatic artery

Portal vein

Capillaries in intestines

Capillaries in kidneys

Capillaries in legs

capillaries. The blood then travels through renal veins to the inferior vena cava.

Hepatic portal circulation, a third part of systemic circulation, involves the digestive tract and liver. *Mesenteric arteries* carry blood from the aorta to the intestines, where water and molecules from digested food enter the capillaries. The blood, which is now enriched with nutrients, travels via the *hepatic portal vein* to the liver, where some nutrients are stored as glycogen. The *hepatic artery* supplies the liver with oxygenated blood. Blood leaves the liver and reaches the inferior vena cava through *hepatic veins.*

43.7 The Lymphatic System

The *lymphatic system* is part of the body's circulatory system. Body fluids are carried in vessels of the lymphatic system as well as in blood vessels. Together the blood vessels and lymph vessels form the body's *vascular,* or vessel, *system.*

Lymph originates from blood plasma and tissue fluid that surrounds all body cells. It provides the medium through which diffusion of nutrients and gases occurs. Each day slightly more fluid filters out of the capillaries than is reabsorbed. Lymph and the valuable proteins it contains are collected in *lymph capillaries,* tiny vessels in almost every organ. The largest of the lymph vessels are *lymph ducts,* which empty into the two *subclavian veins* located in the neck. In this way, fluid and proteins are returned to the bloodstream.

The lymph system also helps protect the body against infection. Tiny bean-shaped organs called **lymph nodes** concentrated in the armpits, neck, and groin filter out such foreign matter as bacteria and viruses from lymph. Lymph tissue is also located in the *tonsils, adenoids, spleen, thymus gland,* digestive tract, and bone marrow. Lymph tissue also produces a type of white blood cell that helps the body fight disease.

If the lymphatic system malfunctions, excessive amounts of fluid collect in the body. This condition is known as *edema.* Generally, edema is a symptom of a more serious physical disorder.

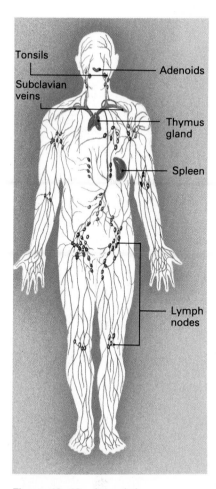

Figure 43–13. Lymph is transported through its own circulatory system. Lymph nodes are concentrated most heavily in the neck, the armpits, and the groin.

Reviewing the Section

1. How is the heartbeat controlled?
2. What is the importance of pulmonary circulation?
3. How is a capillary's structure related to its function?
4. Name the main functions of the lymphatic system.

Defenses Against Disease

- *Distinguish* between non-specific and specific body defenses.
- *Explain* how lymphocytes detect foreign substances.
- *Compare* the functions of B cells and T cells.
- *List* two problems that can occur with the immune system.

Blood and tissue fluid carry nutrients to body cells. These substances are necessary for healthy cells. The blood also carries substances that defend the body against disease.

43.8 Nonspecific Defenses

Some defenses are called *nonspecific defenses* because they operate in the same way against all disease-causing microorganisms. **Among nonspecific defenses are the skin and mucous membranes.** They provide a mechanical barrier against **pathogens,** which are disease-causing agents such as viruses and bacteria. If pathogens do enter the body, a type of white blood cell called a **phagocyte** engulfs and digests them. This process is known as *phagocytosis.* The dead bacteria and white blood cells may become *pus.* The presence of pus indicates an infection.

Virus-infected cells may also release the protein *interferon.* Interferon inactivates attacking viruses by preventing them from reproducing. All viruses stimulate the production of the same type of interferon, and interferon attacks all types of viruses.

43.9 Immune Response

The body also has *specific defenses,* by which it defends itself against specific pathogens. Body defenders constantly circulate in the bloodstream and tissue fluid, tracking down harmful microorganisms and diseased cells. When they locate their prey, they trigger a precisely targeted attack. These defenders are white blood cells called **lymphocytes.** The two main types of lymphocytes are *B cells* and *T cells.* These are complex white blood cells that stop the progression of diseases and infections.

How do B cells and T cells recognize pathogens and foreign substances? Every body cell has molecules on its surface that identify it as "self"—that is, as part of the body. Foreign substances have surface molecules that tag them as "nonself." If a surface molecule contacted by a lymphocyte is a "self" marker, nothing happens, and the lymphocyte moves on. If the molecule is a "nonself" marker, however, the body produces an attack on the foreign substance called an *immune response.* Any molecule that triggers an immune response is an antigen.

When a B cell identifies a "nonself" marker, it carries the pattern for that antigen to a lymph node. The B cell may then become a *plasma cell.* Plasma cells manufacture proteins that exactly fit the "nonself" surface marker of the antigen. These

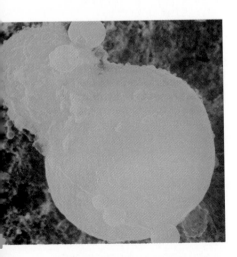

Figure 43–14. This scanning electron photomicrograph shows a small T cell attacking a tumor cell magnified 2,500 times.

proteins are antibodies. Each antibody fits—or combats—just one specific antigen. This type of antibody, called a *circulating antibody,* moves through the body fluids, seeking out the appropriate antigen. When the antigen is located, the antibody hooks on and signals phagocytes to surround and destroy the antigen.

T cells do not produce circulating antibodies. They carry *cellular antibodies* on their surface. The cellular antibodies latch onto an antigen and direct the action of phagocytes. T cells can recognize body cells that have been invaded by cancer and certain viruses. The cancerous cells register as "not quite self," thus allowing T cells to launch a defense.

43.10 Immunity

The body generally requires several days to form antibodies after the first attack by an antigen. Reaction to the first invasion is called the *primary immune response*. Future responses to the same antigen are rapid because of memory cells. **Memory cells** are B cells that carry, or "remember," the antigen pattern.

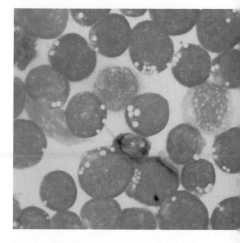

Figure 43–15. B cells like the ones above produce circulating antibodies that move through body fluids in search of a specific antigen.

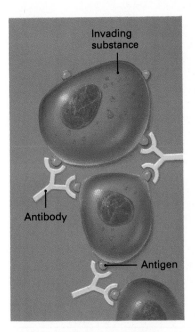

Figure 43–16. Antibodies have two locations at which they can attach to antigens. By attaching to two antigens simultaneously, they cause the antigens to agglutinate.

They produce antibodies immediately if the antigen attacks again. Antibodies produced during such a *secondary immune response* are stronger and last longer than the original antibodies. Memory cells live for years. New memory cells are produced during each response. As a result, the response is faster and stronger each time. This process of warding off disease through antibodies is **immunity.** Immunity prevents a person from getting certain diseases, such as measles or chickenpox, repeatedly.

43.11 Problems with the Immune System

Not all "nonself" markers are harmful. Sometimes the body cannot distinguish between harmful and harmless "nonself" markers. For example, it may fail to recognize certain pollens as harmless. Reactions called *allergies* then occur. Although most allergic reactions are not medically serious, some can be life-threatening. One example is the violent immune response some people have to bee venom.

The body occasionally fails to recognize some body cells as "self" and attacks them as antigens. Such misdirected attacks occur in *autoimmune diseases. Rheumatoid arthritis* is an autoimmune disease affecting joint tissue.

The body may also lose its ability to attack invading microorganisms and diseased cells. This condition is called *immune deficiency*. In its most severe form, immune deficiency is almost always fatal. Its victims suffer from repeated infections and illnesses. Occasionally the deficiency is genetic and is present at birth. More often it develops later in life. *AIDS (acquired immune deficiency syndrome)* is an example of the latter. Scientists suspect that AIDS is caused by a viral attack. However, they do not know how the virus affects the immune response.

Rejection of transplanted organs and tissues is also caused by the body's immune system. Transplanted organs are recognized only as "nonself." Previously, drugs were used to suppress a patient's entire defense system to prevent rejection. The patient then became vulnerable to all diseases. *Cyclosporine*, a recently developed drug, suppresses transplant rejection but does not disrupt other immune functions.

Reviewing the Section

1. Give two examples of nonspecific defenses.
2. How do B cells and T cells differ in function?
3. Describe two problems of the immune system.

Investigation 43: Circulation in Goldfish

Purpose
To locate and observe circulation in the tail of a goldfish

Materials
Petri dish, absorbent cotton, dechlorinated water, two microscope slides, aquarium, goldfish, fish net, compound microscope

Procedure
1. Set up all materials before taking the goldfish out of the aquarium.
2. Place half of the petri dish on your laboratory table. Soak a piece of cotton in dechlorinated water and place the cotton in the petri dish. Place one of the glass slides on the side of the petri dish opposite the cotton.
3. Use a fish net to obtain a goldfish. Following your teacher's instructions on how you are to do this, gently place the head of the goldfish on the cotton in the petri dish. Be sure that the tail is on the glass slide. Place a second piece of wet cotton over the fish being sure that the tail is uncovered. Then place the second slide over the tail.

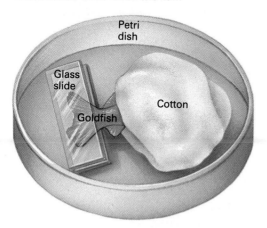

4. Remove the stage clips from the microscope and place them in a safe area. Place the petri dish on the stage of the microscope with the tail over the hole in the stage. Be sure to keep the cotton over the fish cool and moist while making your observations.
5. Focus the microscope on low power and observe the blood flowing through the various types of blood vessels. *Describe the blood cells found in the vessels of the fish's tail.*
6. Draw and label the various kinds of blood vessels. Your drawings should show how the vessels are interconnected. *How can you distinguish arteries from veins in the fish's tail? How can you distinguish capillaries from venules or arterioles in the fish's tail?*
7. Return the goldfish to the aquarium as soon as your observations are complete.

Analyses and Conclusions
1. In which of the blood vessels does the blood appear to throb? Why?
2. In which of the blood vessels does the blood flow at a uniform rate? Why?
3. Describe the movement of blood through a capillary.

Going Further
Study the effects of temperature on blood flow through the fish's tail. Bathe the fish's tail with ice cold water and observe the circulation. Then bathe the fish's tail with warm water and observe the circulation. Describe the effects of these two temperature extremes.

Chapter 43 Review

Summary

The circulatory system consists of the heart, blood and lymph, and a vast system of blood vessels and lymph vessels. It delivers nutrients and oxygen to cells and carries waste materials from cells. The chief carrier in this system is blood, a liquid tissue made of plasma and three types of cells: red blood cells, white blood cells, and platelets.

The heart is a muscular organ that pumps blood throughout the body. It functions as two separate but coordinated pumps, controlled by the sinoatrial node. The right side pumps blood into the pulmonary circulation. The left side pumps blood into the systemic circulation.

The major types of blood vessels are arteries, veins, and capillaries. Arteries carry blood away from the heart. Veins carry blood to the heart. Diffusion of gases and nutrients occurs through capillary walls.

B cells and T cells are two types of lymphocytes that circulate in body fluids. B cells produce circulating antibodies to fight off antigens. T cells carry cellular antibodies and detect cancer and virus-infected cells.

BioTerms

antibody (657)	heart (659)	plasma (655)
antigen (657)	hemoglobin (656)	platelet (656)
aorta (661)	immunity (668)	pulmonary artery (661)
artery (661)	inferior vena cava (661)	pulmonary vein (661)
atrioventricular (AV) node (663)	lymph (665)	red blood cell (656)
atrium (660)	lymph node (665)	Rh factor (658)
blood (655)	lymphocyte (666)	sinoatrial (SA) node (663)
blood type (657)	memory cell (667)	superior vena cava (661)
capillary (663)	pathogen (666)	vein (661)
coagulation (656)	pericardium (659)	ventricle (660)
	phagocyte (666)	white blood cell (656)

BioQuiz   (Write all answers on a separate sheet of paper.)

I. Completion

1. _____ direct the action of B cells and phagocytes.
2. The transfer of blood or blood parts from one person to another person is called _____.
3. The heart is separated into right and left sides by the _____.
4. Impurities are filtered from tissue fluid in the _____.

II. Modified True and False

Mark each statement TRUE or FALSE. If false, change the underlined term to make the statement true.

5. Thromboplastin is the oxygen-carrying molecule in red blood cells.
6. Blood type is determined by the hemoglobin in red blood cells.
7. Clotting could never take place if the body did not have platelets.

III. Multiple Choice

8. The heart receives blood from the body's lower regions via the
   a) pulmonary artery. b) inferior vena cava. c) superior vena cava. d) hepatic portal vein.
9. The pacemaker of the heart is the
   a) coronary sinus. b) memory cell. c) sinoatrial node. d) pericardium.
10. The white blood cells that devour foreign substances are a) phagocytes. b) T cells. c) B cells. d) lymphocytes.
11. Blood vessels with the most elastic walls are a) venules. b) veins. c) capillaries. d) arteries.
12. Antigen A and anti-B antibody are in blood type a) A. b) B. c) AB. d) O.
13. Blood is pumped into the pulmonary artery from the a) right atrium. b) left atrium. c) right ventricle. d) left ventricle.
14. Systemic circulation does not include
    a) coronary circulation. b) hepatic portal circulation. c) renal circulation. d) pulmonary circulation.
15. Circulating antibodies are produced by a) plasma cells. b) phagocytes. c) T cells. d) interferon.

IV. Essay

16. How does tissue fluid differ from plasma?
17. How does immunity occur?
18. What two features permit blood in veins to flow *up* to the heart?
19. What would happen if an Rh-positive woman was pregnant with an Rh-negative child? Explain why.
20. Why is hypertension called the silent killer?

Applying and Extending Concepts

1. High levels of radiation are known to destroy bone marrow. Write a paragraph explaining why a person who has been exposed to such radiation would probably develop anemia.
2. A process called pheresis makes it possible to collect platelets or other blood parts from a donor and then return the rest of the blood to the donor. Contact the Red Cross or a local hospital with a pheresis unit to learn more about the process and the important role of pheresis donors. Report your findings to the class.
3. Vaccines are substances that, when injected into a person, will cause the production of antibodies. Antiserums are injections of antibodies that have been produced by another organism. Immunity induced by vaccines occurs more slowly but lasts longer. Explain why.

Related Readings

Edelson, E., "Cleansing the Blood." *Science 82* 3 (October 1982): 72–77. This article explains the process by which blood components are replaced or treated.

Goodfield, June, "Dr. Coley's Toxins." *Science 84* 4 (April 1984): 68–73. Early experiments into the use of toxins to cure cancer are described.

"How Do Our Bodies Make Blood?" *Science Digest* 90 (March 1982): 103. This interesting article describes the important roles played by body parts such as bone marrow, the kidneys, and the spleen in the human circulatory system.

BIOTECH

Clearing Blocked Arteries

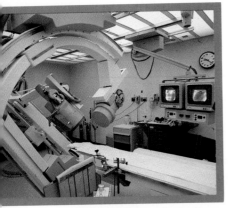

Coronary diagnostic device

Healthy arteries supply oxygen to the heart (top right), but may become blocked (bottom right) and cause a heart attack

Heart transplants and artificial heart implants have repeatedly made front-page headlines. Clearly, medical science has made great progress in the treatment of heart disease. In addition to replacement techniques, exciting developments in technology offer effective treatment before a heart attack occurs.

Coronary artery disease is a condition of reduced supply of oxygen to the heart muscle. The oxygen shortage occurs when cholesterol-containing scar tissue builds up along arterial walls and reduces or prevents blood flow. In severe cases, the oxygen supply is inadequate too long. Then the segment of heart muscle dependent on a blocked artery may die, resulting in a heart attack.

Most cases of coronary artery disease are adequately managed with medication. Today more attempts are being made to treat these conditions through technological innovations.

To help prevent heart attacks, physicians are developing treatments for coronary artery disease. The physicians attempt to improve blood flow and oxygen supply before the heart muscle

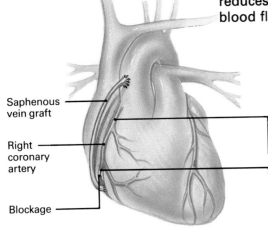

Saphenous vein graft

Right coronary artery

Blockage

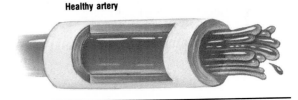

Healthy artery

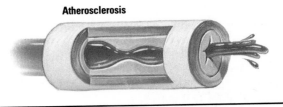

Atherosclerosis

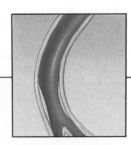

Blocked artery (left) is cleared with balloon catheter (center) to improve blood flow (right)

receives severe damage from a heart attack. An important instrument for such treatment is a slender plastic tube called a *catheter.* X-ray dye can be injected through a catheter to detect the location of artery blockage. Once the blockage is located, one of several methods of treatment can be used. One widely-used surgery involves using a section of leg vein or a small artery from under the breast bone. This vessel is grafted beyond the obstruction so that blood flow can bypass the

New technology is helping to treat and prevent heart attacks.

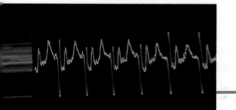

blockage. This technique is called *coronary bypass surgery.*

Another treatment uses a modified catheter to carry a tiny, high pressure balloon into the blocked artery itself. As the catheter is guided through the artery, physicians monitor its position on an X-ray television monitor. When the catheter reaches the blocked artery, the balloon is inflated to a fixed diameter at the precise location of the blockage. The inflated balloon dilates, or stretches, the diameter of the artery and allows more blood to flow through.

Laser technology offers the possibility of yet another type of catheter treatment. Researchers are developing a technique in which miniature lasers will be inserted through catheters in much the same way as balloon catheters. The lasers may be used at the

Computer image of heart

point of obstruction to destroy the cholesterol-containing scar tissue. Then a suction device will remove the resulting gases and debris. One of the major challenges in perfecting this technology is to strictly control the energy level, direction, and penetration of the laser beam, so that it does not damage the artery.

The frontiers of research on coronary artery disease are rapidly expanding. Heart specialists continue to develop new, more effective treatment for the disease, so that one day it may no longer be the leading cause of death in the United States.

44

Respiration and Excretion

Athletes breathing

Introduction

All living things get their energy through a biochemical process that takes place within their cells. Some organisms can produce energy without oxygen. Humans, however, require oxygen. Human bodies are adapted to carry oxygen from the atmosphere to body cells and to eliminate the waste products resulting from the energy-producing process.

The *respiratory system* and the *excretory system* are involved in these functions. Through the respiratory system, oxygen is inhaled and diffused into the blood. In addition, carbon dioxide is diffused from the blood into the lungs and is exhaled. Carbon dioxide is the chief waste product of cellular activity. However, there are other cellular wastes. Those wastes are eliminated by the excretory system.

The Respiratory System

The respiratory system consists of the organs of breathing. However, breathing is only one part of respiration. **Respiration** is the process by which the body takes in oxygen, uses it to produce energy, and then eliminates some waste products of the cellular activity. Three subprocesses are involved in respiration. They are *external respiration, internal respiration,* and *cellular respiration*. In external respiration, or breathing, the body exchanges gases between the atmosphere and the blood. Internal respiration is the diffusion of gases between the blood or tissue fluid and body cells. Cellular respiration is the process by which cells break down glucose molecules in the presence of oxygen to form the energy molecule ATP.

Section Objectives

- *Name* the parts of the respiratory system.
- *Describe* the structure and function of alveoli.
- *Trace* the path of oxygen from the atmosphere to a body cell.
- *List* three ways in which the human body adjusts to high-altitude living.
- *Explain* the operation of the diaphragm and intercostal muscles.

44.1 The Lungs and Breathing

The major breathing organs are two **lungs,** located in the thoracic cavity. The lungs are spongy, cone-shaped, saclike organs. Each lung weighs about 600 g (1.3 lb.). The right lung has three main divisions, or *lobes,* and is slightly larger than the left lung, which has two lobes. Both lungs are encased in a tough membrane that also lines the thoracic cavity. This double membrane, the **pleura,** secretes a lubricating fluid that allows the lungs to move smoothly. Inflammation of the pleura can lead to fluid buildup in the thoracic cavity. This condition is called *pleurisy*.

Breathing begins when the **diaphragm,** the dome-shaped muscle below the chest cavity, contracts and moves downward. The *intercostal muscles* between the ribs also contract, causing

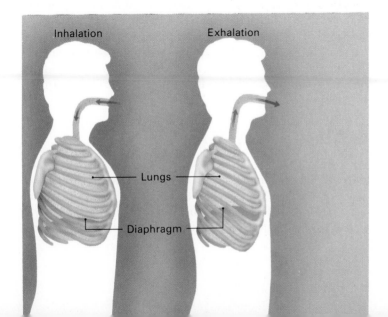

Inhalation

Exhalation

Lungs

Diaphragm

Figure 44–1. The diaphragm and intercostal muscles control the movements of the rib cage and, therefore, breathing.

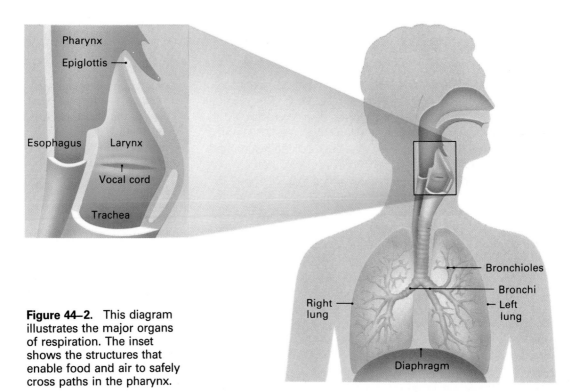

Pharynx

Epiglottis —

Esophagus

Larynx

Vocal cord

Trachea

Bronchioles

Right — lung

Bronchi

Left — lung

Diaphragm

Figure 44–2. This diagram illustrates the major organs of respiration. The inset shows the structures that enable food and air to safely cross paths in the pharynx.

Q/A

Q: *What is a "collapsed" lung?*

A: Air pressure in the chest cavity is normally less than that outside the body. If the pressure increases because air flows into the chest cavity, perhaps because of a puncture wound in the chest wall, the lungs will not fill with air. This condition is called a collapsed lung.

the rib cage to move up and out. Together, these muscle contractions cause the chest cavity to enlarge. When the chest expands, the air pressure in the chest cavity drops. Air pressure outside the body is then greater than that inside the chest cavity. Air then flows into the lungs from outside the body, equalizing the pressure. This part of the breathing process is called *inspiration* or **inhalation.**

When the air pressure has been equalized, it causes the diaphragm and intercostal muscles to relax and return to their normal positions. This in turn reduces the size of the chest cavity. As the size of the chest decreases, the air pressure inside the chest cavity gradually becomes greater than the air pressure outside the body. Air then leaves the lungs, again equalizing the pressure. This part of the breathing process is called *expiration* or **exhalation.**

44.2 The Pathway of Air

Air enters the body through two openings in the nose called **nostrils.** From there the air flows into the **nasal cavities,** two spaces in the nose. The cavities are separated by a cartilage and bone partition called the *septum.* The cavities are lined with mucous tissue that contains many blood vessels. The mucous tissue warms and moistens the incoming air. Moisture must be present for diffusion of gases to take place within the lungs. Cilia and hairs also line the cavities and filter foreign particles from the air. The cilia move constantly, carrying these particles outward toward the nostrils.

Air travels from the nasal cavities into the back side of the *pharynx,* a tube at the rear of the nasal cavities and mouth. The pharynx is a common passageway for both food and air. While air must get into the cartilage-ringed **trachea,** or windpipe, at the front of pharynx, food must get to the esophagus at the back side of the pharynx. Therefore, food and air cross each other's paths. If food entered the air passageway, the person would choke. To ensure that food does not enter the air passageways, the body makes involuntary adjustments. During the process of swallowing, a flap of tissue called the *epiglottis* closes over the *glottis,* or the upper part of the trachea. At the same time, the *soft palate* closes off the nasal cavities. During inhalation, the glottis is open to allow air to enter the trachea.

At the top of the trachea is the *larynx,* or voice box. Two ligaments called **vocal cords** are stretched across the larynx. The larynx is called the voice box because sound is produced when air is forced between the cords. The amount of tension in the cords determines the pitch of a sound. Nine cartilage rings connected by ligaments hold the mucus-lined larynx open during inhalation and against the pressure from food passing through the adjacent esophagus. The largest of the cartilage rings appears as the *Adam's apple* in the throat.

The trachea descends to a point near the middle of the breastbone. There it divides into two branches called **bronchi** (BRAHN ky). Bronchi walls consist of muscle supported by cartilage and are lined with mucus and cilia. The bronchi reach deep into the lungs, subdividing about 25 times into smaller and smaller passageways. The first 10 subdivisions are called *secondary bronchi.* The remaining subdivisions are microscopic-sized tubes called **bronchioles** (BRAHNG kee ohlz). Bronchiole walls consist of smooth muscle and are lined with mucus and cilia. The continuous beating of the cilia in the bronchi and bronchioles carries foreign particles and excess mucus into the pharynx. This material may then be expelled by being swallowed or coughed out.

The smallest bronchioles branch into tiny ducts, which end in clusters of tiny bulges. These bulges are air sacs called **alveoli** (al VEE uh lih). Each lung has more than 300 million alveoli. Each alveolus measures from 0.1 to 0.2 mm (0.004 to 0.008 in.) in diameter. The total surface area provided by the alveoli is estimated at about 70 m^2 (750 sq. ft.).

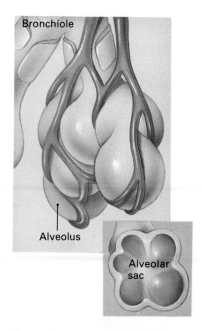

Figure 44–3. Inhaled air travels through the respiratory system to the alveoli, air sacs at the end of tiny bronchioles. Each alveolus is surrounded by blood vessels and is hollow inside (below).

Figure 44–4. The cilia lining the surface of the trachea can be seen magnified 8,000 times in this photomicrograph.

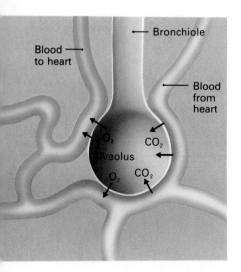

Blood to heart

Bronchiole

Blood from heart

O_2 CO_2

Alveolus

O_2 CO_2

Figure 44–5. Exchange of gases occurs in the alveoli.

44.3 Exchange of Gases

Alveoli are completely surrounded by capillaries. *The actual exchange of gases occurs when oxygen in the air of the alveoli diffuses into the blood in the capillaries. In turn, the carbon dioxide in the blood diffuses into the air of the alveoli.* The epithelial tissue forming the walls of both the alveoli and capillaries is only one cell thick. Together, the walls of an alveolus and an adjacent capillary measure only 0.0004 mm (0.00001576 in.). The oxygen in inhaled air dissolves in the mucus on the lining of the alveoli.

In the blood, most oxygen combines with hemoglobin to form *oxyhemoglobin.* Oxygen from the oxyhemoglobin diffuses into body cells and is used in **metabolism,** the chemical and physical activities within cells. Metabolism includes the building up and breaking down of complex molecules and the releasing of energy during the breakdown. As a result of metabolism, oxygen concentration in the body cells is low, but carbon dioxide concentration is high.

Carbon dioxide, a metabolic byproduct, diffuses from body cells into the blood. Carbon dioxide is transported in the blood in three ways. About 5 percent dissolves in the plasma. About 25 percent enters the red blood cells and combines with hemoglobin. With help from a special enzyme, the remainder—or about 70 percent—combines with water in the red blood cells to form carbonic acid:

$$CO_2 + H_2O \rightarrow H_2CO_3$$
(carbon dioxide)   (water)   (carbonic acid)

Almost immediately, carbonic acid separates into hydrogen ions (H^+), which combine with hemoglobin, and bicarbonate ions (HCO_3^-), which diffuse into the plasma.

$$H_2CO_3 \rightarrow H^+ + HCO_3^-$$
(carbonic acid)   (hydrogen ion)   (bicarbonate ion)

As a result of this chemical process, most carbon dioxide is transported in the plasma as bicarbonate ions.

When blood reaches the lungs, chemical reactions occur that reverse the process, releasing carbon dioxide:

$$H^+ + HCO_3^- \rightarrow H_2CO_3$$
(hydrogen ion)   (bicarbonate ion)   (carbonic acid)
$$\rightarrow CO_2 + H_2O$$
(carbon dioxide)   (water)

The carbon dioxide diffuses from the blood into the lungs. The carbon dioxide is exhaled along with water vapor.

Highlight on Careers: Respiratory Therapist

General Description

A respiratory therapist, or inhalation therapist, gives respiratory treatment and life support aid to patients with breathing problems. These patients may be suffering from asthma, bronchitis, emphysema, or pneumonia. They may be victims of such disasters as fires or automobile accidents, or they may be suffering from complications of surgery. Respiratory therapists work with such people, helping to restore their normal breathing functions and their chance for a physically normal, healthy life.

The therapist may do all or part of the setting up and operating of devices that help patients inhale and exhale. As a patient uses these devices, the therapist monitors the patient's vital signs to see that the treatment is proceeding properly. At the same time, the therapist checks to make sure that all equipment is operating properly.

If the patient does not respond well to treatment, the therapist consults a physician promptly. Under a physician's direction, a respiratory therapist may assist the patient with breathing exercises that are part of the recovery program.

Most respiratory therapists work in hospitals or nursing homes. A small percentage work in rehabilitation institutes, education, or research.

Career Requirements

Many hospitals require certification as a registered respiratory therapist (RRT). Prerequisites for certification include graduation

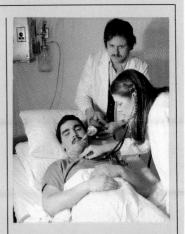

from a two-year college program in respiratory therapy and passing an examination. Helpful personal qualities include the ability to operate mechanical equipment and to work with patients who are often extremely ill.

For Additional Information
American Association for
 Respiratory Therapy
1720 Regal Row
Dallas, TX 75235

44.4 Regulation of Breathing

What determines when and how deeply the body should breathe? Many factors influence the control of breathing, including carbon dioxide and oxygen levels in the blood. *The level of carbon dioxide in the blood plays a vital role in regulating breathing.* Carbon dioxide affects blood acidity. Certain nerve cells are sensitive to changes in blood acidity. These nerves send messages to the *breathing center* at the base of the brain. When the carbon dioxide level in the blood is high, the messages cause the breathing center to trigger a speedup in

▶ Effects of High Altitude

People who move from low altitudes to high altitudes may suffer temporarily from excessive tiredness and sleepiness. The cause of their fatigue is generally attributed to "thin air." More accurately stated, they are suffering from a lack of oxygen.

The higher the altitude, the less oxygen in the air. At sea level, hemoglobin is 97 percent saturated with oxygen. At an altitude of 3,048 m (10,000 ft.), hemoglobin is only 90 percent saturated, and at 6,096 m (20,000 ft.), only 70 percent saturated. The human body begins to feel the effects of reduced oxygen at an altitude of about 1,800 m (1.1 mi.).

Studies of people who normally live at high altitudes provide information on the long-range effects of conditions in which levels of oxygen are low. For example, the Quechua Indians of the Andes, who live above 3,600 m (more than 2 mi.), have developed very large chest and lung capacities. These people also differ in blood composition; they have a higher concentration of red blood cells and hemoglobin than do people who normally live at sea level.

This Indian group also has a higher breathing rate, and the capillaries in their lungs have a greater diameter. Their hearts are much larger, particularly on the right side, and the blood pressure in the lungs is greater than it is in any other part of the circulatory system. However, the heartbeat of the Quechua Indians is slower than that of people living at sea level.

Judging from these facts, scientists theorize that a larger volume of blood is pumped with each beat of the heart in high-altitude dwellers.

breathing rate. Conversely, a low carbon dioxide level reduces the breathing rate. Other messages reach the breathing center from *stretch receptors* in the lungs. When the lungs expand sufficiently, the stretch receptors send messages to the breathing center. The breathing center then sends messages that make the muscles relax. Stretch receptors thus operate as another kind of breathing control mechanism.

Reviewing the Section

1. How do internal and external respiration differ?
2. What role do the diaphragm and intercostal muscles play in breathing?
3. What purposes do mucous tissue and cilia serve in the respiratory system?
4. How is carbon dioxide transported in the blood?
5. How does the exchange of gases occur in the lungs?
6. How is breathing controlled by the body?

The Excretory System

Respiration rids the body of water and carbon dioxide, a waste product of metabolism. Other metabolic wastes, especially nitrogen compounds, are eliminated from the body through the process of **excretion.**

Nitrogen compounds in the form of ammonia are released as the body breaks down excess amino acids. In concentrated form, ammonia is a poison. The body has two processes for making the ammonia less toxic. In one, the ammonia is mixed with great quantities of water. In the other, the ammonia is changed to a less harmful form.

The liver, operating as an excretory organ, combines the ammonia with carbon dioxide to form a less toxic compound called *urea.* Urea enters the blood and circulates throughout the body. Some urea is excreted through the skin in the form of perspiration, which is a mixture of water, minerals, and urea. Most of the urea the body produces, however, is eliminated by the excretory system.

44.5 The Kidneys

The **kidneys** are the major organs of the excretory system. They are two bean-shaped organs, each about 11 cm (4 1/2 in.) long, 6 cm (2 1/2 in.) wide, and 2.5 cm (1 in.) thick. They are located on either side of the spine in back of the abdominal cavity. Their combined weight is less than 0.5 kg (1 lb.). The kidneys are held in position by tough connective tissue and protected by a layer of fatty tissue.

Figure 44–6. The excretory system includes the kidneys, ureters, and urinary bladder (left). Each kidney is made up of three layers (right). The process of filtering wastes from the blood occurs in structures straddling the cortex and medulla.

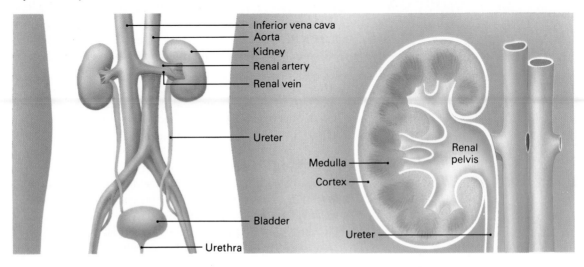

Inferior vena cava
Aorta
Kidney
Renal artery
Renal vein
Ureter
Bladder
Urethra
Medulla
Cortex
Renal pelvis
Ureter

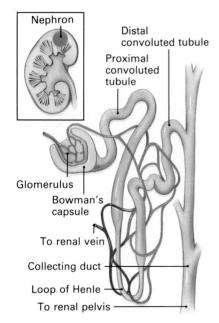

Nephron

Distal convoluted tubule

Proximal convoluted tubule

Glomerulus

Bowman's capsule

To renal vein

Collecting duct

Loop of Henle

To renal pelvis

Figure 44–7. This diagram illustrates the structures that make up the nephron and its surrounding blood vessels.

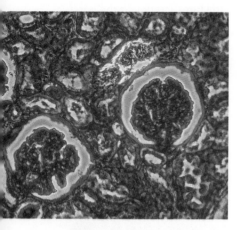

Figure 44–8. This phase contrast microscope image shows several glomeruli magnified 125 times.

The main functions of the kidneys are to remove urea and other wastes, regulate the amount of water in the blood, and adjust the concentrations of various substances in the blood. Thus, the kidneys play a vital role in maintaining *homeostasis,* or balance among elements in the body.

The cross section in Figure 44–6 on page 681 shows the three main sections of a kidney: an outer layer called the **cortex,** a middle layer called the **medulla,** and a central cavity called the **renal pelvis.** Blood enters the kidneys through *renal arteries* and leaves through *renal veins.*

The basic functional unit in each kidney, called a **nephron,** straddles both the cortex and the medulla. Each kidney contains an estimated one million nephrons.

As shown in Figure 44–7, the nephron consists partly of a **glomerulus** (glah MEHR yoo luhs), which is a mass of capillaries that form a tight ball. Each glomerulus is surrounded by a hollow, cup-shaped sac called a **Bowman's capsule.** The glomerulus and Bowman's capsule, located in the cortex, are responsible for filtering wastes from the blood. The Bowman's capsule is the first part of the **renal tubule,** much of which is a coiled tube extending into the medulla and back again. The last part of the renal tubule is called the **collecting duct.** The collecting duct is a straight tube leading to the renal pelvis. Water and mineral composition of the blood are regulated primarily by the coiled part of the renal tubule.

Filtration   The process of removing urea and other wastes from the blood is called **glomerular filtration.** The process starts as blood from the renal arteries flows into the glomerular capillaries. Blood in the glomerulus is under high pressure as it is pumped with great force from the heart into the tiny capillaries. This pressure forces water, urea, glucose, and minerals—a mixture called *filtrate*—into the Bowman's capsule. Red blood cells, white blood cells, and protein molecules do not pass out of the capillaries.

Blood passes through the kidneys at a rate of 1.25 L (1.3 qt.) per minute. In other words, all the blood in the human body passes through the kidneys once every 30 minutes. A total of 170 L (180 qt.) of filtrate is produced daily.

Reabsorption   If the kidneys only filtered the blood, a person would soon die, because along with metabolic wastes, filtration removes glucose, water, and other substances needed for life. However, these materials are returned to the blood in the renal tubule, the second part of the nephron. **Tubular reabsorption** is the process by which these vital materials are returned to the

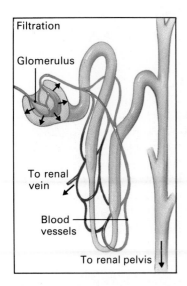
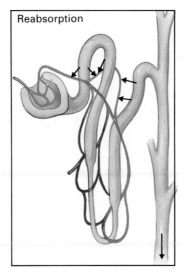
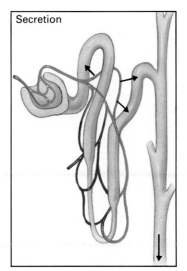

| Filtration | Reabsorption | Secretion |

Glomerulus

To renal vein

Blood vessels

To renal pelvis

blood. Because reabsorption is important to the normal functioning of the body, the kidneys may be more accurately described as organs of regulation rather than excretion.

Reabsorption occurs as materials cross the walls of a renal tubule into a web of surrounding capillaries. Glucose and such chemicals as sodium, potassium, hydrogen, magnesium, and calcium are reabsorbed through active transport. As much as 99 percent of the water in filtrate may return to the blood through osmosis. When blood volume is low, a large amount of water is reabsorbed. When blood volume becomes normal, the rate of osmosis decreases.

In addition to reabsorption, a process called **tubular secretion** occurs in the renal tubule. In the process of tubular secretion, tubule cells actively remove certain substances from the blood and secrete these substances into the filtrate. Penicillin is an example of a substance that is secreted in this way.

Urea, other metabolic wastes, and water that remain in the renal tubule form an amber-colored liquid called **urine.** The urine from several tubules flows into a single collecting duct located in the medulla. In turn, all of the collecting ducts channel urine into the funnel-shaped renal pelvis. From the renal pelvis the urine flows into the urinary system to be removed from the body.

44.6 The Urinary System

From the renal pelvis, urine enters a long, narrow tube called the **ureter.** The ureter from each kidney connects to the **urinary bladder,** a sac of smooth muscle that can hold approximately

Figure 44–9. The kidneys perform three major functions: they remove wastes through filtration (left); return useful materials to the body through reabsorption (center); and remove substances such as penicillin through secretion (right).

Q: *Can a person live with one kidney?*

A: Yes. A person can even live with only part of a kidney. The remaining kidney or kidney part takes over the function of the missing kidney. In other words, the healthy tissue does the work of two kidneys, using what is known as the kidney's reserve capacity.

About 60,000 people in the United States suffer from kidney disease. Each year another 12,000 join that number.

Only about 10 percent of these people will eventually receive a kidney transplant. The rest will be on *hemodialysis,* more commonly known as *dialysis.* A person on dialysis is hooked up to a stationary machine called a hemodialyzer for about five hours every other day. The machine cleans the blood by filtering out waste products through a dialysis membrane. The machine also removes excess fluid

from the body to keep the blood volume at a constant level.

A permanent connective tube that hooks up to the dialyzer is attached to a patient. Once a hookup is made, blood flows from a patient's artery through the tube into the machine and then from the machine into a vein.

A different procedure, called *CAPD (continuous ambulatory peritoneal dialysis),* allows a patient unrestricted movement while undergoing dialysis. With CAPD, plastic tubes are implanted in a patient's abdomen. Through the

tubes, 2 L (2.12 qt.) of a special salt solution are emptied from a plastic bag into the patient's abdominal cavity. The plastic bag can then be rolled up and put in a pocket or purse.

Impurities in the patient's blood diffuse into the salt solution through a membrane lining the abdominal cavity. After about six hours, the patient reconnects the plastic bag and places it below the level of the implanted tubes. By gravity, the waste-containing salt solution drains into the bag. The bag is then discarded. The procedure is done four times a day.

400 to 500 mL (12 to 15 fl. oz.) of urine. When the bladder is full, special nerves in the bladder wall send messages to the brain. The brain's response causes sphincter muscles to relax. This relaxation in turn causes the bladder to contract. Urine is forced from the bladder through the **urethra,** a tube to the outside of the body. The process of expelling urine from the body is called *urination.*

Reviewing the Section

1. What roles do the skin and lungs play in excretion?
2. What is tubular reabsorption?
3. How do the kidneys help maintain homeostasis?
4. Trace the pathway of urine from a renal tubule until it leaves the body.
5. By what two methods are wastes removed from the blood of a person whose kidneys are not functioning? Describe each method.

Investigation 44: Breathing Rate and Volume

Purpose
To determine breathing rate and volume before and after exercise'

Materials
Watch with a second hand; 1,000-ml Florence flask, 100-ml graduated cylinder, glass tubing; two-hole #6 rubber stopper, plastic or rubber tubing

Procedure
1. Make a chart like the one shown.

Measurement of Breathing Rate

Breathing Rate per Minute	Trial 1	Trial 2	Trial 3	Avg.
Resting				
Exercise				

2. Count the number of inhalations you make in one minute while resting. Record your results in the chart.
3. Repeat step 2 two more times. Calculate your average breathing rate and record it in the chart.
4. Do jumping jacks vigorously for two minutes and then count the number of inhalations you make in one minute. Record your results in the chart.
5. Rest quietly until your breathing rate returns to normal. Then repeat step 4 two times and record the results.
6. Rest until your breathing rate is normal.
7. Using the apparatus provided, exhale slowly through the inlet tube.

8. Determine the amount of water in the graduated cylinder and record it on a chart like the one below.

Measurement of Breathing Volume

Volume of Lungs	Trial 1	Trial 2	Trial 3	Avg.
Resting				
Exercise				

9. Replace the water in the flask.
10. Repeat steps 8 and 9 two more times and record the results.
11. Calculate the average displacement volume and record it in the chart.
12. Do 10 jumping jacks and quickly blow into the inlet tube. Measure the amount of water in the graduated cylinder and record this amount on the chart.
13. Rest quietly until your breathing returns to normal. Then repeat step 12.

Analyses and Conclusions
1. How does your normal breathing rate compare to your breathing rate after exercise?
2. How does your average breathing volume before exercise compare to your breathing volume after exercise?
3. What phenomenon accounts for the change in breathing rate and breathing volume?
4. What trigger mechanism causes the increase in respiration?

Going Further
Determine the products of respiration by checking pH, performing a limewater test, and exhaling on a mirror.

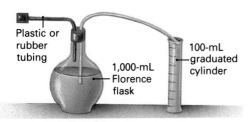

Plastic or rubber tubing

1,000-mL Florence flask

100-mL graduated cylinder

Chapter 44 Review

Summary

Respiration is the process by which gases are exchanged between the atmosphere and body cells and in which cells produce energy from glucose molecules. Air passes through the nostrils, pharynx, larynx, trachea, bronchi, and bronchioles into alveoli. Gases are exchanged between the air in the alveoli and blood in the capillaries surrounding them.

A breathing center in the brain helps control breathing. Inhalation occurs when the diaphragm and intercostal muscles contract, expanding the chest cavity. Air rushes to fill the lungs because air pressure is lower inside the chest than outside the body. When the muscles relax, exhalation occurs.

The major organs of excretion are the kidneys. Their functional units, called nephrons, filter out water, wastes, and other substances from the blood. Most of the water and other substances essential to the body are reabsorbed. Some chemicals are actively removed from the blood by renal tubule cells and secreted into the filtrate. Filtrate that is not reabsorbed forms urine, which collects in the bladder and is then expelled from the body through the urethra.

BioTerms

alveolus (**677**)
Bowman's capsule (**682**)
bronchiole (**677**)
bronchus (**677**)
collecting duct (**682**)
cortex (**682**)
diaphragm (**675**)
excretion (**681**)
exhalation (**676**)
glomerular filtration (**682**)
glomerulus (**682**)

inhalation (**676**)
kidney (**681**)
lung (**675**)
medulla (**682**)
metabolism (**678**)
nasal cavity (**676**)
nephron (**682**)
nostril (**676**)
pleura (**675**)
renal pelvis (**682**)
renal tubule (**682**)

respiration (**675**)
trachea (**677**)
tubular reabsorption (**682**)
tubular secretion (**683**)
ureter (**683**)
urethra (**684**)
urinary bladder (**683**)
urine (**683**)
vocal cord (**677**)

BioQuiz   (Write all answers on a separate sheet of paper.)

I. Completion

1. Soon after the diaphragm and intercostal muscles contract, air pressure in the chest _____ .
2. Formation of urea is an important function of the _____ .
3. Volume and composition of the blood are controlled largely by the process of _____ .

4. _____ in the nasal cavities warms and moistens inhaled air.
5. Increased carbon dioxide in the blood _____ the breathing rate.
6. Nitrogenous wastes in the body come from the breakdown of _____ .
7. The common passageway for food and air is the _____ .

II. Multiple Choice

8. Oxygen passes from the lungs into the blood through walls of the a) bronchi. b) bronchioles. c) alveoli. d) septum.
9. One human adaptation to high-altitude living is a) smaller lung capacity. b) larger lung capacity. c) fewer bronchi. d) additional lung lobes.
10. The functional unit of the kidney is the a) renal tubule. b) glomerulus. c) ureter. d) nephron.
11. Reabsorption takes place in the a) renal tubule. b) renal pelvis. c) glomerulus. d) Bowman's capsule.
12. Drugs are actively removed from the blood to filtrate through a) filtration. b) perspiration. c) tubular secretion. d) respiration.
13. Fluids are forced out of the blood vessels in the nephron during a) tubular reabsorption. b) tubular secretion. c) respiration. d) glomerular filtration.
14. The exchange of oxygen and carbon dioxide between the atmosphere and the blood is a) external respiration. b) internal respiration. c) cellular respiration. d) perspiration.
15. The body eliminates metabolic waste products by means of perspiration, respiration, and a) inhalation. b) secretion. c) urination. d) metabolism.

III. Essay

16. In what way is air pressure related to breathing?
17. How is breathing controlled?
18. How is carbon dioxide carried to the lungs for exhalation?
19. What causes blood to move from the glomerulus to the Bowman's capsule?
20. What respiratory structures are involved in external respiration?

Applying and Extending Concepts

1. Name two ways in which the structure of alveoli helps them perform the function of gas exchange.
2. Use your school or public library to find out about kidney transplants. How successful are transplanted kidneys in taking over the task of excretion? Prepare a report for the class.
3. Firefighters sometimes die as a result of breathing extremely hot air. Use some of the facts you learned in this chapter to suggest the probable cause of death in such cases.
4. Uric acid is a nitrogenous waste resulting from protein metabolism. Trace the pathway of a molecule of uric acid from a cell in the leg where it might be produced to its excretion from the body. Write a paragraph explaining its path and identifying the structures involved.

Related Readings

"Body's Own Filter Material Replaces Kidneys." *Popular Mechanics* 158 (September 1982): 171. This article describes the Continuous Ambulatory Peritoneal System (CAPS) as a replacement for dialysis.

Ingber, D. "Brain Breathing." *Science Digest* 89 (June 1981): 72–75. The author of this interesting article describes breathing as a way to exert control over the involuntary nervous system.

Pollis, R. "Comprehending Kidney Disease." *Science News* 122 (October 2, 1982): 218. This article explains how lower protein consumption can slow renal disease.

45

Nervous Control and Coordination

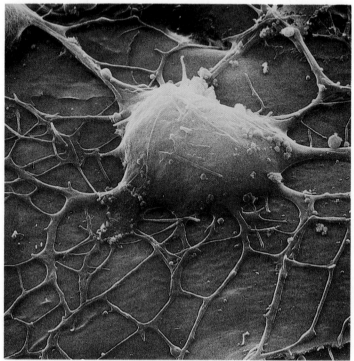
A nerve cell

Introduction

Many living creatures can run faster than human beings can. Some can see better in dim light. Others can hear better, and still others can smell faint odors better at a distance. What, then, has enabled humans to achieve high levels of technology and communication while other species have not? Humans have a highly developed brain that makes it possible to learn, to remember what has been learned, and, most important, to reason.

The brain, however, is just part of the complex *nervous system,* which controls and coordinates essential body functions. The nervous system sends special signals to and receives responses from every organ and tissue of the body. These signals make it possible for you to play the piano, thread a needle, throw a baseball, or just sit and think.

The Nervous System

The nervous system has two main subdivisions. One part, the **central nervous system,** consists of the brain and the spinal cord. The central nervous system receives stimuli from inside and outside the body and then coordinates the body's response. The second part of the nervous system is the **peripheral** (puh RIHF uhr uhl) **nervous system.** It provides the pathways to and from the central nervous system for electrochemical signals called **impulses.**

Three types of body structures are needed for the entire process of picking up stimuli and responding to them. They are *receptors, conductors,* and *effectors.* To understand how these different structures are coordinated, consider what happens when a doorbell rings. First, the ear acts as a receptor that picks up the sound of the ringing bell. A receptor is a cell, group of cells, or organ, that detects a stimulus. The receptor then generates impulses that travel along conductors, or nerve cells. Ultimately, the impulses reach effectors—structures that may react to the original stimulus. In this case, muscles are the effectors. The reaction to the original stimulus is to walk to the door and open it.

45.1 The Neuron

The basic functional unit of the nervous system is the nerve cell, called a **neuron.** Three types of neurons interact in the nervous system. Neurons that receive stimuli and transmit them to the central nervous system are **sensory neurons.** Neurons that carry impulses away from the central nervous system to muscles or glands are **motor neurons.** The third type of neuron, an **interneuron,** links sensory and motor neurons.

Every neuron consists of a **cell body,** which contains the nucleus and cytoplasm, and threadlike extensions of cytoplasm called **nerve fibers.** A neuron has two kinds of nerve fibers. **Dendrites** are fibers that carry impulses from other neurons or receptors toward the cell body. Dendrites are generally short, branched fibers. The second kind of nerve fiber is the **axon,** which carries impulses away from the cell body to other neurons or to effectors. As you can see in Figure 45–1, a neuron has many dendrites but only one axon. Axons are usually longer than dendrites, and any branches axons have exist only at the end of the fiber.

An axon may be wrapped in a fatty insulating layer known as a **myelin sheath.** The sheath is formed by special cells called

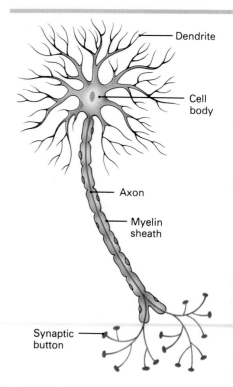

Figure 45–1. Many-branched dendrites carry nerve impulses toward a neuron cell body. A single axon carries impulses away. Axons are relatively long and are branched only at the far end.

Labels: Dendrite, Cell body, Axon, Myelin sheath, Synaptic button

Q: *Do neurons reproduce themselves?*

A: No. The neurons you are born with are all you will ever have. An estimated 50,000 neurons die each day from the time a person is 20 years old. Yet, by the time he or she reaches 50, only about 10 percent have died.

Schwann cells. The sheath supports, insulates, and nourishes the axons. It also helps maintain the chemical balance of the axon. Gaps between the Schwann cells, called the **nodes of Ranvier** (rahn vee AY), occur about every 1 mm (0.04 in.) along the myelinated axons.

Neurons are the largest cells in the body. Some neurons may measure almost 2 m (2.2 yd.). Bundles of nerve fibers, containing hundreds or even thousands of axons, form a **nerve.** Within a nerve, each fiber carries a separate impulse, just as each wire inside a telephone cable can carry a separate phone call at the same time.

45.2 How a Nerve Impulse Travels

Impulses travel not only along the length of a nerve cell but also from cell to cell. Within a neuron the impulse is transmitted electrically. However, chemicals are generally involved in moving the impulse from cell to cell.

The Nerve Impulse   Like all cells, neurons have a certain electrical charge on the inside and outside of their cell membranes. The first part of Figure 45–2 shows the axon of a neuron when the neuron is at its *resting potential*—that is, when it is not carrying an impulse.

The outside of the axon has about 10 times as many sodium (Na^+) ions as the inside. Inside the membrane are negatively charged organic ions and about 30 times as many potassium

Figure 45–2. When an axon is not carrying an impulse, it has many more sodium ions (Na^+) outside than inside (top). An impulse travels along the axon as sodium ions move through the cell membrane into the cell (bottom).

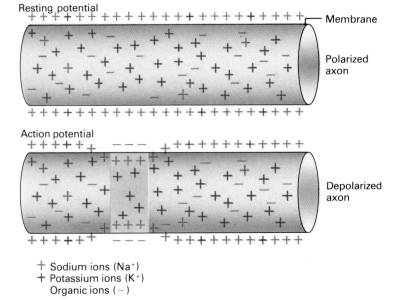

Resting potential

Membrane

Polarized axon

Action potential

Depolarized axon

+ Sodium ions (Na^+)
+ Potassium ions (K^+)
 Organic ions ($-$)

(K^+) ions as outside. The membrane keeps the Na^+ ions outside and the negatively charged organic ions inside. The K^+ ions move in and out of the axon freely. At resting potential, the inside of the cell membrane has a slightly negative charge, and the outside has a slightly positive charge. In this case the cell is said to be *polarized*.

When the nerve fiber is stimulated, its membrane suddenly becomes permeable to Na^+ ions at the place where the stimulation occurs. The negative ions inside the membrane then attract the Na^+ ions. Some Na^+ ions move rapidly to the inside of the cell. The presence of these positively charged ions causes that part of the interior to become more positive than the outside. These electrical changes create an *action potential*, and the neuron is said to be *depolarized*.

The membrane remains permeable to Na^+ ions for only half a millisecond. However, this brief electrical charge is enough to start the action potential moving down the nerve fiber. How does this movement occur? As you see in Figure 45–2, the positively charged ions inside the cell move toward the negatively charged area next to the region of stimulation. The positive ions cause this area to become depolarized and the membrane to become permeable to Na^+. More Na^+ ions then rush inside the membrane, causing that section of the interior to become positive. Again, positively charged ions are attracted to the adjoining negatively charged area, and thus the action potential moves along the nerve fiber.

The rapid change from negative to positive charge within the membrane is an electrical wave called a nerve impulse. *A nerve impulse can be described as the movement of the action potential along a neuron.*

As soon as an impulse passes a section of nerve fiber, the membrane once again becomes permeable only to K^+ ions. The neuron then returns to its resting potential in preparation for the next impulse. The process of returning to resting potential involves an active transport system known as the *sodium-potassium pump*. The sodium-potassium pump carries Na^+ ions to the outside and K^+ ions to the inside of the membrane.

In myelinated axons, the myelin sheath acts as an insulator against electrical impulses. Because of this insulation, the exchange of ions across the membrane takes place only at the nodes of Ranvier, where the sheath is interrupted. This periodic, rather than continuous, exchange results in a leaping of the impulse from node to node. As a result, impulses travel along myelinated axons 50 times faster than they do along unmyelinated axons, sometimes as fast as 100 meters per second (224 miles per hour).

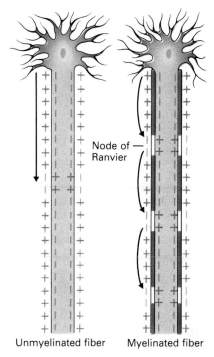

Node of Ranvier

Unmyelinated fiber   Myelinated fiber

Figure 45–3. Nerve impulses travel through every part of an unmyelinated nerve fiber. In a myelinated fiber, the myelin acts as insulation, and the impulse leaps from node to node.

The Synapse   Impulses travel from neuron to neuron, but adjoining neurons generally do not touch one another. Therefore, an impulse must cross from the axon of one neuron to the dendrites of another. This junction is called a **synapse.** An impulse does not ''jump'' across the space, however. In fact, the original impulse ends when it reaches the end of an axon. At that point, however, the impulse causes the release of chemicals that generate new impulses in the next neuron.

Many axon branches terminate in tiny bulblike structures called *terminal buttons,* which contain numerous *synaptic vesicles.* A synaptic vesicle is a tiny sac that holds chemical substances called **neurotransmitters** that stimulate nearby dendrites to start new impulses. A neurotransmitter released into the space, called the *synaptic cleft,* diffuses rapidly to nearby dendrites. There it disturbs the resting potential of the dendrites and so generates new impulses.

An impulse eventually reaches an effector cell, such as a muscle fiber. In this situation, a neurotransmitter is released from motor neurons through *motor end plates,* which are located at the ends of axons near muscle fibers. The neurotransmitter then causes the muscle to contract.

Figure 45–4. When an impulse reaches the end of an axon branch (top, left and right), a neurotransmitter is released and travels toward nearby dendrites, triggering new impulses. Neurotransmitters are also released when an impulse reaches a muscle cell (bottom, left and right), causing the muscle cell to contract.

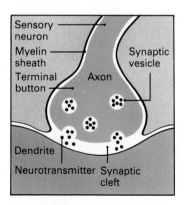

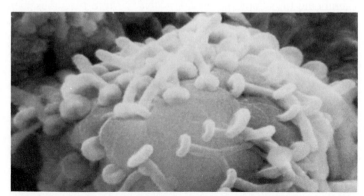

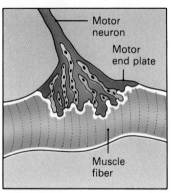

▶ Depression

Depression is a mental disorder that affects an estimated 8 million people in the United States alone. No one knows positively that depression has a biological cause. No biochemical abnormality has yet been discovered in the brain of anyone suffering from depression.

Some researchers, however, are convinced that in many cases neurotransmitters are at fault. Generally, a neurotransmitter is either destroyed by an enzyme or reabsorbed by neurons almost immediately after it enters the synaptic cleft.

Reserpine, a drug used by patients with high blood pressure, causes two neurotransmitters, *serotonin* and *norepinephrine,* to leak into the synaptic cleft. These neurotransmitters are in the part of the brain that controls emotions. Within a few days after reserpine enters a person's body, most of the serotonin and norepinephrine in that part of the brain is destroyed. Patients then often become depressed. As a result, researchers have concluded that a lack of serotonin and norepinephrine in the brain is related to depression.

Drugs now used to combat depression also indicate a possible link between depression and neurotransmitters. One type of drug prevents the reabsorption of serotonin and norepinephrine by neurons. The drug thus causes a buildup of serotonin and norepinephrine in the synaptic cleft. As a result of this buildup, impulses can be generated in the normal manner. The conclusion is that when an abnormally low level of neurotransmitters is brought to a normal level, depression is no longer a problem.

Starting a Nerve Impulse To "fire" a neuron—that is, to get a nerve impulse going in the first place—a stimulus must have a certain level of strength called a *threshold*. If the energy level of a stimulus falls below the threshold, the neuron will not fire. However, a stimulus with an energy level greater than the threshold does not cause a faster or stronger impulse. The neuron either fires or it doesn't, a phenomenon known as the *all-or-none* response. The intensity of a sensation depends on the number of neurons stimulated. After an impulse, the neuron must rest for about one-hundredth of a second. A stimulus, no matter how strong, cannot fire the neuron during this time.

Reviewing the Section

1. How do dendrites and axons differ?
2. Explain how depolarization occurs.
3. What is the function of a neurotransmitter?
4. What is meant by the all-or-none response?

The Central Nervous System

- *Name* the major parts of the brain.
- *Discuss* the role of the cerebrum.
- *List* the functions of the cerebellum and the brain stem.
- *Describe* the structure of the spinal cord.
- *Compare* the functions controlled by the right and left side of the cerebrum.

Impulses travel through the central nervous system, which processes incoming sensory impulses and sends out responding impulses. ***The brain and spinal cord, which make up the central nervous system, each control specific tasks.***

45.3 The Brain

The **brain** is the control center for the human body. Its 100 billion nerve cells not only coordinate and regulate body activities but also enable humans to think. The human brain weighs only about 1.4 kg (3 lb.), but it is the most complex structure on Earth. The surface is *gray matter,* which consists of about 6 million cell bodies and their dendrites packed into each cubic centimeter (0.06 cu. in.). Under the gray matter is *white matter,* formed from myelinated axons.

The brain is composed of three major structures: the *cerebrum* (suh REE bruhm), the *cerebellum* (sehr uh BEHL uhm), and the *brain stem*. Each area seems to control separate functions. However, it is not the independence but rather the interdependence of its parts that makes the brain so effective.

The Cerebrum The **cerebrum** makes up about seven-eighths of the total brain weight. Its two sides, called **cerebral hemispheres,** are joined by a bridge of 200 million nerve fibers. This bridgelike structure between the cerebral hemispheres is known

Figure 45–5. Each hemisphere of the cerebrum (left) can be divided into four lobes. The functional parts of the entire brain (right) are the cerebrum, cerebellum, and the brain stem.

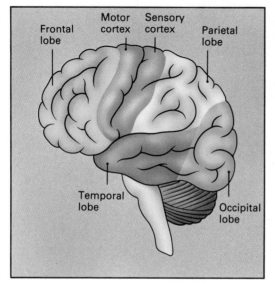

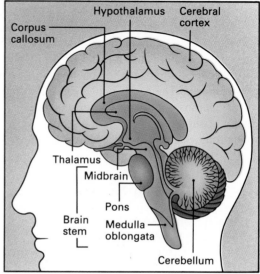

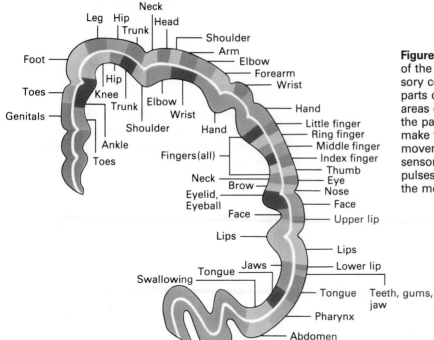

Labels on figure:
Neck · Head · Leg · Hip · Trunk · Shoulder · Arm · Elbow · Forearm · Wrist · Foot · Hip · Knee · Trunk · Elbow · Wrist · Hand · Toes · Shoulder · Hand · Little finger · Ring finger · Genitals · Ankle · Fingers (all) · Middle finger · Index finger · Toes · Thumb · Neck · Brow · Eye · Eyelid, Eyeball · Nose · Face · Face · Upper lip · Lips · Lips · Jaws · Lower lip · Tongue · Swallowing · Tongue · Teeth, gums, jaw · Pharynx · Abdomen

Figure 45–6. Specific areas of the motor cortex and sensory cortex control specific parts of the body. The largest areas of motor cortex control the parts of the body that make the most complex movements. Large areas of sensory cortex receive impulses from body parts with the most sensory receptors.

as the **corpus callosum** (KAWR puhs kuh LOH suhm). Deep grooves mark off four areas on each hemisphere. The four areas are the *frontal, parietal, temporal,* and *occipital* lobes.

The gray matter of the cerebrum is called the **cerebral cortex.** Its main function is to receive sensory impulses from the body and coordinate motor responses to them. Its many ridges and valleys, called *convolutions,* greatly increase the surface area of the brain. As you can see in Figure 45–6, each area on the section called the *motor cortex* controls the movement of muscles in a specific part of the body. Each area of the *sensory cortex* receives impulses from a specific part of the body. The area devoted to each body part is proportional to its sensitivity or motor capability, not to its size. For example, a large area is devoted to the hand, a sensitive area.

Each hemisphere controls the actions and sensations of the opposite side of the body. For example, the left side controls movement of the right hand; the right side controls movement of the left hand. Scientists have discovered that in most people each side also has exclusive control over certain functions.

Several important structures lie within the cerebrum. On each side of the brain is the **thalamus** (THAL uh muhs), a small organ that acts as a relay center for impulses. The thalamus processes incoming sensory impulses before sending them to appropriate parts of the cortex. It also sorts out and combines impulses from the cortex and other areas of the brain. Below the thalamus is the **hypothalamus.** Research has indicated that this

Q/A

Q: *Is brain size related to intelligence?*

A: No. Brain size among human beings is not related to level of intelligence. Previously the number of convolutions was thought to play a role in intelligence, but now even that theory is in doubt.

▶ The Split Brain

Both hemispheres of the cerebrum control similar activities. However, in most people, some activities are controlled by one side or the other. Usually the left side dominates language. The right side dominates spatial perception and musical ability.

These discoveries were made in split-brain research on patients suffering from epilepsy. Surgeons occasionally cut the corpus callosum in epileptic patients to relieve seizures. As a result of this operation, information cannot travel from one side of a patient's brain to the other.

Researchers studied other effects of the operation. In one experiment, an apparatus held the patient's head while different pictures were flashed to each eye. Patients could describe what they saw with the right eye but not what they saw with the left eye. For example, patients were shown the word *heart*. The left eye saw only *he,* and the right eye, only *art.* Patients said they saw *art.* They could not say *he.*

Although patients could not describe what they saw with the left eye, they could make a nonverbal association. For instance, when a patient saw the word *nut* with the left eye, he or she could not say the word *nut.* However, the patient could choose a nut from an assortment of objects. Researchers also found that, in tests involving spatial relationships, the right cortex was far superior to the left. Right-handed patients drew a more accurate picture with the left hand than with the right hand when space perception was involved.

structure controls body temperature, thirst, hunger, salt and water balance, and emotional behavior in general. Near the corpus callosum is a network of neurons called the *limbic system*. The limbic system is thought to translate a person's drives and emotions into actions.

The Cerebellum   The **cerebellum** is located beneath the occipital lobe. The white matter that composes most of the cerebellum is covered by a thin layer of gray matter. The cerebellum coordinates voluntary muscle movements and maintains muscle vigor and body balance. Damage to the cerebellum may result in jerky, awkward movements, although the ability to make the movements is not affected.

The Brain Stem   The **brain stem** contains all the nerves that connect the spinal cord with the cerebrum. The principal divisions of the brain stem, as shown in Figure 45–5 on page 694, are the medulla oblongata, pons, and midbrain. The **medulla oblongata** (mih DUHL uh ahb lawn GAHT uh) is the enlarged portion of the spinal cord that enters the lower skull. It controls breathing, swallowing, digestive processes, and action of the heart and blood vessels. In the medulla, many nerve fibers crisscross. As a result, each hemisphere receives impulses from and sends impulses to the opposite side of the body. The **pons** connects the two hemispheres of the cerebellum and links the cerebellum with the cerebrum. The **midbrain** lies above the pons. It controls responses to sight, such as movements of the eyes and size of the pupils.

A complex network of nerve fibers called the **reticular formation** runs through the brain stem and thalamus. This structure plays an essential role in consciousness, awareness, and sleep. The reticular system activates the rest of the brain when a stimulus is received. However, it first filters every stimulus. For example, people can sleep through loud noises such as traffic sounds but be awakened instantly by the ring of a telephone. Researchers do not know exactly how the reticular formation

Spotlight on Biologists: Candace Pert

Born: New York City, 1946
Degree: Ph.D., Johns Hopkins University

While still a graduate student, American neuroscientist Candace Pert discovered opiate receptors, the sites in the brain that receive pain-deadening drugs or natural chemicals. Her findings affected research in psychiatry as well as in biochemistry.

Pert's discovery of the receptor sites suggested that the body produces its own opiates, which act as painkillers. This hypothesis was confirmed two years after Pert's initial discovery by two British scientists, Hans Kosterlitz and John Hughes.

The opiate receptors are one of many types of receptors in the brain, but the first to be located. Pert's method of detecting brain receptors is to use a radioactive compound to tag a substance that binds to the receptor. She then grinds up brain tissue and analyzes it to see which substances are located in which tissues.

By revealing the makeup and location of all receptors, scientists hope to find chemical cures for many mental illnesses. Pert foresees a time when a photograph will expose all receptor sites and tell which need regulation. Appropriate drugs could then be prescribed.

Pert is now a research scientist at the National Institute of Mental Health. She is currently using her findings and her talent to study the brain's receptor sites for tranquilizers and other drugs that slow down the activities of the nervous system.

Q: *Does the brain need much energy to operate?*

A: Yes. Compared with other organs, the brain needs a tremendous amount of energy. The brain is only 2 percent of the body's total weight, but it uses 20 percent of the body's oxygen and glucose. Because it cannot store either substance, an interruption in the supply of either one is very serious.

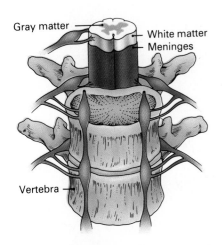

Gray matter — White matter — Meninges — Vertebra

Figure 45–7. The spinal cord is protected by the surrounding meninges and vertebral column.

functions during sleep, but they know that a lack of sleep can seriously affect a person's well-being. A person deprived of sleep becomes quick-tempered, lacks concentration and energy, and is easily distracted. Too little sleep can eventually affect sight and hearing.

Protection of the Brain The brain is protected in three ways. First, the skull helps prevent serious injury from blows to the head. Second, the brain is cushioned inside the skull by **cerebrospinal fluid.** Cerebrospinal fluid is tissue fluid that circulates constantly around the brain and spinal cord. Third, three layers of tissue known collectively as the **meninges** (muh NIHN jeez) protect the surface of the brain. The innermost layer, called the *pia mater*, follows all brain convolutions. Its rich blood supply carries nutrients and oxygen to brain cells and carries waste products away. The middle layer, the *arachnoid*, is a delicate weblike structure. Fluid between the pia mater and arachnoid serves as the pathway for exchange of nutrients and waste products. The outermost layer is a tough fibrous membrane called the *dura mater*.

45.4 The Spinal Cord

The **spinal cord** is a column of nerve tissue extending from the brain through the spinal column. In adults it is about 43 cm (17 in.) long and as thick as a pencil. The spinal cord links the brain with nerves to all parts of the body and controls involuntary movements known as *reflexes*.

Figure 45–7 shows that the center of the cord is filled with gray matter with a cross section shaped somewhat like the letter H. Cell bodies of motor neurons and interneurons are in the gray matter. The cell bodies of sensory neurons form small masses called *ganglia* outside the spinal cord. White matter around the gray matter consists of myelinated axons. Vertebrae, meninges, and cerebrospinal fluid protect the spinal cord.

Reviewing the Section

1. What are the three major sections of the brain?
2. What are the main functions of the cerebral cortex?
3. How might an injury to the cerebellum affect a person's movements?
4. What functions does the reticular formation perform?
5. If the right side of a person's body is paralyzed, what part of the brain may be injured? Explain.

The Peripheral Nervous System

Section Objectives

The peripheral nervous system carries impulses to and from the central nervous system. Twelve pairs of *cranial nerves* and 31 pairs of *spinal nerves* make up the peripheral nervous system. The cranial nerves connect the brain primarily with sense organs, the heart, and other internal organs. The spinal nerves carry impulses between the spinal cord and skeletal muscles.

So far, the nervous system has been presented like a map. Another way of thinking of the nervous system is to focus on what its parts do and not on where they are located. When described in this way, the subdivisions are called the *somatic nervous system* and the *autonomic nervous system*. These systems involve both the peripheral and the central nervous systems.

- *Name* the functional divisions of the nervous system.
- *Trace* the path of a reflex arc.
- *Explain* how the somatic and autonomic nervous systems differ.
- *Compare* the sympathetic and parasympathetic nervous systems.

45.5 The Somatic Nervous System

The **somatic nervous system** transmits impulses to and from skeletal muscles, which are usually under conscious control. For this reason the somatic nervous system is sometimes called the voluntary nervous system. Each pair of spinal nerves has motor and sensory fibers. Each pair carries impulses to and from skeletal muscles in a specific part of the body.

Not all skeletal muscle movements are voluntary. Movements called reflexes are not under conscious control. A reflex pathway, called a **reflex arc,** involves two or three neurons, as shown in Figure 45–8. The simplest arc consists of one sensory and one motor neuron. A three-neuron reflex arc includes an interneuron. The brain is not involved in either arc.

Responses such as dodging a moving object are *conditioned;* they are learned from experience. Reflexes, such as blinking, are *unconditioned.* The advantage of reflexes is speed. Reflexes may take one one-hundredth of a second—much faster than any pathway involving the brain. However, the brain may play a role in other responses to the stimulus. Typically, you quickly withdraw your hand after touching a hot object. Three neurons carry the impulses in the reflex arc, causing the jerk of your hand. Other impulses travel to the brain. At the conscious level, then, you may look at the hot object and gasp, "Ouch."

45.6 The Autonomic Nervous System

The **autonomic nervous system** controls automatic, or involuntary, functions involving glands, internal organs, and other smooth muscle tissue. The autonomic nervous system is divided

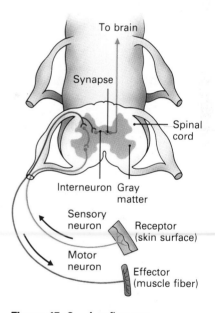

Figure 45–8. A reflex arc includes a sensory neuron, a motor neuron, and in some cases an interneuron. The spinal cord—not the brain— activates the response.

Table 45–1: The Autonomic Nervous System

Anatomic Part	Sympathetic	Parasympathetic
Iris	Dilates pupil	Contracts pupil
Salivary glands	Decreases salivation	Increases salivation
Bronchioles	Relaxes bronchioles	Constricts bronchioles
Heart	Speeds up heartbeat, increases force of contraction	Slows heartbeat, reduces force of contraction
Digestive system	Slows peristalsis, inhibits pancreatic secretions, stimulates release of glucose by liver	Stimulates peristalsis, stimulates pancreatic secretions, promotes production of glycogen by liver
Urinary bladder	Contracts bladder	Relaxes bladder
Circulatory system	Constricts blood vessels of some internal organs, dilates blood vessels of skeletal muscles	Returns constricted blood vessels to normal size
Hair follicles	Contracts muscles around hair follicle; body hair stands erect	Relaxes muscles around the hair follicles

into two systems. The *sympathetic nervous system* enables the body to handle stress through what is called the "fight or flight" response, in which a person either fights or runs away. Fighting and running both require extra energy. Therefore, the sympathetic nervous system causes bodily changes that channel extra glucose and oxygen to skeletal muscles, thus supplying the extra strength needed in an emergency. When stress no longer exists, the *parasympathetic nervous system* is responsible for returning body functions to normal and for maintaining them at that level. For example, the sympathetic nervous system speeds the heart to supply cells more quickly; the parasympathetic nervous system slows the heart.

Reviewing the Section

1. What is the advantage of a reflex?
2. What is the difference between the somatic and the autonomic nervous systems?
3. How does the sympathetic nervous system help the body protect itself?

Investigation 45: Stimulus-Response Performance

Purpose
To measure response time after visual, auditory, and tactile stimuli have been practiced

Materials
Meter stick, watch with a second hand

Procedure
Part A
1. Make a chart like the one shown.

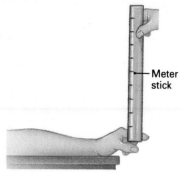

Meter stick

Stimulus-Response Table

Signal	Variable	1	2	3	4	5	6	7	8	9	10
Visual	Distance										
	Time										
Auditory	Distance										
	Time										
Tactile	Distance										
	Time										

2. Place your arm on a desk or table top with your hand resting on the edge. Your thumb and forefinger need to be projecting over the edge and about 2 cm apart.
3. One member of your group will hold the meter stick in a vertical position so that the zero end of the meter stick is halfway between your thumb and forefinger. The holder of the meter stick will drop it without warning you. You are to catch the meter stick as soon as possible after it has been released. Record on the data chart the number of centimeters the meter stick drops before you catch it.
4. Repeat the procedure nine more times.
5. Another member of your group should time how long it takes you to catch the meter stick in seconds. Record the time in the data chart for each of the 10 trials in steps 3 and 4.

Part B
Repeat the procedure in steps 2 and 3 but this time close your eyes and have your partner tell you the moment the meter stick is dropped. Repeat this nine more times and then record the data on the chart.

Part C
Repeat the procedure with a tactile stimulus. That is, have your partner touch you as soon as the meter stick is dropped. Be sure your eyes are closed. Repeat nine more times and record the data as before.

Part D
Exchange places with a member of your group and repeat Parts A, B, and C. Collect data on his/her stimulus-response times and distances.

Analyses and Conclusions
1. How did the auditory response time compare to the other signals?
2. How did the tactile response time compare to the other signals?
3. How did your response time change with practice?
4. Rank the response times from the slowest to the fastest.
5. From these results draw a conclusion about practice with regard to stimulus-response time.

Chapter 45 Review

Summary

The nervous system is made up of the central nervous system and the peripheral nervous system. The basic functional unit of the nervous system is the neuron. Dendrites carry impulses from receptors toward the cell bodies. Axons carry impulses from a cell body to effectors or other neurons. An impulse is an electrochemical message transmitted by changes in the neuron membrane. Neurotransmitters generate new impulses in adjoining neurons or stimulate effectors to act.

The brain consists of the cerebrum, the cerebellum, and the brain stem. Different body functions are controlled by different areas of the brain. Reflexes, for example, are controlled by the spinal cord.

The somatic nervous system controls skeletal muscle movement. The autonomic nervous system includes the sympathetic and parasympathetic systems. The sympathetic system prepares the body to meet stress; the parasympathetic nervous system relaxes the body.

BioTerms

autonomic nervous system (**699**)
axon (**689**)
brain (**694**)
brain stem (**696**)
cell body (**689**)
central nervous system (**689**)
cerebellum (**696**)
cerebral cortex (**695**)
cerebral hemisphere (**694**)
cerebrospinal fluid (**698**)
cerebrum (**694**)
corpus callosum (**694**)

dendrite (**689**)
hypothalamus (**695**)
impulse (**689**)
interneuron (**689**)
medulla oblongata (**696**)
meninges (**698**)
midbrain (**697**)
motor neuron (**689**)
myelin sheath (**689**)
nerve (**690**)
nerve fiber (**689**)
neuron (**689**)
neurotransmitter (**692**)

node of Ranvier (**690**)
peripheral nervous system (**689**)
pons (**697**)
reflex arc (**699**)
reticular formation (**697**)
Schwann cell (**690**)
sensory neuron (**689**)
somatic nervous system (**699**)
spinal cord (**698**)
synapse (**692**)
thalamus (**695**)

BioQuiz   (Write all answers on a separate sheet of paper.)

I. Completion

1. A simple reaction involving only a sensory and motor neuron is a _____.
2. The part of an axon nearest an effector cell is a _____.
3. Depression appears to be linked to a low level of _____.
4. Nerves going to and from the central nervous system constitute the _____.

II. Modified True and False

Mark each statement TRUE or FALSE. If false, change the underlined term to make the statement true.

5. Cerebrospinal fluids come from the synaptic vesicles of axons.
6. The sympathetic nervous system prepares the body for emergencies.
7. The pons is important to sleep.

III. Multiple Choice

8. Muscular coordination is controlled by the a) reticular formation.
b) cerebellum. c) pons. d) midbrain.
9. Impulse transmission is made faster by the a) nodes of Ranvier.
b) cerebrospinal fluid. c) synaptic cleft. d) motor end plates.
10. All impulses to the brain are processed through the a) hypothalamus. b) reflex arc. c) thalamus. d) cerebellum.
11. The three protective coverings of the brain are the a) convolutions.
b) meninges. c) ventricles. d) Schwann cells.
12. A nerve fiber that carries impulses away from the cell body is a(n) a) dendrite.
b) cell membrane. c) nucleus. d) axon.
13. The fatty insulation on some axons is a) myelin. b) neurotransmitter.
c) cytoplasm. d) reticular formation.
14. An impulse is the movement of a(n)
a) resting potential. b) refractory period.
c) synapse. d) action potential.
15. Motor neurons carry impulses to
a) receptors. b) sensory cells.
c) effectors. d) conductors.

IV. Essay

16. What changes within a neuron cause an impulse to travel?
17. How did the split-brain theory arise? Explain what it is.
18. What is a myelin sheath? How does it affect impulse transmission?
19. How do neurotransmitters help impulse transmission?
20. How do reflex arcs differ from other impulse pathways? What is the advantage of a reflex?

Applying and Extending Concepts

1. Write a paragraph describing the pathway of an impulse. Start with a receptor and end with the response made by a skeletal muscle.
2. During the 1970s, scientists discovered that the brain produces its own painkillers called *endorphins*. Use your school or public library to learn more about these substances and how they reduce pain.
3. A driver sees a red light, lifts her right foot from the accelerator, and places it on the brake. Which hemisphere of the brain ordered this action? What other areas of the brain were involved in perceiving the red light and responding to it?
4. Large areas of the motor cortex are involved in controlling the movements of the lips, tongue, and fingers. On the other hand, relatively small areas are devoted to controlling the movements of the arms and shoulders. Suggest a possible explanation for this fact.

Related Readings

Gilling, D. and Brightwell, R. *The Human Brain*. New York: Facts on File Publications, 1982. This book explains the brain's makeup and operation.

McKean, K. "Beaming New Light on the Brain." *Discover* 2 (December 1981): 30–33. The article describes Brain Electrical Activity Mapping (BEAM), a technique that is used to uncover brain dysfunction.

Oldendorf, W. H. and Zabielski, W. "The World Divided: Your Brain's Split Universe." *Science Digest* 90 (January 1982): 56–59. This article explains how the brain hemispheres are linked and how brain damage affects perceptions.

CHAPTER

46

Senses

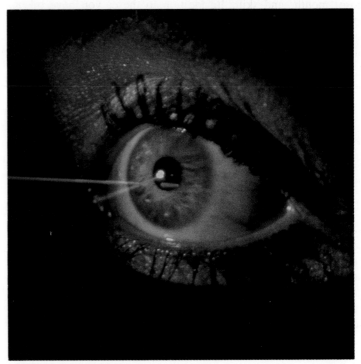

Laser surgery on human eye

Introduction

You know what is going on around you and inside you because of special receptors in your body. Through them, you are able to see beautiful sights, hear wonderful sounds, taste delicious flavors, and smell appealing aromas. You can touch things and feel a variety of sensations. More importantly, however, these receptors help you survive. You can avoid injuries if you can feel, taste, smell, hear, or see potential danger. Without some of the special sense receptors, you could not move well, balance yourself, or judge your position in space.

Each kind of receptor reacts to a specific type of stimulus. The stimuli are transmitted to the central nervous system, which in turn determines the body's response to conditions in the external environment.

Receptors and Sense Organs

Many receptors that enable the body to obtain information from the environment are located in highly specialized organs called *sense organs*. The most familiar sense organs are the eyes, ears, nose, mouth, and skin. In addition to these, you have other sense organs that you may not be aware of. For example, receptors in your ears enable you to keep your balance. All sense organs have specialized receptors for stimuli. Most sense organs have receptors that pick up stimuli from the body's external environment. Other kinds of receptors pick up stimuli from the body's internal environment.

46.1 Types of Receptors

Sense receptors are highly selective. The receptors for taste will not respond to light, no matter how intense it is. The receptors for sight cannot be activated by sound vibrations.

Sense receptors can be classified according to the stimuli that activate them. *Photoreceptors* detect stimuli generated by light. The receptors for taste and smell are triggered by chemicals and are called *chemoreceptors*. *Thermoreceptors* respond to heat or cold, either inside or outside the body. *Pain receptors* generate impulses interpreted as pain. *Mechanoreceptors* respond to mechanical pressures. Such pressures may come from sound vibrations, touch, muscle contractions, or movements of joints. The pressure bends or distorts the part of the sense organ in which the mechanoreceptors are located. **Hair cells,** which have extremely fine projections like cilia, are the most common type of mechanoreceptors.

46.2 Sense Organs

Sense organs act as **transducers**—that is, they transform one form of energy into another form. For example, when light rays strike the inner lining of the eye, they are changed into impulses. These impulses move along a nerve to the brain's visual center where they are interpreted as sight.

Impulses from all sense organs are basically alike. *The way the brain interprets impulses from various sense organs differs.* Impulses from each sense organ travel to a different part of the brain. The impulses from a particular sense organ are interpreted in only one way, according to where they are received in the brain. For example, when the eye receives light signals, it produces impulses that the brain interprets as an image. When

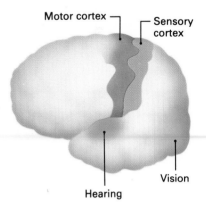

Motor cortex — Sensory cortex

Vision

Hearing

Figure 46–1. The illustration shows a sensory map of the brain. Damage to a specific area will cause problems with the activity associated with that area.

▶ The Other Senses

The human body has many special receptors other than those in the familiar sense organs. While some scientists claim that all these receptors are "senses," others feel they are more accurately described as "controls." Regardless of the term used, the fact remains that these special receptors react to stimuli, though the stimuli are internal rather than external.

The internal controls primarily maintain homeostasis. For example, thirst is triggered by the hypothalamus, which responds to salt concentration in the blood. When the water level in the blood is low, salt becomes more concentrated. When salt concentration is high, the hypothalamus reacts by generating impulses that trigger a thirst sensation. When the water level is high, salt concentration is low, and the body eliminates more water. Similarly, chemicals in the cerebrospinal fluid and a low level of glucose in the blood seem to trigger hunger.

Another type of internal control monitors your skeletal muscles. Muscle spindles, a special type of muscle fiber, are part of skeletal muscle. The spindles contain two types of sensory neurons. One type alerts the central nervous system of a change in the stretch or contraction of a muscle. The other type registers how much stretch is involved. This constant monitoring of muscular contraction helps you maintain posture and keeps your body steady. Joint and tendon receptors work with the muscle spindles. Joint receptors register the angle of ligament movement. Tendon receptors indicate the amount of stretch in the tendons.

the ear receives pressure waves, or sound vibrations, it produces impulses that the brain interprets as sound. The brain never interprets impulses from the eye as sound or impulses from the ear as an image. Even if some other type of energy generates an impulse in a receptor cell, the brain will interpret the impulse exactly as it does all other impulses from that receptor. For example, a blow to the eye may cause you to see an image, even though the impulse was generated not by light but by mechanical pressure.

Reviewing the Section

1. Name three familiar sense organs.
2. How are sense receptors classified?
3. The ears convert one form of energy into another form. What are these two forms of energy?
4. Why might a blow to the eye result in the perception of a visual image?

Vision, Hearing, and Balance

The eyes and ears provide the body with its greatest protection. Because these organs are sensitive to distant stimuli, they can give early warnings about possible dangers.

46.3 The Eyes

The eye is often compared to a camera, but it is more complicated than the most sophisticated camera. Humans have *binocular vision,* the ability to view objects with two eyes. People also have *stereoscopic vision,* the ability to see objects in three dimensions—height, width, and depth. With stereoscopic vision a person can assess the speed of a moving object and determine the distance of an object in space.

Structure of the Eye   The *eye,* or *eyeball,* is an almost perfect sphere with a diameter of about 2.5 cm (1 in.). The eye is protected in a number of ways. A fatty layer within the *orbit,* a socket in the skull, cushions the eyeball. Eyelids and eyelashes also provide protection by preventing foreign particles from entering the front of the eye. If something touches the eyelashes or moves suddenly in front of the eye, the eyelid closes and reopens rapidly in a blinking reflex.

Each eye is moved by three sets of muscles and is lubricated by mucus and tears. The mucus is secreted by the **conjunctiva,** a delicate, blood-rich membrane that lines the inner eyelid and covers the front of the eye. Tears are produced by the **lacrimal gland** near the outer corner of the eye. When the eye closes, the eyelid spreads the mucus and tears, which moisten the eye and help remove foreign particles.

The eye has an outer wall that consists of three layers of tissue: the sclera, the choroid, and the retina. These three layers surround a jellylike substance, the **vitreous** (VIH tree uhs) **humor** that makes up two-thirds of the eyeball.

The **sclera** (SKLIHR uh) is tough, white connective tissue that forms the outermost layer. About 80 percent of the sclera, including the "white" of the eye, is opaque. The remainder is a transparent layer called the **cornea** at the front of the eye.

The **choroid** (KAWR oyd) is the middle, darkly pigmented layer of tissue. It absorbs light and so prevents reflection, which would result in fuzzy images. The choroid contains many blood vessels that nourish the eye. Toward the front of the eye, the choroid forms a colored ring, the **iris,** which gives the eye its color. In the center of the iris is an opening called the **pupil.** In

Section Objectives

- *List* the major parts of the eye and state the function of each part.
- *Contrast* the functions of the rods and the cones.
- *Name* several disorders that can be detected by examining the retina.
- *Trace* a vibration through the ear.
- *Explain* how the inner ear regulates balance.

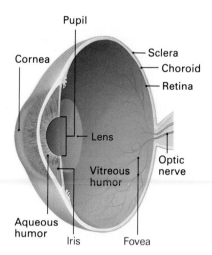

Figure 46–2. The cross-section view of the eye shows both outer and inner structures. The size of the pupil automatically adjusts in response to changes in light.

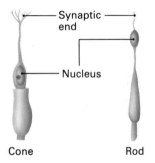

Synaptic end

Nucleus

Cone   Rod

Figure 46–3. Rod and cone photoreceptor cells differ in their response to light stimulation. Without cones, a person would not see colors, only shades of gray.

bright light, one set of muscles in the iris contracts and causes the pupil to become smaller. In dim light, a different set of iris muscles contracts, making the pupil larger.

Behind the pupil is the **lens,** a transparent, curved structure. By changing shape, the lens helps focus images onto receptor cells at the rear of the eye. The curvature of the lens is controlled by *ciliary muscles* attached to the choroid. A clear, watery fluid called the **aqueous** (AY kwee uhs) **humor** fills the space between the lens and cornea. The vitreous humor fills the space behind the lens.

The innermost layer of the eye is the light-sensitive **retina.** The retina contains about 125 million receptors called **rods** and **cones.** *The rods and cones are stimulated by light to generate nerve impulses.* The rods are extremely light-sensitive and can detect various shades of gray even in dim light. However, they cannot distinguish colors, and they produce poorly defined images. The cones detect color, produce sharp images, and are important for seeing in bright light. In a tiny pit at the center of the retina is a concentration of cones. This area, the **fovea** (FOH vee uh), produces the sharpest image.

How You See Light passes through the cornea, aqueous humor, pupil, lens, and vitreous humor on its way to the retina. Impulses generated by the rods and cones travel to the visual center in the occipital lobe of the brain by means of the *optic nerve.* The optic nerve from each eye consists of about 1 million nerve fibers. No rods or cones exist at the point where the optic nerve enters the retina. This area, called the *optic disc,* or blind spot, does not transmit impulses.

▶ What Your Eyes Tell About You

A doctor can learn a great deal about the condition of your entire body by looking into your eyes. Using an *ophthalmoscope,* which has special lenses and a light, he or she can study the optic disc and the blood vessels of the retina.

The retina reveals many disorders that do not directly involve the eyes. For example, high blood pressure can be identified by viewing the blood vessels of the retina. The increased pressure of the blood circulating through these tiny vessels may cause some of them to burst. Diabetes may also cause changes in these blood vessels and in the vitreous humor. Changes in the size and shape of the optic disc may indicate such disorders as glaucoma and brain tumor.

The doctor's ability to see changes within the eyes often permits early diagnosis and treatment of the underlying condition.

Near the base of the brain, half of the nerve fibers from the left eye cross over and join half of the nerve fibers from the right eye. All these fibers go to the right side of the brain. Likewise, half the nerve fibers from the right eye join half from the left eye and go to the left side of the brain. Each side of the brain thus receives images from both eyes. The point at which the partial crossing-over of the fibers occurs is the *optic chiasma*.

Disorders of the Eye Eye disorders affect more than 50 percent of the people in the United States. Among the most common of these disorders are *myopia* and *hyperopia*. In myopia, or nearsightedness, the eyeball is too long from the front to the back. Light focused by the lens falls at a point in front of the retina, resulting in a blurred image of distant objects. In hyperopia, or farsightedness, the eyeball is too short from front to back. Light is focused at a point behind the retina, resulting in a blurred image of close objects. In *astigmatism*, irregularities in the curvature of the cornea result in fuzzy images. Prescribed eyeglasses or contact lenses can correct these conditions.

In a condition called *glaucoma*, the aqueous humor cannot drain into blood vessels around the eye. Because new aqueous humor is constantly produced by the choroid, failure to drain creates excess pressure within the eye. This pressure damages the retina and optic nerve and can lead to blindness.

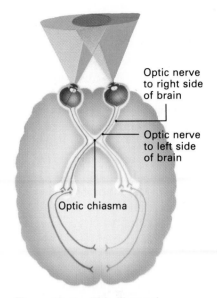

Optic nerve to right side of brain

Optic nerve to left side of brain

Optic chiasma

Figure 46–4. This illustration shows the neural pathways involved in seeing. Because the fields of vision overlap, each side of the brain receives information from each retina.

Nearsighted eye

Farsighted eye

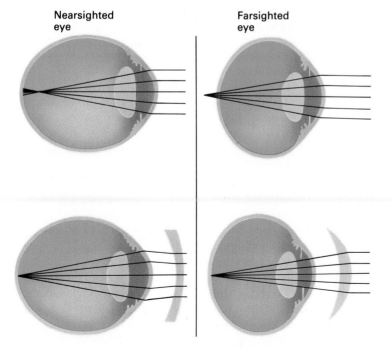

Figure 46–5. A concave lens corrects nearsightedness by spreading the light rays so that they focus on the retina (left). A convex lens bends light rays to correct farsightedness (right).

General Description

An ophthalmologist is a physician who treats diseases and disorders of the eye. An ophthalmologist may prescribe corrective lenses for a vision problem. He or she may also prescribe drugs or perform eye surgery.

Ophthalmologists may work privately or with a group of different specialists in a health maintenance organization (HMO). An HMO is a type of health insurance in which people regularly pay a fixed amount beforehand so they can have ready access to these doctors.

Inventions are constantly changing the practice of ophthalmology. Therefore, continuing education is extremely important.

Career Requirements

An ophthalmologist must complete medical school and take specialty training. The time required to complete college and specialty training is generally 12 years.

All states require that doctors be licensed. To be certified in a specialty, a

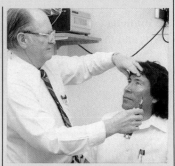

student also takes oral and written exams after completing a residency.

For Additional Information

American Board of
    Ophthalmology
111 Presidential Boulevard
Bala Cynwyd, PA 19004

46.4 The Ears

The eyes have only one function—vision. The ears, however, perform two vital functions—hearing and balance.

Structure of the Ear   The ear is divided into three major sections: the *outer ear,* the *middle ear,* and the *inner ear.* Each region has a specific function.

The outer ear consists of a cartilage flap called the **pinna** and the **auditory canal,** a tube leading to the middle ear. These structures channel sound to the **eardrum,** a tightly stretched membrane between the outer ear and the middle ear. The auditory canal is lined with cilia and special cells that secrete *cerumen,* or earwax. Together, the cilia and earwax clear foreign particles from the auditory canal.

The middle ear lies within an air-filled space called the **tympanic cavity** inside the skull bone. A duct called the **Eustachian** (yoo STAY shuhn) **tube** connects the middle ear to the pharynx. Generally the tube is collapsed, but it opens when you yawn, swallow, cough, or blow your nose. Air pressure between the middle ear and throat is then equalized. Air pressure around you varies with altitude and can change rapidly, as when you ride an elevator or airplane. If the pressure is not equalized, the eardrum can bulge, causing pain and difficulty in hearing. Lying across the middle ear cavity are three tiny bones called the **malleus** (MAL ee uhs), or *hammer;* the **incus** (IHN kuhs), or *anvil;* and the **stapes** (STAY peez), or *stirrup*. The stapes touches a membrane called the **oval window,** located between the middle ear and the inner ear.

The inner ear contains the sensory receptors for hearing and balance. It consists of three main parts: the cochlea, the vestibule, and the semicircular canals. The organ of hearing is within the **cochlea** (KAHK lee uh), a bony, coiled tube filled with fluid and lined with hair cells. A second membrane-covered opening is located in the cochlea below the oval window. Called the **round window,** it maintains a constant pressure within the inner ear. The upper part of the inner ear consists of three **semicircular canals,** which are fluid-filled tubes positioned at right angles to each other. These canals help maintain balance by responding to head movement. A bony chamber called the **vestibule** lies between the semicircular canals and the cochlea.

How You Hear Sound waves are generated when any object vibrates, or moves back and forth, in the air. The human ear can detect sounds between 20 and 20,000 vibrations per second.

Q/A

Q: *How does an animal distinguish one sound from another?*

A: Different hair cells in the cochlea are stimulated by different frequencies. The brain then interprets the sound according to which hair cell an impulse comes from.

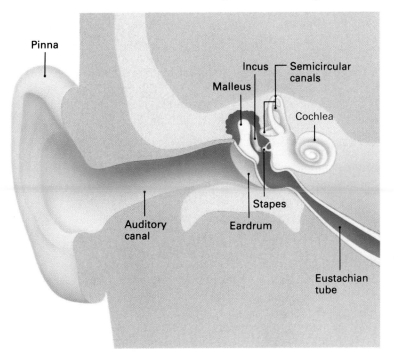

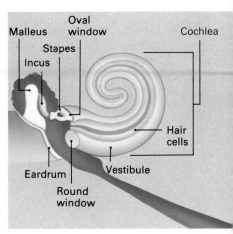

Figure 46–6. Sound waves must pass through the outer, middle, and inner ears (left) for hearing to take place. The cochlea, which contains the receptors for sound, is located in the inner ear (below).

Figure 46–7. This scanning electron micrograph magnifies the inner hair cells of the human ear 1,705 times.

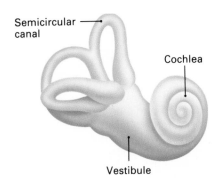

Semicircular canal

Cochlea

Vestibule

Figure 46–8. The organs of balance are located in the inner ear. Stimulation of the hair cells in the saccule and utricle provides information to the brain on changes in the rate as well as the direction of motion.

The vibrations travel through the auditory canal to the eardrum; to the malleus, which touches the eardrum; and then to the incus and the stapes. The stapes touches the oval window. The oval window sets the fluid in the cochlea in motion. Stimulated by the fluid motion, hair cells in the cochlea generate nerve impulses that travel along the *auditory nerve* to the auditory center in the temporal lobe of the brain. Exactly how vibrations are transformed into impulses is not clear.

Hearing loss due to disease or injury of the auditory nerve or cochlea is called *nerve deafness*. It is the most common cause of total and permanent hearing loss. Deafness resulting from interference as vibrations pass to the inner ear is called *conductive deafness*. This condition may be caused by several problems, including excess earwax, infection, swelling and closing of the passage, rupture and inflammation of the eardrum, or immobility of the stapes due to bone overgrowth. Conductive deafness generally can be treated.

How You Balance Yourself Fluid in the semicircular canals flows when you change the angle of your head. A different canal in each ear is affected by any particular movement. For example, a movement to the right causes fluid in the right ear to flow toward the hair cells. As a result, many impulses are sent to the cerebellum from the right ear. At the same time, the movement causes fluid in the left ear to flow away from the hair cells. Few impulses are then sent to the brain from the left ear. The cerebellum interprets the two sets of impulses so you know which way your head is turned.

The **saccule** (SAK yool) and **utricle** (YOO trih kuhl), the two sections of the vestibule, also help with balance. They are lined with hair cells covered by a gelatin-like membrane embedded with grains of limestone. Gravity pulls the limestone down onto the hair cells, causing them to generate impulses. The greater the pull on particular grains, the stronger the impulses. The cerebellum interprets the direction of gravity and lets you know the position of your head.

Reviewing the Section

1. What function does the sclera perform?
2. How do the sensory receptors in the retina differ?
3. How does yawning help relieve ear discomfort in an airplane passenger?
4. What part of the inner ear has sound receptors?
5. How is body balance monitored by the inner ear?

Smell, Taste, and Touch

Smell and taste are closely associated senses. The fact that a stuffy nose makes food seem tasteless demonstrates the close relationship between these senses. Smell and taste seem to operate more simply than sight and hearing, but biologists do not yet know precisely how the receptors for smell and taste discriminate among various chemicals.

The skin, the largest organ of the body, contains several types of receptors. These receptors register touch, pressure, pain, heat, and cold. The receptors for these sensations vary in number and location over the body.

46.5 The Nose

The *nose,* the chief sense organ of smell, contains receptors embedded in mucous membrane. About 50 million of these special cells, called **olfactory receptors,** are located in each nasal passage. Airborne substances dissolve in the mucus that covers the olfactory receptors. The receptors produce nerve impulses that travel through *olfactory nerves* to the *olfactory lobe* in the cerebral cortex.

Some biologists think that the perception of smell occurs when a specialized molecule on the receptor surface reacts with a specific chemical in inhaled air. The reaction generates an impulse that results in a particular smell. Other scientists believe that the outline, or shape, of a molecule is the cause of its particular odor. They think that a molecule of a specific shape fits into an olfactory receptor that will accept only that shape, just as a lock works with one key. These researchers believe that the thousands of odors humans can distinguish are simply combinations of seven basic odors.

46.6 The Tongue

The *tongue* is the major sense organ of taste. The chemical receptors for taste are clusters of sensory hair cells located in the **taste buds.** Each taste bud consists of about 40 receptor and supporting cells and an opening called the *taste pore.* The taste buds lie in bunches called *papillae,* which are visible as the bumps on your tongue. Although most of a person's 10,000 taste buds are on the tongue, a few also exist on the roof of the mouth and in the throat.

Taste buds produce one or a combination of four main taste sensations: sweet, sour, bitter, and salty. A receptor cell may be

Section Objectives

- *State* one theory of how the perception of smell may occur.
- *Explain* what taste buds are and how they may operate.
- *Identify* the various types of sense receptors in the skin.

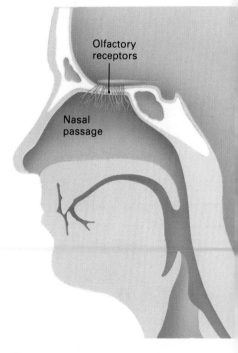

Olfactory receptors

Nasal passage

Figure 46–9. The illustration above shows the location of the olfactory receptors in the human nasal passage.

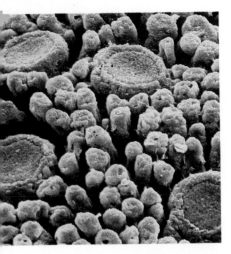

stimulated by only one taste, but most cells are stimulated by two or more tastes. This combination of different tastes may be what produces the wide variety of flavors you enjoy.

Like smell, taste depends upon chemical reactions that take place only in solution. Saliva constantly bathes taste buds, reaching receptor cells through the taste pores. Food molecules also enter the taste pores. The chemical reactions that take place somehow cause the receptor cells to generate nerve impulses. The impulses travel through three different nerves to the taste center in the cerebral cortex. No one knows precisely how taste receptors function. Some researchers think that sensory cells have sites that accept specific chemical molecules. Other researchers think that, as with smell, the shape of a molecule determines its taste. Molecules of a certain shape, they think, activate specific sites on a taste bud to produce one taste.

46.7 The Skin

The skin is considered the organ of touch. It actually contains five distinct senses, most with their own type of receptor. These five senses are *touch, pressure, pain, heat,* and *cold.* Impulses travel from the various sense receptors to different areas of the sensory cortex.

Touch receptors are the ends of certain nerve fibers. Many touch receptors are located at the base of hairs and generate impulses when the hairs move even slightly. However, touch is most sensitive in the fingertips, palms, lips, and other places where hair is not present. Other receptors react to pressure. Some are sensitive to deep pressure and vibration, while others are sensitive to lighter pressure.

Unlike the other senses of touch, the sense of pain has no specialized receptors. Pain receptors are free ends of unmyelinated nerve fibers. Pain appears to stem from a variety of stimuli. Some parts of the body are almost pain-free. Other parts may sense only one type of pain. Sensitivity to pain may be related to other body conditions, such as mental attitude.

Temperature receptors may be either bare nerve endings or specially shaped cells. Different types of receptors detect heat and cold.

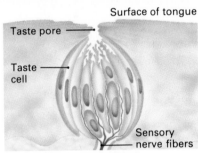

Surface of tongue

Taste pore

Taste cell

Sensory nerve fibers

Figure 46–10. The scanning electron micrograph (top) shows the taste buds on the surface of the human tongue magnified 240 times. The cross section of a human taste bud (bottom) shows the location of taste receptor cells and the nerve fibers that carry the sensations to the brain.

Reviewing the Section

1. What is the function of olfactory receptors?
2. Why is moisture necessary for smell and taste?
3. What types of sense receptors are in the skin?

Investigation 46: Dissecting the Eye

Purpose
To learn about the parts of the eye

Materials

Preserved beef eye, surgical scissors, paper towels, unused single-edge razor blade, dissecting needle, forceps

Procedure
1. **CAUTION: The use of sharp cutting edges is required throughout this investigation. Be very careful as you cut.** Place the eye on several thicknesses of paper towels and study its external structures. Use the illustration of the human eye on page 707 as a reference.
2. After examining the muscles, use the scissors to cut away some of the soft white fat covering the eye. At the rear of the eye, a white stalk (the optic nerve) will begin emerging.
3. Continue removing the fat carefully until you reach the tough outer coating (the sclera).
4. To inspect the eye's internal structures (see Figure 46–2), hold the eye so that the front part is up. With the razor blade, cut through the transparent lower conjunctiva and the cornea, which may appear slightly blue and somewhat opaque. As you make your incision, some of the aqueous humor, a clear, watery substance, may ooze out.
5. Make a second cut through the cornea that is perpendicular to the first one. This will create two triangular flaps; slowly peel them back. Inside is the iris, the colored portion of the eye. The hole in the iris is the pupil. *To which tissue layer of the eye does the iris belong?*
6. Extend the original cuts and pull the flaps back even further. Make an incision from the pupil through the iris and to the side of the eye.
7. Make another incision perpendicular to the last one. Carefully peel back the triangular

flaps as shown in the illustration. Suspensory ligaments that hold the lens in place will be revealed.

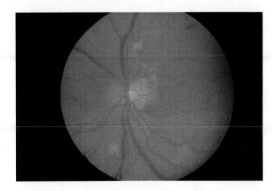

Human retina and optic disc

8. Extend the side incisions so that the eyeball is divided in two halves to expose the vitreous humor, a viscous, jellylike substance.
9. Examine the retina and the fovea centralis. Note the relationship between the latter and the optic nerve. *Why is it that at night, objects in the exact center of vision are least clearly seen?*
10. Wrap all eye parts in paper towels and dispose of everything according to your teacher's instructions.

Analyses and Conclusions
1. Explain how the external eye muscles might work.
2. Explain how the suspensory ligaments could change the thickness of the lens to accommodate farsighted and nearsighted images.

Going Further
- Use your text and other references to study the functions of each part of the eye.
- Make a model to show how eyeglasses correct hyperopia and myopia.

Chapter 46 Review

Summary

The most familiar senses are vision, hearing, smell, taste, and touch. Special receptors receive stimuli in the form of light, chemicals, temperature, vibrations, and pressure. Acting as transducers, the sense organs change the energy from the stimuli into nerve impulses. The impulses are transmitted to the brain, where they are interpreted.

Rod and cone cells in the retina of the eye make sight possible. The ear is the organ of hearing and balance. Sound vibrations are changed into nerve impulses in the cochlea.

Balance is achieved by means of the semicircular canals and the vestibule.

Smell takes place through chemical reactions in the olfactory receptors, and taste is dependent upon chemical receptors in the taste buds. Exactly how the nerve impulses for smell and taste come about is not understood.

The skin contains receptors for touch, pressure, pain, and temperature. These receptors are distributed in varying concentration throughout the body.

BioTerms

aqueous humor (**708**)
auditory canal (**710**)
choroid (**707**)
cochlea (**711**)
cone (**708**)
conjunctiva (**707**)
cornea (**707**)
eardrum (**710**)
Eustachian tube (**710**)
fovea (**708**)
hair cell (**705**)

incus (**711**)
iris (**707**)
lacrimal gland (**707**)
lens (**708**)
malleus (**711**)
olfactory receptor (**713**)
oval window (**711**)
pinna (**710**)
pupil (**707**)
retina (**708**)
rod (**708**)

round window (**711**)
saccule (**712**)
sclera (**707**)
semicircular canal (**711**)
stapes (**711**)
taste bud (**713**)
transducer (**705**)
tympanic cavity (**710**)
utricle (**712**)
vestibule (**711**)
vitreous humor (**707**)

BioQuiz   (Write all answers on a separate sheet of paper.)

I. Completion

1. The three fluid-filled tubes that help a person maintain balance are the

   _____ .

2. Sense organs are _____ because they change one form of energy into another form.

3. The senses of _____ and _____ are dependent upon special receptors that are activated by chemicals.

II. Modified True and False

Mark each statement TRUE or FALSE. If false, change the underlined term to make the statement true.

4. The saccule and utricle are necessary for the sense of balance.

5. The cornea of the eye tells a great deal about one's physical condition.

6. Hair cells are the sensory receptors in the ear.

III. Multiple Choice

7. Air pressure is equalized in the Eustachian tube located in the
   a) semicircular canals. b) middle ear.
   c) vestibule. d) cochlea.
8. Images are focused on the retina by the
   a) sclera. b) pupil. c) lens.
   d) conjunctiva.
9. The specialized sensory cells for seeing color are a) olfactory receptors.
   b) cones. c) rods. d) chemoreceptors.
10. Impulses generated by receptors in the skin travel to the a) cerebellum.
    b) olfactory lobe. c) limbic system.
    d) sensory cortex.
11. Hearing receptors react to stimuli from
    a) chemicals. b) vibrations.
    c) temperature. d) light.
12. A person's sharpest vision is in the
    a) cochlea. b) sclera. c) fovea.
    d) choroid.
13. Taste and smell are impossible without a) mucus. b) ciliary muscles. c) pain receptors. d) cones.
14. No specific organs have been located for the sense of a) balance. b) heat.
    c) pain. d) light pressure.
15. The tissue layer that nourishes the eye is the a) retina. b) sclera. c) malleus.
    d) choroid.

IV. Essay

16. How do muscle spindles operate as sensory receptors?
17. What are the major types of stimuli that activate sense organs? Which sense organs react to each type?
18. How is pressure on the eardrum converted to sound?
19. What controls the amount of light entering the eye? Explain how.
20. Although all nerve impulses are alike, they result in various sensations. How does this occur?

Applying and Extending Concepts

1. Test the skin for sensitivity to various sensations. For example, you might use velvet or sandpaper for touch, a piece of ice for cold, and a metal rod slightly heated in warm water for heat. Test the same skin areas for all sensations and record which areas of the body seem most sensitive to certain stimuli.
2. On a sheet of paper, draw a small X and then a small dot about 7 cm (3 in.) to the right of the X. Cover your left eye, hold the paper about 30 cm (12 in.) from your face, and stare at the X. As you slowly move the paper toward your face, the dot will disappear. Why does this cause the dot to disappear?
3. Smell results from chemicals in vapors that reach the nose from a distance. Taste results from chemicals in the mouth. Which receptors, smell or taste, are more sensitive? Explain your answer.

Related Readings

Bartoshuk, L. "The Separate Worlds of Taste." *Psychology Today* 14 (September 1980): 48–56. This article describes the reasons why people experience different sensations of taste from the same kinds of physical stimuli.

Oldendorf, C. "See Inside Your Eye." *Science Digest* 89 (January–February 1981): 92–93. As the title promises, this article describes how to use a penlight to see the many tiny blood vessels located inside the eye.

CHAPTER

47

Hormonal Control

Firefighters responding quickly under stress

Introduction

Activities within the human body are regulated by two systems, the nervous system and the *endocrine system*. Although both systems control body functions, their methods differ.

The nervous system sends its messengers, called impulses, to specific cells, generally muscle or gland cells. The nervous system acts quickly. Its messages travel rapidly and can change instantly. The response is immediate.

The endocrine system uses chemical messengers. They are widely dispersed to every cell throughout the body. However, only specific target cells—cells equipped with receptors—respond to the messages. The endocrine system generally does not act as quickly as the nervous system. Its messages travel more slowly, but the effects last longer.

The Endocrine Glands

The body contains many *glands*. Glands are cells, groups of cells, or organs that produce and secrete substances. **Exocrine glands,** such as sweat glands and digestive glands, secrete their products through tubes, or ducts. **Endocrine glands,** often called *ductless glands,* release their products directly into the bloodstream. *Endocrine glands produce powerful chemicals called* **hormones,** *which help regulate the activities of body tissues and organs.* Each hormone acts on a specific tissue or organ; that tissue or organ is the hormone's **target.**

47.1 The Thyroid

The **thyroid gland,** located on the trachea, secretes *thyroxine.* Thyroxine controls metabolic activities, including the production of proteins and ATP. Because thyroxine influences protein production, it affects the growth rate of children. This hormone is also necessary for the proper development of the nervous system.

Iodine is necessary for the production of thyroxine. A person needs 1 mg of iodine each week. Eating a moderate amount of iodized salt usually meets that need. Insufficient iodine may cause the thyroid gland to enlarge, a condition called *goiter.* Frequently a person with goiter also suffers from *hypothyroidism,* a lack of thyroxine. The result is a low metabolic rate. In adults the symptoms are low body temperature, sluggishness, weight gain, and excess fluid in the body. In infants hypothyroidism may cause *cretinism.* The effects of cretinism include mental retardation and abnormal bone growth. *Hyperthyroidism,* or an excess of thyroxine, causes a higher-than-normal metabolic rate. The symptoms of hyperthyroidism include weight loss, muscle weakness, excessive sweating, increased heartbeat rate and blood pressure, nervousness, and bulging eyes.

47.2 The Parathyroids

On the back of the thyroid gland are four tiny **parathyroid glands.** They secrete *PTH (parathyroid hormone),* which regulates the levels of calcium ions and phosphate ions in the blood. These minerals are necessary for proper bone development and for normal functioning of muscles and nerve cells. Too little calcium can make nerve cells so unstable that they send impulses without being stimulated. The result is uncontrollable muscle contractions. If muscles remain contracted, a person

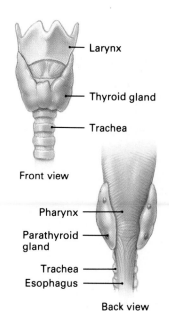

Larynx

Thyroid gland

Trachea

Front view

Pharynx

Parathyroid gland

Trachea

Esophagus

Back view

Figure 47–1. The thyroid and parathyroid glands are located on the trachea.

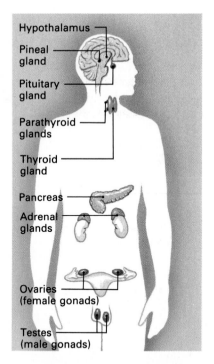

Figure 47–2. The illustration above shows the location of the major endocrine glands in the human body.

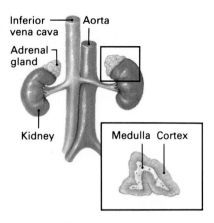

Figure 47–3. The location of the adrenal gland and its two parts (inset) are shown above.

may die because breathing stops. The calcium level sometimes is too high. Nerves and muscles then fail to respond to stimuli. Reflexes are slow, and muscle contractions are weak.

47.3 The Adrenals

An **adrenal** (uh DREE nuhl) **gland** is located on top of each kidney. Each gland functions as two separate endocrine glands. The inner part of the adrenal gland, called the **adrenal medulla,** secretes **epinephrine** (ehp uh NEHF rihn) and **norepinephrine.** These hormones produce the same effects as the sympathetic nervous system. They thus help the body respond to stress. For example, they increase blood pressure and heartbeat and breathing rates, dilate the pupils, and inhibit digestion. They also increase metabolism, sometimes as much as 100 percent.

The outer layer of the adrenal gland is the **adrenal cortex,** which secretes more than 50 hormones. All belong to a group called **corticoids.** Among the corticoids are *aldosterone; hydrocortisone,* also called *cortisol;* and also *androgens.* Aldosterone affects water and salt balance by controlling the reabsorption of sodium and potassium ions in the kidneys. Hydrocortisone controls the breakdown of proteins and fats into glucose, inhibits glucose uptake by cells, and aids in healing. Androgens are sex hormones. They regulate development of secondary sex characteristics.

A lack of corticoids may result in *Addison's disease.* The symptoms of this disease include low blood pressure, darkened skin, dehydration, a low level of sugar and sodium ions in the blood, and a high blood level of potassium ions. A victim will die within a few days if not treated with corticoids. Oversecretion of corticoids may result in *Cushing's disease,* characterized by high blood pressure, fat deposits in the face and back, and accumulation of tissue fluids. Excessive secretion of androgens may result in early sexual development in males and excessive hair and a deep voice in females.

47.4 The Pancreas

The *pancreas* is an exocrine gland that produces digestive enzymes. However, it also has special cells called the **islets of Langerhans** that function as an endocrine gland. They secrete insulin and glucagon. **Insulin** is a hormone that lowers the level of glucose in the blood. It does so by stimulating the uptake of glucose by body cells and the formation of excess glucose into glycogen in the liver and muscles. *Glucagon* triggers the breakdown of glycogen to glucose when the body needs more energy.

► Prostaglandins

Prostaglandins are fatty acids that behave as hormones. Some scientists say they are not hormones, however, because they are produced by almost every body cell, not just by glands.

Prostaglandins appear to regulate the organs and tissues in which they are produced. Some scientists believe that prostaglandins help the brain and endocrine glands to monitor body activities.

Prostaglandins are among the most powerful biological chemicals. They are secreted only when needed and last less than a minute before enzymes break them down. In that time, however, they may affect circulation, digestion, respiration, reproduction, and possibly even nerve control.

Scientists know some effects prostaglandins have, but they still are not sure how these chemicals operate. For example, they know that the cells lining blood vessels produce a prostaglandin called *prostacyclin.* Prostacyclin relaxes blood vessels and suppresses agglutination, or clumping, of platelets. Another prostaglandin, *thromboxane,* is produced by platelets. It constricts blood vessels and promotes agglutination. The two prostaglandins must remain balanced to assure proper blood circulation. If a blood vessel is damaged, prostacyclin production may be reduced in that area. Thromboxane can then cause the blood vessel to constrict and platelets to agglutinate there. Some scientists feel that this effect of thromboxane may be a major cause of strokes and heart attacks.

Prostaglandins make some nerve endings more sensitive to pain. They also increase the fluid and heat in joints. As a result, they contribute to arthritis pain and inflammation. Their role in arthritis has helped explain the mystery surrounding aspirin. For decades aspirin was known to provide relief to people who suffer from arthritis, but no one understood why. Scientists now know that aspirin inhibits the production of some prostaglandins and thus has a soothing effect. Prostaglandin research is also under way in many other areas including kidney function, migraine, cancer, diabetes, and pain.

In the absence of insulin, glucose cannot enter body cells. As a result, the cells use their own proteins and fat for energy. The level of glucose in the blood then becomes abnormally high. This condition, called *diabetes mellitus,* is the third major cause of death in the United States. Without proper treatment it can lead to heart disease, stroke, kidney failure, severe nerve damage, or blindness. Diabetes may also result in infections so severe that limb amputation is necessary.

The two chief forms of diabetes are Type 1, or insulin-dependent, diabetes and Type 2, or non-insulin-dependent, diabetes. In Type 1 diabetes, the islets of Langerhans produce too little or no insulin. Some researchers suspect a virus may be

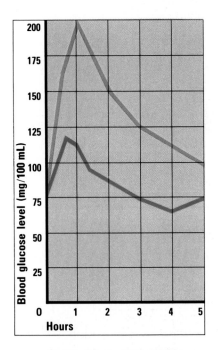

Figure 47–4. This graph compares the levels of glucose in the bloodstreams of a normal individual (blue) and a diabetic (red) during the five-hour period immediately following a typical meal.

involved in Type 1 diabetes. Type 1 usually first appears in people under 20 years of age and can be controlled by strict diet and daily injections of insulin. Approximately 85 percent of all diabetics suffer from Type 2 diabetes. Type 2 generally first appears in people over 40 years of age. These diabetics may have normal or even high levels of insulin, but their bodies cannot use the hormone. The causes of Type 2 diabetes are believed to be a shortage of insulin receptors on body cells or a breakdown of the immune system, which causes the body to become insulin-resistant. Heredity also appears to be a factor in both types of diabetes. Type 2 diabetes can generally be controlled through diet.

Excessive levels of insulin in the blood lead to *hypoglycemia*, a condition in which the level of glucose in the blood drops. Body cells then cannot obtain enough energy. Because brain cells need a constant supply of glucose, a victim may lose consciousness due to the lack of glucose. A diet high in protein and low in carbohydrates can help control hypoglycemia.

47.5 The Gonads

Gonads, the gamete-producing organs of the reproductive system, also produce and secrete hormones. The female gonads secrete estrogens that influence the development of female secondary sex characteristics. Among these are wider hips, enlarged breasts, and rounded body contours. The male gonads produce androgens that stimulate development of the male secondary sex characteristics. These include a deepened voice and enlarged muscles and bones. All these sex hormones play roles in reproduction, which is discussed in Chapter 48.

47.6 The Pituitary

The **pituitary** (pih TOO uh tehr ee) **gland,** located at the base of the brain, is about the size and shape of a kidney bean. It has two major sections, the anterior lobe and the posterior lobe.

The *anterior lobe* produces at least six hormones. Four are **tropic hormones**—that is, hormones that affect the secretions of other glands. Two tropic hormones, *FSH (follicle stimulating hormone)* and *LH (luteinizing hormone),* act on the gonads. These hormones will be discussed further in Chapter 48. The other two tropic hormones are *TSH (thyroid stimulating hormone)* and *ACTH (adrenocorticotropic hormone).* TSH stimulates the thyroid gland to secrete thyroxine, and ACTH affects the adrenal cortex. The anterior lobe also secretes *somatotropin,* or *growth hormone (GH).* Somatotropin has many effects on

Table 47-1: Hormones and Their Functions

Gland	Hormone	Target	Functions
Pituitary, anterior lobe	Growth hormone (GH, somatotropin)	All cells	Maintains protein production, releases fats and glucose
	Thyroid-stimulating hormone (TSH)	Thyroid gland	Stimulates production and secretion of thyroxine
	Adrenocorticotropic hormone (ACTH)	Adrenal cortex	Stimulates production and secretion of corticoids
	Follicle-stimulating hormone (FSH)	Gonads	Plays a role in female monthly cycle, the production of female sex hormones and male gametes
	Luteinizing hormone (LH)	Gonads	Plays a role in female monthly cycle, stimulates production of sex hormones
	Prolactin	Mammary glands	Stimulates growth of gland and production of milk
Hypothalamus	Releasing hormones	Pituitary	Stimulates release of GH, TSH, LH, FSH, ACTH, and prolactin
	Inhibiting hormones	Pituitary	Inhibits release of GH and prolactin
	Oxytocin	Uterus, mammary glands	Stimulates muscle contractions during childbirth, milk release
	Antidiuretic hormone (ADH, vasopressin)	Kidneys	Controls water reabsorption
Thyroid	Thyroxine	All body cells	Stimulates metabolic rate
Parathyroid	Parathyroid hormone (PTH)	Bone	Controls level of calcium ions and potassium ions
Adrenal cortex	Aldosterone	Kidneys	Controls reabsorption of sodium, stimulates excretion of potassium
	Hydrocortisone (cortisol)	Liver, various cells	Inhibits glucose uptake, aids healing, reduces inflammation
	Androgen	Male gonads	Stimulates development of male secondary sex characteristics
Adrenal medulla	Epinephrine, norepinephrine	Various cells	Controls stress reactions: increases heart and breathing rates, raises blood pressure and glucose level, inhibits digestion
Pancreas (islets of Langerhans)	Insulin Glucagon	Liver, muscle Liver	Stimulates glucose uptake Triggers breakdown of glycogen into glucose
Female gonads (ovaries)	Progesterone, estrogen	Female sex organs	Controls female secondary sex characteristic development, female sexual functions
Male gonads (testes)	Testosterone	Male sex organs	Controls development of male gametes and male secondary sex characteristics

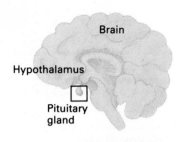

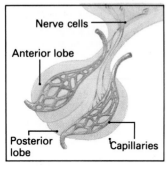

Figure 47–5. The illustration shows the location of the pituitary gland in the brain (top) and a closeup view of the gland (bottom).

metabolism. It stimulates bone and muscle growth and helps control the use of glucose and fatty acids for energy. *Prolactin,* another hormone of the anterior lobe, stimulates the mammary glands to produce milk after the birth of a child.

The posterior lobe of the pituitary does not produce any hormones, but it stores two hormones produced by the hypothalamus. They are *antidiuretic hormone (ADH)* and *oxytocin.* ADH, also called *vasopressin,* keeps the blood volume constant by controlling reabsorption of water in the kidneys. Oxytocin stimulates the contraction of uterine muscles during childbirth and the release of milk from the breasts after childbirth. Prolactin and oxytocin have no known function in males.

Most disorders associated with the pituitary gland involve somatotropin. An excess during childhood results in *gigantism,* or excessive growth. One victim grew to 2.7 m (8 ft. 11 in.). An excess during adulthood results in *acromegaly,* in which the hands, feet, and skull increase in size. Too little somatotropin during childhood results in *dwarfism,* characterized by a short body but otherwise normal proportions and normal mental and sexual development.

47.7 The Hypothalamus

The *hypothalamus,* which is a part of the brain, may be considered the master switchboard of the endocrine system. It links the endocrine system with the nervous system. The nervous system feeds information from the entire body into the hypothalamus. Based on that information, the hypothalamus then sends signals in the form of tropic hormones to stimulate or inhibit hormone secretion by the pituitary gland. At least nine such hormones have been identified. The hormones that stimulate secretions are called *releasing hormones.* Releasing hormones trigger secretion of TSH, GH, LH, FSH, ACTH, and prolactin. Hormones that slow down secretions are *inhibiting hormones.* The hypothalamus secretes inhibitors for GH, TSH, and prolactin. It also produces ADH and oxytocin and signals their release from the posterior pituitary.

Reviewing the Section

1. How do exocrine glands and endocrine glands differ?
2. How does a lack of iodine affect the body?
3. What happens when the body has too little insulin?
4. How does the hypothalamus control much of the endocrine system?

Endocrine System Regulation

The endocrine system and the nervous system together control other body systems. However, the endocrine system also controls itself.

- *Identify* the two general types of hormones.
- *Explain* how hormones affect their targets.
- *Describe* how a negative feedback circuit works.

47.8 Feedback

The endocrine system controls itself through a process called **negative feedback.** This process is similar to the way a thermostat regulates a household furnace. When the temperature falls below the thermostat setting, the furnace switches on and begins producing heat. When the temperature reaches the thermostat setting, the furnace switches off. *Similarly, the level of a hormone in the blood turns its own production off and on.*

Negative feedback controls the thyroxine level in the blood. The hypothalamus plays the role of the thermostat. The hypothalamus has cells that detect the presence of thyroxine in the blood. When the thyroxine level is low, the hypothalamus secretes a releasing hormone that stimulates the pituitary to secrete

Spotlight on Biologists: Irene Duhart Long

Born: Cleveland, Ohio, 1951

Degree: M.D., St. Louis University School of Medicine

American aerospace physician Irene Long studies how the weightless environment of outer space affects medical care of astronauts. Long is chief of the medical operations and human research branch of the National Aeronautics and Space Administration's (NASA) biomedical office. She is part of the team that provides medical care to astronauts in the event of an emergency.

Long and her crew also meet returning astronauts. Information gathered on cardiovascular or endocrine reactions to space travel will help determine future medical care for space travelers.

Long's fascination with air travel began when as a young child she accompanied her father on flying lessons. By the age of nine, she had decided that she wanted to become a NASA physician and work at the Space Center.

Long is one of the highest ranking women at the Center, and the first black to occupy her present position. Her next goal is to one day travel in space as the medical officer aboard the space shuttle.

Q: *How does the body rid itself of hormones?*

A: Excess hormones are in-activated by the liver and kid-neys and may be excreted. A kidney or liver disorder may cause problems from hor-mone buildup in the blood.

TSH. TSH causes the thyroid to secrete thyroxine. When the thyroxine level returns to normal, the hypothalamus stops se-creting the releasing hormone. As a result of this feedback mechanism, the pituitary stops secreting TSH, and the thyroid slows down secretion of thyroxine.

47.9 How Hormones Act

There are two types of hormones: **steroids,** which are fatlike organic compounds, and **protein hormones.** Sex hormones and corticoids are steroids. All others are protein hormones.

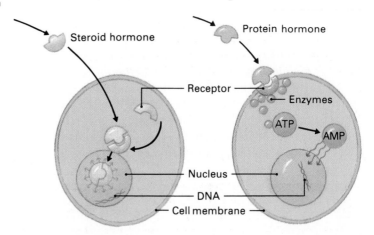

Figure 47–6. Steroid hor-mones (left) act directly within the cell nucleus, after passing through the cell membrane and binding with receptor molecules. Protein hormones (right) combine with receptor molecules at the boundary of the cell and work by activating enzymes inside the cell.

Steroids and protein hormones produce their effects differ-ently. A steroid passes through the target cell membrane. It combines with a receptor molecule and moves into the cell nu-cleus. There it helps determine the manufacture of specific pro-teins. Protein hormones affect their target cells through a two-step procedure called a ''two-messenger'' system. The first messenger, the hormone, combines with a receptor on the target cell membrane. This combination activates an enzyme on the membrane's inside wall. The enzyme helps change ATP into cyclic adenosine monophosphate, or **cyclic AMP.** Cyclic AMP triggers enzymes that bring about changes initiated by the origi-nal hormone. Thus, cyclic AMP is called the *second messenger*.

Reviewing the Section

1. How does negative feedback operate within the endocrine system?
2. What are the two types of hormones?
3. Why is cyclic AMP called the second messenger?

Investigation 47: How the Endocrine System Works

Purpose
To become familiar with the interactive operations of the endocrine system

Materials
Unlined paper, colored pencils, ruler, reference text, textbook

Procedure
1. On a sheet of unlined paper, copy the hormones, endocrine glands, and target organs and cells just as they are arranged at the bottom of this page. Be sure to leave plenty of space between the rows of items and between the items in each row.
2. Use black pencil to draw a box around each endocrine gland.
3. Draw a black triangle around each target that is not an endocrine gland.
4. Draw a black circle around each hormone.
5. Use green pencil to draw an arrow from each endocrine gland to the hormone or hormones it produces.
6. Use blue pencil to draw an arrow from each hormone to its target.
7. Tropic hormones stimulate their target glands to produce hormones. The hypothalamus detects increases in the levels of these hormones through the process of negative feedback. Draw a red arrow from each of these target glands to the hypothalamus.
8. The secretion of two pituitary hormones is regulated by inhibiting hormones as well as releasing hormones. Draw an orange arrow from these hormones to the hypothalamus.

Analyses and Conclusions
1. Study your complete drawing. What does it show you about the interactive operation of the endocrine system?
2. Which endocrine glands might be malfunctioning in an individual with abnormally low levels of hydrocortisone?

Going Further
- Diagram the negative feedback system that regulates the production of thyroxine.
- Study the effects of adrenalin and acetylcholine on heartbeat rate in *Daphnia*.
- Study the effect of testosterone on the secondary sex characteristics of leghorn cockerels.

Hypothalamus

Releasing hormones   Inhibiting hormones

Pituitary

TSH	ACTH	FSH	LH	GH	Prolactin
Thyroid	Adrenals	Gonads	Body cells		Mammary glands
Thyroxine	Aldosterone	Hydrocortisone	Androgens		
Body cells	Kidneys	Liver	Gonads		

Chapter 47 Review

Summary

Hormones secreted by endocrine glands control body functions such as metabolism, water and mineral balance, glucose balance and storage, muscle contraction, impulse transmission, and reproduction. The major endocrine glands are the thyroid, parathyroids, adrenals, pancreas, gonads, pituitary, and hypothalamus.

Hormones act upon specific tissues and organs, or targets. Tropic hormones affect other endocrine glands, causing them to secrete hormones.

The endocrine system controls itself through negative feedback. In the process of negative feedback, the blood level of a hormone causes a gland to either stop or start secretion of that hormone.

Hormones are either steroids or protein hormones. Steroids affect cells by passing through the cell membrane and moving to the nucleus. In the nucleus the steroid influences protein production. Protein hormones attach to receptors on the cell membrane where they activate an enzyme that helps convert ATP to cyclic AMP. Cyclic AMP then stimulates enzymes that bring about the changes initiated by the original hormone.

BioTerms

adrenal cortex **(720)**

adrenal gland **(720)**

adrenal medulla **(720)**

corticoid **(720)**

cyclic AMP **(726)**

endocrine gland **(719)**

epinephrine **(720)**

exocrine gland **(719)**

gonad **(722)**

hormone **(719)**

insulin **(720)**

islets of Langerhans **(720)**

negative feedback **(725)**

norepinephrine **(720)**

parathyroid gland **(719)**

pituitary gland **(722)**

protein hormone **(726)**

steroid **(726)**

target **(719)**

thyroid gland **(719)**

tropic hormone **(722)**

BioQuiz   (Write all answers on a separate sheet of paper.)

I. Completion

1. The endocrine system regulates itself through a ＿＿ system.
2. Deepening of the male voice is caused by the hormones called ＿＿.
3. ＿＿ hormones affect the secretions of other glands.
4. A lack of the hormone ＿＿, which stimulates uptake of glucose, may lead to diabetes mellitus.
5. The ＿＿ secretes hormones that help the body to handle stress.
6. Dwarfism results from too little ＿＿ during childhood.

II. Modified True and False

Mark each statement TRUE or FALSE. If false, change the underlined term to make the statement true.

7. The <u>hypothalamus</u> produces hormones that <u>inhibit or release</u> the secretion of other hormones.
8. The gonads are controlled by secretions of the <u>parathyroids</u>.
9. Protein <u>hormones</u> bring about their effects through cyclic AMP.
10. <u>Prostaglandins</u> act like hormones, but they are produced by almost every body cell, not only by glands.

III. Multiple Choice

11. Steroids affect cells by a) permitting glucose to enter. b) helping determine what proteins are produced. c) forming cyclic AMP. d) regulating mineral balance.
12. All hormones of the adrenal cortex are a) sex hormones. b) tropic hormones. c) growth hormones. d) steroids.
13. Improper functioning of the parathyroid affects a) glucose uptake. b) water balance. c) muscle contractions. d) energy storage.
14. Development of secondary sex characteristics is determined by the a) gonads. b) adrenal medulla.
    c) thyroid. d) parathyroid.
15. The body's water balance is governed by the a) adrenal medulla. b) adrenal cortex. c) gonads. d) thyroid.

IV. Essay

16. What is the difference between Type 1 and Type 2 diabetes?
17. How does the hypothalamus link the nervous and endocrine systems?
18. How does the two-messenger system of hormone activity work?
19. In what way does the hypothalamus regulate the body's water balance?
20. How do tropic hormones affect the endocrine system? Give an example.

Applying and Extending Concepts

1. Give one example of how the endocrine system helps maintain homeostasis.
2. Diagram a feedback loop showing the relationship between the hypothalamus, the pituitary gland, and one other gland of the endocrine system.
3. The parathyroid gland controls the release of calcium from bones into the bloodstream. Occasionally tumors affect the parathyroid gland so that too much calcium is released from the bones. What do you think are some of the results of such an occurrence?
4. Some athletes take steroids to help them build muscles. However, this use of steroids is highly controversial. Use your school or public library to research the current debate concerning steroids. Then write a report explaining why athletes believe that steroids help build muscles. Also, be sure to include information on the negative effects of steroid use.
5. Insulin is a small protein molecule consisting of a chain of 51 amino acids. Taking insulin orally instead of by injection does not help diabetics. Use your knowledge of body processes to explain why diabetics must take insulin by injection and not through insulin pills or medicine.

Related Readings

Ehrenkranz, J. R. L. "A Gland for All Seasons." *Natural History* 92 (June 1983): 18–23. The functions of the pineal gland, which may regulate behavior by seasons, are described.

Rosenfeld, A. "The Reexamination of a Curious Hormone." *Science 81* 2 (November 1981): 20–24. This article discusses the anti-obesity, anti-inflammatory, tumor-inhibiting, and stress-relieving properties of the steroid hormone DHEA.

Shodell, M. "The Prostaglandin Connection." *Science 83* 4 (March 1983): 78–82. Various functions of the hormonelike prostaglandins are described.

Reproduction and Development

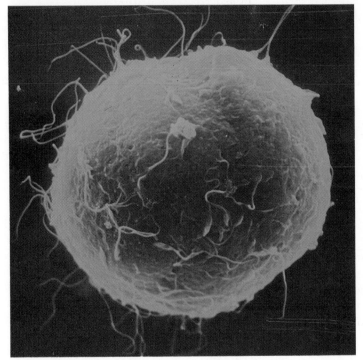

Fertilized egg with sperm attached

Introduction

A child is born. Life is passed from one generation to the next. By the time of birth, an infant has already developed for nine months within its mother. During these nine months, a single fertilized cell has become a complete human being. From the time of conception to the time of birth, the individual has increased in size 2 billion times! It has developed arms, legs, and internal organs to carry out all of life's functions. The development of a human being is a common, but very extraordinary, occurrence. Today, worldwide, about 141 children are born each minute, 8,500 each hour, and 74 million each year.

Like all mammals, humans reproduce sexually. Special structures in the male and female make up the human *reproductive systems.*

The Reproductive Systems

The reproductive organs in males and females are called *go-nads*. Gonads are present at birth. They develop fully during **puberty,** a period during which boys and girls mature physically and sexually. In males puberty generally occurs between the ages of 13 and 15. In females puberty is usually complete by 13 years of age. Following puberty gonads produce sex cells called **gametes** (GAM eets).

48.1 The Male Reproductive System

The male gonads, the **testes,** *are a pair of organs with two primary functions: to produce male gametes, called sperm, and to produce male hormones.* The testes are suspended below the abdomen in an external sac called the *scrotum,* where the temperature is lower than normal body temperature. The lower temperature is necessary for the production of healthy sperm. Inside the testes are hundreds of tightly packed, coiled tubes, called *seminiferous tubules,* where sperm are produced.

Sperm are carried from the seminiferous tubules to a coiled tube, the *epididymis* (ehp uh DIHD ih muhs). From the epididymis sperm move into the *vas deferens,* tightly coiled tubules where sperm are stored. Muscular contractions, called *ejaculation,* force sperm from the vas deferens through the *urethra,* in the *penis,* to the outside of the body.

- *Identify* the structures of the male and female reproductive systems and give their functions.
- *Trace* the pathway of sperm from the site of production to the outside of the male body.
- *Trace* the pathway of an unfertilized egg from its origin to the outside of the female body.
- *List* the components of semen and describe their functions.
- *Describe* the role of hormones in the menstrual cycle.

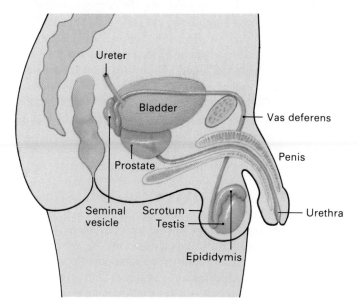

Figure 48–1. The structures of the male human reproductive system are shown here.

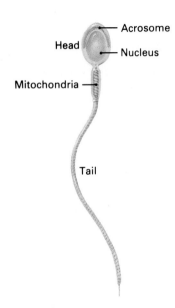

Head — Acrosome
— Nucleus

Mitochondria —

Tail

Figure 48–2. The drawing shows the parts of a human sperm cell.

As sperm are ejaculated, they are mixed with secretions from several glands—the *seminal vesicles, Cowper's glands,* and the *prostate.* The secretions contain nutrients, hormones, and enzymes that provide the sperm with energy and the proper environment. The sperm and the secretions together form a thick, milky liquid called **semen.**

A human sperm cell is extremely tiny. Approximately 4 billion sperm could easily fit into a thimble. Each sperm cell has three parts—the head, the middle region, and the tail. The head is the cell nucleus. A small region at the tip of the head, the *acrosome,* contains enzymes that enable the sperm to penetrate the female gamete. The middle region contains the *mitochondria* that supply the cell's energy. The tail, a long, slender flagellum, uses that energy to move the sperm cell.

During puberty, cells surrounding the seminiferous tubules begin to secrete male hormones, or androgens, the most important of which is **testosterone** (tehs TAHS tuh rohn). Testosterone stimulates development of secondary male sex characteristics. Among these are strong muscles, a deep voice, and body hair. Two tropic hormones secreted by the anterior pituitary are necessary for normal functioning of the testes. *LH (luteinizing hormone)* stimulates the production of testosterone. *FSH (follicle stimulating hormone)* controls the maturation of sperm cells.

48.2 The Female Reproductive System

The female gonads, the **ovaries,** are located in the lower part of the abdominal cavity. These olive-sized organs produce the female gametes, called *ova* or *eggs.* Below the ovaries is the **uterus,** a hollow, thick-walled, muscular organ about the size of a fist. The uterine walls are lined with mucous membrane that contains many glands and blood vessels. An unborn child develops here. A duct, called a **Fallopian tube,** or *oviduct,* extends from each side of the uterus. Fringed projections at the upper end of each Fallopian tube surround each ovary. Cilia lining these projections propel the mature egg to the uterus. A tube called the *vagina* leads from the uterus to the outside of the body. Between the uterus and the vagina is a muscular ring, the *cervix.*

At birth a female has about 2 million immature eggs in her ovaries. Beginning at puberty a single egg matures and is released from an ovary each month. In general, the ovaries alternate in releasing eggs. The left ovary releases an egg one month, and the right ovary releases one the next month. Approximately 400 eggs mature during a female's lifetime. During puberty the ovaries also begin to secrete the female hormone **estrogen** (EHS

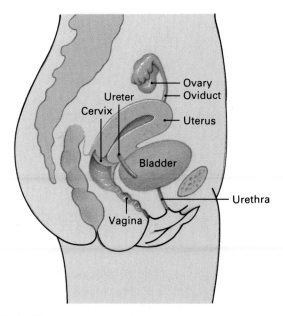

Figure 48–3. The major organs of the female human reproductive system are shown in this illustration.

Ovary
Oviduct
Ureter
Cervix
Uterus
Bladder
Urethra
Vagina

truh juhn). Estrogen causes development of secondary female sex characteristics, such as wide hips, body hair, and enlarged breasts.

48.3 The Menstrual Cycle

The female reproductive system has three primary functions: the production of eggs, the secretion of female sex hormones, and the nourishment and protection of a new individual. Approximately every 28 days, one egg matures and is released from the ovary, and the uterus is prepared to receive it. These activities are controlled by hormones operating on a feedback system. The entire process of ovulation and related changes in the uterus operates on a regular, repeating pattern called the **menstrual** (MEHN stroo wuhl) **cycle.**

The cycle starts with the release of FSH and LH from the anterior pituitary. These hormones trigger the maturing of an egg and its **follicle,** the fluid-filled chamber around the egg. The follicle secretes estrogen, which causes the uterine lining to thicken. Estrogen also triggers an increase in LH. The LH causes the follicle to rupture and release the mature egg from the ovary surface. The process of releasing a mature egg from the ovary is called **ovulation.**

Following ovulation LH stimulates the follicle to enlarge and fill with blood vessels. The follicle is now called the **corpus luteum** (KAWR puhs LOO tee uhm), or "yellow body." It functions temporarily as an endocrine gland, producing the

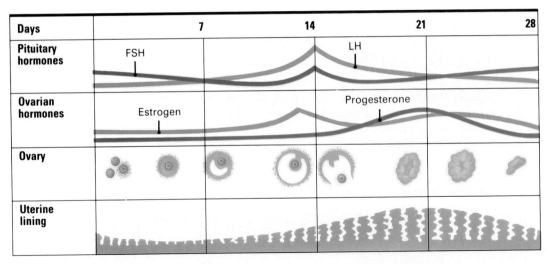

Days		7		14		21		28

Pituitary hormones — FSH, LH

Ovarian hormones — Estrogen, Progesterone

Ovary

Uterine lining

Figure 48–4. The diagram above shows the changes that take place in human females during the 28-day menstrual cycle. The cycle is counted from the first day of the menstrual flow.

ovarian hormones estrogen and **progesterone** (proh JEHS tuh rohn). Progesterone continues the process of thickening the uterine lining in preparation for the uterus to receive a fertilized egg. Both estrogen and progesterone inhibit the production of *GnRH (gonadotropin-releasing hormone)* by the hypothalamus. The lower GnRH level inhibits the production of FSH and LH, thus keeping another follicle from maturing.

If an egg is not fertilized, the corpus luteum disintegrates. The lining of the uterus collapses. Its cells die and are sloughed off through the vagina. This discharge of dead tissue and the unfertilized egg is called **menstruation** (mehn stroo WAY shuhn). The decreased levels of FSH and LH, plus disintegration of the corpus luteum, lead to lower progesterone and estrogen levels. The lower hormone levels trigger GnRH, which stimulates FSH and LH secretion. Thus a new cycle begins.

Males normally produce healthy sperm until about age 80. Females, however, cease to produce mature egg cells after **menopause,** the time at which the menstrual cycle ceases. This period usually begins between the ages 45 and 55. After menopause the pituitary does not secrete LH and follicles do not mature. Therefore, estrogen and progesterone are not produced.

Reviewing the Section

1. What are the chief functions of the male gonads?
2. Describe the formation of the corpus luteum and tell what it does.
3. List the differences in the rate of production of male and female gametes.

Fetal Development and Birth

The fusing of a sperm nucleus and an egg nucleus is called **fertilization.** When an egg is fertilized, menstruation does not occur. Instead, a nine-month **gestation** (jehs TAY shuhn) **period** begins. This period, also called **pregnancy,** is the time during which a fertilized egg develops into a child inside the mother.

48.4 Fertilization

During sexual intercourse semen is ejaculated through the male's penis into the female's vagina. Of some 400 million sperm ejaculated, less than 3,000 sperm get as far as a Fallopian tube and, of these, only about 50 reach the mature egg. In fertilization the head and middle section of the sperm enter the ovum. Only one sperm can fertilize an egg. Changes occur in the egg cell membrane to prevent other sperm from penetrating it.

Since both gametes are haploid, each parent contributes half the normal chromosome number. In human beings the haploid number is 23 chromosomes. When the gametes fuse, the **zygote** (ZY goht), or fertilized egg, has 46 chromosomes.

48.5 Embryonic Development

About 36 hours after fertilization, the zygote begins to divide by mitosis. The zygote does not increase in size as a result of these divisions. This process is called *cleavage*. The zygote continues to divide, forming a hollow ball of cells called the *blastula*. At this stage a major transformation occurs. The inner cell mass develops three layers, which eventually develop into the various organs of the body. This process is called *gastrulation*.

- *Describe* the fertilization process, including *in vitro* fertilization.
- *Define* the term *pregnancy*.
- *Draw* a time line showing the development of an unborn child.
- *Name* the protective embryonic membranes and their functions.
- *Summarize* the process of childbirth.

Figure 48–5. The illustration shows the fertilization of a human egg by sperm. The sperm move up through the oviduct to meet the egg. The egg when fertilized continues to pass down the oviduct and becomes implanted in the lining of the uterus.

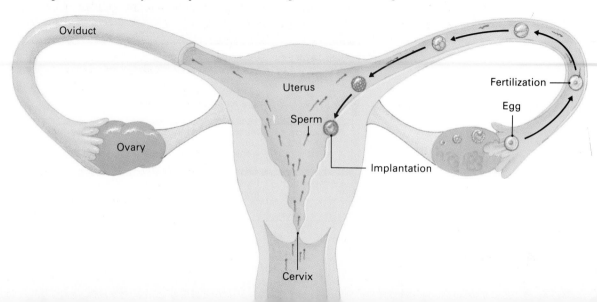

► Hope for the Childless

Blockage of the Fallopian tubes is the chief reason that many women are unable to become pregnant. The blockage prevents sperm from entering the tube and fertilizing a mature egg.

A recently developed technique can overcome this difficulty in a limited number of cases. The technique is known as *in vitro* fertilization (IVF). The term *in vitro* means "in glass" and refers to the fact that the process of fertilization takes place in a laboratory dish rather than in the human body.

The procedure involves taking a mature egg from the woman's body when the egg is ready to be released from the ovary. A physician removes the egg through a small incision in the woman's abdomen. The egg is placed in a laboratory dish to which nutrient has been added to keep the egg healthy. Sperm from the husband is added to the dish. After two days the material in the dish is examined to determine whether the fertilized egg is dividing. If division is taking place, the egg is implanted in the woman's uterus. If implantation is successful, the woman becomes pregnant. Nine months later, if the pregnancy has proceeded normally, the newborn infant joins the ranks of "test tube" babies.

In vitro fertilization has made it possible for a number of childless couples to have babies. However, these successes represent only about 12 percent of the women who undergo IVF. Most failures occur during implantation. The removal of the unfertilized egg from the ovary can cause bleeding or a hormone imbalance. When that happens, the lining of the uterus is not ready to accept the egg at the time of implantation, and pregnancy will not occur.

Table 48–1 shows how these layers develop into different body components, which combine to produce the body's organs. The blastula then moves down the oviduct and into the uterus where it is now called a *blastocyst*. Only one part of the blastocyst will develop into the **embryo.** An unborn child is called an embryo

Table 48–1: Development of Cell Layers

Cell Layer	Becomes
Ectoderm	Epidermis, including hair and nails; nervous system; epithelial tissue of the nose, mouth, and anus; enamel of the teeth
Mesoderm	Muscles, skeleton, the ducts of the excretory and reproductive systems, circulatory system, kidneys, gonads, dermis, connective tissue
Endoderm	Liver, pancreas, digestive tract, respiratory system

during the first eight weeks of its development. Afterward it is referred to as a **fetus.** The outer blastocyst cells, called the *trophoblast,* are not part of the embryo.

Six days after fertilization, the blastocyst attaches itself to the lining of the uterus, a process called *implantation.* From the trophoblast cells, four protective membranes are formed. They are the *yolk sac, allantois, amnion,* and *chorion.* The yolk sac has no yolk, as it does in birds, and serves no major function in the human embryo. The allantois functions in the formation of fetal blood vessels and gives rise to the **umbilical cord.** The umbilical cord, which contains two arteries and one vein, is the chief connection with the mother's uterus and permits the embryo to float in fluid. The amnion forms a sac, which fills with fluid. This *amniotic fluid* cushions the embryo and keeps it moist. The outermost membrane, the chorion, forms fingerlike projections, called *villi,* that attach to the lining of the uterus. Blood vessels from the allantois run through the villi.

Together the chorion and allantois form the **placenta.** The placenta is a mass of tissue that secretes estrogen and progesterone, which help maintain a thickened, blood-enriched uterine lining. The placenta also serves as the point of exchange of substances between mother and fetus. The exchange takes place through the umbilical cord by which the fetus is attached to the placenta. However, the connection between the circulatory systems of the mother and child is not a direct one. The mother's

Figure 48–6. The diagram shows how the exchange of nutrients and waste matter takes place across the placenta and uterine walls into the mother's bloodstream.

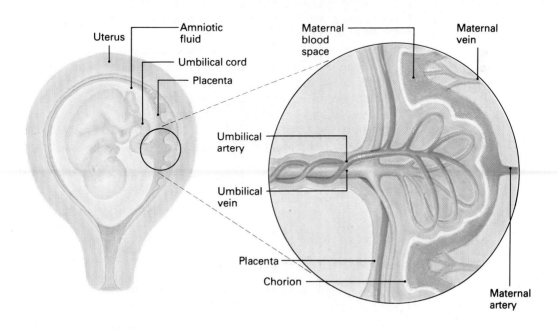

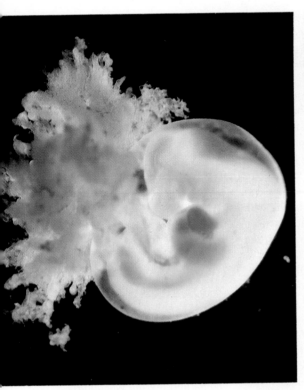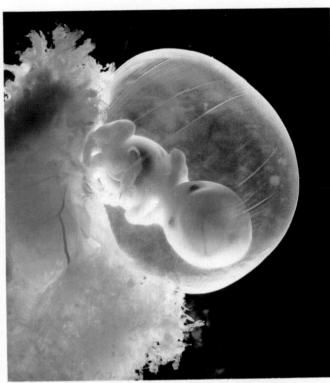

Figure 48–7. The series of photographs above show a human fetus at five stages in its development: (left to right) 5 weeks, 7 weeks, 2 months, 3 months, and 4 months.

body carries out the processes of digestion, excretion, and respiration for the developing child. Food and oxygen pass from the mother's bloodstream into the capillaries within the villi of the placenta. The food and oxygen then pass into adjacent capillaries belonging to the fetus and, thus, into the main bloodstream of the fetus. Waste products pass from the blood of the fetus to the blood of the mother in the same manner.

The nine-month pregnancy is commonly divided into three three-month periods called **trimesters.** During the first trimester, unorganized cells become organized into vital organs. In the second trimester, the organs and body features become more refined. After 25 weeks, development is complete. During the final trimester, most growth in body size takes place. A child born at this time would be premature but, with proper care, it would have a good chance of surviving outside its mother's body.

48.6 Childbirth

After approximately 266 days inside its mother's body, the child is ready for birth. Shortly before birth the mother's body undergoes changes that prepare it for **labor,** the process of literally

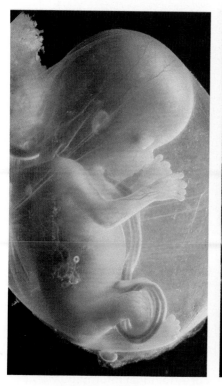 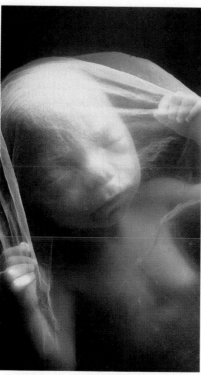 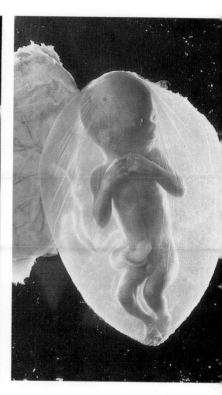

forcing the child out of her body. Contractions of uterine muscles become progressively stronger due to a decrease in progesterone, an increase in estrogen, and the release of the hormone *oxytocin* by the hypothalamus. Oxytocin speeds up contractions of the uterus. The cervix relaxes and widens from about 0.5 cm to 10 cm. As the baby enters the birth canal, the amnion breaks, and amniotic fluid escapes. Muscle contractions become very strong and finally push the infant out of the mother's body. The passing of the child from the uterus into the external environment is called **delivery.**

After delivery the umbilical cord is cut and tied to prevent a loss of blood through the umbilical blood vessels. The child must now support its own life processes. During the gestation period, fetal lungs do not operate. At birth carbon dioxide from cellular activity quickly builds up in the child's blood. This buildup triggers the respiratory center in the brain to take in oxygen, and the child begins breathing on its own.

Fetal blood follows a path that bypasses the lungs. Freshly oxygenated blood flows from the placenta to the heart and deoxygenated fetal blood flows through the heart in a path which leads it back to the placenta where it is oxygenated once again. Two openings in the fetal heart make this circulation possible.

Highlight on Careers: Nurse-Midwife

General Description

A nurse-midwife is a registered nurse who is specially trained to deliver babies. Nurse-midwives always have a contract with one or more physicians who provide medical care if a problem arises.

Most nurse-midwives work in hospitals. Some may deliver babies at home if asked by the parents to do so. Others work in birth centers. The birth center, a relatively new idea, is a delivery room with a homelike setting. Unlike the home, however, the birth center has extra personnel and special equipment ready should some problem arise.

Most nurse-midwives provide prenatal care for the pregnant woman, deliver the baby, and follow the baby's progress for six months. If the baby has a condition that requires medical attention, the nurse-midwife recommends that the infant be taken to a physician.

Career Requirements

The American College of Nurse-Midwives certifies nurse-midwives. To become certified, an applicant must pass a written examination, must be a registered nurse (RN), and must complete a course in midwifery. The course, given by nursing schools, usually takes one to two years. Some states require

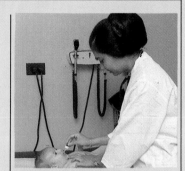

that nurse-midwives be licensed. Nursing schools generally look for nurse-midwife students who are compassionate, concerned about mothers and babies, hard-working, responsible, and self-reliant.

For Additional Information

American College of
   Nurse-Midwives
1000 Vermont Avenue, N.W.
Washington, DC 20005

When a newborn infant begins to breathe on its own, these openings normally close. If either remains open, it must be closed by surgery.

After delivery the uterus continues to contract, expelling fluid, blood, and the placenta with its attached umbilical cord. These materials are called the *afterbirth*. By this time the mother's pituitary gland has begun releasing *prolactin*, a hormone that stimulates her mammary glands to secrete milk.

Reviewing the Section

1. How is a zygote formed?
2. Explain how a fetus receives nutrients and how it rids itself of waste products.
3. What is the function of prolactin?

Investigation 48: Gametes and Embryonic Development

Purpose
To observe egg and sperm cells and the early stages of embryonic development

Materials
Prepared slides of egg and sperm cells, compound microscope, prepared slides of early embryonic development

Procedure
1. Place the slide of the egg cell on the microscope stage. Using low power, focus on the egg cell. Draw and label the parts of the egg cell. *What is the ratio of the size of the nucleus in proportion to the cytoplasm?*
2. Switch to high power and draw the egg cell.
3. Using the slide of the sperm cells, repeat steps 1 and 2. *What is the ratio of the nucleus to the cytoplasm in the sperm cell?*
4. Obtain slides of various stages of embryonic development. The slides might include the following stages: two, four, or eight blastomeres; blastula; early, mid, or late gastrula.
5. Observe each slide under both low and high power of the microscope. Make a drawing of each slide and label according to the information on the slide.
6. Describe each of the embryonic stages you have observed. Use reference books to help you in your descriptions.

Analyses and Conclusions
1. Compare the size of the egg and the sperm cells. Explain the difference in size.
2. How are the shapes of cells related to their functions?
3. What happens to sperm and egg cells after fertilization occurs?
4. The slides you used in this investigation were not of human cells. In what ways might human cells and tissues be similar to those you observed?
5. Explain the process by which fraternal twins and identical twins are formed.

Going Further
Research the beliefs of the animalculists and the ovists. How have views of the function of sperm and eggs changed since that time?

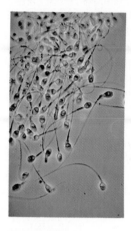
Sperm cells

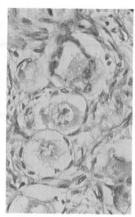
Egg cells

Chapter 48 Review

Summary

Male gonads, called testes, produce sperm. Sperm are the haploid male gametes. Female gonads, called ovaries, usually produce a single, mature, haploid egg each month. These eggs are the female gametes.

The maturing and releasing of an egg follows a pattern called the menstrual cycle. Hormones control the cycle which begins with the maturing of an egg and its follicle. Following ovulation the follicle forms the corpus luteum.

This body secretes hormones that prepare the uterus to receive a fertilized egg.

Fertilization begins a nine-month gestation period. The zygote divides to form a blastocyst. The blastocyst implants itself in the lining of the uterus. Part of the blastocyst becomes the embryo. A pregnancy is divided into trimesters during which unorganized cells first become organized and then develop into distinct tissues and organs.

BioTerms

corpus luteum (733)	gestation period (735)	progesterone (734)
delivery (739)	labor (738)	puberty (731)
embryo (736)	menopause (734)	semen (732)
estrogen (732)	menstrual cycle (733)	testis (731)
Fallopian tube (732)	menstruation (734)	testosterone (732)
fertilization (735)	ovary (732)	trimester (738)
fetus (737)	ovulation (733)	umbilical cord (737)
follicle (733)	placenta (737)	uterus (732)
gamete (731)	pregnancy (735)	zygote (735)

BioQuiz   (Write all answers on a separate sheet of paper.)

I. Completion

1. Because human gametes, which are haploid, have 23 chromosomes, a human zygote, which is diploid, has _____ chromosomes.
2. The period during which gonads mature is called _____ .
3. Ovulation occurs when _____ secretion increases sharply.
4. The fluid, blood, and placenta expelled from the mother's body after childbirth are called _____ .
5. Body organs of a fetus are formed during the _____ trimester.

II. Modified True and False

Mark each statement TRUE or FALSE. If false, change the underlined term to make the statement true.

6. Production of testosterone is controlled by LH.
7. The cervix temporarily becomes an endocrine gland, secreting estrogen and progesterone.
8. Estrogen causes muscles to contract strongly during labor.
9. Sperm is stored in the testes.
10. The progesterone level drops quickly when an egg is not fertilized.

III. Multiple Choice

11. FSH in males a) produces LH.
    b) controls sperm maturation. c) causes muscle growth. d) develops the testes.
12. Transfer of nutrients from the mother's to the fetus's blood occurs in the
    a) ovary. b) yolk sac. c) amnion.
    d) placenta.
13. The uterus is chiefly maintained during pregnancy by a) LH. b) FSH.
    c) progesterone. d) estrogen.
14. Development of female secondary sex characteristics is stimulated by the presence of a) estrogen.
    b) progesterone. c) oxytocin.
    d) testosterone.
15. When LH production in women stops, they experience a) menstruation.
    b) labor. c) pregnancy. d) menopause.

IV. Essay

16. What is the procedure for *in vitro* fertilization?
17. How does a fetus get the oxygen it needs, and how does this system change at birth?
18. What are the three parts of a sperm cell, and what is the function of each?
19. What is the function of the corpus luteum?
20. When a woman reaches menopause, what changes occur in her body?

Applying and Extending Concepts

1. Make a time line showing major events in the development of a fetus. You will need to research this at your school or public library. Separate the time line into trimesters and be sure to include the time at which brain waves are first detectable and when the major organs are formed. If possible, display the time line.
2. The mixing of type A$^-$ blood and type B$^+$ blood causes the blood cells to adhere together in clumps, a process called agglutination. Consider a mother with type A$^-$ blood who is carrying a fetus with type B$^+$ blood. Explain why agglutination does not become a problem as nutrients, oxygen, and waste matter are exchanged between the body of the mother and the body of the fetus.
3. Breast feeding of infants has been found to be effective in preventing a new pregnancy for a short period of time following delivery. According to the prevailing theory, nerve impulses triggered in the breasts of the mother by infant suckling inhibit the release of GnRH. Explain how this process could prevent a female from becoming pregnant.

Related Readings

Block, I. "Sperm Meets Egg." *Science Digest* 89 (April 1981): 96–99. Included in this article on conception are many electron photomicrographs.

Frisch, R. E. "Fatness, Puberty, and Fertility." *Natural History* 89 (November 1980): 16–27. This article explains how stored fat regulates reproductive ability in females.

Gregg, S. "STDs." *Sciquest* 53 (November 1980): 11–15. This article describes treatment of sexually transmitted diseases.

Nilsson, L. *A Child Is Born.* New York: Delacorte Press, 1977. This photographic study includes explanations of the reproductive systems, conception, birth, and breast feeding.

Alcohol, Other Drugs, and Tobacco

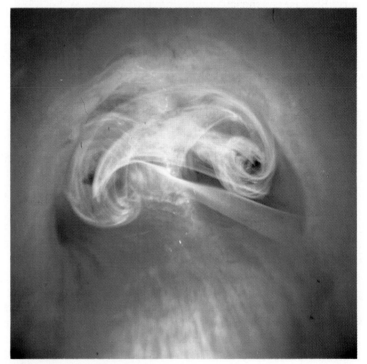

Cigarette smoke entering bronchi

Introduction

For thousands of years, people have known that certain chemical substances called **drugs** have a marked effect on the body and mind. The ancient Incas of South America, for instance, discovered that chewing leaves of the coca plant reduced fatigue and produced a feeling of well-being. In other parts of the world, people found drugs that soothe pain, produce sleep or changes in emotion, and cure disease.

Today people use a huge variety of drugs to prevent and fight disease and to reduce pain. However, a number of drugs are also misused by people seeking escape from personal problems and stressful situations. The effect of such use is almost always harmful, endangering an individual's health and ability to function in society.

Psychoactive Drugs

A **psychoactive drug** is one that affects the central nervous system. Alcohol is an example of such a drug. Many other drugs, ranging from diet pills to LSD, also belong to this category. Because psychoactive drugs produce a false sense of well-being and relief from tension, they are the drugs most commonly misused.

49.1 Drug Abuse

Drug abuse occurs when a person takes excessive amounts of a psychoactive drug for nonmedical reasons. Many psychoactive drugs lend themselves to abuse because when used regularly, they can cause a state of dependence called **addiction.** Psychological addiction occurs when a user has an emotional need for a drug and uses it to maintain a state of well-being. Physical addiction exists when the body needs the drug in order to work properly. A characteristic of addiction is **withdrawal,** a painful reaction when the drug is discontinued. This condition causes severe tremors, severe sweating, and feelings of anxiety, and may progress to physical ailments, such as cramps or muscle pain, nausea, vomiting, and convulsions.

Continued use of a psychoactive drug often leads to **tolerance,** a condition in which larger and larger doses of the drug are needed to produce the desired effect. Drug tolerance is dangerous because taking ever larger doses of a drug may result in a fatal overdose.

49.2 Alcohol

The psychoactive ingredient in beer, wine, and other alcoholic beverages is ethyl alcohol, or ethanol (C_2H_5OH). Of all the types of alcohol, only ethyl alcohol can be consumed with relative safety. *However, ethyl alcohol is a drug—in fact, one of the most widely used and abused of all psychoactive drugs.* People who consume alcohol excessively can cause themselves serious physical harm.

Effects of Alcohol When a glass of beer or wine is consumed, most of the ethyl alcohol it contains is absorbed in the small intestine. However, about 20 percent passes directly into the blood through the walls of the stomach less than two minutes after consumption. From the blood, ethyl alcohol diffuses into tissue fluid and affects every cell in the body.

- *List* the characteristics of drug abuse.
- *Describe* the effect of alcohol on the central nervous system.
- *Summarize* the stages leading to alcoholism and the health problems caused by excessive drinking.
- *Explain* how researchers have been able to diagnose alcoholism.
- *Distinguish* between the effects of hallucinogens, stimulants, and depressants.

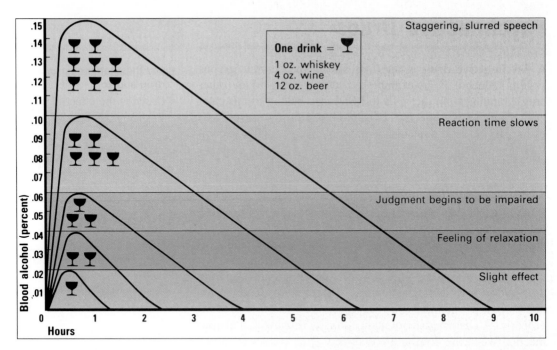

One drink =
1 oz. whiskey
4 oz. wine
12 oz. beer

Staggering, slurred speech

Reaction time slows

Judgment begins to be impaired

Feeling of relaxation

Slight effect

Blood alcohol (percent)

Hours

Figure 49–1. The graph shows the blood alcohol content and the behavioral effects that result after consuming 1, 2, 3, 5, and 8 drinks. Higher levels of alcohol in blood can cause loss of consciousness, coma, and death.

Q/A

Q: *What is fetal alcohol syndrome and what are some of its symptoms?*

A: Babies born of mothers who drink alcohol during pregnancy frequently have a condition called *fetal alcohol syndrome.* Babies with this disorder have small heads, low birth weight, retarded growth, and mental retardation. Many also suffer from heart murmurs, hernias, and urinary tract abnormalities.

Alcohol is oxidized in the liver and in body cells at a fairly constant rate of 29.6 mL (1 oz.) per hour. Oxidation changes an ounce of ethyl alcohol into water, carbon dioxide, and about 200 calories of heat energy. When a person drinks more alcohol than the body can oxidize, the excess alcohol accumulates in the blood and tissue fluid. The excess alcohol acts as a **depressant,** a drug that slows the functions of the central nervous system.

Just a small amount of alcohol affects the cerebral cortex, the part of the brain that controls thought, judgment, and self control. By suppressing these functions, alcohol produces feelings of relaxation and freedom from tension. A higher concentration of alcohol in the blood slows the brain centers governing speech, vision, hearing, coordination, and balance. As a result, the drinker may experience slurred speech, double vision, and staggering. If the concentration of alcohol in the blood reaches high enough levels, it can suppress the brain centers governing breathing and the heartbeat, causing death.

Alcoholism Addiction to alcohol, or **alcoholism,** is the major drug abuse problem in the United States. About 10 million adults and 3 million teenagers are alcoholics. More than 200,000 people die of alcoholism each year. Alcoholism is a form of drug addiction. However, it is a treatable disease.

Alcoholism develops gradually through several stages. During the early stages, an individual drinks to be sociable but soon consumes larger amounts of alcohol and may experience blackouts. The individual continues to function during a blackout, but later cannot remember his or her actions. In advanced stages the individual loses control of the amount of alcohol consumed and eventually becomes intoxicated every day. In addition, a form of mental illness called *alcoholic psychosis* may occur. An individual with alcoholic psychosis becomes confused and may not recognize family members. In extreme cases, the individual may see or hear things that do not exist and develop uncontrollable trembling called **delirium tremens,** or **DTs.**

People who drink excessive amounts of alcohol over long periods of time also develop serious health problems. For example, many alcoholics suffer from a lack of vitamins because they do not eat properly when drinking heavily. A lack of nutrients may cause malnutrition and lead to abnormalities in the heart and circulation and inflammation of the stomach lining. In addition, alcohol is readily converted to energy, preventing the body from breaking down other nutrients, such as sugars, amino acids, and fatty acids. These nutrients are stored as fats in the liver. After several years of heavy drinking, liver cells are filled with fat and begin to die. Once in this condition, the liver may become inflamed, a condition called *alcoholic hepatitis*.

Q/A

Q: *Is alcoholism related to a person's genetic makeup?*

A: Research results are inconclusive, although some alcoholics seem to inherit a tendency for the disease. This tendency may take the form of a difference in the ability to metabolize alcohol.

▶ Finally, a Way to Diagnose Alcoholism

Diagnosing alcoholism before it reaches advanced stages is a difficult problem. One reason is that alcohol is absorbed and oxidized so rapidly that it quickly disappears from the bloodstream. A second reason is that alcoholics generally deny to themselves or anyone else that they are drinking too much. As a result of this denial, doctors facing symptoms similar to those of alcoholism have had no sure way to make an accurate diagnosis.

Recently, however, experiments have indicated that the results of standard blood chemistry tests can be used to pinpoint alcoholics. In one experiment, researchers took blood from test groups of known alcoholics and nonalcoholics. Typical blood tests measure 25 separate chemicals in the blood. Instead of analyzing the test results of these 25 chemicals individually, however, the researchers looked for distinct patterns among the results. By doing a mathematical analysis of many blood chemicals and how they relate to each other, the researchers found one pattern associated with alcoholism and another associated with nonalcoholism. These patterns were then used in analyzing blood tests of other individuals. The researchers found they could identify severe cases of alcoholism 100 percent of the time, less severe cases 94 percent of the time, and nonalcoholics 100 percent of the time.

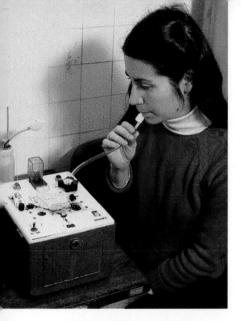

Figure 49–2. Law enforcement officers use breath-analyzing devices like the one shown to determine the level of alcohol in a driver's bloodstream.

Q/A

Q: *How do hallucinogens produce their effects?*

A: All thoughts and feelings are produced by communication between nerve cells of the brain by means of chemicals called neurotransmitters. One of these neurotransmitters is serotonin. Produced in larger than normal quantities, it can induce feelings of great happiness. Many drugs including hallucinogens, which excite the brain, are very similar to serotonin in chemical structure.

If heavy drinking continues, a liver condition called **cirrhosis** may develop. In cirrhosis, functioning liver cells are replaced with useless scar tissue, and the liver gradually shrinks into a small, hard mass. In this form, the liver can no longer eliminate body wastes, produce blood clotting factors, or carry out its other functions.

49.3 Other Psychoactive Drugs

Psychoactive drugs include a variety of substances other than alcohol. The caffeine found in coffee is a psychoactive drug, as are heroin and the drug in marijuana. Some psychoactive drugs, while used in controlled doses to treat pain and disease, cannot be sold to or used by the general public. Abuse of both legal and illegal drugs can lead to many serious health problems.

Hallucinogens Drugs that distort the way the brain translates impulses from the sensory organs are called **hallucinogens** (huh LOO suh nuh jehnz). These distortions may take one of two forms. The brain may alter messages about something real, producing an illusion. The brain also may produce images with no basis in reality called *hallucinations*. Hallucinogens include LSD (lysergic acid diethylamide), mescaline, and peyote.

The effects of hallucinogens depend upon a variety of factors, such as the chemical used, the dosage, and the user's emotional state. Sometimes users see vivid images and have feelings of well-being. Users also claim that hallucinogens make them more creative and perceptive, although research has not supported these claims. However, people using hallucinogens may also experience depression, terror, and fear of dying.

LSD is an especially dangerous hallucinogen because its effects are so unpredictable. LSD may also make users feel that nothing can harm them. In this state of mind, users may take physical risks, such as stepping in front of a car, that lead to injuries or death.

Hallucinogens have a high potential for abuse. They do not create physical dependence but may produce psychological dependence. Hallucinogens do not have any currently acceptable medical use.

Stimulants Certain drugs are called **stimulants** because they stimulate the central nervous system and thereby speed up body processes. One common stimulant is the caffeine found in coffee, tea, cola soft drinks, chocolate, and some diet pills. Amphetamines are a group of strong stimulants often called "speed." Other stimulants include nicotine and cocaine.

Highlight on Careers: Drug Counselor

General Description

A drug counselor helps people overcome drug-abuse problems or drug addiction. Counselors may be professional social workers or psychologists. Some are community workers interested in drug problems.

Many drug counselors are employed by private alcohol treatment centers or hospital-based programs. Others work for state or local government agencies. These agencies offer various programs. In some, clients are first isolated in rural areas away from drug sources. Later they return to city life for counseling. In other programs, counselors may work with groups of drug addicts or abusers rather than with individuals. Special group programs to alleviate or prevent drug abuse are conducted in some big-city schools.

Regardless of the type of program, counselors usually give clients moral support and an opportunity to discuss personal or work problems. The counselor refers medical problems to a physician connected with the treatment program.

Career Requirements

No formal professional requirements exist for drug counseling. However, each state has its own requirements. In some states counselors must earn a certificate. At least one state requires drug counselors to be social workers, while several other states give preference to social workers. A Master's degree in social work is often recommended.

For Additional Information

National Association of
  Drug Abuse Problems
355 Lexington Avenue
New York, NY 10017

The amphetamines are among the most commonly used of all stimulants. These drugs stimulate the cerebral cortex by replacing neurotransmitters at the synapses. Their chief effect is to cause the body to react as if it were in danger. Thus amphetamines speed up metabolism, blood pressure, heartbeat and respiratory rates. They also cause pupils to dilate, reduce the appetite, and produce a feeling of alertness and confidence.

Long-term use of stimulants can be dangerous. Because stimulants produce feelings of energy, users may not realize how exhausted they are. Excessive use of amphetamines for long periods can even cause mental problems resembling schizophrenia, a condition in which a person loses touch with reality. In extreme cases, individuals may become violent.

Cocaine is a stimulant made from the leaves of the coca plant. Cocaine users take cocaine by inhaling it through the nose or by injecting it into a vein. The effects of cocaine resemble those of amphetamines, but cocaine also damages the membranes of the nasal passages. Ulcers often form and holes may penetrate the septum.

Depressants Depressants are drugs that slow down the action of the central nervous system. This group of drugs includes alcohol, calming drugs called *tranquilizers*, and more powerful relaxants called *barbiturates*.

In small doses, depressants decrease awareness and relieve tension and inhibitions. Larger doses are used to treat sleep disorders and anxiety. However, excessive doses of depressants may result in a coma, a condition of deep unconsciousness. Depressants taken in combination with alcohol can be deadly. The depressant effects of both drugs can stop respiration.

Phencyclidine hydrochloride, also known as PCP or angel dust, is another depressant. However, it may also act as a hallucinogen or a stimulant. PCP is an extremely dangerous drug that can cause confusion, delirium, paralysis, and violent behavior. Because the effects of this drug are so serious, it has been prohibited for any human use.

Depressants may cause serious physical and psychological dependence. Withdrawal can be so severe that convulsions and death result.

Inhalants A broad group of chemicals produce psychoactive vapors that can be inhaled. Nitrous oxide, also called "laughing gas," ether, and chloroform all have been used medically as anaesthetics. When inhaled they produce a short-term effect similar to alcoholic intoxication.

Highly toxic are the fumes from a large group of organic solvents. These inexpensive and easily available substances include gasoline, paint thinners, glue, and cleaning fluids. Prolonged use results in irreversible brain and liver damage.

Marijuana Marijuana comes from the dried leaves, flowers, or stems of the Indian hemp plant *Cannabis sativa*. The psychoactive ingredient in marijuana is Δ9–THC (delta-9-tetrahydrocannabinol), which may act as a hallucinogen, stimulant, or depressant. Hashish is made from the dark, sticky resin of the hemp plant. It contains more THC than marijuana and is therefore more powerful.

Marijuana affects users in various ways. In some cases, the drug produces a sense of well-being and enhances the senses of taste, smell, sight, and hearing. If an individual is upset or depressed, however, marijuana may intensify those feelings.

Figure 49–3. Slender, pointed leaves are a distinctive characteristic of the plant *Cannabis sativa,* the source of marijuana and hashish.

While under the influence of marijuana, a person may experience loss of judgment, loss of inhibitions, and distorted vision and hearing. Although marijuana does not seem to cause physical addiction, users may become psychologically addicted. Prolonged use of marijuana can impair short-term memory and muscular coordination.

Narcotics Drugs that dull the senses and relieve pain by depressing the cerebral cortex are called **narcotics.** They also affect the *thalamus,* the body's mood-regulating center. One group of narcotics, called *opiates,* are derived from the opium poppy. Opiates include codeine, morphine, heroin, and opium.

Many opiates are used as pain relievers and cough suppressants. Codeine, for instance, is found in some cough medicines. Morphine, which is a stronger drug than codeine, is used to dull severe pain. Heroin, which is stronger and more addictive than other opiates, is not used in medicine. However, it is the narcotic most widely abused.

Heroin can be inhaled or injected under the skin or into a vein. It induces a feeling of happiness, relieves pain, and affects the body much as a depressant would. Tolerance and addiction develop rapidly. For a heroin addict, withdrawal symptoms appear just a few hours after the last dose and may last for several days. Feelings of anxiety and general discomfort may progress to dilation of the pupils of the eye, diarrhea, and pain in the abdomen, muscles, and joints.

One method of treating heroin addiction is to substitute another narcotic agent, methadone, for the heroin. Methadone, like heroin, is addictive. An advantage is that methadone is taken orally, thus avoiding needle infections. Also, methadone is three to six times longer-lasting than heroin, so that the dosage can be given once a day. Methadone treatment is effective only when administered as part of a multiple program that also includes vocational and psychological guidance.

Reviewing the Section

1. What is withdrawal? How is it related to dependence?
2. How does alcohol affect the central nervous system?
3. How does cirrhosis develop?
4. How have researchers been able to diagnose alcoholism?
5. What is the difference between stimulants and depressants?
6. What is a hallucinogen?
7. What is a narcotic?

- *Identify* several components of tobacco smoke.
- *Describe* some effects of tobacco smoke.
- *Explain* why tobacco smoke may seriously affect non-smokers.
- *Discuss* several diseases caused by or related to smoking.

Tobacco

Like any psychoactive drug that is misused, tobacco poses a serious health threat. Tobacco smoke is a combination of heated gases and suspended particles that contains more than 1,200 poisonous chemicals. One element in tobacco is **nicotine,** a colorless, oily psychoactive drug. *Tars* make up the particles in tobacco smoke. They also form a sticky coating on the lining of the bronchial tubes and can interfere with breathing.

49.4 Effects of Tobacco

Some gases in cigarette smoke, such as hydrogen cyanide, are strong poisons. Carbon monoxide, another of the gases, is thought to contribute to cardiovascular disease among smokers. Carbon monoxide reduces the amount of oxygen carried by hemoglobin, causing shortness of breath. In addition, the action of the cilia that line the respiratory passages may be inhibited for up to eight hours after being exposed to smoke. Constant exposure eventually destroys the cells that produce cilia and mucus in the respiratory tract. A smoker is then more susceptible to respiratory infections.

The carbon monoxide level in the blood returns to normal within eight hours after a cigarette-user stops smoking. The cilia require one to nine months to regrow after smoking stops.

Nicotine has a different effect than carbon monoxide. In small doses it acts as a stimulant and speeds up the transmission of nerve impulses. In high dosages, however, it has an inhibiting effect on nerve impulses. Nicotine also stimulates the adrenal glands. As a result, blood pressure rises, the heartbeat rate increases, and blood vessels constrict.

The smoker is not the only person affected by tobacco smoke. The smoke from a cigarette consists of mainstream smoke, which the smoker breathes in, and sidestream smoke from the tip, which is not drawn through the cigarette. Sidestream smoke contains twice as much tar and nicotine and five times as much carbon monoxide as mainstream smoke. Nonsmokers breathe this smoke into their lungs whenever they are in the vicinity of a smoker.

49.5 Diseases Related to Smoking

Smoking is involved in about 30 percent of all heart disease in the United States. Nicotine increases blood pressure and the heartbeat rate, causing stress on the circulatory system.

Q/A

Q: Is "smokeless" tobacco any less harmful than smoking?

A: Nicotine is also absorbed by the body from chewing tobacco and from dipping snuff. Chewing tobacco and snuff are equally as addictive as tobacco that is smoked, and lead to an increased risk of mouth and throat cancers.

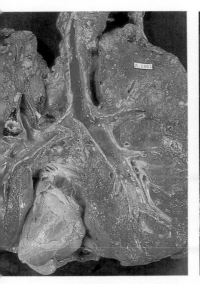 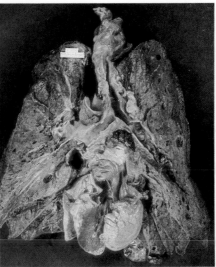 

Smoking is also believed to contribute to 80 percent of all cases of lung cancer. Lung cancer is the leading type of cancer death in the United States, accounting for 120,000 deaths each year. The risk of developing lung cancer is directly related to the length of time a person has been smoking and the number of cigarettes smoked each day. Cancer of the mouth, larynx, esophagus, stomach, and urinary bladder are also more common in smokers than in nonsmokers.

Emphysema and chronic bronchitis are also related to smoking. Chronic bronchitis involves inflammation of the bronchial tubes and an increase in the production of mucus. A chronic cough and breathing difficulties result. Chronic bronchitis is often followed by emphysema. Emphysema is a disease marked by rupture of the alveoli, the tiny sacs in the lungs where gases are exchanged. Because the surface area for gas exchange is reduced, a person suffering from emphysema cannot get rid of carbon dioxide or take in oxygen efficiently. Eighty percent of all emphysema cases are related to smoking.

Figure 49–4. The photo on the left shows a healthy heart and lung, photographed from the back. Compare this with the photos of a lung of a person suffering from emphysema (center) and cancer (right).

Reviewing the Section

1. How does the carbon monoxide in cigarette smoke affect the body?
2. Why might someone living with a smoker suffer more from cigarette smoke than the smoker does?
3. How does nicotine affect the body?
4. What are tobacco tars? Describe what they do.
5. List and describe several diseases related to smoking.

Investigation 49: The Effect of Smoke on Air Passages

Purpose
To observe a simulation of the condensation of substances along the respiratory pathway when cigarettes are smoked

Materials

Cigarettes, matches, glass tubing, rubber tubing, aspirator, rubber top of medicine dropper, scissors, burette clamp, ring stand, cork

Procedure
1. Cut the closed end off the top of a medicine dropper. This makes a holder that has the exact diameter necessary to hold a cigarette.
2. Using a short piece of rubber tubing, connect the other end of the holder to a six-inch piece of glass tubing.
3. Using a second piece of rubber tubing, connect the other end of the glass tubing to an aspirator.
4. Suspend the tubing horizontally by passing the tube through a cork. Hold the cork with a burette clamp at the appropriate level on a ring stand as shown in the illustration.

5. Place a cigarette in the rubber holder.
6. Using the aspirator, draw air through the system. By placing a lighted match at the tip of the cigarette, the cigarette can be "smoked." Gases from the burning cigarette will be drawn through the glass tubing. Two or three cigarettes can be "smoked" this way with the apparatus. *What is the glass tubing simulating?*
7. Observe the material that collects on the inside of the tubing. *Where did it come from?*
8. Disconnect the glass tubing and smell the material inside. *Describe its aroma.*

Analyses and Conclusions
1. What effects could the substances collected in the glass tubing have on the surface cells of human respiratory passageways?
2. How could you determine what chemicals were present in these substances?

Going Further
Study the effects on living cells of the chemical materials collected in the glass tubing.

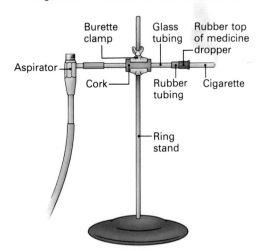

Chapter 49 Review

Summary

Psychoactive drugs are those drugs that affect the central nervous system. These drugs are often abused. Drug abuse may lead to physical dependence, psychological dependence, or both. The body may also build up a drug tolerance. Tolerance is a condition in which the body becomes "used to" a drug. As a result, an increasingly large amount of the drug is required to produce the same psychological effect.

Alcohol is a depressant that slows down body processes. When a person consumes more alcohol than the body can oxidize, the excess alcohol affects the central nervous system. Thought, judgment, and self control are affected first. Increased amounts of alcohol disturb speech, vision, balance, and eventually respiration. Addiction to alcohol, called alcoholism, is a treatable disease. Other psychoactive drugs are hallucinogens, stimulants, inhalants, marijuana, and narcotics. Many of these drugs are illegal because they impair judgment and may make users dangerous to themselves and to others.

Tobacco contains the drug nicotine, plus tars and gases. Tobacco has been linked to cancer, cardiovascular diseases, chronic bronchitis, and emphysema. Smoke affects not only the smoker who inhales mainstream smoke but also anyone nearby who inhales sidestream smoke.

BioTerms

addiction (745)
alcoholism (746)
cirrhosis (748)
delirium tremens (DTs) (747)
depressant (746)

drug (744)
drug abuse (745)
hallucinogen (748)
narcotic (751)
nicotine (752)

psychoactive drug (745)
stimulant (748)
tolerance (745)
withdrawal (745)

BioQuiz   (Write all answers on a separate sheet of paper.)

I. Completion

1. _____ drugs affect the central nervous system.
2. The psychoactive ingredient in _____ can act as a hallucinogen, a stimulant, or a depressant.
3. Excess alcohol in the system acts as a _____.
4. When a user stops taking drugs, he or she is likely to undergo painful symptoms of _____.
5. _____ are drugs that slow down body processes.

II. Modified True and False

Mark each statement TRUE or FALSE. If false, change the underlined term to make the statement true.

6. A person may experience paranoia from taking hallucinogens.
7. Emphysema ruptures the alveoli.
8. Tobacco tars make blood vessels constrict and the heart beat fast.
9. Cola soft drinks contain the stimulant caffeine.
10. Cocaine is the most commonly abused narcotic drug.

III. Multiple Choice

11. Alcohol affects the body rapidly because it is absorbed into the blood from the a) stomach. b) liver. c) large intestine. d) mouth.
12. Drugs that cause the body to react as if it were in danger are a) tranquilizers. b) barbiturates. c) amphetamines. d) opiates.
13. The drug in tobacco is a) nicotine. b) caffeine. c) formaldehyde. d) tar.
14. Severe physical exhaustion may result from prolonged use of a) barbiturates. b) LSD. c) amphetamines. d) heroin.
15. The ingredient in cigarette smoke which may contribute to heart disease is a) nicotine. b) carbon monoxide. c) carbon dioxide. d) tar.

IV. Essay

16. What causes an alcoholic's liver to deteriorate?
17. Why is it dangerous to mix alcohol and barbiturates?
18. Why are smokers susceptible to respiratory infections?
19. How might some people abuse both stimulants and depressants at the same time?
20. What are effects of each of the major categories of drugs?

Applying and Extending Concepts

1. Under the Controlled Substances Act, drugs are classified into five categories. Do research in the library on the factors that determine the category into which a drug is placed. List a few examples for each category.
2. The antidiuretic hormone (ADH) secreted by the hypothalamus controls the amount of water reabsorbed by the kidneys. ADH makes the membranes of the collecting ducts of the kidneys permeable so that fluids return to the bloodstream. Alcohol suppresses ADH secretion. Explain the likely effect of this suppression.
3. Alcohol and caffeine stimulate the secretion of gastrin, a hormone that controls the secretion of hydrochloric acid (HCl) in the stomach. The higher the level of gastrin, the greater the amount of HCl secreted. Explain what happens when a person consumes a large quantity of these drugs.
4. Research the effects of smoking by a pregnant woman on her fetus. Include in a report the effects on birth weight, death rate, and miscarriage.
5. The central nervous system produces secretions called endorphins. Prepare a report for the class on what you can find out about the similarity between narcotics and endorphins.
6. The hemoglobin of a smoker who smokes two packs of cigarettes a day carries 18 percent carbon monoxide, or carboxyhemoglobin. Therefore, hemoglobin cannot carry 100 percent oxygen. Explain how this much smoking affects the heart.

Related Readings

Brecher, E. M. *Licit and Illicit Drugs*. Boston: Little, Brown, 1972. This Consumers Union Report discusses socially acceptable as well as illegal drugs, within a historical framework.

Lehmann, P. "Psyching Out the Athlete's Medicine Chest." *Sciquest* 1 (November 1979): 6–11. This article describes research into the dangers of various drugs often used by athletes.

CHAPTER

50

Introduction to Ecology

A beaver changing its environment

Introduction

Late into the night, a beaver works diligently to fell a tree. It alternately bites the trunk and strips the bark. Finally the sapling falls. The beaver pushes and drags the tree into a nearby stream where it adds the new material to the winter home it has been building for its family.

The story above depicts only one small part of the life of a beaver and the relationship of the animal to its surroundings. The study of the relationships of living things to their environment is called **ecology.** The role of an *ecologist,* a person who studies ecology, is to discover those relationships and how they are affected by changes in the environment. Environmental changes that affect one type of organism usually affect many others as well.

Organization of the Biosphere

Section Objectives

- *List* some effects of abiotic factors on ecosystems.
- *Explain* how populations of organisms are organized into a community.
- *State* the difference between an organism's habitat and its niche.
- *Discuss* the importance of the relationships among members of a community.

Life on Earth extends from the ocean depths to a few kilometers above the land's surface. The great biological drama, the struggle for survival, is played out entirely within this thin and fragile space. Ecologists call this area where life exists the **biosphere.** The biosphere is an extremely complex system and a difficult one to understand. To unravel the mystery of how millions of diverse species live together in the biosphere, ecologists divide it into smaller, more easily studied units. These units are called ecosystems.

50.1 Ecosystems

An **ecosystem** is a physically distinct, self-supporting unit of interacting organisms and their surrounding environment. A forest, for example, is an ecosystem. Its physical boundaries are the areas where trees give way to other types of vegetation. A forest is considered self-supporting because plants change energy from the sun into chemical energy. Forest animals then eat the plants and are eaten by other animals, thus transferring energy from organism to organism. When forest plants and animals die, their bodies are decomposed by microorganisms, and the chemicals released by this decomposition are then reused by other living things. By these four processes—production of chemical energy from sunlight, transfer of energy, decomposition, and reuse of nutrients—an ecosystem sustains itself year after year. A forest is an example of a physically distinct, self-supporting part of the biosphere. Therefore, a forest is considered an ecosystem.

Many kinds of ecosystems exist on Earth. They vary in composition and size. An ecosystem can be as large and complex as a jungle or as small and simple as a tiny patch of garden. Meadows, rivers, and tide pools are examples of ecosystems. Bogs found in the middle of a forest and the waters of mountain lakes are also ecosystems.

An ecosystem consists of two sets of environmental factors: biotic factors and abiotic factors. An ecosystem's **biotic factors** are all of its living organisms. Most ecosystems contain many organisms of many species. The individuals of some species are easy to see, such as a hawk or an oak tree, but many are not visible to the naked eye, such as the microbes that exist everywhere. **Abiotic factors** are the nonliving parts of the ecosystem. These factors include light, soil, water, wind, temperature, and nutrients.

Figure 50–1. The boundary between a forest ecosystem and a sand dune ecosystem is clearly visible in this photograph.

▶ Operation Cat Drop

If you had been in Borneo one day in 1955, you would have seen a strange and surprising sight. The skies over many villages were filled with parachutes dropped by the British Royal Air Force. The parachutes were carrying cats.

The need for "Operation Cat Drop" arose after the World Health Organization attempted to rid the area of malaria. To control the population of malaria-carrying mosquitoes, the villages were sprayed with two potent insecticides. These insecticides wiped out the mosquitoes, but they also created many other problems.

Two of the normal inhabitants of the village huts were cockroaches and lizards called geckoes. The cockroaches, contaminated by the insecticides, were eaten by the geckoes. In turn, the lizards were eaten by cats. As a result, the cats were poisoned by the accumulation of pesticides. Without cats the village rat population increased dramatically. Because rats carry bubonic plague and typhus, the villagers were threatened by these deadly diseases.

Operation Cat Drop restored the cat population. The parachuted cats went to work immediately. Soon the rat population was under control and the disease threat eliminated.

Operation Cat Drop emphasizes the fact that organisms depend on one another for their survival. It reminded scientists that disturbing one part of an ecosystem's biotic community has unexpected and far-reaching effects on its other members.

Figure 50–2. Cactus plants (top) thrive in sunlight so bright it would kill most other plant species. Violets (bottom) grow best on a shady forest floor.

Light and Temperature Light and temperature are two of the most important abiotic factors. Plants such as the barrel cactus grow best in the high-intensity light of the desert. Nettles on a forest floor, however, are adapted to low-intensity light. In addition, the distribution of organisms in a lake or ocean is directly related to how deep light penetrates the water. For instance, some algae live near the surface of a lake where the light is bright, while other kinds of algae live near the bottom in areas of little light.

Temperature is equally important. Some organisms, such as lemmings and dwarf willows, are adapted to the cold arctic wastelands, where the temperatures during the growing season are often no higher than 5°C (41°F). Gila monsters and many other cold-blooded animals live in the desert where summer temperatures are often above 38°C (100°F).

Water and Soil Water and soil are two additional important abiotic factors. Individual organisms differ in their need for water. Many animals, such as the kangaroo rat, live in a desert environment, where water is scarce. Many other animals and plants, however, live in water.

Soil variations are also important. An example is the plant genus *Ammophila*. *Amm* is Latin for "sand"; *philos* is a Greek word meaning "to love." *Ammophila* lives on sand dunes because it requires sandy soil. Other plants, such as basswood trees, require a nutrient-rich soil in which to grow. Earthworms need a soil rich in oxygen, but some bacteria are able to live in soil with no oxygen at all. Chemical variations and soil acidity help determine where an organism will live and how well it will thrive. In addition, the animals themselves affect the composition of the soil, because they add their own organic materials to the soil's original composition.

50.2 Communities and Populations

An ecosystem's biotic factors—the plants, animals, and other organisms in an ecosystem—interact with one another. Because of this, ecologists say that these living things are members of the same biotic **community.** A community is a group of organisms that coexist. *Members of a community form a system of production, consumption, and decomposition.*

To better comprehend communities, ecologists often divide them into smaller units called populations. A **population** is a group of many individuals of a single species that occupy a common area and share common resources. For instance, all the maple trees in a maple-basswood forest are members of the same population. This is also true of all the trout in a mountain lake. All of the populations of a self-sustaining area make up a biological community.

The number of populations within a community varies. A tropical rain forest or a coral reef contains thousands of populations. Other communities, such as deserts, contain relatively few populations.

50.3 Habitats and Niches

Ecosystems are made up of interacting communities, and communities are made up of interacting populations. Likewise, populations are composed of interacting individuals. Each individual organism lives in a specific environment and pursues a specific way of life.

The surroundings in which a particular species can be found are called its **habitat.** For example, the habitat of a California condor is the dry hills in certain parts of southern California. The habitat of a gray whale is the coastal waters of the western Pacific Ocean. An organism may inhabit an entire ecosystem, such as a woodpecker in an oak forest. The oak forest is the

Q/A

Q: *What is the "greenhouse effect" and what causes it?*

A: Light energy passes through the glass of a greenhouse, is absorbed by the plants, and is changed to heat energy. The heat energy cannot pass through the glass so the air in the greenhouse is heated. When sunlight strikes the earth, CO_2 acts like the glass of a greenhouse and prevents the escape of heat.

Figure 50–3. The number of populations inhabiting a community can be extremely large. A coral reef, with its many plant and animal species, is an excellent example.

Figure 50–4. The giant panda has a very limited ecological niche. It lives only in certain areas of China, and its diet consists largely of bamboo shoots.

woodpecker's habitat. However, the habitat of other organisms may be just a small part of the ecosystem, such as that of a certain caterpillar, which lives only on oak leaves.

Some habitats appear to overlap when actually they do not. For example, five different kinds of warblers can live harmoniously on a single spruce tree. One species spends at least half its feeding time on the upper branches and another on the lower branches; another feeds primarily close to the trunk; and the other two feed near the ends of the branches but in different parts of the tree. However, if the habitats of two organisms truly overlap, they may have to compete directly for limited food, water, or space.

The way of life a species pursues within its habitat is called its ecological **niche.** An organism's niche is composed of both biotic and abiotic elements. These include how much moisture it requires, what it eats, where it lives, when and how it reproduces, and other such factors that make up its life. The niche of an armadillo, for example, includes such elements as the necessary food, water, and light. Its niche also includes temperature and space. In short, the total way of life of the armadillo is known as its niche.

Some niches are very broad. Rats, for example, live in houses, sewers, barns, ships, or even in open fields. They eat a varied diet, which may include grain, processed food, garbage, or other animals. They can reproduce when the temperature is hot, cold, or moderate, and may reproduce several times during a year.

In contrast, some niches are narrow. Pandas, for instance, are found in only a few isolated regions in China. They are very sensitive to changes in their environment and will not mate unless conditions for reproduction are perfect. They usually mate just once a year. Their diet consists almost exclusively of bamboo shoots. The panda, unlike the rat, has a very limited ecological niche.

Reviewing the Section

1. Explain why a forest is considered an ecosystem.
2. In what ways does an ecosystem sustain itself year after year?
3. How do the amounts of available light and water affect an ecosystem?
4. What is a community? A population?
5. What is the difference between the habitat of an organism and its niche?

The Flow of Materials

All organisms need certain chemicals in order to live. Chief among these are water, oxygen, carbon, and nitrogen. Organisms use these substances in the molecules that supply their nutrients and make up the materials of their bodies. When organisms die, these chemical materials are returned to the earth and the atmosphere. Other organisms then take up and use these same chemicals. *The continuous movement of chemicals throughout the ecosystem is called recycling.* The pathway through which a chemical substance is recycled is its **biogeochemical cycle.**

Section Objectives

- *Explain* what is meant by a biogeochemical cycle.
- *Describe* how water is recycled throughout a given ecosystem.
- *Show* the pathways taken by oxygen and carbon during recycling.
- *Define* nitrogen fixation, ammonification, nitrification, and denitrification.

50.4 The Water Cycle

The water cycle is an example of a biogeochemical cycle. Heat from the sun evaporates water from oceans, lakes, moist soil, the leaves of plants, and the bodies of animals. Water molecules from these sources are carried into the atmosphere by air currents. There they condense around dust particles and eventually return to Earth as rain, snow, hail, or fog.

Rain replenishes Earth's oceans and lakes. Plants release some of the water from their leaves through *transpiration*. In addition, some water filters down through the soil until it reaches solid rock. By means of seepage and underground streams, this groundwater eventually reaches streams, lakes, and oceans, where much of it is recycled by the process of evaporation.

50.5 The Oxygen-Carbon Cycle

Oxygen and carbon make up much of the body's carbohydrates, proteins, and fats. They are also involved in many chemical reactions. Like water, carbon and oxygen are recycled throughout the ecosystem. Their recycling pathways are so closely related that they are considered part of the same biogeochemical cycle, called the **oxygen-carbon cycle.**

The oxygen-carbon cycle is driven by photosynthesis and respiration. In photosynthesis plants take in carbon dioxide (CO_2). The oxygen is then released into the atmosphere as a gas, and the carbon is used to make glucose. The glucose and carbohydrates made from glucose are used by animals that eat the plants. Organisms release chemical energy from the breakdown of glucose during respiration. After additional chemical changes, some of the products of glucose metabolism become

Q/A

Q: *How much carbon is incorporated into sugar by plants each year?*

A: Photosynthesis produces 50 to 60 billion metric tons (4 to 7 trillion pounds) of sugar annually.

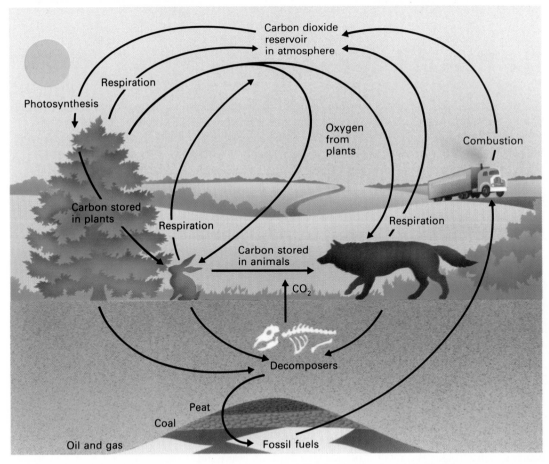

Figure 50–5. Carbon dioxide and oxygen are constantly recycled through plants, animals, and the atmosphere.

part of the animal itself. When these plant eaters are eaten, the carbohydrates are passed on.

During the process of respiration, unused carbon is returned to the air as CO_2 or is released in solid waste. Decomposers break down wastes and dead tissue, releasing more CO_2.

50.6 The Nitrogen Cycle

Nitrogen, a component of proteins and DNA, is also cycled throughout the biosphere. Over 78 percent of the air is nitrogen gas (N_2), but nitrogen in this form is useless to organisms. *Most living organisms use nitrogen only in the form of nitrates (NO_3^-), nitrites (NO_2^-), or ammonia (NH_3).* Atmospheric nitrogen must be converted, or fixed, into usable molecules.

The conversion of nitrogen gas (N_2) to nitrate (NO_3^-) is known as **nitrogen fixation.** Nitrogen fixation is carried out mainly by bacteria found in the roots of *legumes,* such as peas and beans. These bacteria take atmospheric nitrogen (N_2) and change it to nitrates, which the plants can then use. When animals eat these plants, the animals acquire usable nitrogen.

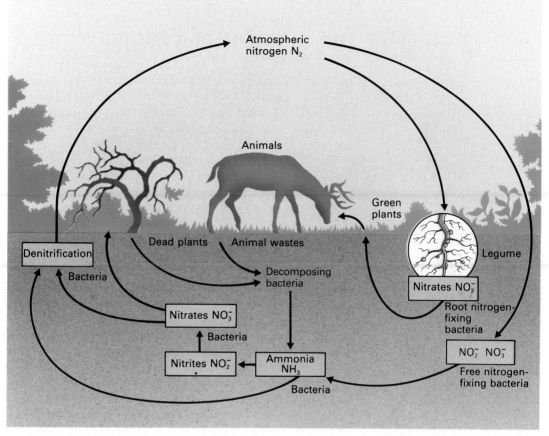

Figure 50–6. Certain bacteria convert atmospheric nitrogen into a form that organisms can use. Other kinds of bacteria recycle nitrogen from living things back into the atmosphere.

When plants and animals excrete waste or die, microorganisms convert their nitrogen compounds to ammonia—a process called **ammonification** (uh moh nuh fih KAY shuhn). Then, because most plants cannot use ammonia, it is converted into nitrate (NO_3^-) in a two-step process called **nitrification** (ny truh fuh KAY shuhn). One group of soil bacteria changes the ammonia to nitrites; another group of bacteria then converts the nitrites into nitrates.

Other bacteria convert ammonia, nitrite, and nitrate back into nitrogen gas—a process known as **denitrification.** Thus nitrogen gas (N_2) is returned to the atmosphere.

Reviewing the Section

1. What is a biogeochemical cycle?
2. How is underground water returned to the water cycle?
3. What are the events in the oxygen-carbon cycle?
4. How does nitrogen fixation make nitrogen available to plants?

- *Name* the trophic levels within an ecosystem.
- *Distinguish* among herbivores, carnivores, omnivores, and scavengers.
- *Discuss* the differences between food chains and food webs.
- *Draw* an energy pyramid, pyramid of mass, and pyramid of numbers for a typical ecosystem.

The Transfer of Energy

Individual organisms within a biotic community survive either by producing food or by feeding on other organisms. Plants, the chief producers, make food by the process known as photosynthesis. This food may be consumed by animals. Food energy thus moves from one organism to another. The process of the transfer of energy is not haphazard, however. It is part of an organized system of energy flow throughout the ecosystem.

50.7 Trophic Levels

A plant is eaten by a rabbit, which in turn is eaten by a coyote. Events such as these take place each day throughout the biosphere. Each of these organisms—plant, rabbit, and coyote—belongs to a distinct level of feeding within the ecosystem. The various levels are referred to as **trophic** (TRAHF ihk) **levels.**

Plants are at the first, or lowest, trophic level in an ecosystem. Plants produce their own food, so ecologists refer to them as *producers.* An animal that eats a plant is acting as a *primary consumer,* also called a *first-order consumer.* An animal that eats primary consumers is called a *secondary consumer,* or *second-order consumer.* An animal that eats a secondary consumer is a *tertiary consumer.* In the example above, the plant is the producer, the rabbit is the primary consumer, and the coyote is the secondary consumer. The particular species found at each trophic level vary from community to community, but the overall pattern remains the same.

Sometimes animals are referred to by the type of food they eat rather than by their trophic level. Animals that eat only plants are called **herbivores.** Animals that eat only animals are **carnivores.** Animals that eat both plants and animals are **omnivores. Scavengers** are animals that feed only on dead organisms. A scavenger may be a herbivore, a carnivore, or an omnivore. Crayfish, which live in ponds and streams, are primarily scavengers. Vultures and certain canines, such as jackals, are also scavengers.

Organisms at all trophic levels die. Decay organisms then act upon their bodies, breaking down tissues into small molecules. These organisms, such as bacteria and fungi, are called **decomposers** because they break down the dead plant and animal tissue and return the nutrients to the soil. In addition, many decomposers feed on materials that would not be considered food at any other trophic level. Among these are animal wastes and the cellulose found in plants.

Figure 50–7. Scavengers such as these jackals feed on the remains of animals left behind by predators.

50.8 Food Chains

As you have seen, organisms at one trophic level feed upon organisms at a lower level. The sequence of one organism feeding upon another at a lower trophic level is called the *food chain*.

Figure 50–8 shows an example of a food chain in the northern Atlantic Ocean. Some types of plankton are producers. These producers are eaten by small herbivorous invertebrates called *copepods*. The copepods are in turn eaten by sand eels, which are in turn eaten by herring. Herring are then caught and ultimately eaten by human beings. This food chain has five levels—plankton, copepod, sand eel, herring, and human.

Most food chains involve no more than four or five trophic levels. In the example above, however, a tapeworm in a human being would add another level.

50.9 Food Webs

Ecosystems almost always contain more than a single food chain. Most food chains also overlap because many organisms eat more than one type of food. As a result, the relationships between various trophic levels are often very complex. In most ecosystems the food chains are intertwined to form a *food web*— a network of interacting food chains.

A relatively simple food web is shown in Figure 50–9. In this particular community, a small meadow, plants are eaten by primary consumers, including mice, rabbits, grasshoppers, and other insects. The grasshopper may be eaten by a secondary consumer, such as a lizard or a bird. The mouse may be eaten by a hawk or a snake. The hawk may also eat the rabbit, the lizard,

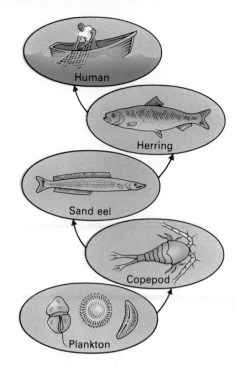

Figure 50–8. Food chains include producers, herbivores, and carnivores. What are the producers in this food chain?

Figure 50–9. Because an organism can represent more than one trophic level, a food web is more complicated than a food chain.

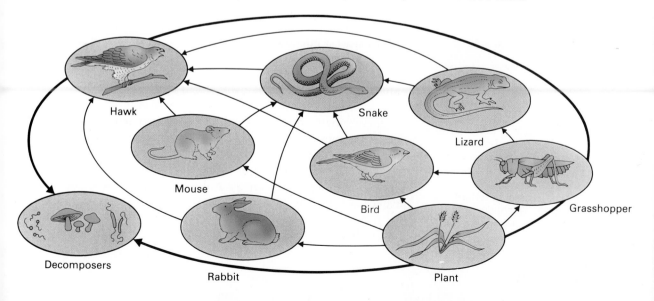

Q: *How much of the sun's energy is actually used by a plant to make food?*

A: Only about 1 percent of the incoming light is used to make sugar. Of the remainder, 19 percent is reradiated, 30 percent is transmitted through the leaf, and 50 percent is used to evaporate water from the leaf's surface.

or the snake. When these organisms die, decomposers act to release carbon, nitrogen, and other essential substances for reuse by other organisms.

In the example described, the hawk is both a secondary and a tertiary consumer. In a food web, a single organism may act at several different trophic levels. When you eat a roast beef sandwich, you are consuming on two different trophic levels at the same time. The bread is composed of wheat, which is made from a producer. The roast beef is made of the muscles of cattle, which are primary consumers. So when you eat a roast beef sandwich, you are acting as both a primary and a secondary consumer.

50.10 Ecological Pyramids

When a head of lettuce grows in the sun, it captures some of the sun's energy. However, all of this energy is not passed on to the rabbit that eats the lettuce. The lettuce plant undergoes metabolism, which uses energy. The amount of energy a plant stores—the amount available to the hungry rabbit—is much less than the plant initially got from sunlight. The bobcat that eats the rabbit will receive only a portion of the energy that the rabbit gained from the lettuce. One reason for this is that the rabbit used some of the energy gained from the plant to meet its own metabolic needs.

The energy available for use by organisms at each trophic level averages only about 10 percent of the preceding level. This means that as much as 90 percent of the energy is used up or lost during metabolism. In a pond ecosystem, for example, for every 1,000 kilocalories of energy taken in as light by the algae, which are the producers, almost 900 kilocalories are lost as heat during metabolism. Only 100 kilocalories are available to the minnows that eat the algae. Bass, the secondary consumers that eat minnows, receive only 10 kilocalories, or only 0.01 of the original energy input. If a human catches and eats the bass, the energy he or she receives is only 1 kilocalorie, or only 0.001 of the energy initially stored by the algae.

To visualize this decrease in available energy in an ecosystem, ecologists create diagrams called **ecological pyramids.** One of the most important of these is the *energy pyramid*. An energy pyramid shows the amount of energy, measured in calories, contained in the bodies of organisms at each trophic level. Figure 50–10 shows an energy pyramid. Notice that at each level, less energy is available.

A *pyramid of biomass* is also useful for visualizing an ecosystem. **Biomass** is the weight of an organism after all the water

Figure 50–10. A pyramid of energy (top), a pyramid of biomass (center), and a pyramid of numbers (bottom) illustrate relationships among trophic levels. From top to bottom, the levels in each pyramid represent tertiary consumers, secondary consumers, primary consumers, and producers.

has been removed. In such a pyramid, the biomass of all the individuals of a trophic level are added together. The pyramid of biomass in Figure 50–10 shows how, in general, biomass tends to decrease along with energy transfer.

Another useful pyramid is the *pyramid of numbers*, also shown in Figure 50–10. A pyramid of numbers is used to show that the number of organisms feeding at each trophic level also tends to decrease.

Reviewing the Section

1. Explain how energy is transferred within an ecosystem.
2. What role do decomposers play in a food chain?
3. How do a food chain and a food web differ?
4. What is the purpose of an ecological pyramid?

Biotic Relationships

- *Tell* why organisms compete in ecosystems.
- *Explain* the importance of predation.
- *Distinguish* among mutualism, commensalism, and predation.
- *Discuss* the difference between circadian and annual rhythms.
- *Explain* the symbiotic relationship between ants and the bull thorn acacia.

All organisms must live side by side with many other organisms. The evolutionary process has shaped ecosystems so that each ecosystem's organisms constantly interact with one another. This interaction may be for the mutual benefit of each organism, or it may result in a struggle for the use of limited natural resources.

50.11 Competition and Predation

In any ecosystem factors such as food, space, and water are found in limited quantities. *Competition* is the struggle among organisms for limited natural resources. When organisms compete with members of their own species, they undergo *intraspecific competition*. When organisms compete with other species, they show *interspecific competition*.

Competition resulted in the evolution of animals with distinct patterns of consumption. Some animals eat only plants; others eat animals. Organisms that eat other organisms are **predators.** The organisms they consume are *prey*. You might think that all of the prey in an area would be wiped out by predators in a short time. Ecologists know, however, that predators and prey coexist in a balanced relationship that keeps the population of each relatively constant year after year.

50.12 Symbiosis

Competition and predation are not the only ways that organisms interact with one another. Organisms also engage in symbiotic relationships. **Symbiosis** (sihm by OH sihs) means "living together." *A symbiotic relationship is a permanent, close relationship between two organisms of different species that benefits at least one of them.* The three types of symbiosis are mutualism, commensalism, and parasitism.

Mutualism **Mutualism** is a symbiotic relationship in which two organisms live together or cooperate with each other for mutual benefit. For example, termites eat wood but are unable to digest its cellulose. The cellulose is digested by protozoa that live in the termite's gut. Termites benefit from the presence of the protozoa by getting food digested, and the protozoa benefit by being protected by the termite's body. In fact, under normal conditions the protozoa would be unable to live outside of the termite's body. The termite-protozoa relationship is mutualistic.

Figure 50–11. An animal's size has nothing to do with whether it eats plants or other animals. The elephant, one of the largest animals, is an herbivore.

Similarly, a lichen is a mutualistic organism composed of an alga and a fungus living intertwined with one another. The alga produces sugars for itself and for the fungus. The fungus receives food and, in turn, covers the alga, thus protecting it from the environment and providing it with moisture.

Commensalism Commensalism is a symbiotic relationship in which one organism benefits and the other is neither helped nor harmed. Commensalism occurs, for example, in tropical rain forests. Large trees often have many small plants called *epiphytes* growing on them. These plants do not damage the supporting tree but, by living high in the air, receive more sunlight than if they grew on the deeply shaded forest floor. Epiphytes get water from rain or vapor. They get minerals from dust and from leaves that fall from their support trees. A commensal relationship exists between the epiphytes and the tree because the epiphytes benefit while the tree is neither harmed nor helped. Many species of orchids are epiphytes.

Parasitism Parasitism is a symbiotic relationship in which one organism benefits and the other is harmed to some extent. In a parasitic relationship, one organism, called a *parasite*, lives in or on another organism. The parasite obtains food from the other organism and harms it in the process. The organism that is harmed is called the *host*.

Mistletoe is a parasitic plant that lives on hosts such as oak and cedar trees. Mistletoe contains chlorophyll and thus is able to make its own food, but it obtains minerals and water from its host. To do this, it invades the host's tissues and causes severe damage.

50.13 Biological Rhythms

Organisms are affected not only by their relationships to other organisms but also by their relationship to the physical environment. Some environmental phenomena, such as day length, seasonal temperature, and phases of the moon, recur at regular intervals. Many organisms respond to these naturally recurring phenomena by undergoing physiological changes called *biological rhythms*.

The pattern of changes that occurs in organisms every 24 hours is called a **circadian** (suhr KAY dee uhn) **rhythm.** *Circa* is Latin for "about"; *dia* is Latin for "day." Sleep and wakefulness in humans is a circadian rhythm. Organisms that are active during the day are called **diurnal.** Organisms active at night are called **nocturnal.**

Figure 50–12. Oak trees are sometimes draped in Spanish moss (top). Both species thrive in this commensal relationship. Mistletoe (bottom) also grows on oaks, but this parasite damages the host.

▶ The Ants and the Acacia

One of the most interesting examples of a mutualistic relationship is the symbiosis of ants and acacia trees in Central America and South America. For a long time, scientists knew that *Acacia cornigera,* the bull thorn acacia, housed large numbers of ants. Careful study of this relationship has revealed the reason.

The acacia tree provides a home for ants in the form of large thorns on its stem. Sometimes 10 or 15 ants can fit into one thorn. The tree also produces special growths on its leaves called *beltian bodies.* Chemical analysis of these structures revealed they are full of glycogen, which is also called animal starch. The plant has a biochemical pathway that produces food that is used by ants.

The acacia derives benefits, too. Ants, protective of their food and housing, attack any predators that attempt to eat the acacia. In addition, the ants destroy other plants that, if allowed to grow, would shade the acacia and cut off its light. Bull thorn acacias are often found in

a clearing in the middle of a dense jungle because the ants have mowed down all the surrounding vegetation.

The internal chemical mechanisms that control circadian rhythms are called *biological clocks*. Although almost all organisms in the plant and animal kingdoms exhibit regular cycles of change, no one knows exactly what a biological clock is.

Another type of biological rhythm is the annual rhythm. An *annual rhythm* is the pattern of a physiological change that occurs once a year. Flowering, egg-laying, mating, and seed germination are examples of annual rhythms. **Hibernation,** the reduction of activity by warm-blooded animals in winter, is based on an annual rhythm. **Estivation** (ehs tuh VAY shun), the reduction of activity by warm-blooded animals in summer, is another change that occurs annually.

Reviewing the Section

1. How do competition and predation differ?
2. How does mutualism differ from commensalism?
3. How do circadian and annual rhythms differ?
4. Differentiate between hibernation and estivation.

Investigation 50: A Decaying Log Community

Purpose
To observe the community found in a decaying log

Materials

Decaying log, hand lens, hatchet, knife, notebook, pencil, field guide

Procedure
1. Locate a suitable log in its natural habitat and describe its location (stream, field, forest, etc.). *What evidence suggests that the log is decayed?*
2. Examine the surface for any organisms that appear to live on or within the log. Use a hand lens to identify smaller organisms not visible to the unaided eye.
3. Record in your notebook the various plants observed. Use the field guide to help identify the organisms by name. Sketch those that you cannot identify. Record whether each species is scarce or plentiful. Record your observations in the form of a table like the one below.

6. Cut away some of the surface of the decaying log with a knife and hatchet. Look for additional organisms living inside, and again record your findings.
7. Continue to cut away at the log until you feel that all organisms living inside have been identified by name or by sketch. Record all findings in your notebook.

Analyses and Conclusions
1. What plants and animals were the most abundant?
2. In what ways do plants and animals depend upon the log for life?
3. Not all organisms living in this community were identified. Explain the reason for this fact.

Going Further
Design a food web for the organisms you identified in the decaying log community. You will need to research this information. You may wish to illustrate your food web on a poster or by making a diorama.

Observations of a Decaying Log Community			
Name of Organism	Sketch of Organism	Scarce	Plentiful

4. On a separate page of your notebook, record the names of the various animals observed. Sketch those you cannot identify by name. Again note which are plentiful and which are scarce.
5. Roll the log over. Look carefully for additional plant and animal forms on the underside of the log. Add these to your notebook.

Chapter 50 Review

Summary

Ecology is the study of the relationship of organisms to their environments. Ecosystems are physically distinct, self-supporting systems of interacting organisms and their environment. Ecosystems are composed of communities of organisms and the abiotic, or nonliving, factors in the environment, such as light, water, soil, and temperature. Within an ecosystem, an organism occupies a physical region, called its habitat. It also has a specific niche, or way of life.

Abiotic factors play an important role in determining where organisms live. Organisms depend upon a continuous supply of recycled nutrients, such as water and carbon. Organisms can be producers, consumers, or decomposers. All organisms occupy a trophic level, and all are members of a specific food chain. Permanent, close relationships between different species are called symbiotic. Organisms respond to regularly reoccurring phenomena in patterns called biological rhythms.

BioTerms

abiotic factors (**759**)
ammonification (**765**)
biogeochemical cycle (**763**)
biomass (**768**)
biosphere (**759**)
biotic factors (**759**)
carnivore (**766**)
circadian rhythm (**771**)
commensalism (**771**)
community (**761**)
decomposer (**766**)

denitrification (**765**)
diurnal (**771**)
ecological pyramid (**768**)
ecology (**758**)
ecosystem (**759**)
estivation (**772**)
habitat (**761**)
herbivore (**766**)
hibernation (**772**)
mutualism (**770**)
niche (**762**)

nitrification (**765**)
nitrogen fixation (**764**)
nocturnal (**771**)
omnivore (**766**)
oxygen-carbon cycle (**763**)
parasitism (**771**)
population (**761**)
predator (**770**)
scavenger (**766**)
symbiosis (**770**)
trophic level (**766**)

BioQuiz (Write all answers on a separate sheet of paper.)

I. Completion

1. An organism's way of life is its _____ .
2. Populations are composed of members of the same _____ .
3. _____ is a symbiotic relationship in which both parties benefit.
4. The _____ is the level at which an organism feeds.
5. Warm-blooded animals undergo _____ in the summer by slowing down their metabolism.

II. Modified True and False

Mark each statement TRUE or FALSE. If false, change the underlined term to make the statement true.

6. The conversion of ammonia to nitrogen gas is called <u>ammonification</u>.
7. Annual rhythms are controlled by <u>internal</u> chemical mechanisms called biological clocks.
8. A group of populations in an area make up a <u>community</u>.

III. Multiple Choice

9. The place an organism lives in is its
   a) rhythm. b) niche. c) habitat.
   d) population.
10. Ammonification is the process of
    converting a) ammonia to nitrate.
    b) ammonia to nitrite. c) nitrogen
    compounds to ammonia. d) atmospheric
    nitrogen to ammonia.
11. Two buffalos seeking the same food is
    an example of a) competition.
    b) parasitism. c) nutrient recycling.
    d) mutualism.
12. Food webs are networks of interacting
    a) habitats. b) food chains. c) primary
    consumers. d) producers.
13. The oxygen-carbon cycle is driven by
    a) gravity and sunlight. b) evaporation
    and precipitation. c) photosynthesis and
    respiration. d) bacteria.
14. Factors such as light, temperature, and
    soil are examples of a) ecosystems.
    b) habitats. c) biotic factors. d) abiotic
    factors.

IV. Essay

15. What two requirements are necessary in
    order to identify an area as an
    ecosystem?
16. How do the three types of symbiotic
    relationships differ?
17. Is it possible for an organism to eat at
    more than one level on a food chain?
    Explain.
18. What chemical transformations take
    place during nitrogen fixation,
    nitrification, and denitrification?
19. How do competition, predation, and
    symbiosis differ?
20. An owl feeds upon a mouse that has
    eaten some seeds. Name the three
    trophic levels represented here.

Applying and Extending Concepts

1. In the grasslands of Africa, many species
   of animals, such as zebras, wildebeests,
   and springboks, coexist in the same area
   even though they all eat plants. Drawing
   on what you know of ecological niches,
   suggest reasons for the harmonious relationships among African plant eaters in
   the face of limited natural resources.
2. Ecologists concerned with the shortage of
   food in the world suggest that people eat
   less meat and more grains and beans. This
   is called "eating lower on the food
   chain." Considering the flow of energy in
   the biosphere, explain why this practice
   makes sense.
3. Ecologists have found that few parasites
   cause the death of their hosts. Those that
   do, say the ecologists, are not as well
   adapted as those that do not. Write a paragraph explaining this statement.

Related Readings

Carson, R. *The Sea Around Us*. New York:
New American Library, 1954. This classic
on the oceans of Earth also shows how
humans are a small but important part of
the biosphere.

Kormondy, E. *Concepts of Ecology*. Engle-
wood Cliffs, NJ: Prentice Hall, 1969. This

book is a standard reference text on the
fundamental concepts presented in this
chapter.

"Why Lions Decide to Be Laid Back." *Science News* 121 (May 8, 1982): 311. This
article interprets competitive behavior
among lions during mating.

51

Succession and Biomes

The border between a sand dune community and a forest community

Introduction

The world is made up of thousands of ecological communities. The sand dunes of the Pacific Ocean, the deep beech forests of New England, and the jungles of Borneo are ecological communities. So are the rugged seashores of Maine and the treeless mountaintops of the Swiss Alps. These communities each have distinct forms of plant and animal life.

Communities are changing all the time. In some communities one group of organisms naturally replaces another group. In other communities natural disasters, such as fires and hurricanes, wipe out the vegetation. This leaves the area bare and ready for exploitation by new organisms. This chapter describes the ways in which communities change and how they form even larger ecological units.

Ecological Succession

You may have walked through a maple forest that was hundreds of years old. If your grandmother and grandfather had walked in the same forest, they probably saw the same kinds of plants and animals that you saw. However, ecologists know that the area was not always a maple forest. A thousand years ago it might have been a meadow or even a shallow lake.

Just as individual organisms grow and change with time, so do the ecological communities of the earth on which we live. A community is a group of different types of organisms that coexist. Such communities replace, or succeed, each other in a predictable, orderly way. The process of replacement is called **ecological succession.**

51.1 The Process of Succession

The complete process of succession may take hundreds, even thousands, of years and always involves a number of intermediate communities. Two hundred years ago a maple forest might have been an oak forest. Three hundred years ago the oak forest might have been a grassland. A thousand years ago the grassland might have been a shallow lake. Each of these intermediate communities is called a **seral community.**

Seral communities replace one another for an interesting reason. *Each seral community alters the physical factors of the area in a way that makes it impossible for the community to regenerate itself.* For example, a pine forest grows and shades the forest floor. New pine seedlings, however, are unable to germinate in the shade created by their parents. Thus, the very existence of a pine forest often ensures that no more pines will grow. However, this shady environment makes it possible for organisms of another seral community to grow, because the new organisms—oaks, for instance—have different requirements.

The replacement of one seral community by another continues until a **climax community** forms. A climax community is a relatively stable, almost permanent, community. A climax community differs from a seral community in that it creates conditions in which its young can regenerate the community.

In any area the kinds of plants and animals that make up a climax community are determined by temperature, soil conditions, rainfall, and other physical factors. In the north central and northeastern United States, maple forests form a common climax community. In parts of Arizona, the saguaro cactus is the climax vegetation; in southern Florida, the sawgrass.

Section Objectives

- *Distinguish* between a seral community and a climax community.
- *Outline* primary and secondary succession.
- *Explain* what takes place during the primary succession of a lake to a forest.
- *Name* the stages in secondary succession by which a vacant field becomes a forest.
- *Explain* the role of weeds in ecological succession.

Figure 51–1. Tall pine trees produce conditions that discourage young pines from growing in their shade. The young trees growing below these tall pines are oaks and sweetgums.

▶ The Ecological Role of Weeds

Most people think of weeds as nothing more than nuisances. Homeowners and farmers wage constant battles against weeds. They spend money on weed killers and buy implements to rid gardens and farms of these unwanted plants. Even though most people regard weeds this way, ecologists know that they are an important component of the ecological environment. Weeds are pioneers.

Weeds, such as dandelions, crabgrass, ragweed, knotweed, and plantain, are plants specifically adapted to quickly colonize soil that has been disturbed. Weeds grow rapidly in soil exposed after landslides or fires. When human beings disturb the soil by farming, gardening, or building roads, weed seeds land in the area. They then germinate, flower, and produce seeds quickly. Four or five generations of weeds grow on the area, altering the physical and chemical conditions. The area can then sustain other, nonweedy kinds of plants.

Weeds often have reproductive systems well adapted to rapid colonization. Many weeds reproduce asexually initially. Since asexual reproduction is more rapid than sexual reproduction, the weeds can spread over an exposed area quickly. The weeds then reproduce sexually, but only after the exposed ground has been colonized and there is no advantage in rapid seed production.

In reality, natural climax communities may never form in an area. It often happens that one of the seral communities remains as the end result of succession. This happens when land is subject to repeated natural occurrences such as fire or grazing that keep the successional process from reaching a climax. *Some communities, in fact, need disasters to sustain themselves.* For example, a prairie that is not periodically burned by wildfires will soon become a forest.

51.2 Primary Succession

Primary succession takes place in areas that have not supported communities before. The development of communities on bare rock, lava flows, sand dunes, and lakes are examples of primary succession. Bare landforms and waters that do not contain living organisms soon give way to seral communities and, ultimately, to a climax community.

From Lake to Forest Imagine a lake that has recently been carved by a glacier out of solid rock. Its water is perfectly clear. This lake is so new that it contains no sediment and no living organisms of any kind.

This lake may remain free of life for a short period of time, but soon dust blows into it, and soil slides in after a heavy rain. In addition, sediments are carried into the lake by the rivers that feed it. Wind-borne spores of bacteria, algae, fungi, and protozoa fall into the lake. These germinate and grow in the sediment that has settled there. These organisms are called **pioneer species** because they are the first to inhabit a community. Pioneer species are adapted to life in low-nutrient conditions. They grow, thrive, and eventually die, adding new nutrients, such as carbon and nitrogen, to the lake. The process of adding nutrients to an ecosystem or community is called **eutrophication** (yoo trahf ih KAY shuhn).

As eutrophication proceeds, the lake can support a few plants such as water lilies and quillworts. These plants take root in the shallow water at the lake's edge. They catch and hold soil. Some animals, such as frogs and insects, find suitable habitats in the plants at the lake's edge.

As the water lilies, quillworts, and insects die and decompose, their remains further eutrophicate the lake. The decaying organic matter collects on the lake's edge and bottom. The lake gradually becomes shallower. Cattails and rushes invade the shallow water. They grow tall and eventually shade out the water lilies.

After a while vegetation grows throughout the lake and the lake becomes a *marsh*—another seral community in primary succession. As the marsh becomes drier, willow trees start to invade its borders. Thus, another seral community, a willow community, forms. Over time other communities, such as a birch forest, a pine forest, and an oak forest, replace one another on the site once occupied by the lake. A climax forest, perhaps one dominated by beech trees, finally forms and primary succession is complete. The process may take as little as 500 years or as long as 10,000 years or more, depending on the initial condition of the area.

Sand Dunes

Another type of primary succession occurs on sand dunes. In parts of Wisconsin, Indiana, and Michigan, for example, sand dunes run parallel to Lake Michigan's shoreline. The dunes nearest the shore were formed recently. The dunes farther back were formed at an earlier time. Thus, as you walk away from the lake across a series of progressively older and older dunes, you are, in effect, walking backward in ecological time. The dunes closest to the lake contain the pioneer and early seral communities. Those dunes located farthest from the lake support the ecological communities that are in the final stages of succession.

Lake during its early stages

Lake has become a marsh

Marsh has become a forest

Figure 51–2. As a lake (top) fills with vegetation, it gradually turns into a marsh (center). The marsh eventually becomes a forest (bottom).

A: Many lichens grow on rocks. They secrete chemicals that cause the rock to break and crack. This breakdown of rock creates soil from which small plants can grow.

Primary succession on sand dunes begins when wind and waves deposit organic matter on the bare sand. Grasses then root in the soil, stabilizing the sand and adding nutrients. Soon the seeds of shrubs carried by wind and water germinate, grow, and replace the grasses.

The shrubs add nutrients to the land. Pine trees then begin to grow. Birds called brown creepers, and other animals that live in association with large trees, begin to inhabit the area. When the pine trees become tall, they shade the ground, killing both the shrubs and their own seedlings. But oak seedlings, which need some shade in which to germinate, begin to flourish. The oaks eventually give way to a climax community dominated by maples. Primary succession on sand dunes takes about 1,000 years.

Highlight on Careers: Soil Scientist

General Description

Soil scientists are important members of the farming community. As experts in the process of soil formation and use, soil scientists help farmers make the best use of the soil.

The duties of a soil scientist vary with the area of specialization and the level of education. Some soil scientists are soil physicists. Others are soil mineralogists, soil microbiologists, or soil meteorologists.

Duties also vary with experience. Beginning soil scientists contact farmers, usually through a county extension service. They study soil samples and collect data on local soil conditions. Using these data, the soil scientist recommends the type and amount of fertilizer that should be added to the field, how irrigation should be developed, and how drainage systems should be designed.

More advanced soil scientists study new developments to help farmers maximize their crop yield. They may conduct research on the relationships between crops, fertilizers, nutrients, and climate.

Opportunities in soil science exist mainly in the public sector. Most soil scientists work at a university or for the government. Others are consultants or perform research for large corporations.

Career Requirements

Extension agents and less advanced soil specialists

need a B.S. degree in soil science. Independent researchers must have a Ph.D. in a soil specialty. These more-advanced soil specialists also need strong computer skills.

For Additional Information
American Society of
  Agronomy
677 South Segoe Road
Madison, WI 53711

51.3 Secondary Succession

When ecological succession takes place in an area where a community once stood rather than in a barren area, the process is called **secondary succession.** *Secondary succession often occurs after natural disasters such as fires, landslides, or floods.*

One example of secondary succession is *old-field* succession, a process by which climax communities are reestablished on abandoned farmland. Early settlers in the northeastern United States cleared land for farming, cutting down extensive oak-hickory forests in some regions. Later, these farms were abandoned. Today oak-hickory forests again grow on the once-farmed land because secondary succession has taken place.

However, when the fields were first abandoned, they did not immediately return to oak-hickory forest. Oak and hickory seedlings need cool shade and moisture to germinate and grow, but the soil in the abandoned fields was sunlit and dry. Only a few weeds, such as crabgrass, grew in the bright, dry conditions. These weeds became the pioneer species in the process of secondary succession.

As generations of pioneer plants died, nutrients were added to the soil. Thus, the crabgrass prepared the ground to support other species of plants. Six to eight years after it had been abandoned, the soil was shaded by shrubs.

Conditions were then favorable for the survival of pine seedlings. Little pines soon grew throughout the field. About 25 years later, the abandoned field became a pine forest community that flourished for the next 50 years.

The deep shade and moist conditions of the pine forest floor, however, were less favorable for pine seedlings than they were for oak and hickory trees. Over the next century these hardwood trees gradually replaced the pines. Eventually, the pine trees died out altogether and a climax community of oak and hickory was again established on the field. Thus, in a period of about 200 years, secondary succession had run its course.

Figure 51–3. An abandoned cornfield (top) is overrun with crabgrass. Five years later (bottom), broom sedge has replaced the crabgrass and is already being crowded out by pines.

Reviewing the Section

1. How are primary and secondary succession different?
2. What conditions usually set the stage for secondary succession?
3. What are the stages of succession on a sand dune?
4. What is the ecological role of weeds?
5. How does secondary succession proceed in an abandoned field in the northeastern United States?

Terrestrial Biomes

- *Tell* what a terrestrial biome is and *name* the major ones.
- *State* how the environment of the tundra affects the kinds of plants and animals that live there.
- *Compare* a grassland, a desert, and a tundra.
- *Describe* the plants that thrive in a tropical rain forest.
- *Tell* the effects of farming on the tropical rain forest.

Climax communities found in similar climates are remarkably alike. A forest in southern Wisconsin, for instance, looks like one in northern Germany where the climate is about the same. Other areas, such as China, also have forests resembling those in Germany and Wisconsin. The species that make up each of these forests may differ, but all are composed of flowering trees that lose their leaves in the fall. These communities are members of a larger ecological unit called a **biome.** A biome is an extensive area of similar climate and vegetation.

The biome's abiotic factors determine what plants and animals live there. The major influences are temperature, light intensity, and the patterns of rainfall, which determine the availability of water. Though scientists disagree about the number of biomes on Earth, the ones that are most generally agreed upon are shown in Figure 51–5.

51.4 Tundra

The **tundra** is a biome of the cold regions like the crest of the Rockies and near the North Pole where the vegetation grows close to the ground. This biome covers a vast expanse of land just south of the permanent ice of the pole.

The tundra is a stark, gently rolling plain with a cold, dry climate. Winters are long and severe. Only 10 to 40 cm (4 to 16 in.) of precipitation fall each year, most of it as snow. Summers in the tundra are cool and very short. Even in the warmest times of the year the surface of the ground thaws only to a depth of about one meter (3 ft.). Below this the ground is permanently frozen. This permanently frozen ground is called *permafrost*.

The most common forms of plant life in the tundra are lichens, mosses, and other small plants that can withstand extreme cold. Most tundra plants, even willow trees, are short. Almost all are less than 10 cm (4 in.) high. The plants are short, an adaptation that helps them to avoid the icy winds.

Polar bears, musk ox, reindeer, and caribou are some of the large animals that live in the tundra. Polar bears are solitary hunters that stalk the ice caps north of the tundra for food. Caribou and reindeer form huge herds that migrate back and forth across the land. They feed mostly on lichens that cling to the soil. Small rodents, called lemmings, live in the tundra together with arctic foxes, arctic hares, and grouse-like ptarmigans. In the summer months, mosquitoes emerge in uncountable millions, making life in the tundra unbearable.

Figure 51–4. The Brooks Mountains region of Alaska presents a scene typical of the tundra. The landscape is bleak, the land is gently rolling, and the vegetation grows close to the ground.

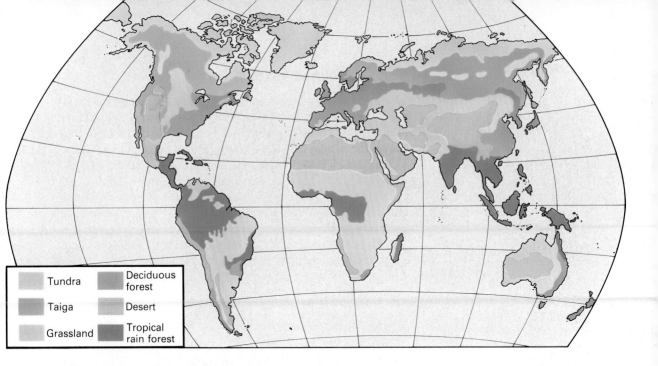

Tundra
Deciduous forest
Taiga
Desert
Grassland
Tropical rain forest

51.5 Boreal Forests

The **boreal forest** is a biome dominated by evergreen trees. It covers the northern reaches of North America, Europe, and the Soviet Union. This biome is also referred to as the *taiga* (TY guh). The taiga's summers are short and warm. The winters are long, snowy, and cold. About 20 to 60 cm (8 to 24 in.) of rain fall annually in this biome, which is also characterized by moist, spongy areas called *bogs*.

The dominant plants in the taiga are cone-bearing trees, such as white and black spruce and balsam fir. They retain their needlelike leaves throughout the harsh winters and reproduce and grow during the short growing season. These *conifers* form vast forests that cover hundreds of thousands of square kilometers. Much of the world's commercially available timber grows in the coniferous forest.

The animal most common to this biome is the moose. The coniferous forest contains so many moose that ecologists jokingly call it the "spruce-moose" biome. Other animals include bears, lynxes, elk, mule deer, and wolves. Smaller animals, such as porcupines, wolverines, hares, bobcats, and many kinds of rodents, also live in this area. During the summer, birds are numerous, particularly warblers.

On a spring evening, one can watch the dancing lights of the aurora borealis. This display of colored lights, often seen in the northern sky, is caused by the interaction of the earth's magnetic field with winds generated by the sun.

Figure 51–5. The worldwide distribution of six major biomes is shown on the map above.

Figure 51–6. The moose, a major inhabitant of the boreal forest, is the largest member of the deer family.

51.6 Deciduous Forests

The **deciduous** (dih SIHJ oo wuhs) **forest,** also called the *temperate forest,* is a biome dominated by deciduous trees. These are trees that shed their leaves each year. The deciduous forest covers the eastern half of the United States from southern Maine to northern Florida. It is also the dominant biome of Europe and eastern China. The deciduous forest receives rainfall year-round, measuring 60 to 100 cm (24 to 40 in.) of precipitation annually. Winters in this biome are long and severe, though not as severe as those of the boreal forest.

Maples, beeches, elms, oaks, hickories, and basswoods grow in many parts of the northern deciduous forest. In the southern hemisphere, other species of deciduous angiosperms replace the typical northern species.

Figure 51–7. Trees that shed their leaves each year make up deciduous forests. In the fall such forests present a brilliant display of colors.

Plants in deciduous forests occupy one of four layers of vertical stratification. Some plants grow on the *forest floor,* the bottom-most level. Most are adapted to low light conditions. Many, however, flourish in the spring before the trees have produced leaves that intercept the light. Plants, such as shrubs and small trees, grow beneath the taller trees in the *understory*. The taller trees form the *canopy* layer where most of the incoming sunlight is captured. Some trees grow above the canopy and form the *emergent* layer.

The animal life of a deciduous forest is varied. White-tailed deer can be found in great numbers in this biome. Other animals include mice, ground squirrels, and other small mammals such as foxes, bears, and wildcats. Salamanders, snakes, lizards, rabbits, and chipmunks also live in the deciduous forest.

51.7 Grasslands

Grasslands—a biome dominated by grasses—occur where the precipitation is between 10 and 60 cm (4 and 24 in.) yearly. *Grasslands make up the largest of the several biomes found in the United States.* They cover much of the central part of the country. They are also found in central Africa and in parts of many other countries. Distinctive kinds of grasslands are given special names: *prairie* in the United States, *steppe* in Russia, *veldt* in Africa, and *pampas* in Argentina.

Spring in the grasslands is warm and wet, but it is usually followed by a scorching dry season. In many grasslands, the winters can be cold and snowy. Where rainfall is moderate, grasslands provide grazing lands for sheep and cattle. Where more moisture is available, grasslands are planted with wheat, oats, barley, and corn. Much of the world's population depends upon these areas for food.

The grasslands of the United States are home for many large animals, such as bison and antelope. Smaller animals, including prairie dogs, coyotes, and badgers, are also characteristic of United States' prairies.

A **savannah** is a special kind of grassland located in a tropical or subtropical area. In a savannah trees are usually scattered throughout the grassy area. Unlike northern grasslands, savannahs remain warm all year. The largest and most famous savannah is the Serengeti Plain of East Africa. Within a single section of the Serengeti live grazing animals such as African buffalos, antelopes, gazelles, elephants, wildebeest, zebras, and giraffes. Numerous carnivores, including lions, cheetahs, wild dogs, and leopards, prey on the various grazing animals. Hyenas, jackals, and vultures are examples of meat-eating scavengers that also prowl the Serengeti.

51.8 Tropical Rain Forests

Tropical rain forests exist in equatorial areas where rainfall is high, sometimes over 500 cm (200 in.) annually. Most of the plants that grow in this biome are flowering trees. Plant growth is extremely rapid. The canopy in the rain forest is very thick, and little light reaches the forest floor. As a result many *epiphytes,* plants that use other plants for support, grow high up on the trees. Vines are also extremely common in a rain forest, as are ferns.

Large numbers of insects and other invertebrates live in tropical rain forests. Suprisingly, far fewer vertebrates live in rain forests than live in other biomes. Birds, monkeys, snakes,

Figure 51–8. The prairies of the midwestern United States are a familiar example of a grassland biome.

Figure 51–9. Bamboo grows abundantly in this tropical rain forest in Hawaii.

▶ Destruction of the Rain Forest

The tropical rain forests of the world cover vast areas of land. Much of South America and much of southeast Asia and central Africa are covered by tropical vegetation. Today, however, rain forests are threatened with destruction by the encroachment of human beings. Destruction of the rain forests may result in increased carbon dioxide in the atmosphere. Destruction of the rain forest may even change the distribution of rainfall throughout the world.

In order to meet the world's growing demand for food, many nations have opened up the rain forests to farming. Huge bulldozers tear down the thick vegetation and foresters cut down the trees. Yet the farms that result from this clearing of the land are often unsuccessful because their soil is unfit to grow many kinds of crops.

The soils in tropical rain forests are extremely low in nutrients. Nutrients that enter the soil are quickly taken up by the many native plants that grow there.

Some tropical soils, called *laterite* soils, are claylike and may become as hard as a brick when exposed to the hot tropical sun. Because the soil has almost no nutrients and because it becomes hard under the heat of the sun, farmland in tropical areas only produces crops for from three to five years. The land must then lie fallow for ten or fifteen years before farmers can grow even one or two more crops.

Because the soil wears out so quickly, farmers will cultivate an area for three to five years and then move on to the next patch of cleared land. Each time they abandon a farm, they leave behind them a legacy of destroyed trees and wildlife as well as barren land.

Q/A

Q: *Do temperate rain forests exist?*

A: Yes. They grow on the Pacific Coast of British Columbia, Washington, Oregon, and California—cool areas of high rainfall. Plants, such as Sitka spruce and redwoods, grow in parts of this small biome.

and lizards are the chief vertebrate representatives. The *arboreal*, or tree-living, habit is common in the rain forest. Even large predators, like jaguars and panthers, spend time in trees.

Tropical rain forests are not jungles. *A jungle is a community within a rain forest.* Jungles grow on the edges of rivers and have very thick, almost impassable, vegetation.

51.9 Deserts

Deserts exist in areas where annual precipitation is less than 20 cm (8 in.). However, even this sparse rainfall is not evenly distributed throughout the year. In deserts in the United States, rain falls primarily in the winter; the rest of the year is often dry. In some places, such as the Sahara desert of northern Africa and

the Gobi desert of Mongolia, rain may fall only once every several years.

Desert plants in these dry environments are adapted to water conservation. Some plants, such as mesquite and creosote bush, grow long root systems that tap water deep in the ground. Other plants, such as cactuses, have shallow root systems. They store water in swollen stems and leaves. These plants have a thick outer covering that keeps the stored water from evaporating. Some desert plants are tiny, fast-growing plants that germinate, flower, produce seeds, and die all within a few weeks, sometimes even within a few days. The life cycles of these plants correspond to the short period when water is available.

The desert supports a wide variety of animal life including insects, reptiles, birds, and small mammals. Many desert animals are active at night. During the heat of the day, these nocturnal animals hide in the crevices of rocks or rest in cool, underground burrows. They venture out only when the evening brings cooler temperatures. Iguanas, Gila monsters, and horned lizards are among the many reptiles found in the deserts of the United States. Kangaroo rats, scorpions, and spiders are also common.

Figure 51–10. The saguaro cactus in the deserts of the southwestern United States and northern Mexico may reach 15 m (50 ft.).

Reviewing the Section

1. How does a biome differ from a community?
2. What characterizes the soil of the tundra?
3. Which two biomes have the lowest annual precipitation? Which has the highest?
4. How does a jungle differ from a tropical rain forest?
5. What two factors make soil in a tropical rain forest unsuitable for many kinds of farming?

Aquatic Biomes

Section Objectives

- *Name* the two kinds of aquatic biomes and tell how they differ.
- *Discuss* the characteristics of the three ecological zones of the marine biome.
- *State* the most important factors that determine the distribution of organisms in a river and in a lake.
- *Explain* why the estuarine habitat is considered a fragile community.

Earth is the water planet. Oceans cover more than three-quarters of the earth's surface. Lakes and ponds dot the land. Rivers and brooks flow from the mountains to the sea. In each of these aquatic habitats, life abounds.

Ecologists divide the watery world into two biomes based upon *salinity*—that is, the amount of salt dissolved in the water. The *marine biome,* with a salt content of about 3.5 percent, includes the world's oceans. The *freshwater biome* comprises bodies of water with little or no salt, such as lakes, ponds, and rivers. A special aquatic community, the *estuary,* includes coastal habitats that are covered with a mixture of fresh and salt water.

51.10 The Marine Biome

The marine biome contains 99.9 percent of the earth's surface water. The distribution of life in this biome is affected by water temperature and by the availability of sunlight and nutrients. These factors vary from area to area.

Organisms in the marine biome live as one of three life forms: plankton, nekton, or benthos. **Plankton** are tiny animals and algae that float near the water's surface. **Nekton** are organisms, such as cod and sharks, that swim freely. **Benthos** are organisms, such as sponges and oysters, that live on the bottom. All of these marine organisms are distributed throughout three distinct ecological zones: the intertidal zone, the neritic zone, and the open sea zone.

The **intertidal zone** is that part of the seashore located between high and low tide. Organisms of the intertidal zone are covered with salt water at high tide and uncovered at low tide.

Figure 51–11. The marine biome is divided into three zones. Not shown in the diagram are the huge mountains that rise from the ocean floor. Many of them are volcanic.

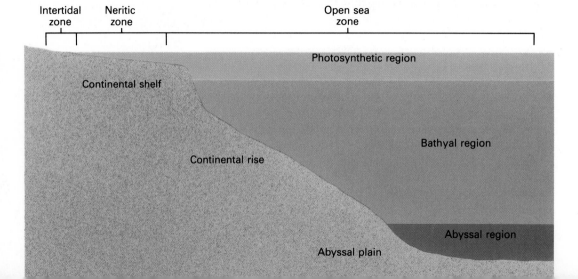

Intertidal zone | Neritic zone | Open sea zone

Photosynthetic region

Continental shelf

Continental rise

Bathyal region

Abyssal region

Abyssal plain

In addition to coping with this dramatic change, these organisms must withstand the powerful pounding of the surf. Despite these apparent hardships, the intertidal zone is crowded with life. Algae, barnacles, snails, and starfish are abundant.

The **neritic zone** is the region of open water close to shore. The water here is shallow compared to the deep seas, and light penetrates to the bottom. Many kinds of plankton live in the neritic zone. They form a plentiful supply of food for the variety of fish including cod and flounder found here.

The largest of the three marine biome zones is the **open sea zone.** The open sea zone has three regions: photosynthetic, bathyal (BATH ee uhl), and abyssal (uh BIHS uhl). The **photosynthetic region** is the area where most of the light penetrates. It extends from the ocean's surface to a depth of about 200 m (660 ft.). Plankton float near the surface in this region. Sharks, whales, large fish, and squid swim below. The **bathyal region** is the region of underwater twilight, extending from 200 m to 2,000 m (660 ft. to 6,600 ft.) deep. Little light reaches this area. It is home to strange organisms. Among these are fish that produce light biochemically through a process called *bioluminescence*. The **abyssal region** lies below 2,000 m (6,600 ft.). Life forms here are adapted to withstand extreme pressure. Dead organisms that sink to the bottom of the ocean provide food for the life forms in this zone.

51.11 The Freshwater Biome

Lakes, ponds, rivers, streams, and brooks are typical habitats within the freshwater biome. The bodies of water that make up the freshwater biome contain relatively little salt. In rivers and streams, the rate of water flow, which determines the dissolved oxygen content, is a major factor in determining the types of organisms existing there. In lakes and ponds, the amount of nutrients is important in determining the forms of life to be found.

Lakes and Ponds   A *pond* is a small, shallow depression in the land filled with fresh water. A *lake* is a larger body of water and may be quite deep. *The nutrient content of ponds and lakes changes over time.* Initially, these bodies of water have few nutrients. Their waters are clean and clear, and they support little life. In this state, lakes and ponds are called **oligotrophic,** (ahl uh goh TRAHF ihk). Lakes or ponds where nutrients are abundant are called *eutrophic* (yoo TRAHF ihk).

Plankton in eutrophic lakes use a tremendous amount of oxygen. Often there is only a little oxygen left over to support other forms of life. Fish such as catfish, carp, and certain bass,

Figure 51–12. Lakes rich in vegetation are also rich in nutrients. Such lakes are called eutrophic.

however, are adapted to life in these oxygen-poor, nutrient-rich waters. On the other hand, oligotrophic lakes such as those found in mountains contain little plankton. As a result, oxygen is plentiful. Fish such as lake trout are found in these oxygen-rich waters.

Rivers and Streams Organisms that live in rivers, streams, brooks, and creeks are adapted to life in flowing water. In slow-flowing streams, algae, mosses, and flowering plants are attached to the bottom of the stream, or its *bed*. Plants, such as water hyacinths, float on the surface. Catfish, frogs, and many invertebrates thrive in this environment.

51.12 Estuaries

An **estuary** is formed where rivers and streams meet the ocean. The salinity of an estuary is intermediate between that of the sea and a river. An estuary may be a mud flat, a salt marsh, or even a mangrove swamp. Most scientists consider estuaries as communities; others think all the world's estuaries make up a special kind of biome.

Changing conditions allow estuaries to support a great diversity of organisms. Fluctuation in water levels and the mixing of salt water and fresh water create a variety of habitats for organisms to live in. Also, estuaries are rich sources of nutrients. Clams, crabs, oysters, barnacles, and a variety of sea worms live in silt deposited by the incoming rivers. Large plants such as marsh grasses, eel grass, and filamentous algae are important producers in the estuary. Minnows and other tiny fish dart through the shallows. Shore birds are plentiful. Many marine animals spend their early lives here.

Individual estuaries are fragile communities. A sudden shift in a river's course can destroy an estuary. A particularly savage storm can overturn sand, uproot plants, and displace attached animals. New estuaries are constantly being formed as others are eliminated by natural catastrophes.

Figure 51–13. This estuary formed by the Delaware River has been designated as a wildlife refuge.

Reviewing the Section

1. What are the two aquatic biomes?
2. How do the salinity of a marine biome, a freshwater biome, and an estuary differ?
3. What are the three ecological zones in an ocean?
4. How do oligotrophic and eutrophic lakes differ?
5. Why is there a great diversity of organisms in estuaries?

Investigation 51: Succession in a Water Community

Purpose
To observe succession in an artificial water community

Materials
Sterilized pond water, dried plant material from the pond, jar with lid, microscope, glass slide, coverslip, medicine dropper, glass-marking pen, notebook

Procedure
1. Write your initials on a clean jar with a glass-marking pen.
2. Fill the jar about one-fourth full with dried plant material.
3. Add enough sterilized pond water to fill the jar about three-fourths full.
4. Place the cap on the jar. Shake the jar thoroughly to mix the plants with the pond water.
5. Copy the table below into your notebook.

Analyses and Conclusions
1. What evidence did you observe indicating that succession has occurred in your water community?
2. Did your observations indicate that the contents of the jar eventually reached the stability of a climax community? Give reasons for your answer.
3. Where did the organisms that appeared in the jar come from?

Going Further
- Design an experiment that will determine what factors in the abiotic community influence the rate of succession. Vary light and temperature. Add various minerals to the water community. Record your observations.
- As succession proceeds in water environments, the amount of oxygen and carbon dioxide in the water changes. Study how scientists measure this change and the effects of free O_2 and dissolved CO_2.

Observations of a Water Community			
Date	Organisms Present	Frequency	Other Observations

6. With a medicine dropper place one drop of the liquid from the jar on a glass slide.
7. Place a coverslip on top of the slide and observe the slide under a microscope. *Is there any evidence of organisms in the liquid on the slide?* Record your observations in the proper columns in the table.
8. Store your jar in a safe place overnight.
9. Observe a drop of water under the microscope every day for one month. Continue to record your observations in the table in your notebook.

Chapter 51 Review

Summary

The biological communities of the world are constantly changing. Many of these changes involve ecological succession, or the replacement of one group of species by another group that is more adapted to life in a given area.

Two types of ecological succession exist. Primary succession takes place where no community existed before. Secondary succession takes place on land that has previously supported communities. During succession the land is occupied by a number of seral communities before a climax community finally becomes established.

Large areas of similar climate and vegetation are called biomes. Tundra, boreal forests, deciduous forests, grasslands, tropical rain forests, and deserts are the major terrestrial biomes. Two biomes make up the aquatic world: the marine and freshwater. Estuaries are special aquatic communities.

BioTerms

abyssal region (**789**)
bathyal region (**789**)
benthos (**788**)
biome (**782**)
boreal forest (**783**)
climax community (**777**)
deciduous forest (**784**)
desert (**786**)
ecological succession (**777**)

estuary (**790**)
eutrophication (**779**)
grassland (**785**)
intertidal zone (**788**)
nekton (**788**)
neritic zone (**789**)
oligotrophic (**789**)
open sea zone (**789**)
photosynthetic region (**789**)

pioneer species (**779**)
plankton (**788**)
primary succession (**778**)
savannah (**785**)
secondary succession (**781**)
seral community (**777**)
tropical rain forest (**785**)
tundra (**782**)

BioQuiz   (Write all answers on a separate sheet of paper.)

I. Completion

1. Most of the world's grain is grown on land that was originally part of the _____ biome.
2. In old-field succession, weeds are known as _____ species because they are the first species to become established.
3. Moose are the most representative animal of the _____ biome.
4. The _____ layer is the uppermost vertical forest layer.
5. The _____ soil of tropical rain forests bakes to a hard clay in the sun.

II. Modified True and False

Mark each statement TRUE or FALSE. If false, change the underlined term to make the statement true.

6. Plankton is found in the <u>benthic</u> region of the open sea zone.
7. Permafrost is the frozen layer of soil in <u>the tundra</u>.
8. In succession on sand dunes, the pine community gives way to the <u>willow</u> community.
9. The Serengeti is an example of a type of grassland called <u>pampas</u>.
10. <u>Nekton</u> swim freely in the sea.

III. Multiple Choice

11. Organisms such as algae and tiny animals that float on the ocean's waters are a) nekton. b) plankton. c) benthos. d) crustaceans.
12. Lakes that get nutrients become a) eutrophic. b) abyssal. c) bathyal. d) pelagic.
13. The Lake Michigan sand dunes farthest from shore are the a) youngest. b) grassiest. c) oldest. d) most fragile.
14. An intermediate stage in succession is a a) climax community. b) primary succession. c) seral community. d) secondary succession.
15. Of all the biomes, the most fragile is the a) estuary. b) boreal forest. c) tundra. d) deciduous forest.

IV. Essay

16. How do climax communities and seral communities differ in regard to stability?
17. What are the three kinds of life forms found in the marine biome and how do they differ?
18. Why is a pine forest community unable to regenerate itself and why is it succeeded by oaks?
19. To what physical conditions must organisms of the intertidal zone be adapted?
20. How does the climate of a boreal forest differ from that of a deciduous forest?

Applying and Extending Concepts

1. The Colorado River, one of the largest rivers in North America, used to flow into the Gulf of California. Now so much water is removed for irrigation that all the water in the river is used up before it gets to the sea. Write a paragraph suggesting how this change in availability of water affects the marine and estuarine life of the Gulf of California.
2. The giant redwood trees of coastal California and Oregon are the world's tallest trees. However, they do not grow in areas of extremely high rainfall. Use your school or public library to research the biology of redwood trees, and write a paper describing how these trees obtain moisture. If possible, illustrate your report with photographs of the redwoods.
3. Maple-basswood forests are the climax communities in inland areas around Lake Michigan's sand dunes. Ecologists have discovered that succession on the dunes themselves usually stabilizes at the oak forest stage and does not progress to a maple-basswood climax community. Write a paragraph stating reasons why this situation might occur.

Related Readings

Boerner, R. "Fire and Nutrient Cycles in Temperate Ecosystems." *Bioscience* 32 (March 1983): 187–192. As its title suggests, this article discusses the impact of fire on the flow of nutrients through a temperate ecosystem.

Horn, H. "Forest Succession." *Scientific American* 232 (May 1975): 90–98. This article discusses how one seral community in the forest modifies the environment and makes it suitable for the emergence of the next community.

Keller, J. "From Yellowthroats to Woodpeckers." *Conservationist* 37 (July/August 1982): 30–35. This article shows how bird life in a given area changes during ecological succession.

Flamingos in Africa

Introduction

In parts of Africa, the sky is often so full of flamingos that it appears pink. In Tennessee and Kentucky, the sky sometimes darkens in the middle of the day when millions of starlings fly overhead. In Malaysia so many fireflies blink on and off at one time that the dark night suddenly becomes bright with biologically produced light.

These phenomena are the result of the activities of enormous biological populations. A *population* is a group of individuals of the same species inhabiting a similar area. Ecologists study the sizes of populations to gain an understanding of how the biological world grows and changes. These studies have a direct impact on our understanding of the human population, its past and future.

Population Growth

Section Objectives

- *Explain* why species do not achieve their biotic potential.
- *Distinguish* between density-independent and density-dependent limiting factors.
- *State* the difference between J-curves, S-curves, and saw-tooth curves.
- *Use* the equation for calculating changes in population size.
- *Summarize* the advantages and disadvantages of high-density populations.

Some populations are made up of less than a dozen individuals. Others are composed of millions, even billions of individuals. Every population has a specific potential for growth and is subject to environmental pressures that control how fast it grows and how large it can be.

52.1 Biotic Potential

A population's **biotic potential** is the rate at which it would produce young if every new individual lived and reproduced at its maximum capacity. *The biotic potential of a population is staggering.* Consider, for example, the reproductive capacity of houseflies. A female housefly usually mates seven times a year and lays about 100 eggs each time. About half of her offspring are females and most produce the same amount of eggs as their mother. If the young were allowed to reproduce unchecked, the housefly population generated by one male and one female would number more than 5 trillion (5,000,000,000,000) in a single year.

Houseflies reproduce relatively rapidly, but even the biotic potential of elephants, which bear young about every two years, is astounding. In 750 years, the descendants of a single pair of elephants would number more than 19 million. In 100,000 years, the descendants of just two elephants would fill the visible universe.

Elephants have been around for more than 100,000 years, but they do not fill the universe. Similarly, the earth is not overrun by houseflies. You need only look around to see that populations do not achieve their biotic potential. What mechanisms control a species' biotic potential and thus limit the size of its populations?

52.2 Carrying Capacity

The environment limits populations from reaching their biotic potential. To understand how the environment controls population growth, it is first necessary to understand how populations grow. Assume that a pair of warblers enters a forest where no other warblers live. At first the increase in the number of warblers is small because there is only one female to produce young. Thus, the population undergoes a period of slow growth called the **lag phase** of growth. As more females are born, they mature, mate, and give birth.

Q/A

Q: *What is the biotic potential of starfish?*

A: If 100 starfish achieved their biotic potential, in 15 generations 10^{79} starfish would exist. The number 10 to the 79th power is 1 with 79 zeroes after it.

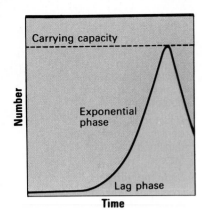

Figure 52–1. The early growth of a new population begins slowly and then explodes, forming a graph shaped like the letter "J."

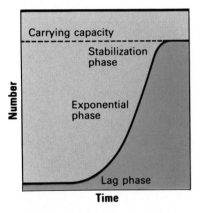

Figure 52–2. Eventually an exploding population begins to stabilize, turning the "J" curve into a curve shaped like an "S."

Soon the population reaches a certain size and **exponential** (eks poh NEHN shuhl) **growth** begins. This is a phase of very rapid growth in which the number of individuals repeatedly doubles in a specific time interval. For example, if a population of 100 warblers lived in the forest this year and the population was growing exponentially, 200 warblers would live in the forest next year. A year later, 400 warblers would live in the forest.

In nature, exponential growth occurs when a new habitat or a new food supply becomes available. For example, the warbler population grew exponentially some time after its arrival in a new habitat. A graph of the lag phase and **exponential phase** is shown in Figure 52–1. Ecologists call this type of population growth curve a **J-curve** because it is shaped like the letter J.

Exponential growth does not continue forever, however. The size of populations soon begins to stabilize, or reaches a **stabilization phase** of growth. This occurs because the **carrying capacity** of the environment is reached. The carrying capacity of an environment is the maximum number of individuals of a species that the environment can support. When carrying capacity is reached, the number of births and deaths becomes about equal. The population size stabilizes.

The growth of the warbler population is shown in Figure 52–2. This type of graph, which includes the lag, exponential, and stabilization phases, is called an **S-curve.**

52.3 Limiting Factors and Density

The environmental factors that stabilize population size and keep species from reaching their biotic potential are called **limiting factors.** The availability of space, food, and nesting materials are some of the limiting factors that probably affected the growth of the warbler population. Similarly, the amount of dissolved oxygen in water and water temperature are limiting factors to rainbow trout populations. The amount of sunlight is a limiting factor for the growth of daffodils. Limiting factors establish the carrying capacity of the land for each population.

Limiting factors may or may not be related to **population density**—that is, the number of individuals in a given area at a specific time. Some limiting factors operate independently of density. Others are density dependent.

Weather, landslides, fires, and floods are examples of **density-independent factors**—that is, factors whose effects are not determined by the density of a population. Imagine a desert where the temperature suddenly drops below 0°C (32°F). Both dense and sparse populations in the desert are equally affected by the freezing temperature.

In contrast **density-dependent factors** operate according to the density of the population. Food, for example, is an important density-dependent limiting factor. The more dense the population, the greater the struggle for limited amounts of food. *When a population's density is low, density-dependent factors are not limiting. As the population becomes more dense, these important factors exert a greater influence on populations.* Water, sunlight, soil nutrients, and availability of space are other density-dependent limiting factors. Stress, accumulation of waste materials, disease, and parasites are also density-dependent factors.

Predators are sometimes density-dependent limiting factors. Predators feed on other organisms and can thereby limit the size of populations. The availability of prey, predator's food, is a limiting factor on populations of predators. Figure 52–4 illustrates graphically a predator-prey relationship. It shows the population sizes of snowshoe hares and lynxes over 100 years. Lynxes prey on hares. When hares are abundant, the population of lynxes rises. As the lynxes eat more hares, hares become more scarce and then some lynxes starve to death. With fewer lynxes, the hare population makes a comeback. Thus, over the years the populations of hares and lynxes rise and fall in response to each other. Figure 52–4 illustrates these fluctuations. This kind of population growth curve, which shows the periodic growth and decline of populations, is called a **saw-tooth curve.** Over time, the populations of both predator and prey remain relatively stable.

52.4 Changes in Population Size

Ecologists are often interested in studying how populations change even after they have become stable. By doing so, ecologists detect growth trends that may be useful in managing a

Figure 52–3. Vegetation on a forest floor is limited to those types of plants that can grow with a minimum of sunlight.

Figure 52–4. Because lynxes eat hares, the populations of lynxes and hares rise and fall together, producing a saw-tooth curve.

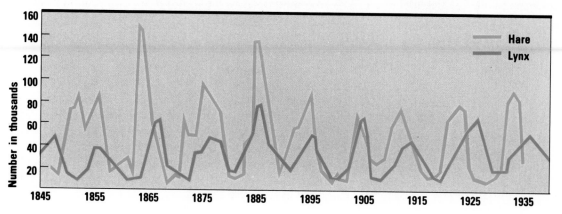

species. For example, knowing the change in the population size of rare whooping cranes enables ecologists to determine if steps must be taken to further ensure their survival. Conversely, if the size of a deer population rises too suddenly, the population may need to be decreased to maintain the ecological balance of an area.

Ecologists determine the change in population size by using the following equation:

Change in population size =
(births + immigrants) − (deaths + emigrants)

This equation can be studied using an example of caribou in the tundra. Consider a herd numbering 10,000 individuals. Assume that in a particular year 400 new caribou were born and 300 old caribou died. As the herd wandered across the icy tundra, 75

Spotlight on Biologists: Akira Okubo

Born: Tokyo, Japan, 1925
Degree: Ph.D., Johns Hopkins University

Oceanographer Akira Okubo began his scientific career researching ocean pollutants. He switched to another field when he became curious about certain behaviors of ocean plankton. Okubo, who moved to the United States in 1958, studies animal dispersal. He wants to know why the animals that form ocean plankton group together in large floating patches rather than dispersing.

Most biologists who study animal populations are interested in changes over time. Okubo is interested in their movements in physical space. He is testing a theory that traveling in large groups is an adaptation that aids in the animals' survival. If the organisms were distributed uniformly over a wide area, he theorizes, the whole group might not find enough food. Those that found food would lack a means of communication with the others.

Okubo also studies a similar phenomenon—swarming—found in mosquito populations. Since it is the male mosquitoes that swarm, he is testing the theory that females can more easily find a mate when the males call attention to themselves by traveling together in a large group. Okubo believes that finding a way to

prevent swarming might be one means of reducing the mosquito population, because it would interfere with mating.

For his work on ocean plankton, Okubo received the 1983 medal of the Oceanographic Society of Japan. He is currently a professor and research scientist at the State University of New York at Stony Brook.

caribou *immigrated*, or joined the herd. Another 200 caribou *emigrated*, or left the herd for another. According to the equation above, then, the change in the size of the population of caribou would be as follows:

(400 births + 75 immigrants) −
(300 deaths + 200 emigrants)

The herd, therefore, lost 25 individuals during that particular year. Ecologists use equations like this to help understand the ebb and flow of population size.

In the above example, 400 births occurred among 10,000 caribou in one year. These data can be used to calculate the population's **birth rate,** the rate at which births occur. To calculate the birth rate, the number of births is divided by the population size and the result multiplied by 100. The rate is expressed as a percent. The birth rate in the above example is $\frac{400}{10,000} \times 100$, or 4 percent.

The **death rate,** the rate at which deaths occur, is calculated the same way. The death rate for caribou is 3 percent. In the above example, the **growth rate,** the rate at which the population is growing or declining, is −0.25 percent. This is arrived at by dividing the population change of 25 by the total population of 10,000 and multiplying this answer by 100. Since there was a population loss, the answer is 0.25. A growth rate of 0 percent would represent a completely stable population. The growth rate of this caribou population is declining but ecologists consider it relatively stable.

Populations may remain relatively stable for a long time. Sometimes, however, a severe famine or a natural catastrophe, such as a landslide or fire, nearly destroys a population. In cases like these, the population undergoes a *population crash*. On the other hand, a population may undergo a *population explosion* where the number of individuals increases dramatically. A population explosion may occur, for example, if a new food supply becomes available or if a predator is somehow removed from the area.

Figure 52–5. Birth and death rates alone do not explain annual fluctuations in the size of a caribou herd. New animals periodically join the herd, while others leave.

Reviewing the Section

1. Give six examples of limiting factors that control biotic potential.
2. What is the difference between density-independent and density-dependent limiting factors?
3. What is the difference between a J-curve and an S-curve?
4. What is meant by the term *growth rate?*

The Human Population

- *State* how the development of agriculture accelerated human population growth.
- *List* the factors that led to the exponential growth of human populations.
- *Explain* what happens to a human population that has reached the land's carrying capacity.
- *Explain* why zero population growth may be difficult to achieve.

The human population has existed for over 500,000 years, a relatively short period in the earth's 4.6 billion year history. In that time, however, the human population has swelled to more that 4.5 billion individuals. How did such a large population arise? What is its future?

52.5 The History of the Human Population

About 20,000 years ago, humans had learned to build fires, construct shelters, and make clothes. The total human population then was about 3 million. This relatively small population was distributed over nearly the entire habitable earth. The sizes of their populations and the life style they adopted have a direct effect on the current size of the human population.

Our ancient ancestors were *hunter-gatherers* whose food consisted of the plants they gathered and the animals they killed. These hunter-gatherers roamed over the land in search of game. They also gathered grain as it ripened in an area, and then moved on when the grain was gone. If food became scarce, people died of starvation and its related diseases. The growth of the human population, therefore, was slow. It was in the lag phase of growth.

About 10,000 years ago, however, humans began to cultivate crops and domesticate animals. **Agriculture increased the carrying capacity of the land.** Agricultural areas provided more

Figure 52–6. Human population growth is in the exponential phase. The effects of medical advances and other human technology have made the "J" curve unusually steep.

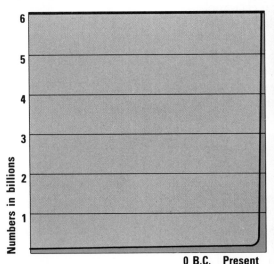

Zero Population Growth

In 1965 the birth rate in Japan was 1.67 percent. By 1980 it was 1.37 percent. In the United States the birth rate in 1965 was 1.94 percent. In 1980 it was 1.62 percent. In both countries the birth rate dropped in 15 years. Yet because the death rate dropped as well, the populations continued to rise and, overall, became more dense.

Because the populations of countries continue to rise, some nations and many private citizens have advocated a concept called *zero population growth*, a stabilized population. Zero population growth can be achieved in nations where immigration equals emigration and where each family has no more than two children. When every family has only two children, new individuals are added to the population at a rate that exactly replaces the individuals that die. The graph shows how the population of the United States would look if every family had just two children. The graph also shows the population growth if each family has three children.

Although the concept of zero population growth will theoretically stabilize a population after two or three generations, the idea is often not well accepted.

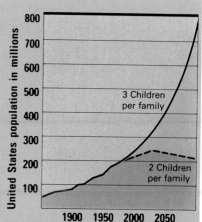

For example, many people see large families as a source of security in their old age. These people are often unwilling to limit the size of their families. Similar cultural factors make the goal of zero population growth a difficult one to achieve.

food than when the same areas were used for hunting. With fields to tend, humans ceased wandering from hunting ground to hunting ground and settled in one place. With a stable food supply and the greater safety offered by a settled community, the human population began to increase.

It took about 500,000 years for the human population to reach 1 billion, which it did in 1840. About that time, advances in medicine began to reduce *infant mortality*, the rate at which infants died, and began to eliminate many contagious diseases. People also began to develop better systems of food production, storage, and distribution.

The population was large enough so that these technological advances enabled the population to enter the exponential phase of growth. Only 90 years later, in 1930, the population had doubled to about 2 billion. Just 45 years after that, in 1975, the human population doubled again, bringing the world population to about 4 billion.

Q: *What is the most densely populated part of the earth?*

A: Macao, a Portuguese province in southern China, has a density of more than 1,700 people per km^2 (44,000 per mi.^{2}).

52.6 The Future of the Human Population

The human population is still in the exponential phase of growth. Because the birth rate exceeds the death rate, over 200,000 new human beings are added to the population each day or about 73 million each year. At the current rate of increase, the population doubles about every 38 years. By 2013, the human population may be close to 8 billion. Many of you reading this book will still be alive in 2051, when there may be as many as 16 billion humans on Earth.

Many factors have contributed to the rise in the number of humans on Earth. Technology, for example, continues to contribute to the population explosion. A hundred years ago it was not unusual to find countries with infant mortality rates over 50 percent. Today, however, infants in even the most impoverished countries have well over an 80 percent chance to live. Furthermore, technological advances have reduced the death rate by eliminating fatal diseases such as smallpox, improving crop yields, and devising new means for distributing food.

The population explosion is also indicated by the changing rates of births and deaths. In underdeveloped countries, people depend on their children for support and security. The infant mortality rate is high because the people lack sufficient food and modern medical care. Parents must have many children to ensure that a few will survive. As a result, birth rates are highest in underdeveloped countries. The simultaneous decline in death rates causes the population of these countries to grow rapidly. In many affluent countries, such as Japan and the United States, the birth rate is actually declining, but because the death rate is declining even faster, their populations continue to grow.

As the population increases, it causes greater competition for limited natural resources. An increasing population is partially caused by problems of hunger and poverty and, in turn, worsens them. Providing for the world's population and controlling its growth are among the serious problems facing the world today.

Reviewing the Section

1. How did the development of agriculture affect the human population?
2. What factors besides agriculture have contributed to the dramatic rise in the human population since 1840?
3. Why has the human population grown in some countries even though the birth rate has decreased?

Investigation 52: Yeast Population Study

Purpose
To discover and graph a comparison of the number of yeast cells present at a given time with the number present five days later

Materials
Graph paper, pencil, ruler, notebook

Procedure
1. Observe the figures below. They show a sample of a yeast population over a five-day period. The lines you see would appear on a special glass slide used for counting yeast cells under a microscope. Assume that each dot represents 1,000 yeast cells.

Day 0

Day 1

Day 2

2. In your notebook, make a table like the one below. Then count the number of yeast cells observed on Day 0 box by box. To avoid duplication, any dots appearing on a line to the right or bottom of a box should be counted as part of that box. Dots appearing on a line at the top or left will not be counted as part of that box.
3. Record your count under Day 0 in your table.
4. Repeat the procedure for days 1 through 5 and record your results in your table.

5. Using the data you have obtained, construct a graph. Use the horizontal axis to plot the time and the vertical axis to plot the number of cells counted.

Analyses and Conclusions
1. Between which days was there the most growth? The least growth?
2. During which day did the yeast population reach its maximum number?
3. What are some reasons why the population decreased after reaching a peak?
4. Name some factors that cause the population growth pattern of humans to differ from that of yeast.

Day 3

Day 4

Day 5

Going Further
You may wish to grow an actual yeast population to observe and count under a microscope. A yeast culture can be grown by mixing yeast, glucose, and distilled water and keeping it in a warm area.

Sample Yeast Population					
Number of Cells Present					
Day					
0	1	2	3	4	5

Chapter 52 Review

Summary

Populations are composed of individuals of the same species living in the same area. The maximum rate of reproduction for a population is its biotic potential. A population allowed to reach its biotic potential would soon cover the earth. The growth of populations is regulated by the environment's limiting factors such as food and space availability.

When a population establishes itself in a new area, it grows slowly at first. The population then enters an exponential phase of growth in which its number doubles at decreasing intervals. Exponential growth ceases when the carrying capacity of the environment is reached. The carrying capacity is determined by limiting factors that restrict population growth. Some limiting factors depend on the population's density; some are independent of density.

Changes in populations are calculated by subtracting the number of deaths and emigrants from the number of births and immigrants. Population crashes and explosions are dramatic changes in the size of populations.

The human population is currently in its exponential phase of growth. The number of humans now doubles every 38 years. Providing for this growing population is one of the major problems facing the modern world.

BioTerms

biotic potential (**795**)	density-independent	lag phase (**795**)
birth rate (**799**)	factor (**796**)	limiting factor (**796**)
carrying capacity (**796**)	exponential growth (**796**)	population density (**796**)
death rate (**799**)	exponential phase (**796**)	saw-tooth curve (**797**)
density-dependent	growth rate (**799**)	S-curve (**796**)
factor (**797**)	J-curve (**796**)	stabilization phase (**796**)

BioQuiz   (Write all answers on a separate sheet of paper.)

I. Completion

1. The largest number of a species the land can support is the land's _____ .
2. A population is composed of individuals of the same _____ .
3. Populations that enter a new habitat where food is abundant may undergo a population _____ .
4. _____ replaced hunting-gathering as the way humans get food.
5. A deadly virus that is transmitted from person to person is a density-_____ limiting factor.

II. Modified True and False

Mark each statement TRUE or FALSE. If false, change the underlined term to make the statement true.

6. The J-curve describes populations during the exponential phase of growth.
7. The human population is in exponential growth partly due to the decline of infant mortality.
8. Contagious diseases are density-independent.
9. The frequency of births in a population is the population's biotic potential.

III. Multiple Choice

10. The slow period at the beginning of the growth of population is the
    a) exponential. b) S-curve. c) lag.
    d) stabilization phase of growth.
11. Fire is a) a density-dependent factor.
    b) a density-independent factor. c) not a limiting factor. d) a factor in the population change equation.
12. Populations that double at regular intervals are in the a) slow stage of growth. b) stabilized phase of growth.
    c) decline of growth. d) exponential phase of growth.
13. Populations in which births and deaths are balanced with immigration and emigration are in a) J-curve growth.
    b) S-curve growth. c) saw-tooth growth.
    d) zero population growth.
14. The maximum rate at which a species reproduces is its a) carrying capacity.
    b) biotic potential. c) limiting factor.
    d) growth rate.
15. If there are 25 births, 15 deaths, 18 immigrations, and 21 emigrations, the change in population size is a) 7. b) 1.
    c) − 1. d) 13.

IV. Essay

16. Why does a long-term relationship between predator and prey lead to a saw-tooth type of population growth curve?
17. What are some of the disadvantages of a population whose density is very low?
18. Why do organisms rarely, if ever, achieve their biotic potential?
19. What are some factors responsible for the current human population explosion?
20. What are some causes of natural population collapses?

Applying and Extending Concepts

1. Since 1900 the birth rate in Pakistan has remained at 4.5 percent. But the population has almost quadrupled. Use the equation describing population change to suggest reasons for this situation.
2. Moose first arrived in Isle Royale, an island in Lake Superior, in 1908. Because no natural predators existed, the moose population soon grew exponentially, only to crash in both 1930 and 1940. In 1948 timber wolves arrived on the island. Since then the moose population has remained between 600 and 1,000 individuals. Explain what you think happened on Isle Royale.
3. Half of the people living in India are less than 15 years old. Using this knowledge of the distribution of the ages of individuals in India's population, write a paragraph predicting the future population growth in India.
4. Imagine that you own a red-snapper fishing business. To ensure yourself the best possible, stable income for your whole life, at what phases in the red-snapper S-curve would you harvest your population of fish? Why would this be the best time for harvesting?

Related Readings

Blake, B. "Threats to Birds of the Sea." *New Scientist* 100 (October 20, 1983): 210–211. Natural population crashes among seabirds are discussed in this article.

Bonner, W. N. *Seals and Man: A Study of Interactions*. Seattle: University of Washington Press, 1984. This text explores the ecology of the seal's world.

BIO*TECH*

Remote Sensing

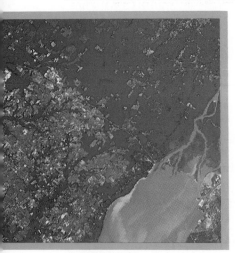

Skylab photo outlines agricultural characteristics of coastal land.

Photographs of Earth taken from an altitude of 435 km (696 mi.) show clearly that the planet is inhabited. Straight lines, rare in nature, break the land surface into geometric figures. Farmland is a mosaic of squares and rectangles, each kind of crop or stage of growth a different hue. Clear-cutting blocks within the dark masses of mature forest vegetation provide evidence of large scale logging operations.

Scientists use **remote sensing** technology to show so clearly how humans have changed the biosphere. Remote sensing is the gathering and recording of information about objects and phenomena from a great distance by systems that are not in contact with them. Tools used in this technology include sensing, recording, transmitting, and information processing equipment as well as high-altitude aircraft, satellites, and spacecraft.

Scientists use remote sensing technology to detect changes in the biosphere.

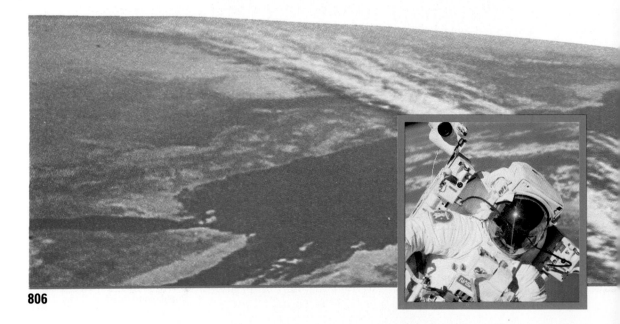

Landsat data indicates types of land cover (far left) and degrees of tree stress (left).

Remote sensing can be as simple as a space-shuttle astronaut taking photographs with a hand-held camera. Most applications, however, are more complex. Satellite scanning systems record the earth's surface by photoelectric means. A scanner aboard a satellite electronically records an image as a series of tiny picture elements, or *pixels.* The information can be stored aboard the satellite on magnetic tape or transmitted to ground receiving stations where the photograph is instantly reconstructed.

The Landsat satellite is one sensing system used to identify types of ground cover. Landsat data can be used to distinguish between hardwood forest, softwood forest, and grassland. The data also distinguish areas where crops are growing normally from areas where growth has been inhibited by disease or drought.

Landsat passes over any given point on Earth only about twice a month. A team of ecologists and physicists has developed a technique for mapping vegetation across a continent on a daily basis, using weather satellite data. They identify the kinds and amounts of vegetation in various sample areas and compare data over time. Using this process, the team produced a map of Africa that graphically reveals the effects of the disastrous drought of 1985. Because the September rainy season never came, vegetation that usually grows amid heavy rains did not appear. The team plans to apply this new technology over a period of years to learn whether, as many ecologists suspect, deforestation and overgrazing of land is the cause of Africa's long drought.

People and the Environment

Boaters in Chicago's Lincoln Park

Introduction

Hundreds of years ago, vast forests covered the eastern part of the United States. Timber wolves stalked herds of elk and deer. The skies teemed with passenger pigeons. Over the years the forests have given way to sprawling cities. The timber wolves, the elk, and the deer have retreated. The passenger pigeon has vanished forever. Wildlife habitats in the city are reduced to small patches of greenery—often referred to as "pocket parks"—which are sandwiched between skyscrapers of concrete and glass.

The settlement of the eastern seaboard vividly demonstrates the tremendous impact that people can have on the environment. As the human population has grown, so has the need to use the resources of this planet wisely.

Conservation of Resources

All the elements of nature, including sunlight, air, water, plants, and animals, are **natural resources.** People depend on natural resources to live, so maintaining the supply is vital. As the human population has grown, the demand for natural resources and the need for their conservation have grown as well. **Conservation** is the careful management, wise use, and protection of natural resources.

Conservation involves both renewable resources and non-renewable resources. **Renewable resources** are those that can be reused or replenished if they are well managed. Air, water, soil, and wildlife are all renewable resources. Soil, for example, can be used over and over to grow crops if it is farmed properly. **Nonrenewable resources,** such as coal and oil, are resources that cannot be reused or replenished. These resources form over thousands or millions of years. Once the supplies are gone, they cannot be replaced.

53.1 Water Conservation

Water is naturally processed for reuse as it moves through a cycle of *precipitation* and evaporation. Precipitation is the part of the cycle in which water falls to the earth as rain, sleet, or snow. The precipitation collects in lakes, rivers, and streams as well as below the ground. The ground water, too, eventually finds its way to these same lakes and rivers. The surface water is then used to supply homes, agriculture, and industry. Some of the water evaporates back into the atmosphere where it again becomes a source of precipitation.

The earth has an abundant supply of water, but over 97 percent is salt water. In addition, the supply of fresh water is unevenly distributed. Thus, some areas have water surpluses, while others face shortages.

Some communities draw their water from natural lakes. Many other communities obtain water from reservoirs that are created by building dams on rivers. Such dams often serve other purposes as well, such as controlling floods and supplying *hydroelectric power.* This is electric power generated by transforming energy produced by water into electrical energy. Conservationists have opposed the construction of certain dams, however, because of the possible harmful ecological effects of the dams. When a reservoir is built behind a dam, for example, it destroys the habitats of the wildlife living there. This destruction can have far-reaching and unexpected effects.

Figure 53–1. Water for drinking and irrigation may come from large reservoirs created by building huge dams to trap the water.

Q: *How much water does an average American use each day?*

A: An average American uses over 304 liters (80 gallons) of water per day.

Many communities obtain their water from wells that pump up ground water. In some communities, the ground water is withdrawn faster than it can be replenished. As the ground water is depleted, the ground surface may begin to sink. In areas of the western United States, the ground water is being depleted by the heavy use of water for irrigation. It is for this reason that many conservationists believe that irrigating arid and semiarid lands may not be practical in the long run. They believe that nothing is gained if irrigation is obtained at such a high cost.

Cities near seacoasts may someday meet their water needs by taking salt out of sea water, a process called **desalination** (dee sal uh NAY shuhn). Currently, however, desalination is too expensive for large-scale use because the process requires so much energy.

In areas where water shortages occur, conservation is a necessity. One method of conserving water is by maintaining the watershed. A **watershed** is the area of land from which water drains into a particular lake or stream. Maintaining plant cover, such as trees, shrubs, and grasses, on watersheds helps prevent rapid runoff of rain water. The water then seeps into the ground and replenishes the ground water supply. To conserve water, industries, farms, and individuals can cut their use of it.

53.2 Soil Conservation

Soil is renewable only if it is managed properly. Soil takes thousands of years to form from disintegrating rock. The soil nearest the surface, the topsoil, is most exposed to wind, water, and the organisms living on or in the soil. The bottom layer, which has been least exposed, most resembles the original rock. In between is a layer of soil containing minerals that have been

Figure 53–2. Steep hillsides are particularly exposed to soil erosion unless steps are taken to protect them from water runoff.

Figure 53–3. Contour plowing (left) protects soil from erosion by producing a pattern of furrows that holds rainwater. Terracing (right) holds water by producing flat steps.

washed down from the topsoil. The topsoil provides the nutrients plants need to grow. Tons of valuable topsoil have been lost to **erosion,** a process in which soil is washed away by water or blown away by wind.

The rate of soil erosion increases when land is stripped of its natural vegetation. Trees, grasses, and other plants reduce erosion by protecting the soil from the direct force of rain and wind. The roots of plants bind the soil and hold it in place. Plants also absorb some rainwater and reduce the runoff rate.

Farmers use various methods to reduce erosion caused by rainwater running down sloping land. **Contour plowing** is a method of plowing across, rather than up and down, a slope. The furrows of plowed soil catch rainwater, allowing it to seep slowly into the ground. In **strip-cropping,** rows of plants, such as grass and clover, are alternated with strips of grain crops. Grass and clover hold water and retard the flow of rainwater better than grain crops do. **Terracing** is a method of converting a hillside into broad, flat steps that help hold water.

Erosion caused by wind especially affects the soil on plains. There, farmers plant rows of trees called **windbreaks** to serve as barriers to the wind.

Another major conservation problem on farmlands is the depletion of soil nutrients. Planting the same crop in a field year after year contributes to a decline in the fertility of the soil. Grain crops, for example, use up the nitrogen in the soil if they are grown in the same field for several years. Farmers maintain soil fertility by **crop rotation,** in which crops are alternated from year to year. The rotation crop is usually alfalfa, soybeans, or other legumes. The advantage of crop rotation comes from a certain kind of bacterium that lives in the roots of legumes. These bacteria restore nitrogen to the soil.

Farmers add plant remains, animal wastes, or chemical fertilizers to their fields. The overuse of chemical fertilizers, however, may prevent bacteria from producing nutrients naturally.

Figure 53-4. A wildlife refuge provides a protected habitat that helps plants and animals avoid extinction.

Figure 53-5. Large numbers of visitors have damaged the natural environment in some national parks.

53.3 Wildlife Conservation

Wildlife has decreased sharply as the human population has grown. During the past few centuries, hundreds of species of animals and countless species of plants have completely died out. Such species are called **extinct species.** The passenger pigeon and the heath hen are but two examples of extinct species. Many more species are in danger of becoming extinct. They are called **endangered species** because they may not survive in the wild unless they are protected. Endangered species include the woolly spider monkey, the blue whale, the Mississippi alligator, and the St. Helena redwood tree.

Extinctions have occurred as long as life has existed on the earth. In earlier times, however, new species have evolved as rapidly as old ones have died out. Within the last few centuries, organisms have become extinct too rapidly for new species to develop. People and their policies are responsible for this rapid rate of extinction.

One cause of animal extinctions is uncontrolled hunting, fishing, and trapping. Animals may be hunted not only for their meat, hides, and other products, but also for sale to zoos, researchers, or pet traders. The passage and enforcement of hunting and fishing laws has protected many species. Even so, illegal hunting, called *poaching,* threatens many animals, including the Siberian tiger and the black rhinoceros.

A primary threat to wildlife today is the destruction of habitats, which occurs when land is cleared for homes, farms, or other developments. Such developments threaten not only individual species but also entire ecosystems. To save wildlife in varied ecosystems, government and private organizations

▶ The Comeback of the Brown Pelican

The brown pelican is a fish-eating coastal bird that nests along the Atlantic, Pacific, and Gulf shores of the Americas. More than 50,000 brown pelicans once nested in Texas and Louisiana alone, but the population there dropped to almost zero in the early 1960s. By the late 1960s, California's pelican population had also declined and was not reproducing normally. In 1970 only one bird hatched at Anacapa Island, a major breeding colony. The brown pelican was placed on the United States list of endangered species in 1973.

Biologists searched for the causes of the pelican's decline. They found that the drop in the Texas population occurred after large numbers of fish died in the Mississippi River delta. The fish had been poisoned by a pesticide called *endrin*. Thus, the pelican's food

supply had been cut off. Since the cause was traced to endrin, the use of that pesticide has decreased. However, the pelican may never thrive again in Texas and Louisiana because of decreased food supplies, oil spills, and the loss of habitat.

In California biologists linked the pesticide DDT with the decline of the brown pelican. The pelican's eggs contained some

traces of DDT. The eggshells were so thin that they broke while the eggs were laid or during incubation. Investigators found that a chemical company had been dumping waste DDT into the Los Angeles sewer system. The DDT contaminated the fish in the oceans and the pelicans ate many fish. The birds thus accumulated large amounts of DDT in their bodies. The process by which a substance becomes more concentrated in animals higher in the food chain is called *biological magnification*.

After DDT was banned in 1972, the brown pelican began to recover in California. In the early 1980s, about 2,000 breeding pairs nested on Anacapa Island. Biologists continue to watch the bird closely, and they believe that the brown pelican is making a strong comeback.

have worked to establish preserves and refuges. National parks also provide homes for wildlife. However, because some parks are visited by so many tourists, the ecology of the parks is being disrupted as well.

Many of the larger zoos are attempting to contribute to wildlife conservation. Through *captive breeding programs*, scientists working in zoos hope to save such endangered species as the cheetah and the black lemur. In these programs zoologists try to create the conditions needed for the animals to mate.

Figure 53–6. A nuclear power plant includes huge towers for cooling the water used to remove waste heat from the reactor.

53.4 Energy Conservation

About 95 percent of the energy used in the world comes from oil, coal, and natural gas. Because these substances are the fossilized remains of prehistoric plants and animals, they are called **fossil fuels.** Fossil fuels are a nonrenewable resource. The present supply took millions of years to form and cannot be replaced. As the supply diminishes, extracting the fuels becomes more difficult and more expensive. *Finding alternate sources of energy to replace fossil fuels is essential.*

At one time many people believed that nuclear energy offered the cleanest and least expensive solution to the world's ever-increasing energy needs. Nuclear energy is produced in two ways: nuclear fission and nuclear fusion. **Nuclear fission** results from the splitting of atomic nuclei, a process that releases usable heat energy. The nuclear power plants in operation today use fission. **Nuclear fusion** involves fusing two hydrogen atoms to form helium. Scientists have not found a practical way to control the release of fusion energy. If a method is found, fusion could supply most of the world's energy needs.

The first nuclear power plant went into operation in the United States in 1957. Since that time the use of nuclear power to replace fossil fuels has not significantly increased. More than 25 years later, only about 4 percent of the nation's energy is provided by nuclear power. Worldwide, nuclear energy accounts for only about 1 percent of the energy supply. The use of nuclear power has been held back for several reasons. There has been much public concern over the dangers of radiation emission and the safe disposal of hazardous wastes. Also, the cost and the complexity of building a nuclear power plant has proved to be much greater than once thought.

Other energy sources now being researched include the harnessing of water, wind, steam, and the energy of the sun. Water power provides about 2 percent of the world's energy supply. A significant increase in the use of water power is unlikely, however, because many suitable sites already have hydroelectric plants.

Wind power has been used for centuries to pump water and to grind grain. More recently large windmills have been used to generate electricity. Wind power is limited, however, because few areas have a steady enough supply of wind.

Geothermal (jee oh THER muhl) **power** can be harnessed from steam created below the earth's surface. The steam is created when water is warmed by hot rocks. Beds of hot rock below the earth's surface provide huge reservoirs of heat. In some cases steam is naturally formed by ground water that flows

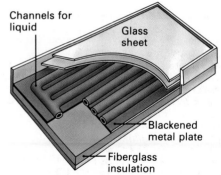

Channels for liquid

Glass sheet

Blackened metal plate

Fiberglass insulation

Figure 53–7. Large solar collectors are an obvious feature of a home that uses solar energy (left). Each individual solar collector (right) is designed to capture maximum sunlight.

over the rocks. This steam can be tapped by drilling and then directed to drive steam turbines. Where underground steam does not naturally occur, it can be created by injecting water into hot rocks. However, geothermal power is limited by the number of natural sites that lie near the earth's surface.

Solar energy, or energy derived from sunlight, is a safe and unlimited source of energy. Figure 53–7 shows how solar collectors absorb the sun's heat for home use. Sunlight can also be collected by devices that convert it into electricity. The use of solar energy, however, is not yet practical in areas where the sunlight is limited.

Wood, once the world's chief fuel, is still the main source of energy in many less developed countries. Growing populations in those areas have greatly reduced forest resources. If the present rate of use is not checked, these resources will one day disappear.

A small amount of electric power is produced by burning trash and other solid wastes. In addition, some treatment plants convert sewage into *methane,* a clean-burning gas. Methane has been used to power small combustion engines.

Until some of these alternate energy sources can be developed on a large scale, people must conserve fossil fuels.

Reviewing the Section

1. What objections have been raised to each of the following water supply methods: dams, irrigation, and desalination?
2. How does natural vegetation work to slow down the rate of soil erosion?
3. Explain how DDT harmed the brown pelican.
4. Why should world leaders be concerned with locating other sources of energy than fossil fuels?

Pollution of the Environment

Section Objectives

- *Distinguish* between biodegradable and nonbiodegradable wastes.
- *Name* the main sources of air and water pollution.
- *Define* the terms *eutrophication* and *thermal pollution.*
- *Identify* factors that contribute to the problem of solid wastes disposal and state one solution.

While conservationists labor to maintain the world's natural resources, other factors are at work that are destructive to these resources. Contamination of the environment with waste products and other impurities is called **pollution.**

Any kind of waste product is a *pollutant*. People have always produced waste products. Before industrial cities were developed, however, these pollutants were not released in huge amounts nor concentrated in small areas. In addition, the types of waste products formerly produced could be broken down naturally into harmless substances by microorganisms and reused by other organisms. Such waste products are **biodegradable** (by oh dih GRAY duh buhl). Today, many discarded products are made of metal or of newer materials, such as plastics, which are often **nonbiodegradable.** This means that they cannot be broken down by microorganisms. Because of the large quantity and types of pollutants that the industrialized nations produce, pollution is now a major problem facing the modern world. Pollution endangers human society, nonhuman organisms, and the entire physical environment.

53.5 Air Pollution

In large cities, breathing the air may be harmful to your health. It can cause headaches and burning eyes. It may contribute to the development of such diseases as lung cancer and emphysema. Air pollutants can stunt the growth of plants and eventually kill them. The pollutants in the air can even eat away at the outside surfaces of buildings.

Air pollution consists of solid or liquid particles called **particulate matter** and of various gases. The particulate matter includes dust, smoke, ashes, asbestos, and tiny particles of lead and other heavy metals. The gases of air pollution include carbon monoxide, nitrogen oxides, sulfur oxides, hydrocarbons, and excessive amounts of carbon dioxide. *The greatest source of air pollution is the burning of fossil fuels to produce electricity.* Other air pollutants are produced by the burning of fuel to drive automobiles, to heat buildings, and to supply power for industries.

Carbon monoxide, a serious air pollutant in cities, combines with hemoglobin and reduces the oxygen carrying capacity of the blood. As a result the heart and lungs must work harder to supply the tissues with oxygen when there are high levels of carbon monoxide in the air.

Figure 53–8. Factories, power plants, and automobile exhausts all contribute to the pollutant called smog, which hangs over many large cities.

Figure 53-9. Metrorail in Miami, Florida provides a means of transportation that does not pollute the atmosphere.

Air pollution is greatest in densely populated, industrialized cities. In some cities, air pollution may take the form of **smog,** a combination of smoke, gas, and fog. Smog worsens when a cool air mass settles under a warm air mass. This condition, called **thermal inversion,** is the opposite of the normal air layer arrangement. In thermal inversion, pollutants become trapped in the cool air layer which would ordinarily warm up, rise, and carry the pollutants away. If this condition persists for several days, the pollutants build up. Breathing the air may then become a major health risk.

Smog may contain large amounts of sulfur oxides from fossil fuels. These react with water vapor to form sulfuric acid. Sulfuric acid in the atmosphere falls to the ground as **acid rain,** which erodes buildings, damages crops, and kills fish and other aquatic organisms.

To reduce air pollution, governments have passed laws regulating the release of various air pollutants. Industries are required to install antipollution devices in factories. Automobile makers in the United States have been asked to install pollution control devices in cars. Cars are now designed to run on lead-free gasoline because it burns cleaner than leaded gasoline. Residents of cities have been encouraged to make increased use of public transportation as a means of reducing the number of automobiles on the road. They have also been encouraged to use non-polluting energy sources.

Q/A

Q: *What is the difference between "brown air" cities and "gray air" cities?*

A: "Brown air" cities are enveloped in a brownish haze of nitrogen dioxide. This haze forms when nitrous oxide released from vehicle exhausts combines with oxygen in the air. "Gray air" cities are cities in which the main source of pollution is the burning of coal and oil. This colors the air gray.

53.6 Water Pollution

The main sources of water pollution are sewer systems, industries, and farms. Pollutants from these sources contaminate lakes, streams, oceans, and ground water.

Most urban sewage systems remove poisonous substances before releasing sewage into lakes or streams. However, although the sewage is treated, it still contains chemicals, such as phosphates and nitrates, which are nutrients for aquatic plants.

Highlight on Careers: Science Writer

General Description

Science writers report on science for a general audience. They may report on a current news item, such as an international conference on the effects of acid rain. They may explain ongoing research in an area such as reproduction in an endangered species. Science writers may specialize in one field, such as ecology, oceanography, medicine, or physics, or they may cover a wide variety of topics.

Science writers produce materials for textbooks, magazines, television, or radio. Some work in public information departments of government agencies, such as the Environmental Protection Agency, or private groups, such as the National Wildlife Federation.

Writers begin their work by gathering information, chiefly through library research or personal interviews. They generally prepare a first draft and then revise it to make it clear, accurate, and interesting. Many science writers find it highly satisfying to take a concept that readers might find difficult and explain it clearly. This may involve giving the necessary background information to make the explanation clear.

Career Requirements

There are no set requirements for a career as a science writer. Some science writers have a bachelor's degree in journalism or English, having taken some additional science courses. An increasing number of universities offer special courses in science writing, usually within a journalism department. Other science writers have a college education in biology or another science and acquire on-the-job experience in writing. Some agencies or organizations may hire only writers with an advanced degree in a science.

For Additional Information
National Association of
   Science Writers, Inc.
P.O. Box 294
Greenlawn, NY 11740

Algae thrive on these nutrients. Later, as the algae die and decay, they use up the water's oxygen supply. Other plants and animals cannot survive in the water without oxygen. This nutrient enrichment process, or eutrophication, changes the ecology of a lake or stream and often results in the death of certain fish species.

Industries dump lead, mercury, and other chemical wastes into streams and lakes. These chemicals have poisoned not only the organisms living in these lakes and streams but also people and other animals that have eaten contaminated fish.

Industries sometimes draw cold water from a lake or stream, use it for cooling purposes, and release it back into its source. Nuclear power plants are the principal users of this cooling method. When the water is used for cooling, the water itself becomes heated. This form of pollution, called **thermal pollution,** kills plants and animals that normally live in cooler water. Oil from accidental spills and from oil industry operations is a primary pollutant of ocean waters. When released near beaches, the oil sometimes coats and kills marine animals and shore birds.

Farms contribute to water pollution through the use of chemical fertilizers and weed killers as well as through the use of chemicals, called **pesticides.** Pesticides are used to control insect pests. These products seep into the ground water and are washed into lakes and streams. Certain harmful pesticides are now banned in some nations.

53.7 Solid Waste Disposal

Solid wastes include paper products, bottles, cans, old appliances, and all the other trash that people throw away. Each year people discard billions of tons of solid wastes. These wastes accumulate in city dumps and litter streets. The disposal of solid wastes is a growing problem.

One solution to the problem of solid wastes is to recycle wastes. Products made of paper, iron, steel, glass, and aluminum can be treated in ways that make them reusable. Wasted food, grass, leaves, and other organic debris can be made into fertilizer. Some solid wastes can be burned to produce energy. The remaining wastes can be used to build up low-lying land. Wastes that are sandwiched between layers of earth create a **landfill.**

Old patterns of carelessly using and wasting resources will not work in the future. Instead, new patterns of carefully using and reusing resources will soon become essential. Not until then will people have truly learned how to live in harmony with the environment.

Figure 53–10. Oil spills from tankers and from industrial operations located near the ocean are a major threat to shorebirds and other forms of wildlife.

Reviewing the Section

1. What are the main components of air pollution?
2. What chemicals in treated sewage contribute to eutrophication?
3. What effect does the burning of fossil fuels have on cities and on rural areas?
4. Name four solutions to the problem of disposing of solid wastes.

Investigation 53: Acidity and Alkalinity of Water

Purpose
To determine the acidity and alkalinity of water samples

Materials
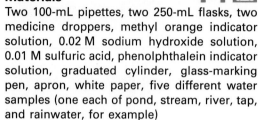
Two 100-mL pipettes, two 250-mL flasks, two medicine droppers, methyl orange indicator solution, 0.02 M sodium hydroxide solution, 0.01 M sulfuric acid, phenolphthalein indicator solution, graduated cylinder, glass-marking pen, apron, white paper, five different water samples (one each of pond, stream, river, tap, and rainwater, for example)

Procedure
1. With a glass-marking pen, label one flask "A" and use it to test for an acid. The other flask should be labeled "B" and used to test for a base.
2. In your notebook, make a table like the one below.
3. To test for an acid, pour 50 mL of the first water sample into a graduated cylinder and gently transfer the water to flask A.
4. Add three drops of phenolphthalein solution.
5. Carefully draw 50 mL of 0.02 M sodium hydroxide into a pipette.
6. Add one drop at a time into the beaker until the solution becomes pale pink. Record the number of milliliters used in Table 1.
7. Multiply the number of milliliters of sodium hydroxide used by 20 to get the parts per million (ppm) of acid.
8. Record your results in Table 1.
9. To test for a base, measure 100 mL of the same water sample into a graduated cylinder and pour gently into flask B.
10. Add five drops of phenolphthalein solution.
11. If the solution becomes colored, go to step 12. If it does not change color, go to step 14.
12. Carefully draw 50 mL of 0.01 M sulfuric acid into a pipette.

13. Add one drop at a time to the beaker until the color disappears. Then record the milliliters of acid used in Table 1.
14. Add five drops of methyl orange indicator to the flask.
15. If the solution becomes yellow, add 0.01 M sulfuric acid until the solution turns pale pink. Record the milliliters of acid used in Table 1.
16. Total alkalinity in ppm is the sum of the milliliters of acid required in steps 13 and 15 multiplied by 10.
17. Clean all glassware thoroughly and repeat the procedure for all water samples. Record the results in your notebook in the form of a table like the one below.

PPM of Acid and Base in Water

Water Sample	Flask A		Flask B		
	mL from step 6	PPM Acid	mL from step 13	mL from step 15	PPM Base

Analyses and Conclusions
1. Which water samples are the most polluted? How is this determined?
2. What type of ion is more prevalent in acid rain, H^+ or OH^-?
3. How does air pollution affect the natural waters of the earth?

Going Further
You may wish to use a water testing kit to test the same samples to determine whether or not the results are the same as the results previously obtained.

Chapter 53 Review

Summary

As the human population has grown, the demand for natural resources has also grown. Resources such as water, air, land, and wildlife, though renewable, need careful management to maintain the supply. Fossil fuels and other nonrenewable resources are diminishing. Alternate sources of energy must be developed for future use.

Pollution of the environment creates a serious threat to our precious natural resources.

People need to develop better methods of dealing with the pollution resulting from the burning of fuel, the dumping of sewage, and the disposal of solid wastes.

Conservation—the careful management, wise use, and protection of natural resources—is essential for the future. Through conservation, people can maintain the supply and the quality of this planet's natural resources.

BioTerms

acid rain (**817**)
biodegradable (**816**)
conservation (**809**)
contour plowing (**811**)
crop rotation (**811**)
desalination (**810**)
endangered species (**812**)
erosion (**811**)
extinct species (**812**)
fossil fuels (**814**)

geothermal power (**814**)
landfill (**819**)
natural resources (**809**)
nonbiodegradable (**816**)
nonrenewable
   resources (**809**)
nuclear fission (**814**)
nuclear fusion (**814**)
particulate matter (**816**)
pesticides (**819**)

pollution (**816**)
renewable resources (**809**)
smog (**817**)
solar energy (**815**)
strip-cropping (**811**)
terracing (**811**)
thermal inversion (**817**)
thermal pollution (**819**)
watershed (**810**)
windbreak (**811**)

BioQuiz   (Write all answers on a separate sheet of paper.)

I. Completion

1. Excessive irrigation of arid lands, particularly in the western states, has depleted the _____ .
2. _____ can be reused or replenished if soundly managed.
3. Alternating the planting of grains and _____ prevents the depletion of nutrients in the soil.
4. Decaying algae in a lake reduces the lake's supply of _____ .
5. Planting windbreaks is a method of preventing _____ .

II. Modified True and False

Mark each statement TRUE or FALSE. If false, change the underlined term to make the statement true.

6. Nuclear fusion results from the splitting of atomic nuclei.
7. Fossil fuels are an example of renewable resources.
8. Endangered species of plants and animals cannot survive in the wild without human protection.
9. Such waste products as plastics and aluminum are biodegradable.

III. Multiple Choice

10. Conservationists sometimes oppose dam construction because the dams may
a) cause pollution. b) introduce foreign species. c) deplete ground water. d) destroy animal habitats.
11. Nutrient enrichment of water plants through addition of chemicals to sewage is an example of a) desalination. b) eutrophication. c) thermal inversion. d) thermal pollution.
12. Soil erosion can be reduced by a) irrigation. b) desalination. c) terracing. d) crop rotation.
13. Waste products made of paper or glass are best disposed of by a) burning. b) dumping at sea. c) recycling. d) using as landfill.
14. Air pollutants are produced chiefly by a) thermal inversion. b) burning of fuel. c) use of pesticides. d) erosion.

IV. Essay

15. What is the cause and possible effect of thermal inversion?
16. What causes acid rain and what are its effects?
17. How can soil erosion on sloping land be reduced?
18. How does the maintenance of watersheds help conserve water?
19. What are the limitations to the use of nuclear power, solar energy, and geothermal power?
20. Why is conservation needed?

Applying and Extending Concepts

1. Choose an endangered animal or plant. Prepare a report for the class describing the species, the threats to its survival, and its current status.
2. Cogeneration is a method of heating buildings using heat that results as a byproduct of the generation of electricity. Use your library to research cogeneration as an alternative energy source.
3. One of the world's greatest art treasures, the Parthenon in Athens, Greece, is threatened by the effects of acid rain. Why is this a greater problem now than it has ever been before? Research the causes, and the steps Greek conservationists are taking to protect this monument.
4. Goats and pigs were introduced to the Galapagos Islands by sailors who hoped those animals would breed and be a source of meat in the future. What do you suppose the effect was on the native wildlife population of the Galapagos? If necessary, use your library to help with your answer.

Related Readings

Ehrlich, P. and A. Erlich. *Extinction: The Causes and Consequences of the Disappearance of Species.* New York: Random House, 1981. This book details the reasons behind species endangerment and the effects of the loss of a species.

Epstein, S. S., L. Brown, and C. Pope. *Hazardous Waste in America.* San Francisco: Sierra Club, 1982. This analysis of toxic wastes explores ways to protect the environment from these pollutants.

Rahn, Kenneth A. "Who's Polluting the Arctic?" *Natural History* 93 (May 1984): 30–38. This article describes the eight-year investigation that led to the conclusion that industrial pollution from the Soviet Union and other European nations is causing the arctic haze.

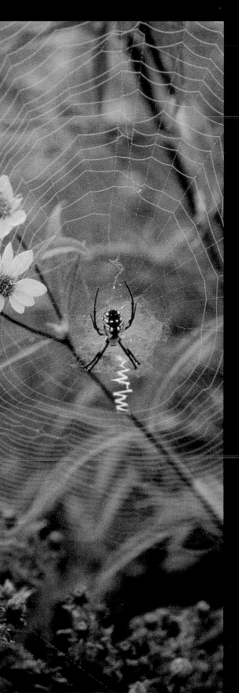

Reference Section

Five-Kingdom Classification of Organisms

Kingdom Monera

Prokaryotes (cells lack a true nucleus and membrane-bound organelles); mostly unicellular; some occur in filaments or clusters.

Phylum Schizophyta: Bacteria; about 2,500 species including eubacteria (true bacteria), rickettsias, mycoplasmas, and spirochetes; mostly heterotrophs; some photosynthetic and chemosynthetic autotrophs; reproduction usually asexual by binary fission.

Phylum Cyanophyta: Blue-green algae or cyanobacteria; about 200 species of photosynthetic autotrophs with chlorophyll *a* and accessory pigments; no chloroplasts; mostly filamentous; some unicellular; reproduction asexual by binary fission or fragmentation.

Phylum Prochlorophyta: Protosynthetic autotrophs; contain chlorophyll *a* and *b*, xanthophylls, and carotenes.

Kingdom Protista

Diverse group of unicellular and simple multicellular eukaryotes (cells have a true nucleus and membrane-bound organelles).

Phylum Euglenophyta: Euglenoids; about 800 species of unicellular, photosynthetic/heterotrophic organisms with chlorophyll *a* and *b*; usually having a single flagellum; reproduction asexual.

Phylum Mastigophora: Flagellates; about 2,500 species; mostly parasitic; includes *Trypanosoma* and *Trichonympha*.

Phylum Sarcodina: Sarcodines; about 11,500 species that move by means of pseudopodia; includes amoebas.

Phylum Ciliophora: Ciliates; about 7,200 species; locomotion by cilia, or sessile; includes paramecia and stentors.

Phylum Sporozoa: Sporozoans; about 6,000 species of nonmotile parasites; includes *Plasmodia,* the cause of malaria.

Phylum Chrysophyta: Golden algae; about 12,000 species of photosynthetic autotrophs with chlorophylls *a* and *c* and carotenes, xanthophylls, and fucoxanthins; most are unicellular and aquatic; includes diatoms.

Phylum Pyrrophyta: Fire algae; about 1,100 photosynthetic species with chlorophylls *a* and *c* and xanthophyll; unicellular; major component of marine phytoplankton; includes dinoflagellates.

Phylum Chlorophyta: Green algae; about 7,000 photosynthetic species with chlorophylls *a* and *b* and carotenoids; includes unicellular, colonial, and multicellular species; probable ancestor of modern land plants.

Phylum Phaeophyta: Brown algae; about 1,500 photosynthetic species with chlorophylls *a* and *c* and fucoxanthin; includes kelps.

Phylum Rhodophyta: Red algae; about 4,000 photosynthetic species with chlorophylls *a* and *d*, carotenes, and phycobilins; includes filamentous, multicellular seaweeds.

Kingdom Fungi

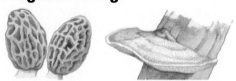

Eukaryotic heterotrophs that obtain food by absorption; includes saprophytes and parasites; most are multicellular, composed of intertwined filaments (hyphae); cell wall of most species contains chitin.

Phylum Myxomycophyta: Slime molds; about 600 species; body consists of multinucleate plasmodium that creeps by amoeboid movement; plasmodium separates into funguslike sporangia during reproduction; forms spores.

Phylum Eumycophyta: True fungi; about 81,500 species; mostly filamentous with chitinous cell walls; reproduction may include sexual and asexual stages.

Class Oomycetes: Aquatic fungi; some motile cells in certain stages of life cycle; approximately 200 species.

Class Zygomycetes: Terrestrial molds; about 600 species; reproduce asexually by spores and sexually by the fusion of nuclei from the tips of different mating strains.

Class Basidiomycetes: Club fungi; about 25,000 species of terrestrial fungi that produce spores on basidia; includes mushrooms, rusts, and smuts.

Class Ascomycetes: Sac fungi; about 30,000 terrestrial and aquatic species; spores form in an ascus (little sac) which results from the sexual combination of two gametes or strains; includes yeasts that reproduce asexually by budding, powdery mildews, morels, and truffles.

Class Deuteromycetes: Imperfect fungi; about 25,000 species that either do not reproduce sexually or have sexual life cycles that are not fully understood; includes *Penicillium*.

Kingdom Plantae

Multicellular, eukaryotic autotrophs that carry out photosynthesis in chloroplasts; chlorophyll *a* is the photoreactive pigment; chlorophyll *b* and various carotenoids serve as accessory pigments; primarily terrestrial; cell walls contain cellulose; body has distinct tissues; life cycle of alternating sporophyte and gametophyte generations.

Phylum Bryophyta: Bryophytes; about 15,600 species that lack vascular tissues and true roots, stems, and leaves; obtain nutrients by osmosis and diffusion; found chiefly in moist habitats; sperm must swim to eggs; gametophyte is dominant form in life cycle.

Class Muscopsida: Mosses; about 9,500 species with small, leafy gametophytes; sporophyte nonphotosynthetic, attached to and dependent upon gametophyte.

Class Hepaticopsida: Liverworts; about 6,000 extremely tiny species; gametophytes generally leafy, with scaly upper surface; sporophyte nonphotosynthetic, attached to and dependent upon gametophyte.

Class Antherocerotopsida: Hornworts; about 100 species; gametophyte not differentiated into roots, stems, or leaves; stomata on sporophyte; sporophyte grows from basal meristem.

All of the following plant phyla are tracheophytes with true roots, stems, and leaves in the dominant sporophyte; gametophyte is greatly reduced.

Phylum Psilophyta: Whisk ferns; a few species of seedless plants lacking roots and leaves; sperm must swim to eggs.

Phylum Sphenophyta: Horsetails; about 15 species of seedless plants with hollow, siliceous stems; sperm must swim to eggs.

Phylum Lycophyta: Club mosses; about 1,000 diverse species of seedless plants with leafy sporophytes; some have only one type of spore; others produce microspores and megaspores; sperm must swim to eggs.

Phylum Pterophyta: Ferns; about 12,000 diverse species of seedless plants; sporophyte bears haploid spores on underside of fronds; spores germinate into free-living gametophytes; sperm must swim to eggs.

Phylum Cycadophyta: Cycads; about 100 species of palmlike plants with slow cambial growth; gymnosperms.

Phylum Ginkgophyta: Ginkgo; one species only; fan-shaped leaves; gymnosperms; separate sexes.

Phylum Gnetophyta: Seed plants similar to angiosperms; about 70 species; motile sperm; xylem with vessels.

Phylum Coniferophyta: Conifers; about 550 species of gymnosperms; gametophyte much reduced: male pollen is dispersed by wind; female gametophytes remain on seed cone, where fertilization and seed development take place; leaves needlelike or scale-like; most species are evergreens.

Phylum Anthophyta: Angiosperms; about 235,000 species of plants that produce enclosed seeds; reproductive structures are flowers; mature seeds are enclosed in fruits; gametophyte is much reduced in size; double fertilization involves two sperm nuclei.

Class Monocotyledones: Monocots; angiosperms that produce seeds with one cotyledon (seed leaf); flower parts usually in threes; leaves with parallel veins; almost all are herbaceous.

Class Dicotyledones: Dicots; angiosperms that produce seeds with two cotyledons; flower parts usually in fours or fives; leaves usually with nonparallel veins; herbaceous and woody plants.

Kingdom Animalia

Multicellular, eukaryotic heterotrophs that obtain food by ingestion; most are motile; reproduction is predominantly sexual.

Phylum Porifera: Sponges; about 5,000 aquatic, mostly marine, species of asymmetrical animals that lack distinct tissues and organs; body consists of two layers supported by a stiff skeleton; sponges are aquatic, mostly marine, and sessile; reproduction sexual or asexual.

Class Calcarea: Simple shallow-water sponges; calcium carbonate spicules; includes *Grantia*.

Class Hexactinella: Deep water sponges; silica spicules; includes Venus's flower basket.

Class Demospongiae: Large sponge; spicules of spongin and silica; freshwater and marine species; includes bath sponges and finger sponges.

Class Sclerospongiae: Spicules of calcium carbonate, silica, and spongin.

Phylum Coelenterata (Cnidaria): Coelenterates; about 9,000 aquatic species; radially symmetrical; digestive cavity with only one opening; two body layers separated by jellylike mesoglea; tentacles armed with stinging cells; two body forms: vase-shaped polyp and bell-shaped medusa; live singly or in colonies often made up of specialized individuals; reproduction sexual or asexual.

Class Hydrozoa: Hydras and related animals; freshwater hydras occur singly with only polyp form; other hydrozoans are colonial and have a life cycle with both polyps and medusae; reproduction sexual or asexual.

Class Scyphozoa: Jellyfish; marine coelenterates; dominant medusa form reproduces sexually; free-swimming larvae become sessile polyps that form new medusae asexually.

Class Anthozoa: Sea anemones, corals, and related "flower animals;" marine coelenterates with no medusa stage; sea anemones are solitary; corals are colonial organisms with either an internal or external skeleton.

Phylum Ctenophora: Sea walnut and comb jellies; about 90 species; gelatinous marine animals with eight bands of cilia; often bioluminescent.

Phylum Platyhelminthes: Flatworms; about 13,000 species; bilaterally symmetrical with three germ layers; digestive cavity has only one opening; no circulatory or respiratory systems; no coelom or pseudocoelom.

Class Turbellaria: Planarians and related free-living, carnivorous flatworms; simple nervous system with sense receptors; sexual reproduction by mutual fertilization; asexual reproduction by fission.

Class Trematoda: Flukes; parasites covered by protective cuticle; most are endoparasites; may have a complex life cycle involving more than one host.

Class Cestoda: Tapeworms; parasitic species that live as adults in the intestines of vertebrates; no digestive system; food absorbed through body surfaces.

Phylum Nematoda: Roundworms; about 12,000 mostly parasitic species; tubular and bilaterally symmetrical body form; digestive tract has two openings; pseudocoelom; reproduction is sexual.

Phylum Nematomorpha: Horsehair worms; bodies very slender, up to one meter (3.3 ft.) long; adults are free-living; larva parasitic.

Phylum Acanthocephala: Spiny-headed worms; about 500 species; parasitic with no digestive tract; head with recurved spines for attachment to host.

Phylum Rotifera: Rotifers or "wheel" animals; wormlike or spherical; complete digestive tract; crown of cilia on anterior end resembling a wheel.

Phylum Bryozoa: "Moss" animals; microscopic, aquatic organisms that form branching colonies; U-shaped row of ciliated tentacles for feeding (lophophore).

Phylum Brachiopoda: Lamp shells; about 250 species, 30,000 extinct; two shells, one dorsal and one ventral; adults sessile; feed by lophophore.

Phylum Mollusca: Mollusks; about 47,000 species of soft-bodied animals with a true coelom, a head-foot, a visceral mass, and a mantle; most are aquatic; many have one or more shells.

Class Pelecypoda: Bivalves; mollusks with two shells and a hatchet-shaped foot; no head; open circulatory system; many species sessile; includes clams, scallops, and oysters.

Class Gastropoda: Snails and slugs; aquatic and terrestrial mollusks with a locomotor belly-foot; open circulatory system; adults have an asymmetrical body twisted by torsion; most have a coiled shell.

Class Cephalopoda: Octopuses, squids, and nautiluses; marine mollusks with a large head and a foot divided into tentacles; closed circulatory system; highly developed nervous system; shell may be internal, external, or absent.

Class Scaphopoda: Tooth shells; about 350 species; marine mollusks with tubular shells.

Class Polyplacophora (Amphineura): Chitons; closely resemble ancestral mollusk; eight dorsal shell plates with reduced head and elongated body.

Phylum Annelida: Segmented worms; about 9,000 species; body has a true coelom; longitudinal and circular muscles; fairly complex circulatory, respiratory, and nervous systems; elimination by nephridia.

Class Oligochaeta: Earthworms and related species; earthworms have paired setae, a closed circulatory system, and a cerebral ganglion; hermaphroditic; sexual reproduction by mutual fertilization.

Class Hirudinea: Leeches; mainly freshwater annelids with posterior and anterior suckers; either free-living or parasitic; reproduction sexual, usually by mutual fertilization of hermaphrodites.

Class Polychaeta: Marine worms; body with many bristles; about 6,000 species; includes sandworms.

Phylum Arthropoda: Arthropods; at least 1 million species; segmented body; paired, jointed appendages; exoskeleton; open circulatory system; complex nervous system with two ventral nerve cords and a brain.

Class Merostomata: Horseshoe crabs; four species; aquatic with fangs and book gills.

Class Arachnida: Arachnids; about 57,000 species; body has two divisions, eight legs, and paired chelicerae and pedipalps; respiration by tracheae, book lungs, or both; includes spiders, ticks, and mites.

Class Crustacea: Crustaceans; about 25,000 mostly aquatic species; paired mandibles and thoracic appendages; usually two pairs of maxillae; respiration by gills; includes lobsters and barnacles.

Class Chilopoda: Centipedes; about 3,000 terrestrial species; mandibles; many body segments, each with one pair of legs.

Class Diplopoda: Millipedes; about 7,500 terrestrial species; mandibles; many body segments, each with two pairs of legs.

Class Insecta: Insects; over 750,000 terrestrial and freshwater species; three body divisions; three pairs of legs; head has antennae, compound eyes, and mouthparts; thorax has legs and, in many species, wings; most species undergo incomplete or complete metamorphosis; respiration by tracheae.

Order Protura: Proturans; piercing-sucking or chewing mouthparts; lacking wings, antennae, and eyes; front legs adapted as organs of touch; no metamorphosis.

Order Thysanura: Bristletails and silverfish; chewing mouthparts; two or three tails; wingless; no metamorphosis.

Order Collembola: Springtails; chewing mouthparts; wingless; compound eyes reduced or absent; no metamorphosis.

Order Ephemeroptera: Mayflies; chewing mouthparts that are reduced; membranous wings (usually two pairs); nymphs are aquatic; adults do not feed at all during their brief existence; incomplete metamorphosis.

Order Odonata: Dragonflies and damsel flies; two pairs of transparent wings; chewing mouthparts; not able to walk; incomplete metamorphosis.

Order Plecoptera: Stone flies; chewing mouthparts which are reduced in many species; two pairs of membranous wings; nymphs are aquatic; incomplete metamorphosis.

Order Orthoptera: Crickets and grasshoppers; two pairs of wings or wingless; chewing mouthparts; incomplete metamorphosis.

Order Dermaptera: Earwigs; chewing mouthparts; two pairs of wings or wingless; incomplete metamorphosis.

Order Embioptera: Web spinners; chewing mouthparts; wingless with the exception of some males; spin silk with special organs on forelegs; incomplete metamorphosis.

Order Isoptera: Termites; social insects that have chewing mouthparts; only the reproductive males and females have wings; incomplete metamorphosis.

Order Mallophaga: Chewing lice; wingless; chewing mouthparts; parasites of mammals and birds; incomplete metamorphosis.

Order Anoplura: Sucking lice; wingless; piercing-sucking mouthparts; parasites of mammals; incomplete metamorphosis.

Order Corrodentia: Bark lice, book lice; chewing mouthparts; wingless or two pairs of membranous wings; incomplete metamorphosis.

Order Hemiptera: Bugs; piercing-sucking mouthparts; two pairs of wings or wingless; incomplete metamorphosis.

Order Homoptera: Aphids, scale insects, and cicadas; wingless or winged (one or two pairs); piercing-sucking mouthparts; incomplete or complete metamorphosis.

Order Thysanoptera: Thrips; piercing-sucking mouthparts; wingless or two pairs of wings with long hairs attached to them; incomplete or complete metamorphosis.

Order Mecoptera: Scorpion flies; chewing mouthparts; wingless or two pairs of membranous wings; in some males, tip of abdomen is curved, resembling the tail of a scorpion; complete metamorphosis.

Order Neuroptera: Helgramites (dobson fly); Larva aquatic with large mandibles.

Order Trichoptera: Caddis flies; chewing mouthparts which are reduced; two pairs of wings; larvae are aquatic; adults do not feed extensively; complete metamorphosis.

Order Lepidoptera: Moths and butterflies; two pairs of broad, scaly wings; tubelike, sucking mouthparts; complete metamorphosis.

Order Diptera: Flies and mosquitoes; transparent front wings; hind wings reduced to knobby balancing organs; lapping and piercing mouthparts; complete metamorphosis.

Order Siphonaptera: Fleas; wingless; flattened body; piercing-sucking mouthparts; parasites of birds and mammals; complete metamorphosis.

Order Coleoptera: Beetles; winged (two pairs of wings, front pair serves as horny cover for membranous hind pair) or wingless; chewing mouthparts; complete metamorphosis.

Order Strepsiptera: Strepsipterans; chewing mouthparts which are reduced; females are wingless; males have two pairs of wings, with the front pair greatly reduced; complete metamorphosis.

Order Hymenoptera: Ants, bees, wasps; winged (two pairs that interlock in flight) or wingless; chewing or lapping mouthparts; some social species; complete metamorphosis.

Phylum Echinodermata: Echinoderms; about 6,000 marine species; calcium endoskeleton, tube feet, and a water-vascular system; adults radially symmetrical; includes starfish, sand dollars, and sea urchins.

Class Crinoidea: Sea lily; five rays; tube feet without suckers; most have stalklike body.

Class Asteroidea: Starfish; five rays with two rows of tube feet in each ray; eyespots.

Class Ophiuroidea: Brittle stars; five slender, delicate rays.

Class Echinoidea: Sea urchin, sand dollar; rays lacking; body spherical or oval.

Class Holothuroidea: Sea cucumber; long, thick body with tentacles; no rays.

Phylum Hemichordata: Acorn worms; about 80 species; hydrostatic skeleton similar to starfish's water vascular system; dorsal and ventral nerve cords, dorsal cord hollow in some species.

Phylum Chordata: Chordates; about 43,000 species that at some stage have a notochord, dorsal nerve cord, pharyngeal gill slits, and tail.

Subphylum Urochordata: Tunicates; about 1,300 marine species; adults sacklike; usually sessile; larvae free-swimming.

Subphylum Cephalochordata: Lancelets; about 30 species of thin, fishlike marine animals; basic chordate features throughout life.

Subphylum Vertebrata: Vertebrates; about 41,700 species; long or cartilaginous endoskeleton includes vertebrae; characterized by distinct cephalization and a closed circulatory system; excretion by kidneys; most have two sets of paired appendages.

Class Agnatha: Jawless fishes; about 60 species of lampreys and hagfishes; have suckerlike mouths; scaleless, cylindrical bodies without appendages; skeleton of cartilage; two-chambered heart.

Class Chondrichthyes: Sharks, skates and rays; about 625 mostly marine species of jawed fishes; cartilaginous skeleton; teeth replaced continuously; skin covered with tooth-derived scales; eggs fertilized internally; two-chambered heart.

Class Osteichthyes: Bony fishes; about 20,000 species of jawed fishes; bony skeletons; found in marine and freshwater habitats; body has bone-derived scales, an air-filled swim bladder, and a two-chambered heart; fertilization external in most species.

Class Amphibia: Amphibians; about 2,500 species characterized by a gill-breathing larval stage and a lung-breathing adult stage; adults have moist skin and three-chambered heart with double circulatory system; eggs usually laid in water.

Order Anura: Frogs and toads; adults are tailless with rear legs adapted to jumping; fertilization external, in water.

Order Urodela: Salamanders; adults have tails and four legs of about equal size; fertilization internal.

Order Apoda: Caecilians; legless, usually blind amphibians that live underground.

Class Reptilia: Reptiles; about 6,000 terrestrial species; fertilization occurs internally; embryo protected by a shelled, amniotic egg; skin is waterproof, covered with horny plates or overlapping scales; double circulatory system.

Order Rhyncocephalia: Tuatara; primitive four-legged species; skin covered by plates; three-chambered heart.

Order Chelonia: Turtles and tortoises; shelled, toothless species with four limbs and a three-chambered heart.

Order Squamata: Lizards and snakes; reptiles with overlapping scales and a three-chambered heart; most lizards have four legs; snakes are legless.

Order Crocodilia: Crocodiles, alligators, and related species; broad-bodied, water-dwelling carnivores with powerful tails and jaws; skin covered by bony plates; four-chambered heart.

Class Aves: Birds; about 8,600 species; winged vertebrates with feathers; endothermic (warm-blooded); double circulation and a four-chambered heart; fertilization internal; amniotic eggs usually incubated in nest; respiratory system with lungs and air sacs.

Order Rheiformes: Rheas; large, flightless South American birds.

Order Struthioniformes: Ostriches; large, flightless African birds.

Order Procellariiformes: Albatrosses, shearwaters, petrels, and other species of related seabirds with tubelike nostrils.

Order Sphenisciformes: Penguins; flightless swimming birds with wings modified as paddles.

Order Gaviiformes: Loons; diving birds with three webbed toes.

Order Pelecaniformes: Pelicans, boobies, and related water birds with four-toed, webbed feet.

Order Ciconiiformes: Storks, herons, and related long-legged wading birds.

Order Gruiformes: Cranes, coots, limpkins, rails, gallinules, and related species.

Order Charodriiformes: Auks, gulls, sandpipers, and related shore and water birds.

Order Anseriformes: Ducks, geese, swans, screamers; short-legged water birds.

Order Falconiformes: Falcons, eagles, hawks, condors, and related diurnal birds of prey.

Order Galliformes: Chickens, turkeys, pheasants, grouse, and related fowl-like birds.

Order Columbiformes: Pigeons, doves, sandgrouse, and related species.

Order Psittaciformes: Parrots; hook-billed birds that eat seeds, fruit, or nectar.

Order Cuculiformes: Cuckoos, roadrunners, and related species.

Order Strigiformes: Owls; nocturnal birds of prey.

Order Caprimulgiformes: Goat-suckers, whippoorwill, night hawk, and related species.

Order Apodiformes: Swifts and hummingbirds; weak-legged birds with strong wings.

Order Coraciiformes: Kingfisher; fishing birds.

Order Piciformes: Woodpeckers, toucans, barbets, and related birds.

Order Passeriformes: Perching birds, including all songbirds.

Class Mammalia: Mammals; about 4,500 species; young are nourished by mother's milk and receive extensive parental care; endothermic and covered by hair; double circulation; four-chambered heart; respiration by lungs; highly developed, complex brain; fertilization internal.

Order Monotremata: Monotremes, spiny anteaters and duck-billed platypus; the only mammals that lay eggs; incomplete control of body temperature; milk secreted from abdominal sweat glands.

Order Marsupialia: Marsupials, including kangaroos, wombats, and opossums; young are born in an immature state and develop further attached to a nipple in the mother's marsupium (pouch).

The following orders are placental mammals in which young undergo substantial prenatal development within the uterus. During gestation, exchange of nutrients, gases, and wastes occurs through the placenta.

Order Insectivora: Insectivores; small, chiefly nocturnal mammals that feed largely on insects.

Order Chiroptera: Bats; the only flying mammals.

Order Artiodactyla: Hoofed mammals with two or four toes; large herbivores including deer, pigs, cattle, sheep, goats, camels, antelopes, giraffes, and hippopotamuses; most are ruminants.

Order Perissodactyla: Hoofed mammals with one or three toes; large herbivores including horses, tapirs, and rhinoceroses.

Order Proboscidea: Elephants; enormous herbivores with trunk.

Order Carnivora: Carnivores; bears, cats, dogs, seals, and related predators.

Order Cetacea: Whales, porpoises, and dolphins; marine mammals with streamlined body; forelimbs adapted as flippers; no hindlimbs; nostril (blowhole) at top of head.

Order Sirenia: Sea cows; aquatic herbivores with short, paddlelike forelimbs and no hindlimbs.

Order Rodentia: Rodents; small, plant-eating mammals; one upper and one lower curving, chisel-like incisors that grow continuously.

Order Lagomorpha: Rabbits, hares, and pikas; rodentlike mammals with two pairs of upper and one pair of lower incisors.

Order Edentata: Armadillos, anteaters, and tree sloths; insect eaters with few or no teeth.

Order Primates: Primates, including lemurs, monkeys, apes, and humans; mammals with features adapted to arboreal life.

Laboratory Procedures

Using a Compound Light Microscope

Parts of the Compound Light Microscope
- The *eyepiece* magnifies the image $10\times$.
- The *low-power objective* magnifies the image $10\times$.
- The *high-power objective* magnifies the image either $40\times$ or $43\times$.
- The *revolving nosepiece* holds the objectives and can be turned to change from one magnification to the other.
- The *body tube* maintains the correct distance between eyepiece and objectives.
- The *coarse adjustment* moves the body tube up and down to allow focusing of the image.
- The *fine adjustment* moves the body tube slightly to bring the image into sharper focus.
- The *stage* supports a slide.
- *Stage clips* secure the slide in position for viewing.
- The *diaphragm* controls the amount of light coming through the stage.
- The *light source* provides light for viewing the slide.
- The *arm* supports the body tube.
- The *base* supports the microscope.

Proper Use of the Compound Light Microscope
1. Carry the microscope to your lab table using both hands, one beneath the base and the other hand holding the arm of the microscope. Hold the microscope close to your body.
2. Place the microscope on the lab table, at least 5 cm (2 in.) in from the edge of the table.
3. Check to see what type of light source the microscope has. If the microscope has a lamp, plug it in, making sure that the cord is out of the way. If the microscope has a mirror, adjust it to reflect light through the

Compound light microscope

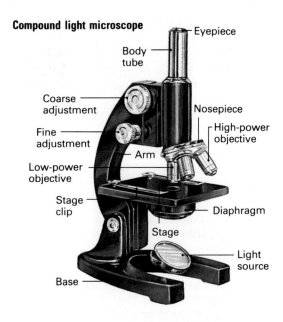

hole in the stage. **CAUTION: If your microscope has a mirror, do not use direct sunlight as a light source. Direct sunlight can damage your eyes.**

4. Adjust the revolving nosepiece so that the low-power objective is in line with the body tube.
5. Place a prepared slide over the hole in the stage and secure the slide with stage clips.
6. Look through the eyepiece and move the diaphragm to adjust the amount of light coming through the stage.
7. Now, look at the stage from eye level, and slowly turn the coarse adjustment to lower the objective until it almost touches the slide. Do not allow the objective to touch the slide.
8. While looking through the eyepiece, turn the coarse adjustment to raise the objective until the image is in focus. *Never focus objectives downward*. Use the fine adjustment to sharpen the focus. Keep both eyes open while viewing a slide.

9. Make sure that the image is exactly in the center of your field of vision. Then, switch to the high-power objective. Focus the image with the fine adjustment. *Never use the coarse adjustment at high power.*
10. When you are finished using the microscope, remove the slide. Clean the eyepiece and objectives with lens paper and return the microscope to its storage area.

Making a Wet Mount

1. Use lens paper to clean a glass slide and coverslip.
2. Place the specimen you wish to observe in the center of the slide.
3. Using a medicine dropper, place one drop of water on the specimen.
4. Position the coverslip so that it is at the edge of the drop of water and at a 45° angle to the slide. Make sure that the water runs along the edge of the coverslip.
5. Lower the coverslip slowly to avoid trapping air bubbles.
6. As the water evaporates from the slide, add another drop of water by placing the tip of the medicine dropper next to the edge of the coverslip. (Use this technique also when adding stains or solutions to a wet mount.) If you have added too much water, remove the excess by using the corner of a paper towel as a blotter. Do not lift the coverslip to add or remove water.

Wet mount

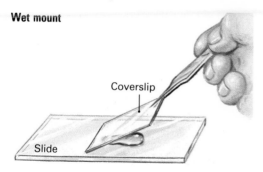

Coverslip

Slide

Lighting a Bunsen Burner

1. Before lighting the burner, observe the locations of fire extinguishers, fire blankets, and sand buckets. Wear safety goggles, gloves, and an apron. Tie back long hair and roll up long sleeves.
2. Turn the gas full on by using the valve at the laboratory gas outlet.

Gas valve
Gas line
Hottest part of flame
Barrel
Air port
Tubing
Gas adjustment valve
Base

3. Make a spark with a striker. If you are using a match, hold it slightly to the side of the opening in the barrel.
4. Adjust the air ports until you can clearly see an inner cone within the flame.
5. Adjust the gas flow for the desired flame height by using the gas adjustment valve either on the burner or at the gas outlet. **CAUTION: If the burner is not operating properly, the flame may burn inside the base of the barrel. Carbon monoxide, an odorless gas, is released from this type of flame. Should this situation occur, turn off the gas at the laboratory gas valve immediately. Do not touch the barrel of the burner.** After the barrel has cooled, partially close the air ports before relighting the burner.

Using a Triple-beam Balance

1. Make sure the balance is on a level surface. Use the leveling screws at the bottom of the balance to make any necessary adjustments.
2. Place all the counterweights at zero. The pointer should be at zero. If it is not, adjust the balancing knob until the pointer rests at zero.
3. Place the object you wish to measure on the pan. **CAUTION: Do not place hot objects or chemicals directly on the balance pan as they can damage its surface.**
4. Move the largest counterweight along the beam to the right until it is at the last notch that does not tip the balance. Follow the same procedure with the next largest counterweight. Then, move the smallest counterweight until the pointer rests at zero.
5. Total the readings on all beams to determine the mass of the object.
6. When weighing crystals or powders, use a filter paper. First weigh the paper, then add the crystals and powders and reweigh. The actual weight is the total less the weight of the paper. When weighing liquids, first weigh the empty container, then the liquid and container.

Triple-beam balance

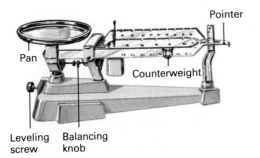

Measuring Volume in a Graduate

1. Set the graduate (graduated cylinder) on a level surface.
2. Carefully pour the liquid you wish to measure into the cylinder. Notice that the surface of the liquid has a lens-shaped curve, the *meniscus*.
3. With the surface of the liquid at eye level, read the measurement at the bottom of the meniscus.

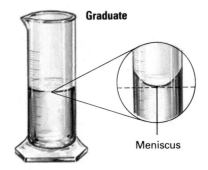

Graduate

Meniscus

Metric Conversion Table

Metric Units		Converting Metric to English	Converting English to Metric
Length			
kilometer (km)	= 1,000 m	1 km = 0.62 mile	1 mile = 1.609 km
meter (m)	= 100 cm	1 m = 1.09 yards	1 yard = 0.914 m
		= 3.28 feet	1 foot = 0.305 m
centimeter (cm)	= 0.01 m	1 cm = 0.394 inch	= 30.5 cm
millimeter (mm)	= 0.001 m	1 mm = 0.039 inch	1 inch = 2.54 cm
micrometer (μm)	= 0.000001 m		
nanometer (nm)	= 0.000000001 m		
Area			
square kilometer (km^2)	= 100 hectares	1 km^2 = 0.3861 square mile	1 square mile = 2.590 km^2
hectare (ha)	= 10,000 m^2	1 ha = 2.471 acres	1 acre = 0.4047 ha
square meter (m^2)	= 10,000 cm^2	1 m^2 = 1.1960 square yards	1 square yard = 0.8361 m^2
			1 square foot = 0.0929 m^2
square centimeter (cm^2)	= 100 mm^2	1 cm^2 = 0.155 square inch	1 square inch = 6.4516 cm^2
Mass			
kilogram (kg)	= 1,000 g	1 kg = 2.205 pounds	1 pound = 0.4536 kg
gram (g)	= 1,000 mg	1 g = 0.0353 ounce	1 ounce = 28.35 g
milligram (mg)	= 0.001 g		
microgram (μg)	= 0.000001 g		
Volume of Solids			
1 cubic meter (m^3)	= 1,000,000 cm^3	1 m^3 = 1.3080 cubic yards	1 cubic yard = 0.7646 m^3
		= 35.315 cubic feet	1 cubic foot = 0.0283 m^3
1 cubic centimeter (cm^3)	= 1,000 mm^3	1 cm^3 = 0.0610 cubic inch	1 cubic inch = 16.387 cm^3
Volume of Liquids			
kiloliter (kL)	= 1,000 L	1 kL = 264.17 gallons	1 gallon = 3.785 L
liter (L)	= 1,000 mL	1 L = 1.06 quarts	1 quart = 0.94 L
milliliter (mL)	= 0.001 L	1 mL = 0.034 fluid ounce	1 pint = 0.47 L
microliter (μL)	= 0.000001 L		1 fluid ounce = 29.57 mL

Temperature Conversion

The top of the thermometer is marked off in degrees Fahrenheit (°F.). To read the corresponding temperature in degrees Celsius (°C), look at the bottom side of the thermometer. For example, 50°F. is the same temperature as 10°C. You may also use the formulas at the right to make conversions.

Conversion of Fahrenheit to Celsius:

$$°C = \tfrac{5}{9} (°F - 32)$$

Conversion of Celsius to Fahrenheit:

$$°F = \tfrac{9}{5} °C + 32$$

°F.

°C

Freezing point of water   Boiling point of water

Safety Guidelines

Participating in biology laboratory investigations should be an enjoyable experience as well as a learning experience. You can ensure both learning and enjoyment of the experience by making the laboratory a safe place in which to work. Carelessness, apathy, and showing off are the major sources of laboratory accidents. It is, therefore, important that you follow safety procedures at all times. If an accident should occur, you should know exactly where to locate emergency equipment. Good safety practice means being responsible for your fellow students' safety as well as your own.

You will be expected to practice the following safety measures whenever you are in the laboratory:

1. **Preparation** Study your laboratory assignment in advance. Before beginning your investigation, ask your teacher to explain any procedures you do not understand.

2. **Chemicals and Other Dangerous Substances** When carelessly handled, chemicals can be dangerous. *Never taste chemicals or place them near your eyes.* Do not use mouth suction when using a pipette to transfer chemicals. Use a suction bulb instead.

   Never pour water into a strong acid or base. The mixture produces heat. Sometimes the heat causes splattering. To keep the mixture cool, pour the acid or base slowly into the water.

   If any solution is spilled on a work surface, wash it off at once with plenty of water. When noting the odor of chemical substances, wave the fumes toward your nose with your hand rather than putting your nose close to the source of the odor.

   Never eat in the laboratory. Counters and glassware may contain substances that can contaminate food. Do not use flammable substances near flame. Handle toxic substances in a well-ventilated area or under a ventilation hood. Do not place flammable chemicals in a refrigerator. Sparks from the refrigeration unit can ignite these substances or their fumes.

3. **Heat** Whenever possible use an electric hot plate instead of an open flame. If you must use an open flame, shield it with a wire screen with a ceramic fiber center. When heating chemicals in a test tube, do not point the test tube toward anyone. Keep combustible materials away from heat sources.

4. **Electricity** Be cautious around electrical wiring. When using a microscope with a lamp, do not place its cord where it can cause someone to trip and fall. Do not let cords hang loose over a table edge in a way that permits equipment to fall if the cord is tugged. Do not use equipment with frayed cords.

5. **Knives** Use knives, razor blades, and other sharp instruments with extreme care. Do not use double-edged razor blades in the laboratory.

6. **Glassware** Examine all glassware before heating. Glass containers for heating should be made of boro-silicate glass or some other heat-resistant material. Never use cracked or chipped glassware. Never force glass tubing into rubber stoppers. Broken glassware should be swept up immediately, never picked up with the fingers. Broken glassware should be discarded in a special container, never into a sink.

7. **Bacterial Cultures** Return all bacterial and other cultures to your teacher for proper disposal. Wash and disinfect all glassware and other equipment that has

come in contact with the culture. Follow any additional instructions your teacher may give you regarding specific cultures.

8. **Dissection** Never dissect a specimen while holding it in your hand. Place all specimens in dissecting pans before beginning dissection.

9. **Eye Safety** Wear goggles when handling acids or bases, using an open flame, or performing any activity that could harm the eyes. If a solution is splashed into the eyes, wash the eyes with plenty of water and notify your instructor at once. Never use reflected sunlight to illuminate a microscope. This practice is dangerous to the eyes.

10. **Safety Equipment** Know the location of all safety equipment. This includes fire extinguishers, fire blankets, first-aid kits, eyewash fountains, and emergency showers. Note the location of the nearest telephone. Take responsibility for your fellow students and report all accidents and emergencies to your teacher immediately.

11. **First Aid** In case of severe bleeding, apply pressure or a compress directly to the wounded area and see that the injured student reports immediately to the school nurse or a physician.

Minor burns caused by heat should be treated by applying ice. Immerse the burn in cold water if ice is not available. Treat acid burns by applying sodium bicarbonate (baking soda). Treat burns caused by bases with boric acid. Any burn, regardless of cause, should be reported to your teacher and referred to the school nurse or a physician.

In case of fainting, place the fainting victim's head lower than the rest of the body and see that the person has fresh air. Report to your teacher immediately.

In case of poisoning, report to your teacher at once. Try to determine the poisoning agent if possible.

12. **Unauthorized Experiments** Do not perform any experiment that has not been assigned by your teacher. Never work alone in the laboratory.

13. **Neatness** Keep work areas free of all unnecessary books and papers. Tie back long, loose hair and button or roll up loose sleeves when working with chemicals or near an open flame.

14. **Cleanup** Wash your hands immediately after handling bacteria or other hazardous materials. Before leaving the laboratory, clean up all work areas. Put away all equipment and supplies. Make sure water, gas, burners, and electric hot plates are turned off.

Remember at all times that a laboratory is a safe place only if you regard laboratory work as serious work.

The instructions for your laboratory investigations will include cautionary statements where necessary. In addition you will find that the following safety symbols appear whenever a procedure requires extra caution:

 Laboratory apron

 Sharp/pointed object

 Goggles

 Flame/heat

 Dangerous chemical

 Biohazard—disease-causing organisms

Key Discoveries in Biology

500 B.C.-1 B.C.

About 500 B.C.
Greek philosopher Alcmaeon performs pioneer studies in human anatomy.

About 420 B.C.
Greek philosopher Hippocrates establishes the scientific basis of medicine.

About 350 B.C.
Greek philosopher Aristotle develops a systematic classification of animals.

Hippocrates

About 300 B.C.
Greek philosopher Theophrastus creates the first scientific classification of plants.

About 75 B.C.
Roman philosopher Lucretius applies the atomistic theory to living things and offers a particulate theory of heredity.

A.D. 1-A.D. 899

About A.D. 50
Roman naturalist Pliny the Elder compiles an encyclopedia of natural history.

About 150
Galen, a Greco-Roman physician, contributes medical theories based on the dissection of apes.

1700s

1753-58 Swedish naturalist Carolus Linnaeus establishes a modern taxonomy.

1779 Dutch naturalist Jan Ingenhousz shows that sunlight is the source of energy in photosynthesis.

1796 English physician Edward Jenner introduces vaccination.

1800s

1817 French scientist Georges Cuvier pioneers the field of comparative anatomy.

1838-39 German biologists Matthias Schleiden and Theodor Schwann establish cell theory.

1855 German physician Rudolf Virchow launches the science of cellular pathology.

1859 British naturalist Charles Darwin proposes his theory of evolution.

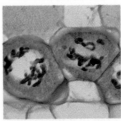
Meiosis

1862 French scientist Louis Pasteur disproves abiogenesis.

1865 Austrian monk Gregor Mendel establishes the basic laws of heredity.

1892 German biologist August Weismann proposes the germ-plasm (gamete) theory of inheritance.

1940s

1941 American researchers George Beadle and Edward Tatum propose the one gene one enzyme theory.

1944-52 Experiments by Americans Oswald Avery (1944) and Alfred Hershey and Martha Chase (1952) prove that DNA is the genetic chemical.

1946-61 American biochemist Melvin Calvin elucidates the process of carbon fixation during photosynthesis.

1947 American geneticist Barbara McClintock proposes that genes can move between chromosome sites.

1950s

1953 American James Watson and Briton Francis Crick discover the molecular structure of DNA.

1956 American molecular biologist Arthur Kornberg synthesizes DNA in the laboratory.

1960s

1961 Americans Marshall Nirenberg and Severo Ochoa break the genetic code of DNA and messenger RNA.

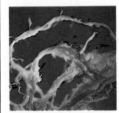
DNA strands

The key discoveries given here are only a limited selection over 2,500 years and are not meant to be a definitive list. The number of women scientists has increased dramatically over the past ten years.

900-1499

About 900
Persian physician Rhazes authors an encyclopedia of practical medicine.

About 1030
Avicenna (Ibn Sina), a Persian philosopher-physician, completes the *Canon of Medicine,* which codifies medical knowledge for 500 years.

1500s

1543 Flemish physician Andreas Vesalius publishes the first scientific textbook on human anatomy.

Drawing by Vesalius

1600s

1628 English physician William Harvey accurately describes the human circulatory system.

1665 English scientist Robert Hooke observes cell structure in the bark of the cork oak.

1676 Dutch microscopist Anton van Leeuwenhoek first sees one-celled organisms.

Hooke's microscope

1900-1919

1900 Austrian-American scientist Karl Landsteiner discovers ABO-blood types in humans.

1900 Mendel's work is independently rediscovered by three scientists—Erich Tschermak von Seysenegg, Hugo de Vries, and Carl Correns.

1902 American cytologist Walter Sutton proposes that genes are carried on the chromosomes.

1910-15 American geneticist T. H. Morgan discovers linkage and crossing over, establishing that genes exist as linear units on the chromosomes.

1920-1939

1928 British biologist Alexander Fleming discovers penicillin.

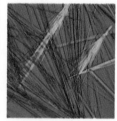

Penicillin crystals

1937 German biochemist Sir Hans Krebs discovers the citric acid cycle, the key event in cellular respiration.

1937 American geneticist Theodosius Dobzhansky initiates the synthesis of genetic and evolutionary theory.

1970s

1973 American molecular biologists Stanley Cohen and Herbert Boyer discover how to induce gene splicing.

1977 American scientists Rosalyn Yalow, Roger Guillemin, and Andrew Schally win the Nobel prize for tracking radioactive hormone in the body.

1980s

1982 American molecular biologist Robert A. Weinberg discovers an oncogene—a dormant gene capable of causing cancer when activated—in human cells.

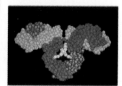

Immunoglobulin

1984 Three immunologists share the Nobel prize: Niels Jerne, in England, for his theories on how the immune system works; and Briton Cesar Milstein and Swiss Georges Kohler for their creation of monoclonal antibodies.

Pharmaceutical lab

837

Glossary

Pronunciation Key

Symbol	As In	Phonetic Respelling	Symbol	As In	Phonetic Respelling
a	bat	a (bat)	ô	dog	aw (dawg)
ā	face	ay (fays)	oi	foil	oy (foyl)
ã	careful	ai (CAIR fuhl)	ou	mountain	ow (MOWN tuhn)
ä	argue	ah (AHR gyoo)	s	sit	s (siht)
ch	chapel	ch (CHAP uhl)	sh	sheep	sh (sheep)
e	test	eh (tehst)	u	love	uh (luhv)
e	eat	ee (eet)	u̇	pull	u (pull)
	ski	ee (skee)	ü	mule	oo (myool)
ėr	fern	ur (furn)	zh	treasure	zh (TREHZ uhr)
i	bit	ih (biht)	ə	medal	uh (MEHD uhl)
ī	ripe	y (ryp)		effect	uh (uh FEHKT)
	idea	eye (eye DE uh)		serious	uh (SIHR ee uhs)
k	card	k (kahrd)		onion	uh (UHN yuhn)
o	lock	ah (lahk)		talent	uh (TAL uhnt)
ō	over	oh (OH vuhr)			

A

abdomen in arachnids, second section of the body, which holds most of the organs (491)

abdominal cavity in mammals, part of the coelom that contains the digestive, reproductive, and excretory organs (612)

abiogenesis (ay by oh JEHN uh sihs) belief that some organisms form from nonliving materials (28)

abiotic factor nonliving part of an ecosystem (759)

abscission shedding of a plant's leaves (395)

absorption in the body, movement of nutrient molecules into blood vessels or other vessels during digestion (644)

abyssal (uh BIHS uhl) **region** area of the open sea below a depth of 2,000 m (1.24 mi.) (789)

acetyl CoA organic substance formed during aerobic respiration resulting from the combination of an acetyl group and coenzyme A (126)

acid compound that releases hydrogen ions in water (68)

acid rain polluting rain, a result of sulfur oxides from fossil fuels that react with water vapor in the environment to form sulfuric acid (817)

activation energy energy required to start a chemical reaction (59)

active immunity resistance to disease resulting from the reproduction of antibodies (289)

active site place on an enzyme where it bonds with its substrate (72)

active transport carrier transport in which energy is used to move substances against the concentration gradient (106)

adaptation a trait in an organism that makes it better suited to its environment (9, 244)

adaptive radiation process by which members of a species adapt to a variety of habitats (250)

addiction state of dependence on a drug (745)

adenine (AD uh neen) carbon-nitrogen compound, one of the four bases of DNA (182)

adrenal (uh DREE nuhl) **cortex** outer layer of the adrenal gland (720)

adrenal gland organ on each kidney that increases metabolism and helps the body react to stress (720)

adrenal medulla inner part of the adrenal gland (720)

aerobic (ehr OH bihk) **respiration** cellular process of breaking down pyruvic acid in the presence of oxygen (122)

air bladder air-filled structure that permits certain brown algae to carry on photosynthesis (336)

air sac in birds, one of several cavities that takes in air, making the animal lighter (577)

albumin nutritious protein that makes up the white of an egg (578–579)

alcoholic fermentation process in which hydrogen and pyruvic acid formed during glycolysis combine and produce ethyl alcohol (123)

alcoholism addiction to alcohol (746)

algae (AL jee) autotrophic protists (328)

algal bloom uncontrolled growth of algae (309, 332)

alimentary canal passageway from mouth to anus (644)

allantois (uh LAN tuh wihs) in reptiles, birds, and mammals, sac that stores nitrogenous wastes produced by the developing embryo (556)

allele (uh LEEL) either member of a pair of genes that determines a single trait (156)

alternation of generations life cycle involving alternating haploid (gametophyte) and diploid (sporophyte) stages (335)

altricial characteristic of a young bird dependent upon its parents for its care and feeding (579)

alveolus (al VEE uh luhs) tiny air sac in the lungs where the exchange of gases occurs (677)

amino acid organic compound, the chief component of protein (71)

ammonification (uh moh nuh fih KAY shuhn) process by which microorganisms convert plant or animal wastes or dead organisms into ammonia (765)

amniocentesis (am nee oh sehn TEE sihs) removal and examination of amniotic fluid to test for genetic disorders in a fetus (208)

amnion membrane that surrounds an embryo and forms a chamber filled with a saline fluid (555)

amniotic egg fluid-filled sac enclosed by a protective porous shell (555)

amoebocyte (uh MEE buh syt) in a sponge, amoebalike cell that carries nutrients to the inner layer and takes away waste matter (445)

amoeboid movement characteristic creeping movement of amoeba, caused by the pseudopodia (319)

amplexus grasping of the female frog by the male, which stimulates her to release eggs (547)

anaerobic (an ehr OH bihk) that which does not require oxygen (122)

anaerobic respiration process of breaking down glucose to form pyruvic acid, and breaking down pyruvic acid to form ethyl alcohol or lactic acid (123)

anal fin stabilizing and maneuvering appendage on the ventral surface of a fish (530)

analogous (uh NAL uh guhs) **structure** body part similar to another in function but not structure (236)

anal pore in paramecia, opening through which undigested wastes leave the cell (322)

anaphase third stage of mitosis, during which chromatids separate and pull apart (141)

angiosperm (AN jee uh spuhrm) flowering plant that produces seeds in fruit and generally loses its leaves annually (374)

antenna in many arthropods, a sensory organ (494)

antennule in crayfish and other arthropods, one of a pair of two appendages that maintain balance, touch, and taste (494)

anterior pertaining to the front end (444)

anther in a flower, oblong portion of a filament, in which pollen is produced (401)

antheridium (an thuh RIHD ee uhm) male reproductive structure of a moss (360)

antibiotic chemical capable of inhibiting the growth of some bacteria (306)

antibody protein that attacks, or neutralizes, the antigen that triggered its production (289, 657)

anticodon (AN tee koh dahn) sequence of three bases on transfer RNA (189)

antigen (AN tuh juhn) foreign molecule that stimulates an organism to produce antibodies (289, 657)

anus posterior opening through which solid wastes leave the body (462)

aorta major artery that carries blood from the heart (508, 661)

appendage any movable extension of the body (489)

appendicular (ap uhn DIHK yuh luhr) **skeleton** vertebrate support system comprised of the pectoral girdle, the pelvic girdle, and their associated limbs (521, 623)

applied genetics implementation of genetic knowledge (216)

applied science fields of study in which scientific findings are put to use (27)

aqueous (AY kwee uhs) **humor** clear, watery fluid between the lens and cornea of the eye (708)

archegonium (ahr kuh GOH nee uhm) female reproductive structure of a moss (360)

artery vessel that carries blood away from the heart (661)

ascospore in sac fungi, haploid spore that forms in the ascus (348)

ascus in some fungi, small sac in which spores form (347)

asexual reproduction production of offspring from one parent (135)

aster spindle fibers that radiate out from a centriole, during mitosis in animal cells (140)

atom building block of all matter (12, 48)

atomic number number of protons in the nucleus of one atom of a specific element (50)

ATP adenosine triphosphate, molecule that allows food energy to be delivered where needed (113)

atrioventricular (ay tree oh vehn TRIHK yuh luhr) **node** also AV node; in vertebrates, region of muscle in the heart that transmits the heartbeat to the ventricles (663)

atrium (AY tree uhm) one of two upper sections of the heart (660)

auditory canal tube to the middle ear (710)

Australopithecus genus name of the oldest known hominid (257)

autonomic nervous system in vertebrates, system that controls involuntary responses (699)

autosome chromosome that does not determine the sex of the individual (172)

autotroph (AWT uh trahf) organism that can use inorganic molecules to produce organic food molecules (123, 275)

auxin hormone that tends to stimulate the elongation of plant cells (419)

axial (AK see uhl) **skeleton** vertebrate support system comprised of the backbone, bones of the skull, and ribcage (521, 623)

axon long cytoplasmic fiber in a neuron that carries impulses away from the cell body (689)

B

bacillus (buh SIHL uhs) rod-shaped bacterium (299)

bacteriophage (bak TIHR ee uh fayj) virus that invades bacteria (284)

bacterium simple, one-celled organism (296)

barb one of many small projections that comprise the soft, flexible part of a feather (574)

barbule in a feather, one of many small lateral projections that link barbs together (574)

base compound that releases hydroxide ions in solution or accepts hydrogen ions (68)

basidiospore in a mushroom, structure formed when the basidium nucleus undergoes meiosis (347)

basidium in some fungi, club-shaped microscopic structure on which spores are formed (346)

bathyal (BATH ee uhl) **region** area of the open sea from 200 m (656 ft.) to 2,000 m (6,562 ft.) deep (789)

benthos in the marine biome, organisms that are attached to some surface (788)

bilateral symmetry arrangement of an organism's body parts so that one-half of the body is an apparent mirror image of the other half (444)

binary fission simplest form of asexual reproduction in which one parent cell splits in two, forming two identical daughter cells (136, 320)

binomial nomenclature in taxonomy, naming organisms by the genus and species (272)

biodegradable (by oh dih GRAY duh buhl) characteristic of a product that can be broken down naturally (816)

biogenesis (by oh JEHN uh sihs) theory stating that all living things arise from other living things (28)

biogeochemical cycle pathway through which a chemical substance is recycled (763)

biology the study of living things (2)

bioluminescence production of light by living things (332)

biomass weight of an organism after all the water has been removed (768–769)

biome large geographic area of similar climate and life forms (16, 782)

biomedical engineering design and development of artificial body parts (616)

biosphere all the life-supporting environments on earth and the organisms in them (17, 759)

biostereometrics study of biological form and function in three dimensions (64)

biosynthesis formation of chemical compounds by the cells of living organisms (340)

biosystematics study of the evolution of species (277)

biotelemetry measurement of living things at a distance (586)

biotic factor living organism in an ecosystem (759)

biotic potential rate at which a population would produce young if every new individual lived and reproduced at its maximum capacity (795)

bipedal walking on two legs (257–258)

birth rate rate at which births occur in a population (799)

bivalve class of mollusks with two hinged shells and a muscular foot (474)

bladderworm undeveloped worm inside a tapeworm cyst (460)

blade broad part of a leaf, containing most of the photosynthetic cells (393)

blastula (BLAS choo luh) ball of cells that forms from repeated cleavages of a zygote (461)

blood fluid that serves as the chief carrier of the body's transport system (655)

blood type classification of an individual's blood according to the presence of an antigen on red blood cells (657)

bolus moist, soft ball created when saliva acts on food in the mouth (645)

book lung in arachnids, respiratory structure that absorbs oxygen through the spiracles (491)

boreal forest biome characterized by short summers, long winters, and an abundance of evergreen trees and wildlife (783)

botany the study of plants (20)

Bowman's capsule in the kidney, a hollow, cup-shaped sac that surrounds the glomerulus (682)

brain control center for the vertebrate body (694)

brain stem portion of the brain where the nerves connect the spinal cord with the cerebrum (696)

bristle in birds, short, hairlike feather near the nostril, which filters dust (574)

bronchiole (BRAHNG kee ohl) one of many microscopic tubes that branch off from the bronchi (677)

bronchus (BRAHN kuhs) one of the two divisions of the trachea leading to the lungs (677)

brood to guard eggs until they hatch (478)

bryophyte (BRY uh fyt) plant that lacks a vascular system and gets food and nutrients by osmosis and diffusion (357)

bud small, protective structure that holds meristematic cells and embryonic plant materials along a stem (392)

budding type of asexual reproduction in which a cell divides into two cells of unequal size; the smaller cell pinches off to become a new individual (137)

button tight mass of hyphae that comprise the first stage of a mushroom's development (346)

C

Calorie amount of heat energy needed to raise the temperature of one kilogram of water 1°C (643)

camouflage defense mechanism in which an organism blends into its surroundings (513)

cap fruiting body of a mushroom (346)

capillary smallest blood vessel where the exchange of nutrients and waste products takes place (663)

capsid outer protein coat that comprises most of the body of a virus (284)

capsule protective layer of slime surrounding the outer cell wall of some infectious bacteria (300)

carapace in some arthropods, protective outer shield or exoskeleton (494); in turtles and tortoises, dorsal half of shell (561)

carbohydrate one of the various molecules containing carbon, hydrogen, and oxygen that supplies most of the body's energy needs (69, 640)

cardiac muscle group of specialized cells that provides for all movements of the heart (628)

cardiac sphincter (SFINGK tuhr) muscular valve that prevents food from reentering the esophagus (646)

carnivore animal that eats only animals (766)

carotene orange pigment present in most green plants (116)

carrier individual who possesses a recessive allele but does not express it (205)

carrier molecule protein molecule that transports large molecules across a membrane (106)

carrying capacity maximum number of individuals in a species that the environment can support (796)

cast mold formed when a material fills the impression created by a decomposed organism (232)

catalyst substance that starts or speeds up a chemical reaction (72)

caudal fin vertical expansion of a fish's tail, used for stabilization (526)

cecum in some mammals, intestinal pouch that contains cellulose-digesting microorganisms (602)

cell functional unit, or building block, of all organisms; smallest unit that can carry on the activities of life (3)

cell body main structure of a nerve cell, containing the nucleus, cytoplasm, and nerve fibers (689)

cell cycle four-stage sequence of cell growth and division between the beginning of one mitosis and the beginning of the next (139)

cell fusion process by which cells from two different kinds of plants are joined (221)

cell membrane outer layer of lipids and proteins that protects a cell, gives it shape, and regulates what enters and leaves the cell (85)

cell plate in plants, structure formed during cytokinesis; a new cell wall forms on both sides of the cell plate (142)

cell theory a three-part explanation stating that all organisms are composed of cells; cells are the basic units of structure and function in organisms; and all cells come from preexisting cells (80)

cellular respiration process of breaking the chemical bonds of organic food molecules and releasing energy that can be used by the cell (114)

cell wall a membrane that helps to protect and support the cells of green plants, algae, fungi, and some bacteria (85)

central nervous system in vertebrates, the brain and spinal cord (689)

centrifugation (sehn trihf yuh GAY shuhn) spinning process that separates cell parts for study (41)

centriole (SEHN tree ohl) small, dark, cylindrical body located outside the nucleus of animal cells and used during cell division (90–91)

centromere (SEHN truh mihr) point at which two chromatids are joined (138)

cephalization (sehf uh lih ZAY shuhn) adaptation in which an organism's neural and sensory organs are concentrated in the anterior end (444)

cephalothorax (sehf uh luh THAWR aks) in invertebrates such as arachnids, a fused head and chest area to which legs are attached (491)

cereal grain small, one-seeded fruit of grasses (431)

cerebellum part of the vertebrate brain that controls movement and muscular coordination (546, 696)

cerebral cortex the gray matter of the cerebrum, which receives sensory impulses from the body and coordinates motor responses to them (695)

cerebral hemisphere either half of the cerebrum; each controls actions and sensations on the opposite side of the body (694–695)

cerebrospinal fluid protective fluid that circulates around the brain and spinal cord (698)

cerebrum control center of the brain (546, 694–695)

cetacean (sih TAY shee uhn) a whale, porpoise, dolphin or related aquatic mammal (600)

chelicera (kuh LIHS uh ruh) in arachnids, poisonous fang or clawlike appendage (491)

cheliped (KEE luh pehd) in crustaceans, one of the second pair of appendages on the thorax part of the cephalothorax (494)

chemical bond force that holds atoms together in any molecule or compound (56)

chemical equation symbols that describe what happens in a chemical reaction (60)

chemical property characteristic that describes how a substance acts when it combines with other substances to form different kinds of matter (47)

chemical reaction process of breaking existing chemical bonds and forming new ones (59)

chemosynthesis breakdown of inorganic chemicals (300)

chitin (KYT uhn) in fungi, hard, water-insoluble substance in the cell walls (343); in arthropods, protein-carbohydrate compound that forms the exoskeleton (489)

chlorophyll (KLAWR uh fihl) green pigment in plants, necessary for photosynthesis (116)

chloroplast plastid that stores chlorophyll (89)

chorion (KOHR ree ahn) in reptiles and birds, outermost membrane that lines an egg shell; in mammals, the membrane that attaches to the uterus wall (556)

chorionic villi biopsy technique involving the examination of a tissue sample from the fetal membrane to detect genetic disorders (210)

choroid (KAWR oyd) dark layer of tissue in the middle of the vertebrate eye (707)

chromatid (KROH muh tihd) strandlike structure that, in pairs, comprises a chromosome during mitosis (138)

chromatin diffused material carrying hereditary information about a cell (84)

chromoplast in plant cells, plastid that stores pigments for the colors red, orange, and yellow (89)

chromosome rod- or rope-shaped body, composed of DNA and proteins, that carries hereditary information (84, 151)

chromosome map graphic device that shows where genes are located on a chromosome (176)

chromosome mutation change involving many genes (191)

chromosome theory scientific theory stating that hereditary factors, or genes, are carried on chromosomes (169)

chrysalis protective covering from which a pupa emerges as a butterfly (510)

chyme semifluid mass of food that passes from the stomach to the small intestine (647)

cilia (SIHL ee uh) tiny, hairlike projections whose wavelike motion helps move substances along a surface (91, 316, 612)

circadian (suhr KAY dee uhn) **rhythm** a pattern of changes that occurs in organisms every 24 hours (771)

circulatory system group of organs that transports materials throughout the body and removes cellular wastes (614)

cirrhosis condition in which liver cells are destroyed and the organ ceases to function (748)

citric acid six-carbon molecule formed during aerobic respiration by the transfer of the acetyl group of acetyl CoA to a four-carbon molecule (126)

cleavage process by which a zygote divides into halves (461)

climax community stable community whose characteristics allow for its own regeneration (777)

clitellum in earthworms, enlarged segment that secretes a mucous substance that holds together mating earthworms (465)

cloaca cavity that receives waste materials from the kidney, the urinary bladder, and the sex organs, and passes them out (462, 543)

clone organism or group of cells developed from one parent and genetically identical to it (220)

cloning production of organisms with identical genes (220)

clutch group of eggs in a nest (580)

cnidocyte (NYD uh syt) stinging cell in coelenterates (448)

coagulation (koh ag yoo LAY shuhn) clotting of blood (656)

coccus (KAHK suhs) sphere-shaped bacterium (299)

cochlea (KAHK lee uh) in the ear of mammals, bony, coiled tube filled with fluid and lined with hair cells (711)

cocoon in metamorphosis, protective covering around the pupa (510)

codominance genetic situation in which neither allele is completely dominant or recessive (163)

codon (KOH dahn) sequence of three bases in RNA that codes for a specific amino acid (187)

coelenterate (sih LEHN tuh rayt) baglike invertebrate animal with tentacles, including hydra, jellyfish, sea anemones and corals (448)

coelom (SEE luhm) large, central body cavity that contains the vital organs (460, 521, 612)

collar cell in a sponge, cell that digests food (445)

collecting duct part of the renal tubule that regulates the water and mineral composition of blood (682)

colloid mixture in which the suspended particles are smaller than those in an ordinary suspension but larger than those of a solute in a solution (54)

colon last portion of the digestive tract, where absorption of water, minerals, and vitamins occurs (650)

colonial algae organisms made up of algal cells held together by a jelly-like substance or cytoplasm (333)

colonial organism individual cells that live in groups and resemble a multicellular organism (14)

colony large group of bacteria descended from a single bacterium (302)

colorblindness inability to distinguish certain colors (205)

commensalism symbiotic relationship in which one organism benefits and the other is unaffected (771)

common name in taxonomy, name given to an organism by people of an area (271–272)

community collection of interacting populations that live in the same area (15, 761)

comparative biochemistry study of molecules that make up living things (235)

comparative embryology study of embryos of different species (236)

complete flower one that contains all the essential and non-essential parts (402)

compound two or more elements that are chemically combined (51)

compound eye in arthropods, eye that has more than one lens (496)

compound leaf leaf with several separate parts attached to an extension of the petiole (393)

compound light microscope instrument that uses light and an ocular and an objective lens to magnify objects (37)

concentration amount of solute dissolved in a given amount of solvent (53)

concentration gradient difference in concentration from the highest to the lowest number of molecules in a substance (100)

condensation reaction chemical reaction in which water is produced (69)

cone in the vertebrate eye, receptor cell that can detect color and produce sharp images (708)

conjugation transfer or exchange of genetic material between organisms (304)

conjunctiva delicate, blood-rich membrane that lines the inner eyelid and covers the front of the eye (707)

connective tissue tissue that joins, supports, and protects other types of tissue (613)

conservation the careful management, wise use, and protection of natural resources (809)

contour feather in birds, one of many smooth, sleek feathers that cover the head, body, and wings (574)

contour plowing method of reducing erosion by plowing across a slope (811)

contractile (kuhn TRAK tuhl) **vacuole** in unicellular organisms, organelle that expels excess water (104, 319)

control group in an experiment, the group that is not exposed to the variable being tested (26)

controlled breeding process of selecting individuals with desired traits to produce the next generation (217)

controlled experiment one in which extraneous conditions are held constant (26)

convergent evolution process by which unrelated species become more alike (251, 597)

cork cambium meristematic tissue that produces bark (387)

cornea transparent layer of tissue that covers the front of the eye (707)

corpus callosum (KAWR puhs kuh LOH suhm) in the human body, bridgelike structure of 200 million nerve fibers that joins the cerebral hemispheres (694–695)

corpus luteum (KAWR puhs LOO tee uhm) enlarged, blood-rich follicle that produces and releases hormones that prepare the uterus for a fertilized egg (733–734)

cortex in plants, rigid outer layer of cells surrounding the pith (384); in the body, outer layer of an organ, such as the kidney (682)

corticoid any of a number of hormones secreted by the adrenal cortex (720)

cotyledon (kaht uhl EED uhn) embryonic leaf enclosed in a seed (374)

covalent bond electrical attraction between two atoms that share electrons (56)

cranial cavity space that houses the brain (612)

critical dark period amount of darkness a photoperiodic plant must have before it flowers (418)

Cro-Magnon kind of human that lived about 35,000 years ago and belonged to the same species as modern humans (262)

crop storage chamber in many birds and invertebrates through which food passes from the esophagus to the gizzard (466)

crop rotation method of conserving soil nutrients by alternating crops from season to season (432, 811)

crossing over process by which alleles exchange places on a chromosome (175)

cross-pollination reproductive process involving the sex cells of two plants (153)

cuticle (KYOOT ih kuhl) in plants, protective outer coating (355); in parasites such as flukes, thick protective coating that prevents digestion by the host (458)

cyclic AMP form of adenosine monophosphate that acts as a ''second messenger'' for many vertebrate hormones (726)

cyst (sihst) in some protists, thickened outer structure that forms when conditions are unfavorable for survival (320); in some parasites, thick-walled structure formed around the larva (459)

cytokinesis (syt oh kih NEE sihs) process during which cytoplasm divides (142)

cytokinin hormone that stimulates cell division in plants (421)

cytoplasm (SYT uh plaz uhm) material between the nucleus and outer boundary of a cell (83)

cytoplasmic streaming constant motion of cytoplasm (86)

cytosine (SYT uh seen) carbon-nitrogen compound, one of the four bases of DNA (182)

D

dark reactions second phase of photosynthesis, in which glucose is formed (117)

data facts collected during an experiment (26)

day-neutral plant one that does not flower in response to the duration of light (418)

death rate rate at which deaths occur in a population (799)

decay element stable substance into which a radioactive isotope eventually breaks down (232)

deciduous (dih SIHJ oo wuhs) **forest** biome characterized by long winters, constant rainfall, trees that shed leaves annually, and much wildlife (784)

decomposer microorganism that breaks down dead tissue and returns the nutrients to the soil (766)

delirium tremens (DTs) in extreme alcoholism, state marked by hallucination and uncontrollable trembling (747)

delivery expulsion of a fetus into the external environment (739)

dendrite branched, cytoplasmic fiber in a neuron that carries impulses toward the cell body (689)

denitrification process in which ammonia, nitrite, and nitrate become nitrogen gas (765)

density-dependent factor environmental element whose effects are determined by population density (797)

density-independent factor environmental element whose effects are not determined by population density (796)

deoxyribose (dee ahk sih RY bohs) five-carbon sugar present in DNA (183)

depressant drug that slows the functions of the central nervous system (746)

dermis thick, inner layer of the skin (632)

desalination (dee sal uh NAY shuhn) process by which salt is removed from salt water (810)

desert biome characterized by very low rainfall, plant life adapted to dry conditions, and nocturnal animals (787)

development series of changes an organism undergoes as it matures (7)

diaphragm (DY uh fram) dome-shaped sheet of muscle that divides the coelom (589, 612, 675)

diatom (DY uh tahm) unicellular golden alga with a silica shell (332)

dicot plant with two cotyledons (376)

differentiation specialization of cells (383)

diffusion movement of molecules of a substance from areas of higher concentration of that substance to areas of lower concentration (99)

digestion process by which food is changed into a form that the body can use (644)

digestive system group of organs that ingest food, break it down, absorb the nutrients, and eliminate solid wastes (614)

digestive tract tubelike passageway from mouth to anus, where digestion occurs (644)

dihybrid cross genetic cross involving two pairs of alleles that determine two separate traits (161)

dinoflagellate unicellular fire alga with a hard, stiff cell wall (332)

diploid (DIHP loyd) the basic chromosome number (2n); two from each pair of homologous chromosomes (143)

directional selection natural selection in a species that proceeds in a given direction (247)

disaccharide molecule formed by the condensation of two monosaccharides (69)

disruptive selection type of natural selection that favors the extremes of a trait (247)

diurnal characteristic of organisms that are active during the day (771)

divergent evolution process by which related organisms become less alike (250)

DNA deoxyribonucleic (dee AHK sih ry boh noo klee ihk) acid, one of two forms of nucleic acid; records instructions for cellular activity and transmits them from generation to generation (73, 181)

dominant gene that masks the other gene in a pair (154)

dormancy in plants, condition marked by minimal growth and low metabolism (419)

dorsal pertaining to the upper side of an organism (444)

dorsal fin in fishes, one of two stabilizing appendages that extend along the back (526)

double fertilization process in flowering plants in which two kinds of fertilization occur, forming both a zygote and an endosperm (405)

down feather in birds, soft, fluffy feather located under a contour feather (574)

Down syndrome genetic disorder in individuals who have an extra chromosome 21 (207)

drug chemical substance that has a marked effect on the body or mind (744)

drug abuse taking excessive amounts of a psychoactive drug for nonmedical reasons (745)

duodenum first part of the small intestine (543)

E

eardrum tightly stretched membrane between the outer and middle ear (710)

echolocation in bats, locating of objects by the reflection of sound waves (591)

ecological pyramid visual display of the decrease in available energy in an ecosystem (768)

ecological succession predictable, orderly replacement of communities in an ecosystem (777)

ecology study of the relationships of living things to their environment (758)

ecosystem distinct, self-supporting unit of interacting organisms and their environment (16, 759)

ectoderm outermost layer of body tissue (445)

ectoplasm in an amoeba, clear, thin layer of cytoplasm found between the endoplasm and the cell membrane (319)

egg tooth in baby birds, sharp structure on the beak tip, used in pecking out of the shell (579)

electron negatively charged subatomic particle (49)

electron microscope instrument that creates enlarged images with a beam of electrons (39)

electron transport chain series of reactions in which energized electrons move from one molecule to another, each time releasing some energy (118)

element substance that cannot be changed into a simpler substance by chemical means (12, 49)

embryo immature form of an organism (374, 596); in humans, the unborn child during its first eight weeks of development (736–737)

embryo sac female gametophyte of a flowering plant (404)

endangered species species of plant or animal that may become extinct unless it is protected (812)

endergonic (ehn duhr GAHN ihk) **reaction** chemical reaction that uses more energy than it releases (60)

endocrine gland group of cells whose secretions are released directly into the bloodstream (719)

endocrine (EHN duh krihn) **system** group of organs that help control body functions through the secretion of hormones (615)

endocytosis (ehn doh sy TOH sihs) bulk transport of substances into a cell (107)

endoderm innermost layer of body tissue (445)

endodermis in plants, innermost layer of cells in the root cortex, which regulates the intake of minerals and other substances (386)

endoplasm in an amoeba, thick, grainy cytoplasm that comprises most of the cell (319)

endoplasmic reticulum (ehn duh PLAZ mihk rih TIHK yuh luhm) also ER; series of canals or channels that serves as a route for materials from cytoplasm to nucleus (86)

endoskeleton internal support system made of bone, cartilage, or a combination of the two (480, 521)

endosperm nutritive tissue that provides a temporary food source for a new plant (405)

endospore cell formed by a bacterium to survive harsh conditions (303)

energy power needed to carry on life activities (5)

energy level region of space around the nucleus of an atom where electrons are located (55)

entomologist scientist who studies insects (504)

environment everything in an organism's surroundings that affects it in any way (9)

enzyme protein that acts as a catalyst (72, 639)

epicotyl part of an embryonic stem above the cotyledons (409)

epidermis in plants and animals, the protective, outermost layers of cells (383, 632)

epiglottis (ehp uh GLAHT ihs) flap of tissue that closes off the trachea (645–646)

epinephrine (ehp uh NEHF rihn) hormone secreted by the adrenal gland (720)

epiphyseal plate in long bones, region at either end where growth occurs (625)

epithelial (ehp uh THEE lee uhl) **tissue** tissue that covers all internal and external body surfaces (612)

epoch (EHP uhk) division of time within an era (233)

equator midplane of a cell (141)

equilibrium state of balance (103)

era one of four spans into which scientists divide time since life began (233)

erosion the removal of soil from an area by water or wind (811)

esophagus (ih SAHF uh guhs) muscular tube leading from the mouth to the stomach (466, 646)

estivation in organisms such as frogs, period of inactivity during the summer (546)

estrogen (EHS truh juhn) hormone that stimulates development of certain female characteristics (733)

estrus in mammals, period of fertility in the female (592)

estuary place where a river or stream meets an ocean (790)

euglenoid movement wormlike method of locomotion typical of *Euglena* (318)

eukaryote (yoo KAR ee oht) cell with a nucleus (92)

Eustachian (yoo STAY shuhn) **tube** duct that connects the middle ear to the pharynx and equalizes air pressure (710)

eutrophication (yoo trahf uh KAY shuhn) the process of adding nutrients to an ecosystem or community, resulting in a decrease in oxygen (779)

evolution change in living things over time (228)

excretion process of eliminating metabolic wastes (681)

excretory system group of organs that remove cellular waste from the blood and maintain the body's fluid and chemical balance (615)

excurrent siphon in bivalves, tube through which water is expelled (474)

exergonic (ehk sur GAHN ihk) **reaction** chemical reaction that releases more energy than it uses up (59–60)

exhalation emptying the lungs of air (676)

exocrine gland group of cells that produces and secretes a substance through tubes, or ducts (719)

exocytosis (ehk soh sy TOH sihs) bulk transport of substances out of a cell (107)

exoskeleton an exterior skeleton (489)

experiment test of a hypothesis (26)

experimental group group exposed to the variable being tested (26)

exponential (ehks poh NEHN shuhl) **growth** period of rapid growth in which the number of individuals repeatedly doubles in a given time period (796)

exponential phase period during which a population undergoes very rapid growth (796)

extensor muscle that straightens a joint (631)

extinct species kind of animal or plant that has disappeared (812)

eyepiece in a microscope, that portion of the instrument closest to the eye of the user (37)

eyespot light-sensitive structure in euglenas and planarians (318)

F

F₁ first filial (FIHL ee uhl) generation; in genetics, the first group of offspring from the crossing of two plants or animals (153)

F₂ second filial generation; offspring produced by the self-pollinating first filial generation (153)

facilitated diffusion form of carrier transport in which substances move from areas of high to low concentration, requiring no energy expenditure (106)

facultative anaerobe bacterium that can live with or without oxygen (302)

Fallopian tube in females, the duct from an ovary to the uterus (732)

fang sharp tooth connected to a poison gland (566)

fat molecule composed of hydrogen, carbon, and oxygen that provides a highly concentrated source of energy (641)

fat body in frogs, structure on the ovary or testis that stores fat as nutrition for developing gametes (547)

feces solid wastes (458, 650)

fermentation anaerobic process that breaks down pyruvic acid into ethyl alcohol or lactic acid (123)

fertilization process in which a male and a female sex cell combine (137, 735)

fetoscopy (fee TAH skuh pee) microscopic examination of a developing fetus in utero to detect physical abnormalities (210)

fetus in humans, unborn child from after the first eight weeks until its birth (737)

fiber substance that aids digestion by stimulating the muscles of the digestive tract (641)

fibrous root in monocots, one of many secondary roots that forms a network over a wide area (385)

filament in a flower, thin, stemlike part of a stamen (401)

filoplume in birds, short, thin feather that looks like hair; also called a *pin feather* (574)

first polar body smaller of the cells resulting from meiosis I in female animals (145)

flagella (fluh JEHL uh) long, hairlike structures on the surface of a cell that aid in locomotion (91, 300, 316)

flame cell excretory cell in planarians (458)

flexor muscle that bends a joint (631)

follicle in birds, one of many small sacs in the skin from which feathers grow (574); fluid-filled chamber around an egg before it is released from the ovary (733)

forestry business of cultivating trees to provide fuel, lumber, and other wood products (435)

fossil any preserved part or trace of an organism that once lived (231)

fossil fuels substances that consist of fossil remains of prehistoric plants and animals (814)

fovea (FOH vee uh) area at the center of the retina, where a concentration of cones produces the sharpest image (708)

fragmentation kind of asexual reproduction in which part of an organism breaks off and grows on its own (309, 334)

fraternal twins offspring that develop from two eggs in the mother that are fertilized by two different sperm (200)

frond leaf of a fern (371)

fruit ripened ovary of a flowering plant (407)

fruiting body visible part of a fungus, containing the spore-producing structures (343)

fungus nonmotile organism that obtains food by decomposing organic matter (343)

G

gall bladder small sac where bile is stored (648)

gamete (GAM eet) sex cell (137, 151, 731)

gametophyte (guh MEET uh fyt) in algae and plants, the haploid generation that produces gametes (334, 358)

ganglion (GAN glee uhn) concentration of nerve cells (457)

gastric caeca (SEE kuh) in an insect, pocket of the stomach that secretes enzymes necessary for digestion (508)

gastrula (GAS troo luh) in animal development, indentation of blastula cells that gives rise to differentiated tissue (461)

gemmae (JEHM ee) spores produced asexually by some liverworts (361)

gemmule (JEHM yool) asexually produced, food-filled ball of amoebocytes that becomes a sponge (447)

gene unit of hereditary information (151)

gene frequency measure of the relative occurrence of a given allele in a population (245)

gene linkage situation in which two or more genes occur on the same chromosome (174)

gene mutation change in a single gene (191)

gene pool alleles of all the genes in all of the individuals in a population (245)

gene therapy technique for replacing a defective gene (214)

generative cell smaller of two cells in a pollen grain (403)

genetic counseling informing couples of their chances of passing on a harmful genetic trait (208)

genetic drift change in gene frequency of a small population due to the effects of random mating (248)

genetic engineering process of transferring DNA segments from one organism into the DNA of another species (221)

genetic equilibrium relative stability of the genetic makeup of a population (246)

genetic isolation absence of genetic exchange between populations because of geographical separation or other factors that prevent reproduction (249)

genetics study of heredity (151)

genotype (JEE nuh typ) pairs of alleles in the cells of an organism (156)

genus in taxonomy, group of similar species (271)

geologist scientist who studies the physical nature and history of the earth (233)

geothermal (jee oh THER muhl) **power** energy created by steam below or near the earth's surface (814)

geotropism plant's response to gravity (422)

germinate in a plant seed, to resume growth (408)

germ layer any of the three layers of cells in most multicellular animals (460)

germ mutation change in a reproductive cell (193)

gestation (jehs TAY shuhn) **period** in mammals, time in which young develop in the uterus (598, 735)

gibberellin hormone that stimulates rapid growth (421)

gill in mushrooms, thin sheet of tissue under the cap (346); in aquatic animals, respiratory organ (474)

gill slits in chordates, openings in the throat region (519)

gizzard digestive chamber in which organic matter is crushed (466)

glomerular filtration in vertebrates, removal of urea and other wastes from the body (682)

glomerulus (glah MEHR yoo luhs) in the kidney, mass of capillaries forming a tight ball within a nephron (682)

glycolysis (gly KAHL uh sihs) process of breaking a glucose molecule in half to form two molecules of pyruvic acid; the first step in cellular respiration (122)

Golgi (GOHL jee) **body** in a cell, area for storage and packaging of chemicals (88)

gonad gamete-producing organ; also produce and secrete hormones (722)

grana tiny disklike sacs in a chloroplast where photosynthesis begins (116)

grassland biome characterized by wet springs, dry summers, and many types of grasses (785)

green gland in most crustaceans, excretory organ located at the base of the antennae (495)

Green Revolution program to introduce high-yield crops into poor agricultural regions (433)

ground tissue in plants, relatively unspecialized tissue that cushions and protects vascular tissue (384)

growth rate rate at which a population changes (799)

guanine (GWAH neen) carbon-nitrogen compound, one of the four bases of DNA (182)

guard cell one of two kidney-shaped cells that regulate the passage of water through a stoma in a leaf (394)

gullet in paramecia, chamber beneath the mouth pore where food is stored (322)

gymnosperm (JIHM nuh spuhrm) type of plant that produces seeds in cones and generally keeps its leaves all year (374)

H

habitat surroundings of a particular species (761)

hair cell sense receptor in which fine, hairlike projections respond to mechanical pressure (705)

hair follicle small fold of epidermis where hair is manufactured (633)

half-life the time it takes for half the isotopes in radioactive material to decay (232)

hallucinogen (huh LOO suh nuh jehn) drug that distorts sensory impressions (748)

haploid number half the diploid number; one from each pair (n) of homologous chromosomes (143)

hard palate bony plate in the roof of the mouth (645)

Hardy-Weinberg principle principle stating that the frequency of alleles in a population stays the same unless altered by some external factor (246)

Haversian (huh VUR shuhn) **canal** small channel in the compact bone containing blood vessels that nourish the osteocytes (625)

head-foot in mollusks, one of three main body parts, consisting of mouth, sensory organs, foot, and other motor organs (473)

heart muscular organ that pumps blood to all parts of the body (659)

hemoglobin (HEE muh gloh bihn) molecule that carries oxygen in the blood (656)

hemophilia (hee muh FIHL ee uh) hereditary disorder that prevents normal blood clotting (205)

hemotoxin poisonous protein that destroys red blood cells and breaks down vessel walls (566)

herbaceous plant monocot with a typically green stem lacking secondary growth (377)

herbivore animal that eats only plants (766)

heredity the passing of traits from parents to offspring (150)

hermaphrodite (huhr MAF ruh dyt) organism that can produce both eggs and sperm (446)

heterotroph (HEHT uhr uh trahf) organism that depends on other organisms for food (123, 275)

heterozygous (heht uhr oh ZY guhs) characteristic of an individual with a dominant and a recessive gene for a trait (154)

hibernation in some animals, period of severely reduced activity in winter (546)

hindgut posterior part of the digestive tract; an insect's intestine (508)

holdfast rootlike structure that anchors many brown algae (336)

homeostasis (hoh mee oh STAY sihs) self-adjusting balance between life functions, the environment, and an organism's activities (11)

home range in mammals, area over which an animal travels during normal activities (592–593)

Homo erectus first hominid species of the genus *Homo* (260)

homologous (hoh MAHL uh guhs) **chromosome** one of two like chromosomes (143)

homologous structure body part with the same basic structure as that of another organism, suggesting common ancestry (236)

Homo sapiens species name of modern humans (261)

homozygous (hoh moh ZY guhs) characteristic of an individual with identical genes for a trait (154)

homozygous dominant individual having two dominant genes for a certain trait (154)

homozygous recessive individual having two recessive genes for a certain trait (154)

hormone internally produced chemical that regulates the functions of tissues and organs (419, 719)

host cell cell in which a virus reproduces (283)

Huntington disease hereditary disease of the nervous system, involving loss of muscle control, mental deterioration, and eventual death (202)

hybrid in genetics, individual with a dominant and a recessive gene (154)

hybridization mating of two different species, breeds, or varieties (219)

hybrid vigor improvement in quality resulting from hybridization (219)

hydrolysis breaking down complex molecules by the combination with molecules of water (71)

hydrotropism movement of a plant's roots toward water (423)

hypertonic solution solution in which the concentration of the solutes outside a cell is greater than that inside it (103)

hypha (HY fuh) one of many filaments that comprise the body of a fungus (343)

hypocotyl part of an embryonic stem above the radicle and below the cotyledons (409)

hypothalamus structure in the vertebrate brain that controls body temperature, thirst, hunger, salt and water balance, and emotional behavior (695–696)

hypothesis (hy PAHTH uh sihs) statement that can be tested (25)

hypotonic solution one in which the concentration of solutes outside a cell is lower than that inside it (103)

I

identical twins two offspring that develop from a single fertilized egg that separates into two halves early in development (200)

ileum second part of the small intestine (543)

immunity natural resistance to disease through the production of antibodies (289, 668)

imperfect flower one that has the reproductive parts of only one sex (402)

imprint impression in a developing rock formed by fossils of soft body structures (231)

impulse electrochemical signal in the central nervous system (689)

inbreeding mating of genetically similar individuals (218)

incomplete dominance situation in which neither allele is dominant or recessive (163)

incomplete flower one that lacks one or more of the essential or nonessential parts (402)

incubate to warm bird eggs to promote development of the young (580)

incurrent pore in sponges, one of many openings in the outer layer, through which water enters (445)

incurrent siphon in bivalves, tube through which water enters the organism (474)

incus (IHN kuhs) in mammals, one of three tiny bones in the middle ear (711)

inferior vena cava (VEE nuh KAY vuh) vessel that brings blood from the lower part of the body to the heart (661)

inhalation process of filling the lungs with air (676)

inherited passed down from generation to generation (9)

inorganic compound one not made by living things (67)

insecticide chemical used to kill insects (513)

insertion point at which a muscle is attached to the moving bone (631)

insulin substance secreted by the pancreas that controls the level of glucose in the blood (543, 720)

integument enclosing layer of an organism, such as the skin of the human body (632)

integumentary system group of organs that form a protective outer layer (615)

interbreed to mate within a population (14)

interferon (ihn tuhr FIHR ahn) protein that interferes with viral replication (290)

interneuron nerve cell that links sensory and motor nerve cells (689)

interphase time between the formation of a cell by mitosis and that cell's next mitosis (138)

intertidal zone that part of the seashore located between high and low tide (788)

invertebrate animal without a backbone (444)

ion atom that has gained or lost electrons (49)

ionic bond electrical attraction between a positive and a negative ion (56)

iris colored ring of tissue that gives the eye its color (707)

irrigation process of watering crops by means other than natural rainfall (433)

irritability ability to respond to stimuli (8)

islets of Langerhans cells in the pancreas that secrete insulin and glucagon (720)

isomers compounds with the same molecular formula but a different arrangement of atoms (69)

isotonic (eye suh TAHN ihk) **solution** one in which the concentration of solutes outside a cell is the same as that inside the cell (103)

isotope (EYE suh tohp) atom of an element with different number of neutrons than other atoms of the same element (52)

J

Jacobson's organ two small hollows in the roof of a snake's mouth sensitive to taste and smell (564)

J-curve pattern formed by the combination of exponential growth and lag phases in a population (796)

joint point at which bones meet (626)

K

karyotype picture of paired human chromosomes arranged by size; used to identify chromosomal abnormalities (206, 273)

keel in birds, high, narrow ridge, at the bottom of the sternum, to which breast muscles are attached (575)

keratin protein that makes skin hard and waterproof (556)

kidney in vertebrates, one of a pair of bean-shaped excretory organs (681)

kinetic energy energy of motion (55)

Klinefelter syndrome genetic disorder in males with an extra X chromosome (207)

Krebs cycle series of chemical reactions, beginning and ending with citric acid, that occur during one stage of aerobic respiration (126)

L

labium in insects, lower lip, which presses food against the jaws (506)

labor process during which uterine muscles contract to expel a fetus (739)

labrum in insects, upper lip, which holds the insect's food (506)

lacrimal gland structure near the outer corner of the eye that produces tears (707)

lacteal tiny vessel in the villi that absorbs fats (649)

lactic acid fermentation process in which pyruvic acid and hydrogen formed during glycolysis combine and produce lactic acid (124)

lag phase period of slow growth in a population (795)

landfill layer of wastes sandwiched between layers of earth (819)

large intestine last portion of the digestive tract, where absorption of water, minerals, and vitamins occurs (543, 650)

larva (LAHR vuh) immature form of an organism that is very different from the adult organism (447)

larynx upper part of the trachea, where the vocal cords are located (646)

lateral line system in a fish, system of fluid-filled canals that allows the animal to detect movement and to locate objects (526)

law of conservation principle stating that matter cannot be created or destroyed (60)

leaf base place on the stem where the petiole is attached (393)

leg scale one of many small plates of keratin that cover a bird's leg (576)

legume plant that produces seeds in pods (432)

lens transparent, curved structure in the eye that helps focus images (708)

leucoplast (LOO kuh plast) in plant cells, colorless plastid that stores food (89)

lichen (LY kuhn) organism consisting of a fungus and an alga living symbiotically (344)

life span average length of life (7)

ligament strong band of connective tissue on the bones of a movable joint (626)

light reactions first phase of photosynthesis, in which light energy is trapped and materials required in the next phase are formed (117)

lignin (LIHG nihn) complex carbohydrate that combines with cellulose to support plant tissues and facilitate photosynthesis (355)

limiting factor environmental element that stabilizes population size and keeps species from reaching their biotic potential (796)

linkage group genes that occur together on a chromosome (174)

lipid one of a group of substances insoluble in water; includes fats, oils, and waxes (71)

liver largest internal organ; producer of bile (647–648)

lobe-finned fish one of a group of bony fishes having dorsal and pectoral fins with large, fleshy bases supported by leglike bones (529)

long-day plant one that produces flowers in summer, when nights are shorter than the plant's critical dark period (418)

lung one of a pair of spongy, saclike organs where breathing occurs (675)

lungfish one of a group of bony fishes with lungs as well as gills (529)

lymph tissue fluid that surrounds all body cells (665)

lymphatic system part of the circulatory system that collects fluid from tissue and returns it to the blood (614)

lymph node one of many tiny bean-shaped organs that filter foreign matter from lymph (665)

lymphocyte type of white blood cell that defends the body against disease and infections (666)

lysogenic (ly suh JEHN ihk) **cycle** period of inactivity that occurs in some viruses after invading a host cell (287)

lysosome (LY suh sohm) in a cell, organelle that digests large particles (90)

lytic (LIHT ihk) **cycle** process in which a virus kills its host cell (284)

M

macroevolution sudden appearance of a distinct species (252)

macronucleus in paramecia, the larger nucleus, which directs all functions but reproduction (321–322)

magnification apparent increase in an object's size (37)

malleus (MAL ee uhs), one of three tiny bones in the middle ear (711)

Malpighian (mal PIHG ee uhn) **tubule** in arachnids, one of many small tubes of the excretory system, which remove excess nitrogen from blood (492)

mammary gland in mammals, modified sweat gland that secretes milk used to feed young (589)

mandible in crustaceans, insects, and myriapods, one of a pair of chewing jaws (494)

mantle in mollusks, thin membrane that surrounds the visceral mass (473)

marrow tissue inside some bones that produces red and white blood cells (623)

marsupial mammal that has a pouch (595)

mass measure of the amount of matter in an object (47)

mass number number of protons plus the number of neutrons in an atom (50)

mass selection raising many plants or animals and selecting the best in each generation for further breeding (218)

mating strain one of two genetically different molds that reproduce sexually (345)

matrix (MAY trihks) thick, nonliving material in which connective tissue cells are embedded (613)

matter anything that takes up space (47)

maxilla in crustaceans and some other arthropods, one of the first or second pair of mouthparts behind the mandible (494)

maxilliped in crustaceans, one of the first pair of appendages on the thorax part of the cephalothorax (494)

medulla inner layer of an organ such as the kidney (682)

medulla oblongata (mih DUHL uh ahb lawn GAHT uh) in vertebrates, enlarged portion of the spinal cord that carries out most automatic nervous responses (546, 696–697)

medusa (muh DOO suh) bell-shaped body of some coelenterates (448)

meiosis (my OH sihs) type of nuclear division in which the chromosome number is halved (143)

meninges (muh NIHN jeez) three layers of tissue that surround and protect the surface of the brain (698)

memory cell white blood cell that ''remembers'' an antigen pattern and produces antibodies (667–668)

menopause in females, period during which the ovaries cease production of mature eggs (734)

menstrual cycle repeating pattern of ovulation and related changes in the female (733)

menstruation in females, discharging of dead tissue and an unfertilized egg from the uterus (734)

meristem (MEHR uh stehm) specific areas in plants at which cells divide (383)

mesentery thin, tough membrane that holds the small intestine in place (543)

mesoderm third layer of cells, encased between the ectoderm and the endoderm of bilaterally symmetrical animals (460)

mesoglea (mehz uh GLEE uh) in coelenterates, jellylike substance separating endoderm and ectoderm (448)

mesophyll middle portion of a leaf (394)

mesothorax in insects, the second segment of the three-part thorax (507)

messenger RNA also mRNA; type of RNA that carries sequences of nucleotides that code for protein from the nucleus to the ribosomes (186)

metabolism (muh TAB uh lihz uhm) the sum of all the chemical reactions within cells or organisms (5–6, 678)

metamorphosis series of marked changes by which an immature organism becomes an adult (509)

metaphase second phase of mitosis, during which the chromosomes line up along the cell's midplane (140–141)

metathorax in insects, third segment of the three-part thorax (507)

microdissection kind of surgery in which biologists remove or add structures to cells (41)

microevolution slow, gradual changes in species (252)

micrometer unit of measurement equaling one-millionth of a meter (0.000039 in.) (38)

micronucleus in paramecia, the smaller nucleus, which controls reproduction (321)

microtome (MY kruh tohm) instrument that slices a specimen for microscopic viewing (40)

microtubule (my kroh TOOB yool) hollow cylinder of protein that supports and shapes a cell (90)

microvillus one of many extensions of epithelial tissue on the villi, where absorption occurs (648)

midbrain structure in the brain that controls responses to sight (697)

midbrain tectum in birds, center of vision (578)

middle lamella (luh MEHL uh) in plants, layer formed between two newly formed adjacent cell walls (86)

midgut stomach of an insect (508)

migration regular, seasonal movement of organisms (248, 533, 581)

milt in fishes, fluid containing sperm (533)

mimicry the resemblance of one organism to that of a more powerful one, which acts as a defense mechanism (512)

mineral one of many inorganic substances that form an important part of living tissue (643)

mitochondrion (myt uh KAHN dree uhn) in a cell, organelle that releases energy from nutrients (88)

mitosis (my TOH sihs) nuclear division that results in the replication and division of the parent cell into two identical daughter cells (138)

mixture mingled molecules not chemically combined (53)

mold depression in a rock shaped by fossils of hard body parts (231–232)

molecular formula notation of the number of atoms in a molecule of any given element (52)

molecular weight sum of the mass numbers of all the atoms in a given molecule (53)

molecule smallest particle of a compound or element that can have a stable, independent existence (12, 52)

molting process by which some animals periodically shed the outer layer of skin (489, 563)

monocot plant with one cotyledon (376)

monohybrid cross genetic cross involving one trait (160)

monomer basic unit of most organic molecules (69)

monosaccharide sugar molecule that cannot be broken down into smaller organic molecules; the foundation of all carbohydrates (69)

monosomy (MAHN uh so mee) condition in which an individual has 45 chromosomes (207)

monotreme mammal that lays eggs (595)

motor neuron nerve cell that carries impulses away from the central nervous system (689)

mouth breathing in frogs, using vessels in the mouth to diffuse oxygen into the blood (544)

mouth pore in paramecia, opening that channels food to the endoplasm (322)

mucus gland in frogs, one of many skin glands that secrete a slimy substance to prevent drying out (540)

multicellular (muhl tih SEHL yoo luhr) **organism** organism composed of more than one cell (3–4)

multiple alleles three or more alleles of a gene that can occur at one location on a chromosome and affect a given trait (203)

multiple fission nuclear division in amoebic cysts (320)

muscle organ made up of tissue that can contract (627)

muscle fiber long tapering cell capable of movement (627)

muscle tissue tissue with the ability to contract and thus produce movement (613)

muscular system group of organs that work with bones to move the body (614)

mutagen anything that increases the rate of mutations in cells (194)

mutant organism in which a genetic or chromosomal change is expressed (191)

mutation change in a gene or chromosome (191)

mutualism symbiotic relationship in which organisms interact for mutual benefit (770)

mycelium (my SEE lee uhm) mass of intertwined hyphae that form the body of a fungus (343)

mycoplasma (my koh PLAZ muh) moneran that lacks a cell wall (298)

myelin sheath fatty insulating layer that supports and helps maintain the chemical balance of the axon (689–690)

myofibril (my oh FY bruhl) one of many protein threads that comprise muscle fibers (627)

myriapod (MIHR ee uh pahd) collective name for centipedes and millipedes (490)

N

narcotic drug that dulls the senses and relieves pain by depressing the cerebral cortex (751)

nasal cavity one of a pair of spaces in the nose, where air is warmed and moistened to aid respiration (676–677)

nastic movement plant movement unrelated to the direction of an external stimulus (423)

natural resources all the elements of nature (809)

natural selection process proposed by Charles Darwin whereby the environment acts to preserve, or select, fit individuals (244)

Neanderthal kind of human that lived between 130,000 and 35,000 years ago (261)

negative feedback process by which the endocrine system regulates itself; when the hormone level falls below a certain point the gland is stimulated to secrete more of the hormone (725)

negative phototropism response in which an organism moves away from light (320)

nekton in the marine biome, animals that swim freely (788)

nematocyst (neh MAT uh sihst) coiled stinger in the tentacles of coelenterates (448)

neoteny development in which an organism retains larval characteristics but achieves sexual maturity and reproduces (550)

nephridia (neh FRIHD ee uh) in many invertebrates, pair of ciliated tubes on each segment, which eliminate liquid wastes (467)

nephron in reptiles, birds, and mammals, the functional unit of the kidney (682)

neritic zone region of open water near the shore (789)

nerve collection of bundles of fibers that carry impulses to and from parts of the body (690)

nerve cord in chordates, hollow, tubular cord along the back of the body above the notochord (519)

nerve fiber threadlike extension of cytoplasm in the nerve cell that carries impulses to and from the cell (689)

nervous system group of organs that monitor the environment and control body activities (615)

nervous tissue tissue that transmits messages (613)

neuron basic functional unit of the nervous system (689)

neurotoxin poisonous protein that attacks the nervous system, paralyzing neurons (566)

neurotransmitter chemical substance that transmits an impulse from one neuron to another (692)

neutron subatomic particle with no electrical charge (48)

niche way of life a species pursues within its habitat (762)

nicotine psychoactive drug in tobacco (752)

nictitating (NIHK tuh tay ting) **membrane** in frogs and some other vertebrates, third eyelid that keeps the eyeball moist (541)

nitrification (ny truh fuh KAY shuhn) two-step process by which ammonia becomes nitrate (765)

nitrogen fixation process by which atmospheric nitrogen is incorporated into compounds (305, 764)

nocturnal characteristic of organisms that are active at night (771)

node of Ranvier (rahn vee AY) gap between the cells of the myelin sheath (690)

nonbiodegradable characteristic of a product that cannot be broken down by microorganisms (816)

nondisjunction (nahn dihs JUHNGK shuhn) failure of a chromosome pair to separate during meiosis or mitosis (207)

nonrenewable resources any elements of nature that cannot be reused or replenished (809)

nonvascular plant bryophyte; plant that lacks vascular tissue (357)

norepinephrine (nor ehp uh NEHF rihn) hormone secreted by the adrenal gland (720)

nostril one of a pair of openings in the nose, through which air enters the body (676)

notochord in chordates, long, firm rod along the back of the body (519)

nuclear envelope in a cell, double layer of lipids and proteins between the nucleus and cytoplasm (84)

nuclear fission splitting of atomic nuclei, which releases heat energy (814)

nuclear fusion joining of two atoms to form helium (814)

nuclear pore opening in the nuclear envelope that allows passage of materials (84)

nucleic acid organic compound that carries all instructions for cellular activity (73)

nucleolus (noo KLEE uh luhs) spherical body of DNA, RNA, and proteins in a nucleus (84)

nucleotide unit of DNA comprised of a nitrogen-carrying base, a sugar molecule, and a phosphate group (73, 183)

nucleus in atoms, central core (48); in cells, control center, containing genetic information (79)

nutrient one of many substances needed for body growth and maintenance (639)

O

objective in a compound light microscope, the lens set near the specimen (37)

obligate aerobe bacterium that needs oxygen to live (302)

obligate anaerobe bacterium that cannot live in the presence of oxygen (302)

ocular in a compound light microscope, the lens set near the viewer's eye (37)

olfactory organ organ that monitors smell (532)

olfactory receptor cell in the nasal passage that produces impulses of smell (713)

oligotrophic (ahl uh goh TRAHF ihk) characteristic of a lake or pond that can support little life (789)

omnivore animal that eats both plant and animal materials (600, 766)

oogamy (oh AHG uh mee) differentiation of gametes into distinctly male and female forms (346)

open sea zone in the marine biome, three-leveled area of a large body of water (789)

operculum (oh PUHR kyoo luhm) in some snails, flat plate on the side of the foot, used to close off the shell (477); in fishes, protective flap of tissue covering the gills (530)

opposable thumb in primates, a thumb that can be positioned opposite the fingers to grasp objects (609)

optic lobes in birds, portion of the midbrain tectum that interprets visual impulses (578)

oral disc round, sucking mouth of a lamprey (523)

oral groove in paramecia, channel that receives food particles (322)

organ structure composed of a number of tissues that work together to perform a specific task (13)

organelle structure that performs a specific function within a living cell (12, 83)

organic compound compound made by organisms, containing carbon (67)

organism complete, individual living thing (3)

origin point at which a muscle is attached to an anchoring bone (631)

osculum (AHS kyuh luhm) large opening in a sponge, through which water exits (445)

osmosis (ahz MOH suhs) diffusion of water through a membrane (102)

ossification (ahs uh fuh KAY shuhn) in embryonic development, process in which cartilage is replaced by bone (625)

osteocyte (AHS tee uh syt) one of many living cells that make up bone (625)

ostia in crustaceans, any of three pairs of pores through which blood enters the heart (495)

oval window membrane between the middle and inner ear (711)

ovary in plants, swollen base of the pistil (402); in animals, egg-producing reproductive organ (449, 547, 732)

oviduct long tube through which eggs travel from the ovary (578)

oviparous (oh VIHP uhr uhs) characteristic of an animal that lays eggs that hatch outside the mother's body (564)

ovipositor in a female insect, one of a pair of pointed organs used to deposit fertilized eggs (508)

ovoviviparous characteristic of an animal that hatches eggs inside the mother's body and gives birth to live young (564)

ovulation release of an egg from the ovary (733)

ovule in a flower, that portion of the ovary in which eggs are produced (402)

ovum female sex cell (137)

oxygen-carbon cycle interrelated recycling pathways of photosynthesis and respiration (763)

P

P parental generation; in genetics, first pair of plants or animals in a study (152)

pair bond mating relationship of a male and female bird for the duration of a reproductive season (580)

paleontologist (pay lee ahn TAHL uh jihst) scientist who looks for and studies fossils (233)

pancreas in vertebrates, organ that secretes digestive enzymes and insulin and glucagon hormones (543, 647–648)

parapodia (pa ruh POHD ee uh) in polychaetes, a pair of appendages that assist in locomotion (468)

parasite organism that benefits from a close association with another organism and harms it in the process (304)

parasitism symbiotic relationship in which one organism benefits and the other is harmed (315, 771)

parathyroid gland any of four glands on the back of the thyroid gland that regulate levels of calcium ions and phosphate ions in the blood (719)

parenchyma (puh REHN kih muh) soft, spongy cells found in the center of roots and stems (384)

parietal (puh RY uh tuhl) **eye** in some reptiles, a third eye, located on top of the head; believed to be used for temperature control (560)

particulate matter solid or liquid particles in air pollution (816)

passive immunity resistance to disease acquired from the antibodies from another immune person or animal (290)

pathogen disease-causing microorganism (288, 666)

pectin jellylike substance that holds together layers of algal cells (329)

pectoral fin in fishes, one of a pair of stabilizing appendages just behind the head (526)

pectoral (PEHK tuhr uhl) **girdle** in vertebrates, wide, flattened surface that connects the upper limbs to the axial skeleton (521)

pedigree record that shows how a trait is inherited over several generations (199)

pedipalp (PEHD ih palp) in arachnids, one of two appendages used in chewing and in sensory perception (491)

pellicle in many protists, flexible layer of protein strips under the cell membrane (318)

pelvic fin one of a pair of stabilizing appendages on the underside of a fish (526)

pelvic girdle in vertebrates, wide, flattened surface that connects the lower limbs to the axial skeleton (521)

penicillin antibiotic that inhibits the growth of bacterial cell walls (306)

peptide bond chemical bond between amino acids (72)

perfect flower one that has both male and female reproductive parts (402)

pericardium (pehr uh KAHR dee uhm) tough, protective sac that surrounds the heart (659)

pericycle outer layer of meristematic cells around the vascular cylinder, responsible for secondary root growth (386)

periosteum (pehr ih AHS tee uhm) protective membrane that covers bones and helps connect them to muscle (625)

peripheral (puh RIHF uhr uhl) **nervous system** part of the nervous system that provides pathways for impulses to and from the central nervous system (689)

peristalsis (pehr uh STAWL sihs) wavelike motion that moves food along the digestive tract (646)

pesticide chemical used to control insect pests (819)

petal leaf that grows between the sepals and the essential flower parts; collectively, forms the corolla (402)

petiole (PEHT ee ohl) leaf stalk that supports the blade (393)

petrified fossil one created by minerals (232)

PGAL phosphoglyceraldehyde; an organic compound formed during the second phase of photosynthesis (120)

phagocyte white blood cell that engulfs and digests a disease-causing agent (666)

phagocytosis (fag oh sy TOH sihs) movement of solids or large particles into a cell (107)

pharynx in some invertebrates, muscular organ that ingests and partially digests food and also emits waste (458); in vertebrates, passageway for food and air (645)

phase contrast microscope one that uses the interference of bent and unbent light rays to show the structure of a cell (37–38)

phenotype (FEE nuh typ) an expressed trait (156)

pheromone chemical secreted by an insect that influences the behavior of other insects (512)

phloem (FLOH ehm) vascular tissue that carries sugar and other products of photosynthesis from the leaves to the rest of the plant (367)

photon unit of light energy (115)

photoperiodism (foht oh PIHR ee uhd ihz uhm) response of plants to periods of light and dark (418)

photosynthesis process by which green plants convert the energy from sunlight into chemical energy (114)

photosynthetic region area of the open sea from the surface to about 200 m (656 ft.) (789)

phototropism response of a plant to the direction of its light source (422)

pH scale means of measuring the relative concentration of hydrogen ions in a substance (68)

phycobilin (FY koh by luhn) accessory pigment found in blue-green algae (307–308)

physical anthropologist (an thruh PAHL uh jihst) scientist who studies human fossils (233)

physical property feature of a substance that can be determined without changing its basic makeup (47)

pigment substance that absorbs light (116)

pilus (PIHL uhs) one of many short, thin extensions that bacteria use to attach themselves to food or oxygen sources or to other bacteria (300)

pin feather in birds, small, dark, immature feather with a scaly covering (574)

pinna flap of cartilage that forms the outer ear (710)

pinocytosis (pihn oh sy TOH sihs) movement of liquids and small particles into a cell (108)

pioneer species first organisms to inhabit a community (779)

pistil female part of a flower (401)

pith ground tissue in the center of roots and stems (384)

pituitary (pih TOO uh tehr ee) **gland** organ at the base of the brain that secretes hormones that regulate a variety of functions (722–723)

PKU phenylketonuria (fehn ihl keet uh NYOOR ee uh); biochemical disorder produced by two recessive alleles (202)

placenta in mammals, mass of tissue that helps maintain the uterine wall and serves as the point of exchange between mother and fetus (595, 737)

placental mammals animals whose young remain in the mother's body until ready to maintain life independently (595)

placoid scale in fishes such as sharks, one of many cone-shaped structures that cover the skin (526)

plankton (PLANK tuhn) small organisms that float near the surface of the ocean (329, 788)

plasma straw-colored, nonliving part of blood (655)

plasmid (PLAZ mihd) small circle of DNA in some bacteria (222, 300)

plasmodium brightly colored, jellylike cytoplasm that comprises the body of a slime mold (349)

plastid in plant cells, organelle that stores food or contains pigment (89)

plastron ventral half of a turtle or tortoise shell (561)

platelet in mammals, cell fragment that aids in blood clotting; also called thrombocyte (656)

pleura in mammals, tough membrane around the lungs that secretes a lubricating fluid (675)

point mutation change in a single gene (191)

polar compound one that has opposite charges at either end of its molecules (67)

polar nuclei two central nuclei in a megaspore during egg cell formation (404)

pollen spores produced by a flower (401)

pollen tube in a flowering plant, tube that grows through the stigma during fertilization (405)

pollination in flowering plants, transferring pollen from the anther to the stigma (405)

pollution contamination of the environment with waste products and other impurities (816)

polydactyly (pahl ih DAK tih lee) condition of having extra fingers or toes (202)

polygenic determined by several genes (204)

polymer complex molecule produced by condensation of many monomers (69)

polyp vase-shaped body of some coelenterates (448)

polyploidy (PAHL ih ploy dee) condition in which an organism has more than two complete sets of chromosomes (220)

polysaccharide carbohydrate that consists of a long chain of monosaccharides (70)

pons structure in the human brain that connects the cerebral hemispheres and links the cerebellum with the cerebrum (697)

population individuals of a single species that occupy a common area and share resources (14, 245, 761)

population density number of individuals in a given area (796)

population genetics study of the number and kind of genes in a population (245)

population sampling genetic study of a small, randomly selected group and projecting the results to the whole population (200)

positive phototropism movement of an organism toward light (318)

posterior pertaining to the hind end of an organism (444)

potential energy energy an object possesses because of its position or composition (55)

precocial characteristic of a young bird that leaves the nest within a few hours after birth (579)

predator organism that kills and eats other organisms (478, 770)

preen gland in birds, gland that secretes oil to waterproof feathers (574)

pregnancy in mammals, timespan during which young develop within the mother (598, 735)

primary growth lengthening of roots and stems (383)

primary root first root of a young plant (385)

primary succession development of a community where none existed before (778)

principle of dominance principle stating that one gene in a pair may prevent the other from being expressed (153)

principle of independent assortment principle stating that pairs of genes segregate independently during gamete formation (156)

principle of segregation principle stating that members of each pair of genes separate when gametes form (155)

probability likelihood that an event will occur (157)

product rule principle stating that the probability of independent events occurring together is the product of the probabilities of each event occurring alone (158)

progesterone (proh JEHS tuh rohn) female hormone that helps prepare the uterus for implantation (734)

proglottid (proh GLAHT ihd) one of several segments that comprise the body of a tapeworm (459–460)

prokaryote (proh KAR ee oht) cell without a true nucleus (92)

prophage (PROH fayj) viral nucleic acid that attaches itself to the DNA of the host cell (287)

prophase first stage of mitosis: chromosomes coil up; centrioles, spindle fibers, and an aster form; and centrioles move to opposite ends of the cells (139)

prosthesis (prahs THEE sihs) artificial body part (616)

protein basic building material of all living things (71, 639)

protein hormone one that combines with a receptor cell on the target cell membrane, activating an enzyme that helps change ATP into cyclic AMP (726)

protein synthesis process by which proteins are made from amino acids (186)

prothallus gametophyte generation of a fern (373)

prothorax in insects, first segment of the three-part thorax (507)

proton positively charged subatomic particle (48)

protonema (proht uh NEE muh) in mosses, horizontal photosynthetic filament, produced by a spore (359)

protozoa (proht uh ZOH uh) complex unicellular heterotrophs (314)

pseudocoelom (SOO doh SEE luhm) false body cavity in a roundworm (460–461)

pseudopodia (soo duh POH dee uh) footlike projections of some protozoa, used in locomotion (319)

psychoactive drug chemical substance that affects the central nervous system (745)

puberty period of physical and sexual maturation (731)

pulmonary artery blood vessel leading from the heart to the lungs (661)

pulmonary circulation blood moving to and from the lungs (545)

pulmonary vein blood vessel that brings blood from the lungs to the heart (661)

Punnett square grid, or chart that shows all possible gene combinations for a cross (159)

pupa in metamorphosis, inactive stage between the larval and adult stages (510)

pupil opening in the center of the iris where light enters the eye (707–708)

purebred individual with identical gene pairs (154)

pure science fields of study that involve the search for new knowledge (27)

pyloric sphincter muscular valve that controls the passage of food out of the stomach (646)

pyrenoid (py REE noyd) small protein body that stores starch in a chloroplast (334)

pyruvic (py ROO vihk) **acid** substance formed during the first stage of cellular respiration (122)

Q

quill hollow section of a feather's rachis (574)

quill feather large feather, located on a bird's wing or tail (574)

R

rachis (RAY kihs) cylinder that runs down the center of a feather (574)

radial symmetry arrangement of body parts around a central point (443)

radiating canal in paramecia, one of many pipelike structures that collect excess water (322)

radicle embryonic root of a plant (408–409)

radioactive isotope atom with an unstable nucleus whose known, constant rate of decay can be used to determine the age of a fossil (232)

radula (RAJ oo luh) in some mollusks, toothed organ that tears or scrapes loose bits of food (473)

Ramapithecus species of small, monkeylike primates that lived between 17 million and 8 million years ago (257)

ray-finned fish one of a group of bony fishes with fins supported by a number of long bones, or rays (529)

RDP ribulose diphosphate; five-carbon sugar that is the major ingredient in the second phase of photosynthesis (120)

receptacle base of a flower (402)

receptor site location on a cell wall and on the tail of a virus at which a chemical bond forms for the purpose of adsorption (285)

recessive a gene that is hidden by another gene in a pair (154)

recombinant DNA new strand of DNA formed when the DNA of two different species is combined (221)

rectum muscular opening of the large intestine, which controls the elimination of feces (650)

red blood cell that part of the blood that transports respiratory gases; also called erythrocyte or red corpuscle (656)

red tides toxic growths of dinoflagellates (332)

reflex arc pathway traveled by impulses involved in involuntary movements (699)

regenerate to grow a new part to replace one that is lost (446)

regeneration a type of asexual reproduction in which a complete animal develops from a body part (137)

remote sensing gathering and recording of information about objects and phenomena from a great distance by systems that are not in contact with them (806)

renal artery vessel that carries blood to the kidney (547)

renal pelvis central cavity of the kidney (682)

renal tubule structure in the kidney responsible for filtering wastes from the blood and for regulating the water and mineral composition of the blood (682)

renewable resources any elements of nature that can be reused or replenished (809)

replication (rehp luh KAY shuhn) duplication of a chromosome (136, 184)

reproduction process of producing offspring (8)

reproductive system a group of organs that provide a means of producing offspring (615)

resolution increase in the visible detail of an object (37)

respiration process in which the body takes in oxygen, uses it to produce energy, and eliminates some waste products (675)

respiratory system group of organs that take in oxygen and eliminate carbon dioxide and water (614)

response any behavior of an organism that results from a stimulus (8)

reticular formation complex network of nerve fibers in the brain stem and thalamus that regulates reactions during consciousness and sleep (697)

retina in the vertebrate eye, innermost, light-sensitive layer (708)

Rh factor antigen present in the blood of about 85 percent of the U.S. population (658)

rhizoid in molds, short extension of a stolon that anchors the mold to its food supply (345); in bryophytes, rootlike structure that anchors the plant and absorbs water but does not channel it to the plant (358)

rhizome in some vascular plants, underground stem that anchors plant and absorbs water and nutrients from soil (369)

ribose five-carbon sugar present in RNA (186)

ribosomal RNA also rRNA; type of RNA present in ribosomes; helps bind messenger RNA and transfer RNA during protein synthesis (186)

ribosome (RY buh sohm) in a cell, tiny knoblike organelle in which protein is manufactured (87)

rickettsia (rih KEHT see uh) type of bacterium that can live only inside other cells (298)

RNA ribonucleic (ry boh noo KLEE ihk) acid, one of two forms of nucleic acid; "reads" instructions for cellular activity and carries them out (73, 186)

rod in the vertebrate eye, extremely light-sensitive receptor cell that generates a nerve impulse (708)

root cap layer of protective cells around the apical meristem (386)

root crop any of the various edible roots and underground stems (432)

root hair tiny outgrowth that increases the surface area of a root (385)

round window membrane-covered opening in the cochlea that maintains pressure in the inner ear (711)

ruminant mammal with a four-chambered stomach (599)

S

saccule (SAK yool) hair-lined section of the vestibule that helps maintain balance (712)

saliva glandular secretion, made up of mucus and the enzyme ptyalin, which moistens food in the mouth (644)

salivary glands three pairs of glands in the mouth that produce saliva to aid digestion (508, 644)

saprophyte (SAP ruh fyt) organism that feeds on dead organisms (302)

sarcomere (SAHR koh mihr) one of many tiny units that comprise the bands in striated muscle (627)

savannah grassland located in a tropical or subtropical area (785)

saw-tooth curve pattern formed by the periodic growth and decline of a population (797)

scale in fishes, one of many thin, overlapping segments covering the skin (531)

scanning electron microscope (SEM) type of microscope that sends a beam of electrons across the specimen from left to right (39)

scavenger animal that feeds only on dead organisms (766)

Schwann cell one of many specialized cells that comprise the myelin sheath (689–690)

science body of factual knowledge that exists about the world and the method of study used to arrive at that knowledge (24)

scientific method a logical, organized method of study through which scientists establish principles (25)

scientific name in taxonomy, short, standard name for an organism (272)

scientific principle best current explanation of how a part of nature works (27)

sclera (SKLIHR uh) in the vertebrate eye, tough, elastic, connective tissue that forms the outermost layer (707)

scolex (SKOH lehks) head of a tapeworm (459)

S-curve pattern formed by the combination of lag, exponential, and stabilization phases (796)

scute (SKYOOT) in turtles and tortoises, one of several plates covering the shell (561)

secondary growth cumulative growth of a plant's xylem and phloem tissues (377)

secondary succession establishment of a community in an area formerly inhabited by a different community (781)

second polar body in meiosis II, the smaller of the two cells resulting from the unequal cell division that occurs in female animals (145)

sedimentary formed in layers (231)

seed coat hard, protective covering of a seed embryo (374)

selectively permeable characteristic of materials that allow only certain substances to pass through them (102)

self-pollination reproductive process in which fertilization is carried on in a single plant (151)

semen thick, milky liquid containing sperm and glandular secretions (732)

semicircular canal one of three fluid-filled tubes in the inner ear that help maintain balance (711)

sensory neuron nerve cell that receives a stimulus and transmits it to the central nervous system (689)

sensory seta in arachnids, one of many tiny hairs covering the organism, which detect pressure and movement (492)

sepal one of the leaflike structures that grows out from the base of a flower and encloses the bud before it blooms; collectively, they form the calyx (402)

seral community temporary community that alters the environment, preventing its own regeneration (777)

sessile characteristic of an organism that remains anchored in one place for most of its life (321, 442)

setae in an earthworm, pairs of bristles that aid in locomotion (465)

sex chromosome chromosome that determines the sex of an individual (172)

sex-influenced trait one generally associated with one sex but produced by genes carried on autosomes (205–206)

sex-linked trait trait that is determined by alleles carried only on a sex chromosome (173)

sexual reproduction formation of a new individual from the union of two cells (135)

short-day plant one that flowers during spring or fall, when nights are as long as or longer than its critical dark period (418)

sickle-cell disease biochemical disorder caused by recessive alleles involving the distortion of red blood cells into sickle shapes and eventual destruction of vital organs (202–203)

sieve-tube member phloem cell through which food passes freely (384)

simple eye in arachnids, structure that detects light but cannot form images (491)

simple leaf one with a single, undivided blade (393)

simple microscope single lens, or curved piece of glass, that magnifies objects (37)

sinoatrial (sy noh AY tree uhl) **node** also SA node; in vertebrates, small region of muscle that triggers the heartbeat (663)

skeletal muscle group of specialized cells that moves bones (627)

skeletal system group of organs that move, support, and protect the body (614)

slime mold organism with a mixture of traits found in protozoa and fungi (345)

small intestine three-sectioned tube where most chemical digestion and absorption occurs (543, 647)

smog air pollution made up of smoke, gas, and fog (817)

smooth muscle group of spindle-shaped cells that provides for the involuntary movements of digestive, respiratory, and circulatory organs (627)

soft palate muscle tissue located behind the hard palate (645)

solar energy energy derived from sunlight (815)

solute substance that dissolves in a solvent (53, 99)

solution uniform mixture of the particles of one substance in another (53, 99)

solvent substance that can dissolve another substance (53, 99)

somatic (soh MAT ihk) **cell** any body cell except one that gives rise to gametes (138)

somatic mutation change in a body cell (193)

somatic nervous system system that transmits impulses to and from skeletal muscles (699)

sori clusters of sporangia (372)

spawn in fishes, to shed eggs (533)

speciation formation of a new species (250)

species biological group whose members resemble one another and interbreed to produce fertile offspring (8, 241)

sperm in sexual reproduction, male sex cell (137)

spherical symmetry round arrangement of body parts, having no back, front, left side, or right side (443)

spicule (SPIHK yool) one of many interlocking spikes of calcium or silicon that form the skeletons of some sponges (446)

spinal cord column of nerve tissue from the brain through the spinal column, which links the brain with nerves to all parts of the body and controls involuntary movements (698)

spindle fiber temporary microtubule formed to help move chromosomes during cell division (90)

spinneret in arachnids, web-spinning organ (492)

spiracle (SPY ruh kuhl) in arachnids, slit in the exoskeleton that regulates the flow of air to the tracheae (491); in some fishes, opening on top of the head, through which water enters the body and is moved to the gills (527)

spirillus (spy RIHL uhs) corkscrew-shaped bacterium (299)

spirochete (SPY ruh keet) type of bacterium characterized by its large size (298)

spongin flexible protein that forms the skeleton of some sponges (446)

spontaneous generation unsupported theory that some organisms form from nonliving materials (28)

sporangiophore in molds, specialized hypha that produces spores (345)

sporangium in molds, case or structure that produces spores (345)

spore asexual reproductive cell (136)

sporophyll in club mosses, narrow clublike structure on which spores are produced (370)

sporophyte (SPAWR uh fyt) organism with diploid (2n) chromosomes that produces spores (n) by meiosis (334, 358)

stabilization phase period when population size levels off (796)

stabilizing selection natural selection that reduces variation by eliminating extremes (247)

stain dye that colors certain tissues for easier viewing under a microscope (40)

stamen male part of a flower (401)

stapes (STAY peez) one of three tiny bones in the middle ear (711)

states of matter descriptions of molecular arrangement in matter: solid, liquid, or gas (47)

statistics (stuh TIHS tihks) mathematical method of evaluating numerical data (26)

stereomicroscope type of microscope with ocular and objective lenses for each eye (37)

sternum breast bone (575)

steroid a lipid that helps determine the synthesis of certain proteins (726)

stigma in flowers, tip of the style, which produces a sticky substance that traps pollen (402)

stimulant drug that causes the central nervous system to speed up body processes (748)

stimulus any condition to which an organism can react (8)

stipe in mushrooms, stemlike structure that supports the button (346)

stolon in molds, hypha that branches out along the surface of the food supply (345)

stomach organ where mechanical and chemical digestion take place (646)

stomata (stoh MAH tah) in plants, small pores that allow gas exchange (355)

strip-cropping method of reducing erosion by planting alternating rows of grass or clover between grain crops (811)

stroma dense fluid between the grana and the outer membranes of a chloroplast (116)

structural formula figure that shows how atoms are bonded in a molecule (70)

style in flowers, slender middle portion of a pistil (402)

subatomic particle component of an atom (48)

substrate molecule upon which an enzyme acts (72)

superior vena cava (VEE nuh KAY vuh) blood vessel that brings blood from the upper regions of the body to the heart (661)

suspension temporary mixture of particles (54)

swarm large mass of insects (506)

swim bladder in bony fishes, gas-filled organ in the coelom that acts as a float (531)

swimmeret in crustaceans, one of five pairs of appendages used in swimming (494)

symbiosis (sihm by OH sihs) permanent, close relationship between two organisms of different species that benefits at least one of them (304, 315, 770)

symmetry arrangement of body parts around a center point or line (443)

synapse place where an impulse crosses from the axon of one neuron to the dendrite of another (692)

synapsis (sih NAP sihs) in meiosis, process during which the homologous chromosomes come together (143)

synovial (sih NOH vee uhl) **fluid** lubricant of body joints (626)

syrinx (SIHR ihnks) pair of vibrating membranes at the bottom of a bird's trachea, which produces a bird's call (577)

system group of organs that cooperate in a series of related functions (13)

systemic circulation blood moving between the heart and the rest of the body other than the lungs (545)

T

tadpole larval form of a frog (548)

talon sharp, curving claw of a bird (573)

taproot mature, thick primary root of many dicots (385)

target specific tissue or organ upon which a given hormone acts (719)

taste bud cluster of specialized hair cells in the tongue that act as chemical receptors for taste (713)

taxonomy science of grouping organisms on the basis of their similarities (271)

telophase the fourth stage of mitosis, during which a nuclear envelope forms around each set of chromosomes (141–142)

telson in some crustaceans, middle part of the tail, used in locomotion (494)

temperate phage bacteriophage that undergoes a long period of inactivity in a host cell (287)

tendon band of inelastic connective tissue that attaches muscle to bone (627)

tentacle long, flexible appendage (448)

terracing method of converting a hillside into broad, flat steps to prevent water loss (811)

territoriality practice of claiming and defending a particular area against other members of the same species (592)

testcross procedure that determines the genotype of an individual whose phenotype is dominant (159)

testis sperm-producing organ (449, 547, 731)

testosterone (tehs TAHS tuh rohn) hormone that stimulates development of certain male characteristics (732)

tetrad (TEHT rad) in meiosis, structure formed when chromosomes intertwine (144)

thalamus (THAL uh muhs) two-part organ at either side of the vertebrate brain that acts as a relay center for impulses (695)

thallus unspecialized, multicellular body of algae (329)

theory a scientific explanation of known facts (27)

thermal inversion condition in which pollutants are trapped in cool air under a warm air mass (817)

thermal pollution killing aquatic life by raising water temperature (819)

thigmotropism plant's curving response to contact with a solid object (423)

thoracic (thaw RAS ihk) **cavity** portion of the coelom above the diaphragm that contains the heart, lungs, and esophagus (612)

thymine (THY meen) carbon-nitrogen compound, one of the four bases of DNA (182)

thyroid gland in vertebrates, organ on the trachea that controls metabolic activities, including the production of protein and ATP (719)

tissue group of similar cells that perform a common function (13)

tolerance condition in which increasingly larger doses of a drug are needed to produce the same effect (745)

torsion in gastropods, twisting of the body during larval development (477)

trachea (TRAY kee uh) in some terrestrial arthropods, tube that carries air directly to the cells (491); in terrestrial vertebrates, the windpipe (677)

tracheid (TRAY kee ihd) long, tapered conducting cell found in gymnosperms (383)

tracheophyte (TRAY kee uh fyt) plant that gets food and nutrients from a vascular system (367)

trait characteristic that is transferred from one generation to the next (9)

transcription process by which messenger RNA is copied from DNA molecules (186)

transducer structure that transforms one form of energy into another (705)

transduction (trans DUHK shuhn) transference by a virus of genetic information between host cells (287)

transfer RNA also tRNA; type of RNA that picks up individual amino acids in the cytoplasm and carries them to the ribosomes (186)

transformation transfer of genetic material from a dead bacterium to a live one (304)

translation process during which ribosomes attach to messenger RNA to form protein (188)

translocation transfer of food produced by photosynthesis from the leaves to the rest of the plant (393)

transmission electron microscope (TEM) type of microscope that sends a beam of electrons through the specimen (39)

transpiration water loss in plants (395)

trichocyst (TRIHK uh sihst) in paramecia, tiny poisonous hair discharged by the organism under attack (323)

trimester one of three three-month periods in human pregnancy (738)

trisomy (try SOH mee) condition in which an individual has an extra chromosome (207)

trochophore (TRAHK uh fawr) mollusk larva (473)

trophic (TRAHF ihk) **level** feeding level within an ecosystem (766)

tropical rain forest biome characterized by heavy rainfall, rapid plant growth, and large numbers of insects and other invertebrates (785–786)

tropic hormone chemical substance that affects the secretions of other glands (722)

tropism response in which an organism moves towards or away from the stimulus (422)

true fungus larger of the two main groups of fungi (345)

tube cell larger cell in a pollen grain (403)

tube foot in echinoderms, one of several hollow cylinders tipped with suckers, on the underside of the arms (480)

tubular reabsorption kidney's return of glucose, water, and other vital materials to the blood (682–683)

tubular secretion process by which kidneys remove certain substances from the blood (683)

tundra biome characterized by bleak terrain, little vegetation, and a cold, dry climate (782)

turgor (TUHR guhr) pressure that builds in a plant cell as a result of osmosis (104)

Turner syndrome genetic disorder in individuals with only one X, and no other, sex chromosomes (207)

tympanic cavity air-filled space inside the skull bone that houses the middle ear (710)

tympanic membrane in frogs, circular structure located behind the eye, which receives sounds and covers the internal ear (541)

tympanum in insects, thin, flexible membrane covering the organ of hearing; an eardrum (507)

U

ultrasound testing technique using sound waves to detect possible abnormalities in a fetus (209–210)

umbilical cord in mammals, blood-rich connector of the fetus to the uterine wall (737)

ungulate land mammal that has hoofs (599)

unicellular (yoo nuh SEHL yoo luhr) **organism** organism composed of a single cell (3)

univalve class of mollusks with a single shell (476)

uracil (YOOR uh sihl) pyrimidine base present in RNA (186)

ureter long, narrow tube that transports urine from the kidney to the urinary bladder (683)

urethra tube through which urine is expelled from the body (684)

urinary bladder sac of smooth muscle that collects and holds urine until it is expelled (683–684)

urinary system that part of the excretory system that eliminates liquid wastes from the body (615)

urine amber-colored liquid consisting of urea, water, and other metabolic wastes (683)

uropod (YOOR uh pahd) in crustaceans, fused pair of appendages that functions as a flipper (494)

uterus in mammals, female reproductive organ that houses a developing fetus (597, 732)

utricle (YOO trih kuhl) in the ear, hair-lined section of the vestibule that helps maintain balance (712)

V

vaccine solution of weakened viruses that stimulates the production of antibodies (290)

vacuole in a cell, nonliving, bubblelike storage structure (89)

vane in a feather, structure comprised of several barbs hooked together by barbules (574)

variable condition that varies in an experiment (26)

variation set of differences (10)

variety subdivision of a species (275)

vascular bundle grouping of vascular tissue in young plant stems (389)

vascular cambium inner layer of meristematic cells in the vascular cylinder, responsible for xylem and phloem production (386)

vascular cylinder center of a root, made up of xylem, phloem, and meristematic cells, responsible for growth (386)

vascular plant one that gets food and nutrients through a network of vessels (357)

vascular system network of interconnected tubes and vessels that carry water, nutrients, and the products of photosynthesis throughout the plant (367)

vascular tissue in plants, internal system of interconnected tubes and vessels that transport food and nutrients (357)

vegetative body parts of the plant—roots, stems, and leaves—that carry out photosynthesis and normal growth (374)

vegetative propagation type of asexual reproduction found in plants, in which offspring separate from the parent plant to become individual plants (137, 410)

vein vessel that carries blood to the heart (661)

ventral pertaining to the lower side of an organism (444)

ventricle (VEHN trih kuhl) one of two lower sections of the heart (660)

vernalization (vuhr nuhl eye ZAY shuhn) period of chilling required before a seed can germinate (419)

vertebra one of the bony parts of a vertebral column (521)

vertebral column backbone of a vertebrate (521)

vertebrate animal with a backbone or spine (444)

vessel member short, tubelike conducting cell found in angiosperms (383)

vestibule bony chamber between the semicircular canals and the cochlea of the ear (711)

vestigial (vehs TIHJ ee uhl) **structure** body part reduced in size and with no apparent function (236)

villus (VIHL uhs) one of many tiny projections in the mucous lining of the small intestine, where absorption occurs (648)

virion (VY ree ahn) new complete virus particle formed by viral nucleic acid and protein during a virus's reproductive cycle (286)

viroid (VY royd) smallest known disease-causing agent, consisting of a single strand of RNA (283)

virulence ability of a virus to cause disease (289)

virus microscopic life form that reproduces only inside a living cell (282)

visceral mass in mollusks, digestive, excretory, and reproductive organs (473)

visible spectrum array of colors formed when white light passes through a prism (115)

vital stain dye that highlights structures in living tissues (40)

vitamin organic substance that assists enzymes (641)

vitreous (VIHT ree uhs) **humor** in the vertebrate eye, clear fluid behind the lens (707)

viviparous characteristic of an animal that gives birth to live young which develop within the mother's body (589)

vocal cord one of a pair of ligaments stretched across the larynx, where sound originates (677)

vomerine teeth in amphibians, two teeth in the roof of the mouth that aid in holding prey (543)

W

walking leg in crustaceans, any one of the four pairs of appendages used for locomotion (494)

watershed area of land from which water drains into a particular lake or stream (810)

water-vascular system in echinoderms, method of locomotion using water pressure to create a ''skeleton'' for muscles to contract against (481)

white blood cell that part of the blood that fights viruses, bacteria, and other foreign organisms (656)

windbreak row of trees that serves as a wind barrier to prevent soil erosion (811)

withdrawal painful reaction to the cessation of drug use (745)

woody plant plant in which xylem and phloem produce cumulative layers of tissue (377)

X

xanthophyll (ZAN thuh fihl) yellow pigment present in most green plants (116)

xylem (ZY luhm) vascular tissue that transports water and minerals absorbed by the roots to the aboveground portion of the plant (367)

Y

yolk large amount of food stored in egg cells for a developing embryo (556)

yolk sac food container attached to the abdomen of the embryo (555–556)

Z

zoology study of animals (20)

zooplankton (zoo uh PLANK tuhn) marine protozoa and other microscopic heterotrophs (315)

zygospore thickened, protective cell wall of a zygote (334)

zygote (ZY goht) in sexual reproduction, cell formed when a sperm and an ovum combine (137, 461, 735)

Index

Boldface numbers refer to an illustration on that page.

Computers, 36, 41–42, 277, 617
Concentration, 53–54
Concentration gradient, 100, 103, 106–107
Condensation reactions, 69, 71, 72
Condor, California, 572
Conductive deafness, 712
Conductors, 689
Cones, of eye, 609, 708, **708**
Cone scales, 374
Conifers, **374,** 374–375, 395, 783
Conjugation, 304, **322,** 322–323, 334, **334**
Connective tissue, 612, 613, **613,** 632, 659
Conservation, law of, 60
Conservation of natural resources, 809–815
Constipation, 650
Constrictors, 564, **564,** 565
Consumers, 766–768
Contour plowing, 811, **811**
Contractile vacuoles, 104, 319, 322
Controlled breeding, 217–219
Convergent evolution, **251,** 251–252, 597
Convolutions, on brain, 695
Copepods, 767
Coral reefs, 452
Corals, 442, 448, 451–452
Cork cambium, 387, 390
Corn, 431–432
Cornea, 707, 709
Corpus callosum, 694–695, 696
Corpus luteum, 733–734
Corolla, 402
Coronary circulation, 664
Coronary sinus, 664
Cortex
  of kidney, 682
  plant, 384, 386
Corticoids, 720, 726
Cortisol, 720
Cotyledons, 374, 376–377, 408–409
Cotylosaurs, 556, **556**
Courtship, in birds, 580
Cousteau, Jacques-Yves, 451, **451**
Covalent bonds, 56, 57, **57**
Cowper's glands, 732
Crabs, 490, **490,** 494, 496, **496**
Cramp, 631
Cranial cavity, 259, 612
Cranial nerves, 699
Crayfish, 490, 494–496, **495**
Cretinism, 719
Crick, Francis H. C., 182–185
Crime laboratory technician, 58, **58**
Critical dark period, 418
Crocodiles, 559, 560, 561, **561**

Crocodilia, 560, 561
Cro-Magnons, 262, **262**
Crop rotation, 432–433, 811
Cross breeding, 433, 434
Crossing over, 174–176, **175**
Cross-pollination, 152–153, 405
Crustaceans (Crustacea), 490, 494–496
Cuboidal epithelium, 612, **612**
Cushing's disease, 720
Cuticle
  in plants, 355–357, **357,** 367
  in worms, 458
Cutin, 393, **394**
Cuttings, 411
Cuttlefish, 474, 478, 479
Cycads, 374, 375–376, **376**
Cyclic AMP, 726
Cyclosporine, 616, 668
Cyst, 320, 459, 464
Cytochrome c, 235
Cytokinesis, 139–141, **140, 141,** 142, **142,** 145
Cytokinins, 419, 421
Cytolysis, 104
Cytoplasm, 83, **83,** 86–91. *See also* Cytokinesis
Cytoplasmic inclusions, 90
Cytoplasmic streaming, 86, *table* 93
Cytosine, **182,** 182–185

D

Dark reactions, 117, 119–120, **120, 121**
Darwin, Charles, 240, **241,** 242–245, 250, 273
Day-neutral plants, 418
DDT, 572, 813
Death rate, 799, 801, 802
Decay elements, 232
Deciduous forests, 784, **784**
Deciduous trees, 395
Decomposers, 764, 766, 768
Defense reaction, 323
Dehydration synthesis. *See* Condensation reactions
Deletions, 192, 193, **193**
Delirium tremens, 747
Dendrites, 689, **689,** 692, 694
Denitrification, 765
Deoxyribonucleic acid. *See* DNA
Deoxyribose, 183
Depressants, 746, 750
Depression, 693
Dermis, 632, 633, **633**
Desalination, 810
Deserts, 786–787, **787**
Development of organisms, 6–7, **7**
Diabetes mellitus, 223, 721–722

Dialysis, 684
Diaphragm, 589
  in humans, 612, **675,** 675–676
Diastolic pressure, 662
Diatomaceous earth, 332
Diatoms, 332, **332**
Dicots, 376–377, **377**
  leaves in, **393**
  root system in, 385, **385**
  seedling development in, 409, **409**
  vascular bundles in, 389, **389,** 390
Dietitian, 649, **649**
Differentiation, 383
Diffusion, 99–100, **100,** 106
  facilitated, 106
  through cell membrane, 102–105
Digestive system
  of birds, 576, **577**
  of bony fishes, 531
  of crayfish, 495
  of earthworms, 466
  of frogs, **542,** 542–543
  human, 614, 644–650, **645,** *table* 700
  of planarians, 458
  of vertebrates, *table* 522
Digestive tract. *See* Alimentary canal
Digitalis, 436, **436**
Dihybrid cross, **161,** 161–163
Dinoflagellates, 332
Dinosaurs, 554, 556–558, **558,** 571, 589
Diphtheria, 305–306
Diplococci, 299
Diplodocus, 556–557
Diploid number, 143, 334
Directional selection, 246–247, **247**
Disaccharides, 69–70, **70,** 640, 648
Disease. *See also* Parasitism
  and alcoholism, 747–748
  and bacteria, 305–306, 666
  defenses against, 666–668
  and smoking, 752–753
  and viruses, 288–290, 666
Disruptive selection, 247, **247**
Dissociation, 53, **53,** 57, 67–68
Diurnal organisms, 771
Divergent evolution, **250,** 250–251
DNA, 73, 84, 181–190
  of bacteria, 181–182, 222–223, 300, 303
  discovery of, **181–185,** 181–185
  and meiosis, 143, 144, 185
  and mitosis, 138, 184
  and polyploidy, 220–221
  of prokaryotes, 92
  and protein synthesis, 186–190, **188–189**
  recombinant, 214–215, 221–223, 290

of planarians, 458
of spiders, 492
of vertebrates, *table* 522
Excurrent siphon, 474, **475**
Exergonic reactions, **59,** 59–60
Exobiologist, 15, **15**
Exocrine glands, 719
Exocytosis, 107, 108, 319–320
Exoskeleton, 489
Experiment, 26
Exponential phase of growth, 796, **796, 800,** 801, 802
Extensor, 631
Extinct species, 812
Eyes
    compound, 496, 503, **506,** 507
    human, 705–706, **707,** 707–709
    of primates, 609
Eyespot, 318, **318,** 457

F

F_1. *See* First filial generation
F_2. *See* Second filial generation
Facilitated diffusion, 106
Factors, 153, 156. *See also* Genes
Facultative anaerobes, 302
Falcon, Peregrine, **572**
Fallopian tubes, 732, 735, 736
Family, as classification, 274
Farsightedness, 709, **709**
Fat bodies, 547
Fats
    digestion of, 648, 649
    as nutrients, 639, 640, 641, **641**
Fat-soluble vitamins, 641–643
Fat tissue, 613, **613**
Fatty acids, 71
Feathers, of birds, 571, 572, 574–575, **575**
Female reproductive system, 732–734, **733, 734**
Femur, 507, 543, 576, **624,** 625
Fermentation, 123–125, **124,** 348
Ferns, **371,** 371–373, **373**
Fertilization, 137
    in flowering plants, 405
    in humans, 735, **735,** 736, 737
Fetoscopy, 210
Fetus, 737–739, **738, 739**
Feulgen, Robert, 183
Fibers, plants in production of, 436
Fibrillation, 663
Fibrin, 657, 664
Fibrinogen, 657
Fibrous proteins, 72
Fibrous root system, 385
Fibula, 576
Fiddlehead, 372
Filament, 401

Filter feeders, 474
Filtrate, 682, 683
Fire algae, 329, 331, 332
First filial generation, 153, **153**
First-order consumers. *See* Primary consumers
Fishes, 518–534
    bony, 520, **528,** 528–534, **529, 530, 531**
    cartilaginous, 520, **525,** 525–527, **527**
    jawless, 520, **523,** 523–524
Flagella, 91, **91,** *table* 93, 300, 316. *See also* Flagellates
Flagellates (Mastigophora), 316, 317–318
Flame cells, 458
Flat bones, 624, 625
Flatworms, 457–461, **459**
Fleming, Alexander, 306
Flemming, Walther, 169
Flexor, 631
Flowering plants. *See* Angiosperms
Flowers
    and gamete formation, 403–404
    kinds of, 402–403
    and photoperiodism, 418
    pollination of, 405–406
    structure of, 377, **401,** 401–402, **402**
Fluid mosaic model, 102, **102**
Flukes (Trematoda), 457, 458–459, **459**
Follicle. *See also* Hair follicles
    of feather, 574
    of ovum, 733
Follicle-stimulating hormone. *See* FSH
Fontanels, 625
Food, 638–643
    and air passageway, 677
    basic groups of, 639, **639**
    plants as, 430–434
    and population growth, 796, 797
Food chains, 767, **767**
Food-storing roots, 410, 411
Food vacuole, 319
Food webs, **767,** 767–768
Foramen magnum, 259
Forestry, 435, **435**
Forests
    boreal, 783, **783**
    deciduous, 784, **784**
    and ecological succession, 777–781, **779**
    as ecosystems, 759
    tropical rain, 771, **785,** 785–786
Forewings, 507
Fossil fuels, 370, 814, 816

Fossils, 64, **64,** 231–233, **231,** 234, 257, **258, 260, 261,** 262, **262,** 268–269, 473
Foxglove, 436, **436**
Fragmentation, 309, 334, 361, 447
Franklin, Rosalind, 183
Fraternal twins, 200, **200**
Freshwater biome, 788, 789–790
Frogs, **540,** 540–548, **541**
    life cycle of, 547–548, **548**
    structure of, 541–547, **542, 543, 545, 546, 547**
Frontal lobe, 695
Fructose, 69, **69,** 640
Fruit fly, in genetics, 172–176
Fruits, 377, 407, **407,** 418
FSH, 722, *table* 723, 724, 732–734
Fucoxanthins, 331, 335
Fungi, 342–349, **343, 345, 347, 349**
    classification of, *table* 275, 276, *table* 276, 824–825
    as decomposers, 766
    in lichens, 344, 771

G

Galactose, 640
Galapagos Islands, 242, 243, 250
Gall bladder, 648
Gametes
    and alternation of generations, 334–335
    of angiosperms, 403–404
    chromosome number in, 143, 207
    germ mutations in, 193
    human, 731, 735
Gametophyte, 334–335. *See also* Alternation of generations
Gamma globulin, 290
Ganglia, 457, 508, 698
Gases, 47–48, 49
    exchange of, in blood, 678, **678**
Gastropods (Gastropoda), 476–478
Gastrula, 461
Gastrulation, 735
Gavials, 561, **561**
Geckos, 562, **562**
Gemmae, 361
Gemmules, 447
Gene frequency, 245–246
Gene mutation, 191
Gene pool, 245. *See also* Genetic equilibrium
Gene therapy, **214,** 214–215, **215**
Genes, 151, 153. *See also* Alleles; Genetics; Mutations
    behavior of, **174,** 174–176, **175**
    and chromosomes, relationship between, 169, 171, 172–173
    nucleotide sequence in, 184, 190

and trait inheritance, 202–207
variations produced by, 244
Genetic counseling, 208, 209
Genetic counselor, 209, **209**
Genetic disorders
detection of, 205–210, **206**
therapy for, 214–215
Genetic drift, 248–249
Genetic engineering, 214–215, 221–223, 433, 434
Genetic equilibrium, 246–249
Genetic isolation, 249, **249,** 250
Genetic recombination, 244, 304
Genetics, *table* 20, 151
applied, 216–223
chemical bases of, 180–194
and chromosomes, 168–176
and evolution, 245–249
human, 198–210
Mendel's principles of, 151–156, 161–164
and probability, 157–160
Genotype, 156
Genotypic ratio, 158, 160–163
Genus, 271, 272, 274
Geographical isolation, 249
Geologic time scale, 233, *table* 234
Geothermal power, 814–815
Geotropism, 422–423
Germination, 408–409, **409**
Germ layers, 460–461
Germ mutations, 193
Gestation, 598, 735–738, 739
GH. *See* Somatotropin
Giant tube worm (Pogonophora), 468
Gibberellins, 419, 421
Gibbons, 603, **603**
Gigantism, 724
Gila monster, 563
Gills
of clams, 474
of crayfish, 495
of fishes, 523, 525, 530, 531, **531,** 539
of mushrooms, 346
Gill slits, 519, **519, 527**
Ginkgoes, 374, 376, **376**
Giraffes, and evolution, 241, 246–247, **247**
Gizzard, 466, 576
Glands, 719–726, 732
Glaucoma, 709
Gliding joints, 626
Globular proteins, 72
Glomerular filtration, 682
Glomerulus, 682, **682**
Glottis, 677
Glucose, **69,** 69–71
in blood, 720–722

breakdown of, 122–128, 763–764
in foods, 640
production of, 71, 114, 115–121
Glycerol, 71
Glycogen, 70–71
Glycolysis, 122–123, **124,** 125, 128, **128**
Gnetophyta, 374
GnRH, 734
Goiter, 719
Golden algae (Chrysophyta), 329, 331–332
Goldson, Alfred Lloyd, 136, **136**
Golgi bodies, 88, **88,** 90, 93, 142
Gonadotropin-releasing hormone, 734
Gonads, 722, 731, 732
Gorillas, 603, **603**
Grafting, 219, 411
Grain amaranth, 434
Gram stain, 303
Grana, 116, **117**
Grasshoppers (Orthoptera), **506,** 506–508, **507,** 509
Grasslands, 785
Gray matter, 694, 695, 696, 698
Green algae, **328,** 331, 333–335, **334, 335,** 354–356, 367
Green glands, 495
Green Revolution, 433
Griffith, Frank, 181–182, 304
Ground tissue, 383, 384, 389
Growth
of organisms, **6,** 6–7
of plants, 416–421, **417**
of populations, 795–799
Growth hormone. *See* Somatotropin
Growth rate of population, 799
Growth rings, 390, 391
Guanine, 90, **182,** 182–185
Guard cells, 355, 394
Guayule, 436–437
Guttation, 418, **418**
Gymnosperms, 374–376, **375**

H _____

Habitats, 761–762
destruction of, **19,** 812–813
Hagfishes, 523, 524, **524**
Hair, human, 632, 633–634, 714
Hair cells, 705, 711–712, **712,** 713
Hair follicles, 633–634, *table* 700
Hallucinogens, 748, 750
Hammer. *See* Malleus
Haploid number, 143–145, 334
Hard palate, 645
Hardy, Godfrey, 245–246
Hardy-Weinberg equilibrium. *See* Genetic equilibrium
Hardy-Weinberg principle, 245–246

Hares, 602
Haversian canals, 625, **625**
Head-footed mollusks. *See* Cephalopods
Hearing, 705–706, 710–712
Heart. *See also* Circulatory system; Heart, human
in birds, 577
in fishes, 531
in frogs, 545, **545**
in grasshoppers, 508
in mammals, 590
in reptiles, 559
Heart, human, 659–660
and autonomic nervous system, *table* 700
cardiac muscle in, 628
and circulation, **661,** 661–665, **664**
fetal, 739–740
structure of, **659,** 659–660
Heart attack, 662, 664
Heartworms, 463
Heat. *See also* Temperature
and chemical reactions, 59–60
sensing of, 705, 714
Heat capacity of water, 67
Helix shape of DNA molecule, 183, 184
Helmont, Jean van, 28
Hemodialysis, 684
Hemoglobin, 72, 202–203, 656, 678, 680
Hemophilia, 205
Hemotoxins, 566
Hepatic portal circulation, 665
Herbaceous plants, 377, 389–390
Herbivores, 766, 767, **770**
mammal, **595,** 599
reptile, 556
Heredity, 150. *See also* Genetics
Hermaphrodites, 446, 458, 460, 467, 524
Heroin, 751
Herpes simplex, 289
Hershey, Alfred, 182
Heterotrophs, 123, 275, 300, 315
Heterozygous individuals, 154, 199, 219
Hibernation, 546, 594, 772
Hinge joints, 626
Holdfast, 336
Homeostasis, 10–11, 682, 706
Home range, 592–593
Hominids, 257
Homo erectus, 260, **260**
Homo habilis, 259
Homologous chromosomes, 143–144, **169,** 169–171, 175

Koalas, **191,** 596
Komodo dragon, 563
Krebs cycle, **126,** 126–127, 128, **128**

L

Labor, 738–739
Laboratory technician, 42, **42**
Lacrimal gland, 707
Lactase, 648
Lacteals, 649
Lactic acid fermentation, 124–125
Lactose, 70
Lagomorphs (Lagomorpha), 602
Lag phase, 795, **796,** 800
Lakes, 788, **789,** 789–790
Lamarck, Jean Baptiste de, 241, **241,**
    242
Laminaria, 336
Laminarin, 335
Lampreys, **523,** 523–524
Lancelets, 519, **519**
Land
    animal adaptation to, 539
    plant adaptation to, **355,** 355–356
Landfill, 819
Landsat, 806–807
Landsteiner, Karl, 203
Language, 611
Large intestine, **542,** 543, **645,** 650
Larva
    of fishes, 524
    of flukes, 459, **459**
    of insects, **509,** 509–510, 512
    of sponges, 447
    trochophore, 473, **474,** 475
Larynx, 646, **676,** 677
Lateral bugs, 392, **392,** 419
Lateral line system, 526, **526,** 532
Laterite soils, 786
Layering, 412
Leaf base, 393, **393**
Leakey, Louis, 259, 264
Leakey, Mary Nicol, 258, 259, 264,
    264
Leakey, Richard, 258, 259, 264
Leaves, 393–396
    function of, 367, 382, 393
    and geotropism, 422–423, **423**
    structure of, **393,** 393–394, **394**
    and vegetative propagation, 410,
        411, **411**
    and water loss, 395, **395**
Leeuwenhoek, Anton van, 38
Leg scales, of birds, 576
Legumes, 432, **432,** 764
Lemaitre, Georges, 229
Lemurs, 603
Lens of eye, **707,** 708
Lethal mutations, 191, 192

Leucoplasts, 89
Leukocytes. *See* White blood cells
Levene, Phoebus A., 182, 183
Lewin, Roger, 308
LH, 722, *table* 723, 724, 732, 733,
    734
Lichens, 344, **344,** 771
Life, 2
    change in forms of, 231–236
    origin of, 229–230
Life span, 7–8, **8**
Ligaments, 626
Light
    as abiotic factor, 759, 760, **760,**
        782
    and photosynthesis, 115–116, **116**
    and plant growth, 417–418, 422,
        422
Light microscope, **37,** 37–39, 40
Light reactions, 117–119, **118, 121**
Lignin, 86, 355, 367
Limbic system, 696
Limbs
    of amphibians, 548
    of reptiles, 559
Limiting factors, 796–797
Limnologist, 333, **333**
Linen, 436
Linnaeus, Carolus, 271–273, 275
Lipids, 69, 71, **71,** 102, **102**
Liquids, 47–48
    diffusion in, 99–100
    molecular movement in, 99
Live-bearers, 534
Liver, 107, 531, 543
    in humans, 647–648, **648,** 649,
        681, 746, 747–748
Liverworts, 358, 361, **361**
Lizards, 560, 562–563
Lobe-finned fishes, 529, **529,** 538
Lobsters, 490, 494, 496, **496**
Long, Irene Duhart, 725, **725**
Long bones, 624, 625, **625**
Long-day plants, 418
Longitudinal binary fission, 318
Loose connective tissue, 613, **613**
LSD, 745, 748
Lucy. See *Australopithecus*
Lung cancer, 753, **753**
Lungfishes, 529, **529,** 546
Lungs, 529. *See also* Respiratory
        system
    of amphibians, 539
    book, 491–492, **492**
    human, 675–680, **676, 677,** 753
Lunula, 634
Luteinizing hormone. *See* LH
Lycopodium, 370, **370**
Lymph, 613, 665, **665**

Lymphatic system, human, 614, 665,
    665
Lymph nodes, 665, **665,** 666
Lymphocytes, 666–668
Lysosomes, 90, 108
Lysogenic cycle, **286,** 287, 289
Lytic cycle, 284–287, **286,** 289

M

McCarty, Maclyn, 181–182
McClintock, Barbara, 192, **192**
MacLeod, Colin, 181–182
Macroevolution, 252
Macronucleus, 321–323, **322**
Maltase, **72,** 648
Magnification, 37–39, 830
Malaria, 323–324, 436
Maleic hydrazine, 421
Male reproductive system, **731,** 731–
    732
Malleus, 711, 712
Malpighian tubules, 492, 497, 508
Malt sugar, 69–70
Maltose, 69–70
Mammals, 520, 588–603
    characteristics of, 378, 589–594,
        595
    classification of, 595–603, *table*
        601, 829
    origin of, 378, 589
Mammary glands, 589
Mammoths, **589,** 589
Manatees, 602, **602**
Mandibles, 494, 497, 503, 506
Mangroves, 387–388
Margulis, Lynn, 301, **301**
Marijuana, 750–751, **750**
Marine biologists, 486–487, **487,**
    498, **498,** 534, **534**
Marine biome, **788,** 788–789
Marrow, 623, 625, **626,** 656, 665
Marshes, 779, **779**
Marsupials, 595, 596–597
Mass, 47
Mass number, 50
Mass selection, 217, 218, 219
Mastodons, 599
Mating strains, 345
Matrix, 613, **613**
Matter, properties of, **47,** 47–48
Mechanoreceptors, 705
Medulla oblongata, 546, **546,** 696–697
Medusa, 448–451, **449, 450**
Megaspores, 371, 403–404, **404**
Meiosis, 143–145, **144, 145**
    chromosome changes in, 143–145,
        169–170, 174, 175
    and mitosis, comparison between,
        table 143

nondisjunction in, 207
and segregation principle, 155
in sporophyte 334, 358
Meischer, Friedrich, 183
Melanin, 632–633
Memory cells, 667–668
Mendel, Gregor, 151–157, 159, 161, 163, 174, 245
Meninges, 698, **698**
Menopause, 734
Menstrual cycle, 733–734, **734**
Meristems, 383. *See also* Apical meristems
Mesenteric arteries, 665
Mesentery, 543
Mesoderm, 460–461
Mesoglea, 448
Mesophyll, 394, **394**
Mesophytes, 418
Mesozoic era, *table* 234, 556
Messenger RNA, 186–190, 286
Metabolism, 5–6, 678
energy loss during, 768
Metamorphosis
in fishes, 524
in frogs, 548
in insects, 509–510
Metaphase, 139, **140,** 140–141, 144
Methadone, 751
Methanogens, 302
Microbiology, *table* 20, 38
Microdissection, 41, **41**
Microevolution, 252
Micrometer, 38
Micronucleus, 321–323
Microscope, 36–40
and discovery of cells, 79, **79**
electron, 39–40, 79, 86, 275
light, **37,** 37–39, 40
Microspores, 371, 403, **404**
Microtome, 40, **40**
Microtubules, 90–91, **91**
Microvilli, **648,** 648–649
Midbrain, 696, 697
Midbrain tectum, 578
Middle ear, 710–711, **711**
Middle lamella, 86, **86**
Migrations, 248, 533, 581–582, 593–594
Miller, Stanley, 230
Millipedes (Diplopoda), 490, **497,** 497–498
Mimicry, 251–252, **512,** 512–513
Minerals, and nutrition, 639, *table* 642, 643
Mistletoe, 771, **771**
Mites, 490, 491, 493, **493**
Mitochondria, **88,** 88–89, 92, 125, 126, 132–133, 732

Mitosis, 138–142, **140, 141,** *table* 143
Mitral valve, 660, 662
Mixtures, 53–54
Molars, 595, 609, 611, 644, **644**
Molds, 276, 340–341, **345,** 345–346
Molecular formula, 52–53
Molecular genetics, 180
Molecular weight, 53
Molecules, 12, 52–53
carrier, 106–107
chemical bonds in, 56–58
movement of, 99–100
Mollusks (Mollusca), 472–479, **473, 474, 475, 477, 478, 479**
Molting, 489, **489,** 563, 563–564, **564**
Monera, *table* 275, 275–276, *table* 276, **297,** 297–298, **298**
Monkeys, 603, **603,** 609, 611
Monocots, **376,** 376–377
leaves in, **393**
root system in, 385, **385**
seedling development in, 409, **409**
vascular bundles in, 389, **389**
Monohybrid cross, 160, **160**
Monomers, 69, 71
Monosaccharides, 69, 640, 648
Monosomy, 207
Monotremes, 595–596, **596,** 597
Moose, 783, **783**
Morgan, Thomas Hunt, 172–173, 174–175, 180
Morphine, 436
Mosquito, *Anopheles*, 323–324
Mosses (Muscopsida), 358, **358, 359,** 359–361, **360**
Motile organisms, 300, 316
Motor cortex, 695, **695**
Motor end plates, 692
Motor nerve fibers, 629
Motor neurons, 689, 698, 699, **699**
Mouth, human, 644–645, **645,** 705, 713–714
Mouth breathing, 544
Movable joints, 626
mRNA. *See* Messenger RNA
Mucous membranes, 289, 666, 676–677
Mucus, 644, 647
Mucus glands, 540, 541
Mud puppies, 550, **550**
Multicellular organisms, 3–4, **4**
Multiple alleles, 203–204
Multiple fission, 320
Multiple fruits, 407
Muscles
function of, 627
movement of, **629,** 629–631

Muscle spindles, 706
Muscle tissue, 612, 613, **627,** 627–628, 632
Muscular dystrophy, 631
Muscular system
in birds, 575–576
human, 614, 627–631
in vertebrates, *table* 522
Museum curator, 274, **274**
Mushrooms, **343, 346,** 346–347, **347**
Mutations, 191–194, **193**
in bacteria, 304
and genetic equilibrium, 249
and genetic variation, 244
Mutualism, 304–305, 315, 770–771, 772
Myasthenia gravis, 631
Mycelium, 343, **343,** 344, 347
Mycoplasmas, 80, 298, 299, 300
Myelin sheath, 689–690, 691, **691**
Myofibrils, 627
Myopia, 709, **709**
Myosin filaments, 629–631
Myriapods, 490, 497–498

N

Nails, 632, 634
Narcotics, 751
Nasal cavities, 676–677
Nastic movements, 423–424
Natural resources, 809–815
Natural selection, 244, 246–247, 250, 252
Nautilus, 478, 479, **479**
Neanderthals, 261, **261,** 262
Nearsightedness, 709, **709**
Nectar guides, 406
Needham, John, 30, **30**
Negative feedback, 725–726
Negative phototropism, 320
Nekton, 788
Nematocyst, 448, 449, **449**
Nematodes. *See* Roundworms
Neoteny, 550
Nephron, 682, **682**
Neritic zone, 789
Nerve cells. *See* Neurons
Nerve cord, 519, **519**
Nerve deafness, 712
Nerve fibers, 689, 690
in brain, 694, 697
in optic nerve, 708–709, **709**
Nerve impulses, 689, 690–693, **690–693**
Nerve net, 449
Nervous system
of birds, 578
of bony fishes, 532

Unsaturated fat, 641, **641**
Uracil, 186, 187
Urea, 681–682, 683
Ureter, **681**, 683
Urethra, 684, 731
Uric acid, 578
Urinary bladder, **681**, 683–684, *table* 700
Urinary system, human, 615, 683–684
Urine, 683–684
Utah Arm, 617
Uterus, 597
    in humans, 732, 733–734, 736–740, **737**

V _____

Vaccines, 289–290
Vacuoles, **89,** 89–90
    contractile, 104, 319, 322
    and bulk transport, 107, 108
Vagina, 732, 734, 735
Variable, 26, 29
Variation(s), 10–11
    origin of, 244
Vascular bundles, 389, **389,** 394
Vascular cambium, 386, 387, 390, 391
Vascular cylinder, 386
Vascular plants. *See* Tracheophytes
Vascular system, 367
    in humans, 665
Vascular tissue, 357, 383–384, 389
Vas deferens, 731, **731**
Vasopressin, 724
Vectors, 215
Vegetative body, 374
Vegetative propagation, 137, **410,** 410–411
Veins, 394, 531
    of human circulatory system, **660,** 661, 663–665, **663, 664**
Venom, 565–566
Ventral surface, 444, **444**
Ventricles, 523, 545, 577
    of human heart, 660–663, **661**
Venus' flytrap, 396, **396,** 423
Vernalization 419, 422
Vertebrae, 521, 543
    in humans, 623, 698
Vertebral column, 521
    in humans, 623, **624, 698**
Vertebrates, 444, 518–522, 612
    characteristics of, **520,** 520–522, **521**
    and echinoderms, resemblance between, 480
    systems of, 522, *table* 522
Vessel members, 383

Vestibule, 711
Vestigial structures, 236
Vibrissae, 599
Villi, **648,** 648–649, 737, 738
Virchow, Rudolph, 80
Viroids, 283
Virulence, 289
Viruses, 282–290
    and cell theory, 80
    characteristics of, 283–284, **284**
    and disease, 288–290, **289, 290,** *table* 291, 294–295, 666, 668
    and gene therapy, 214–215
    reproductive cycle of, 284–287, **286**
    transduction by, 287, **287**
Visceral mass, 473
Visible spectrum, 115
Vision, 705, 706, 707–709
Vitamins, 72–73, 639, 641–643, *table* 642
Viviparous mammals, 589
Vocal cords, 542, 677
Voice box, 646, 677
Voluntary muscles. *See* Skeletal muscle
Voluntary nervous system, 699
Volvox, 333–334, **334**
Vomerine teeth, 543

W _____

Walruses, 600
Warm-blooded animals, 557, 590
Wasps, 511, 512, **512**
Water
    as abiotic factor, 759, 760–761
    and bodily functions, 639, 643
    in cells, movement of, 99, 102–105
    conservation of, **809,** 809–810
    molecule of, 52, 57
    movement to land from, 539
    and photosynthesis, 115, 117
    and plant growth, 417, 418
    in plants, movement of, 395
    pollution of, 817–819, **819**
    as solvent, 53, 67
Water cycle, 763
Water molds (Oomycetes), 346, **346**
Watershed, 810
Water-soluble vitamins, 641–643
Water-vascular system, 481
Watson, James D., 182–185
Web of life, 19
Weeds, ecological role of, 778
Weinberg, Wilhelm, 245–246
Weismann, August, 169, 244
Welwitschia, **36,** 375, 376
Went, Frits W., 420

Whales, 251, **251,** 586–587, **600,** 600–602
Whisk ferns (Psilophyta), 369, **369**
White blood cells, 306, 655, **655,** 656, 665, 666–668
White corpuscles. *See* White blood cells
White matter, 694, 696, 698
Whittaker, Robert, 321, **321**
Wildlife biologist, 544, **544**
Wildlife conservation, **812,** 812–813
Wilkins, Maurice, 183
Wilson, Edward Osborne, 504, **504**
Windbreaks, 811
Withdrawal, 745
Wood thrush, 582
Woody plants, 377
    buds of, 392–393
    stems of, **390,** 390–391
Worms, 456–468

X _____

Xanthophylls, 116, 307–308, 331, 332
X chromosome, 172–173
Xerophytes, 418
Xylem, 367, 383, 386, **386,** 387, 390, **390, 394,** 395

Y _____

Y chromosome, 172–173
Yeasts, 348
Yolk, 555–556
Yolk sac, 555–556, 737

Z _____

Zero population growth, 801
Zinjanthropus, 264
Zookeeper, 562, **562**
Zoology, 20
Zooplankton, 315
Zygospore, 334, 345
Zygote, 137, **137,** 461
    and alternation of generations, 334
    chromosome number in, 170, 207
    of flowering plants, 405
    human, 735

Credits

The positions of illustrations and photographs are indicated by the following abbreviations: (t) top, (b) bottom, (l) left, (r) right, (c) center, (ins) inset.

Cover: Closeup of agave, Joseph Holmes; (*front inset*) Bengal tiger, Ron Garrison, © Zoological Society of San Diego; (*back inset*) Bengal tiger running, Zig Leszczynski, Animals Animals

Frontispiece: Comstock

Illustrations:
Scott Thorn Barrows 222, 614, 615, 628 (r), 644, 645, 646, 648, 659, 660, 661, 663, 664, 665, 668, 672, 673, 719, 720 (t) **Sally Bensusen** 409 (c) **Leon Bishop** 21, 33, 43, 95, 109, 129, 146, 195, 230, 237, 253, 291, 397, 413, 425, 438, 469, 610, 612, 613, 720 (b1), 724, 726, 731, 732, 733, 735, 737 **Howard Friedman** 37, 39, 83, 86, 88 (tl), 89, 92, 94, 117, 284, 286, 287, 290, 297, 300, 304, 332, 334, 335, 343, 345, 347, 355, 357, 358, 360, 492, 503, 506, 507, 509, 511, 512 **Biruta Ackerbergs Hansen** 102, 107, 108, 367, 368, 373, 375, 383, 385, 386, 387, 389, 390, 392, 394, 395, 458, 459, 460, 461, 465, 466, 473, 474, 475, 481, 495, 519, 526, 529, 530, 531, 532, 533 **Jean Cassels Helmer** 106, 258, 259, 260, 261, 262, 276, 377, 378, 391, 393, 401, 402, 404, 407, 408, 409 (br), 410, 420, 431, 443, 444, 446, 448, 449, 450, 505, 542, 543, 545, 546, 547, 548, 561 (tr), 573, 577, 578, 601 **Deirdre A. McConathy** 88 (bl) **Teri J. McDermott** 215, 497, 499, 521 (l), 535, 551, 619, 629 (c), 669, 677, 682, 683, 685, 689, 701, 803 **Leonard E. Morgan** 6, 13, 67, 71 (t, b), 72 (t), 80, 87, 113, 114 (b), 118, 120, 124, 126, 127 (t), 128, 183, 184, 185, 187, 188–189, 208, 229, 232, 521 (r), 624, 625, 626, 627, 628 (tl), 629 (tr), 631, 633, 675, 676, 678, 681, 705, 707, 708, 709, 711, 712, 713, 714 **Sarah Forbes Woodward** 8, 29, 30, 31, 48, 49 (tl, tc, tr), 53, 56 (t, b), 57, 65, 99, 100, 103, 104, 121, 123, 136, 137, 138, 139, 140–141, 144–145, 145 (r), 151, 152, 153, 155, 156, 159, 160, 161, 163, 169, 172, 173, 174, 175, 181, 193, 199, 200, 204, 206, 207, 213, 236, 247, 251, 263, 269, 295, 316, 318, 319, 322, 323, 324, 331, 428, 486, 520, 555, 556, 558, 561 (br), 565, 566, 589, 590, 591, 595, 599, 600, 603, 641, 657, 658, 690, 691, 692, 694, 695, 698, 699, 722, 734, 746, 764, 765, 767, 768, 779, 783, 788, 796, 797, 800, 801, 815

Photographs
Unit 1: Page 1 E. R. Degginger; **2** Carol Hughes, Bruce Coleman; **3** (l) A. Pasieka, Taurus Photos; (c) J. Gennard/L. Grillone, Photo Researchers; (r) Eric Grave, Photo Researchers; **4** (l) M. Abbey, Photo Researchers; (c) James Bell, Photo Researchers; (r) L. L. T. Rhodes, Animals Animals; **6** Petit Formal and Guigoz, Photo Researchers/Science Source; **7** (l) Danny Brass, Photo Researchers; (r) Chace, National Audubon Society, Photo Researchers; **9** (t) Charles Summers, Bruce Coleman; (b) Wayne Lankinen, Bruce Coleman; **10** Bruno Zehnder, Peter Arnold; **13** Manfred Kage; **14** Bruno Zehnder, Peter Arnold; **15** NASA; **16** (l) Harold Hoffman, Photo Researchers; (tr) Keith Gunnar, Bruce Coleman; (br) Russ Kinne, Photo Researchers; **17** NASA; **18** Mantis Wildlife Films, Oxford Scientific Films; **19** David Austen, Stock Boston; **24** Grapes Michaud, Photo Researchers; **25** (b) W. E. Ruth, Bruce Coleman; (t) L. West, Bruce Coleman; (ins) Eric Simmons, Stock Boston; **26** Bruce Powell; **28** Culver Pictures; **32** C. Johnson, GammaLiaison; **36** P. Dayanandan, Photo Researchers; **38** Carl Zeiss Inc.; **39** (l) James Somers, Taurus Photos; (c) BioPhotos Associates, Photo Researchers; (r) Ann Smith, Photo Researchers; **40** Lester Bergman and Associates; **41** Jon Gordan, Phototake; **42** Beth Ullman, Taurus Photos; **46** Stephen Dalton, Photo Researchers; **47** Steven Fuller; **49** Manfred Kage, Peter Arnold; **54** David Austen, Stock Boston; **55** Craig Aurness, West Light; **56** E. R. Degginger, Bruce Coleman; **58** Bruce Powell; **59** (b) Jeremy Ross, Photo Researchers; (ins) Bruce Powell; **64** (tl) Terri Sherman/John Kinnamon, University of Colorado; (bl, br) Washington University; **65** (tc, tr) Gower Medical Publishing Ltd.; (bl) CNRI, Vision International; **66** Yata Haneda; **68** Rick Browne, Picture Group

Unit 2: 77 Runk/Schoenberger, Grant Heilman; **78** Manfred Kage, Peter Arnold; **79** (t) Brown Brothers; (b) Bettman Archive; **82** (r) BioPhotos Associates, Photo Researchers; (l) Eric Grave, Photo Researchers; **84** Don Fawcett, Photo Researchers; **85** Rob Stepney, Photo Researchers; **87** Don Fawcett, Photo Researchers; **88** Abbott Labs; **89** Jeremy Burgess, Photo Researchers; **91** (l) D. Fawcett/D. Phillips, Photo Researchers; (c) Manfred Kage, Peter Arnold; (tr) Omikron, Photo Researchers; (br) D. Fawcett/D. Philips, Photo Researchers; **98** Richard Feldman, National Institute of Health; **100–101** Bruce Powell; **104** Michael Sheetz, University of Connecticut; **105** Gower Medical Publishing; **107** Eric Grave, Photo Researchers; **112** Henry Ausloos, Earth Scenes; **116** David Parker, Photo Researchers; **117** BioPhotos Associates, Photo Researchers; **119** USDA, Photo Researchers; **132** (tl) Manfred Kage, Peter Arnold; (cl) CNRI, Vision International

(bl) David Scharf; (br) International Scientific Instruments Inc., Milpitas, CA; **133** (tl) Richard Carter, University of Colorado; (tr) Terri Sherman/John Kinnamon, University of Colorado; (bc) Massachusetts General Hospital; **134** Keith Porter, Photo Researchers; **136** U.S. News; **140–141** Philip Coleman; **142** Manfred Kage, Peter Arnold

Unit 3: 149 Ira Block, The Image Bank; **150, 157** Bruce Powell; **162** John Cowell, Grant Heilman; **164** (l) Bruce Powell; (r) Dennis Purse, Photo Researchers; **168** Russ Kinne, Photo Researchers; **170** Milton Mann, Cameramann International; **176** Manfred Kage, Peter Arnold; **180** Phillip and Phylis Morrison, Charles and Ray Eames, Scientific American Books, *Powers of Ten;* **183** J. D. Watson, *The Double Helix,* C. S. H. Labs; **191** (t) Tom McHugh, Photo Researchers; (b) Grant Heilman, Grant Heilman; **192** Nik Kleinberg, Picture Group; **198** Lee Balterman, Gartman Agency; **201** Enrico Ferorelli, DOT; **202** (t) Owen Franken, Stock Boston; (b) Daniel Mass, University of Chicago Medical Center; **203** (l) Bill Longcore, Photo Researchers; (r) Sklar/Peiper, Photo Researchers; **204** Peter Vandermark, Stock Boston; **205** Russ Kinne, Photo Researchers; **207** Rosaria Baldino; **209** Bruce Powell; **210** (l) Bill Gallery, Stock Boston; **211** Jim Ballard; **214** (tl) Robert Langridge/Dan McCoy, Rainbow; (bl) Dan McCoy, Rainbow; (br) Mark Godfrey, Archive Pictures Inc.; **215** (tr) Richard Feldman, National Institute of Health (br) Hank Morgan, Photo Researchers; **216** Margot Conte, Animals Animals; **217** Ray Hillstrom, Root Resources; **218** (l) Stark Nurseries, Louisiana, MO; (r) Hoffman-LaRoche; **219** E. R. Degginger; **220** The Land, Epcot Center, Walt Disney World Resort Center; **221** (l) David Evans/William Sharp, DNA Plant Technology Corp.; (r) Stark Brothers Nurseries; **223** H. Potter/D. Dressler, *Life Magazine,* © 1980, Time Inc.

Unit 4: 227 James Carmichael Jr., Nature Photographers; **228** R. J. Dufour/Rice University, Hansen Planetarium; **231** (l) Frank Carpenter, Harvard University; (c) Runk/Schoenberger, Grant Heilman; (r) Field Museum of Chicago; **233** Lee Balterman, Gartman Agency; **235** Richard Feldman, National Institute of Health; **240** E. S. Ross; **241** (t) Mansell Collection; (b) Harry Ransom Humanities Research Gernsheim Collection, University of Texas at Austin; **242** (t) Mickey Gibson, Tom Stack and Associates; (b) Kenneth W. Fiak, Bruce Coleman; **243** (l) Russ Kinne, Photo Researchers; (r) Alan Root, Bruce Coleman; **244** Walter Chandoha, Gartman Agency; **245** Milton and Joan Mann, Gartman Agency; **248** Donald Dietz, Stock Boston; **249** (t) F. Erize, Bruce Coleman; (b) Thomas Friedmann, Photo Researchers; **250** (l) David Fritts, Animals Animals; (r) Richard Kolar, Animals Animals; **252** American Museum of Natural History; **256** Margo Crabtree/AAAS; **257** Cleveland Museum of Natural History; **258, 260** Margo Crabtree/AAAS; **261** Margo Crabtree, Musee de l'homme; **262** (l) Tom McHugh, Photo Researchers; (r) Margo Crabtree, Musee de l'homme; **264** John Reader; **268** (tl) Donna Rona, Bruce Coleman; (bl) Manfred Kage, Peter Arnold; **269** (t) Alvin Upitis, The Image Bank; (b, br) Geological Society of America, LaMont Doherty CLIMAP Project; **270** Alan Brown, Photo Design; **271** E. Duscher, Bruce Coleman; **272** Yale University Library; **273** R. E. Ferrell, University of Pittsburgh; **274** James P. Rowan, Gartman Agency

Unit 5: 281 E. H. Cook, Photo Researchers/Science Photo Library; **282** Ny Carlsberg Glyptotek; **284** (l) National Cancer Institute, Vision International Source; (c) Lee D. Simon, Photo Researchers; (r) Biology Media, Photo Researchers; (bl) Lee D. Simon, Photo Researchers; **285** C. Steele Perkins, Magnum; **288** Russ Kinne, Photo Researchers; (ins) Biology Media, Photo Researchers; **289** Pam Hasegawa, Taurus Photos; **294** Keystone Press; **295** (tl) Dan McCoy, Rainbow; (tc, bl) U.S. Department of Health and Human Services/Centers for Disease Control; **296** Manfred Kage, Peter Arnold; **298** (t) Eric Grave, Photo Research; (r) Peter Parks, Oxford Scientific Films; **299** (l) Gower Medical Publishing; (r) Tony Brain, Photo Researchers; **300** Judith Hoeniger, University of Wisconsin; **301** Dan McCoy, Rainbow; **302** Bruno Zehnder, Peter Arnold; **303** (r, l) E. R. Degginger; **304** E. L. Wollman, Institute for Cancer Research; **305** Walter Dawn, Oxford Scientific Films; **307** (bl) James Somers, Taurus Photos; (r) Walter Dawn; **309** Daniel Brody, Stock Boston; **310** Martin Rotker, Taurus Photos

Unit 6: 313 John Shaw, Tom Stack and Associates; **314** M. I. Walker, Photo Researchers; **315** (t, b) Gary Grimes, Taurus Photos; **317** Eric Grave, Photo Researchers; **318** Spike Walker, Photo Researchers; **319** Manfred Kage, Peter Arnold; **320** Eric Grave; **321** (t) Manfred Kage, Peter Arnold; (b) Cornell University Biological Sciences; **328** Oxford Scientific Films, Animals Animals; **329** (t) Alfred Pasieka; (b) Jeff Foote, Tom Stack and Associates; **330, 332** W. H. Hodge, Peter Arnold; **333** E. R. Degginger, Bruce Coleman; **334** Manfred Kage, Peter Arnold; **336** (t) Walt Anderson, Tom Stack and Associates; (b) Walter Dawn; **340** (t, c, bl) Lederle Laboratories; (br) H. J. Phaff, University of California at

Davis; **341** (tl) CNRI, Vision International; (tc) Manfred Kage, Peter Arnold; (br) Ralph Morse, Scientific American; **342** W. H. Hodge, Peter Arnold; **344** (l) Gwen Fidler, Tom Stack and Associates; (r) Eric Grave; **346** (l) Angelina Lax, Photo Researchers; (tr) D. Smiley, Peter Arnold; (br) Noble Proctor, Photo Researchers; **348** (l) M. P. Kahl, Photo Researchers; (r) Jim Strawser, Grant Heilman; **349** (t) Walter Dawn; (b) Gwen Fidler, Tom Stack and Associates

Unit 7: 353 Kathleen Norris Cook; **354** Stephen Dalton, Photo Researchers; **356** Bill Ross, West Light; **357** E. S. Ross; **358** Milton Rand, Tom Stack and Associates; **361** (t) John Gerlach, Tom Stack and Associates; (b) E. S. Ross; **362** Story Litchfield, Stock Boston; **363** John Shaw, Bruce Coleman; **366** Kathleen Norris Cook; **369** (l) R. C. Simpson, Tom Stack and Associates; (r) Stephen J. Kraseman, Photo Researchers; **370** Rod Plank, Tom Stack and Associates; **371** (t) E. S. Ross; (bl) Bruce Coleman; (br) W. H. Hodge, Peter Arnold; **372** New York State College at Cornell University; **374** Cindy McIntyre, West Stock Inc.; **376** (l) E. S. Ross; (c) Terry Ashley, Tom Stack and Associates; (r) E. S. Ross; **382** George H. Harrison, Grant Heilman; **384** (l, r) BioPhotos Associates, Photo Researchers; **387** (l, r) Walter Chandoha; **388** John Kohout, Root Resources; **394** (t) B. E. Juniper, Oxford Botany School; (b) P. Dayanandan, Photo Researchers; **396** (l) E. S. Ross; (c) Hans Pfletschinger, Peter Arnold; (r) J. H. Robinson, Photo Researchers; **400** David Cavagnaro; **402** (c) Walter Chandoha; **403** (l) Joan Nowicke, Smithsonian Institution; (c) Allen Rokach, New York Botanical Gardens; (r) A. J. Belling, Photo Researchers; **405** G. I. Bernard, Earth Scenes, Oxford Scientific Films; **406** (l, r) E. S. Ross; (c) Alan Pitcairn, Grant Heilman; **411** Jerome Wexler, Photo Researchers; **412** E. R. Degginger; **416** E. S. Ross; **417** Walter Chandoha; **418** Walter Dawn; **421** Walter Chandoha; **422** Grant Heilman, Grant Heilman; **423** (l) Runk/Schoenberger, Grant Heilman; (r) B. E. Juniper, Oxford Botany School; **424** E. R. Degginger; **428** (l) Chuck O'Rear, West Light; (r) Tom Petroff, Petroff Photography; **429** (bl) Bill Ross, West Light; (tl) The Land, Epcot Center, Walt Disney World Vacation Resort; (tr) NASA, Photo Researchers/Science Source; (c) Plant Genetics, Inc.; **430** Craig Aurness, West Light; **432** Philip Harrington, Peter Arnold; **433** Fred Ward, Black Star; **434** Plant Genetics Inc.; **435** Menschenfreund, Taurus Photos; **436** (t) Michael Philip Manheim, Photo Researchers; (b) Zig Leszczynski, Photo Researchers; **437** Ted Streshisky, *Science Year*, © 1983 World Book Inc.

Unit 8: 441 E. R. Degginger, Animals Animals; **442** A. Power, Bruce Coleman; **443** (l) Robert Lee, Photo Researchers; (r) Dave Woodward, Taurus Photos; **444** Steve Solum, Bruce Coleman; **445** Jeff Rotman; **450** Maria and Rod Borland, Bruce Coleman; **451** (t) Parrot-Sygma, Sygma; (b) R. N. Mariscal, Bruce Coleman; **452** L. Isy-Schwart, The Image Bank; **456** R. L. Sefton, Bruce Coleman; **459** (t) E. R. Degginger; (b) Manfred Kage, Peter Arnold; **462** Martin M. Rotker, Taurus Photos; **463** Alfred Pasieka, Taurus Photos; **464** National Academy of Sciences; **466** Runk/Schoenberger, Grant Heilman; **467** Hans Pfletschinger, Peter Arnold; **472** Neil G. McDaniel, Tom Stack and Associates; **474, 475** James Carmichael Jr., Nature Photographers; **476** National Fisheries Center, U.S. Fish and Wildlife; **477** Y. Momatiuk, Photo Researchers; **478** Fred Bavendam, Peter Arnold; **479** (t) Jeff Rotman; (c) James Carmichael Jr., Nature Photographers; **480** (l) Brian Parker, Tom Stack and Associates; (c) E. R. Degginger; (r) Jeff Rotman, Peter Arnold; **482** Carl Roessler, Tom Stack and Associates; **483** Gary Milburn, Tom Stack and Associates; **486** (tl) Neil McDaniel, Tom Stack and Associates; (bl) Bill Tronca, Tom Stack and Associates; (bc) Gary Milburn, Tom Stack and Associates; **487** (t) Dan McCoy, Rainbow; (tr, br) Christopher Swann; **488** J. H. Robinson, Photo Researchers; **489** Hans Pfletschinger, Peter Arnold; **490** Townsend P. Dickenson, Photo Researchers; **491** Hans Pfletschinger, Peter Arnold; (t) E. R. Degginger; **493** (c) David Scharf; **495** E. R. Degginger; **496** (t) David Scharf; (b) Christopher Crowley, Tom Stack and Associates; **497** (t) E. S. Ross; (b) E. R. Degginger; **498** Dave Woodward, Taurus Photos; **502** E. S. Ross; **504** Rick Friedman, Black Star; **507** James Carmichael Jr., Bruce Coleman; **512** (t) E. S. Ross; (b) Hans Pfletschinger, Peter Arnold

Unit 9: 517 Charles Krebs; **518** Allen Morgan, Peter Arnold; **523** (l) Heather Angel, BioFotos; (r) Oxford Scientific Films, Animals Animals; **524** Heather Angel, BioFotos; **525** Steve Martin, Tom Stack and Associates; **527** (l) Zig Leszczynski, Animals Animals; (r) Jane Shaw, Root Resources; **528** (l) Zig Leszczynski, Animals Animals; (c) Gary Milburn, Tom Stack and Associates; (r) Ron and Valerie Taylor, Bruce Coleman; **534** University of Maryland; **538** George Porter, Photo Researchers; **540** (t) Stephen Dalton, Photo Researchers; (b) Zig Leszczynski, Animals Animals; **541** A. B. Joyce, Photo Researchers; **542** Runk/Schoenberger, Grant Heilman; **544** Cary Wolinsky, Stock Boston; **549** (l) Jack Dermid; (r) Zig Leszczynski, Animals Animals; **550** (t) Jack Dermid; (b) Tom McHugh, Photo Researchers; **554** A. Nelson, Tom Stack and Associates; **555** Zig Leszczynski, Animals Animals; **559** Zig Leszczynski, Animals Animals; **560** John Markham, Bruce Coleman; **561** Zig Leszczynski, Animals Animals; **562** (t) Gary Milburn, Tom Stack and Associates; (b) James Rowan, Gartman Agency;

563 John Gerlach, Tom Stack and Associates; **564** (t) Zig Leszczynski, Animals Animals; (bl) James Carmichael Jr., The Image Bank; (br) Michael Fogden, Bruce Coleman; **566** Zig Leszczynski, Animals Animals; **570** Wayne Lankinen, Bruce Coleman; **571** American Museum of Natural History; **572** E. R. Degginger; **574** Patti Murray, Animals Animals; **575** David Cavagnaro; **579** (l) Photri and Gartman Agency; (c, tr, br) Runk/Schoenberger, Grant Heilman; **580** Renee Purse, Photo Researchers; **581** Kim Steele, West Stock; **582** James L. Gulledge, Laboratory of Ornithology, Cornell University; **586** (tl) Bruce Penrod, New York State Department of Environmental Conservation; (bc, br) Jeff Foote, Bruce Coleman; **587** (t) Sam Ridgway, U.S. Navy; (bl, br) Larry McNease, Rockefeller Wildlife Center; **588** M. P. Kahl, Photo Researchers; **592** (t) Gregory G. Dimnian, Photo Researchers; (b) Tom Bean, Tom Stack and Associates; **593** Louis Trusty, Animals Animals; **594** Richard Farnell, Animals Animals; **596** (l) Alan Roberts; (r) F. Prenzel; **598** (t) Elizabeth Weiland, Photo Researchers; (b) Stephen Dalton, Animals Animals; **602** Fred Barendam, Peter Arnold

Unit 10: 607 Focus West; **608** Dan McCoy, Rainbow; **609** Peter B. Kaplan, Photo Researchers; **616** Dr. P. Galletti, Brown University; **617** National Center for Rehabilitation Engineering; **618** Call, Gamma-Liaison; **622** Bill Ross, West Light; **627** Eric Grave, Photo Researchers; **630** David Madison, Bruce Coleman; **638** Manfred Kage, Peter Arnold; **639** Bruce Powell; **649** Joseph Sterling, Click/Chicago; **654** Manfred Kage, Photo Researchers; **655** K. R. Porter, Photo Researchers; **656** James White, University of Minnesota; **657** Manfred Kage, Peter Arnold; **666** A. Liepins, Photo Researchers; **667** (t) American Red Cross; (b) Neil, Photo Researchers/Science Source; **672** (tl) Alvin Upitis, The Image Bank; **673** (tr) Merrell Wood, The Image Bank; (bl) Photo Researchers/Science Source; Photo Researchers; **674** David Madison, Bruce Coleman; **677** Manfred Kage, Peter Arnold; **679** Richard Wood, Taurus Photos; **682** Odyseaus, Peter Arnold; **688** Manfred Kage, Peter Arnold; **692** Ed Reschke, Peter Arnold; **697** Henry Grossman; **704** Alexander Tsiaras, Photo Researchers/Science Source; **710** Eric Kroll, Taurus Photos; **712** G. Bredberg, Photo Researchers; **714** Omikron, Photo Researchers; **715** Don Wong, Photo Researchers; **718** Craig Aurness, West Light; **725** NASA; **730** P. Sundstrom, Gamma-Liaison; **738** (l, r) **739** (l, c) Lennart Nilsson, *Behold Man*, Little, Brown, and Co.; (r) Lennart Nilsson, *A Child Is Born*, Dell Publishing Co.; **740** Gerry Souter, Click/Chicago; **741** (l) John Walsh, Photo Researchers/Science Photo Library; (r) Martin Rotker, Taurus Photos; **744** Lennart Nilsson, *Behold Man*, Little, Brown, and Co.; **748** A. Duncan, Taurus Photos; **749** David Barnes; **750** E. R. Degginger; **753** Martin Rotker, Taurus Photos

Unit 11: 757 L. Isy-Schwart, The Image Bank; **758** Leonard Lee Rue III, Tom Stack and Associates; **759** Michael Fogdon, Oxford Scientific Films, Earth Sciences; **760** (t) Stephen Kraseman, Peter Arnold; (b) John Macgregor, Peter Arnold; **761** L. Isy-Schwart, The Image Bank; **762** Tom McHugh, Photo Researchers; **766** Mitch Reardon, Photo Researchers; **769** Reynolds Aluminum Company; **770** G. Ziesler, Peter Arnold; **771** (t) W. H. Hodge, Peter Arnold; (b) Alan Roberts; **772** Betty Kubis, Root Resources; **776** Kevin Schafer, Tom Stack and Associates; **777** Grant Heilman, Grant Heilman; **780** Walter Dawn; **781** (t, b) Jack Dermid; **782** David Fritts, Animals Animals; **783** Richard Smith, Tom Stack and Associates; **784** Isaac Gelb, Grant Heilman; **785** (t) Brian Parker, Tom Stack and Associates; (b) Phil Degginger, Click/Chicago; **786** Andrew Holbrooke, Black Star; **787** Katherine Thomas, Taurus Photos; **789** R. Massa; **790** Larry Lefever, Grant Heilman; **794** M. Philip Kahl, Tom Stack and Associates; **797** Grant Heilman, Grant Heilman; **798** Marine Science Research Center, State University of New York at Stony Brook; **799** Steven Kaufman, Peter Arnold; **800** Bonnie Freer; Peter Arnold; **806** (tl) NASA; (ins) NASA; **800–807** (b) NASA, Photo Researchers/Science Source; **807** (tl, tc) NASA, Photo Researchers/Science Source; **808** Lee Balterman, Gartman Agency; **809** Charles Kennard, Stock Boston; **810** Peter Menzel, Stock Boston; **811** (t) Cary Wolinsky, Stock Boston; (b) Steven Kaufman, Peter Arnold; **812** (t) Jack Dermid; (b) Jerome Wycoff; **813** Jack Dermid; **814** David Falconer, West Stock; **815** Jonathon Wright, Bruce Coleman; **816** Tom Stack, Tom Stack and Associates; **817** John Lopinot, Black Star; **818** George Kufrin, Click/Chicago; **819** Henri Bureau, Sygma; **823** Lynn Stone, The Image Bank; **836** (t) Historical Pictures Service, Chicago; (c) Philip Harris Biological Ltd., Photo Researchers; (bl) Philip Harrington, Peter Arnold; (br) Photo Researchers/Science Source; **837** (tl, tr) Historical Pictures Service, Chicago; (c) Gower Medical Publishing Ltd.; (bl) Dan McCoy, Rainbow; (br) Hank Morgan, Rainbow